# Advanced Interconnects for ULSI Technology

# Advanced Interconnects for ULSI Technology

Edited by

Mikhail R. Baklanov, Paul S. Ho and Ehrenfried Zschech

WILEY

A John Wiley & Sons, Ltd., Publication

This edition first published 2012


*Registered Office*
John Wiley & Sons, Ltd, The Atrium, Southern Gate, Chichester, West Sussex, PO19 8SQ, United Kingdom

For details of our global editorial offices, for customer services and for information about how to apply for permission to reuse the copyright material in this book please see our website at www.wiley.com.



*Library of Congress Cataloging-in-Publication Data*

Advanced interconnects for ULSI technology / [edited by] Mikhail R. Baklanov, Paul S. Ho, Ehrenfried Zschech.
p. cm.
Includes bibliographical references and index.
ISBN 978-0-470-66254-0 (hardback)
1. Interconnects (Integrated circuit technology). 2. Integrated circuits–Ultra large scale integration.
I. Baklanov, Mikhail. II. Ho, P. S. III. Zschech, Ehrenfried.
TK7874.53.A39 2012
621.39′5–dc23
2011038787

A catalogue record for this book is available from the British Library.

Print ISBN: 9780470662540

Set in 10/12pt Times by SPi Publisher Services, Pondicherry, India
Printed and bound in Singapore by Markono Print Media Pte Ltd

1 2012

# Contents

# About the Editors

## MIKHAIL R. BAKLANOV

Dr Mikhail R. Baklanov is a Principal Scientist at IMEC. He graduated from the Faculty of Natural Sciences at Novosibirsk State University in 1971 and joined the Institute of Semiconductor Physics of the Siberian Branch of the Russian Academy of Sciences (Novosibirsk). He received his PhD degree (candidate of science) in Physical Chemistry in 1978 and the Doctor of Science degree in 1991. From 1991 to 1995 he headed a Laboratory at the Institute of Semiconductor Physics. He joined the Interuniversity Microelectronics Centre in Belgium (IMEC) in 1995 as a Visiting Professor. In 2000 and 2003 he worked as R&D Manager at XPEQT (Switzerland/Belgium). Since 2003 Mikhail Baklanov has been a Principal Scientist at IMEC. His current interest is related to low-*k* dielectric films for interconnects in advanced technology nodes and new materials for nanoelectronics. He is serving as a member of Organizing Committees of several international conferences. He has edited and contributed to several books and published more than 200 papers in peer reviewed journals, 24 patents and more than 40 invited presentations at International Conferences.

## PAUL S. HO

Dr Paul S. Ho is the Director of the Laboratory for Interconnect and Packaging at the University of Texas at Austin. He received his BS degree in Mechanical Engineering from National Chengkung University, MS degree in physics from the National Tsinghua University, both in Taiwan, and PhD degree in Physics from Rensselaer Polytechnic Institute. He joined the Materials Science and Engineering Department at Cornell University in 1966 and became an Associate Professor in 1972. In 1972, he joined the IBM T.J. Watson Research Center and became Senior Manager of the Interface Science Department in 1985. In 1991, he joined the faculty at the University of Texas at Austin and was appointed the Cockrell Family Regents Chair in Materials Science and Engineering. His current research is in the areas of materials science and reliability for interconnect and packaging for microelectronics. He has edited several books and published extensively in the area of thin films and materials science for microelectronics. He has received several technical awards, including the Callinan Award from the Electrochemical Society in 2000 and the University Research Award from the Semiconductor Industry Association in 2007. He is a Fellow of the American Physical Society, the American Vacuum Society and the Institute of Electrical and Electronics Engineering.

## EHRENFRIED ZSCHECH

Dr Ehrenfried Zschech is working at the Fraunhofer Institute for Nondestructive Testing (IZFP). He was Senior Manager of the Center for Complex Analysis at GLOBALFOUNDRIES in Dresden. He joined Advanced Micro Devices in 1997. His responsibilities include the analytical support for process control and technology development, as well as physical failure analysis. He received his diploma degree in solid-state physics and his Dr rer. nat. degree from Dresden University of Technology. After having spent four years as a project leader in the field of metal physics and reliability of microelectronics interconnects at the Research Institute of Non-Ferrous Metals in Freiberg, he was appointed as a university teacher for ceramic materials at Freiberg University of Technology. In 1992, he joined the development department at Airbus in Bremen. There he managed the metal physics group and worked on laser joining metallurgy of light metals. His current research interests are in the areas of thin film materials compatibility, structure and materials analysis and physical failure analysis in integrated circuit applications. He has published three books and more than 100 papers in scientific journals in the areas of solid-state physics and materials science.

# List of Contributors

**Oliver Aubel**
GLOBALFOUNDRIES, Dresden
Module One LLC & Co. KG,
Wilschdorfer Landstrasse 101,
01109 Dresden, Germany

**Mikhail R. Baklanov**
IMEC, Kapeldreef 75, B-3001 Leuven,
Belgium

**Sridhar Balakrishnan**
Intel Corp., Hillsboro, OR 97124, USA

**Gautam Banerjee**
Air Products and Chemicals, Inc.,
Allentown, PA 18195, USA

**Wim Bogaerts**
Ghent University – IMEC,
Department of Information Technology,
Ghent, Belgium

**Ruth Brain**
Intel Corp., Hillsboro, OR 97124, USA

**Jean-François de Marneffe**
IMEC, Kapeldreef 75, B-3001 Leuven,
Belgium

**Alain Diebold**
College of Nanoscale Science and
Engineering at the University at Albany,
Albany, New York, USA

**Valery M. Dubin**
NANO3D SYSTEMS LLC, Portland,
Oregon, USA

**Geraud Dubois**
Hybrid Polymeric Materials Group,
IBM Almaden Research Center,
650 Harry Road, K-17/E-1, San Jose,
CA95120, USA

**Laurent Favennec**
STMicroelectronics, 850 rue Jean Monnet,
38921 Crolles, France

**Olivier Gourhant**
STMicroelectronics, 850 rue Jean Monnet,
38921 Crolles, France

**Meike Hauschildt**
GLOBALFOUNDRIES, Dresden
Module One LLC & Co. KG,
Wilschdorfer Landstrasse 101,
01109 Dresden, Germany

**Michael Hecker**
Center for Complex Analysis,
GLOBALFOUNDRIES, Dresden
Module One LLC & Co. KG,
Wilschdorfer Landstrasse 101,
01109 Dresden, Germany

**Paul S. Ho**
Lab for Interconnect and Packaging,
The University of Texas at Austin, UT-PRC
10100 Burnet Road, Bldg 160,
Mail Code R8650, Austin, TX 78758, USA

**Chao-Kun Hu**
IBM Research Division,
T.J. Watson Research Center,
Yorktown Heights, NY 10598, USA

**Huai Huang**
Lab for Interconnect and Packaging,
The University of Texas at Austin, UT-PRC
10100 Burnet Road, Bldg 160,
Mail Code R8650, Austin, TX 78758, USA

**René Hübner**
Fraunhofer Institute for Non-Destructive
Testing IZFP, Dresden Branch,
Maria-Reiche-Strasse 2, 01109 Dresden,
Germany

**Vincent Jousseaume**
CEA-LETI, MINatec Campus,
17 rue des Martyrs, 38054 Grenoble Cedex 9,
France

**Akio Kawabata**
MIRAI-Selete, 10-1
Morinosato-Wakamiya,
Atsugi 243-0197,
Japan

**Takamaro Kikkawa**
Research Institute for Nanodevices
and Bio Systems, Hiroshima University,
1-4-2 Kagamiyama, Higashi-hiroshima,
Hiroshima 739-8527, Japan

**Els Kesters**
IMEC, Kapeldreef 75, B-3001 Leuven,
Belgium

**John U. Knickerbocker**
IBM Research Division,
Thomas J. Watson Research Center,
Yorktown Heights, NY 10598, USA

**Lay Wai Kong**
College of Nanoscale Science and
Engineering at the University at Albany,
Albany, New York, USA

**Quoc Toan Le**
IMEC, Kapeldreef 75, B-3001 Leuven,
Belgium

**Han Li**
IBM Research Division,
Thomas J. Watson Research Center,
Yorktown Heights, NY 10598, USA

**Sven Niese**
Fraunhofer Institute for Non-Destructive
Testing IZFP, Dresden Branch,
Maria-Reiche-Strasse 2, 01109 Dresden,
Germany

**Mizuhisa Nihei**
MIRAI-Selete, 10-1 Morinosato-Wakamiya,
Atsugi 243-0197, Japan

**Ennis T. Ogawa**
Broadcom Corporation, Irvine,
CA 92617, USA

**Motonobu Sato**
MIRAI-Selete, 10-1 Morinosato-Wakamiya,
Atsugi 243-0197, Japan

**Shintaro Sato**
MIRAI-Selete, 10-1 Morinosato-Wakamiya,
Atsugi 243-0197, Japan

**Denis Shamiryan**
GLOBALFOUNDRIES, Dresden
Module One LLC & Co. KG,
Wilschdorfer Landstrasse 101,
01109 Dresden, Germany

**Hualiang Shi**
INTEL Corporation, Chandler, Arizona,
USA

**Herbert Struyf**
IMEC, Kapeldreef 75, B-3001 Leuven,
Belgium

**Kris Vanstreels**
IMEC, Kapeldreef 75, B-3001 Leuven,
Belgium

**Guy Vereecke**
IMEC, Kapeldreef 75, B-3001 Leuven, Belgium

**Joost J. Vlassak**
School of Engineering and Applied Sciences, Harvard University, Cambridge, MA 02138, USA

**Willi Volksen**
Advanced Organic Materials Group, IBM Almaden Research Center, 650 Harry Road, K-17/E-1, San Jose, CA95120, USA

**Aziz Zenasni**
CEA-LETI, MINatec Campus, 17 rue des Martyrs, 38054 Grenoble Cedex 9, France

**Lijuan Zhang**
IBM System and Technology Group, Hopewell Junction, NY 12533, USA

**Larry Zhao**
Intel Corp., Hillsboro, OR 97124, USA

**Ehrenfried Zschech**
Fraunhofer Institute for Non-Destructive Testing IZFP, Dresden Branch, Maria-Reiche-Strasse 2, 01109 Dresden, Germany

# Preface

Since the 1960s, the density of transistors in integrated circuit (IC) chip has been doubling every 1.5 years. This progression is known as Moore's law, after Gordon Moore, one of the early IC pioneers and founders of Intel Corporation. The semiconductor industry primarily focused on improvements related to the speed of individual transistors and enhanced performance through scaling – by squeezing more transistors into a single IC device. By the early 1990s, the distances between IC components became incredibly small, and the relative effect of the interconnect delay became a greater portion of the overall signal propagation delay. The total resistance ($R$) of the interconnect structure became a significant factor affecting chip performance. At the same time, the capacitance ($C$) between the wires was increasing proportionally to the decreasing spacing between the wires. Both of these factors significantly increased the RC delay of IC circuits. The microelectronics community realized the need to improve the interconnect delay by making changes to the materials used for wires and the materials used to insulate the wires.

Metallic conductivity and resistance to electromigration of bulk copper (Cu) was known to be better than aluminum (Al), which historically has been the dominant interconnect material. Using a lower resistivity interconnect material like Cu decreases interconnect RC delay, which, in turn, increases the IC speed. The transition to Cu as a conductor is one of the most significant changes in semiconductor manufacturing history. The first working microprocessor using copper was made by IBM in 1997 and the process was used in high-volume manufacturing in IBM's facilities from 1998. The next step was related to the creation of the AMD/Motorola strategic alliance in the development of Cu interconnects in July 1998. AMD built a manufacturing plant in Dresden with the purpose of producing copper-only chips in volume IC manufacturing.

Replacement of Al by Cu was an enormous obstacle for the semiconductor industry, since Al is deposited over the entire wafer surface and then patterned by reactive ion etching (RIE). Cu cannot be patterned by RIE because of the very low vapor pressure of the reaction products, and a new process had to be developed. As a result, the damascene process[1] has emerged as the industry standard. The dielectric layer is deposited and patterned first before the metal deposition. Then copper should fill a patterned dielectric by superfilling, which refers to the effect where higher deposition rates are achieved in the bottom of trenches than on the sides, resulting in void-free and seamless filling of trenches and vias with high aspect ratio features, playing a pivotal role in the success of this technology.

However, the introduction of copper was not sufficient for the necessary reduction of RC delay. The 1994 National Technology Roadmap for Semiconductors (NTRS) – the US predecessor of the International Technology Roadmap for Semiconductors (ITRS) – stated

[1]The word damascene is derived from the city Damascus in Syria, where the damascene process was originally developed for jewelry manufacturing.

that materials with a lower dielectric constant would be needed for wire insulation as the feature sizes of IC devices became smaller. The NTRS projected that within 10 years the industry should be able to achieve a standard dielectric constant of less than 1.5 in their production interconnect material. However, the real situation has been much more complicated. After several revisions, the latest edition of ITRS, low-*k* materials with only $k = 2.5$, is expected to be integrated in 2012. The problem is that porous low-*k* materials are generally soft, mechanically weak, and did not adhere well to silicon or metal wires. Further, porous low-*k* materials did not withstand conventional processing (i.e. they degrade during the plasma and chemical processing, crack or delaminate).

Several different types of low-*k* materials were considered as candidates for the reduction of the dielectric constant. Relatively fast developed organosilicate (OSG)-based ultra-low-*k* materials, like XLK, LKD, NCS, etc., met huge challenges during their integration. Low-*k* dielectrics based on organic polymers have low polarizability and, therefore, they are able to provide the lowest *k* value without requiring the introduction of porosity. However, significant efforts to integrate organic materials into ICs have also not been sufficiently successful. In addition to poor mechanical and thermal properties, the most important problems were related to the relatively high coefficient of thermal expansion (CTE) in comparison with other components of integrated circuits.

Additional challenges met during the introduction of copper and low-*k* dielectrics have been related to the high diffusivity of copper ions. To avoid copper diffusion into the dielectric significant effort was made on the development of Cu diffusion barriers. These are conductive barriers isolating Cu and low-*k* in one metallization level and dielectric barriers deposited after chemical mechanical planarization and before starting the next metallization level. The major requirements to the barrier layers are their density and amorphous nature and they do not have to contain any groups oxidizing copper to avoid formation of Cu ions. Many potential barriers have been developed so far and some of them will be analyzed in this book.

However, the efforts of scientists and engineers during the last 15 years have made the Cu/low-*k* technology mature. OSG low-*k* materials with $k > 2.5$ are already at the stage of integration into ICs. The advantage of OSG materials is the similarity of their chemical properties to traditional $SiO_2$, which makes it possible to use traditional technological equipment and chemistries during the integration. Plasma enhanced chemical vapor deposition (PECVD) has been the most important low-*k* deposition method for low-*k* materials with a dielectric constant $> 2.3$. The integrated structures need careful analysis of their reliability as reliability related challenges become more and more important.

This book presents an in-depth overview of the present status, novel developments and new materials and approaches for advanced interconnect technology. It covers a broad range of topics, from physical principles to design, fabrication, characterization and application of new materials for nano-interconnects. This book is intended for postgraduate level students, PhD students and industrial researchers, to enable them to gain an insight into this important area of research.

The chapters included in this book are divided into four separate sections:

Section I, 'Low-*k* Materials', includes four chapters. Chapter 1, 'Low-k Materials: Recent Advances', discusses the present state-of-the-art of low-*k* materials, technological challenges observed during their integration and the present tendencies and methods needed for further scaling and improvement of intrinsic properties of low-*k* materials. Chapter 2, 'Ultra-Low-*k* by CVD: Deposition and Curing', describes presently popular PECVD low-*k* materials with a special emphasis on mechanisms of chemical reactions occurring during the deposition and UV curing. Chapters 3 and 4, 'Plasma Processing of Low-*k* Dielectrics' and 'Wet Clean Applications in Porous Low-*k* Patterning Processes', give an overview of plasma and wet processing developed for integration of low-*k* materials. Special attention is devoted to the reactions mechanisms and degradation of low-*k* materials during exposure in oxidative ambient.

Section II, 'Conductive Layers and Barriers', includes two chapters related to recent progress in copper electroplating for on-chip metallization (Chapter 5) and detailed analysis of Cu diffusion barriers (Chapter 6). The analysis in Chapter 6 is mainly related to the conductive barrier.

Section III, 'Integration and Reliability' includes five chapters: 'Integration and Electrical Properties' (Chapter 7), 'Chemical Mechanical Planarization for Cu–Low-*k* Integration' (Chapter 8), 'Scaling and Microstructure Effects on Electromigration Reliability for Cu Interconnects' (Chapter 9), 'Mechanical Reliability of Low-*k* Dielectrics' (Chapter 10) and 'Electrical Breakdown in Advanced Interconnect Dielectrics' (Chapter 11). This section provides extensive analysis of Cu/low-*k* integration and the possible challenges. Two chapters in this section are related to reliability problems due to electrical characteristics. In addition, the chapters related to CMP and mechanical reliability have also been included in this section.

Section IV, 'New Approaches', includes four chapters. Chapter 12 provides an extensive description on 3D interconnects. It contains two parts: a general overview of this technology and the advanced microscopy used for characterization of 3D interconnects. The last three chapters give overviews of exploratory research of alternative approaches for interconnects. They are related to application of carbon nanotubes as conductors (Chapter 13), optical interconnects (Chapter 14) and wireless interchip interconnects (Chapter 15).

All the chapters included in this book are written by leading experts in fields of materials science and integration of advanced interconnects.

MIKHAIL R. BAKLANOV
*IMEC, Belgium*

PAUL S. HO
*The University of Texas at Austin, USA*

EHRENFRIED ZSCHECH
*Fraunhofer Institute for Nondestructive Testing,*
*Dresden, Germany*

# Abbreviations

| | |
|---|---|
| 1MS | methylsilane |
| 2MS | dimethylsilane |
| 3MS | trimethylsilane |
| 4MS | tetramethylsilane |
| AAS | atomic absorption spectrometry |
| AES | Auger electron spectroscopy |
| AFM | atomic force microscopy |
| ALD | atomic layer deposition |
| APD | avalanche photodetector |
| ARXPS | angle-resolved XPS |
| ATR-FTIR | attenuated total reflectance FTIR |
| ATRP | a-terpinene |
| AWG | arrayed waveguide grating |
| BCHD | bicyclohexadiene |
| BDV | distribution of breakdown voltage |
| BEOL | back-end-of-line |
| BMO | butadiene monoxide |
| BTASE | bis(trialkoxysilyl)ethane |
| BTASM | bis(trialkoxysilyl)methane |
| BTESEN | bis(trialkoxysilyl)ethene |
| BTMSM | bistrimethylsilyl methane |
| BTS | bias-temperature stress |
| C4 | controlled-collapsed-chip connection |
| CBED | convergent beam electron diffraction |
| CCP | capacitively coupled plasma |
| CD | critical dimensions |
| CDO | carbon-doped oxide |
| CHO | cyclohexene oxide |
| CMOS | complementary metal oxide semiconductor |
| CMP | chemical–mechanical polishing |
| CP | carbosilane precursor |
| CPI | chip packaging interaction |
| CPO | cyclopentene oxide |
| CTE | coefficient of thermal expansion |
| CVD | chemical vapour deposition |
| DD | dual damascene |
| DEMS | diethoxymethylsilane |
| DFB | distributed feedback (laser) |

| | |
|---|---|
| DMCPS | decamethylcyclopentasiloxane |
| DMSO | dimethyl sulfoxide |
| DSP | downstream plasma |
| *E* | Young's modulus |
| EB | electron beam |
| EBSD | electron backscatter diffraction |
| ED | electron diffraction |
| EDXS | energy dispersive X-ray spectroscopy |
| EELS | electron energy loss spectroscopy |
| EFTEM | energy-filtered TEM |
| ELK | extreme low-*k* |
| ELP | electroless plating |
| EM | electromigration |
| EP | ellipsometric porosimetry |
| ERR | energy release rate |
| FACVD | filament-assisted chemical vapour deposition |
| FD | framework density |
| FEOL | front-end-of-line |
| FIB | focused ion beam |
| FM | frequency modulation |
| FOUP | front opening universal pod |
| FP | Fabry–Perot (cavity) |
| FTIR | Fourier transform infrared (spectroscopy) |
| *G* | fracture resistance |
| GB | grain boundaries |
| GDOES | glow discharge optical emission spectroscopy |
| GISAXS | grazing incidence small-angle X-ray scattering |
| *H* | hardness |
| HBPCSO | hyperbranched polycarbosiloxane |
| HMDS | hexamethyldisilazane |
| HPC | high-performance computing |
| HPLC | high-performance liquid chromatography |
| HSSL | high suppression strength levelers |
| I/O | input–output |
| ICP | inductively coupled plasma |
| iCVD | initiated CVD |
| ILD | interlayer dielectric |
| ITRS | International Technology Roadmap for Semiconductors |
| KD | Kikuchi diffraction |
| LSV | linear sweep voltammograms |
| Me | methyl |
| MEA | monoethanolamine |
| MEL | zeolite with a two-dimensional 10-ring pore structure |
| MEMS | microelectromechanical systems |
| MFI | zeolite with a two-dimensional 10-ring pore structure and pore size 5.5 Å |

| | |
|---|---|
| MMR | material removal rate |
| MNL | molecular nanolayers (self-assembled) |
| MP | methyl-2-pyrrolidone |
| MPS | 3-mercaptopropylsulfonate |
| MSSQ | methylsilsesquioxane |
| MTMS | methyltrimethoxysilane |
| MZI | Mach–Zehnder interferometer |
| NBD | norbornadiene |
| NBE | norbornene |
| NBECVD | ncutral-bcam-cnhanccd CVD |
| NBED | nanobeam electron diffraction |
| NMR | nuclear magnetic resonance |
| OF PSZ | organic functionalized PSZ |
| OnoC | optical network on a chip |
| OOK | on–off keying (modulation format) |
| OPC | optical proximity correction |
| OPL | organic planarization layer |
| ORP | oxidation–reduction potential |
| OSG | organo-silicate glass |
| PALS | positronium annihilation lifetime spectroscopy |
| PC | propylene carbonate |
| PCBO | post-CMP bake-out |
| PCG | planar curved grating |
| PEALD | plasma-enhanced atomic layer deposition |
| PEBO | post-etch bake-out |
| PECVD | plasma-enhanced chemical vapour deposition |
| PEG | polyethylene glycol |
| PID | plasma-induced damage |
| PPG | polypropylene glycol |
| PR | photoresist |
| PS | pitch splitting (of a metal layer) |
| PSZ | pure silica zeolites |
| PVD | physical vapor deposition |
| RIE | reactive ion etching |
| SAED | selected area electron diffraction |
| $ScCO_2$ | supercritical $CO_2$ |
| SEM | scanning electron microscopy |
| SIMS | secondary ion mass spectroscopy |
| SIV | stress-induced voiding |
| SOI | silicon on insulator |
| SP | spacer patterning |
| SPS | bis(3-sulfopropyl) disulfide |
| TDDB | time-dependent dielectric breakdown |
| TDESC | 1,3,5-tris(diethoxysila)cyclohexane |
| TDS | thermal desorption spectroscopy |
| TE | transverse electric field (polarization) |

| | |
|---|---|
| TEM | transmission electron microscopy |
| TEOS | tetraethyl-*ortho*-silicate |
| TGA | thermogravimetric analysis |
| TIA | transimpedance amplifier |
| TM | transverse magnetic field (polarization) |
| TMAH | tetramethylammonium hydroxide |
| TMCTS | tetramethylcyclotetrasiloxane |
| TMOS | tetramethyl-*ortho*-silicate |
| TPAOH | tetrapropylammonium hydroxide |
| TSV | through-silicon via |
| TVMOS | trivinylmethoxysilane |
| TVS | triangular voltage sweep |
| ULK | ultra-low-*k* |
| UV | ultraviolet |
| UVSE | UV spectroscopic ellipsometry |
| V3D3 | trimethyltrivinylcyclotrisiloxane |
| V4D4 | tetravinyltetramethylcylclotetrasiloxane |
| VCSEL | vertical cavity surface emitting laser |
| VTMOS | vinyltrimethoxysilane |
| VTMS | vinyltrimethylsilane |
| VUV | vacuum ultraviolet |
| WDM | wavelength division multiplexing |
| XPS | X-ray photoelectron spectroscopy |
| XRD | X-ray diffraction |
| XRF | X-ray fluorescence |
| XRR | X-ray reflectivity |
| ZLK | zeolite-inspired low-*k* |

# Section I

## Low-*k* Materials

# 1

# Low-$k$ Materials: Recent Advances

*Geraud Dubois and Willi Volksen*

*Hybrid Polymeric Materials Group, IBM Almaden Research Center, San Jose, California, USA*

## 1.1 Introduction

From a historical point of view, the search for new low-$k$ and ultra-low-$k$ materials has always been dictated by industrial needs, resulting in a strong connection between fundamental research and technology. Although a variety of potential candidates has been reported in the literature during the past decade, integration is the deciding factor driving the selection of the most promising materials for a given technology node [1]. For instance, the industry flirted briefly with organic polymers [2–7]. Besides the anticipated thermal stability challenges, this class of materials, known to be tough and crack-resistant, had other issues, such as softness, large coefficient of thermal expansions (CTE) and delamination. While most of these problems were mitigated by material reengineering and chip redesign, solutions occurred too late, relegating the organic polymers to possible hybrid build applications [8]. For the new low-$k$ materials, it was expected that their electrical and mechanical properties would be comparable to those of silicon dioxide [9, 10], the insulating material of reference for the quite conservative semiconductor industry. Chemical modification of the silicon network, first by introduction of fluorine and eventually by the addition of carbon, was then adopted as the lower risk path to introducing low-$k$ materials in the back-end-of-the-line (BEOL) [1]. At that time, silicates and

*Advanced Interconnects for ULSI Technology*, First Edition. Edited by Mikhail R. Baklanov, Paul S. Ho and Ehrenfried Zschech.

organosilicates emerged as the dominant candidates. They can be deposited by both spin-on and chemical vapor deposition (CVD) processes. However, the final thin-film properties are in large part dictated by the chemical structure of the deposited film, a result of both the precursor or resin chemistry and the deposition process. From before 1997 until now (Figure 1.1), plasma-enhanced chemical vapor deposition (PECVD) has been the method of choice for depositing silicon dioxide ($SiO_2$), fluorine-doped oxides (F-$SiO_2$), carbon-doped oxides (SiCOH: elementally descriptive but not representing the stoichiometry) and porous carbon-doped oxides (p-SiCOH). As illustrated in Figure 1.1, PECVD SiCOH materials were successfully implemented in IBM microprocessors: at the 90 nm ($k = 3.0$) and 65 nm ($k = 2.7$) technology nodes in 2004 and 2006, respectively. PECVD materials containing additional porosity (p-SiCOH) appeared for the first time in high-volume manufacturing in 2008 (Figure 1.1).

Interestingly, the real IBM technology node/dielectric constant relationship differs significantly from the targets proposed by the NTRS in 1997. This shows the danger of predicting dielectric targets based solely on engineering designs when new materials are not readily available. The challenges in designing dielectric insulators that meet all the BEOL requirements (electrical, thermal and mechanical) have been the source of many publications over the last 15 years [1]. Among them, many excellent reviews have been published, addressing the different aspects of these dielectric materials: physical properties [7, 11–13], integration requirements and challenges [14, 15], characterization [9, 16–19] and chemistry [1]. For advanced and future technologies, the addition of porosity at levels necessary to obtain dielectric constants of 2.4 and beyond has exacerbated already known integration issues [14]. In particular, the processing induced damage and the decrease in material mechanical properties present serious concerns for the reliability of these advanced structures. In this chapter, we will first define the main integration challenges currently encountered. Then we will discuss proposed solutions to these issues by both alternative processing approaches and development of advanced novel materials.

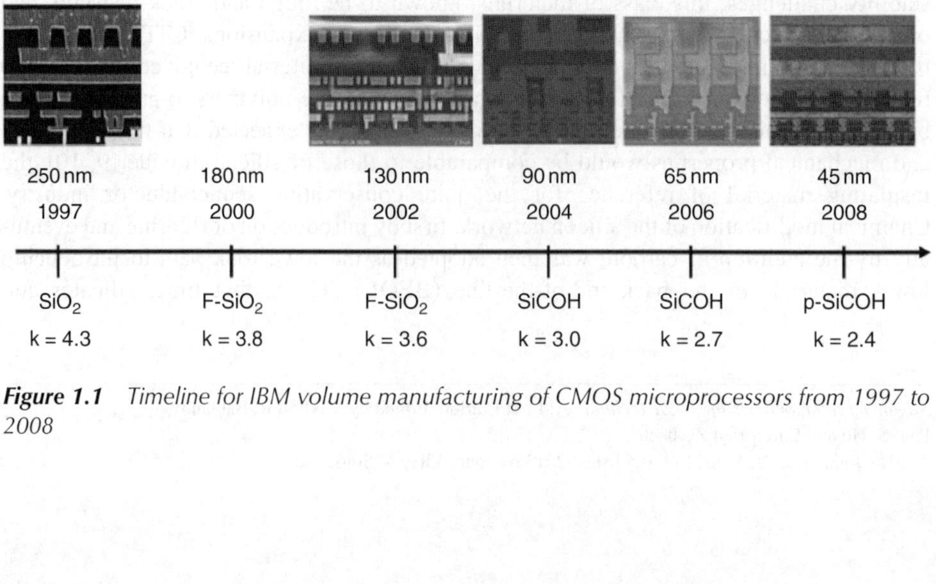


*Figure 1.1* *Timeline for IBM volume manufacturing of CMOS microprocessors from 1997 to 2008*

## 1.2 Integration Challenges

The design of low-*k* dielectric materials with the requisite properties needed for implementation into modern semiconductor manufacturing processes is an extremely challenging task. In large part, this is a consequence of trying to adapt $SiO_2$-based processes to carbon containing low-*k* dielectric materials with or without additional porosity. Whereas the introduction of carbon into the inorganic network can provide an initial lowering of the dielectric constant, addition of porosity is required to achieve ultra-low-*k* (ULK) performance.

The introduction of carbon and the addition of porosity serve merely in the process of lowering the dielectric constant and nothing more. However, along with this single beneficial attribute come two major drawbacks. Firstly, the increased, accessible surface area of the porous network constitutes a new weakness with regard to plasma-induced damage and wet chemistry processes. Secondly, the decrease in network connectivity as compared to silica and the reduction of density strongly impact the dielectric insulator mechanical properties.

The comparative relationship of the interlayer-dielectric (ILD) mechanical properties and ILD process-induced damage as a function of porosity and materials chemistry is presented in Figure 1.2. It should be mentioned that these trends are qualitative in nature only. Nevertheless, they clearly demonstrate the ever-growing challenges in successfully building advanced semiconductor structures with increased performance and good reliability.

The process-induced damage is primarily reflected in deterioration of the dielectric constant, leakage current and breakdown voltage. It can also affect the mechanical properties of the integrated structure, although these are more dependent on the initial materials characteristics. The Young's modulus (*E*), hardness (*H*) and adhesive/cohesive fracture

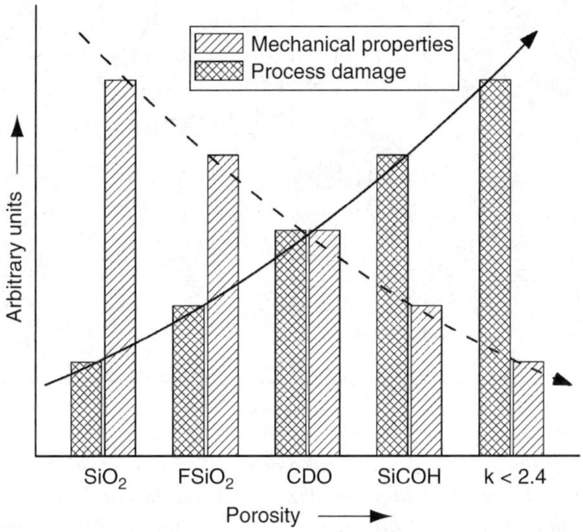


***Figure 1.2*** *Comparative relationship of ILD mechanical properties and ILD process-induced damage as a function of porosity and materials chemistry*

resistance ($G$) of future ultra-low-$k$ materials have reached a level where the mechanical integrity of the device is of major concern.

### 1.2.1 Process-Induced Damage

In order to identify and better visualize key integration issues as encountered in a typical dual damascene (DD) 'via-first' approach [20], the reader is referred to Figure 1.3. Here the critical integration areas are arranged in alphabetical order, following a sequence that might be encountered in a typical process flow. For the sake of simplification, the integration processes are further broken down into pre- and post-metallization steps.

In the pre-metallization portion of the 'via-first' integration approach, the first problem areas encountered are a result of the introduction of new interfaces. As with many interfaces, adhesion at such boundary layers is of tremendous importance. The first interface results from the ILD deposition on a capping layer, which serves to protect the underlying copper metal (see Figure 1.3 (a)). Capping layers composed of $SiN_x$, $SiC_xN_y$ and $SiC_x$ are commonly employed and present surface chemistries considerably different from those based on $SiO_x$ [21–25]. Coupled with the fact that adhesive properties scale inversely with porosity [26, 27], this combination may lead to severe delamination during chemical–mechanical polishing (CMP) and chip–package interaction (CPI). To prevent such a catastrophic failure, the capping layer surfaces are modified by plasma treatment

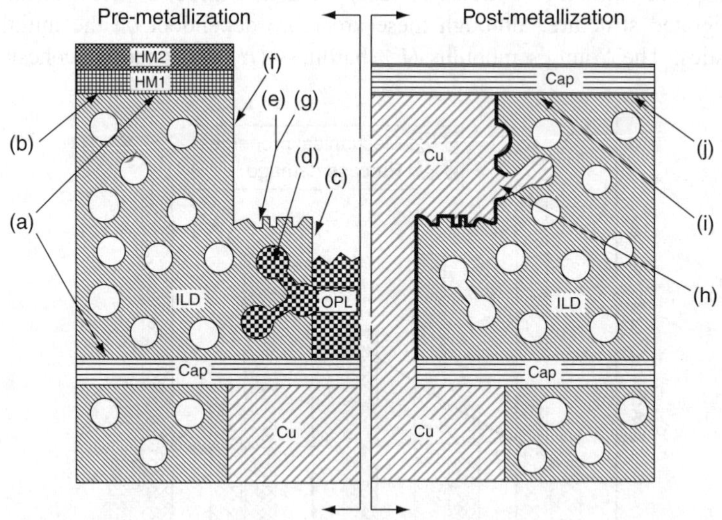


*Figure 1.3 Simple, two-layer dual-damascene cross-section illustrating various process-related problem areas before and after metallization: (a) adhesion failure, (b) ILD plasma damage, (c) sidewall ILD damage from via-etch and PR strip, (d) OPL penetration during via-fill, (e) LBR and pitting from uneven etch front, (f) sidewall ILD damage from line etch and PR strip, (g) exacerbated LBR and pitting due to cap-open, (h) discontinuous barrier layer due to large, interconnected pores, (i) ILD damage from CMP, (j) Cu pre-clean/cap deposition plasma damage*

or, as in the case of CVD-deposited porous ILDs, by employing a gradient layer at the interface [28]. Dense ILD is deposited at the immediate cap/ILD interface, ensuring good adhesion by eliminating the effect of porosity and then continuously increasing the porosity of the ILD until the target $k$ is reached. Other new interfaces are created during hardmask deposition, which usually consists of two layers, HM1 and HM2, *vide infra* [29–31]. The function of the hardmask is to prevent direct contact of the photoresist (PR) with the ILD in order to eliminate the absorption of basic species, such as atmospheric amines, by the highly porous ILD, which leads to poisoning of the highly sensitive chemically amplified photoresists [32–34]. An additional function of the hardmask is to allow for PR rework.

The PR rework process typically involves the use of $O_2$ plasmas, requiring the use of $SiO_2$ as a hardmask (HM2). However, since $SiO_2$ plasma deposition processes severely damage porous, low-$k$ ILDs, an intermediate layer (HM1) is deposited prior to the $SiO_2$ layer. For HM1, plasma deposition conditions are selected that are considerably less damaging to the porous ILD, yet provide sufficient adhesion during ensuing CMP processes (Figure 1.3 (a) and (b)) [35].

Next, via-structures are lithographically patterned and etched into the porous ILD employing $C_xF_y$-based plasmas [36]. These etch processes, initially developed for $SiO_2$ or other dense materials, tend to damage the via-sidewalls (Figure 1.3 (c)) [37–39]. By carefully balancing the polymerization and etch reactions of the plasma process, less damaging etch processes are possible. However, plasma damage to the via-sidewall is not limited to the etch step. In fact, the majority of the sidewall damage can be attributed to the strip (ash) of the remaining photoresist. Whereas the etch plasma damage to the via-sidewall is primarily confined to the surface, a PR plasma strip may induce damage that extends far into the bulk of the ILD and is a function of the overall porosity, pore size and pore interconnectivity.

Following patterning of the via-structure, the line level has to be lithographically defined. In order to allow for uniform coverage of the photoresist on top of the via-structures, the topography has to be planarized. This leveling of the topography is achieved in the so-called via-fill step employing a low molecular weight organic planarizing layer (OPL) [40–42]. The low molecular weight allows for high solids content formulations with correspondingly low viscosities, providing for the ability to fill narrow spaces and to planarize. This ability to fill narrow spaces may present a drawback for porous ILDs with larger and interconnected pores. Here, the OPL may actually penetrate into the ILD (Figure 1.3 (d)). Furthermore, since OPL polymers are generally polydisperse, the penetration problem is exacerbated by the lowest molecular weight species from the low end of the molecular weight distribution. The issue of OPL penetration is just starting to emerge as $k$ values are driven to ever lower values. Hopefully, solutions to this problem can be found in a new generation of OPL candidates with properly chosen molecular weight characteristics, by designing high-porosity materials with small, noninterconnected pores or by pore sealing approaches.

Once the structure has been planarized using an OPL, deposition of a low-temperature plasma-deposited oxide layer follows, which allows rework of the PR deposited in the next step. PR deposition is then followed by lithographic patterning and etching the line levels into the ILD, leaving a small plug of OPL at the bottom of the via-structure. This protects the bottom capping layer and is subsequently removed during the PR strip.

However, since the line does not extend all the way to the bottom of the ILD layer, line bottom roughness (LBR) due to a nonhomogeneous etch front may be observed (see Figure 1.3 (e)) [43].

The LBR can then be further exacerbated along with additional sidewall damage to both the line and via, as shown in Figure 1.3 (c), (e) and (f), from the PR strip/OPL clean-out step, which typically requires oxidative plasma conditions. LBR appears to be strongly affected by ILD porosity and pore interconnectivity, even leading to pitting of the line bottom at high porosity levels.

Once the line/via-structures have been defined, another etch step is required to open the capping layer, allowing for the eventual connection of the new line/via level to the underlying metal level. The nature of the cap material dictates plasma etch conditions of this cap-open step, which are even more aggressive, thus exposing the line bottom to further damage (see Figure 1.3 (g)) [44]. In order to mitigate the plasma damage and to remove any plasma residues from the various exposed surfaces, wet-clean processes using dilute HF (DHF) are often employed [45–47]. This treatment ensures good contact between the via and the underlying copper metal as well as removing hydrophilic sidewall damage resulting from the various process plasmas. In the case where sidewall damage was very severe, the wet-clean process can potentially remove so much material as to significantly change the critical dimensions of the line/via-structures.

After the line/via-structures have been defined and the via bottom has been opened to the underlying metal layer, metallization is performed. The metallization portion of the process consists of two steps:

1. barrier layer deposition and
2. Cu plating.

The barrier is generally a bilayer consisting of a metal nitride and a barrier metal to protect the ILD from Cu migration/penetration and typical examples are TaN/Ta and TiN/Ti. The function of the metal nitride is to provide excellent adhesion to the ILD, whereas the barrier metal provides an optimum interface to the Cu. For maximum effectiveness, this barrier layer needs to be continuous and without defects and pinholes. For highly porous ILDs, this requires the use of relatively thick barrier layers to seal off the pores, preventing Cu intrusion into the ILD, as shown in Figure 1.3 (h) [48, 49]. However, as line and via dimensions shrink in accordance with future technology nodes, thinner barrier layers are needed to maintain or improve electrical performance. This requirement presents a major challenge for highly porous materials and may, in fact, limit the extendibility of porous ILDs to dielectric constants $>2.0$. Next, copper is deposited into the patterned dual-damascene structure and the Cu overburden including the bilayer hardmask is polished back to the ILD using chemical mechanical polishing (CMP). Here, high mechanical forces and the chemical nature of the CMP slurry may adversely affect the stack in the form of adhesive failure at the various interfaces, cohesive failure of the ILD [50–52] and adversely impacting the electrical properties [53] (see Figure 1.3 (i)). Proper selection of the CMP slurries in addition to low down-force CMP processes can significantly mitigate these problems.

Finally, the last step in the dual-damascene process involves the deposition of a capping layer to protect the oxidatively sensitive Cu metallurgy, so that the entire process sequence can be repeated over and over again.

### 1.2.2 Mechanical Properties

For traditional PECVD and spin-on materials, the mechanical strength of ultra-low-*k* films decreases as the dielectric constant is reduced. Consequently, their stiffness and resistance to cohesive or adhesive fracture is strongly affected. In addition, the number of ULK levels usually increases from one technology node to the next to minimize the BEOL $RC_{delay}$. On the other hand, the stiffness of controlled-collapsed-chip-connection (C4) solders has increased with the switch to lead-free materials and ceramic carriers have been replaced by organic substrates (with a higher coefficient of thermal expansion (CTE) and lower Young's modulus). Altogether, more thermomechanical stress is applied to the BEOL levels whereas the overall fracture resistance of the structure has been reduced. This conflicting trend in chip–package materials is summarized in Table 1.1 [54, 55].

It is very interesting to note in Table 1.1 that the preferred evolution of BEOL and packaging characteristics is the mirror image to what is currently developed in the semiconductor industry. It is then not surprising that critical reliability issues have been encountered during chip–package interaction (CPI) [56, 57]. Indeed, the thermal deformation of the packaging structure is directly coupled into the BEOL interconnects, creating large stresses at the ultra-low-*k* levels that can induce materials fracture and/or delamination [58]. A good example of such a failure is presented in Figure 1.4 and is known as the 'white bump' issue. These 'white bumps' are observed by acoustic wave spectroscopy (Figure 1.4(a)) after the chip join reflow processing once the flip-chip ball interconnects have solidified [58]. The presence of these 'white spots' indicates that fracture (adhesive or cohesive) occurred

***Table 1.1*** *Conflicting trend in chip–package material*

| | BEOL | | Packaging | | Substrate flexibility |
|---|---|---|---|---|---|
| | ULK Young's modulus | ULK cohesive strength | ULK levels | C4 stiffness | |
| Preferred | ↑ | ↑ | ↓ | ↓ | ↓ |
| Reality | ↓ | ↓ | ↑ | ↑ | ↑ |

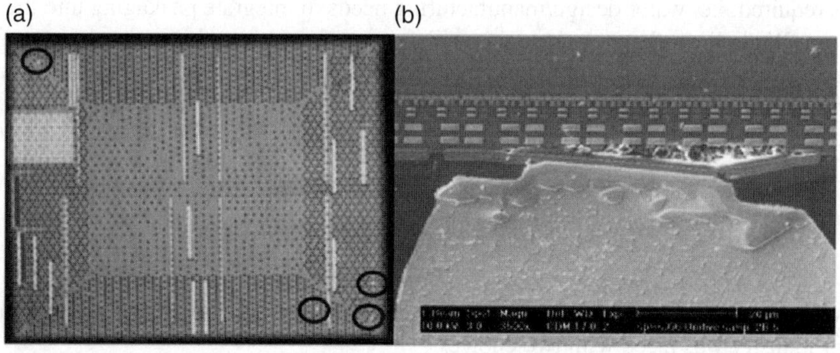

***Figure 1.4*** *(a) Acoustic wave spectroscopy image of white bumps. (b) Cross-section of a white bump*

underneath the solder ball, usually at the ULK levels (Figure 1.4(b)). It has also been observed that these 'white bumps' were generated during the thermal testing of packaged dies (microprocessors must pass a thousand cycles from sub 0 °C to greater than 100 °C).

Since this failure mechanism is only observed after packaging and thermal testing the cycle of trial and error can be long and expensive. At the same time, new materials are also introduced in the BEOL and at the packaging level for each technology node. Consequently, the identification of the main factors involved with CPI failures in Cu/low-*k* interconnects is a daunting task. In that regard, finite element based modeling coupled with fracture mechanics has been of great help in pinpointing these determining factors. Early on, it was found using three-dimensional (3D) finite element analysis based on a multilevel submodeling approach that for a stand-alone chip, the energy release rate (ERR) induced during processing was lower than the critical ERR for interfacial delamination obtained from a four-point bend test [59]. After the die is assembled in a flip-chip package, a significant increase of this ERR was obtained for the interconnect structure interfaces parallel to the die surface. In a two-metal layer interconnect structure, the ERR was shown to increase rapidly when the modulus of the ILD is lower than 10 GPa [60]. In a more complex structure, the interface exposed to the highest ERR varies as a function of the materials (TEOS, low-*k* and ULK) used at each interconnect level [60]. For a full interconnect structure based on the 65 nm technology node, the ERR increased with increasing crack length. A similar observation has been made by different authors, confirming that the crack feeds from the energy supplied by the package until a steady state is reached [61].

The packaging challenges for ultra-low dielectric materials may be summarized as follows [62]:

- die separation process = dicing,
- underfill selection for ULK and bump protection,
- lead-free bump, with higher stress on ULK levels and
- fine-pitch bump application.

If not addressed, the mechanical issues for BEOL interconnect and packaging will lead to more and more difficulties in successfully passing reliability testing when moving from one technology node to the next. Traditionally seen as two independent sectors, BEOL and packaging must now face the same challenge and a common effort of these two communities is required, i.e. wafer design/manufacturing needs to integrate packaging into the total solution. We believe that opportunities will emerge from a combination of new gentler integration processes, toughening post-treatments, development of stiffer and more crack-resistant advanced ULK materials, implementation of more compliant packaging materials and novel BEOL and packaging designs [55, 62].

## 1.3 Processing Approaches to Existing Integration Issues

Considering all the integration challenges described in the previous section, the semiconductor industry was faced with two choices:

- develop new processing/engineering solutions or
- design advanced dielectric materials.

Obviously, for a conservative industry, the former represented the more prudent and readily implementable approach. Therefore, much of the initial attention was focused on this area.

In general, these processing/engineering solutions can be divided into three distinct categories:

1. Post-deposition treatments to enhance the physical properties of highly porous dielectric materials.
2. Prevention or repair of plasma-induced processing damage.
3. Use of multilayer structures comprised of an organic dielectric and an organosilicate.

### 1.3.1 Post-deposition Treatments

The deposition of organosilicate-based dielectric materials, as practiced in the semiconductor industry, typically involves a thermally activated process either during or immediately following the deposition step. The thermal portion of this process step is dictated by the need to decompose the thermally labile pore generator (porogen) and to ensure optimum network formation. Both requirements are favored by high curing temperatures, typically in excess of 400 °C. Network formation and hence mechanical properties greatly benefit from high curing temperatures, as shown by laser spike annealing studies of spin-on organosilicate films [63]. Keeping in mind that for microelectronic applications, temperatures of 400–425 °C are considered an upper limit and curing times are only on the order of several minutes, dielectric films are most likely to be undercured and network formation is not optimized. Although this may not have presented a problem for dense dielectric materials, it is significantly more critical for highly porous coatings with commensurately lower modulus and hardness values. In order to address these weaknesses, thermally assisted e-beam and ultraviolet (UV) curing processes were auditioned. The goal was to facilitate the complete porogen removal at temperatures ≤400 °C and to boost mechanical properties as a result of the known hardening effect of these treatments [64–69].

Of the two radiation processes, e-beam treatments were investigated first, but when it was found that UV irradiation was nearly as effective in promoting complete porogen decomposition and mechanical properties enhancement, the former was dropped due to perceived detrimental effects to the front-end devices [70]. Overall, both processes are similar in that optimum results are generally obtained at 400 °C with exposures typically ranging from 3 to 10 minutes [71–73], the primary difference being that e-beam exposure appears to be more directional as compared to broadband UV exposure [67]. Furthermore, both treatments are equally effective in promoting porogen decomposition at 400 °C [67, 68]. For example, XPS data obtained for porogen containing PECVD organosilicate films as a function of UV exposure at 400 °C indicates a rapid decrease in the C/Si ratio during the first 10 minutes of UV exposure, followed by a significantly lower rate of carbon loss [74]. The initial rapid carbon loss was attributed to porogen decomposition, while the slower change in C concentration was associated with other carbon depleting changes, *vide infra*. However, completeness of the porogen decomposition is not always readily assessable and the presence of porogen residues can persist even after an extended thermally assisted UV cure [75].

For the remainder of this discussion the main focus will be on UV post-deposition treatment, unless there is a specific call for e-beam examples. In addition, primary consideration will be given to broadband UV radiation ($\lambda = 200$–$400$ nm) over monochromatic UV radiation ($\lambda < 200$ nm) due to the fact that the former is better suited for porogen removal and mechanical enhancement [76], less damaging to the dielectric material [77], less prone to standing wave effects [78, 79] and the preferred spectral range of presently used manufacturing tools.

Spectroscopic examination by Fourier transform infrared (FTIR) and $^{29}$Si-NMR of organosilicates exposed to thermally assisted radiation treatments provides a relatively consistent picture of the structural changes taking place. The main observed structural change is an increase in the network connectivity via additional condensation reactions. This is most clearly demonstrated when comparing the $^{29}$Si-NMR spectra of pristine and UV-treated films, which exhibit an increase in *Q* (Si bonded to four oxygens) and *T* (Si bonded to three oxygens) moieties of 7% and 4%, respectively, after UV exposure [72]. The increase in *T* and *Q* occurs at the expense of *D* (Si bonded to two oxygens) and the complete disappearance of *M* (Si bonded to one oxygen) structural units. In addition, there is a decrease in *T* and *D* structural units with pendant silanol groups. The overall effect is an increase in the degree of condensation, i.e. the degree of cross-linking or network connectivity. Similar and consistent observations were reported by other groups [67, 80]. Complimentary FTIR studies provided additional information and subtleties not apparent from the nuclear magnetic resonance (NMR) experiments. Thus, radiation treated films exhibit a definite change in the Si–O–Si bond arrangement reflected in the 1200–1000 cm$^{-1}$ spectral range. There is a definite decrease in the 1140 cm$^{-1}$ absorption (assigned to the Si–O–Si cage network structure or large bond angle structure) and a concomitant increase in the absorption at 1040 cm$^{-1}$ (assigned to the Si–O–Si small bond angle structure and suboxide). It is not certain whether this structural rearrangement in the Si–O–Si backbone structure results in an overall increase in network connectivity, but the more stable (140–144° bond angle) bonding configuration is expected to increase the overall rigidity of the network and hence Young's modulus. Along with these major structural changes, there are smaller observed differences, such as small decreases in the Si–$CH_3$ absorptions at 2975 cm$^{-1}$, 2910 cm$^{-1}$ and 1275 cm$^{-1}$ and the appearance of Si–H absorptions at 2240 and 2180 cm$^{-1}$. Based on these observations it is possible to postulate a series of potential reactions that would account for the spectroscopic data as shown in Figure 1.5. Of these reactions, those represented by pathways (a) to (c) are consistent with experimental observations, with pathway (a) being the most dominant and the other two playing a lesser role. Pathways (d) and (e) lead to carbon-bridged Si-network structures. Although reactions (d) and (e) cannot be totally ruled out, there is a strong possibility that these may arise from residual porogen fragments left in the organosilicate matrix [67, 81]. Finally, reaction pathway (f) has been proposed based on the observed production of Si–H bonds in both UV and e-beam treated organosilicate samples [67, 68, 81]. This reaction is more predominant in e-beam treated samples. However, the relative amount of Si–H formation during UV or e-beam exposure is hard to quantify since this moiety is rather reactive and could be readily consumed by reaction pathway (c). Nevertheless, the presence of residual Si–H bonds is not desirable and presents another reason why broadband UV exposure, with little or no Si–H formation, is preferred.

With an understanding of the underlying structural changes accompanying UV or e-beam exposure of organosilicates, it is now possible to elucidate the observed changes in

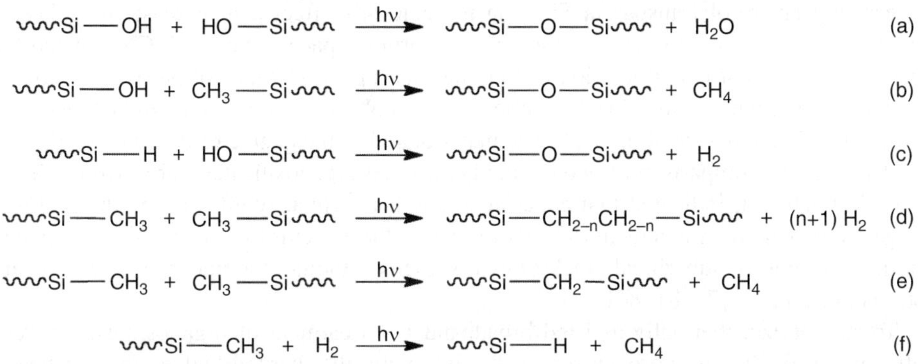


***Figure 1.5*** *Possible structural modifications iinduced by e-beam or UV exposure of organosilicate materials*

mechanical properties. In general, broadband UV exposure at 400 °C and exposure times on the order of less than 10 minutes lead to a dramatic improvement in Young's modulus (40–50 % increase) while typically slightly lowering the dielectric constant [68, 71, 72, 79]. However, treatment of multistack structures appears less effective and dictates the treatment of individual layers for optimum results [71]. As shown in Figure 1.5, reactions (a) to (c) increase network connectivity, increasing $E$ while consuming pendant Si–OH groups, which leads to a decrease in $k$ due to lower polarizability of the matrix and less propensity to absorb atmospheric moisture. Futhermore, the organosilicate bond rearrangement of UV exposed samples contributes to additional stiffening of the matrix and a concomitant increase in $E$. This contribution, although not obvious, was elegantly demonstrated by comparing a PECVD organosilicate resin, resin A, with a thermally cured spin-on organosilicate resin, resin B [82]. Resin A, prior to UV exposure, was characterized by low network connectivity, i.e. the presence of significant Si–OH groups. UV cure of this resin exhibited a significant increase in $Q$ groups of the Si-NMR spectrum and little or no Si–O–Si bond rearrangement as reflected in the FTIR spectrum. In contrast, resin B showed high network connectivity, as reflected by the lack of $M$ and $D$ groups in the Si-NMR spectrum. This was not completely unexpected since this resin had been subjected to an initial thermal cure. UV exposure of resin B yielded little or no increase in $Q$ groups, but a noticeable decrease in the large-angle Si–O–Si structure was evident from the FTIR spectrum. Given that both materials exhibited a similar improvement in Young's modulus, the connection between mechanical enhancement by an increase in network connectivity and/or Si–O–Si bond rearrangement could be made. Furthermore, a decrease of Poisson's ratio as a function of UV cure for organosilicate films suggests that increases in the shear modulus dominate over increases in the bulk and/or compressive modulus [82].

Besides the improvement in the stiffness of UV-treated organosilicate materials, the impact of UV curing on the fracture energy was also studied. In this connection, it was shown that while adhesive fracture energies at the top interface can improve by as much as 200 %, cohesive fracture energies and bottom adhesive fracture energies improve at best by 20 %. This suggests a depth dependence of the UV cure. Indeed, studies using monochromatic UV radiation, where the depth dependence is significantly more severe, indicate UV light interference, which forms a standing wave during the curing process [78, 79]. While underlying layers of transparent SiN can mitigate this effect and increase the adhesive fracture energy at the bottom interface, UV absorbing spacers such as SiCN negated the beneficial effects of UV irradiation at both interfaces [83]. The lack or presence of contributions by thermally assisted UV treatment on the cohesive strength may in fact be due to differences in the nature of the dielectric material and/or the degree of initial network connectivity. Thus, comparison of two distinctly different organosilicates, subjected to identical UV treatment, indicated that cohesive strength is primarily affected by the Si–O–Si large-angle bond rearrangement [82]. Furthermore, there seems to be no significant change in the threshold or subcritical crack growth velocity although the underlying reasons are still being debated [73, 81, 84].

To summarize, thermally assisted broadband UV treatment of organosilicates is very effective in facilitating porogen removal, minimizing the thermal budget and increasing Young's modulus. This enhancement of the mechanical properties is a prerequisite in being able to go to higher porosities (lower $k$ values) and in turn extending the technology to future device generations. It should be noted that along with the beneficial effects of such post-deposition treatments on the dielectric insulator there might also be less understood implications with respect to underlying structures, such as capping layers, etc. Obviously, more studies are necessary in this area.

### 1.3.2 Prevention or Repair of Plasma-Induced Processing Damage

As the dielectric constant of the ILD is pushed to lower values, larger and highly interconnected pores are created, raising the following concerns:

- access of reactive plasmas deeper into the ILD,
- OPL penetration into the ILD and
- noncontinuous barrier layer coverage at the sidewall leading to metal intrusion.

Of course, the majority of these problems could be addressed by generating the porosity after the DD structure build is completed. This has been proposed in a 'solids first' or 'porosity last' approach [85–89]. In these schemes the porosity is created either after etch (PEBO) or after Cu CMP (PCBO), the latter requiring the use of a permeable cap. Furthermore, the PCBO approach is more desirable based on the fact that it protects from plasma damage, minimizes liner discontinuities and eliminates metal intrusion. Either way, considerable improvements in reliability and downscaling of the metal barrier could be realized. However, these advantages are in the least partially negated by large ILD shrinkage (~ 10 %) upon porogen burn-out. As a result, this approach has not been pursued any further.

Another alternative is to seal the surface of the porous interfaces. This approach addresses barrier metal coverage and prevention of metal intrusion, but has no effect in mitigating plasma damage. Nevertheless, two potential pathways have been explored. The first pathway interposes a thin, continuous layer between the porous ILD and the barrier metal [90, 91]. This material could be deposited by various means and both organic and inorganic candidates have been proposed [25, 48, 49, 92–95]. Furthermore, since this material contributes to the overall dielectric constant, candidates with relatively low $k$ values are preferred.

Potential drawbacks to this approach present themselves in the introduction of new material interfaces and the need to remove the additional material at the via bottom. Thus, the former could give rise to interfacial adhesion issues, while the latter could adversely affect the cap-open step to establish metal contact.

The second pathway to pore sealing is by plasma modification. Proper selection of a suitable plasma may densify the outermost surface of the via and line sidewalls, enabling deposition of thin, continuous barrier layers [96–101]. As it turns out, this approach is feasible for microporous dielectric materials, i.e. small pores with little or no interconnectivity, but fails for mesoporous materials. Due to the high interconnectivity of mesoporous materials, plasma modification is not merely restricted to the surface but penetrates deep into the bulk of the material. Here dielectric damage appears to dominate with only limited pore-sealing success.

Whereas these two modifications address the issue of smooth sidewalls for optimum barrier metal deposition, they do not consider the plasma damage to the ILD from etch and strip processes. Although strip (resist removal) damage can be minimized with the use of hardmasks (the lithographic pattern is first transferred to the hardmask, which allows a PR strip without exposing the ILD) [29, 30, 102–105], plasma damage arising from the actual ILD etch cannot be neglected. This damage manifests itself in the form of oxidative degradation of the ILD and as a result produces silica-based oxides and hydroxides at the exposed surfaces. The presence of the latter is particularly detrimental to the dielectric constant due to its propensity to absorb moisture. In addition, the effect of the oxidative ILD damage on the mechanical properties is still uncertain.

Given that plasma damage is hard to prevent, the alternative is to repair this damage. In this respect, two techniques have emerged. The first technique tries to make use of a hydrogen plasma treatment, which converts the damaged layer to a denser layer of Si–H rich material [106–109]. In addition, the porous matrix material may experience rearrangement of the Si bonds to include formation of $Si(CH_3)_3$, $SiOCH_3$ and Si–C bonds, while also suffering from a reduction in the Si–$CH_3$ bonds [110, 111]. Of course, the extent of plasma damage is dependent on the specific plasma conditions [112].

The second process utilizes reactive silylating agents to convert hydrophilic Si–OH groups into hydrophobic derivatives [113–119]. This reaction is best accomplished with highly reactive chlorosilanes, but the corrosive nature of the by-product HCl and the potential to introduce chloride ions makes them poor candidates. Much more suitable silylating agents are silylamines, such as hexamethyldisilazane and dialkylaminotrimethylsilane, which only generate nonhalogen containing by-products. Here, the difficulty lies in the delivery of these reagents in order to maximize the penetration into the porous dielectric resin. Although solvent mediated treatments have been proposed, only supercritical $CO_2$ (SCO2) or gas-phase treatments are of practical value [120–122]. Gas-phase silylation is particularly attractive since this restoration process uses a neat reagent at elevated temperatures, greatly amplifying the reactivity of the nonhalogen containing candidates. However, even in the best of situations, the silylation process cannot fully restore the damage ILD to its previous virgin state [113].

### 1.3.3 Multilayer Structures

Finally, plasma damage can be mitigated by the presence of high carbon content materials. This observation was utilized in hybrid integration approaches, which utilize a two-layer ILD comprised of an organosilicate layer at the via level and an organic polymer layer at the trench level [8, 123–125]. Although this approach limits plasma damage mainly to the via level (top and sidewall surfaces), it complicates the integration approach. Thus, since most organic polymers, e.g. polyarylenes or poly(arylene ethers), are deposited by spin-on techniques and the majority of organosilicate materials are prepared by PECVD, two distinctly different deposition tool sets would be required. Furthermore, the introduction of new interfaces, as mentioned previously for pore sealing using thin liners, may significantly affect interfacial adhesion. One novel integration approach combines both hybrid integration (useful for minimizing ILD damage at the trench level) and a new concept referred to as 'effective porogen control' (ECB) [126]. The latter is achieved by limiting the amount of porogen decomposition of the porous SiCOH layer, i.e. the via level, to less than 70 %. The remaining porogen effectively increases the carbon content of this layer protecting the via level. The remaining porogen is partially removed during the plasma processing steps (trench level), followed by a thermal cure after a complete DD build.

## 1.4 Material Advances to Overcome Current Limitations

In the search for improved mechanical properties of highly porous materials, two candidates have emerged over the past few years. The first type, strictly spin-on, is based on the superior mechanical properties of crystalline zeolites. The second type of material relies on maintaining a highly connected network by introducing carbon as a bridging group in between the silicon atoms. Initially developed for spin-on materials, the latter strategy has been also recently adopted by the PECVD community. Carbon-bridged CVD precursors have appeared to replace current materials for the 32 nm technology node and beyond.

### 1.4.1 Silica Zeolites

Zeolites (derived from the Greek words *zeo* (to boil) and *lithos* (stone)) are microporous crystalline materials that consist of open aluminosilicate frameworks derived from $[SiO_4]^{4-}$ and $[AlO_4]^{5-}$ tetrahedra linked to form cages, channels or cavities of various sizes [127]. The negatively charged frameworks are balanced by positively charged cations of appropriate size located at various positions in the lattice. Because of their well-defined micropores and the presence of acidic sites inside the pores, zeolites have been successfully employed for a diverse range of applications: ion exchange, separations and catalysis (oil refining, petrochemistry, organic synthesis) [128]. While pure-silica zeolites, silicalite-1 (similar to ZSM-5) [129] and silicalite-2 (similar to ZSM-11) [130], were synthesized for the first time at the end of the 1970s, it is not until recently that these materials have received some attention from semiconductor manufacturers as potential low-*k* dielectrics. Since silicon is tetravalent, pure-silica zeolites (PSZs) do not contain framework charges and consequently no cations, which would be detrimental for electronic applications, are present. Due to their dense crystalline structure, better mechanical properties are expected [131], and since their pore size (<2 nm) is significantly smaller than integrated circuit (IC) features, the problem of electrical breakdown should be mitigated by the absence of any randomly occurring large pores.

The first films made of silica zeolites were obtained following two strategies [132]:

(a) in situ crystallization in a Teflon-lined Parr autoclave (165 °C, 2h) of a solution composed of 0.32 TPAOH/TEOS/165 $H_2O$ and
(b) spin-coating of a silicalite nanocrystals containing solution [132–134].

Independent of the method used to grow silicalite films, a post-synthesis thermal treatment at 450 °C for 2 hours is applied to remove the tetrapropylammonium hydroxide (TPAOH) template.

In the case of in situ crystallized films, thicknesses vary from 200 to 500 nm, and chemical mechanical polishing (CMP) needs to be performed initially to smooth the surface. These films possess excellent mechanical properties with $E$ = 30–40 GPa (nanoindentation) for dielectric constants ranging from 2.7 to 3.1. The effect of water adsorption on the $k$ value of silicalite films was examined by exposing the sample to air at 60% relative humidity. The $k$ value increases from 2.7 to 3.3 after an exposure time of 30 h ($k$ = 3.5 after several days). This observation is in good agreement with an ellipsometric porosimetry (EP) study using water as an adsorbent, reporting that the in situ crystallized film's hydrophilicity increases with aging [135].

On the other hand, silicalite spin-on films present lower dielectric constants ($k$ = 1.8–2.1) due to the additional porosity gained from internanocrystal packing voids [132]. A uniform interparticle pore size of 17 nm and an interparticle pore volume of 0.40 $cm^3$ $g^{-1}$ were measured by $N_2$ adsorption experiments performed on bulk samples. While beneficial in terms of the dielectric constant, the presence of larger mesopores is not only of concern for practical applications but also affects the adhesive properties of these films, leading to failure during CMP [136]. To address this issue, a brief secondary growth of silicalite nanocrystals can be applied, resulting in the loss of the interstitial porosity, an increase of the dielectric constant to a value of 3.0 and restoration of the adhesive properties observed for the in situ crystallized films. This second silicalite growth can be avoided by using an amorphous silica containing silicalite nanocrystals solution. Here, the mesopore size is reduced from 17 to 2.3–2.6 nm, resulting in a film with better mechanical properties ($E$ = 16–18 GPa by nanoindentation) and a dielectric constant of 2.3 [137]. Positronium annihilation fifetime spectroscopy (PALS) was used in this case to confirm the presence of micropores (0.55 nm) due to the zeolite framework, in addition to open/interconnected mesopores (2.3–2.6 nm) [138]. Unfortunately the film is quite hydrophilic, as shown by an increase in the dielectric constant from 2.3 to 3.9 within 1 h of exposure to ambient air at 50–60% relative humidity. To render the film hydrophobic, vapor-phase silylation was conducted. After silylation, a dielectric constant of 2.1 was obtained; this value changes slightly over time upon exposure to air. Interestingly, this post-curing treatment step can be eliminated if the silylation is performed during the zeolite synthesis and/or during calcination of the template [139].

In order to achieve a tunable $k$ value in an even lower range on the MFI (zeolite with a two-dimensional 10-ring pore structure and pore size of 5.5Å) low-$k$ material, incorporation of nanosized voids using γ-cyclodextrin as a porogen was explored [140]. By introducing 5–15 wt % of γ-cyclodextrin, controllable interparticle porosity ranging from 2.71 to 3.29 nm was created. The dielectric constant decreases in relation to the amount of porogen used. A value of 1.8 is obtained for a 15 wt % porogen loading, while the mechanical properties for this film remain excellent ($E$ = 14.3 GPa by nanoindentation). Unfortunately, all of the cyclodextrin containing films required heating to 450 °C for 9 h to ensure full decomposition of the porogens (cyclodextrin and TPAOH), making this process unsuitable for integration.

Because all the films above are composed of a mixture of amorphous silica and silicalite nanocrystals, not only the overall ratio of one silica component versus the other (in other words the total level of film crystallinity) [141] but also the nature of the zeolite (MFI or MEL, zeolites with a two-dimensional 10-ring pore structure) [142] influence the dielectric properties of the material. Using a two-stage hydrothermal synthesis method, the yield of nanocrystals for a given particle size can be increased in the solution used for film preparation. A direct effect is observed on the dielectric constant of the films with higher crystallinity, leading to lower *k* values [141, 142]. At the same relative level of crystallinity, changing the nature of the zeolite nanocrystals from MFI to MEL also results in a lower dielectric constant. The authors point out that a lower *k* value is expected because MEL has a lower framework density (FD) than MFI [142]. Since different sizes of nanocrystals are used in this study and the total porous volume of the films is different, it is not clear whether this result can be directly correlated to the nature of the zeolite or if it is governed by other parameters such as the size of the nanocrystals, mesoporous volume and number of residual silanols. Without silylation, and independent of the level of crystallinity, the synthesized films are still highly hydrophilic.

While this strategy initially looks very attractive, it is worth noting that the longer the second crystallization stage (i.e. higher crystallinity), the lower the mechanical properties [143, 144] and the bigger the interstitial voids [143, 145]. The mechanical properties decrease with increasing crystallinity because there is less amorphous silica to ensure the homogeneity of the film and also because the interstitial volume and size increase as well [143]. Regarding the last point, it was clearly shown that for silicalite films the pore size and distribution evolve as a function of the crystallization time [143, 145]. For crystallization times less than 3–4 days, pores at 0.55 nm (from silicalite) and at 2–5 nm are detected. At longer crystallization times, additional pores in the range of tens of nanometers are formed [143, 145]. Furthermore, it has been demonstrated that an increase of crystallinity is accompanied by a growth of the nanocrystal size and a reduction in the relative amount of amorphous silica [146]. It is most likely that these two parameters have a direct effect on the size of the voids. In that regard, Liu *et al.* have developed an 'evaporation-assisted two-stage synthesis that apparently limits the formation of 40–80 nm zeolite nanoparticles in solution leading to low-*k* films with improved properties' [147]. They have reported that small MEL nanocrystals (e.g. 14 nm) could be obtained in a 63 % yield using this strategy. Unfortunately, spin-coating of the above formulation leads to low-*k* films containing pores as wide as 50 nm, as evidenced by a comparative study conducted by a different group [148].

Remarkably, it was recently discovered that crystallization can also occur during the annealing of the spin-on films in a new process called 'on-wafer crystallization' [149]. The formation of big particles in solution is prevented by the absence of any hydrothermal treatment. After anneal and silylation, the film has a dielectric constant of 1.8, a Young's modulus of 16.8 GPa and an average pore size diameter centered around 6 nm. While smaller, the pores are still too big to envision a trouble-free integration of these materials.

Another approach to decrease the pore size diameter of PSZ low-*k* films was developed by Eslava *et al.* [150]. They have shown that UV curing (5 min at 425 °C under an inert atmosphere) of silica zeolite as-synthesized films leads to the formation of mesopores in the 2–8 nm range. For comparison, the calcined film only contains much larger pores in the 10–50 nm range. In addition, the UV-assisted curing not only increases the hydrophobicity of the porous zeolite film by condensation of silanols but also through grafting of the desorbing organic template fragment, an unexpected result. The film has a dielectric constant of 2.16 and a Young's modulus of 10.7 GPa. While the mesopores are again too big for application as a low-*k* dielectric material, UV curing appears as a novel path to the design of PSZ films with the desired mechanical, electrical and pore size distribution properties. In a more recent publication, post-deposition UV curing was combined with two silylation steps using 1,3,5,7-tetramethylcyclotetrasiloxane (TMCTS) [151]. The film modulus improved from 4.2 to 10 GPa after the complete treatment, for a dielectric constant $k = 2.18$. It is worth noting that no pore size distribution was reported in this paper [151].

A different strategy to address the water adsorption in PSZ films is to synthesize organic-functionalized PSZ MFI films (OF PSZ) [152]. Incorporation of Si Me groups in the zeolite framework was achieved through addition of $MeSi(OMe)_3$ to the synthesis solution of MFI nanoparticles. Upon exposure to ambient air for 1 h, the dielectric constant of the OF PSZ film increased from 2.25 to 2.74 (20 %), compared to a 70 % increase for the nonfunctionalized film (10 % after silylation). Other organic groups, different from Si–Me, 3,3,3-trifluoropropyltrimethoxysilane (C6) or 1*H*,1*H*,2*H*,2*H*-perfluorooctyltriethoxysilane (C14), have also been used to functionalize PSZ MFI and PSZ MEL [153]. In this case, the quantity of water adsorbed, while reduced, is still significant. The amount of organic groups incorporated, initially low to preserve the crystalline structure, is reduced even further after calcination due to partial decomposition, possibly explaining why higher hydrophobicity is not achieved. Interestingly, OF PSZ films have also been mixed with an organosilicate (50 % MTES/50 % TEOS) containing F127, a triblock porogen of the Pluronics® family [154]. No post-calcination silylation is necessary when the amount of OF PSZ is kept below 20 %. However, the Young's modulus value and the pore size distribution reported in this paper [154] are questionable.

In the case of OF PSZ films, it has been clearly shown that the introduction of Si–Me groups significantly increases the hydrophobicity of the film but also affects its level of crystallinity [155]. The authors mimicked the synthesis of silicalite using the following solution composition: $0.5MTMS/0.5TEOS/5H_2O/0.2TPAOH$ [155]. The film is amorphous and has been named ZLK (zeolite-inspired low-*k*). This film is microporous (no pores bigger than 2 nm) and has a dielectric constant of 2.2 that only increases by 0.5 % upon exposure to ambient air. The Young's modulus was measured by nanoindentation and is in the 6–7 GPa range.

As of today, the major drawbacks associated with the preparation of zeolite films, such as, big nanoparticles, hydrophilicity and the presence of big mesopores, have not been fully solved. For these reasons the use of silicalite films as future low-*k* materials in semiconductor manufacturing does not seem realistic at this time. An exception are ZLK films, which exhibit the expected properties to warrant some integration work. Nevertheless, their properties are similar to those reported for oxycarbosilane materials [1], which are prepared by nonhydrothermal synthesis and are easily scalable.

### 1.4.2 Hybrid Organic–Inorganic: Oxycarbosilanes

The observation that incorporation of organic moieties into sol-gel glasses can reduce their brittleness [156–160] has drawn the attention of the semiconductor industry. However, in order to be viable candidates for application in the BEOL, the silicon precursors to these

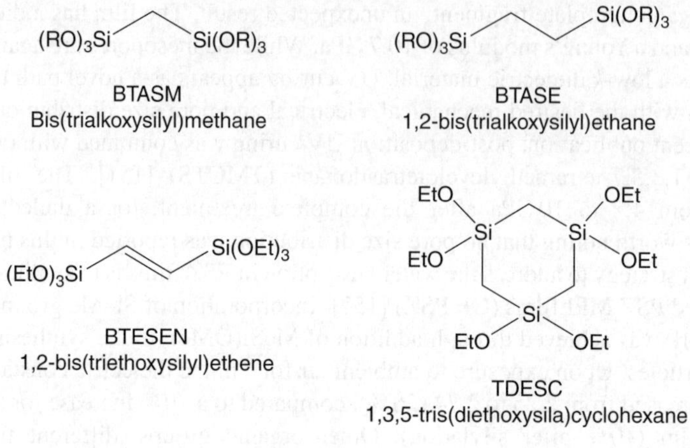


***Figure 1.6*** *Carbon-bridged alkoxysilane precursors used in the preparation of spin-on porous low-k films (R = OMe, OEt)*

hybrid materials should be low cost, commercially available or readily prepared on a large scale. For these reasons, the choice has been limited to simple alkoxysilane molecules containing a carbon bridge between the silicon atoms (Figure 1.6). These hybrid organic–inorganic materials are known in the literature under different names: bridged polysilsesquioxanes [161, 162], oxycarbosilanes [1, 7, 163–165] and carbosiloxanes [166]. Similarly to traditional organosilicates, they are stable up to 425–450 °C in an inert atmosphere and incorporation of a certain level of porosity is necessary to achieve *k* values below 2.4. Although the physical properties of the porous films are in part dependent on the type of porogen used, the chemical nature of the hybrid network is more important [1]. In this regard, the mechanical properties vary significantly for materials prepared in the presence of a coprecursor (e.g. MTMS) rather than from the single monomer.

Different authors have studied the properties of porous thin films prepared from MTMS/BTASM [167] or MTMS/BTASE [168–171] mixtures with a porogen. Unfortunately, full characterization, i.e. dielectric constant, Young's modulus and pore size distribution, was not reported for a given composition. Nevertheless, it was observed that at a dielectric constant of ~2.3, the Young's modulus doubles when the MTMS/BTASE ratio is increased from 100:10 to 100:90 [169]. This result is in good agreement with a study on neat films showing the significant impact on the Young's modulus of adding 10 to 100 mol % of BTASE to MTMS [172]. Still, the addition of porosity to MTMS/BTASE copolymers leads to an important decrease of the mechanical properties, even at low porogen loadings. This is due to the power law decay of the modulus as a function of porosity, typically observed for organosilicates [1].

BTASM, BTASE, BTESEN and TDESC have also been mixed with TEOS (or TMOS) and a surfactant to produce periodic mesoporous organosilicate low-*k* films [13, 173–175]. Materials with a dielectric constant in the 2.0 range were easily accessible using the above strategy. However, the question remains whether a controlled and reproducible level of pore organization over a very large surface is possible in a manufacturing environment. Indeed, differences in pore organization across the wafer would result in material properties variability and lack of etch uniformity. As of today, no integration data have been reported for these materials and therefore their potential is still difficult to assess.

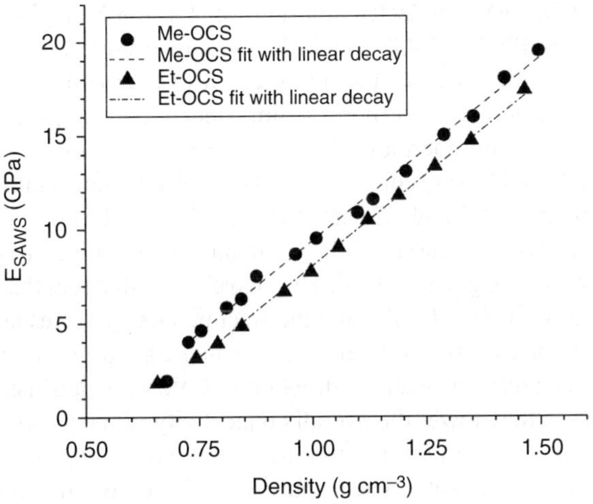


***Figure 1.7*** *Young's modulus for Me-OCS and Et-OCS as a function of density*

Our strategy to prepare hybrid organic–inorganic low-*k* films focused on two aspects: (a) compatibility of the sol–gel chemistry and solution deposition with semiconductor requirements and (b) use of oxycarbosilane precursors (BTASM or BTASE) in their neat form. Compared to previous studies, we have developed a process that is compatible with current integration requirements, i.e. the surfactant has to be nonionic and thermally decomposed (extraction by solvent washing was avoided), high-quality films without defects can be spin-coated on 8 in and 12 in wafers (low boiling alcohols cannot be used as a solvent), halogenated acids like HCl need to be replaced by $HNO_3$ to prevent chloride ion contamination and, finally, sol solutions should be stable for at least one month at room temperature. Regarding the sol–gel chemistry, we have demonstrated that hydrolysis conditions for oxycarbosilane precursors need to be precisely controlled in order to obtain high-quality films with the targeted electrical and mechanical properties. Under acidic conditions, a stoichiometric amount of water or higher is preferred to favor the formation of higher molecular weight polymers and consequently eliminate volatilization of low molecular weight oligomers and cyclics during the baking stage. Furthermore, the stability of 30 wt % oxycarbosilane solutions in 1-methoxy-2-propanol containing a poly(alkylene oxide)-based porogen, water and a catalytic acid was studied over time until the onset of gel formation. At room temperature the prepared formulations were stable for at least one month before a gel is formed. Under optimized conditions, we also found that the poly(alkylene oxide) porogens used needed to contain an ethylene oxide chain of at least 10 units to ensure good compatibility of the polymer with the oxycarbosilane matrix. The mechanical properties for oxycarbosilane films of varying porosity generated using either Pluronic 123® ($-(CH_2-CH_2-O)_{20}-(CH_2-CH-(CH_3)-CH_2)_{70}-(CH_2-CH_2-O)_{20}-$) or Brij 35® ($C_{12}H_{25}(O-CH_2-CH_2)_{23}OH$) were also investigated and are presented in Figure 1.7. Me-OCS and Et-OCS are used here to designate the thermally cured films obtained from BTASM and BTASE, respectively. The Young's modulus decreases linearly with decreasing density (i.e. increasing porosity), which is a unique characteristic of coadditive-free oxycarbosilanes [164, 165]. This result contrasts with SSQ–porogen mixtures where the decay in modulus is exponential. At a dielectric constant of 2.0, the Young's modulus values by SAWS are ~ 5 and ~ 6 GPa for Et-OCS and Me-OCS, respectively.

The modulus values obtained for thermally cured oxycarbosilane films are by far the largest of any we have observed for porous films generated using the sacrificial porogen route. Similarly, the fracture energies of the oxycarbosilane films are also considerably higher than their CDO and MSSQ counterparts. At a dielectric constant of 2.0, *G* (cohesive) is 2.8 J $m^{-2}$ as compared to 1.5 and 1.3 J $m^{-2}$ for both CDO and MSSQ [164]. Interestingly, recent computational modeling has established fundamental scaling laws for the elastic and fracture properties of hybrid glasses [176]. They have found that for ethane-bridged glasses, the bulk modulus scales as $(p - 0.6)^{1.92}$ and the fracture energy scales as $(p - 0.55)$, with $p$ being the network connectivity. In other words, while Si-Me groups provide significant benefits such as plasma resistance and hydrophobicity, they are detrimental to the mechanical properties because they lower the overall connectivity of the network.

Another strategy to prepare porous carbon-bridging containing materials was recently developed by Rathore *et al.* [166]. Their approach is based on the sol–gel processing of hyperbranched polycarbosiloxanes (HBPCSO). These polymers contain different silicon microenvironments corresponding to $[O_{4-x}Si(CH_2)_x]$, with $x = 1$–4. This constitutes a major difference from Me-OCS where the nominal structure is only composed of $[O_3SiCH_2]$ units. Films with dielectric constants ranging from 3.0 to ~ 1.8 were obtained using Pluronic 123® as a porogen. Over this entire range, Young's moduli obtained for HBPCSO films are largely superior to the MSSQ samples (prepared under similar conditions), but slightly inferior to Me-OCS samples. This observation confirms that silicates containing carbon-bridging groups belong to a different class of materials when considering the mechanical properties of porous low-*k* dielectrics.

Finally, PECVD films prepared from carbon-bridged precursors have also recently emerged to replace current materials at the 32 nm technology node and below [177, 178]. While the generic formula is still proprietary, it is believed that these precursors are based on a methane-bridged (MB) structure. They are typically used in combination with DEMS (diethoxymethylsilane) and BCHD (bicyclohexadiene) to prepare porous films using UV curing and which exhibit *k* values ranging from 2.2 – 2.4 [177, 178]. Interestingly, films prepared from the MB precursor had slightly lower mechanical properties but their porosity was significantly modified. Indeed, at the same dielectric constant, the proportion of ultramicropores (< 10 Å) versus supermicropores (10–20 Å) increases for MB-containing films. These films present a higher resistance to the plasma-induced damage observed during nanofabrication [179]. Knowing that the MB-containing films also possess a higher carbon content, we believed that their superior resistance to plasma-induced damage is the combination of this characteristic and their remarkable porous structure.

## 1.5 Conclusion

It is definitely an interesting time in the field of low-*k* materials as materials chemistry and process engineering need to come together to facilitate the successful implementation of $k < 2.4$ dielectrics in future technology nodes. Obviously, similar progress is also required for other BEOL materials such as dielectric caps, hardmasks, liners, metallurgy, etc.

The main challenges for ultra-low-*k* materials are associated with their usually poor mechanical properties and their highly interconnected porous network. The introduction of UV curing in manufacturing has permitted the first generation of $k = 2.4$ PECVD materials to achieve sufficient mechanical properties to be integrated. Beyond this dielectric constant, it is anticipated that new PECVD precursors will have to be implemented in order to satisfy the BEOL mechanical requirements. In that regard, spin-on silicalites, ZLK and oxycarbosilanes offer a significant advantage as their mechanical properties are equivalent or superior to manufactured PECVD $k = 2.4$ materials, at a $k < 2.2$. Among them, oxycarbosilane materials seem to be the most advanced in terms of development. Nevertheless, successful integration at the most aggressive ground rules needs to be demonstrated for these spin-on candidates to attract the full attention of the semiconductor industry. Electrical and mechanical reliability of these structures will be under scrutiny as the effect of plasma-induced damage and pre-clean treatments will become worse with increasing porosity. Toward this goal, it is incumbent that gentler plasma processes and process damage repair need to play a significant role. In the future, the solution might also arise from the combination of these new materials with new and innovative integration schemes [180].

## References


[1] Volksen, W., Miller, R.D. and Dubois, G. (2010) Low dielectric constant materials. *Chem. Rev.*, **110**, 56–110.

[2] Goto, K., Akiike, T., Konno, K., Shiba, T., Patz, M., Takahashi, M., Inoue, Y. and Matsubara, M. (2002) Thermally stable polyarylenes with low dielectric constant: direction towards the lowest limit of dielectrics. *J. Photopolym. Sci. Technol.*, **15**, 223–229.

[3] Maier, G. (2001) Low dielectric constant polymers for microelectronics. *Prog. Polym. Sci.*, **26**, 3–65.

[4] Miwa, T. (2001) Polyimides in microelectronics applications. *J. Photopolym. Sci. Technol.*, **14**, 29–32.

[5] Treichel, H., Ruhl, G., Ansmann, P., Wurl, R., Muller, C. and Dietlmeier, M. (1998) Low dielectric constant materials for interlayer dielectric. *Microelectron. Engng*, **40**, 1–19.

[6] Treichel, H., Withers, B., Ruhl, G., Ansmann, P., Wurl, R., Muller, C., Dietlmeier, M. and Maier, G. (1999) Low dielectric constant materials for interlayer dielectrics, in *Handbook of Low and High Dielectric Constant Materials and Their Applications* (ed. Nalwa, H.S.), vol. 1, Academic Press, San Diego, California, pp. 1–71.

[7] Dubois, G., Volksen, W. and Miller, R.D. (2007) Spin-on dielectric materials, in *Dielectric Films for Advanced Microelectronics* (eds Baklanov, M., Maex, K. and Green, M.), John Wiley & Sons, Inc., New York.

[8] Nakamura, T. and Nakashima, A. (2004) Robust multilevel interconnects with a nano-clustering porous low-*k* ($k < 2.3$). Proceedings of the 7th IEEE International Interconnect Technology Conference, Burlingame, California, pp. 175–177.

[9] Homma, T. (1998) Low dielectric constant materials and methods for interlayer dielectric films in ultralarge-scale integrated circuit multilevel interconnections. *Mater. Sci. Engng, R*, **R23**, 243–285.

[10] Kirk-Othmer (1992) *Kirk-Othmer Encyclopedia of Chemical Technology*, 3rd edn, vol. 20, John Wiley & Sons, Inc., New York, pp. 748–825.

[11] Grill, A. (2007) Low and ultralow dielectric constant films prepared by plasma-enhanced chemical vapor deposition, in *Dielectric Films for Advanced Microelectronics* (eds Baklanov, M., Maex, K. and Green, M.), John Wiley & Sons, Inc., New York.

[12] Grill, A. (2009) Porous pSICOH ultralow-*k* dielectrics for chip interconnects prepared by PECVD. *Annu. Rev. Mater. Sci.*, **39**, 49–69.

[13] Hatton, B.D., Landskron, K., Hunks, W.J., Bennett, M.R., Shukaris, D., Perovic, D.D. and Ozin, G.A. (2006) Materials chemistry for low-*k* materials. *Mater. Today*, **9**, 22–31.

[14] Hoofman, R.J.O.M., Nguyen, V.H., Arnal, V., Broekaart, M., Gosset, L.G., Besling, W.F.A., Fayolle, M. and Iacopi, F. (2007) Integration of low-*k* dielectric films in damascene processes, in *Dielectric Films for Advanced Microelectronics* (eds Baklanov, M., Maex, K. and Green, M.), John Wiley & Sons, Inc., New York.

[15] Morgen, M., Ryan, E.T., Zhao, J.-H., Hu, C., Cho, T. and Ho, P.S. (2000) Low dielectric constant materials for ULSI interconnects. *Annu. Rev. Mater. Sci.*, **30**, 645–680.

[16] Baklanov, M.R. (2007) Porosity of low dielectric materials. Ellipsometric porosimetry, in *Dielectric Films for Advanced Microelectronics* (eds Baklanov, M., Maex, K. and Green, M.), John Wiley & Sons, Inc., New York.

[17] Gidley, D.W., Peng, H.-g. and Vallery, R. (2007) Porosity of low dielectric materials. Positron annihilation spectroscopy, in *Dielectric Films for Advanced Microelectronics* (eds Baklanov, M., Maex, K. and Green, M.), John Wiley & Sons, Inc., New York.

[18] Maex, K., Baklanov, M.R., Shamiryan, D., Iacopi, F., Brongersma, S.H. and Yanovitskaya, Z.S. (2003) Low dielectric constant materials for microelectronics. *J. Appl. Phys.*, **93**, 8793–8841.

[19] Soles, C.L., Lee, H.-J., Vogt, B.D., Lin, E.K. and Wu, W.-L. (2007) Porosity of low dielectric materials. Structure characterization of nanoporous interlevel dielectric thin films with X-ray and neutron radiation, in *Dielectric Films for Advanced Microelectronics* (eds Baklanov, M., Maex, K. and Green, M.), John Wiley & Sons, Inc., New York.

[20] Wolf, S. (2002) *Silicon Processing for the VLSI Era*, vol. 4, *Deep-Submicron Process Technology*, Lattice Press, Sunset Beach, California.

[21] Chiang, C.-C. Chen, M.-C., Li, L.-J., Wu, Z.-C., Jang, S.-M. and Liang, M.-S. (2004) Physical and barrier properties of amorphous silicon-oxycarbide deposited by PECVD from octamethylcyclotetrasiloxane. *J. Electrochem. Soc.*, **151**, G612–G617.

[22] Chiang, C.-C., Ko, I.H. and Chen, M.-C. (2003) Leakage and breakdown mechanisms in Cu damascene with a bilayer-structured a-SiCN/a-SiC dielectric barrier. Proceedings of the 6th IEEE International Interconnect Technology Conference, Burlingame, California, pp. 201–203.

[23] Chiang, C.-C., Ko, I.H., Chen, M.-C., Wu, Z.-C., Lu, Y.-C., Jang, S.-M. and Liang, M.-S. (2004) Improvement in leakage current and breakdown field of Cu-comb capacitor using a silicon oxycarbide dielectric barrier. *J. Electrochem. Soc.*, **151**, G606–G611.

[24] Martin, J., Filipiak, S., Stephens, T., Huang, F., Aminpur, M., Mueller, J., Demircan, E., Zhao, L., Werking, J., Goldberg, C., Park, S., Sparks, T. and Esber, C. (2002) Integration of SiCN as a low *k* etch stop and Cu passivation in a high performance Cu/low *k* interconnect. Proceedings of the 5th IEEE International Interconnect Technology Conference, Burlingame, California, pp. 42–44.

[25] Tsui, T.Y., Willecke, R. and McKerrow, A.J. (2003) Effects of silicon carbide composition on dielectric barrier voltage ramp and TDDB reliability performance. Proceedings of the 6th IEEE International Interconnect Technology Conference, Burlingame, California, pp. 45–47.

[26] Guyer, E.P., Patz, M. and Dauskardt, R.H. (2006) Fracture of nanoporous methyl silsesquioxane thin-film glasses. *J. Mater. Res.*, **21**, 882–894.

[27] Ong, M.D., Jousseaume, V., Maitrejean, S. and Dauskardt, R.H. (2006) Fracture properties of porous MSSQ films: impact of porogen loading and burnout. Proceedings of the Materials Research Society Symposium, San Francisco, California, pp. 15–20.

[28] Grill, A., Edelstein, D., Lane, M., Patel, V., Gates, S., Restaino, D. and Molis, S. (2008) Interface engineering for high interfacial strength between SiCOH and porous SiCOH interconnect dielectrics and diffusion caps. *J. Appl. Phys.*, **103**, 054104/054101–054104/054106.

[29] Assous, M., Simon, J., Broussous, L., Bourlot, C., Fayolle, M., Louveau, O., Roman, A., Tabouret, E., Feldis, H., Louis, D. and Torres, J. (2003) Porous dielectric dual damascene patterning issues for 65-nm node: Can architecture bring a solution? Proceedings of the 6th IEEE International Interconnect Technology Conference, Burlingame, California, pp. 97–99.

[30] Hiroi, M., Ohtake, H., Saito, S., Onodera, T., Furutake, N., Harada, Y. and Hayashi, Y. (2001) Dual hard mask process for low-*k* porous organosilica dielectric in copper dual damscene interconnect fabrication. Proceedings of the 4th IEEE International Interconnect Technolnology Conference, Burlingame, California, p. 295.

[31] Kinoshita, K., Tada, M., Usami, T., Hiroi, M., Tonegawa, T., Shiba, K., Onodera, T., Tagami, M., Saitoh, S. and Hayashi, Y. (2000) Process design methodology for via-shape-controlled, copper dual-damascene interconnects in low-*k* organic film. Technical Digest – International Electron Devices Meeting, pp. 257–260.

[32] Fayolle, M., Torres, J., Passemard, G., Fusalba, F., Fanget, G., Louis, D., Assous, M., Louveau, O., Rivoire, M., Haxaire, K., Mourier, M., Maitrejean, S., Besson, P., Broussous, L., Arnaud, L. and Feldis, H. (2002) Integration of Cu/SiOC in Cu dual damascene interconnect for 0.1-mm technology. *Microelectron. Engng*, **64**, 35–42.

[33] Li, H.Y., Su, Y.J., Tsang, C.F., Sohan, S.M., Bliznetsov, V. and Zhang, L. (2004) Process improvement of 0.13 mm Cu/low *k* (Black Diamond) dual damascene interconnection. *Microelectron. Reliab.*, **45**, 1134–1143.

[34] Lin, S., Jin, C., Lui, L., Tsai, M., Daniels, M., Gonzalez, A., Wetzel, J.T., Monnig, K.A., Winebarger, P.A., Jang, S.M., Yu, D. and Liang, M.S. (2001) Low-*k* dielectrics characterization for damscene integration. Proceedings of the 4th IEEE International Interconnect Technology Conference, Burlingame, California, p. 146.

[35] Mosig, K., Jacobs, T., Brennan, K., Rasco, M., Wolf, J. and Augur, R. (2002) Integration challenges of porous ultra low-*k* spin-on dielectrics. *Microelectron. Engng*, **64**, 11–24.

[36] Schwartz, G.C. (2006) Methods/principles of deposition and etching of thin-films, in *Handbook of Semiconductor Interconnection Technology*, 2nd edn (eds Schwartz, G.C. and Srikrishnan, K.V.), CRC Press, Boca Raton, Florida.

[37] Baklanov, M.R., Le, Q.T., Kesters, E., Iacopi, F., van Aelst, J., Struyf, H., Boullart, W., Vanhaelemeersch, S. and Maex, K. (2004) Challenges of clean/strip processing for Cu/low-*k* technology. Proceedings of the 7th IEEE International Interconnect Technology Conference, Burlingame, California, pp. 187–189.

[38] Furukawa, Y., Patz, M., Kokubo, T. and Snijders, J.H.M. (2003) Material modification of the patterned wafer during dry etching and strip determined by XPS. *Microelectron. Engng*, **70**, 267–273.

[39] Furukawa, Y., Wolters, R., Roosen, H., Snijders, J.H.M. and Hoofman, R. (2004) Etch and strip induced material modification of porous low-*k* ($k = 2.2$) dielectric. *Microelectron. Engng*, **76**, 25–31.

[40] Huang, R. (2004) Via fill properties of organic BARCs in dual-damascene application. Proceedings of SPIE – International Society of Optical Engineering, Santa Clara, California, pp. 711–717.

[41] Hussein, M., Sivakumar, S., Brain, R., Beattie, B., Nguyen, P. and Fradkin, M. (2002) A novel approach to dual damascene patterning. Proceedings of the 5th IEEE International Interconnect Technology Conference, Burlingame, California, pp. 18–20.

[42] Lamb, J.E. and Shao, X. (2002) *Fill material for dual damascene processes*, US Patent 6,391,472 B1.

[43] Ryan, E.T., Labelle, C., Nitta, S., Fuller, N.C.M., Bonilla, G., McCullough, K., Taft, C., Lin, H., Simon, A., Simonyi, E., Malone, K., Sankarapandian, M., Dunn, D., Zaitz, M.A., Cohen, S., Klymko, N., Moon, B.K., Li, Z., Li, S., Yan, Y., Liu, J. and Ho, P.S. (2005) Nanoporous materials integration into advanced microprocessors. Proceedings of the Material Research Society Symposium, San Francisco, California, pp. 147–158.

[44] Yang, H.-L., Zhang, F., Nelson, K., Tseng, J.M., Forster, J., Sunddarrajan, A., Bhatnagar, A., Kumar, N. and Gopalraja, P. (2008) Off-angular deposition compensation for PVD selective re-sputtering process. Proceedings of the Material Research Society Symposium, San Francisco, California, Paper 1079-N1003-1005.

[45] Louis, D., Beverina, A., Arvet, C., Lajoinie, E., Peyne, C., Holmes, D. and Maloney, D. (2001) Cleaning status on low-*k* dielectric in advanced VLSI interconnect: characterisation and principal issues. *Microelectron. Engng*, **57–58**, 621–627.

[46] Louis, D., Beverina, A., Arvet, C., Lajoinie, E., Peyne, C., Holmes, D., Maloney, D., Lee, S. and Lee, W.M. (2000) Cleaning process strategies compatible with low-*k* dielectric and copper state of the art, evolution and perspectives. Proceedings of the 3rd IEEE International Interconnect Technology Conference, Burlingame, California, p. 250.

[47] Louis, D., Peyne, C., Lajoinie, E., Vallesi, B., Holmes, D., Maloney, D. and Lee, S. (1999) Improved post etch cleaning for low-*k* and copper integration for 0.18 mm technology. *Microelectron. Engng*, **46**, 307–310.

[48] Iacopi, F., Tokei, Z., Le, Q.T., Shamiryan, D., Conard, T., Brijs, B., Kreissig, U., Van Hove, M. and Maex, K. (2002) Factors affecting an efficient sealing of porous low-*k* dielectrics by physical vapor deposition Ta(N) thin films. *J. Appl. Phys.*, **92**, 1548–1554.
[49] Iacopi, F., Zistl, C., Jehoul, C., Tokei, Z., Le, Q.T., Das, A., Sullivan, C., Prokopowicz, G., Gronbeck, D., Gallagher, M., Calvert, J. and Maex, K. (2002) Dependence of the minimal PVD Ta(N) sealing thickness on the porosity of ZirkonTM LK dielectric films. *Microelectron. Engng*, **64**, 351–360.
[50] Kondo, S., Tokitoh, S., Yoon, B.U., Namiki, A., Sone, A., Ohashi, N., Misawa, K., Sone, S., Shin, H.J., Yoshie, T., Yoneda, K., Shimada, M., Ogawa, S., Matsumoto, I. and Kobayashi, N. (2003) Low-pressure CMP for reliable porous low-*k*/Cu integration. Proceedings of the 6th IEEE International Interconnect Technology Conference, Burlingame, California, pp. 86–88.
[51] Leduc, P., Farjot, T., Savoye, M., Demas, A.-C., Maitrejean, S. and Passemard, G. (2006) Dependence of CMP-induced delamination on number of low-*k* dielectric films stacked. *Microelectron. Engng*, **83**, 2072–2076.
[52] Leduc, P., Farjot, T., Savoye, M., Demas, A.-C., Maitrejean, S. and Passemard, G. (2006) CMP-induced peeling in multi-level ultra low-*k*/Cu interconnects. Proceedings of the Material Research Society Symposium, San Francisco, California, pp. 225–230.
[53] Satyanarayana, S., McGowan, R., White, B. and Hosali, S. (2005) Damage mechanisms in porous low-*k* integration. *Semicond. Int.*, **28**, 63–72.
[54] Brofman, P. (2009) IBM's Packaging Technology Roadmap and the 'collaboratory' approach to advanced packaging development. Proceedings of the International Conference on Electronics Packaging, ICEP 2009, Kyoto, Japan, pp. 1–6.
[55] Dubois, G. (2010) Chip–package interaction from a low-*k* materials perspective. Proceedings of the International Conference on Electronics Packaging, ICEP 2010, Hokkaido, Japan, pp. 448–453.
[56] Gallois-Garreignot, S., Fiori, V. and Nelias, D. (2010) Fracture phenomena induced by front-end/back-end interactions: dedicated failure analysis and numerical developments. *Microelectron. Reliab.*, **50**, 75–85.
[57] Van Driel, W.D. (2007) Facing the challenge of designing for Cu/low-*k* reliability. *Microelectron. Engng*, **47**, 1969–1974.
[58] Susko, R.A., Daubenspeck, T.H., Wassick, T.A., Sullivan, T.D., Sauter, W. and Cincotta, J. (2009) Solder bump electromigration and CPI challenges in low-*k* devices. *ECS Trans.*, **16**, 51–60.
[59] Wang, G., Ho, P.S. and Groothuis, S. (2005) Chip-packaging interaction: a critical concern for Cu/low *k* packaging. *Microelectron. Reliab.*, **45**, 1079–1093.
[60] Uchibori, C.J., Lee, M., Zhang, X., Ho, P.S. and Nakamura, T. (2009) Impact of Cu/low-*k* interconnect design on chip package interaction in flip chip package. Proceedings of Stress-Induced Phenomena in Metallization: 10th International Workshop, Austin, Texas, pp. 185–196.
[61] Liu, X.H., Shaw, T.M., Lane, T.W., Liniger, E.G., Herbst, B.W. and Questad, D.L. (2007) Chip–package-interaction modeling of ultra low-*k*/copper back end of line. Proceedings of the 10th IEEE International Interconnect Technology Conference, Burlingame, California, pp. 13–15.
[62] Appelt, B.K., Chen, W.T. and Lai, Y. (2010) Chip–package interactions at advanced silicon nodes. Proceedings of the International Conference on Electronics Packaging, ICEP 2010, Hokkaido, Japan, pp. 38–42.
[63] Volksen, W., Dubois, G., Kellock, A., Magbitang, T.P., Miller, R.D., Miller, D., Cohen, S., Simonyi, E.E., Ramirez, L., Markle, D., Chen, S., Zhou, S., Wang, X. and Wang, Y. (2008) Mechanical enhancement of low-*k* organosilicates by laser spike annealing. *J. Electrochem. Soc.*, **155**, G224–G230.
[64] Gage, D.M., Peng, L., Stebbins, J., Yim, K.S., Al-Bayati, A., Demos, A. and Dauskardt, R.H. (2010) Effects of e-beam curing on glass structure and mechanical properties of nanoporous organosilicate thin films. *Int. J. Mater. Res.*, **101**, 228–235.
[65] Goto, K., Hashii, S., Matsumoto, M., Miura, N., Furusawa, T., Matsuura, M., Ohsaki, A., Ohara, N., Tsuji, N. and Matsushita, K. (2006) Film characterization and integration of UV-cured ultra low-*k* for 45 nm node Cu/low-*k* interconnects. Proceedings of the Advances in Metals Conference 2005, Colorado Springs, Colorado and Tokyo, Japan, pp. 277–283.

[66] Ito, F., Takeuchi, T. and Hayashi, Y. (2006) Improvement of mechanical properties of porous SiOCH films by post-cure treatments. Proceedings of the Advances in Metals Conference 2005, Colorado Springs, Colorado and Tokyo, Japan, pp. 291–296.

[67] Jousseaume, V., Zenasni, A., Favennec, L., Gerbaud, G., Bardet, M., Simon, J.P. and Humbert, A. (2007) Comparison between E-beam and ultraviolet curing to perform porous a-SiOC:H. *J. Electrochem. Soc.*, **154**, G103–G109.

[68] Nakao, S.-I., Ushio, J., Ohno, T., Hamada, T., Kamigaki, Y., Kato, M., Yoneda, K., Kondo, S. and Kobayashi, N. (2006) UV/EB cure mechanism for porous PECVD/SOD low-*k* SiCOH materials. Proceedings of the IEEE International Interconnect Technology Conference, Burlingame, California, 5–7 June 2006, pp. 66–68.

[69] Yoneda, K., Kato, K., Kondo, S., Kobayashi, N., Matsuki, N., Matsushita, K., Ohara, N., Fukazawa, A. and Kimura, T. (2005) Impacts of UV cure for reliable porous PECVD SiOC integration. Proceedings of the IEEE International Interconnect Technology Conference, Burlingame, California, 5–7 June 2005, pp. 220–222.

[70] Mehta, S., Dimitrakopoulos, C., Augur, R., Gambino, J., Chou, A., Hook, T., Linder, B., Tseng, W., Bolam, R., Harmon, D., Massey, D., Gates, S. and Nye, H. (2006) Evaluation of device damage from e-beam curing of ultra low-*k* BEOL dielectrics. Proceedings of the Advances in Metals Conference 2005, Colorado Springs, Colorado and Tokyo, Japan, pp. 361–367.

[71] Chapelon, L.L., Vitiello, J., Gonchond, J.P., Barbier, D. and Torres, J. (2006) UV curing effects on mechanical and electrical performances of a PECVD non-porogen porous SiOC:H films (in *k* [2.2–2.4] range) for 45 nm node and below. *Microelectron. Engng*, **83**, 2136–2141.

[72] Iacopi, F., Travaly, Y., Eyckens, B., Waldfried, C., Abell, T., Guyer, E.P., Gage, D.M., Dauskardt, R.H., Sajavaara, T., Houthoofd, K., Grobet, P., Jacobs, P. and Maex, K. (2006) Short-ranged structural rearrangement and enhancement of mechanical properties of organosilicate glasses induced by ultraviolet radiation, *J. Appl. Phys.*, **99**, 053511/053511–053511/053517.

[73] Smith, R.S., Tsui, T.Y. and Ho, P.S. (2009) Effects of ultraviolet radiation on ultra-low-dielectric constant thin film fracture properties. *J. Mater. Res.*, **24**, 2795–2801.

[74] Smith, R.S., Tsui, T. and Ho, P.S. (2007) The effect of ultraviolet light curing on the material and fracture properties of a *k*~2.5 low-*k* dielectric. Proceedings of the Materials Research Society Symposium, pp. 27–31.

[75] Marsik, P., Verdonck, P., De Roest, D. and Baklanov, M.R. (2010) Porogen residues detection in optical properties of low-*k* dielectrics cured by ultraviolet radiation. *Thin Solid Films*, **518**, 4266–4272.

[76] Berry, I., Waldfried, C. and Durr, K. (2007) Requirements and constraints on optimizing UV processing of low-*k* dielectrics. Proceedings of the Materials Research Society Symposium, p. B02–01.

[77] Martinez, E., Rochat, N., Guedj, C., Licitra, C., Imbert, G. and Le Friec, Y. (2006) Influence of electron-beam and ultraviolet treatments on low-*k* porous dielectrics. *J. Appl. Phys.*, **100**, 124106/124101–124106/124105.

[78] Kim, T.-S., Tsuji, N., Kemeling, N., Matsushita, K., Chumakov, D., Geisler, H., Zschech, E. and Dauskardt, R.H. (2008) Depth dependence of ultraviolet curing of organosilicate low-*k* thin films. *J. Appl. Phys.*, **103**, 064108/064101–064108/064108.

[79] Kim, T.-S., Tsuji, N., Matsushita, K., Kobayashi, N., Chumakov, D., Geisler, H., Zschech, E. and Dauskardt, R.H. (2008) Tuning depth profiles of organosilicate films with ultraviolet curing. *J. Appl. Phys.*, **104**, 074113/074111–074113/074116.

[80] Gage, D.M., Guyer, E.P., Stebbins, J., Cui, Z., Al-Bayati, A., Demos, A., MacWilliams, K.P. and Dauskardt, R.H. (2006) UV curing effects on glass structure and mechanical properties of organosilicate low-*k* thin films. Proceedings of the IEEE International Interconnect Technology Conference, Burlingame, California, 5–7 June 2006, pp. 149–151.

[81] Gage, D.M., Stebbins, J.F., Peng, L., Cui, Z., Al-Bayati, A., MacWilliams, K.P., M'Saad, H. and Dauskardt, R.H. (2008) Effects of UV cure on glass structure and fracture properties of nanoporous carbon-doped oxide thin films. *J. Appl. Phys.*, **104**, 043513/043511–043513/043518.

[82] Iacopi, F., Beyer, G., Travaly, Y., Waldfried, C., Gage, D.M., Dauskardt, R.H., Houthoofd, K., Jacobs, P., Adriaensens, P., Schulze, K., Schulz, S.E., List, S. and Carlotti, G. (2007) Thermomechanical properties of thin organosilicate glass films treated with ultraviolet-assisted cure. *Acta Mater.*, **55**, 1407–1414.

[83] Kim, T.-S., Chumakov, D., Zschech, E. and Dauskardt, R.H. (2009) Tailoring UV cure depth profiles for optimal mechanical properties of organosilicate thin films. *Appl. Phys. Lett.*, **95**, 071902/071901–071902/071903.

[84] Lin, Y., Xiang, Y., Tsui, T.Y. and Vlassak, J.J. (2008) PECVD low-permittivity organosilicate glass coatings: adhesion, fracture and mechanical properties. *Acta Mater.*, **56**, 4932–4943.

[85] Calvert, J.M. and Gallagher, M.K. (2003) A new approach to ultralow-*k* dielectrics. *Semicond. Int.*, **26**, 56–58, 60.

[86] Dalton, T.J., Greco, S.E., Hedrick, J.C., Nitta, S.V., Purushothaman, S., Rodbell, K.P. and Rosenberg, R. (2002) *Method for forming a porous dielectric layer in a semiconductor device and device formed*, Patent 6,451,712 B1.

[87] Fayolle, M., Jousseaume, V., Assous, M., Tabouret, E., le Cornec, C., Haumesser, P.H., Leduc, P., Feldis, H., Louveau, O., Passemard, G. and Fusalba, F. (2004) Cu/ULK integration using a post integration porogen removal approach. Proceedings of the 7th IEEE International Interconnect Technology Conference, Burlingame, California, pp. 208–210.

[88] Gallagher, M.K. and You, Y. (2002) *Electronic device manufacture*, Patent 6,667,147 B2.

[89] Jousseaume, V., Favennec, L., Zenasni, A. and Passemard, G. (2006) Plasma-enhanced-chemical-vapor-deposited ultralow *k* for a postintegration porogen removal approach. *Appl. Phys. Lett.*, **88**, 182908/182901–182908/182903.

[90] Zielinski, E.M., Russell, S.W., List, R.S., Wilson, A.M., Jin, C., Newton, K.J., Lu, J.P., Hurd, T., Hsu, W.Y., Cordasco, V., Gopikanth, M., Korthuis, V., Lee, W., Cerny, G., Russell, N.M., Smith, P.B., O'Brien, S. and Havemann, R.H. (1997) Damascene integration of copper and ultralow-*k* xerogel for high performance interconnects. Technical Digest – International Electron Devices Meeting, pp. 936–938.

[91] Jin, C., Luttmer, J.D., Smith, D.M. and Ramos, T.A. (1997) Nanoporous silica as an ultralow-*k* dielectric. *MRS Bull.*, **22**, 39–42.

[92] Bonitz, J., Schulz, S.E. and Gessner, T. (2004) Ultra thin CVD TiN layers as diffusion barrier films on porous low-*k* materials. *Microelectron. Engng*, **76**, 82–88.

[93] Furuya, A., Soda, E., Yoneda, K., Yoshie, T., Okamura, H., Shimada, M., Ohtsuka, N. and Ogawa, S. (2004) Pore-sealing by etch-byproduct followed by ALD-Ta adhesion layer for Cu/porous low-*k* interconnects. Proceedings of the 7th IEEE International Interconnect Technology Conference, Burlingame, California, pp. 39–41.

[94] Mourier, T., Jousseaume, V., Fusalba, F., Lecornec, C., Maury, P., Passemard, G., Haumesser, P.H., Maitrejean, S., Cordeau, M., Pantel, R., Pierre, F., Fayolle, M. and Feldis, H. (2003) Porous low *k* pore sealing process study for 65 nm and below technologies. Proceedings of the 6th IEEE International Interconnect Technology Conference, Burlingame, California, pp. 245–247.

[95] Tada, M., Harada, Y., Tamura, T., Inoue, N., Ito, F., Yoshiki, M., Ohtake, H., Narihiro, M., Tagami, M., Ueki, M., Hijioka, K., Abe, M., Takeuchi, T., Saito, S., Onodera, T., Furutake, N., Arai, K., Fujii, K. and Hayashi, Y. (2003) A 65-nm-node, Cu interconnect technology using porous SiOCH film ($k = 2.5$) covered with ultra-thin, low-*k* pore seal ($k = 2.7$). Proceedings of Technical Digest – International Electron Devices Meeting, pp. 845–848.

[96] Abell, T. and Maex, K. (2004) Damage minimized plasma pore sealing of microporous low-*k* dielectrics. *Microelectron. Engng*, **76**, 16–19.

[97] Iacopi, F., Baklanov, M.R., Sleeckx, E., Conard, T., Bender, H., Meynen, H. and Maex, K. (2002) Properties of porous HSQ-based films capped by plasma-enhanced chemical vapor deposition dielectric layers. *J. Vac. Sci. Technol., B*, **20**, 109–115.

[98] White, B., Book, G., Hautala, J. and Tabat, M. (2004) Ultra low-*k* integration solutions using GCIB processing. Proceedings of the 7th IEEE International Interconnect Technology Conferance, Burlingame, California, pp. 193–195.

[99] Wu, Z.C., Lu, Y.C., Chiang, C.C., Chen, M.C., Chen, B.T., Wang, G.J., Chen, Y.T., Huang, J.L., Jang, S.M. and Liang, M.S. (2002) Advanced metal barrier free Cu damascene interconnects with PECVD silicon carbide barriers for 90/65-nm BEOL technology. Technical Digest – International Electron Devices Meeting, pp. 595–598.

[100] Aimadeddine, M., Arnal, V., Roy, D., Farcy, A., David, T., Chevolleau, T., Posseme, N., Vitiello, J., Chapelon, L.L., Guedj, C., Brechet, Y., Volpi, F. and Torres, J. (2006) Effect of $CH_4$

plasma on porous dielectric modification and pore sealing for advanced interconnect technology nodes. Proceedings of the IEEE International Interconnect Technology Conference, Burlingame, California, pp. 81–83.
[101] Cui, H., Carter, R.J., Moore, D.L., Peng, H.-G., Gidley, D.W. and Burke, P.A. (2005) Impact of reductive $N_2/H_2$ plasma on porous low-dielectric constant SiCOH thin films. *J. Appl. Phys.*, **97**, 113302/113301–113302/113308.
[102] Arnal, V., Hoofman, R.J.O.M., Assous, M., Bancken, P.H.L., Broekaart, M., Brun, P., Casanova, N., Chapelon, L.L., Chevolleau, T., Cowache, C., Daamen, R., Farcy, A., Fayolle, M., Feldis, H., Furukawa, Y., Goldberg, C., Gosset, L.G., Guedj, C., Haxaire, K., Hinsinger, O., Josse, E., Jullian, S., Louveau, O., Michelon, J., Posseme, N., Rivoire, M., Roman, A., Vandeweyer, T., Verheijden, G.J.A.M. and Torres, J. (2004) Process optimisation and dual damascene integration of porous CVD SiOC dielectric at 2.4 and 2.2 *k*-values for 45 nm CMOS technology. Proceedings of the 7th IEEE International Interconnect Technology Conference, Burlingame, California, pp. 202–204.
[103] Furusawa, T., Machida, S., Ryuzaki, D., Sameshima, K., Ishida, T., Ishikawa, K., Miura, N., Konishi, N., Saito, T. and Yamaguchi, H. (2003) Novel dissoluble hardmask for damage-less Cu/low-*k* interconnect fabrication. Proceedings of the 6th IEEE International Interconnect Technology Conference, San Francisco, California, pp. 195–197.
[104] Hinsinger, O., Fox, R., Sabouret, E., Goldberg, C., Verove, C., Besling, W., Brun, P., Josse, E., Monget, C., Belmont, O., Van Hassel, J., Sharma, B.G., Jacquemin, J.P., Vannier, P.,A. Humbert, A.,D. Bunel, D.,R. Gonella, R.,E. Mastromatteo, E.,D. Reber, D., Farcy, A., Mueller, J., Christie, P., Nguyen, V.H., Cregut, C. and Berger, T. (2004) Demonstration of an extendable and industrial 300 mm BEOL integration for the 65-nm technology node. Technical Digest – International Electron Devices Meeting, pp. 317–320.
[105] Struyf, H., Hendrickx, D., Van Olmen, J., Iacopi, F., Richard, O., Travaly, Y., Van Hove, M., Boullart, W. and Vanhaelemeersch, S. (2005) Low-damage damascene patterning of SiOC(H) low-*k* dielectrics. Proceedings of the IEEE International Interconnect Technology Conference, Burlingame, California, pp. 30–32.
[106] Chang, T.C., Liu, P.T., Shih, F.Y. and Sze, S.M. (1999) Effects of hydrogen on electrical and chemical properties of low-*k* hydrogen silsesquioxane as an intermetal dielectric for nonetchback processes. *Electrochem. Solid-State Lett.*, **2**, 390–392.
[107] Kim, Y.-H., Kim, H.J., Kim, J.Y. and Lee, Y. (2002) Passivation effect on low-Î° SiOC dielectrics by $H_2$ plasma treatment. *J. Korean Phys. Soc.*, **40**, 94–98.
[108] Shieh, J.-M., Tsai, K.-C., Dai, B.-T., Wu, Y.-C. and Wu, Y.-H. (2002) Reduction of etching plasma damage on low dielectric constant fluorinated amorphous carbon films by multiple $H_2$ plasma treatment. *J. Vac. Sci. Technol., B*, **20**, 1476–1481.
[109] Xu, K., Vereecke, G., Kesters, E., Le, Q.T., Lux, M., Henry, S.A., Kraus, H., Archer, L., Mertens, P.W., Kovacs, F., Dalmer, M., Gaulhofer, E., Luo, S.J., Han, Q.Y. and Berry, I. (2007) Modifications of porous low-*k* by plasma treatments and wet cleaning. *ECS Trans.*, **11**, 409–416.
[110] Grill, A. and Patel, V. (2004) Interaction of hydrogen plasma with extreme low-*k* SiCOH dielectrics. *J. Electrochem. Soc.*, **151**, F133–F134.
[111] Grill, A., Sternhagen, V., Neumayer, D. and Patel, V. (2005) Hydrogen plasma effects on ultralow-*k* porous SiCOH dielectrics. *J. Appl. Phys.*, **98**, 074502/074501–074502/074507.
[112] Berry, I., Shiota, A., Han, Q., Waldfried, C., Sekiguchi, M. and Escorcia, O. (2003) $H_2$/He plasma for photoresist stripping over porous MSQ low-*k* films. Proceedings of the Electrochem. Society, pp. 202–210.
[113] Bhanap, A., Korolev, B., Nitta, S., Purushothaman, S., Bonilla, G. and Ryan, E.T. (2007) A carbon-restoring silylation process for low-*k* dielectric repair. *Solid State Technol.*, **50**, 79–80, 82, 84, 86.
[114] Mor, Y.S., Chang, T.C., Liu, P.T., Tsai, T.M., Chen, C.W., Yan, S.T., Chu, C.J., Wu, W.F., Pan, F.M., Lur, W. and Sze, S.M. (2002) Effective repair to ultra-low-*k* dielectric material (*k* approx. 2.0) by hexamethyldisilazane treatment. *J. Vac. Sci. Technol., B*, **20**, 1334–1338.
[115] Orozco-Teran, R.A., Gorman, B.P., Mueller, D.W., Baklanov, M.R. and Reidy, R.F. (2004) Effect of silylation on triethoxyfluorosilane xerogel films by means of atmospheric pressure drying. *Thin Solid Films*, **471**, 145–153.

[116] Rajagopalan, T., Lahlouh, B., Chari, I., Othman, M.T., Biswas, N., Toma, D. and Gangopadhyay, S. (2008) Hexamethyldisilazane vapor treatment of plasma damaged nanoporous methylsilsesquioxane films: structural and electrical characteristics. *Thin Solid Films*, **516**, 3399–3404.
[117] Singh, A.P., Gandhi, D.D., Singh, B., Simonyi, E., Liniger, E.G., Nitta, S.V., Lane, M.W. and Ramanath, G. (2009) Pore orientation and silylation effects on mesoporous silica film properties. *Appl. Phys. Lett.*, **94**, 093502/093501–093502/093503.
[118] Chen, J.Y., Pan, F.M., Cho, A.T., Chao, K.J., Tsai, T.G., Wu, B.W., Yang, C.M. and Chang, L. (2003) Microstructure and mechanical properties of surfactant templated nanoporous silica thin films: effect of methylsilylation. *J. Electrochem. Soc.*, **150**, F123–F127.
[119] Liu, J., Kim, W., Bao, J., Shi, H., Baek, W. and Ho, P.S. (2007) Restoration and pore sealing of plasma damaged porous organosilicate low *k* dielectrics with phenyl containing agents. *J. Vac. Sci. Technol., B*, **25**, 906–912.
[120] Jung, J.M., Kwon, H.S., Lee, W.-K., Choi, B.-C., Kim, H.G. and Lim, K.T. (2010) Repair of plasma-damaged p-SiOCH dielectric films in supercritical $CO_2$. *Microelectron. Engng*, **87**, 1680–1684.
[121] Oszinda, T., Schaller, M. and Schulz, S.E. (2010) Chemical repair of plasma damaged porous ultra low-Î° SiOCH film using a vapor phase process. *ECS Trans.*, **25**, 19–30.
[122] Gorman, B.P., Orozco-Teran, R.A., Zhang, Z., Matz, P.D., Mueller, D.W. and Reidy, R.F. (2004) Rapid repair of plasma ash damage in low-*k* dielectrics using supercritical $CO_2$. *J. Vac. Sci. Technol., B: Microelectron. Nanometer Struct. – Process., Measmt, Phenom.*, **22**, 1210–1212.
[123] Arakawa, S., Mizuno, I., Ohoka, Y., Nagahata, K., Tabuchi, K., Kanamura, R. and Kadomura, S. (2006) Breakthrough integration of 32 nm-node Cu/ultra low-*k* SiOC ($k = 2.0$) interconnects by using advanced pore-sealing and low-*k* hard mask technologies. Proceedings of the IEEE International Interconnect Technology Conference, Burlingame, California, pp. 210–212.
[124] Usui, T., Tsumura, K., Nasu, H., Hayashi, Y., Minamihaba, G., Toyoda, H., Sawada, H., Ito, S., Miyajima, H., Watanabe, K., Shimada, M., Kojima, A., Uozumi, Y. and Shibata, H. (2006) High performance ultra low-*k* ($k = 2.0/k_{eff} = 2.4$)/Cu dual-damascene interconnect technology with self-formed $MnSi_xO_y$ barrier layer for 32 nm-node. Proceedings of the IEEE International Interconnect Technology Conference, Burlingame, California, pp. 216–218.
[125] Watanabe, T., Hayashi, Y., Tomizawa, H., Usui, T., Gawase, A., Shimada, M., Watanabe, K. and Shibata, H. (2008) Robust BEOL process integration with ultra low-*k* ($k = 2.0$) dielectric and self-formed $MnO_x$ barrier technology for 32 nm-node and beyond. Proceedings of the IEEE International Interconnect Technology Conference, Burlingame, California, pp. 208–210.
[126] Kagawa, Y., Enomoto, Y., Shimayama, T., Kameshima, T., Okamoto, M., Kawashima, H., Yamada, A., Hasegawa, T., Akiyama, K., Masuda, H., Miyajima, H., Shibata, H. and Kadomura, S. (2006) Robust 45-nm node Cu/ULK interconnects using effective porogen control. Proceedings of the IEEE International Interconnect Technology Conf., Burlingame, California, pp. 207–209.
[127] Davis, M.E. and Lobo, R.F. (1992) Zeolite and molecular sieve synthesis. *Chem. Mater.*, **4**, 756–768.
[128] Corma, A. (1997) From microporous to mesoporous molecular sieve materials and their use in catalysis. *Chem. Rev.*, **97**, 2373–2419.
[129] Flanigen, E.M., Bennett, J.M., Grose, R.W., Cohen, J.P., Patton, R.L., Kirchner, R.M. and Smith, J.V. (1978) Silicalite, a new hydrophobic crystalline silica molecular sieve. *Nature (London)*, **271**, 512–516.
[130] Bibby, D.M., Milestone, N.B. and Aldridge, L.P. (1979) Silicalite-2, a silica analog of the aluminosilicate zeolite ZSM-11. *Nature (London)*, **280**, 664–665.
[131] Li, Z., Johnson, M.C., Sun, M., Ryan, E.T., Earl, D.J., Maichen, W., Martin, J.I., Li, S., Lew, C.M., Wang, J., Deem, M.W., Davis, M.E. and Yan, Y. (2006) Mechanical and dielectric properties of pure-silica-zeolite low-*k* materials. *Angew. Chem., Int. Ed.*, **45**, 6329–6332.
[132] Z. Wang, Z.,H. Wang, H.,A. Mitra, A.,L. Huang, L. and Y. Yan, Y. (2001) Pure-silica zeolite low-*k* dielectric thin films. *Adv. Mater.*, **13**, 746–749.

[133] Gaynor, J.F. (2001) *Dielectric layer including silicalite crystals and binder and method for producing same for microelectronic circuits*. Patent 6,329,062.
[134] Gaynor, J.F. (2003) *Dielectric layer including silicalite crystals and binder and method for producing same for microelectronic circuits*. Patent 6,566,243.
[135] Eslava, S., Baklanov, M.R., Kirschhock, C.E.A., Iacopi, F., Aldea, S., Maex, K. and Martens, J.A. (2007) Characterization of a molecular sieve coating using ellipsometric porosimetry. *Langmuir*, **23**, 12811–12816.
[136] Hu, L., Wang, J., Li, Z., Li, S. and Yan, Y. (2006) Interfacial adhesion of nanoporous zeolite thin films. *J. Mater. Res.*, **21**, 505–511.
[137] Wang, Z., Mitra, A., Wang, H., Huang, L. and Yan, Y. (2001) Pure silica zeolite films as low-*k* dielectrics by spin-on of nanoparticle suspensions. *Adv. Mater.*, **13**, 1463–1466.
[138] Li, S., Sun, J., Li, Z., Peng, H., Gidley, D., Ryan, E.T. and Yan, Y. (2004) Evaluation of pore structure in pure silica zeolite MFI low-*k* thin films using positronium annihilation lifetime spectroscopy. *J. Phys. Chem. B*, **108**, 11689–11692.
[139] Lew, C.M., Liu, Y., Day, B., Kloster, G.M., Tiznado, H., Sun, M., Zaera, F., Wang, J. and Yan, Y. (2009) Hydrofluoric-acid-resistant and hydrophobic pure-silica-zeolite MEL low-dielectric-constant films. *Langmuir*, **25**, 5039–5044.
[140] Li, S., Li, Z. and Yan, Y. (2003) Ultra-low-*k* pure-silica zeolite MFI films using cyclodextrin as porogen. *Adv. Mater.*, **15**, 1528–1531.
[141] Li, Z., Li, S., Luo, H. and Yan, Y. (2004) Effects of crystallinity in spin-on pure-silica-zeolite MFI low-dielectric-constant films. *Adv. Funct. Mater.*, **14**, 1019–1024.
[142] Li, Z., Lew, C.M., Li, S., Medina, D.I. and Yan, Y. (2005) Pure-silica-zeolite MEL low-*k* films from nanoparticle suspensions. *J. Phys. Chem. B*, **109**, 8652–8658.
[143] Eslava, S., Kirschhock, C.E.A., Aldea, S., Baklanov, M.R., Iacopi, F., Maex, K. and Martens, J.A. (2009) Characterization of spin-on zeolite films prepared from Silicalite-1 nanoparticle suspensions. *Microporous Mesoporous Mater.*, **118**, 458–466.
[144] Johnson, M., Li, Z., Wang, J. and Yan, Y. (2007) Mechanical characterization of zeolite low dielectric constant thin films by nanoindentation. *Thin Solid Films*, **515**, 3164–3170.
[145] Eslava, S., Baklanov, M.R., Neimark, A.V., Iacopi, F., Kirschhock, C.E.A., Maex, K. and Martens, J.A. (2008) Evidence of large voids in pure-silica-zeolite low-*k* dielectrics synthesized by spin-on of nanoparticle suspensions. *Adv. Mater.*, **20**, 3110–3116.
[146] Davis, T.M., Drews, T.O., Ramanan, H., He, C., Dong, J., Schnablegger, H., Katsoulakis, M.A., Kokkoli, E., McCormick, A.V., Penn, R.L. and Tsapatsis, M. (2006) Mechanistic principles of nanoparticle evolution to zeolite crystals. *Nat. Mater.*, **5**, 400–408.
[147] Liu, Y., Sun, M., Lew, C.M., Wang, J. and Yan, Y. (2008) MEL-type pure-silica zeolite nanocrystals prepared by an evaporation-assisted two-stage synthesis method as ultra-low-*k* materials. *Adv. Funct. Mater.*, **18**, 1732–1738.
[148] Eslava, S., Seo, J.W., Kirschhock, C.E.A., Baklanov, M.R., Maex, K. and Martens, J.A. (2010) Comment on 'MEL-type pure-silica zeolite nanocrystals prepared by an evaporation-assisted two-stage synthesis method as ultra-low-*k* materials'. *Adv. Funct. Mater.*, **20**, 2377–2379.
[149] Liu, Y., Lew, C.M., Sun, M., Cai, R., Wang, J., Kloster, G., Boyanov, B. and Yan, Y. (2009) On-wafer crystallization of ultralow-*k* pure silica zeolite films. *Angew. Chem., Int. Ed.*, **48**, 4777–4780, S4777/4771.
[150] Eslava, S., Iacopi, F., Baklanov, M.R., Kirschhock, C.E.A., Maex, K. and Martens, J.A. (2007) Ultraviolet-assisted curing of polycrystalline pure-silica zeolites: hydrophobization, functionalization, and cross-linking of grains. *J. Am. Chem. Soc.*, **129**, 9288–9289.
[151] Yoshino, T., Ohnuki, N., Hata, N. and Seino, Y. (2009) Young's modulus enhancement of mesoporous pure-silica-zeolite low-dielectric-constant films by ultraviolet and silylation treatments. *Jpn. J. Appl. Phys.*, **48**, 050210/050211–050210/050213.
[152] Li, S., Li, Z., Medina, D., Lew, C. and Yan, Y. (2005) Organic-functionalized pure-silica-zeolite MFI low-*k* films. *Chem. Mater.*, **17**, 1851–1854.
[153] Lew, C.M., Li, Z., Li, S., Hwang, S.-J., Liu, Y., Medina, D.I., Sun, M., Wang, J., Davis, M.E. and Yan, Y. (2008) Pure-silica-zeolite MFI and MEL low-dielectric-constant films with fluoro-organic functionalization. *Adv. Funct. Mater.*, **18**, 3454–3460.

[154] Zhu, Y., Mueller, T.E. and Lercher, J.A. (2008) Single step preparation of novel hydrophobic composite films for low-*k* applications. *Adv. Funct. Mater.*, **18**, 3427–3433.
[155] Eslava, S., Urrutia, J., Busawon, A.N., Baklanov, M.R., Iacopi, F., Aldea, S., Maex, K., Martens, J.A. and Kirschhock, C.E.A. (2008) Zeolite-inspired low-*k* dielectrics overcoming limitations of zeolite films. *J. Am. Chem. Soc.*, **130**, 17528–17536.
[156] Hobson, S.T. and Shea, K.J. (1997) Bridged bisimide polysilsesquioxane xerogels: new hybrid organic–inorganic materials. *Chem. Mater.*, **9**, 616–623.
[157] Huang, H.H., Orler, B. and Wilkes, G.L. (1985) Ceramers: hybrid materials incorporating polymeric/oligomeric species with inorganic glasses by a sol-gel process. 2. Effect of acid content on the final properties. *Polym. Bull.*, **14**, 557–564.
[158] Huang, H.H., Orler, B. and Wilkes, G.L. (1987) Structure-property behavior of new hybrid materials incorporating oligomeric species into sol-gel glasses. 3. Effect of acid content, tetraethoxysilane content, and molecular weight of poly(dimethylsiloxane). *Macromolecules*, **20**, 1322–1330.
[159] Sharp, K.G. (1998) Inorganic/organic hybrid materials. *Adv. Mater.*, **10**, 1243–1248.
[160] Wen, J. and Wilkes, G.L. (1996) Organic/inorganic hybrid network materials by the sol-gel approach. *Chem. Mater.*, **8**, 1667–1681.
[161] Loy, D.A. and Shea, K.J. (1995) Bridged polysilsesquioxanes. Highly porous hybrid organic–inorganic materials. *Chem. Rev.*, **95**, 1431–1442.
[162] Shea, K.J. and Loy, D.A. (2001) Bridged polysilsesquioxanes. Molecular-engineered hybrid organic–inorganic materials. *Chem. Mater.*, **13**, 3306–3319.
[163] Dubois, G., Magbitang, T., Volksen, W., Simonyi, E. and Miller, R.D. (2005) New spin-on oxycarbosilane low-*k* dielectric materials with exceptional mechanical properties. Proceedings of the IEEE International Interconnect Technology Conference, Burlingame, California, 5–7 June 2005, pp. 226–228.
[164] Dubois, G., Volksen, W., Magbitang, T., Miller, R.D., Gage, D.M. and Dauskardt, R.H. (2007) Molecular network reinforcement of sol-gel glasses. *Adv. Mater.*, **19**, 3989–3994.
[165] Dubois, G., Volksen, W., Magbitang, T., Sherwood, M.H., Miller, R.D., Gage, D.M. and Dauskardt, R.H. (2008) Superior mechanical properties of dense and porous organic/inorganic hybrid thin films. *J. Sol-Gel Sci. Technol.*, **48**, 187–193.
[166] Rathore, J.S., Interrante, L.V. and Dubois, G. (2008) Ultra low-*k* films derived from hyperbranched polycarbosilanes (HBPCS). *Adv. Funct. Mater.*, **18**, 4022–4028.
[167] Choi, B.-G., Kim, B.R., Moon, M.-S., Kang, J.-W. and Ko, M.-J. (2005) New carbon-bridged hybrid polymers for low-*k* materials. *Mater. Res. Soc. Symp. Proc.*, **863**, 121–126.
[168] Char, K., Cha, B.J. and Kim, S. (2004) Material issues for nanoporous ultra low-*k* dielectrics. Proceedings of the 7th IEEE International Interconnect Technology Conference, pp. 219–221.
[169] Kim, B.R., Son, J.M. and Ko, M.J. (2007) The effect of porogen on physical properties in MTMS-BTMSE spin-on organosilicates. *J. Mater. Sci.*, **42**, 5381–5390.
[170] Kim, S., Char, K., Hahn, J., Lee, J.-K., Yoon, D.Y., Rhee, H.-W. and Jin, M.Y. (2007) Effect of pore generating materials on the electrical and mechanical properties of porous low-*k* films. *Macromol. Res.*, **15**, 1–4.
[171] Toivola, Y., Kim, S., Cook, R.F., Char, K., Lee, J.-K., Yoon, D.Y., Rhee, H.-W., Kim, S.Y. and Jin, M.Y. (2004) Organosilicate spin-on glasses. *J. Electrochem. Soc.*, **151**, F45–F53.
[172] Ro, H.W., Char, K., Jeon, E,-c., Kim, H.-J., Kwon, D., Lee, H.-J., Lee, J.-K., Rhee, H.-W., Soles, C.L. and Yoon, D.Y. (2007) High-modulus spin-on organosilicate glasses for nanoporous applications. *Adv. Mater.*, **19**, 705–710.
[173] Hatton, B.D., Landskron, K., Whitnall, W., Perovic, D.D. and Ozin, G.A. (2005) Spin-coated periodic mesoporous organosilica thin films – towards a new generation of low-dielectric-constant materials. *Adv. Funct. Mater.*, **15**, 823–829.
[174] Landskron, K., Hatton, B.D., Perovic, D.D. and Ozin, G.A. (2003) Periodic mesoporous organosilicas containing interconnected $[Si(CH_2)]_3$ rings. *Science*, **302**, 266–269 (2003).
[175] Lu, Y., Fan, H., Doke, N., Loy, D.A., Assink, R.A., LaVan, D.A. and Brinker, C.J. (2000) Evaporation-induced self-assembly of hybrid bridged silsesquioxane film and particulate mesophases with integral organic functionality. *J. Am. Chem. Soc.*, **122**, 5258–5261.

[176] Oliver, M.S., Dubois, G., Sherwood, M., Gage, D.M. and Dauskardt, R.H. (2010) Molecular origins of the mechanical behavior of hybrid glasses. *Adv. Funct. Mater.*, **20**, 2884–2892.
[177] Gates, S.M., Dubois, G., Ryan, E.T., Grill, A., Liu, M. and Gidley, D. (2009) Adjusting the skeleton and pore structure of porous SiCOH dielectrics. *J. Electrochem. Soc.*, **156**, G156–G162.
[178] Gates, S.M., Grill, A., Dimitrakopoulos, C., Patel, V., Chen, S.T., Spooner, T., Ryan, E.T., Cohen, S.A., Simonyi, E., Liniger, E., Ostrovski, Y. and Bhatia, R. (2009) Integration compatible porous SiCOH dielectrics from 45 to 22 nm. Proceedings of the Advances in Metals Conference 2008, pp. 531–536.
[179] Ryan, E.T., Gates, S.M., Grill, A., Molis, S., Flaitz, P., Arnold, J., Sankarapandian, M., Cohen, S.A., Ostrovski, Y. and Dimitrakopoulos, C. (2008) Property modifications of nanoporous pSiCOH dielectrics to enhance resistance to plasma-induced damage. *J. Appl. Phys.*, **104**, 094109/094101–094109/094107.
[180] Frot, T., Volksen, W., Purushothaman, S., Bruce, R.L., and Dubois, G. (2011) Application of the protection/deprotection strategy to the science of porous materials, *Adv. Mater.*, **23**, 2828.

# 2

# Ultra-Low-*k* by CVD: Deposition and Curing

*Vincent Jousseaume[1], Aziz Zenasni[1], Olivier Gourhant[2], Laurent Favennec[2] and Mikhail R. Baklanov[3]*

[1]*CEA-LETI, Grenoble, France*
[2]*STMicroelectronics, Crolles, France*
[3]*IMEC, Leuven, Belgium*

## 2.1 Introduction

Microelectronics is a field in constant evolution that requires colossal research and development efforts. Integrated circuit interconnections are not immune to this trend and major developments have been made in recent years in this subject. Aluminium and silicon oxide, which were still used in the most advanced technologies at the end of the last century, have gradually given way to copper and low-*k* materials. The introduction of new materials was justified by the gain in performance (lower resistivity of Cu and a lower dielectric constant for low-*k*), even if new technical problems appear in their integration. The International Technology Roadmap for Semiconductors (ITRS) is a privileged witness of this development [1]. However, when focusing on the predictions made by experts of the organization regarding the evolution of the dielectric constant of the interlayer dielectric (ILD) as a function of a technology node, we can understand how the difficulty to achieve and integrate materials with very low permittivity was underestimated. For instance, the ITRS predicted in 2001 that low-*k* with a dielectric constant of 1.9 would be necessary for the 45 nm technology node in

*Advanced Interconnects for ULSI Technology*, First Edition. Edited by Mikhail R. Baklanov, Paul S. Ho and Ehrenfried Zschech.

production in 2010. In retrospect, it appears that these estimations were too ambitious. However, they have supported active researches in the field of low-*k* dielectrics, particularly in the last 10 years. This history of low-*k* dielectrics is well described in several reviews that are now referenced in this area [2–4]. The first major step used to reduce the dielectric constant of the ILD was the introduction of carbon (through $CH_3$ groups) in a silica matrix. The introduction of these carbon groups connected to Si can both reduce the density of the material (and thus *k*) and the total polarizability (Si–C bonds are less polarizable than Si–O). A new generation of SiOCH thin films was developed, the dielectric constant being lowered from 4.2 for a PECVD oxide to $k = 2.9$ for SiOCH, which is integrated in production for technologies at the node at 90 nm. However, this strategy of carbon incorporation in a silica matrix shows its limits in terms of *k* reduction. Depending on the precursors used, deposition conditions and finally the carbon content in the layer, the dielectric constant can be lowered to 2.7 [5]. Then a second breakthrough was reached by introducing porosity in the insulating matrix. The objective here was to decrease the film density further, leading to a decrease of the dielectric constant. The initial work was done using wet processes by taking benefits of advances in the field of sol–gel [6], chemists being the first to demonstrate the feasibility of mesoporous thin films. During this era, our scientific community has worked on two major deposition techniques: spin-coating (materials are also called spin-on materials) and plasma-enhanced chemical vapour deposition (PECVD). The first technique is more simple, less expensive to implement and can benefit from the engineering of molecules and polymers made by chemists. The second is a standard in the field of deposition of dielectrics in microelectronics and is often preferred (wrongly or rightly) by manufacturers. The example of ultra-low-*k* (ULK) (dielectric permittivity <2.5) is a particularly interesting example of the back and forth movement between the two communities (sol–gel and PECVD). The early works on ultra-low-*k* were initiated mainly on materials deposited by spin-coating: JSR 5109 from JSR Micro and Zirkon LK from Shipley (then Rohm and Haas) were references in terms of ULK [2]. They have helped to provide evidence of the different issues of integrating porous dielectrics [7–9] and have allowed characterization techniques to be developed for mesoporous thin films. In parallel, different approaches to achieve porous SiOCH thin films have been studied in order to transfer the concepts demonstrated by wet approaches to PECVD [10]. The so-called 'porogen approach' consisting of the introduction and subsequent removal of an organic sacrificial phase in a dielectric matrix has become the principle approach used to achieve porous dielectrics for insulation in advanced interconnects.

This chapter's main goal is to describe what are key points for producing an ultra-low-*k* thin film using a PECVD technique. We will first focus on the porogen approach, which is now the most common approach used throughout the world. The first part will be dedicated to the understanding of the precursor's choice and deposition conditions that are necessary to succeed in this approach. A second part will be devoted to the thermal treatment assisted by ultraviolet, which is now the post-deposition treatment reference for the realization of an ultra-low-*k*. The impact of deposition and post-treatment processes on the porosity, electrical properties and mechanical properties will be described. Finally, we will focus on the main challenges that this approach is facing in terms of thin-film realization to improve and extend it for future generations: lower *k*, higher mechanical properties, impact of organic residues and other drawbacks. Finally, alternative ways and CVD techniques of achieving highly porous materials and specific material development for a better integration compatibility will be presented.

## 2.2 Porogen Approach by PECVD

### 2.2.1 Precursors and Deposition Conditions

Among the different techniques allowing the deposition of porous SiOCH dielectrics, the plasma-enhanced chemical vapour deposition porogen approach is the most used. This approach, also known as the 'subtractive process' [2, 11], is the one currently performed in high-volume manufacturing for the 45 and 32 nm technology nodes. In this process, the film is deposited from the decomposition of (at least) two precursors in the plasma (Figure 2.1). One of them is a pure organic molecule referred to as a 'porogen'. The second molecule consists of silicon atoms and organic radicals and is called the 'matrix' precursor. In the first step of the porogen approach, both molecules (matrix and porogen precursors) are transformed into species that eventually lead to the formation of a 'hybrid' film composed of an organosilicate-based matrix enclosing organic inclusions. Then, during an appropriate post-deposition treatment such as thermal annealing, the organic phase, mostly consisting of the porogen molecule fragments, is removed. At this point, the film should become porous and has ultra-low-*k* properties.

In 2001, Grill and colleagues were the first to present results on a porous SiOCH using a PECVD porogen approach (using tetramethyl-cyclotetrasiloxane (TMCTS) as the matrix precursor) [12]. Afterwards, other research groups successfully achieved this porous SiOCH (Table 2.1): Vrtis *et al.* using diethoxy-methyl-silane (DEMS) [13], Bilodeau *et al.* using dimethyl-dioxiranyl-silane and diethoxy-methyl-oxiranyl-silane [14], Burkey and Gleason using trimethyl-trivinyl-cyclotrisiloxane [15] and Favennec *et al.* using decamethyl-cyclopentasiloxane (DMCPS) and diethoxy-methyl-silane [16]. These different works have shown that cyclic and linear organo-silicate precursors were suitable to perform on porous material without a clear understanding of the impact of the matrix precursor structure on the success of the porogen approach (see Table 2.1).

For the porogen precursor, any hydrocarbon source with sufficient volatility to be delivered as a gas to the PECVD reactor could be, in principle, used as a porogen [11]. In the literature, many of them were tested and studied, such as unsaturated cycles like terpinene (such as a-terpinene or ATRP) or norbornene [17–20] or molecules with strained rings like cycloalkene oxide [16, 21].

This approach also requires the use of a post-deposition treatment (thermal-based at temperatures usually ≥ 400 °C) in order to remove the organic phase and create porosity. This step also induces some chemicals changes in the skeleton structure and is accompanied by film shrinkage. However, whether done by a purely thermal treatment or assisted

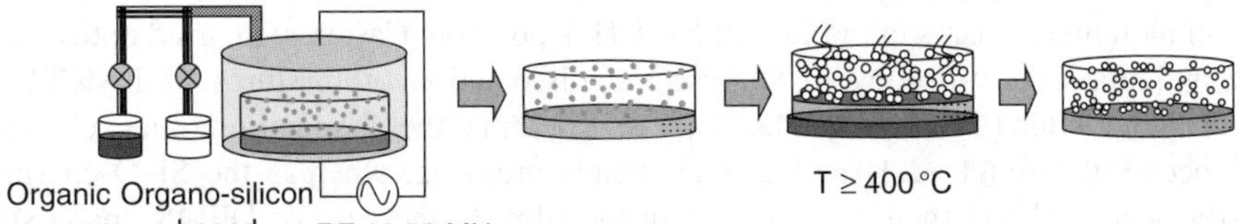


*Figure 2.1* Schematic view of the PECVD porogen approach

***Table 2.1*** *Examples of matrix and porogen precursors*

| | | |
|---|---|---|
| Diethoxy-methyl-silane (DEMS) | Tetramethyl-cyclotetrasiloxane (TMCTS) | Decamethyl-cyclopentasiloxane (DMCPS) |
| 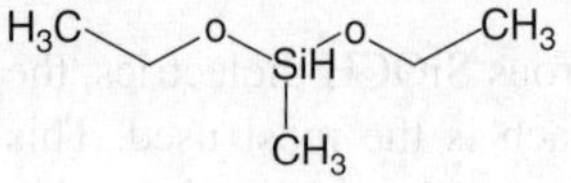 | 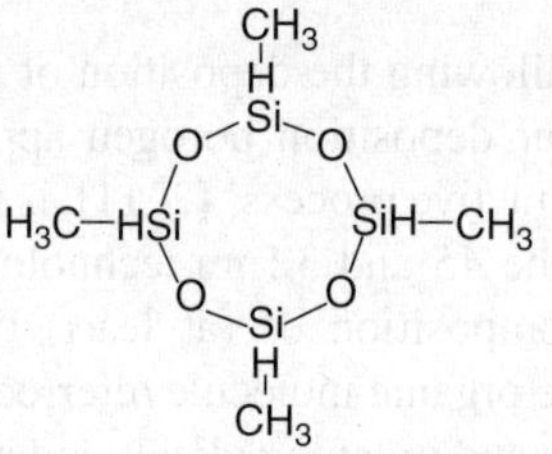 | 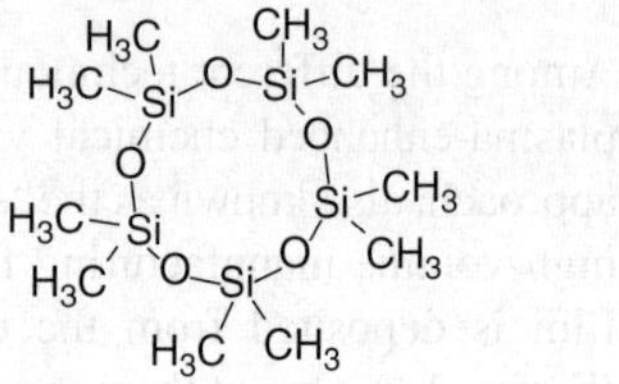 |
| Diethoxy-methyl-oxiranyl-silane | Dimethyl-dioxiranyl-silane | Trimethyl-trivinyl-cyclotrisiloxane (V3D3) |
| 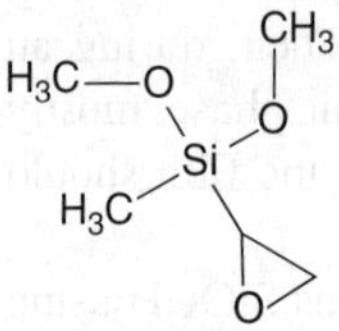 | 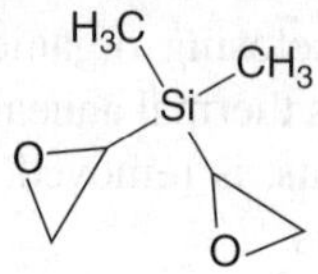 | 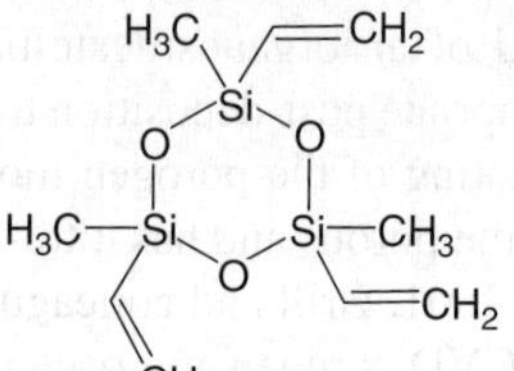 |
| Norbornadiene (NBD) | Norbornene (NBE) | α-Terpinene (ATRP) |
|  | 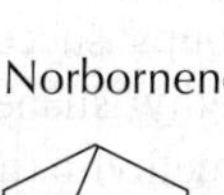 | 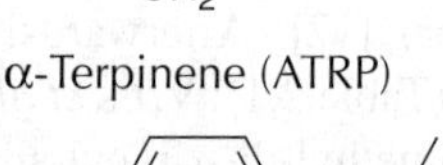 |
| Cyclopentene oxide (CPO) | Cyclohexene oxide (CHO) | Butadiene monoxide (BMO) |
| 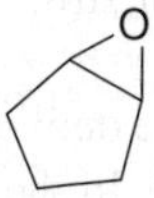 | 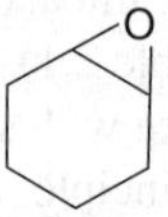 | 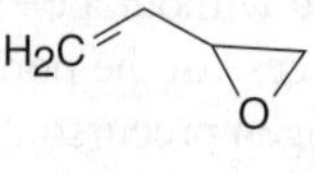 |

by UV radiation, the final annealing does not systematically lead to the realization of a porous material. The PECVD porogen approach has often proved to be a difficult one.

Firstly, matrix and porogen precursors should be chemically compatible. Moreover, the plasma condition should be optimized in order to prevent excessive dissociation of the skeleton precursor and to produce an SiOCH film with *k* close to 3, but should be able to dissociate the porogen precursor. The dissociation of the precursors in the plasma is a complex phenomenon and only a few studies are devoted to the understanding of molecule dissociation mechanisms in the case of SiOCH deposition. Castex *et al.* used optical emission and mass spectroscopy to analyse the dissociation of mixtures of DMCPS and cyclohexene oxide (CHO) [22]. The study revealed that the dissociation of a cyclic molecule occurred through C–H and Si–$CH_3$ bonds breaking, whereas the Si–O–Si ring is mostly preserved and then incorporated in the film. In the case of DEMS, many single silicon-atom-bearing species are produced. The study also found that CHO dissociates into H, $C_xH_y$ and CO radicals and that the epoxy moiety is believed to help the CHO molecule fragmentation because it is not found in hybrid films. They concluded that choosing a

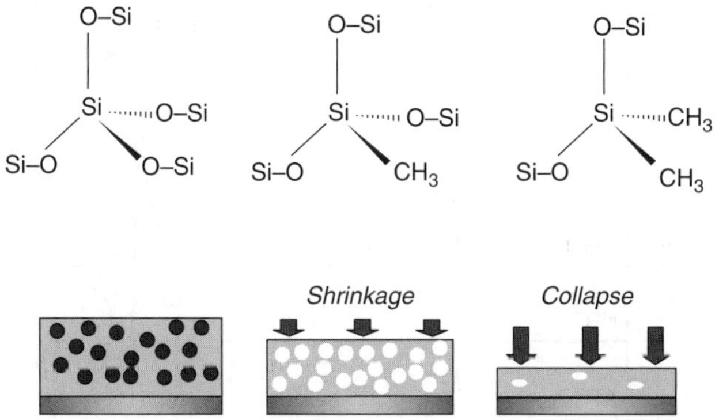


***Figure 2.2*** *Different –Si–O– configurations and the difference between shrinkage and collapse*

porogen and precursor that have high reactivity (such as an epoxy ring strain) in the case of porogen and a close dissociation energy threshold between the porogen precursor and the matrix one seems to succeed better in realizing porous ULK [23].

Secondly, in order to succeed in the PECVD porogen approach, Burkey and Gleason have shown that the mechanical properties of the film are a key point [24]. The matrix should be strong enough to avoid collapse after removal of the porogen species and should also be minimally bonded to the porogen species to avoid forming dangling bonds or other defect sites. Favennec *et al.* have studied the influence of the SiOCH matrix (composition and structure) and the porogen loading in the hybrid material [25]. They have evidenced a close link between the shrinkage phenomenon observed during the porogen removal treatment (thermal treatment at 450 °C) and the composition of the matrix structure. The higher the carbon content is, the lower the ability to prevent the film collapse. In terms of structure, the SiOCH constituted by silicon bonded to two methyl (Me) groups (or more) are not able to prevent significant film shrinkage; the presence of $O_3$–Si–$Me_1$, rather than $O_2$–Si–$Me_2$, appears more favourable to prevent a high shrinkage during porogen removal (Figure 2.2). For example, for hybrid materials deposited using DMCPS, silicon is mostly bounded to two methyl groups. Then, during porogen removal treatment, the film shrinkage 'corresponds' to the porogen removal rate in volume; there is no porosity generation [25]. This behaviour can be discussed using the percolation of rigidity arguments, this being first used for organosilicon thin films by Burkey and Gleason [26]. Networks below the percolation threshold are more deformable than those above the percolation point and the maximization of $O_3$–Si– (with potentially also $O_4$–Si links) should be favoured to avoid the film collapse. In hybrid deposited using DEMS, there is only a small part of $O_2$–Si–$Me_2$ in the film structure, leading to a minimal shrinkage [25]. Then the carbon content decrease in the matrix and the presence of $–SiO_4$ are both favourable to prevent a too-high shrinkage during porogen removal, even though they both increase the $k$ value. Reducing Si–Me in the matrix of the hybrid film (and increasing the Si–O–Si ratio of the skeleton) appears important to prevent high shrinkage. There is a balance to find between the minimization of the shrinkage (by reinforcing the mechanical strength) and the $k$ value (by including enough carbon) of the final porous material. The authors concluded that DEMS (which originally contains silicon bonded to only one methyl) is more appropriate than

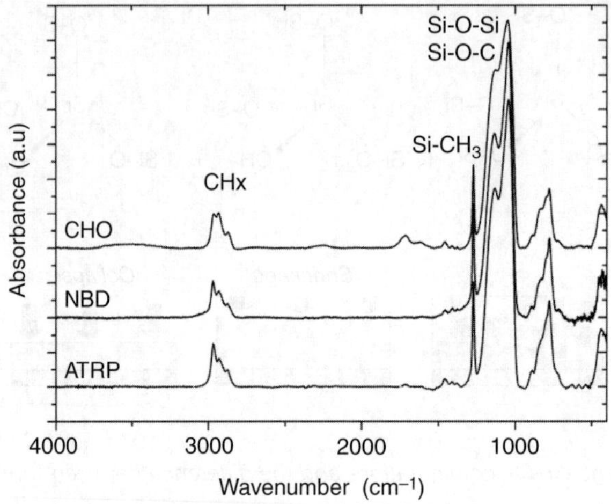


***Figure 2.3*** *FTIR of different hybrid films using DEMS as the matrix precursor (with CHO, ATRP or NBD as porogen)*

DMCPS (which contains silicon atoms bonded to two methyl groups) as a matrix precursor to deposit SiOCH. Finally, addition of a small amount of oxygen in the plasma gas feed allows an SiOCH material to be created by minimizing the unfavourable $O_2$–Si–$Me_2$ configuration [25]. Then the shrinkage is reduced for the same porogen loading. However, the addition of a large amount of oxygen in the plasma is highly unfavourable for the successful incorporation of the porogen species [24, 25]. A limited flow of oxidant should then be used.

The efficiency of porogen incorporation in the as-deposited SiOCH film is also strongly dependent on the deposition conditions (in particular to the ratio $R$ = (porogen precursor flow rate/precursor matrix flow rate)) and on the matrix precursor used. The incorporation is favoured at a lower deposition temperature and low plasma power (pressure has a minimal impact on the incorporation but allows variations of the cross-linking of the skeleton). In a first approach, the amount of porogen in the hybrid film can be evaluated thanks to FTIR analysis (see examples in Figure 2.3) through the evaluation of a parameter $P$:

$$P = \text{Area}_{2800-3200\,\text{cm}^{-1}} / \text{Area}_{980-1350\,\text{cm}^{-1}} \approx [\text{CH}_x]/[\text{SiOSi}]$$

This parameter can be seen as an indicator of the porogen species amount in each hybrid film.

As shown in Figure 2.4, $P$ evolves quasi-linearly as a function of the porogen loading in plasma gas feed, meaning that the porogen content in hybrid film can be efficiently tuned [27, 28]. The same trend is obtained for different substrate temperatures in the range 200 to 300 °C in the case of DEMS-NBD chemistry [18]. However, even if a deposition at a lower temperature allows the porogen content to be increased in the hybrid film, it leads to the formation of a partially cross-linked SiOCH skeleton. This induces higher film shrinkage during the curing process. A high deposition temperature could be a solution but the nature

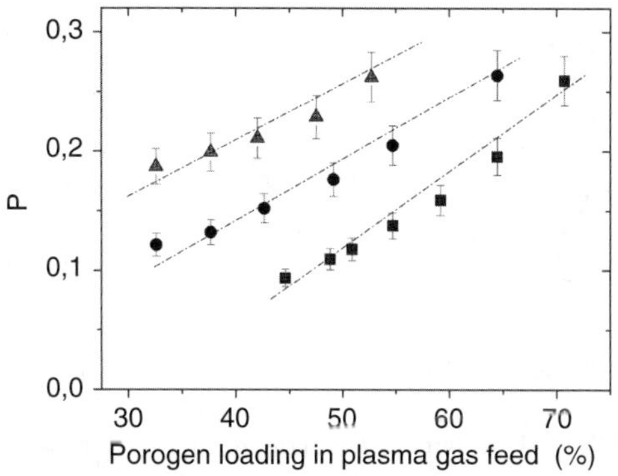


***Figure 2.4*** *Evolution of the parameter P as a function of the porogen loading in plasma gas feed and for different deposition temperatures: triangles = 225 °C, circles = 260 °C and squares = 300 °C. The parameteCr P is an indicator of the ratio C/Si* $\left(P = \text{Area}_{2800-3200\,\text{cm}^{-1}} / \text{Area}_{980-1350\,\text{cm}^{-1}} \approx [\text{CH}_x]/[\text{SiOSi}]\right)$

of incorporated organic species changes at high temperature, which makes the extraction of porogen difficult. At this point, a balance has to be found.

In addition, all porogens do not result in the same efficiency. When comparing the incorporation of butadiene monoxide (BMO) and cyclopenteneoxide (CPO) with TMCTS, Grill and Patel have found that at a very high ratio *R*(BMO) = 83.3 is required to obtain a dielectric constant of 2.27, whereas a lower dielectric constant of 2.07 is obtained for smaller *R*(CPO) = 55.6 (assuming that mass flows are properly calibrated) [21]. Moreover, high *R* values are required to obtain very low *k* values in films deposited with the cyclic TMCTS, whereas much smaller *R* values are required to obtain similar *k* values for films deposited from DEMS [21]. Grill and Patel concluded that the high ratio of porogen to SiOCH skeleton precursor that was required to obtain ULK films based on TMCTS makes this skeleton precursor unattractive for commercial applications. Bilodeau *et al.* found that relatively small changes in the chemical structure of the porogen can result in large changes in incorporation and removal efficiencies [19]. They have shown, for instance, that bicyclic precursors have the highest incorporation and removal efficiencies.

At the time of writing this article and even if this information is still difficult to obtain, it appears that the chemistry based on DEMS and ATRP or NBD are the ones that have the greatest success from an industrial point of view.

### 2.2.2 Mystery Still Unsolved: From Porogens to Pores

Several questions are still unsolved regarding the deposition of the hybrid film (SiOCH matrix + porogens). One of them concerns the exact link between the porogen and the final pore size. In order to test the hypothesis that pore size was related to the molecular volume of the porogen precursor, O'Neill *et al.* have determined pore sizes from both positronium annihilation lifetime spectroscopy (PALS) and ellipsometric porosimetry (EP). They have compared results to the molecular volume of the porogen molecule estimated from its

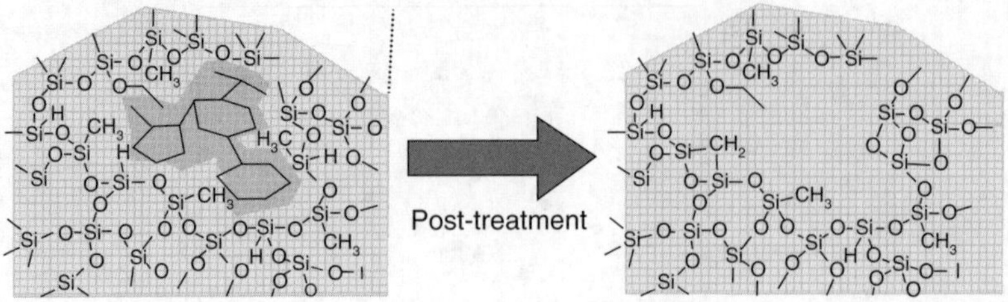


***Figure 2.5*** *Schematic view of the porogen incorporation (ATRP) by O'Neill* et al. [29]

molecular weight and liquid density [29]. The number of porogen molecules per pore ranges from 4 to 10 by this estimation. These results did not show any correlation of pore size with either the molecular volume of the porogen or with its chemical nature (i.e. the degree of unsaturation, boiling point, etc.). They concluded that the pores must be formed by a species larger than one porogen precursor molecule (more likely clusters of molecules, see Figure 2.5), which could be created either in the plasma phase, on the film surface or even in the bulk of the film during the curing stage (akin to an Ostwald ripening effect).

More recently, Jousseaume *et al.* have observed that a grazing incidence small-angle X-ray scattering (GISAXS) pattern is already present in hybrid materials deposited by PECVD [30]. One possible interpretation is that this signal is due to the porogen phase separation from the SiOCH skeleton, this type of hump being always observed in phase separation with high volume fractions [31]. The mean size deduced from this experiment was much smaller than 1 nm but the authors mentioned that a reliable value could not be deduced owing to the insufficient energy range explored.

Moreover, the exact way by which the porogen is bounded (or not) to the SiOCH skeleton remains unsettled. O'Neill *et al.* suggested that even though the porogen nature seems to have little influence on the pore size, it has a major impact on the film properties because it modifies the structure of the organosilicate network produced [29]. For example, when a deuterated porogen is used with DEMS, extensive hydrogen and deuterium scrambling is observed between the porogen and the inorganic network [32]. Nevertheless, the real link between the porogen and the matrix still remains unclear!

## 2.3 UV Curing

As mentioned previously, the porogen approach needs the use of a post-deposition treatment in order to strengthen the film structure and remove the organic phase to create porosity. This step appears to be crucial for the realization of porous ultra-low-*k* with accurate mechanical properties and a low dielectric constant.

### 2.3.1 General Overview of Curing

Several methods can be used to cure the as-deposited SiOCH films in order to remove the porogen, make them porous and reduce their dielectric constants. These techniques include thermal annealing, curing assisted by an electron beam (EB), curing assisted by ultraviolet (UV) radiation, $H_2$ plasma and supercritical $CO_2$ [12, 17, 25, 33–38].

A thermal annealing of 4h at 400 °C in a helium atmosphere has been initially used by Grill and colleagues in order to remove the porogen and produce porous films [12, 27]. However, at this temperature and time, this annealing is not enough to produce stable porous ULK films [12, 25, 27]. In particular, thermal treatment does not enhance sufficiently the volumic concentration of Si–O–Si bonds and then the film exhibits poor mechanical properties. Actually, thermal annealing even at extreme conditions (high temperature and for a long time; up to 12h at 450 °C in Reference [18]) let some uncondensed silanols bonds (isolated SiOH) that are not participating in the cross-linking mechanisms [18].

Moreover, thermal curing, especially for highly cross-linked as-deposited films, is not able to remove all porogens. Some residues are still present and can lead to the deterioration of the low-*k* film properties. Then, curing by thermal annealing requires long times at temperatures at the upper limit of acceptability for back-end-of-the-line (BEOL) processes and produces porous films with uncompleted cross-linking and, therefore, insufficient mechanical properties.

To circumvent these limitations, thermal treatment is assisted by additional energetic sources. EB and UV curing remained the most promising ones because they offered the opportunity to process at temperatures of 400 °C for short durations (a few minutes) and to achieve porous films with better mechanical properties [17, 34]. Many similarities have been evidenced with these two treatments: both remove efficiently the porogen from the deposited films but they can also react with some skeleton bonds (such as Si–$CH_3$, Si–H and (Si–OH)), leading to the increase of the Si–O–Si cross-links concentration [17]. However, even if an EB cure is still used, damages that this treatment leads to certain types of active devices led some manufacturers to abandon it in favour to UV treatment [39, 40]. Since then, most of the attention has been focused on the use of UV-assisted curing.

### 2.3.2 UV Curing Mechanisms

Recent studies highlighted the fact that the curing process cannot be considered any more as an isolated process from the deposition step especially for the mechanical properties enhancement. The couple (deposition and curing) should be taken into account in the porous film process efficiency [33]. Since UV has been adopted as the major process in curing ultra-low-*k* material, numerous studies have been devoted to the fundamental understanding of the mechanisms behind the transformation of organosilicate films through UV exposure [33, 34]. It is also well admitted that UV has a lot of benefits to remove porogens species efficiently. Figure 2.6 displays the transformation occurring during the UV curing (broadband UV lamp of 200–500 nm). FTIR spectra show a depletion of carbon content with the continuous decrease of the large band between 2900 and 3100 $cm^{-1}$ linked to porogen organic species and a monotonic decrease of Si–$CH_3$ peaks at 1275 $cm^{-1}$ attributed to the carbon linked to the matrix.

Besides, when UV cured, a small peak attributed to H–$SiO_3$ appears at 895 $cm^{-1}$. Moreover, a rearrangement of the Si–O–Si structure can be observed. All these modifications are indicators of the porogen removal and the mechanical properties enhancement.

Indeed, UV curing is a suitable process to enhance the mechanical properties. In SiOCH porous films, the Si–O–Si network constitutes the skeleton that sustains the structure of the film. Mechanical properties enhancement would be logically related to the Si–O–Si

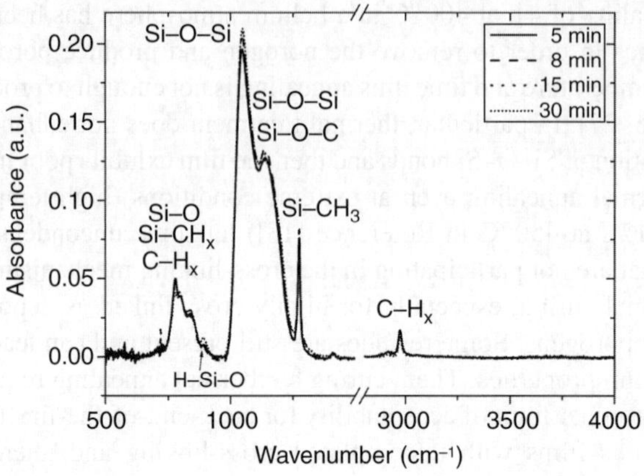


*Figure 2.6* FTIR spectra of SiOCH cured films for different UV curing durations

structure. Some works have claimed the ability of energetic radiation to improve the skeleton cross-linking, initiated by silanol condensation [41]. The increase in the amount of Si–O–Si could result from a condensation reaction of two silanol groups

$$\text{Si}-\text{OH}+\text{HO}-\text{Si}\rightarrow\text{Si}-\text{O}-\text{Si}+\text{H}_2\text{O} \tag{2.1}$$

Reaction (2.1), well known in organosilicate films, is thermally activated and needs enough Si–OH bonds to lead to a closeness of these groups, making the chemical reaction easier. This cross-linking mechanism induces a decrease of the distance between the two atoms of silicon, which leads finally to a film shrinkage. Thus, an aspect that has to be taken into account in the curing process is the deposition conditions and, more specifically, the hydroxyls content –OH. A low deposition temperature favours the appearance of Si–OH bonds into the films [18, 21]. During curing, the condensation of abundant Si–OH results in Si–O–Si formation, as estimated by an increase of the FTIR signal. In this case, the mechanical properties can be enhanced thanks to a better cross-linking [17]. At a high deposition temperature (typically $T > 250\,°\text{C}$), there is no evidence of the presence of a high amount of Si–OH in the film, meaning the absence of a significant condensation phenomenon. Reaction (2.1) does not allow an explanation of the improvement of the thin-film strength (and consequently of the mechanical properties), which is observed after UV curing of these films.

There are some works that have dealt with the Si–O–Si cross-linking with UV in order to obtain a unified explanation of the UV photon effect and how they can improve the mechanical properties [33, 34, 41, 42]. Although the total phenomena involved in this mechanism are still unknown, some advances have been accomplished and give enlightenment on what really occurs within the film. In this field, two factors have been identified as crucial: UV photon wavelength and Si–OH content.

UV photons are absorbed by the material and more specifically by its compounds (depending on their own specific absorption). Thus, UV photons have enough energy to embrittle the film structure through a selective photodissociation. Depending on the UV source energy the photonic radiation would target different structures (Figure 2.7). Two cases have been studied in general: large broadband UV with $\lambda > 200\,\text{nm}$ and short

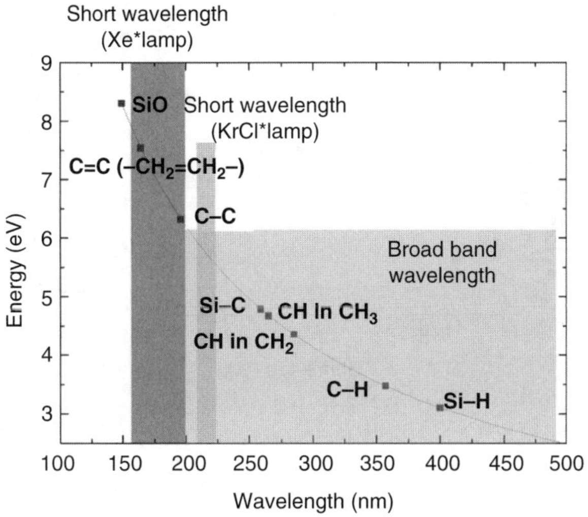


***Figure 2.7*** *Bond dissociation energies of different structures in the SiOCH material* [43]

wavelength with $\lambda < 200\,\text{nm}$. A large broadband UV source emits a light from 200 to 500 nm. As displayed in Figure 2.7, a bond energy of C–H in $CH_3$ (270–285 nm) matches well with an emitting UV bulb. UV energy is, then, efficient to break those C–H groups from the porogen and skeleton to create the desired porosity.

On the one hand, the main studies have maintained that broadband UV serves only as a breaking energy source that cannot be considered to favour any silanols condensation [33]. Indeed, a structural rearrangement of the existing silica bonds into a small-angle Si–O–Si configuration can be initiated by UV radiation at wavelengths above the silica absorption energy ($E_d \sim 155\,\text{nm}$). In the UV broadband configuration, there is neither rearrangement occurring in the Si–O–Si network nor a condensation of silanols.

Zenasni *et al.* have shown that, in the case of hybrid films with a low amount of silanols, the increase of the concentration of Si–O–Si bonds, previously adopted as cross-link dependent [33], is not due to a higher Si–O–Si bond number, but mainly involves a densification of the bulk structure under UV irradiation, leading to an enhancement of the siloxanes volumic concentration. This densification leads to the enhancement of the mechanical properties (Figure 2.8) [33].

When sufficient Si–OH are created during deposition, they can cross-link to form Si–O–Si bonds thanks to thermal activation. In this case, an appropriate UV wavelength can eventually compensate for the lack of Si–OH in the initial SiOCH structure.

Additionally, more recently Gourhant *et al.* have highlighted original mechanisms of condensation involving Si–O–C bonds to form Si–O–Si bonds [18]. This kind of transformation has a substantially smaller energy barrier and is exothermic [44]. As a result, the cross-linking mechanism of Si–O–C structures has been proposed based on the FTIR and NMR analyses:

$$-\text{Si}-\text{O}-\text{CH}_2-\text{CH}_3+\text{SiCH}_3 \rightarrow \text{Si}-\text{O}\bullet + -\text{Si}\bullet + \text{CH}_3-\text{CH}_2-\text{CH}_3 \rightarrow \text{Si}-\text{O}-\text{Si}+\text{C}_x\text{H}_y \quad (2.2)$$

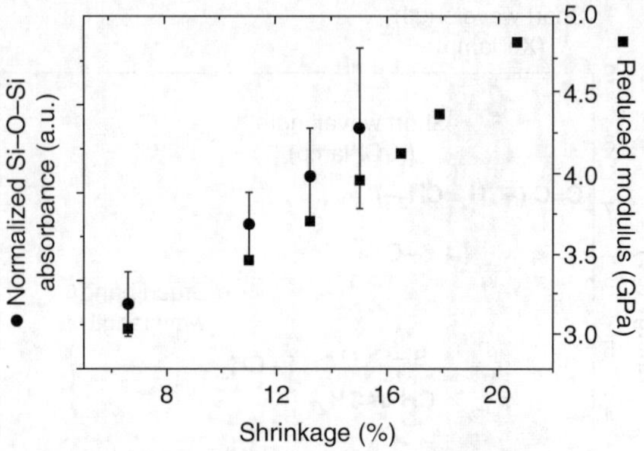


*Figure 2.8* Normalized Si–O–Si absorbance and reduced modulus as a function of thickness shrinkage

Other works mentioned that structure rearrangement would also involve Si–$CH_2$–Si link creation (carbon bridging bonds) resulting from $SiCH_3$ transformation [44]. This kinetic energy would involve different reactions, such as the following:

$$-Si-CH_3 + H_3C-Si- \rightarrow -Si-(CH_2)_2-Si- + H_2 \nearrow \quad (2.3)$$

$$-Si-CH_3 + H_3C-Si- \rightarrow -Si-CH_2-Si- + CH_4 \nearrow \quad (2.4)$$

However, by associating FTIR and NMR analyses, it has been shown that creation of Si–$(CH_2)_n$–Si does not seem to be created during UV curing [18, 45].

Using a short wavelength, the dependence of the wavelength of UV light on porogen desorption and mechanical strength improvement was investigated in SiOCH films. Several lamps have been tested with a specific look at their impact on photolysis of substructures like Si–$CH_3$, Si–O and C–H. On all the peaks evaluated, it can be seen that the impact with a more energetic wavelength is more pronounced. Porogen is effectively removed by UV light since C–C bonds in SiOCH films are broken by vacuum ultraviolet (VUV) irradiation (C–C bonding energy ~196 nm [43]). It was also observed that irradiation at 172 nm decreases the Si–$CH_3$ content (not observed at 185 and 222 nm). The nature of this threshold was studied by Prager *et al.* [46]. From quantum-chemical calculations on the model substances octamethyl-and tetramethylcyclotetrasiloxane as well as on hexamethyl- and tetramethyldisiloxane, they have shown that a threshold wavelength exists for the excitation of the molecule into the first excited singlet state of 190 and 198 nm as well as of 189 and 192 nm, respectively. After excitation and intersystem crossing in an excited triplet state, the scission of the Si–$CH_3$ bond may occur, gaining an energy benefit of around 50 kcal $mol^{-1}$ (see Figure 2.9). The calculated Si–$CH_3$ bond dissociation energy (about 90 kcal $mol^{-1}$) is in agreement with the experimental data given in Reference [47]. These findings reveal the presumption that only photons with $\lambda < 200$ nm can generate Si-centred radicals that subsequently attract protons from neighbouring methyl groups. This is in good agreement with the behaviour of the H–SiO peak after irradiation with 172 or 222 nm photons. Moreover, it opens the pathway for cross-linking via Si–Si, Si–O–Si and Si–$CH_2$–Si bonds.

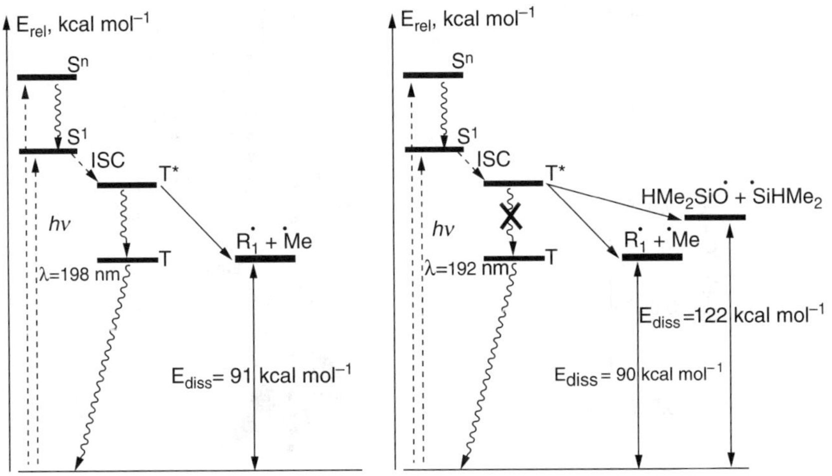


***Figure 2.9*** *Results of quantum-chemical calculations on tetramethylcyclotetrasiloxane (left) and tetramethyldisiloxane (right)*

The formation of this kind of defect (Si-centred radicals) has also been demonstrated by Nakao *et al.* [48] and confirmed by Tajima *et al.* [44] and Seo *et al.* [49] as the following:

$$-\mathrm{Si}-\mathrm{CH_3}+\mathrm{eV}\rightarrow -\mathrm{Si}\bullet +\mathrm{CH_2(H)} \qquad (2.5)$$

This defect can easily interact with hydrogen and oxygen atoms to form Si–H and Si–OH bonds. This is the reason why during UV curing, 172 nm light forms more Si–H bonds than broadband light with $\lambda > 200$ nm. This allows explaining observed anticorrelation between concentration of Si–$CH_3$ and Si–H groups (Figure 2.10) [50]. In the case of formation of Si–H bonds, this also leads to generation of less free volumes than Si–$CH_3$ and contributes to shrinkage of the film.

Even though effective on the degradation of porogen, the high energetic radiation could also break Si–O bonds, which could induce a rearrangement of the structure (densification) towards a higher mechanical strength [51, 52]. The opening of the Si–O–Si structure, coupling with the strong carbon depletion (the lack of hydrophobicity of the film), may also induce the appearance of hydroxyls –OH. Another reason might be related to the influence of residual oxygen in the UV vacuum chamber, which could be excited by $\lambda < 200$ nm radiation. Therefore, an optimal balance between improvement of mechanical properties and hydrophilization of a low-*k* value must be found. Recent works have, for instance, proposed a multistep curing process consisting of VUV and deep ultraviolet irradiation [49, 53].

UV also has a lot of benefits in that it removes porogens species efficiently. For porogen removal, UV curing enhances the photolysis of organic species leading to the porogen removal and porosity creation. During UV irradiation, the porosity is created through a combination of the photonic breaking effect (ultraviolet irradiation action) and thermally activated degradation of porogen. Thanks to this double effect, the porogen is easily removed from the film after a few minutes of UV curing, whereas more than an hour was required when the film was only thermally cured. Figure 2.11 shows an example of the dielectric constant evolution

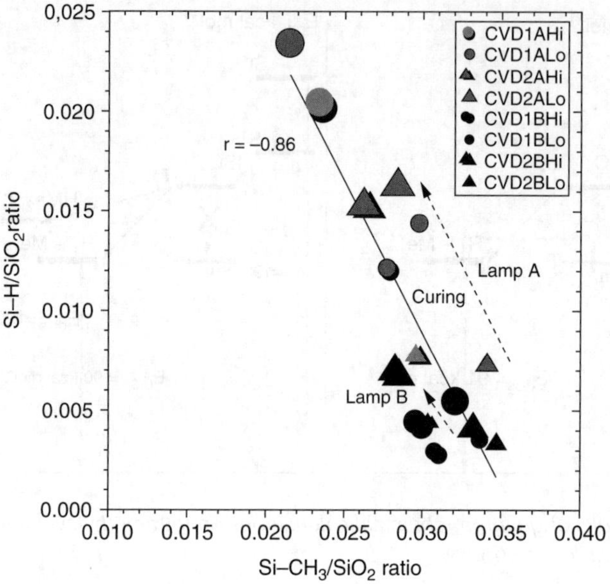


***Figure 2.10*** *Anticorrelation between FTIR measured concentrations of Si–CH$_3$ and Si–H groups observed after curing near a monochromatic lamp with $\lambda = 172$ nm (A) and broadband lamp with $\lambda > 200$ nm (B)*

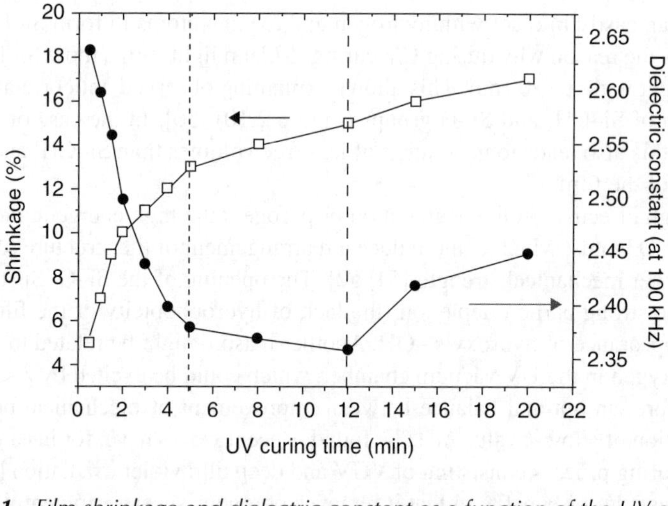


***Figure 2.11*** *Film shrinkage and dielectric constant as a function of the UV curing time*

versus the curing time. The porogen removal occurs on the primary steps of curing (during the first 5 min for broadband UV). This results in a fast increase of shrinkage up to 12 %, whereas the *k* falls from 2.7 to 2.4 after 5 min of curing. A thickness change occurring during porogen removal does not significantly impact porosity creation. Although the bulk material should be impacted by shrinkage (free volume loss), this effect is counterbalanced by porogen removal

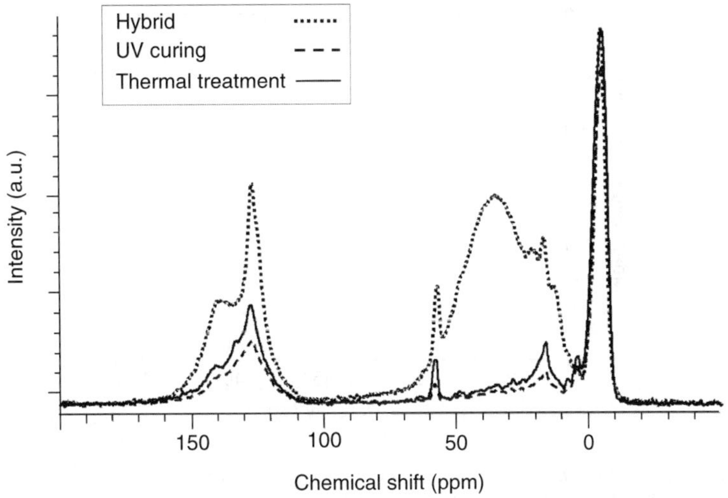


***Figure 2.12*** *$^{13}C$ NMR chemical shift spectra for the as-deposited, thermally cured and UV-treated SiOCH films*

(potential porosity creation), leading to a lower film density. Indeed, the porosity evaluation through ellipsometric porosimetry measurement shows a continuous increase of porosity created within the film as a function of the UV-assisted thermal curing time. The dielectric constant decreases with the UV curing time, which is mainly driven by the porosity increase. Consequently, even if the film shrinkage should involve film densification, the porogen removal compensates this structure contraction by the pore creation.

Figure 2.12 illustrates the benefit brought by UV curing for porogen removal. A better efficiency in terms of porogen removal is observed on the UV cured spectrum in comparison to thermal curing with the quasi-disappearance of porogen-related structures ($-CH_2O$, $-CH_2-$ and $-CH_3$). The peaks of $Si-CH_3$ slightly decreased, meaning that there was a substantial effect of UV on the photolysis of methyl groups.

In conclusion, all works conducted to highlight the role of an UV wavelength during curing are facing complex chemistry occurring in the films. Although some tendencies could be observed, a final conclusion is a long way from being established. Because most of the modifications are expected to be based on the generation of silicon and/or carbon defects, more dangling bond studies should be supported and linked to a microstructure in order to obtain a clear view of the effects of UV photons on the SiOCH films [54, 55].

## 2.4 Impact of Curing on Structure and Physical Properties: Benefits of UV Curing

### 2.4.1 Porosity

To investigate the impact of the curing on the porous film characteristics, it is essential to understand properly what kind of modification such treatment can induce. Figure 2.13 shows the pore size evaluated as a function of the total porosity using ellipsometric

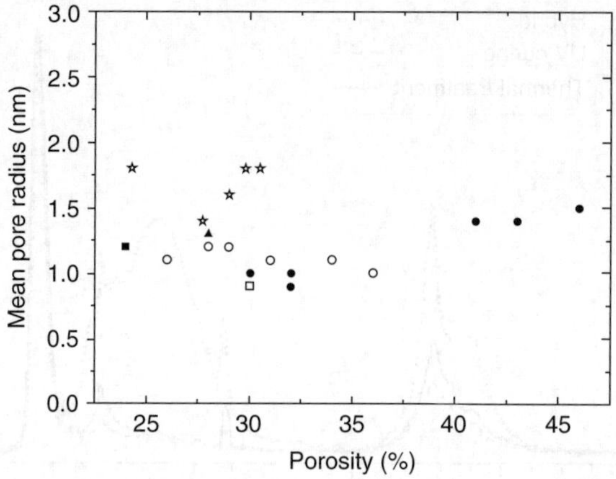


*Figure 2.13* *Mean pore sizes obtained by EP versus porosity:* ▲ *and* ■ *DEMS-ATRP e-beam and UV cured respectively* [17]*;* ○ *DEMS-NBD UV cured* [18], ★ *TMCTS-porogen thermally cured* [56] *and* ● *matrix-porogen UV cured* [53] □ PECVD 3MS-porogen (trimethylsilane) UV cured

porosimetry for several chemistries, deposition conditions and different post-treatments. As pointed out by many authors, the porogen approach by PECVD often allows small pores to be obtained [17, 18, 53, 56]. A slight increase of the pore size is noticed for high porogen loading, although the radius remains usually below 2 nm [53]. However, it is important to note that such small radii are at the bottom limit of the Kelvin model used to deduce the mean pore size [57]. In addition, it would be significant to compare other characteristics such as pore size distribution and interconnectivity between pores. However, a systematic comparison remains difficult because these data are more dependent on the characterization method [56, 58, 59]. Measurement sensitivity with respect to pore size and interconnectivity can lead to significant differences in apparent porosity levels for a given material using different porosimetry techniques.

### 2.4.2 Chemical Structure and Mechanical Properties

The introduction of porosity into an amorphous matrix induces strong modifications of elastic properties of thin films. Figure 2.14 shows the variations of the reduced modulus, evaluated by nanoindentation, versus the porosity for different cured films. As expected, a sharp decrease of the elastic properties is observed when more porosity is created. Several authors have shown that this behaviour is compatible with a model typically used for foam [18, 60]. Indeed, classical foam mechanical models predict a linear relation between the modulus of porous materials and their square density [61], with the assumption that the material skeleton remains unchanged. However, this model fails to explain the difference between mechanical properties observed on materials cured with different post-treatments, even though they have the same density (porosity) [60].

A typical example is presented in Figure 2.14, which evidences that, at equivalent porosity, a thermal curing leads in general to porous films with weaker mechanical properties than those obtained for UV cured films. This result can only be explained by taking into account the chemical structure of the porous film.

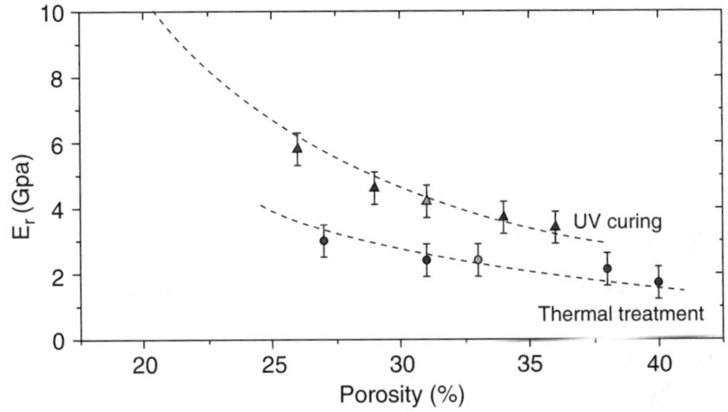


***Figure 2.14*** *Reduced modulus ($E_r$) obtained by nanoindentation as a function of the porosity for DEMS-NBD after UV or thermal curing* [18]

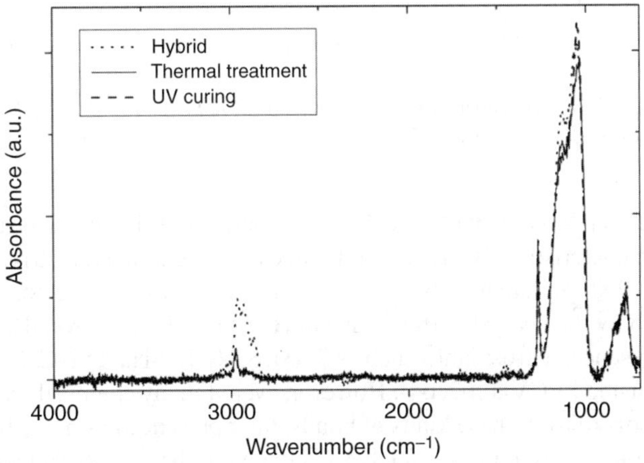


***Figure 2.15*** *FTIR spectra of different porous a-SiOCH (k = 2.4–2.5). The FTIR spectrum of the hybrid film is also shown*

To do so, it is important to have a clear view of the film structure and to understand better what kind of role the SiOCH skeleton can play in the physical properties (elastic and also electrical). Even though various chemistries, different deposition conditions and curing techniques are used to perform porous SiOCH material using the porogen approach, many similarities exist between all of them in terms of composition and skeleton structure. Figure 2.15 shows FTIR spectra of porous SiOCH films performed by PECVD using the porogen approach. A first-order analysis shows that these spectra are quite similar and the skeletons of these materials (all porous and having dielectric constants close to 2.4–2.5) are mainly constituted by an Si–O–Si network with methyl groups bounded to silicon.

Both porous spectra present the same kind of contributions that have been intensively discussed in the literature [43, 62, 63]. The strong peak between 1040 and 1055 $cm^{-1}$ with its shoulder at about 1142 $cm^{-1}$ is mainly due to Si–O–Si bonds, but can also be potentially attributed to Si–O–C vibration, which usually occurs between 1100 and 1140 $cm^{-1}$. The sharp

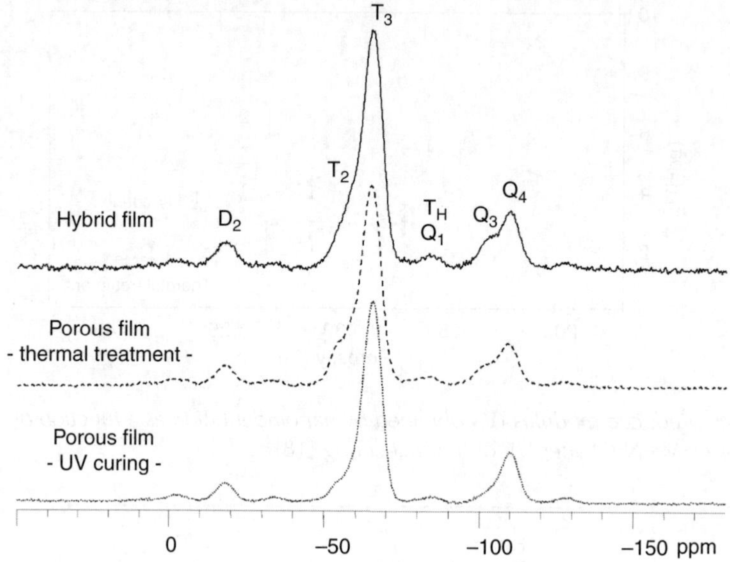


***Figure 2.16*** *$^{29}$Si NMR spectra of different porous a-SiOCH (k = 2.2–2.3). The NMR spectrum of the hybrid film is shown for comparison*

peak around 1275 cm$^{-1}$ is attributed to Si–$CH_3$ bonds, and the small peaks at 2975 and 2880 cm$^{-1}$ are characteristic of C–H stretching in $CH_3$ present in the matrix. The bonds centred around 800–900 cm$^{-1}$ are mainly due to $Si(CH_3)_n$. Few Si–H can be observed in PECVD film (symbolized by the peak at 900 cm$^{-1}$ due to H–SiO) [17, 62]. Several authors have also suspected the presence of cross-links such as Si–Si, Si–$CH_2$–Si and Si–$CH_2$–O–Si in SiOCH films deposited using PECVD [62, 64]. However, even if many chemical reactions may take place in the plasma to form these kinds of bonds, their presence has never been evidenced.

High-resolution solid-state NMR analysis is known to be very complementary to FTIR experiments to characterize the structural properties of thin films and to obtain information on the chemical environment of the observed atoms.

Figure 2.16 shows an example of $^{29}$Si NMR spectra performed on porous samples deposited using a porogen approach by PECVD after different types of curing. Several papers have already discussed such spectra and have proposed peak assignations [17, 42, 45, 51, 65]. The band between –120 and –80 ppm corresponds to Si bonded to four O atoms ($SiO_4$ denoted as Q); the band between –75 and –40 ppm can be correlated to Si bonded to three O atoms ($O_3$–Si–R, denoted as T); the band between –25 and –5 ppm is due to Si bonded to two O atoms ($O_2$–Si–R′, denoted as D) and the band between 5 and 15 ppm corresponds to Si bonded to only one O atom ($O_1$–Si–R′R″, denoted as M). R, R′ and R″ can be alkyl groups of different sizes. The number corresponds to the –OSi number linked to the Si atoms. The same contributions are observed for all the studied porous films, the main differences being in the T and Q environments (especially $T_2$ and $Q_3$), which are linked to film cross-linking. Many works confirm that, whatever the deposition technique used, the porous film skeleton is mainly constituted of an Si–O–Si bond network with $CH_3$ that are directly bounded to silicon. The first major difference between all different porous SiOCH is the amount of terminal bonds

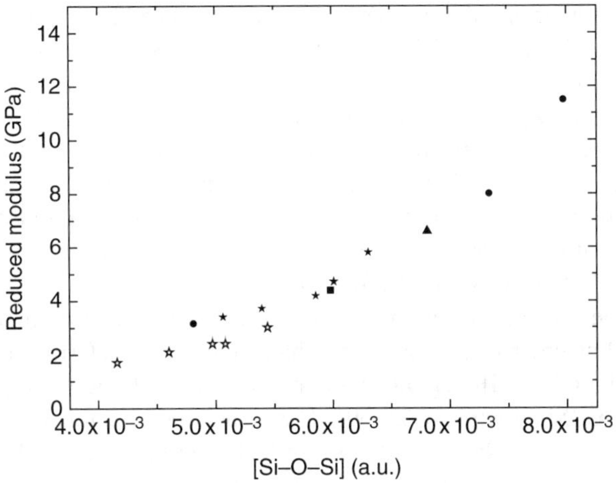


***Figure 2.17*** *Variation of the reduced modulus obtained by nanoindentation as a function of the volumic concentration of Si–O–Si bonds (obtained from FTIR analyses)* ([17], [18] *and* [67]*)*

(such as Si–$CH_3$, SiH and $SiOC_xH_y$). This quantity is directly related to the concentration of Si–O–Si bonds in the films. The other difference comes from the diversity in the Si–O–Si bonding environment. For instance, depending of the deposition conditions, porous SiOCH films deposited using OMCTS as the matrix precursor and cyclopentene oxide as the porogen can present various structures such as $D_2$, $M_1$ and MH, even after porogen removal [66].

As shown previously, the final skeleton structure can be tuned depending on the post-treatment used to cure the film. This phenomenon occurs by a modification of the concentration of Si–O–Si bonds (by cross-linking or densification) and a reorganization of the Si–O–Si bonding environment. The increase of Si–O–Si bond concentration can result from a cross-linking improvement due to better silanol condensation. Nevertheless, as discussed in the previous section, it has been shown that, in specific cases (where hydroxyl groups are not sufficiently incorporated), this increase can only be explained by film densification [33, 34].

The difference in skeleton cross-linking observed with different deposition techniques or post-treatments allow an explanation of the differences observed in terms of elastic properties. In a first approach, Ciaramella *et al.* have proposed a way to plot the elastic properties versus the volumic concentration of Si–O–Si bonds in the film evaluated thanks to FTIR spectra ([Si–O–Si] = area of the Si–O–Si band/film thickness; see Figure 2.17) [60]. It is worth noticing that the measurements of elastic parameters of a thin porous film are very difficult to obtain and the lack of harmonization across research centres makes it difficult to get any realistic material reference and ensure a faithful comparison between mechanical results. We only have compared, in this section, data from the same source.

This graph shows a good correlation between the concentration of Si–O–Si bonds and the elastic properties. This result has been confirmed by different groups [67–69], whereas Jousseaume *et al.* have highlighted that this approach can be extended to other SiOCH films (for different deposition techniques including spin-coating, various post-treatment, different porosity rates) [10].

It is also worth remarking that this concentration is related both to the film density and the skeleton chemical structure. This means that the Si–O–Si bond concentration (whatever results from cross-linking or densification phenomena) is the primary key parameter for the elastic properties of SiOCH thin films. Then, the enhanced mechanical properties observed after optimized post-treatments are mainly due to a higher concentration of cross-links. The improvement of the cross-link concentration is the key point indicating enhancement of the mechanical properties and the correlation between elastic properties and Si–O–Si bonds is possible only if cross-linking is done preferentially through Si–O–Si bonds (and not through Si–$CH_2$–Si or C–C bonds, for example).

Other approaches have also been proposed in order to take better account the impact of the Si–O–Si bonding environment on the mechanical properties. Grill and others have suggested that the Si–O–Si FTIR region should be decomposed into three peaks assigned to three types of Si–O–Si structures [62, 69]. They have operated this decomposition based on the FTIR signature of silicon oxide, which has been taken as a reference. Indeed, in fully relaxed stoichiometric thermal silicon oxides, the bonding angle is reported to be 144° with an FTIR absorption around 1080 $cm^{-1}$ and for low-temperature stoichiometric silicon oxides, the FTIR stretching frequency has been shown to decrease from 1080 to 1060 $cm^{-1}$ as the Si–O–Si angle decreases. The first peak of their decomposition, centred at 1135 $cm^{-1}$, was attributed by Grill and Neumayer to larger angle Si–O–Si bonds in cage-like structures with a bond angle of approximately 150° [62]. The peak at 1063 $cm^{-1}$ was assigned to the stretching of smaller angle Si–O–Si bonds in a network-like structure. A last peak close to 1023 $cm^{-1}$ was assigned to stretching of an even smaller Si–O–Si bond angle, such as might be encountered in a networked silicon suboxide. Thus, the Si–O–Si asymmetric stretching band frequency shifts are attributed to changes in the Si–O–Si bonding angle. Based on their analysis, they have observed that the elastic modulus increases while the mechanical stress decreases with increasing fraction of network oxide [70]. Huang *et al.* have also explained the increase of mechanical strength of porous SiOCH film after UV curing by the effect of Si–O–Si suboxide structures and the transformation of broken Si–O–Si cage-like bonds into an Si–O–Si network structure [69]. However, these approaches display only a partial explanation because they do not take into account the concentration of bonds, which must be proportional to the elastic properties. Iacopi *et al.* have proposed that, in addition to the increase in the number of cross-links, a structural rearrangement of the existing silica bonds into a small angle Si-O-Si configuration could be initiated by UV irradiation if large angle Si–O–Si bonds and structural imperfections are present in the matrix [42].

However, more recently, Gourhant *et al.* have emphasized that for hybrid SiOCH films deposited by PECVD with DEMS as the matrix precursor, Si–O–C bonds could be at the origin of a large part of the reorganization observed in the Si–O–Si band (their contribution being overlapped in the FTIR Si–O–Si band) [18]. They have suggested that the decrease at the shoulder centred at 1150 $cm^{-1}$ can be explained by the disappearance of Si–O–C bonds. Moreover, they have evidenced a typical cross-linking mechanism involving Si–O–C bonds in the creation of Si–O–Si bonds, this behaviour being promoted during UV curing. This later result tends to complicate the decomposition of the Si–O–Si FTIR peak and leads to an overestimation of the Si–O–Si bonds.

The elastic properties improvement discussed here is observed both after UV curing using a broadband UV lamp and for single wavelength UV curing. However, a specific mechanical

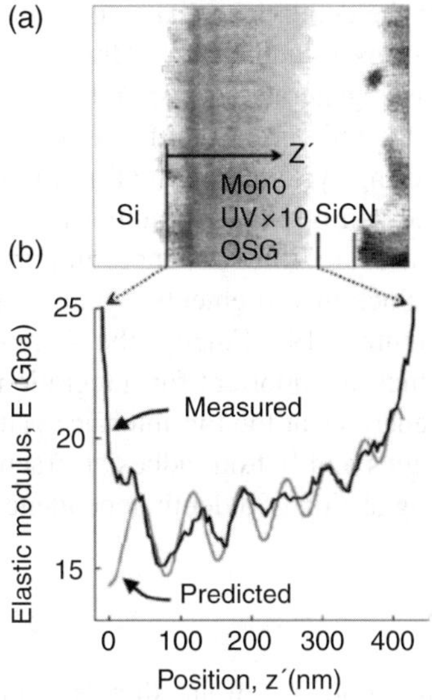


***Figure 2.18*** *(a) The frequency modulation (FM) signal image of a cross-section of a specimen containing a monochromatic UV × 10 organosilicate film. (b) An elastic modulus depth profile of the monochromatic UV × 10 film measured by FM-AFM is compared with one predicted using the standing wave simulation*

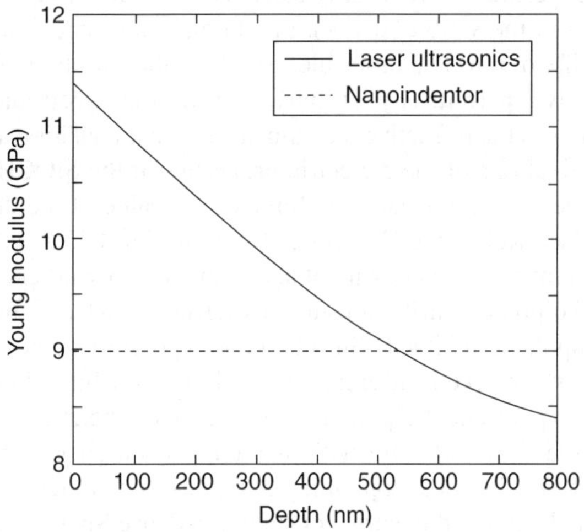


***Figure 2.19*** *Young's modulus of UV-cured CVD SiOCH film measured by nanoidentation and laser ultrasonics*

behaviour is observed using a monochromatic bulb due to the absorption spectra of organosilicate films. Indeed, the absorption spectra depend on the carbon concentration and type of porogen and sharply increases with decreasing wavelength [52]. This leads to a different depth dependence of the UV curing. For instance, studies using near monochromatic

(172 nm) UV radiation, where the depth dependence is significantly more severe, indicate UV light interference, which forms a standing wave during the curing process [52, 71, 72].

In general, the standing waves generate corresponding oscillation of Young's modulus (Figure 2.18), but this problem is only important for a monochromatic light source with a single lamp. In the case of a curing system with light of different wavelengths and with an appropriate design (for instance an optimized number and location of UV lamps), such gradients are not observed. Therefore, it is very more important to be able to measure the gradient of Young's modulus. Such measurements can be done by using picosecond ultrasonic interferometry [73] (Figure 2.19). Finally, the formation of these gradients also affects the fracture energy, which is important for integration. For instance, it was shown that, while adhesive fracture energies at the top interface can be improved by as much as 200 %, cohesive fracture energies and bottom adhesive fracture energies are improved at best by 20 %, which is clear suggestion of a depth dependence of the UV curing [71, 72].

### 2.4.3 Electrical Properties

The first objective to introduce porosity in an SiOCH matrix is to lower the dielectric constant. Minimizing this parameter must be first operated by the identification of film components that affect the electrical properties. In an insulating material, the dielectric constant is the parameter that describes the modification of the charges distribution when the material is put in an electric field. To obtain the static relative permittivity of the material ($k = \varepsilon_r = 1 + \varepsilon_e + \varepsilon_{ion} + \varepsilon_{dip}$) and to take into account the total polarization (electronic, ionic and dipolar), experiments are usually carried out at a low frequency ($f \leq 1$ MHz). At a high frequency (from the visible to UV), the dipolar and ionic contributions are weak because of the inability of ions and molecules to follow the variation of the electrical field. Therefore, the relative permittivity is only due to the electronic polarization and $\varepsilon_r = 1 + \varepsilon_e = n^2$, where $n$ is the refractive index measured in the visible range. There are only a few detailed investigations of the dielectric properties of the SiOCH material in the literature [74–77]. In general, information is limited to a value of the dielectric constant at 100 kHz or 1 MHz (see Reference [78] for a discussion on CV measurements) and to a value of leakage current and breakdown voltage. Kim and co-workers initially studied the electronic, ionic and dipolar contributions to the dielectric constant and found evidence of a strong dipolar component in SiOCH films that could possibly come from Si–$CH_x$ groups [75]. This constitutes a marked difference with $SiO_2$ for which dipolar polarization is quasi-null (there is no permanent dipole in $SiO_2$ due to tetrahedral symmetry). They have shown that the (dipolar + ionic) contributions are always higher than the electronic part except in the case of SiOCH films containing a high carbon content (%C > 50 %). Gonon *et al.* have studied in detail the dielectric constant of a dense SiOCH films as a function of the frequency and temperature and have shown that the method used to incorporate carbon into the film has a huge impact on the ionic and dipolar parts of the permittivity [74]. They have also confirmed an (ionic + dipolar) contribution of the same order of magnitude as the electronic part of the permittivity ($\varepsilon_e = 0.93$, $\varepsilon_{ion} = 0.53$ and $\varepsilon_{dip} = 0.37$). More recently, Gourhant *et al.* have extended this investigation for several porous SiOCH films and show that the electronic part of the permittivity is always higher than the others [77]. They have obtained an (ionic + dipolar) contribution close to the electronic part of the permittivity ($\varepsilon_e = 0.75$, $\varepsilon_{ion} = 0.38$ and $\varepsilon_{dip} = 0.24$ for a sample with $k = 2.3$). Moreover, they have

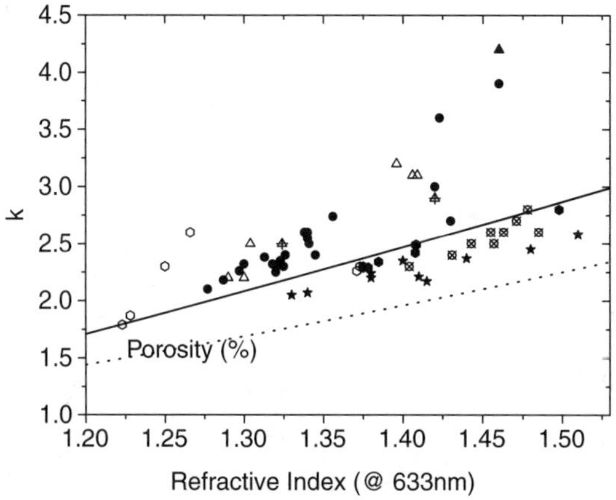


*Figure 2.20* *Variation of the dielectric constant of different SiOCH versus the refractive index measured at 633 nm. The full line corresponds to the sum of the vacuum dielectric constant and the electronic constituent of the dielectric constant* ($n^2 = 1 + \varepsilon_e$)*:* ● [17, 18, 67], Δ [19], ≑ [79], ▲ *$SiO_2$ PECVD,* ★ [21, 27], ⊠ [15], ○ [53], ⬢ [80]

observed that the ionic contribution of the dielectric constant decreases continuously with the porosity, this contribution being mainly correlated with the concentration of Si–O–Si bonds in the film. In addition, the dipolar contribution remains unchanged for different porogen loadings (except for highly loaded samples). Similar contributions to the dielectric constant are observed when the sample is either UV cured or thermally treated [77]. Figure 2.20 shows the evolution of the dielectric constant versus the refractive index (at 633 nm) for different SiOCH films. The dot line corresponds to the sum of the vacuum permittivity and the contribution of the electronic polarization ($n^2 = 1 + \varepsilon_e$). The solid curve is deduced from Gourhant's works and shows that dipolar and ionic contributions are significant in SiOCH films. The material close to this line corresponds to thin films containing very few SiOH bonds (unfavourable species because they are very easily polarizable) and thus close to the ideal cross-linking state. It seems very unlikely to have points far beneath this limit (for a SiOCH skeleton) even if many experiments have reported such results in the literature. The lack of unique methodology as a reference in measuring the dielectric constant could be responsible for this disparity in results.

Different authors have also investigated the leakage current of porous SiOCH and have found that the electrical conduction in these dielectrics is controlled mainly by the Poole–Frenkel mechanism, which is independent of porosity [81, 82]. The breakdown field usually decreases with increasing degree of porosity, and for UV-cured films deposited from DEMS + NBD, it decreased from >7 MV cm$^{-1}$ for a $k = 2.4$ film to 6 MV cm$^{-1}$ for a $k = 2.2$ film [11].

## 2.5 Limit/Issues with the Porogen Approach

As pointed out in the previous section, the porogen approach suffers from several issues, especially at high porogen loading. The following section reports the main challenges that have to be faced.

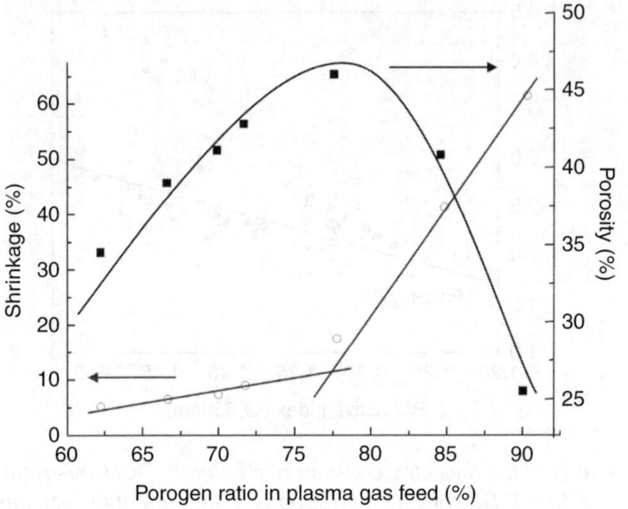


*Figure 2.21* *Film shrinkage and porosity of PECVD films as function of the porogen loading (after UV curing at 400 °C)*

### 2.5.1 Porosity Creation Limit

The porogen is introduced as a sacrificial specie in the hybrid film with the aim to convert all organic sites in the final porosity using an appropriate curing process. Nevertheless, this ideal case is far from what can be possibly done. The study of the porogen conversion into porosity indicates the existence of a threshold. Below this limit, the increase of porogen loading leads to more porosity. Above a porogen loading limit, no more porosity is created due to a significant increase of film shrinkage (as shown in Figure 2.21). This shrinkage becomes the major phenomenon that limits porosity generation [11, 18, 25]. Favennec *et al.* have shown that the film shrinkage can be correlated to the weakness of the skeleton. Indeed, for a high porogen loading, the weakness is not due to the local structure but mainly to the too-small quantity of matrix in the film [25]. When a lot of porogen is incorporated, there is not enough Si–O–Si structure, which mainly constitutes the matrix skeleton. When porogen is degraded during curing, the porous material has a lower Si–O–Si concentration than a threshold and the skeleton is not sufficiently strong to prevent collapse. This threshold delimited a particular behaviour, which is displayed as U-shape evolution of the dielectric constant versus the porogen loading (Figure 2.22).

The reinforcement of the skeleton structure by additional 'useful matter' can be obtained by tuning the deposition conditions and the post-treatment used to cross-link the film. The maximum of porosity is then also a function of these processes [18]. A higher deposition temperature (below porogen degradation) has been used to strengthen the hybrid film structure (up to 300 °C for DEMS + NBD chemistry) [18]). In addition, as already mentioned in the first section, an appropriate selection of the matrix precursor and of the deposition condition (especially by addition of oxygen during the hybrid deposition) limits the film shrinkage during porogen removal [25]. Gourhant has shown that porous ULK presenting a dielectric constant of 2.1 (using thermal treatment) and 2.3 (using UV curing)

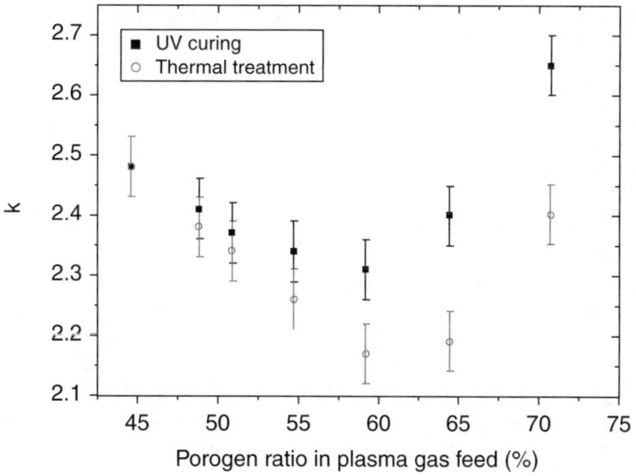


***Figure 2.22*** *Evolution of the dielectric constant versus the porogen ration in plasma gas feed*

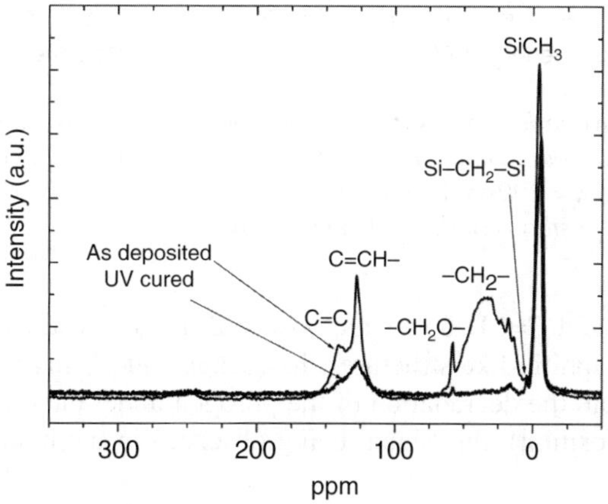


***Figure 2.23*** *$^{13}$C NMR spectra of as deposited and porous ULK films (after UV curing)*

can be achieved [84]. Grill and others have also reported the realization of porous film with *k* close to 2 after thermal curing [21, 56]. More recently, Urbanowicz *et al.* have shown the possibility of preparing porous films with a porosity higher than 40 % using an appropriate curing strategy; *k* close to 1.8 has been measured [53].

### 2.5.2 Porogen Residues

Even if the possible existence of porogen residues after curing has already been reported, their effects on the low-*k* properties have been studied only recently. Jousseaume *et al.* have highlighted the presence of C=C bonds in nuclear magnetic resonance spectra after e-beam or UV curing [17]. Gates *et al.* have also reported their formation after thermal curing [66]. Indeed, Figure 2.23 shows the $^{13}$C NMR spectra of hybrid and porous

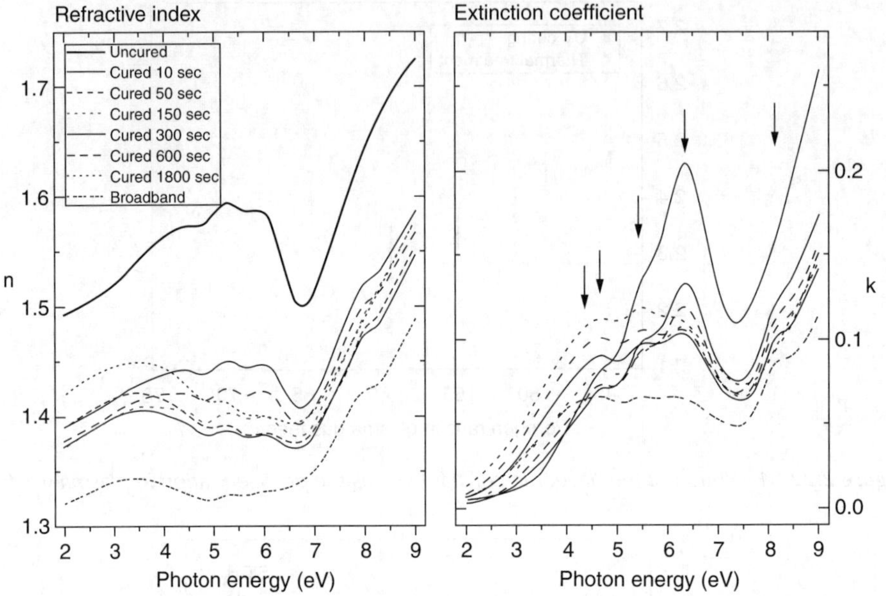


***Figure 2.24*** *Refractive index and extinction coefficient of uncured (solid thick black line) and UV-cured low-k dielectric films. The optical function of the sample cured by a broadband lamp with $\lambda = 200\,nm$ is plotted for comparison (dash-dot-dot line). The arrows show the positions of the G–L peaks* [83]

SiOCH films. C=C and C=CH bonds are observed in the spectra with a chemical shift corresponding to graphitic-like structures. Jousseaume *et al.* suggested that such structures could arise from the degradation of the porogen under energetic conditions in the plasma (during deposition), this signal being observed in the hybrid film and not in a matrix film [17].

These contributions also remain present in the porous film, indicating that the curing treatment does not completely remove these structures. The hypothesis that porogen species can also be partially converted by UV light into nonvolatile graphitized-carbon residues is possible, but probably not important (it is, for instance, interesting to notice that usually these types of graphitic-like residues are not observed in spin-on films after UV curing [45]).

Although some progress has been made on porogen residues investigations, their detection remains uneasy. Indeed, their quantitative evaluation cannot be performed by Fourier transform infrared spectrometry because it has a limited sensitivity to amorphous carbon, such as C=C and C–C bonds. NMR is a powerful technique used to highlight the presence of residues but it remains costly and difficult to set [45]. Recently, Marsik *et al.* have developed a versatile method based on UV spectroscopic ellipsometry (UVSE) in order to estimate accurately the amount of residues [83]. They have found that the optical properties of the low-*k* films between 2 and 9 eV are mostly sensitive to the presence of porogen and its removal during UV curing. Figure 2.24 shows a typical example of optical functions of various UV cured samples.

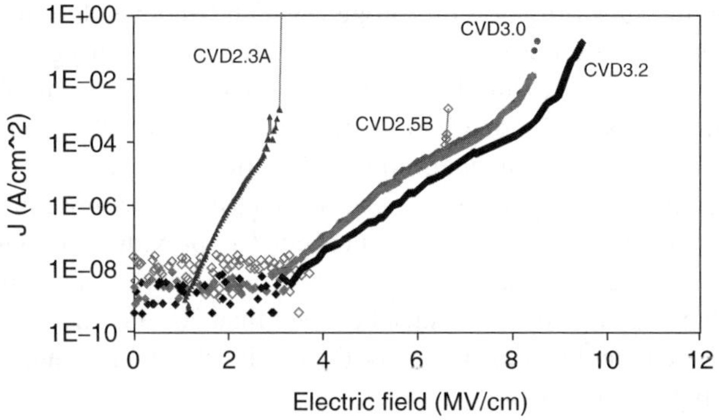


***Figure 2.25*** *CVD2.3A – CVD low-k with* k = *2.3 cured with lamp A (172 nm); CVD2.5B – CVD low-k with* k = *2.5 cured with lamp B (>200 nm); CVD3.0 and CVD3.2 are nonporogen films, not UV cured* [85]

The figure emphasizes that the band at 4.5 eV (in the extinction coefficient curve) is due to a porogen residue: $\pi$–$\pi^*$ electronic transitions between the $sp^2$ carbon orbital in the ultraviolet. A following of this band versus the UV curing time shows clearly that few porogens are still kept within the materials and do not disappear, even for long curing times. This carbon fraction could thus be considered as a porogen residue in the low-*k* material. The amount and the nature of residues are closely dependent on porogen chemistry and composition, deposition and curing conditions.

These porogen residues can seriously impact physical properties of porous thin film. For instance, Gourhant has shown that the presence of porogen residues leads to an increase of the electronic and dipolar contribution of the permittivity [84]. Indeed, the porogen residues are carbon bonds containing graphitic-like phases (C=C–H). These structures hold a permanent dipolar moment and nonlocalized valence electrons. This would explain the negative effect of porogen residues on the electrical properties of porous SiOCH films. Moreover, these graphitic-like structures are more conductive than the SiOCH skeleton. Baklanov *et al.* have shown a clear correlation between ‘porogen residue’ and leakage current [85]. As an illustration, Figure 2.25 shows leakage current that shifts to a high value with more porogen residues (such as CVD $k$ = 2.3 films). The breakdown field is also affected by the presence of these residues.

Moreover, the presence of porogen residues also impacts the elastic properties of porous SiOCH films. For instance, in the case of very high porogen loading (>60 %), the mechanical strength is lower than those simulated by the foam mechanical model [18]. Actually, even though these graphitic structures are part of the film skeleton, their contribution to the film strength and to the elastic properties are worse than that of Si–O–Si bonds. This result has been confirmed by Urbanowicz *et al.*, who have shown that excellent mechanical properties are obtained for ‘porogen-residues-free’ highly porous films [53].

In conclusion, these graphitic structures, which are detrimental from the electrical point of view, should also be avoided in order to have the optimal Young’s modulus/porosity ratio. Some ideas to limit the formation of porogen residues are reported in the literature. Gourhant has shown that a too-high deposition temperature is not appropriate because it favours the graphitic transformation of porogen, which becomes difficult to extract [84]. This leads to an incomplete porogen removal, even for a long curing time. Marsik *et al.* have also reported that a broadband lamp with $\lambda > 200\,\text{nm}$ is more suitable as it produces less residues [83]. Finally, Urbanowicz *et al.* have used both a hydrogen plasma treatment and UV curing in order to combine their effects [53]. In this case, hydrogen plasma is first performed to degrade the porogen species and make easier their subsequent extraction by conventional UV curing. This process has succeeded in providing porous materials with undetectable porogen residues and maximum porosity (46 %). The limitation of this approach is that the targeted thickness must be lower than 200 nm. Indeed, when the film is thicker, hydrogen treatment can also densify the top layer, which induces inhomogeneity and prevents the porogen extraction.

## 2.6 Future of CVD Low-*k*

As mentioned before, the creation of porosity and the enhancement of mechanical properties have both to be addressed. However, porosity means free volume and low density while mechanical strength requires bond concentration and film densification. Then a trade-off is necessary. This is one of the main challenges that is being investigated by researchers and manufacturers. One way consists in using new precursors that allow enhancement of the skeleton of porous SiOCH while maintaining a low *k*. Other approaches (apart from the porogen one) to perform porous thin films by CVD techniques are also being investigated. Finally, we discuss the optimization of SiOCH thin films in order to sustain the other processes used during the integration (etching and CMP, for example).

### 2.6.1 New Matrix Precursor

Recent theoretical studies were performed by Tajima *et al.* in order to explore the appropriate chemical structures that allow the enhancement of the elastic properties of low-*k* films. Using atomistic simulations, they have predicted that PECVD SiOCH films containing methane, ethane or propane bridging units should deliver a significantly higher modulus [86]. They concluded that SiOCH films with small dielectric constants can have markedly improved mechanical strengths without adversely affecting the small *k*, if the hydrocarbon components form cross-links instead of a network terminal. This concept makes sense since the SiOCH structure is based on the Si–O–Si network with terminal carbon structures; a limitation of these terminal bonds will lead to improved mechanical properties.

Dubois *et al.* first demonstrated experimentally that organosilicates with improved mechanical properties can be obtained at high levels of porosity using bridged oxycarbosilane precursors in a sol–gel approach (Figure 2.26) [87]. They have shown that the Young's modulus of these carbon-bridged spin-on organosilicates films remain higher when compared to traditional MSQ materials (Figure 2.27). They have also shown that this elastic modulus decreases linearly with increasing porosity (i.e. decreasing density), which seems a specific characteristic of porous silicates incorporating organic bridges [87].

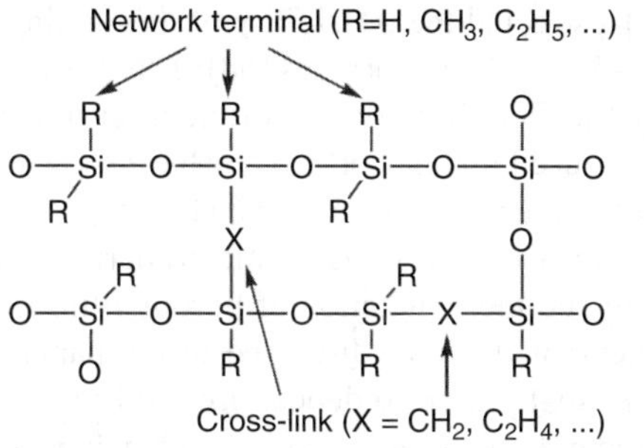


***Figure 2.26*** *Hydrocarbon groups at the network terminal and those in the polymer network (cross-links) of SiOCH films [86]*

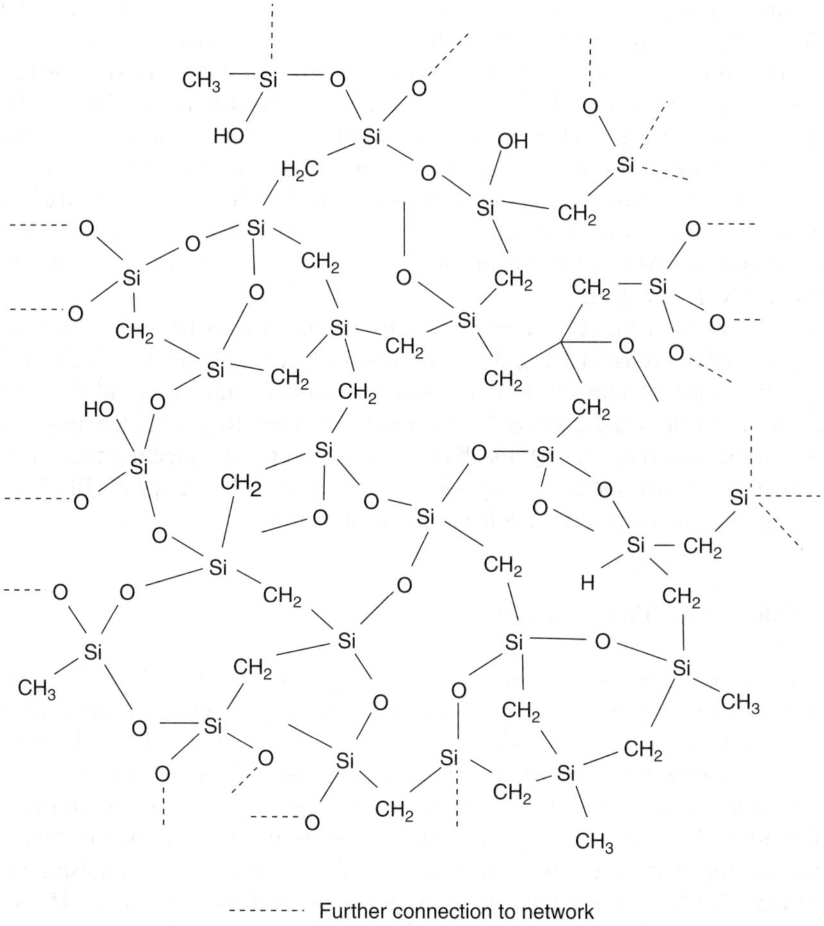


***Figure 2.27*** *Illustration of a portion of the fully condensed network structure (ideal case) of oxycarbosilane-based materials [4]*

Shinriki *et al.* have investigated the possibility of developing this concept by PECVD using SiOCH films with –Si–$C_2H_4$–Si– bridges [88]. The films were deposited by PECVD using trivinylmethoxysilane (TVMOS) or vinyltrimethoxysilane (VTMOS) as the matrix precursor (no porogen was used in this work). They have shown that organic bridges were successfully incorporated in the thin films (using FTIR), especially in the case of TVMOS. Unfortunately, no clear improvement of the elastic properties was proven. More recently, Gates *et al.* have investigated the realization of porous films deposited by addition of a specific carbosilane precursor (CP, not disclosed in the paper) containing–Si–$CH_2$–Si– groups to the precursors usually used to deposit hybrid films (with DEMS as the matrix precursor and NBD as the porogen precursor) [20]. They have performed porous SiOCH material exhibiting a *k* value between 2.2 and 2.4 after UV curing. The presence of Si–C–Si bonds has been observed using FTIR. They have also demonstrated that the addition of the CP precursor does not significantly change the dielectric constant while it permits the control of the film porosity characteristics. However, CP introduction has not proved any improvement of the mechanical properties. Worse, it exhibits films presenting Young's moduli that are slightly inferior to the reference without CP. Therefore, the proven benefit of carbon bridging bonds to the mechanical properties in spin-on oxycarbosilanes has not been yet evidenced for PECVD material. Additional experiments have to be conducted in order to make a reliable advantage to the use of such a kind of structure. As discussed previously, PECVD, unlike spin-on, involves more complex chemistry reactions in the plasma phase. All attempts to control the PECVD film microstructure that would enable a better improvement of targeted properties (mechanical and electrical) should be pursued. Moreover, the impact of the UV curing on these organic bridges remains poorly documented.

In conclusion, the benefit targeted by adding organic bridges between Si atoms (such as –Si–$CH_2$–Si) in porous SiOCH thin films deposited by PECVD still remains a challenge. However, this approach remains very attractive in order to obtain porous SiOCH thin films with ultra-low-*k* values and improved mechanical properties. Regarding the progress operated on spin-on materials, this approach has to be supported by selecting the right conditions (chemistry, deposition and curing parameters) to succeed. Once again, PECVD could make a huge step forward following the progress of spin-on.

### 2.6.2 Other Deposition Strategies

The porogen approach, using a couple (or more) of precursors, is not the only way to produce porous SiOCH films by PECVD. There are other ways where only one kind of molecule is used to deposit ultra-low-*k* films. However, as pointed out by Grill in a recent review [11], most of these studies used a thermal curing to degrade and remove organic fractions from the film, making it (more) porous. This method can be considered as subtractive-based, where sacrificial components are originated from the monomer itself. For instance, the team of Shi-Woo Rhee [89] has deposited a porous film using vinyltrimethylsilane (VTMS) in a mixture with oxygen. After a thermal curing at 450 °C, they obtained a dielectric constant as low as 2.0. However, the *k* value seems incompatible with the refractive index reported (higher than 1.40) and, thus, such results have to be considered very carefully.

Navamathavan *et al.* have used the same approach with a single precursor (bistrimethylsylylmethane or BTMSM) mixed with oxygen [90]. The best *k* values are reported equal to 2.38 after thermal annealing at 500 °C. In addition, Wu and Gleason have shown that films deposited from mixtures containing methylsilane (1MS), dimethylsilane (2MS) or trimethylsilane (3MS) with oxygen and argon can allow dielectric constants of 2.4 to 2.6 to be achieved after annealing at 550 °C [91]. A similar result has been obtained by Kim *et al.* using bistrimethylsilylmethane (BTMSM), an Si–Me rich precursor. Films deposited at 30 °C with oxygen and annealed at 500 °C exhibit a *k* value of 2.44 [92]. Even though most of these low-*k* values have been reported for films prepared from a single precursor, they have also been processed at very high temperatures, usually >500 °C.

Another approach to perform porous low-*k* by PECVD was initially proposed based on the incorporation of siloxane rings into the thin film, the idea being to introduce microporosity in the material through cyclic structure. Tada *et al.* have used a cyclic siloxane (V3D3), where the methyl group on each silicon atom can be replaced by longer alkyl chains [93]. These monomers can be deposited by PECVD at 350 °C to generate porous SiCOH with *k* <2.5. No porogen addition and/or post-deposition annealing were necessary to achieve this low dielectric constant. Moreover, in addition to the vinyl groups, longer alkyl chains on the silicon concurrently lead to films with *k* values as low as 2.3. However, there is still an inconsistency between the low dielectric constant and the refractive index of such types of films (also known as a molecular pore tack (MPS), SiOCH [94]), the refractive index being nearly 1.46. As mentioned before, the deposition of decamethylcyclopentasiloxane (DMCPS) by PECVD has been tested in order to introduce large siloxane cages in the deposited thin film. In that case, even if FTIR analysis shows that soft plasma conditions keep the siloxane rings intact within the film, no porosity creation has been evidenced [22, 25]. The dielectric constant cannot be lowered below 2.7. Thus, the ability of this approach in lowering the dielectric constant must be considered carefully.

A last possible approach to perform highly porous SiOCH is a 'foaming technique', which allows porosity to be created by the use of gas nucleation in a polymer matrix [95]. This technique, which is used, in general, to produce macropores in thick polymers, was also reported for thin SiOCH films [67, 96]. It consists in depositing an SiOCH film by PECVD using an organosilane precursor (tetramethylsilane (4MS) for instance) with oxygen at room temperature. The film deposition is followed by a hydrogen plasma treatment at 400 °C (Figure 2.28).

The post-treatment induces simultaneously a film thickness increase (unlike in the case of the porogen approach), a strong carbon depletion (with a decrease of Si–$CH_3$ bonds after

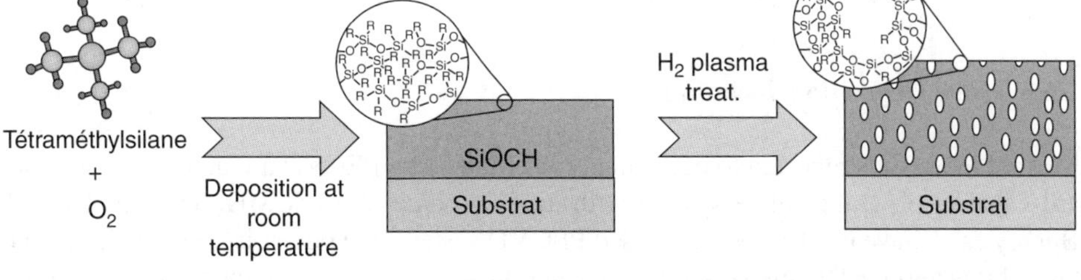


***Figure 2.28*** *Example of foaming process by PECVD*

*Figure 2.29* *Transmission electron microscope/electron energy-loss spectroscopy of a porous thin film obtained by a foaming process. A dense $SiO_x$ crust is clearly detected at the top surface (Si not shown)*

plasma curing) and a film density decrease and allows the formation of a porous layer [97]. Previous works have tried to explain the porosity formation by invoking skeleton restructuration mechanisms through a combination of effects due to the thermal curing and $H_2$ plasma [98]. It was suggested that the process might consist of the rupture of Si–$CH_3$ groups in favour of a less bulky structure (such as Si–H bonds) without any network collapse. However, Jousseaume *et al.* have shown that the right mechanism involved in this case is probably closer to a foaming phenomenon [67]. Indeed, in a polymer foaming process, the film should be saturated by a gas in such a way that appropriate temperature curing allows the gas to coalescence and the formation of a porous layer [95]. In this specific plasma approach, hydrogen reacts with methyl (from Si–$CH_3$) during the post-deposition hydrogen plasma treatment to form methane. In the same time, a dense '$SiO_2$-like' crust is formed at the top of the SiOCH thin film (see Figure 2.29), preventing the gas from escaping through the SiOCH films. Then it can coalesce and allows the formation of nanopores, even for thin films (for thicknesses lower than 200 nm). The porosity can be adjusted as a function of the curing treatment condition. A $k$ value as low as 2.2 has been obtained with porosity higher than 40 % and mean pore diameter close to 3 nm [67].

Nevertheless, in the case of this foaming approach, the porous thin films exhibit porosity and composition gradients: the bulk film is highly porous and the film is only slightly porous or even dense close to the interfaces. This gradient constitutes one of the main drawbacks of this approach. Some solutions were recently proposed in order to avoid the presence of a gradient by intentionally depositing a dense top layer and by using more uniform post-treatment (such as UV curing) [97]. By combining the foaming technique with a porogen approach, it is claimed that a highly porous dielectric (porosity >50 %) by PECVD can be performed.

### 2.6.3 New Deposition Techniques

Most of the results reported to perform porous SiOCH by PECVD are done using a standard capacitive parallel plate system with an RF source at 13.56 MHz [12–15, 18, 25]. Burkey *et al.* have used a specific pulsed PECVD system (the power is pulsed), the advantage of this approach being potentially a better preservation of the precursor structure [24]. However, PECVD forms films from precursors that are highly dissociated due to collisions

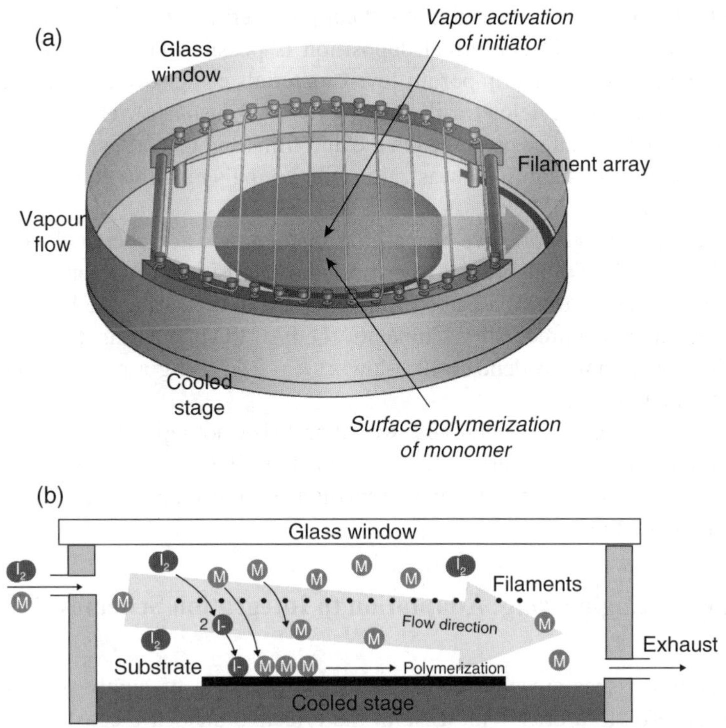


***Figure 2.30*** *Principle of initiated CVD (iCVD) [100]*

in the plasma, and UV and charged particles continuously irradiate the substrate surface. The plasma, even at low powers, provides such an abundance of energy to the monomer species in the reactor that the desired precursor functionalities are mostly not retained in their entirety in the deposited film. In order to preserve the precursor structure better, new deposition techniques have been investigated. Yasuhara *et al.* have developed a neutral-beam-enhanced CVD (NBECVD) in which the precursors are directly injected into the process chamber and adsorbed on to the substrate surface (the substrate temperature is kept at –20 °C in order to absorb the precursor efficiently on the substrate surface) [99]. Then, an accelerated nonreactive Ar neutral beam is used to bombard the surface without charged particles or UV photons. The random dissociation of these precursors in the gas phase can thus be nearly eliminated, and low-*k* films seem to be deposited with structures corresponding to those of the precursors. Gleason and her team have promoted an original CVD process that allows performing polymer thin films by an iCVD method (initiated CVD) [100]. This technique avoids plasma excitation and damage to the growing film as it provides only enough thermal energy to cleave the easily activated thermally labile bond in the initiator species, while keeping the monomer structure intact. A typical iCVD process uses a resistively heated filament array controlled accurately at temperatures ranging from 200 to 550 °C. The filament temperature is used to decompose the initiator and pyrolize it to form radicals (see Figure 2.30). The radicals and the monomer species migrate to a temperature-controlled cooled substrate, typically at room temperature, and adsorb on to the surface to undergo free radical polymerization, forming a polymer film. Using this concept, highly conformal deposition is possible [101]. Trujillo *et al.* have applied this technique to perform porous low-*k* materials using several chemistries (V3D3 and tetravinyltetramethylcylclotetrasiloxane (V4D4) for instance). They have shown that the precursor structure can be preserved in the thin films and that, after a thermal annealing at 410 °C, dielectric constants as low as 2.15 (with Young's modulus equal to 3.2 GPa) can be obtained [102].

More recently Faguet *et al.* have proposed a filament-assisted chemical vapour deposition technique (FACVD) in order to perform the deposition of porous SiOCH [103]. FACVD is based on the use of a heated zone adjacent to the surface to be coated to fragment the input chemical precursors. Unlike iCVD, FACVD processing does not require an initiator. They were able to deposit ultra-low-*k* ($k = 2.2$) using a porogen approach and DEMS as the matrix precursor.

These advanced deposition techniques would probably not replace the classical PECVD process, which is usually preferred for the deposition of ILD in microelectronics, but it opens a new way, for example, for the deposition of conformal porous low *k* for three-dimensional integration.

## 2.7 Material Engineering: Adaptation to Integration Schemes

Many integration issues are observed with the introduction of highly porous materials, leading, in general, to a drastic increase in the effective dielectric constant in integrated structures [104–108]. In particular, because the porosity is open and interconnected, reactive gases and chemistries (used during etching, resist removal, cleaning and metal deposition) can easily penetrate into the porous structure and damage its intrinsic properties.

To overcome these material degradations, usual integration strategies have settled these issues by the insertion of new processes that induce an increase in the effective dielectric constant and the process step numbers (use of a dielectric hard mask, pore sealing treatment, repair, etc.). Moreover, as explained in a previous section, the introduction of nanopores in a dielectric matrix drastically decreases the film mechanical properties, which could have tragic consequences during the chemical mechanical polishing (CMP) and the packaging steps.

One possible way to limit the impact of plasma-induced damage (PID) consists in optimizing the deposition and curing conditions of the ULK. A challenge is to improve plasma damage resistance without compromising its mechanical strength and dielectric constant. Ryan *et al.* have shown that an increased content of Si–$CH_3$ can allow an improvement of plasma damage [109]. Furasawa *et al.* have optimized the UV curing in order to limit the impact of an $NH_3$ plasma treatment [110]. In all cases, a trade-off should be found in order not to decrease too much the amount of Si–O–Si and to keep good mechanical properties. However, these approaches that consist in modifying the skeleton structure still have limitations.

To do otherwise, an original integration scheme (known as 'solid first') has been proposed by Calvert and Gallagher [111]. It allows the solid phase processing of a spin-on porous low-*k* material. As shown in Figure 2.31, this approach uses a hybrid material made of a methylsilsesquioxane matrix and labile organic nanoparticles (porogens). Critical process steps such

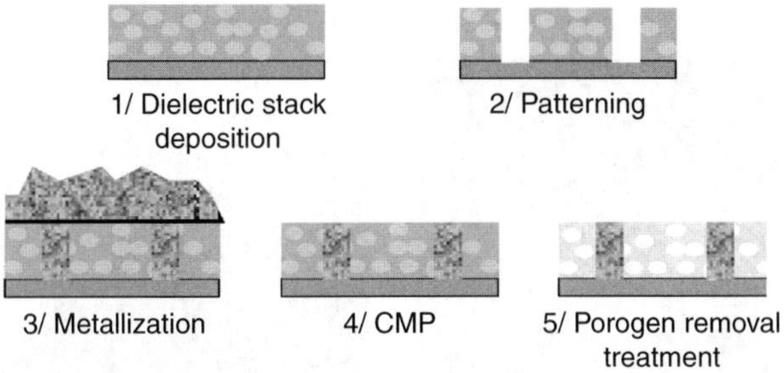


***Figure 2.31*** *Solid first approach integration scheme*

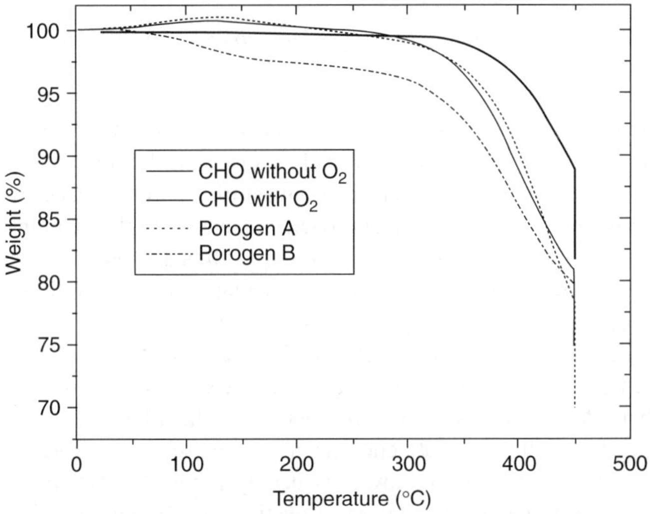


***Figure 2.32*** *TGA curves of hybrid films. The hybrid deposited using CHO and $O_2$ in the gas feed is compared to the hybrid deposited without $O_2$ and to hybrids deposited using other porogens*

as etching, resist removal, metal deposition and CMP are performed on a nonporous film and the porosity is created after integration by using an appropriate porogen removal treatment.

This integration strategy is able to solve major issues of porous dielectric integration, but requires the development of a specific hybrid film with defined characteristics. In particular, the porogen thermal stability should be improved towards a high temperature (typically above 325 °C) to allow the use of conventional BEOL integration processes [112]. Obviously, this should be operated without any porogen degradation. Afterwards, the porogen removal treatment should allow the creation of porosity with minimum film shrinkage at the standard curing temperature (typically 400 °C). Favennec *et al.* have studied the thermal stability of different hybrid films in order to use this 'solid first' integration scheme with PECVD materials (using DEMS as the matrix precursor) [113]. Figure 2.32 shows the thermogravimetric analysis (TGA) of hybrid films deposited in appropriate conditions with different porogens. They have shown that cured films deposited using DEMS and CHO

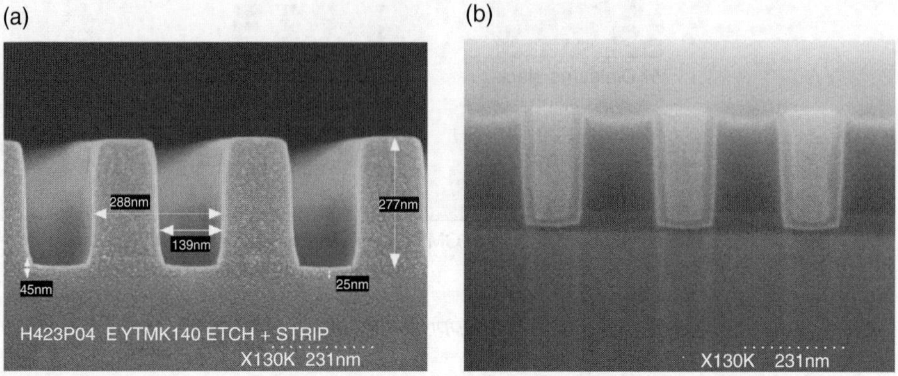


***Figure 2.33*** *SEM pictures after etching (a) and after integration (b) using a post-integration porogen removal (the hybrid (nonporous) film was deposited using DEMS and CHO)*

mixture still behave as a dense film at temperatures close to 380 °C, without significant porogen degradation. A $k = 2.3$ material was then successfully integrated without a hard-mask on one metal level (Figure 2.33) using conventional BEOL processes [113, 114]. To conclude, this post-integration porogen approach can overcome the main integration issues encountered with porous ULK integrations. Moreover, this approach is extendible to next-generation nodes: a lower final $k$ value only means a larger porogen ratio in the composite material, the integrated material being always nonporous.

This post-integration porogen approach could be a competitive solution for advanced nodes when porosity becomes too important and the pore size is close to (or larger than) 2 nm. However, further improvements will probably be required in order to limit the film shrinkage impact occurring during the UV treatment. The integration scheme should also be adapted in order to allow complete porogen removal, especially for dual-damascene structures. The recent works on the post porosity plasma protection which consists of filling the ULK porosity with a degradable organic polymer prior to integration constitute a promising improvement in this direction [115].

## 2.8 Conclusion

Low dielectric constant material has attracted a lot of attention over the last decade. One major breakthrough was the insertion of porosity into an SiOCH thin film using plasma-enhanced chemical vapour deposition. The porogen approach is now the most common technique used to deposit porous SiOCH by a CVD method. This approach needs the use of a post-treatment, which is usually assisted by UV. Thanks to the different developments performed, very low dielectric constants can be obtained (as low as 2.0). Moreover, major progress has been also made in the improvement of specific material properties (especially mechanical), which are essential for reliable integration. Several improvements are now under investigation to fit more closely with the requirements of advanced interconnects integration: skeletons containing Si–C–Si bonds, appropriate treatment to eliminate graphitic residues and reinforcement of the skeletons in order to limit plasma-induced damage.

Of course, equipment configuration, deposition conditions, chemistry composition, film stability, plasma and beam-surface interactions, cross-linking mechanisms and transformation kinetics are just the first steps in the chain that lead to development of porous low-*k* materials suitable for implementation into integrated circuit interconnects. Many other works were done in order to optimize each process step and integration scheme to preserve the quality of the dielectric in the final device. Some manufacturers have succeeded in introducing ultra-low-*k* ($k = 2.5$) materials in production. However, the difficulties encountered in integrating ULK in 32 nm technologies and below are such that the decrease of the dielectric constant seems less relevant today. Depending on integration schemes, direct CMP can be required, which leads to the use of a capping layer on top of the ULK (for instance dense SiOCH in Reference [116]). Interface layers are also required in order to improve the reliability of interconnects [117]. The dielectric barrier used (usually SiCNH or SiCH with $k >> 3$) also contributes to increasing the overall dielectric constant of the dielectric stack. Another critical step is the packaging, which remains particularly challenging with highly porous dielectrics. These different examples illustrate how difficult it is to integrate highly porous materials and point out that works are still necessary to successfully transfer them into production.

## References


[1] International Technology Roadmap for Semiconductors (ITRS), http://www.itrs.net.

[2] Maex, K., Baklanov, M.R., Shamiryan, D., Iacopi, F., Brongersma, S.H. and Yanovitskaya, Z.S. (2003) Low dielectric constant materials for microelectronics. *J. Appl. Phys.*, **93**, 8793–8841.

[3] Morgen, M., Ryan, E.T., Zhao, J.H., Hu, C., Cho, T.H. and Ho, P.S. (2000) Low dielectric constant materials for ULSI interconnects, *Annual Rev. Mat. Sci.*, 30, 645–680.

[4] Volksen, W., Miller, R.D. and Dubois, G. (2010) Low dielectric constant materials. *Chem. Rev.*, **110**, 56–110.

[5] McGahay, V. (2010) Porous dielectrics in microelectronic wiring applications. *Materials*, **3**, 536–562.

[6] Hedrick, J.L., Miller, R.D., Hawker, C.J., Carter, K.R., Volksen, W., Yoon, D.Y. and Trollsas, M. (1998) Templating nanoporosity in thin film dielectric insulators. *Adv. Mater.*, **10**, 1049.

[7] Fayolle, M., Passemard, G., Louveau, O., Fusalba, F. and Cluzel, J. (2003) Challenges of back end of the line for sub 65 nm generation. *Microelectron. Engng*, **70**, 255–266.

[8] Malhouitre, S., Jehoul, C., Van Aelst, J., Struyf, H., Brongersma, S., Carbonell, L., Vos, I., Beyer, G., Van Hove, M., Gronbeck, D., Gallagher, M., Calvert, J. and Maex, K. (2003) Single damascene integration of porous Zirkon™ version 1 low-*k* dielectric films. *Microelectron. Engng*, **70**, 302–307.

[9] Das, A., Kokubo, T., Furukawa, Y., Struyf, H., Vos, I., Sijmus, B., Iacopi, F., Van Aelst, J., Le, Q.T., Carbonell, L., Brongersma, S., Maenhoudt, M., Tokei, Z., Vervoort, I., Sleeckx, E., Stucchi, M., Schaekers, M., Boullart, W., Rosseel, E., Van Hove, M., Vanhaelemeersch, S., Shiota, A. and Maex, K. (2002) Characterisation and integration feasibility of JSR's low-*k* dielectric LKD-5109. *Microelectron. Engng*, **64**, 23–33.

[10] Jousseaume, V., Le Cornec, Ch., Ciaramella, F., Favennec, L., Zenasni, A., Simon, G., Simon, J.P., Gerbaud, G. and Passemard, G. (2006) PECVD versus spin-on to perform porous ULK for advanced interconnects: chemical composition, porosity and mechanical behavior. Materials Research Society Symposium Proceedings, vol. 914, pp. F04–F06.

[11] Grill, A. (2009) Porous pSiCOH ultralow-*k* dielectrics for chip interconnects prepared by PECVD. *Annual Rev. Mater. Res.*, **39**, 49–69.

[12] Grill, A. and Patel, V. (2001) Ultralow-*k* dielectrics prepared by plasma-enhanced chemical vapor deposition. *Appl. Phys. Lett.*, **79**, 803–805.

[13] Vrtis, R.N., O'Neill, M.L., Vincent, J.L., Lukas, A.S., Peterson, B.K., Bitner, M.D. and Karwacki, E.J. (2003) Plasma enhanced chemical vapor deposition of porous organosilicate

glass ILD films with $k < 2.4$. Materials Research Society Symposium Proceedings, vol. 766, p. 259.

[14] Bilodeau, S.M., Borovik, A.S., Ebbing, A.A., Vestyck, D.J., Xu, C., Roeder, J.F. and Baum, T.H. (2004) Chemical routes to improved mechanical properties of PECVD low $k$ thin films. Materials Research Society Symposium Proceedings, vol. 812, pp. F6.18.1–F6.18.6.

[15] Burkey, D.D. and Gleason, K.K. (2004) Temperature-resolved Fourier transform infrared study of condensation reactions and porogen decomposition in hybrid organosilicon porogen films. *J. Vac. Sci. Technol. A*, **22**, 61–70.

[16] Favennec, L., Jousseaume, V., Rouessac, V., Durand, J. and Passemard, G. (2005) Ultra low $k$ PECVD porogen approach: matrix precursors comparison and porogen removal treatment study. Materials Research Society Symposium Proceedings, vol. 863, pp. 49–54.

[17] Jousseaume, V., Zenasni, A., Favennec, L., Gerbaud, G., Bardet, M., Simon, J.P. and Humbert, A. (2007) Comparison between e-beam and ultraviolet curing to perform porous SiOC:H. *J. Electrochem. Soc.*, **154**, 1.

[18] Gourhant, O., Gerbaud, G., Zenasni, A., Favennec, L., Gonon, P. and Jousseaume, V. (2010) Crosslinking of porous SiOCH films involving Si–O–C bonds: impact of deposition and curing. *J. Appl. Phys.*, **108**, 124105.

[19] Bilodeau, S.M., Chen, P., Giannetto, W., Xu, C., Hunks, W., Baum, T.H. and Roeder, J.F. (2006) Porogen precursors for ULK ($k < 2.2$) PECVD dielectrics. Proceedings of the Advanced Metallization Conference, vol. 367.

[20] Gates, S.M., Dubois, G., Ryan, E.T., Grill, A., Liu, M. and Gidley, D. (2009) Adjusting the skeleton and pore structure of porous SiCOH dielectrics, *J. Electrochem. Soc.*, **156**, G156–G162.

[21] Grill, A. and Patel, V. (2008) Ultralow dielectric constant pSiCOH films prepared with tetramethylcyclotetrasiloxane as skeleton precursor. *J. Appl. Phys.*, **104**, 024113.

[22] Castex, A., Favennec, L., Jousseaume, V., Bruat, J., Deval, J., Remiat, B., Passemard, G. and Pons, M. (2005) Study of plasma mechanism of hybrid a-SiOC:H low-$k$ film deposition from decamethylcyclopentasiloxane and cyclohexene oxide. *Microelectron. Engng*, **82**, 416.

[23] Castex, A., Jousseaume, V., Deval, J., Bruat, J., Favennec, L. and Passemard, G. (2008) Ultra low $k$ films by using a plasma enhanced chemical vapor deposition porogen approach: study of the precursor reaction mechanisms, *J. Vac. Sci. Technol. A*, **26**, 1343–1354.

[24] Burkey, D.D. and Gleason, K.K. (2004) Organosilicon thin films deposited from cyclic and acyclic precursors using water as an oxidant. *J. Electrochemical Soc.*, **151**, F105–F112.

[25] Favennec, L., Jousseaume, V., Gerbaud, G., Zenasni, A. and Passemard, G. (2007) Ultralow-$k$ using a plasma enhanced chemical vapor deposition porogen approach: matrix structure and porogen loading influences. *J. Appl. Phys.*, 102, 064107.

[26] Burkey, D.D. and Gleason, K. (2003) Structure and mechanical properties of thin films deposited from 1,3,5-trimethyl-1,3,5-trivinylcyclotrisiloxane and water. *J. Appl. Phys.*, 93, 5143–5150.

[27] Grill, A. (2003) Plasma enhanced chemical vapor deposited SiCOH dielectrics: from low-$k$ to extreme low-$k$ interconnect materials. *J. Appl. Phys.*, **93**, 1785–1790.

[28] Gourhant, O., Jousseaume, V., Favennec, L., Zenasni, A., Maury, P., Mage, L., Gonon, P. and Vincent, G. (2007) Extendibility of the PECVD porogen approach for ULK materials. Materials Research Society Symposium Proceedings, vol. 990, pp. B06–B06.

[29] O'Neill, M.L., Haas, M.K., Peterson, B.K., Vrtis, R.N., Weigel, S.J., Wu, D., Bitner, M.D. and Karwacki, E.J. (2006) Impact of pore size and morphology of porous organosilicate glasses on integrated circuit manufacturing. Materials Research Society Symposium Proceedings, vol. 914, p. 3.

[30] Jousseaume, V., Gourhant, O., Zenasni, A., Maret, M. and Simon, J.-P. (2009) Grazing incidence small angle X-ray scattering study of the structure of nanoporous ultralow-$k$ dielectrics prepared by plasma enhanced chemical vapor deposition. *Appl. Phys. Lett.*, **95**, 022901.

[31] J.P. Simon, J.P.,V. Jousseaume, V. and G. Rolland, G. (2007) GISAXS study of porous dielectrics used in advanced microelectronic interconnections. *J. Appl. Cryst.*, **40**, s363–s366.

[32] Hurley, P.T., Du, L.-S., McDaniel, P.L., Peterson, B.K., Weigel, S.J., Haas, M.K., Vrtis, R.N., Sinatore, D., Bitner, M.D., Theodorou, K.E. and O'Neill, M.L. (2007) Formation of porous organosilicate glasses produced by PECVD and UV treatment. Materials Research Society Symposium Proceedings, vol. 990, p, 3.

[33] Zenasni, A., Jousseaume, V., Holliger, P., Favennec, L., Gourhant, O., Maury, P. and Gerbaud, G. (2007) The role of ultraviolet radiation during ultralow *k* films curing: strengthening mechanisms and sacrificial porogen removal. *J. Appl. Phys.*, **102**, 094107.
[34] Nakao, S.I., Ushio, J., Ohno, T., Hamada, T., Kamigaki, Y., Kato, M., Yoneda, K., Kondo, S. and Kobayashi, N. (2006) UV/EB cure mechanism for porous PECVD/SOD low-*k* SiCOH materials. Proceedings of the IEEE Interconnect Technology Conference, p. 66.
[35] Jousseaume, V., Payet, M., Maury, P. and Passemard, G. (2006) $H_2$ plasma as porogen removal treatment to perform porous ULK. Meeting of the Abstract Electrochemical Society, vol. 601, p. 370.
[36] Zenasni, A., Jousseaume, V., Gourhant, O., Favennec, L. and Maury, P. (2008) Porosity generation using hydrogen plasma assisted thermal curing for ultra low-*k* material. *Microelectron. Engng*, **85**, 2102.
[37] Lubguban, J.A., Sun, J., Rajagopalan, T., Lahlouh, B., Simon, S.L. and Gangopadhyay, S. (2002) Supercritical carbon dioxide extraction to produce low-*k* plasma enhanced chemical vapour deposited dielectric films. *Appl. Phys. Lett.*, **81**, 4407–4409.
[38] Jousseaume, V., Beynet, J., Rémiat, B., Gaillard, F., Lumia, G., Didier, G., Fournel, B., Rebiscoul, D., Louis, D. and Passemard, G. (2005) Total porogen extraction by supercritical carbon dioxide to produce porous ULK thin film. Proceedings of the Advanced Metallization Conference, pp. 317–323.
[39] Mehta, S., Dimitrakopoulos, C., Augur, R., Gambino, J., Cjou, A. et al. (2006) Evaluation of device damage from e-beam curing of ultra low-*k* inter-level BEOL dielectrics. Proceedings of the Advanced Metallization Conference, pp. 361–367.
[40] Mickler, E., Lin, C.-T., Krishnan, A.T., Jin, C. and Jain, M. (2004) A charge damage study using an electron beam low *k* treatment. Proceedings of the 7th International Interconnect Technology Conference, pp. 190–192.
[41] Ito, F., Takeuchi, T. and Hayashi, Y. (2004) Improvement of mechanical properties of porous SiOCH films by post-cure treatments. Proceedings of the Advanced Metallization Conference, pp. 32–33.
[42] Iacopi, F., Travaly, Y., Eyckens, B., Walfried, C., Abell, T., Guyer, E.P., Gage, D.M., Dauskardt, R.H., Sajavaara, T., Houthoofd, K., Grobet, P., Jacobs, P. and Maex, K. (2006) Short-ranged structural rearrangement and enhancement of mechanical properties of organosilicate glasses induced by ultraviolet radiation. *J. Appl. Phys.*, **99**, 053511.
[43] Lide, D.R. (ed.) (2004) *Handbook of Chemistry and Physics*, 85th edn, CRC, Boca Raton, Florida, pp. 9–52.
[44] Tajima, N., Ohashi, Y., Nagano, S., Xu, Y., Matsumoto, S., Kada, T. and Ohno, T. (2009) Theoretical analyses of chemical reactions for forming hydrocarbon-bridged SiCOH low-*k* films in PECVD processes. Proceedings of the International Interconnect Technology Conference, p. 116.
[45] Gerbaud, G., Hediger, S., Bardet, M., Favennec, L., Zenasni, A., Beynet, J., Gourhant, O. and Jousseaume, V. (2009) Spin-coated and PECVD low dielectric constant porous organosilicate films studied by 1D and 2D solid-state NMR. *Phys. Chem. Chem. Phys.*, **11**, 9729–9737.
[46] Prager, L., Marsik, P., Wennrich, L., Baklanov, M.R., Naumov, S., Pistol, L., Schneider, D., Gerlach, J., Verdonck, P. and Buchmeiser, M. (2008) Effect of pressure on efficiency of UV curing of CVD-derived low-*k* material at different wavelengths. *Microelectron. Engng*, **85**, 2094–2097.
[47] Chatchilialoglu, C. (2004) *Organosilanes in Radical Chemistry*, John Wiley & Sons, Ltd, Chichester.
[48] Nakao, S., Kamigaki, Y., Ushio, J., Hamada, T., Ohno, T., Kato, M., Yoneda, K., Kondo, S. and Kobayashi, N. (2007) Paramagnetic defect spin centers in porous SiOCH film investigated using electron spin resonance. *Jpn. J. Appl. Phys.*, **46**, 3351–3353.
[49] Seo, K., Oka, Y., Nomura, K., Tsutsue, M., Kobori, E., Goto, K., Mizukami, Y., Ohtsuka, T., Tsukamoto, K., Matsumoto, S. and Ueda, T. (2009) New multi-step UV curing process for porogen-based porous SiOC. Proceedings of the International Interconnect Technology Conference, pp. 101–103.
[50] Marsik, P., Urbanowicz, A., Verdonck, P., Ferchichi, K., DeRoest, D., Prager, L. and Baklanov, M.R. (2009) Engineering of chemical and physical properties of low-*k* materials by different wavelengths of UV light. Proceedings of the Advanced Metallization Conference 2008, pp. 543–549.

[51] Matsuura, M., Goto, K., Miura, N., Hashii, S. and Asai, K. (2006) Film characterization of ultra low-*k* dielectrics modified by UV curing with different wavelength bands. Materials Research Society Symposium Proceedings, vol. 914, pp. F01–F06.

[52] Eslava, S., Eymery, G., Marsik, P., Iacopi, F., Kirschhock, C.E.A., Maex, K., Martens, J.A., Baklanov, M.R. (2008) Optical property changes in low-*k* films upon ultraviolet-assisted curing. *J. Electrochem. Soc.*, **155**, G115–G120.

[53] Urbanowicz, A.M., Vanstreels, K., Verdonck, P., Shamiryan, D., De Gendt S.,and Baklanov, M.R. (2010) Improving mechanical robustness of ultralow-*k* SiOCH plasma enhanced chemical vapor deposition glasses by controlled porogen decomposition prior to UV-hardening. *J. Appl. Phys.*, **107**, 104122.

[54] Tanbara, K. and Kamigaki, Y. (2010) Paramagnetic defect generation and microstructure change in porous low-*k* SiOCH films with vacuum baking. *J. Electrochem. Soc.*, **157**, G95–G99.

[55] Bittel, B.C., Lenahan, P.M. and King, S.W. (2010) Ultraviolet radiation effects on paramagnetic defects in low-*k* dielectrics for ultralarge scale integrated circuit interconnects. *Appl. Phys. Lett.*, **97**, 063506.

[56] Grill, A., Patel, V., Rodbell, K.P., Huang, E., Baklanov, M.R., Mogilnikov, K.P., Toney, M. and Kim, H.C. (2003) Porosity in plasma enhanced chemical vapor deposited SiCOH dielectrics: a comparative study. *J. Appl. Phys.*, **94**, 3427–3435.

[57] Rouquerol, J., Avnir, D., Fairbridge, C.W., Everett, D.H., Haynes, J.H., Pernicone, N., Ramsay, J.D.F., Sing, K.S.W. and Unger, K.K. (1994) Recommendations for the characterization of porous solids. *Pure Appl. Chem.*, **66**, 1739.

[58] Baklanov, M.R. and Mogilnikov, K.P. (2002) Non destructive characterization of porous low *k* dielectric films. *Mircoelectron. Engng*, **64**, 335–349.

[59] Jousseaume, V., Rolland, G., Babonneau, D. and Simon, J.P. (2007) Structural study of nanoporous ultra low-*k* dielectrics using complementary techniques: ellipsometric porosimetry, X-ray reflectivity and grazing incidence small-angle X-ray scattering. *Appl. Surf. Sci.*, **254-2**, 473–479.

[60] Ciaramella, F., Jousseaume, V., Maitrejean, S., Verdier, M., Remiat, B., Zenasni, A. and Passemard, G. (2006) Crosslinking impact of mesoporous MSQ films used in microelectronic interconnections on mechanical properties. *Thin Solid Films*, **495**, 124–129.

[61] Gibson, L.J. and Ashby, M.F. (1997) *Cellular Solids: Structure and Properties*, 2nd edn, Cambridge University Press.

[62] Grill, A. and Neumayer, D.A. (2003) Structure of low dielectric constant to extreme low dielectric constant SiCOH films: Fourier transform infrared spectroscopy characterization. *J. Appl. Phys.*, **94**, 6697–6707.

[63] Rau, C. and Kulish, W. (1994) Mechanisms of plasma polymerization of various silico-organic monomers. *Thin Solid Films*, **249**, 28.

[64] Jun, X., Yang, C.S., Jang, H.R. and Choi, C.K. (2003) Chemical structure evolution of SiOCH films with low dielectric constant during PECVD and postannealing. *J. Electrochemical Soc.*, **150**, F206.

[65] Mabboux, P.-Y. and Gleason, K.K. (2005) Chemical bonding structure of low dielectric constant Si:O:C:H films characterized by solid-state NMR. *J. Electrochemical Soc.*, **152**, F7–F13.

[66] Gates, S.M., Neumayer, D.A., Sherwood, M.H., Grill, A., Wang, X. and Sankarapandian, M. (2007) Preparation and structure of porous dielectrics by plasma enhanced chemical vapour deposition, *J. Appl. Phys.*, **101**, 094103.

[67] Jousseaume, V., Favennec, L., Zenasni, A. and Gourhant, O. (2007) Porous ultra low *k* deposited by PECVD: from deposition to material properties. *Surf. Coat. Technol.*, **201**, 9248.

[68] Lin, Y., Xiang, Y., Tsui, T. and Vlassak, J. PECVD low-permittivity organosilicate glass coatings: adhesion, fracture and mechanical properties. *Acta Materialia*, **56**, 4932–4943.

[69] Huang, C.H., Huang, H.L., Hung, C.I., Wang, N.F., Wang, Y.H. and Houng, M.P. (2008) Bond structure in porous SiOCH low-*k* film fabricated by ultraviolet irradiation. *J. Appl. Phys.*, **47**, 1532–1535.

[70] Grill, A., Edelstein, D., Restaino, D., Lane, M., Gates, S., Liniger, E., Shaw, T., Liu, X.H., Klaus, D., Patel, V., Cohen, S., Simonyi, E., Klymko, N., Lane, S., Ida, K., Vogt, S., Van Kleeck,

T., Davis, C., Ono, M., Nogami, T. and Ivers, T. (2004) Optimization of SiCOH dielectrics for integration in a 90 nm CMOS technology. Proceedings of the 7th IEEE International Interconnect Technology Conference, pp. 54–56.
[71] Kim, T.-S., Tsuji, N., Kemeling, N., Matsushita, K., Chumakov, D., Geisler, H., Zschech, E. and Dauskardt, R.H. (2008) Depth dependence of ultraviolet curing of organosilicate low-*k* thin films. *J. Appl. Phys.*, **103**, 064108.
[72] Kim, T.-S., Tsuji, N., Matsushita, K., Kobayashi, N., Chumakov, D., Geisler, H., Zschech, E. and Dauskardt, R.H. (2008) Tuning depth profiles of organosilicate films with ultraviolet curing. *J. Appl. Phys.*, **104**, 074113.
[73] Mechri, C., Ruello, P., Breteau, J.M., Baklanov, M.R., Verdonck, P. and Gusev, V. (2009) Depth-profiling of elastic inhomogeneities in transparent nanoporous low-*k* materials by picosecond ultrasonic interferometry. *Appl. Phys. Lett.*, **95**, 091907.
[74] Gonon, P., Sylvestre, A., Meynen, H. and Van Cotthem, L. (2003) Permittivity and conductivity of low-dielectric-constant SiOC:H films deposited by plasma-enhanced chemical vapour deposition. *J. Electrochem. Soc.*, **150**, F47–F52.
[75] Kim, J.Y., Hwang, M.S., Kim, Y.-S., Kim, H.J. and Lee, Y. (2001) Origin of low dielectric constant of carbon-incorporated silicon oxide film deposited by plasma enhanced chemical vapor deposition. *J. Appl. Phys.*, **90**, 2469–2473.
[76] C.S. Kim, C.S. and H.D. Jeong, H.D. (2008) What originates the dielectric permittivity of silicate–silsesquioxane hybrid thin films? *J. Phys. Chem. B*, **112**, 16257–16260.
[77] O. Gourhant, O.,V. Jousseaume, V.,L. Favennec, L.,A. Zenasni, A.,P. Gonon, P. and G. Vincent, G. (2009) Porous ultra-low *k* dielectrics deposited by PECVD: impact of the elaboration process on the structural properties and on the different contributions of the dielectric constant. Proceedings of the Advanced Metallization Conference 2008, San Diego, California, pp. 537–542.
[78] Ciofi, I., Baklanov, M.R., Tokei, Z. and Beyer, G.P. (2010) Capacitance measurements and *k*-value extractions of low-*k* films. *Microelectron. Engng*, **87**, 2391–2406.
[79] Zheng, Y., Zhu, W., Huang, T.F., Nemani, S., Yim, K.S., Amer, H., Xia, L.Q. and M'Saad, H. (2004) Plasma enhanced CVD low-*k* Black Diamond™ formation process for 65 nm technology node. Proceedings of the Advanced Metallization Conference, pp. 543–547.
[80] Verdonck, P., De Roest, D., Kaneko, S., Caluwaerts, R., Tsuji, N., Matsushita, K., Kemeling, N., Travaly, Y., Sprey, H., Schaekers, M. and Beyer, G. (2007) Characterization and optimization of porogen-based PECVD deposited extreme low-*k* materials as a function of UV-cure time. *Surf. Coat. Technol.*, **201**, 9264.
[81] Guedj, C., Portier, X., Mondon, F., Arnal, V., Guillaumond, J.F., Arnaud, L., Barnes, J.P., Jousseaume, V., Roule, A., Maitrejean, S., Chapelon, L.L., Reimbold, G., Torres, J. and Passemard, G. (2005) Bias-stress-induced evolution of the dielectric properties of porous ULK/copper advanced interconnects. *Microelectronic Engng*, **80**, 345–348.
[82] Aimadeddine, M., Arnal, V., Volpi, F. and Torres, J. (2008) Impact of pore density on conduction mechanisms in SiOCH based interconnects. Proceedings of the Advanced Metallization Conference 2007, pp. 427–432.
[83] Marsik, P., Verdonck, P., De Roest, D. and Baklanov, M.R. (2010) Porogen residues detection in optical properties of low-*k* dielectrics cured by ultraviolet radiation. *Thin Solid Films*, **518**, 4266–4272.
[84] Gourhant, O. (2008) Élaboration et caractérisation de matériaux à très faible constante diélectrique de type a-SiOCH élaborés par PECVD: application aux interconnexions des circuits intégrés. PhD thesis, INP Grenoble, France.
[85] Baklanov, M.R., Zhao, L., Van Besien, E. and Pantouvak, M. (2011) Effect of porogen residue on electrical characteristics of ultra low-*k* materials. *Microelectron. Engng*, **88**, 990–993.
[86] Tajima, N., Ohno, T., Hamada, T., Yoneda, K., Kondo, S., Kobayashi, N., Shinriki, M., Inaishi, Y., Miyazawa, K., Sakota, K., Hasaka, S. and Inoue, M. (2007) Carbon-doped silicon oxide films with hydrocarbon network bonds for low-*k* dielectrics: theoretical investigations. *Jpn. J. Appl. Phys.*, **46**, 5970–5974.
[87] Dubois, G., Magbitang, T., Volksen, W., Simonyi, E. and Miller, R.D. (2005) New spin-on oxycarbosilane low-*k* dielectric materials with exceptional mechanical properties. Proceedings of the IEEE International Interconnect Technology Conference, pp. 226–228.

[88] Shinriki, M., Tajima, N., Hamada, T., Ohno, T., Yoneda, K., Kondo, S., Ogawa, S., Miyazawa, K., Inaishi, Y., Sakoda, K., Hasaka, S. and Inoue, M. (2007) Proceedings of the Advanced Metallization Conference, pp. 339–344.

[89] Kwak, S.-K., Jeong, K.-H. and Rhee, S.-W. (2004) Nanocomposite low-*k* SiCOH films by direct PECVD using vinyltrimethylsilane. *J. Electrochem. Soc.*, **151**, F11–F16.

[90] Navamathavan, R., Oh, K.S., Chang, S.Y., Kim, S.H., Jang, Y.J.,A.S. Jung, A.S., Lee, H.J., Lee, K.M. and Choi, C.K. (2006) Electrical properties of low-dielectric-constant SiOC(–H) films prepared by plasma-enhanced chemical vapor deposition from methyltriethoxysilane and $O_2$. *Jpn. J. Appl. Phys.*, **45**, 8435–8439.

[91] Wu, Q. and Gleason, K.K. (2003) Plasma-enhanced chemical vapor deposition of low-*k* dielectric films using methylsilane, dimethylsilane, and trimethylsilane precursors, *J. Vac. Sci. Technol. A*, **21**, 388.

[92] Kim, Y.-H., Lee, S.-K. and Kim, H.J.J. (2000) Low-*k* Si–O–C–H composite films prepared by plasma enhanced chemical vapor deposition using bis-trimethylsilylmethane precursor. *J. Vac. Sci. Technol. A*, **18**, 1216.

[93] Tada, M., Yamamoto, H., Ito, F., Takeuchi, T., Furutake, N. and Hayashi, Y. (2007) Chemical structure effects of ring-type siloxane precursor on properties of plasma-polymerized porous SiCOH films. *J. Electrochem. Soc.*, **154**, D354–D361.

[94] Ueki, M., Tagami, M., Ito, F., Kume, I., Yamamoto, H., Kawahara, J., Inoue, N., Hijioka, K., Takeuchi, T., Saito, S., Onodera, T., Furutake, N., Okada, N. and Hayashi, Y. (2008) High performance Cu interconnects with damage-less full molecular-pore-stack (MPS) SiOCH for 32 nm-node LSIs and beyond. Proceedings of the IEDM, pp. 1–4.

[95] Eaves, D. (2004) *Handbook of Plastic Foams*, ChemTec Publishing.

[96] Giles, K. (2005) *Low dielectric constant layers*, US Patent, 6,963,137 B2, 8 November 2005.

[97] Jousseaume, V. (2007) *Procédé de préparation de couches minces de matériaux diélectriques nanoporeux*, Patent FR2918997, 20 July 2007.

[98] Chapelon, L.L., Arnal, V., Broekaart, M., Gosset, L.G., Vitiello, J. and Torres, J. (2004) Characterization and integration of a CVD porous SiOCH ($k<2.5$) with enhanced mechanical properties for 65 nm MOS interconnects and below. *Microelectron. Engng*, **76**, 1.

[99] Yasuhara, S., Chung, J., Tajima, K., Yano, H., Kadomura, S., Yoshimaru, M., Matsunaga, N., Kubota, T., Ohtake, H. and Samukawa, S. (2009) Structure-designable method to form super low-*k* SiOC film ($k=2.2$) by neutral-beam-enhanced chemical vapour deposition. *J. Phys. D: Appl. Phys.*, **42**, 055208.

[100] Sreenivasan, R. and Gleason, K.K. (2009) Overview of strategies for the CVD of organic films and functional polymer layers. *Chem. Vap. Deposition*, **15**, 77–90.

[101] Gupta, M. and Gleason, K.K. (2009) Surface modification of high aspect ratio structures with fluoropolymer coatings using chemical vapor deposition, *Thin Solid Films*, **517**, 3547–3550.

[102] Trujillo, N.J., Wu, Q. and Gleason, K.K. (2010) Ultralow dielectric constant tetravinyltetramethylcyclotetrasiloxane films deposited by initiated chemical vapor deposition (iCVD). *Adv. Funct. Mater.*, **20**, 607–616.

[103] Faguet, J., Lee, E., Liu, J., Brcka, J. and Akiyama, O. (2009) Novel dielectric deposition technology for advanced interconnect with air gap. Proceedings of the International Interconnect Technology Conference, pp. 35–37.

[104] Arnal, V., Farcy, A., Aimadeddine, M., Jousseaume, V., Gosset, L.G., Guillan, J., Assous, M., Favennec, L., Zenasni, A., David, T., Hamioud, K., Chapelon, L.-L., Jourdan, N., Vanypre, T., Mourier, T., Chausse, P., Maitrejean, S., Guedj, C. and Torres, J. (2007) Materials and processes for high signal propagation performance and reliable 32 nm node BEOL. Proceedings of the IEEE International Interconnect Technology Conference, pp. 1–3.

[105] Hoofman, R., Michelon, J., Bancken, P., Daamen, R., Verheijden, G., Arnal, V., Hinsinger, O., Gosset, L., Humbert, A., Besling, W., Goldberg, C., Fox, R., Michaelson, L., Guedj, C., Guillaumond, J.F., Jousseaume, V., Arnaud, L., Gravesteijn, D., Torres, J. and Passemard, G. (2005) Reliability challenges accompanied with interconnect downscaling and ultra low-*k* dielectrics. Proceedings of the International Interconnect Technology Conference, pp. 85–87.

[106] Hoofman, R.J.O.M., Verheijden, G.J.A.M., Michelon, J., Iacopi, F., Travaly, Y., Baklanov, M.R., Tőkei, Zs. and Beyer, G.P. (2005) Challenges in the implementation of low-*k* dielectrics in the back-end of line. *Microelectron. Engng*, **80**, 337–344.

[107] McGahay, V., Bonilla, G., Chen, F., Christiansen, C., Cohen, S., Cullinan-Scholl, M., Demarest, J., Dunn, D., Engel, B., Fitzsimmons, J., Gill, J., Grunow, S., Herbst, B., Hichri, H., Ida, K., Klymko, N., Kiene, M., Labelle, C., Lee, T., Liniger, E., Liu, X.H., Madan, A., Malone, K., Martin, J., McLaughlin, P.V., Minami, P., Molis, S., Muzzy, C., Nguyen, S., Patel, J.C., Restaino, D., Sakamoto, A., Shaw, T.M., Shimooka, Y., Shobha, H., Simonyi, E., Widodo, J., Grill, A., Hannon, R., Lane, M., Nye, H., Spooner, T., Wisnieff, R. and T. Ivers, T. (2006) 65 nm Cu integration and interconnect reliability in low stress $k = 2.75$ SiCOH. Proceedings of the International Interconnect Technology Conference, pp. 9–11.

[108] Balakrishnan, R., Brain, R. and Zhao, L., Chapter 7 in this book.

[109] Ryan, E.T., Gates, S.M., Grill, A., Molis, S., Flaitz, P., Arnold, J., Sankarapandian, M., Cohen, S.A., Ostrovski, Y. and Dimitrakopoulos, C. (2008) Property modifications of nanoporous pSiCOH dielectrics to enhance resistance to plasma-induced damage. *J. Appl. Phys.*, **104**, 094109.

[110] Furusawa, T., Miura, N., Matsumoto, M., Goto, K., Hashii, S., Fujiwara, Y., Yoshikawa, K., Yonekura, K., Asano, Y., Ichiki, T., Kawanabe, N., Matsuzawa, T. and Matsuura, M. (2005) W-hardened high-modulus CVD-ULK material for 45-nm node Cu/low-*k* interconnects with homogeneous dielectric structures. Proceedings of the International Interconnect Technology Conference, pp. 45–47.

[111] Calvert, J.M. and Gallagher, M.K. (2003) A new approach to ultralow-*k* dielectrics. *Semicond. Int.*, **26**, 56.

[112] Jousseaume, V., Assous, M., Zenasni, A., Maitrejean, S., Rémiat, B., Leduc, P., Trouvé, H., Le Cornec, Ch., Fayolle, M., Roule, A., Ciaramella, F., Bouchu, D., David, T., Roman, A., Scevola, D., Morel, T., Rebiscoul, D., Prokopowicz, G., Jackman, M., Guedj, C., Louis, D., Gallagher, M. and Passemard, G. (2005) Cu/ULK integration for 45 nm node and below using an improved hybrid material with conventionnal BEOL processing and a late porogen removal. Proceedings of the IEEE International Interconnect Technology Conference, pp. 60–62.

[113] Favennec, L., Jousseaume, V., Zenasni, A., Assous, M., David, T. and Passemard, G. (2007) PECVD ultra low *k* material appropriate to the post-integration porogen removal approach. Proceedings of the Advanced Metallization Conference 2006 – Asian Session, Tokyo, Japan, pp. 345–350.

[114] Jousseaume, V., Favennec, L., Zenasni, A. and Passemard, G. (2006) Plasma-enhanced-chemical-vapor-deposited ultra low *k* for a post-integration porogen removal approach. *Appl. Phys. Lett.*, **88**, 182908.

[115] Frot, T., Volksen, W., Magbitang, T., Miller, D., Purushothaman, S., Lofaro, M., Bruce, R., and Dubois, G. (2011) Post Porosity Plasma Protection : a new approach to integrate $k \leq 2.2$ porous ULK materials, Proceedings of the IEEE International Interconnect Technology Conference, pp. 1–3.

[116] Gall, S., Euvrard, C., Shhun, S., Maitrejean, S., Assous, M., Haumesser, P.-H. and Rivoire, M. (2008) Investigation of ULK ($k = 2.5$) damage by direct CMP process for 45 nm technology node, in Proceedings of the Advanced Metallization Conference 2007. *Mater. Res. Soc. Conf. Proc.*, **23**, 115–121.

[117] Grill, A., Edelstein, D., Lane, M., Patel, V., Gates, S., Restaino, D. and Molis, S. (2008) Interface engineering for high interfacial strength between SiCOH and porous SiCOH interconnect dielectrics and diffusion caps. *J. Appl. Phys.*, **103**, 054104.

# 3

# Plasma Processing of Low-*k* Dielectrics

*Hualiang Shi*[1], *Denis Shamiryan*[2], *Jean-François de Marneffe*[3], *Huai Huang*[4], *Paul S. Ho*[4] *and Mikhail R. Baklanov*[3]

[1]*INTEL Corporation, Chandler, Arizona, USA*
[2]*GLOBALFOUNDRIES, Dresden, Germany*
[3]*IMEC, Leuven, Belgium*
[4]*University of Texas at Austin, Texas, USA*

## 3.1 Introduction

Plasma technology is widely used in semiconductor processing. It performs etching, deposition, cleaning and surface treatment of different materials used in integrated circuit (IC) manufacturing. Besides the many advantages of plasma processing, a significant drawback is that it can be harmful to some materials. Plasma is a rather aggressive medium in the presence of vacuum ultraviolet (VUV) light, energetic ions, electrons and highly reactive radicals. In addition, low-*k* materials are sensitive to chemical modification and are usually porous, which increases the penetration depth of reactive species and radiation. As a result, plasma damage is the main obstacle to a successful integration of low-*k* dielectrics into an IC. The main consequence of plasma damage is a modification of the chemical composition of the low-*k* dielectric, although other detrimental effects, like densification, could also be observed.

*Advanced Interconnects for ULSI Technology*, First Edition. Edited by Mikhail R. Baklanov, Paul S. Ho and Ehrenfried Zschech.

The chemical composition of low-*k* dielectrics is tailored to make them hydrophobic. Plasma-induced chemical changes can turn the low-*k* dielectric hydrophilic. With a dielectric constant $k = 80$, water, when absorbed in the porous structure of the low-*k* dielectrics, increases the total *k* value beyond that of $SiO_2$ (~4.5), making the material totally useless for interconnect applications. Adsorbed water also deteriorates other electrical properties of the low-*k* dielectrics, e.g. increasing leakage current and decreasing reliability. If plasma processing converts the low-*k* dielectric (or part of it) into a hydrophilic material, such processing is unacceptable and must be avoided. Therefore, when low-*k* dielectrics are exposed during processing, plasmas must comply with two general requirements: perform the designed task (etch, strip or cleaning) and do not damage the low-*k* [1, 2].

There are several plasma processing steps encountered by low-*k* materials: deposition of dielectric masks on top of the low-*k*, etching vias or trenches, removal (stripping) of photoresist, deposition of diffusion barrier and post-CMP cleaning [3, 4]. Additional steps might include plasma treatments, for instance to improve the adhesion of deposited layers to the low-*k* [5]. There are many challenges in etching low-*k* materials. Some of them are inherent to plasma etching, like etch rate uniformity across a wafer or profile control, while some are more specific to a particular low-*k* and patterning scheme, like using a hardmask (dielectric or metal), avoiding micromasking and etch stop. Photoresist stripping after dry etching is known to be one of the most harmful plasma steps for low-*k* materials. It originates from the fact that strip chemistries are targeted to remove organic polymer while modern low-*k* dielectrics rely on hydrocarbon groups for hydrophobicity. As a result, the same chemistry usually removes photoresist as well as hydrophobic groups (usually $-CH_3$) from the low-*k* dielectrics, making them hydrophilic (in other words, inflicts severe damage).

There are three main routes to cope with the plasma damage issue: avoid, reduce or restore. One can avoid plasma damage by avoiding plasma exposure (e.g. use nonplasma deposition or cleaning methods, or protect the low-*k* from plasma exposure); the damaging effect could be reduced by using more plasma-resistant low-*k* or less-damaging plasma chemistry and, finally, plasma damage could be restored (to some extent) after the plasma processing.

This chapter will first describe the plasma etch equipments used for the patterning of low-*k* materials and the methods used for analyzing the performances of the etch process, in terms of plasma damage and morphological inspections (Sections 3.2 and 3.3). Then, the basics of plasma interaction with low-*k* films will be covered, including patterning strategies, etch chemistries and mechanisms (Section 3.4). The biggest part of this chapter will describe the mechanism of plasma damage for porous SiOCH materials, with a focus on the effect of ion bombardment, radical diffusion and VUV radiation, including some approaches allowing the reduction of plasma damage (Section 3.5). Finally, we present techniques used for the restoration of properties of low-k dielectrics (Section 3.6).

## 3.2 Materials and Equipment

Two approaches can be followed to reduce the *k* value of interconnect dielectrics: reduce their polarizability by the use of low-polar bonds (like C–C, C–H, $Si-CH_3$, etc.) and/or reduce their density by means of porosity. Modern low-*k* materials can be divided into two major classes: organic polymers that mostly make use of low-polar bonds, but might also be porous, and hybrid silica-based porous materials that, besides Si–O bonds, contain Si–$CH_3$ bonds. The latter bonds should be preserved during plasma processing in order to keep the material hydrophobic. Detailed reviews of low-*k* materials can be found in the literature [1, 6]. Hybrid low-*k* dielectrics that are being currently investigated have *k* values ranging from 2.0 to 3.0 (with a continuous trend to decrease it), porosity of 30–50% and a pore size of 2–3 nm.

There are three main plasma reactor types that are used in plasma etching: inductively coupled plasma (ICP), capacitively coupled plasma (CCP) and remote or downstream plasma (DSP). ICP sources have a higher plasma density ($10^{11}$–$10^{12}$ electrons/$cm^3$) than CCP ($10^9$–$10^{10}$ electrons/$cm^3$) and are mostly used for conductor etching (metals or silicon), while CCP reactors are used for a dielectric etch [7]. DSP sources are used for cleaning and resist stripping and do not allow anisotropic etching.

In CCP reactors, plasma density and ion energy are interdependent. In CCP, the energy is transferred through capacitive coupling. Due to the capacitor separating the power generator from the electrode, a self-bias build-up between the plasma and the electrode, which accelerates ions toward the surface (the sheath). By increasing the power in order to increase the plasma density, the self-bias increases as well and, in turn, increases the ion energy. As a result, plasma density and ion energy are coupled and cannot be changed separately. This issue could be partly alleviated by using higher frequencies: above 50 MHz the coupling between plasma density and ion energy becomes less pronounced [8]. Another characteristic of the CCP is that power is coupled through the sheath, meaning that the energy transfer is usually poor and the species are not easily dissociated. In addition, the top electrode is usually made of Si, which captures free fluorine from the plasma and enhances the C/F ratio. All this favors the presence of high molecular weight $C_xF_y$ radicals and ions and, together with the intrinsic high ion energy (up to several hundreds of eV), CCP discharges are able to etch high-bond energy materials (Si–O) with good selectivity toward Si and $Si_3N_4$.

In ICP reactors, plasma density and ion energy are controlled separately. The energy is transferred to electrons through inductive coupling with a coil separated from the plasma by a quartz or sapphire window. The oscillating magnetic field induces an oscillating current in the plasma and, contrary to the CCP case, penetrates the bulk of the plasma directly with little power coupling through the sheaths. This allows a very efficient energy transfer and creates a highly dissociated and dense plasma. This results also in a low self-bias (between 10 ~ 15 and a few tens of eV), which does not depend on the plasma source power. As a result, plasma density can be easily increased without increasing ion energy. In order to control ion energy, a second generator is capacitively coupled to the substrate. Due to the deficiency in large molecular weight radicals, ICP reactors lack passivating species required by a typical dielectric etch. They are very attractive for a conductor etch (semiconductors or metals), but precise control of ion energy is required to achieve highly selective etching (e.g. for a gate etch). It should be mentioned that pulsed ICP sources are becoming increasingly popular [9], due to their ability to decrease the average ion energy down to the eV range, and providing additional chemistry tuning by playing with the variable decay rate of radicals during the afterglow period.

From the above considerations, CCP reactors are more suitable for low-*k* etch; however, many attempts to etch low-*k* in ICP are made, especially when precise control of ion energy is required.

## 3.3 Process Results Characterization

After the plasma processing, material characterization aims at two objectives: first, make sure that the plasma processing reached the desired morphological characteristics (trenches are etched with the specified depth, shape and uniformity, photoresist is fully removed, etc.); second, the low-*k* must retain its original properties (no water adsorption, no deterioration of the *k* value and the mechanical properties). The former assessment is usually performed by means of scanning electron microscopy (SEM) or transmission electron microscopy (TEM). Both those methods provide direct observation of the plasma processing results (feature shapes, the presence of photoresist or etch residues, etc.) but they are destructive (if done in a cross-section mode) and time consuming (especially TEM). Scatterometry [10] can be used for post-etch morphological evaluation of patterned structures; it is a fast and nondestructive technique, but indirect, and relies on fitting of the measured spectra to libraries. Top-down SEM can be used for measurements of critical dimensions (CD) after low-*k* patterning and defectivity studies over the full wafer surface.

The characterization of plasma-induced damage is more complicated. Although it is relatively easy to characterize unpatterned wafers (vertical damage), the characterization of patterned structures (horizontal damage, normal to the patterns sidewalls) is not straightforward.

Fourier transmission infrared (FTIR) spectroscopy is the most widely used method of low-*k* damage assessment since it is fast, nondestructive and allows direct observation of adsorbed water as well as changes of chemical structure of the bulk low-*k*. An example of FTIR spectra of an SiOCH low-*k* film before and after plasma exposure is shown in Figure 3.1. A broad peak can be observed around 3000–3600 $cm^{-1}$, which is caused by adsorbed water in the porous low-*k* films. However, FTIR is a bulk technique that requires film thicknesses to be measured above ± 100 nm in order to maximize the signal/noise ratio. Attenuated total reflection FTIR (ATR-FTIR) is sometimes used because it allows a significant decrease of the Si substrate contribution. FTIR is not accurate enough to measure the physicochemical modifications that may appear in the top surface, over depths around 0–15 nm. For that purpose, X-ray photoelectron spectroscopy (XPS) allows a clear signal coming from the top 10 nm of the surface to be obtained (composition and chemical bonds), but attention must be paid to carbon contamination from the ambient (usually a depth profiling is performed). Spectroscopic ellipsometry is another technique commonly used to characterize plasma damage of unpatterned low-*k* films. Usually, the refractive index is regarded as an indicator of plasma damage. The increase of the refractive index indicates a densification of the film (decreasing porosity), water adsorption or a combination of both. Care should be taken when the low-*k* film is not fully modified by plasma; e.g. the top layer is carbon depleted or denser, while underneath the material retains its original properties. In that case a double-layer model should be used, which is more complex and less accurate. More detailed information about plasma damage could be obtained by ellipsometric porosimetry (EP) [12], where porosity, pore size distribution and the proportion of adsorbed water can be measured [13]. EP allows discrimination between densification (decrease of porosity and pore size distribution, measured with a solvent as a pore-filling agent) and water adsorption (by using water vapor as a pore-filling agent). Another advantage of EP is its ability to measure pore sealing. A delay in solvent penetration into a porous film indicates sealing of the top surface [14].

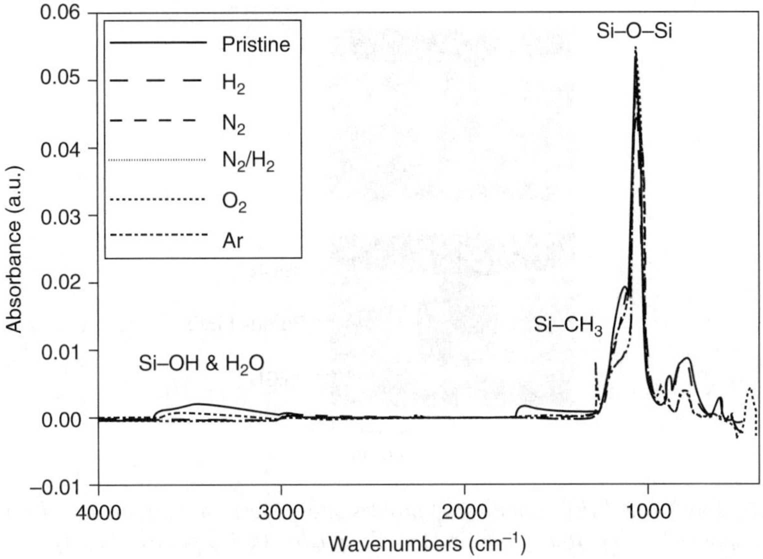


***Figure 3.1*** *FTIR spectra of a low-k film after Ar, $O_2$, $N_2$, $N_2/H_2$ and $H_2$ plasma treatments at 30 sccm, 30 mtorr and 150 W for 20 min in an Oxford RIE chamber. Reprinted with permission from* [11]. *Copyright 2008, American Vacuum Society*

Other commonly used methods for the evaluation of plasma damage on unpatterned wafers include secondary ion mass spectrometry (SIMS), which measures the depth profile of various elements present in the film (mainly used for carbon depletion); thermal desorption spectroscopy (TDS), where the evaporation of the adsorbed water at elevated temperatures is measured; X-ray reflectivity (XRR), which measures top surface density and can, for example, reveal a densified layer that seals the surface (if such sealing is measured by EP). The number of techniques that allow the characterization of the plasma damage on patterned structures is rather limited. The simplest method, called 'HF decoration', is based on the fact that a pristine low-*k* (SiOCH) is usually not attacked by HF, which is the opposite to that of a damaged film where C-depletion turns the material hydrophilic and $SiO_2$-like. As a result, after dipping into an HF solution, cross-sectional SEM inspection will indicate which part of the low-*k* film is converted to an $SiO_2$-like material (see Figure 3.2). Although simple and intuitive, this method has the disadvantage of low sensitivity and limited reproducibility, especially for narrow features where damage can be restrained to only a few nanometers.

A more advanced technique of plasma damage characterization on patterned structures is energy-filtered transmission electron microscopy (EFTEM), where elements on the image can be mapped by filtering electrons with a certain energy (based on principles of electron energy loss spectroscopy). EFTEM visualizes plasma damage (e.g. carbon depletion) with nanometer resolution, but it is time consuming (complicated sample preparation) and destructive. Recent techniques do allow a fast and nondestructive characterization of plasma damage on patterned structures, e.g. scatterometry, microwave impedance measurements or microbeam IR spectroscopy. Scatterometry may take into account, besides the shape of the evaluated structure, its optical properties [16]. A model can be built where the

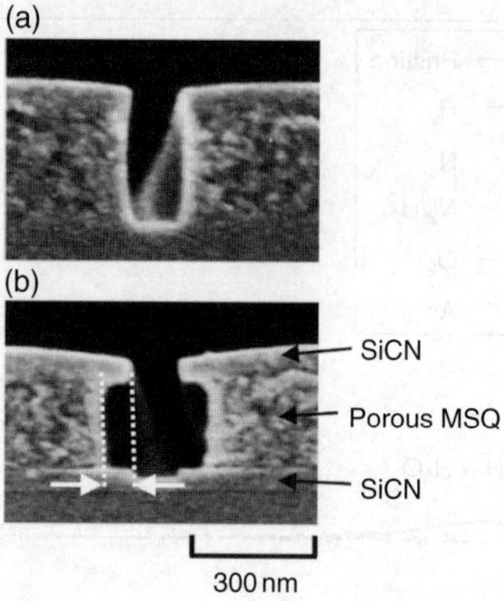


*Figure 3.2* *Low-*k *(methylsilsesquioxane) profiles after $O_2$ ash (a) without and (b) with HF dipping. Reprinted with permission from* [15]. *Copyright 2005, Japanese Society of Applied Physics*

damaged sidewalls show different optical properties from the bulk low-k material. Then, fitting the model to the measured spectra, the depth of the sidewall damage can be determined. The disadvantage of this method is the extensive modeling and the need to know the exact optical properties of the damaged film. Ellipsometric porosimetry can be combined with scatterometry, giving rise to scatterometric porosimetry, which allows the dynamic measurement of sidewall permeation [17]. Microwave impedance measurements reveal changes of the impedance of the patterned structures caused by plasma damage [18]. Since microwaves are absorbed by metals, the technique cannot measure structures that are covered by metal, e.g. after metallization or after patterning by means of a metal hardmask. Another interesting technique is proposed by H. Seki *et al.* [19]. By aligning the electrical vector of the IR electromagnetic wave perpendicular to the metal lines, the IR light is no longer absorbed since the electron motion is restricted by the line boundaries. As a result, it is possible to estimate damage not only after patterning but also after full processing, including Cu metallization.

The final test for the plasma damage is electrical measurements of the fully processed interconnect. It gives the final answer if the RC delay is reduced plus other important electrical data (e.g. leakage current and reliability), but cannot pinpoint any particular step that might have caused damage to the low-*k* film. There are simplified electrical tests, which do not include the complete process flow and allow identification of the influence of various process steps. A good example is a planar capacitor structure (p-cap), where a low-*k* film is deposited in a window in $SiO_2$ (to exclude edge effects), exposed to the studied process step, followed by barrier deposition, Cu electroplating and Cu CMP in order to allow electrical measurements, as shown in Figure 3.3. The metal passivation layer is used to encapsulate the structure and provide a probing location, preventing mechanical damage to the low-*k* material from the probing needle.

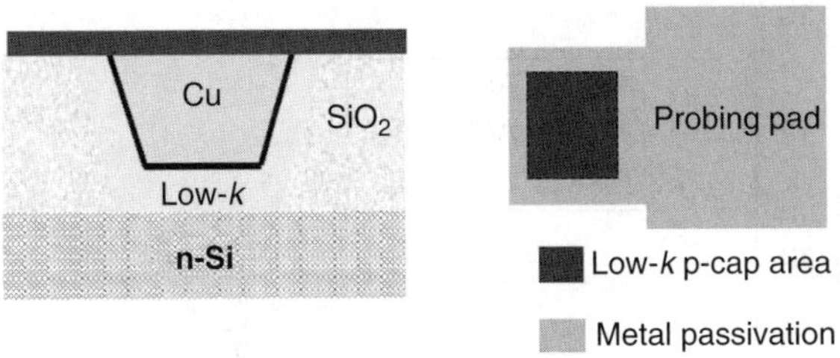


***Figure 3.3*** *Planar capacitor test structure used for evaluation of the electrical characteristics of low-k films* [20]

## 3.4 Interaction of Low-*k* Dielectrics with Plasma

Due to the presence of many energetically and chemically active species, plasma is a very aggressive medium for low-*k* material that may cause severe degradation of its original properties. In semiconductor processing, the low-*k* material is exposed to plasmas during patterning, namely etching and photoresist removal (resist strip or ash).

### 3.4.1 Low-*k* Etch Chemistries

There are two types of low-*k* materials, which require different etch approaches: organic polymers and hybrid Si-based materials (SiOCH), also called organic silicon glass or carbon-doped oxide. Organic polymers can be etched by oxygen- or hydrogen-based plasma by forming volatile $CO_2$ or $C_xH_y$. By addition of $N_2$ ($H_2/N_2$ plasma), the etch rate is higher than in $H_2$/He and $H_2$/Ar discharges because of the formation of volatile HCN molecules.

Due to the high reactivity of oxygen, $O_2$-based plasmas do not provide anisotropic profiles (vertical sidewalls) – oxygen reacts with polymers even without ion bombardment. Hydrogen-based chemistries (like $N_2/H_2$ or He/$H_2$) are less aggressive and can therefore be successfully used for the patterning of organic polymers [21]. The main issue in organic polymer etching is the so-called 'graphitization' – the removal of lighter hydrocarbon fractions by ion bombardment, which leaves behind graphite-like carbon [22].

The etch of SiOCH materials requires fluorine since Si can be removed only by forming volatile $SiF_x$ compounds (other halides are not widely used for low-*k* etching due to their low etch rate, compatibility with the etch equipment and metal corrosion aspects). The chemistry of choice is fluoro(hydro)carbons $C_xH_yF_z$ (with or without hydrogen). Fluorine is able to remove both Si (by forming $SiF_4$) and carbon (by forming volatile fluorocarbons like $COF_2$). However, the etch process is efficient only in the presence of ion bombardment, due to the strength of the Si–O and Si–C bonds. SiOCH materials are usually not etched in downstream $C_xH_yF_z$ plasma. The etch rate can be drastically increased by a small oxygen addition, but this may damage low-*k* materials, especially if these are highly porous [23]. Another important parameter that should be tuned in the etch chemistry is the polymerization rate, commonly described by the C/F ratio. A too weak polymerization

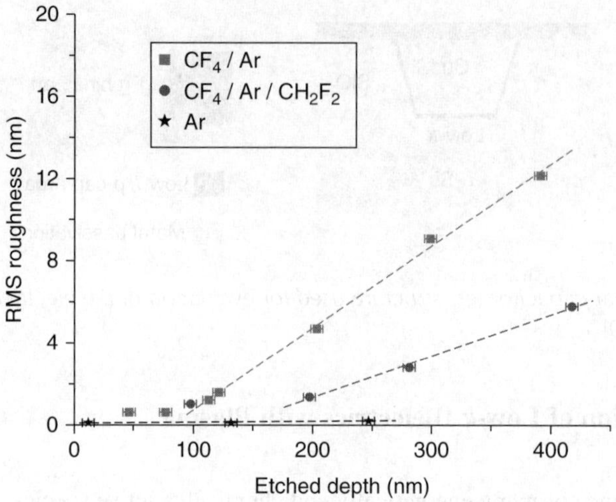


***Figure 3.4*** *Root mean square (RMS) roughness as a function of the etched depth for porous SiOCH (25 %) using a $CF_4$/Ar, a $CF_4$/Ar/$CH_2F_2$ or a pure Ar plasma. Reprinted with permission from* [24]. *Copyright 2010, American Institute of Physics*

will result in bowed profiles (fluorine will react with the sidewalls) and fluorine diffusion into the low-*k* material, leading to plasma damage and increased roughness [24].

Figure 3.4 demonstrates that the roughness increases with the decrease of the polymerization character of a fluorocarbon gas mixture. Too strong polymerization will result in sloped profiles or even etch stop, and possibly cause heavy residues and post-etch cleaning issues. Indeed, plasma-deposited polymers may be difficult to remove, especially when a metal hardmask is used, and these polymers could contain metals. When the polymerization and the etch rates are well balanced, the polymer film passivates the sidewalls without affecting the vertical etch rate, resulting in a vertical etch profile. It can also seal the porous surface and prevent the penetration of unwanted chemicals into the bulk of the porous low-*k* material. The polymerization rate can be controlled in different ways. One approach is to mix low- and high-polymerizing gases (e.g. $CF_4$ and $CHF_3$ or $CH_2F_2$) in the correct proportion. The advantage of this approach is that there are no agents other than fluorocarbons that interact with the low-*k* material, which simplifies the low-*k* plasma interactions. The drawback is that balancing the polymerization with the $CF_4/C_xH_yF_z$ ratio does not give a lot of freedom in process optimization. Another approach to control the polymerization rate is by the use of additives that will prevent polymer deposition, e.g. oxygen or nitrogen. Oxygen easily breaks polymers reacting with carbon and hydrogen and even a small fraction of oxygen in the plasma significantly reduces polymerization and sidewall protection. However, the addition of oxygen has drawbacks: first, it reduces selectivity over organic masks (if present). Second, oxygen radicals cause severe plasma damage by removing –$CH_3$ groups from the low-*k* structure. Nitrogen can also be used for polymerization control since it reacts with carbon, forming C–N bonds and volatile HCN molecules. Much larger quantities of nitrogen (usually 100 times more than $O_2$) are needed to achieve a comparable reduction in the polymerization rate. Nitrogen is also detrimental for low-*k* materials in some conditions [2].

### 3.4.2 Patterning Strategies and Masking Materials

Two masks can be used for pattern transfer into low-*k* material: photoresist (or softmask) and inorganic materials (or hardmask). The photoresist mask is the simplest since it does not require the deposition and removal of any additional layers. In the case of a hardmask, first the hard mask is deposited on top of the low-*k* material and then the photoresist is coated and lithographically patterned on top of the hardmask. The plasma etch process is divided into three steps: first the hardmask opening, second the photoresist strip and third the low-*k* material is etched using the hardmask as a protection. Although the resist mask-based patterning is preferred from a cost perspective, it has a number of disadvantages. For organic polymers, the resist mask is excluded since both the material to be etched and the mask belong to the same class of materials: organic polymers. Therefore, plasma that etches the low-*k* material will also etch the resist mask. As a consequence, first, the selectivity of the low-*k* etching over the resist will be low and, second, during the post-etch strip, the low-*k* material will be removed as fast as the photoresist. Therefore, for the pattering of the organic polymers, a hardmask is an absolute requirement.

For hybrid silica-based materials, the resist mask can be an option. Due to the different natures of the resist mask and the SiOCH (actually not completely different since SiOCH also contains some hydrocarbons), there are plasma conditions that may etch SiOCH selectively over the resist mask. A resist strip should be optimized in such a way that it does not damage the low-*k* material, which could be a challenge since SiOCH materials rely on $-CH_3$ groups in order to remain hydrophobic, while strip plasma is designed to remove hydrocarbons. In technology nodes older than 65 nm, the dimensions of the structures were large enough to tolerate some limited sidewall damage caused by the resist strip, and therefore such patterning has been successful. However, the continuous decrease of critical dimensions (CD) brings two new challenges: first, the resist thickness is significantly decreased in order to avoid pattern collapse; second, the narrower lines in low-*k* no longer allow any sidewall damage. As a result, a hardmask is the only viable option used to pattern SiOCH low-*k* materials for modern technology nodes.

There are two major types of hardmask: dielectric and metal. A dielectric hardmask could be Si-based ($SiO_2$, $Si_3N_4$ or SiC) or carbon-based (amorphous carbon). Metal hardmasks usually consist of metal nitrides: TaN or, more popularly, TiN. Si-based masks are best suited for organic polymer low-*k* material because it remains unaffected by the plasma that etch the low-*k* ($H_2$- or $O_2$-based) material. As a result, very high selectivity could be achieved that allows relatively thin masks to be used. For SiOCH materials, Si-based hardmasks are more difficult to implement since both the low-*k* material and the mask contain Si, and it is difficult to reach a high selectivity. The low selectivity imposes a thicker hardmask, but this increases the effective *k* value of the structures if left on top of the low-*k* material after patterning. Again, the removal of the Si-based hardmask poses a problem since it has a composition similar to the low-*k*.

Recent solutions for SiOCH hardmasks include the metal hardmask and the amorphous carbon hardmask. In the case of the metal hardmask (e.g. TiN) a high selectivity is achieved since the main Ti etch product – $TiF_4$ – has a relatively low volatility (boiling point at atmospheric pressure 284 °C) as compared to $SiF_4$ (boiling point at atmospheric pressure −86 °C). From that perspective, TaN is less preferred since its main etch product $TaF_5$ is more volatile

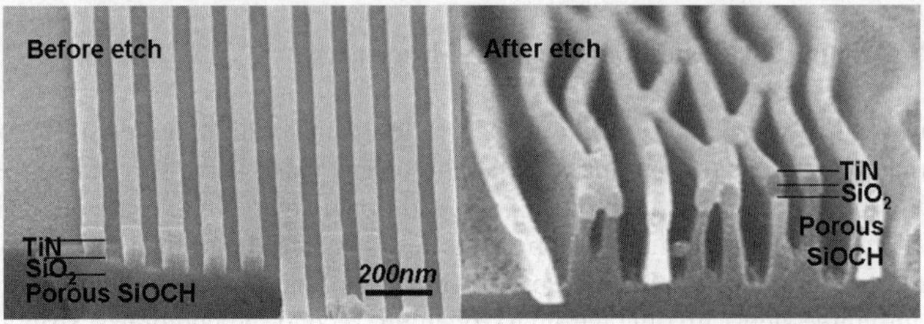


*Figure 3.5* Tilted SEM view of a porous SiOCH/$SiO_2$/TiN stack (left) after etching of the TiN and (right) after etching of the porous SiOCH. Reprinted with permission from [25]. Copyright 2007, American Institute of Physics

than $TiF_4$. Besides selectivity aspects, the metal hardmask does not contribute to the increase of the effective *k* value because it is removed at CMP after metal filling, and can be used as a diffusion barrier for subsequent Cu deposition. The disadvantages of the metal hardmask are the formation of metal-containing etch residues that are difficult to remove, the formation of micromasking caused by the low-volatile Ti etch by-products and the appearance of mechanical stress in the metal strips, which can cause severe distortion (so-called 'wiggling' or 'undulation'; see Figure 3.5) of a relatively mechanically weak low-*k* material when narrow lines with a high aspect ratio are patterned [25]. In order to compensate for the many issues brought by metal hardmasks, new materials are being investigated, such as boron-based compounds [26], which introduce less stress on the microstructures, generate $BF_x$ by-products that are more volatile than $TiF_x$ and are amorphous (opposite to TiN).

The amorphous carbon mask [27, 28] is a relatively new concept in low-*k* patterning, which uses carbon-based material deposited by CVD or spin-on. It shows good etch resistance (better than photoresist) but may require ashing, which is based on carbon removal and, therefore, is potentially damaging for the low-*k* materials. However, recently investigated He/$H_2$ downstream plasma at elevated temperature might be used for amorphous carbon hardmask removal without low-*k* damage.

### 3.4.3 Etch Mechanisms

By scaling down the feature dimensions, plasma damage, introduced by etching, becomes more and more important. In order to understand the etch damage, it is necessary to first understand the etch mechanisms. It is known that the etch of $SiO_2$ in fluorocarbon plasma is governed by a thin fluorocarbon layer that grows on the surface [29, 30]. The thickness and composition of this layer is governed by plasma conditions (pressure, power, composition) as well as by the composition of the material itself – oxygen that is released during $SiO_2$ etch reacts with the fluorocarbon layer on the surface. When $SiO_2$ is replaced by a low-*k* material (SiOCH) the formation of the layer will be influenced by the material composition and, as will be shown later, by the presence of porosity. Indeed, the less oxygen present in a dielectric, the thicker the fluorocarbon film formed on top of the dielectric during plasma etching, as illustrated by Figure 3.6, which shows that the layer becomes thicker with the addition of a polymerizing gas ($C_4F_6$ or $CH_2F_2$).

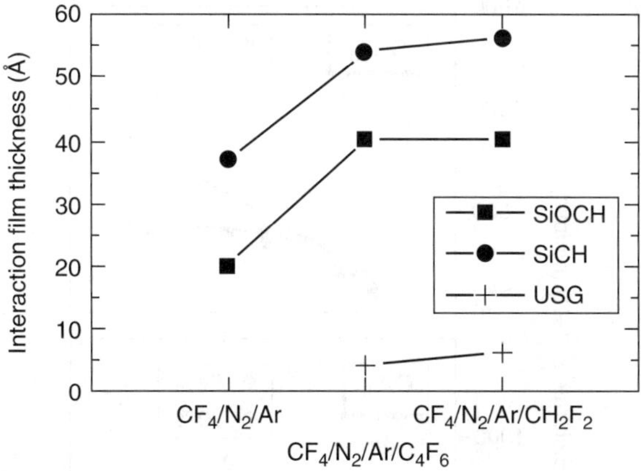


***Figure 3.6*** *Perturbed layer thickness (determined by XPS) on SiOCH, SiCH and $SiO_2$ in fluorocarbon-based plasma versus high polymerizing gas addition. Reprinted with permission from* [31]. *Copyright 2003, American Vacuum Society*

The situation becomes more complicated in the presence of porosity. Assuming that the porosity has no impact on the etch mechanism, the etch rate should scale according to the density of the material, following a simple law [32]:

$$ER_{norm} = (1 - \Pi)ER \tag{3.1}$$

where $ER_{norm}$ is the normalized etch rate, $ER$ the etch rate and $\Pi$ is the porosity.

Figure 3.7 shows the normalized etch rates for $SiO_2$ and Xerogel (porous $SiO_2$, 30% porosity) in a relatively low polymerizing plasma ($CHF_3$) and a more polymerizing one ($C_4F_8$). For $C_4F_8$, the scaled etch rates of the Xerogel and the $SiO_2$ do overlap, indicating a similar etch mechanism – formation of a fluorocarbon polymer layer on top of the etched surface. However, for less polymerizing plasma ($CHF_3$), at a high bias voltage the Xerogel is etched faster than predicted by the scaling rule (3.1). This observation suggests a different etch mechanism in this range of bias.

During the etch of a porous low-*k* material in a fluorocarbon plasma, fluorine diffuses into the porous material causing a higher etch rate, an increased roughness [24] and damage. Comparing porous and nonporous SiOCH low-*k* materials after etch by a $CF_4$/$CH_2F_2$/Ar plasma, Posseme *et al.* [33] found that the reactive layer formed during a porous low-*k* etch is a mixed layer composed of C, Si, O and F atoms, which is uniform throughout the depth probed by XPS at various take-off angles. This is opposite to the dense SiOCH, for which the fluorocarbon film is localized at the top surface. They concluded that there is diffusion of the fluoropolymer through the pores into the porous SiOCH. The etch process is a trade-off between the fluorocarbon radical diffusion rate through the mixed layer formed at the surface and the chemical sputtering rate of the mixed layer. In order to estimate the damage inflicted by fluorine, ATR-FTIR spectroscopy was applied to samples etched in different fluorine-based plasma, including $SF_6$, which is not polymerizing at all and, therefore, should result in a higher permeation of fluorine. It was found that the greatest damage (reduction of the Si–$CH_3$ peak) is produced in a less polymerizing plasma (Ar) and, especially, $SF_6$ (see Figure 3.8).

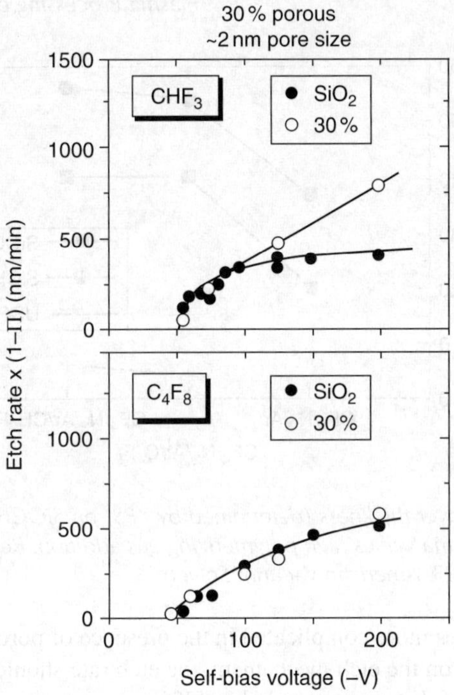


*Figure 3.7* *Blanket etch rates of 30 % porous xerogel (open dots) in $CHF_3$ and $C_4F_8$ after correction for porosity. The $SiO_2$ etch rates (solid dots) are shown for comparison. The source power and pressure were fixed at 1400 W and 10 mtorr, respectively. Reprinted with permission from* [31]. *Copyright 2000, American Vacuum Society*

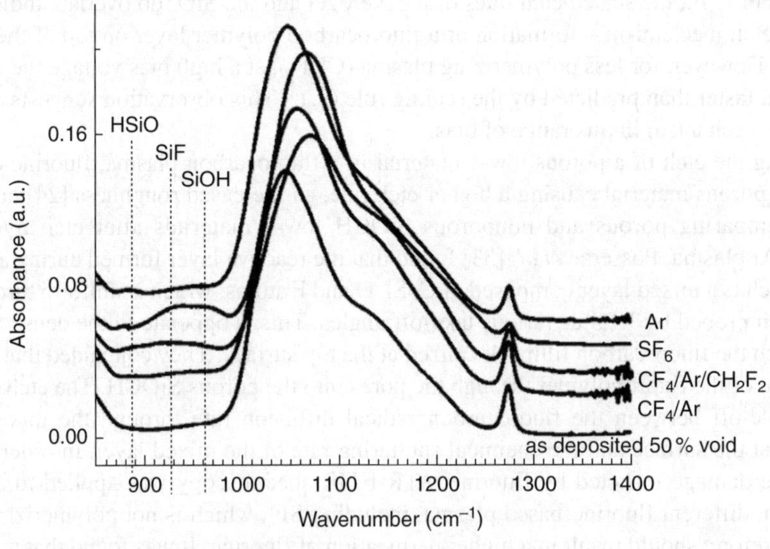


*Figure 3.8* *ATR spectra comparison of porous SiOCH (50 % porosity) as deposited and after partial etching in a fluorocarbon-based plasma. The Si–$CH_3$ peak is located at 1275 cm$^{-1}$. Reprinted with permission from* [33]. *Copyright 2004, American Vacuum Society*

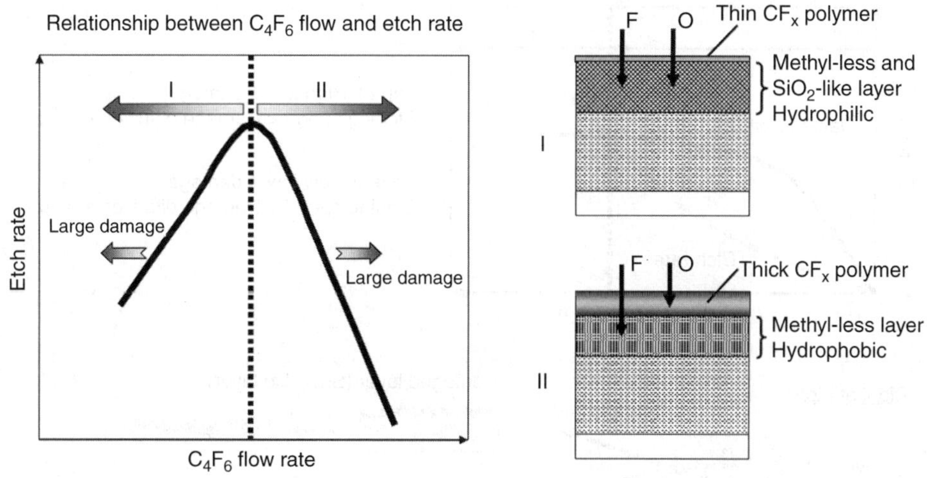


***Figure 3.9*** *Etching damage model for $C_4F_6/O_2/Ar$ chemistry. Reprinted with permission from [34]. Copyright 2008, Japanese Society of Applied Physics*

It can be concluded that a more polymerizing plasma will deposit a protective fluorocarbon layer that protects the porous low-*k* from carbon depletion. The relation between polymerization and damage, however, is not so straightforward. Iba *et al.* [34] showed that damage of porous nanoclustering silica (NCS), after etching in $C_4F_6/O_2/Ar$ plasma, has a non-monotonous dependence on the polymerizing character of the discharge. At low $C_4F_6$ flows (low polymerization rate) the damage is inflicted by O radicals diffusing through a thin fluorocarbon film. At high $C_4F_6$ flow the damage is related to fluorine radicals, diffusing into the NCS, as illustrated in Figure 3.9. In this case, it can be presumed that the fluorocarbon film acts as a fluorine source.

All the results discussed above were obtained on unpatterned wafers, in other words, on horizontal surfaces, exposed to the plasma. Although important for the understanding of etch and damage mechanisms, horizontal surfaces are not of the greatest concern, since the damaged layer is removed as the etch progresses. The situation is different for feature sidewalls, i.e. vertical surfaces. There, the damage will remain after etch, increasing the *k* value of the dielectric between two adjacent trenches. In this case several factors should be taken into account, as illustrated in Figure 3.10. Low-*k* damage is the result of a competition between the etch rate and the diffusion rate of active species ($O^*$, fluorine) causing the damage. The etch rate is driven by the thickness and composition of the fluoropolymer that appears on the low-*k* surface during the etch. For polymerizing discharges, due to the absence of ion bombardment, a $CF_x$ protective layer will form on the vertical sidewalls, which will slow down the horizontal etch rate, therefore making the resulting process very sensitive to diffusion of the damaging species. In order to obtain a nondamaging process, a high etch rate should be met and the amount of radical diffusion must be suppressed as much as possible. This is only possible by carefully balancing the plasma polymerization propensity, i.e. to find conditions where simultaneously the radical density is low and/or sidewall polymers have a structure/composition that seals the porous low-*k* surface against radical diffusion.

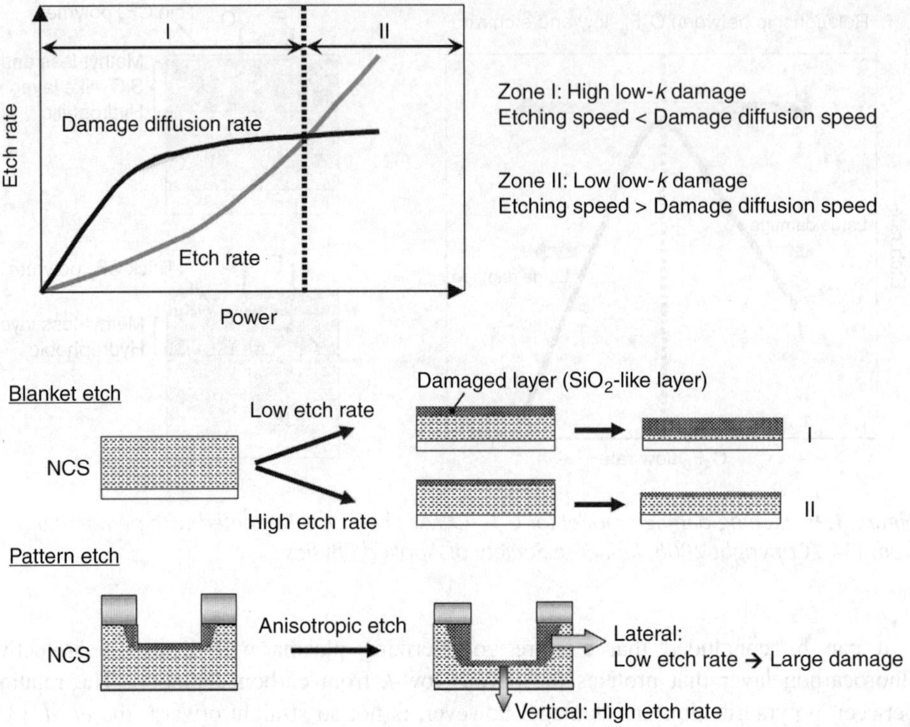


***Figure 3.10*** *Etching damage model taking into account etch and damage diffusion rates. Reprinted with permission from* [34]. *Copyright 2008, Japanese Society of Applied Physics*

In summary, many parameters must be taken into account in order to achieve a good low-*k* etch process, i.e. nondamaging and with a good morphology. Not only the etch profile and uniformity should be optimized, but also the plasma lateral damage should be minimized. A low-damage etch process could be realized by avoiding damaging agents (like O or N radicals) and controlling the F diffusion into the porous film. The F diffusion can be reduced by creating an adequate protective layer (reducing the diffusivity) and by increasing the etch rate (as compared to the diffusion rate).

## 3.5 Mechanisms of Plasma Damage

The damage formation in low-*k* dielectrics is a complex phenomenon involving both physical and chemical effects; the former arises from ion bombardment (energy distribution and flux for each ionic species) while the latter involves chemical reactions between the plasma and the low-*k* constituents or is induced by plasma radiation (photochemistry). The increase of the dielectric constant is directly correlated to changes in the bonding configuration, the formation of a carbon-depleted layer, film shrinkage and surface densification. During processing, the plasma constituents (ions, photons and radicals) contribute and interact in a complex manner to induce damage.

In this section, we review the results of studies aiming at understanding the mechanisms of damage formation induced by oxidative plasmas to porous low-*k* dielectrics. The first part summarizes the results from studies using various shading methods that allow discrimination of the individual roles of ions, photons and radicals. The damage induced by oxidative plasmas such as $O_2$ and $CO_2$ was found to be proportional to the oxygen radical concentration and enhanced by VUV photon radiation. Ion bombardment induces surface densification that blocks the diffusion of oxygen radicals. The results of the studies also indicate that the UV and VUV photons alone can also lead to surface damage.

In the second part, the mechanism of damage formation by oxidative plasma is analyzed by considering the kinetics of plasma interaction with porous low-*k* dielectrics, i.e. the kinetics of radical diffusion, reaction and surface recombination in the pore structure of low-*k* dielectrics. In order to study these chemical effects, several analytical techniques were used, including spectroscopic ellipsometry (SE), FTIR and XPS to examine the chemical change and correlate to the dielectric loss measured by the CV method. In the high ion energy region, damage is induced by momentum transfer between the ion and the structure of the low-*k* dielectric. This interaction, leading to surface densification, can be described by the 'sputtering yield model' that will be presented.

### 3.5.1 Gap Structure Studies

To understand the effect of the plasma process to a low-*k* material, it is essential to study the specific role of each plasma constituent (ions, photons and radicals). For this purpose, there are two general approaches. On the one hand, three separate sources for radicals, ions and light are assembled each next to the other (downstream plasma for radicals, ion beam generator, UV–VUV lamps). When focused on the sample surface, this setup simulates the individual or the combined effect of the plasma constituents [11, 35–37]. On the other hand, experiments can be performed in regular plasma chambers, by using specially designed masks (windows) that allow the selection of ions and/or photons within a certain range of energy and wavelength. In a typical design, the gap spacing between the mask and the low-*k* film surface can be varied in order to control the diffusion length and the density of radicals underneath the mask [38–42].

The gap structure used by Shi *et al.* [11, 41] is shown in Figure 3.11. It contains two parallel Si spacers and a top optical window that is used as a mask. When an Si substrate with a low-*k* dielectric layer on top is inserted into the tunnel, a gap is formed between the low-*k* dielectric layer and the optical mask. The gap structure can also be mounted directly on top of a standard Si wafer coated with a low-*k* film. The structure was loaded into a commercial plasma chamber equipped with a 13.56 MHz inductive coupled plasma (ICP) source and a 13.56 MHz radio frequency (RF) bias source at the bottom electrode. Plasma ions, photons and radicals are driven to impinge on to the low-*k* dielectrics outside the masked area, while underneath the mask only photons and radicals can react with the low-*k* layer. Depending on the cut-off wavelength of the optical mask, photons within different energy ranges can transmit through the optical mask and impinge on the low-*k* surface. Three kinds of mask materials ($MgF_2$, fused silica or

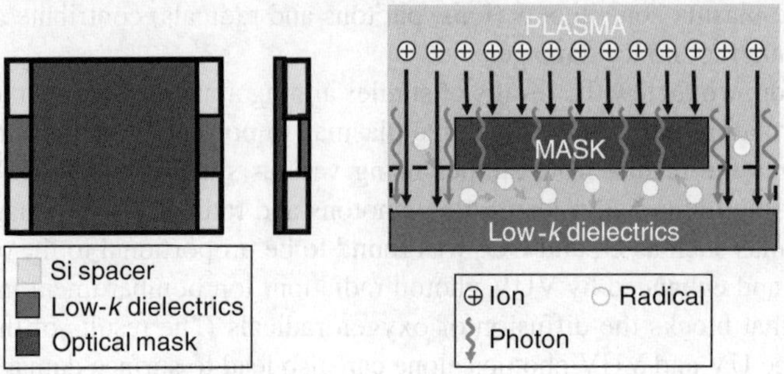


*Figure 3.11* Schematic drawing of gap structure: (left) top view and side view, (right) detailed side view [41] (see color Plate 1)

*Table 3.1* Properties of optical masks: $E_g$ = electronic bandgap

| Optical mask | $E_g$ (eV) | $\lambda_{cut\text{-}off}$ (nm) | Photons underneath mask |
|---|---|---|---|
| $MgF_2$ | 11.8 | 120 | VUV, UV, Vis, IR |
| Fused silica | 7.8 | 185 | UV, Vis, IR |
| Si | 1.1 | 1500 | IR |

quartz, and Si) were used to build the gap structure, yielding a range of photon energies as summarized in Table 3.1. The photon intensity can be tuned by varying the ICP source power and the chamber pressure. Through both open sides of the gap structure radicals diffuse into the channel and react with the low-*k* surface. By changing the gap height and plasma operating conditions, the radical density in the gap structure can be adjusted. At the same time, the RF bias power at the bottom electrode can be adjusted to vary the ion energy in the study.

A similar system was used by Uchida *et al.* [39] and Takeda *et al.* [42]. In their setup, parts of the wafer are covered with two $MgF_2$, two quartz and one Si windows.They were able to separate different factors: only VUV radiation using the $MgF_2$ and UV radiation using the quartz window when these are positioned directly on top of the wafer. When raised by 0.7 mm above the wafer surface, $MgF_2$ and quartz windows allowed the observation of the concomitant effects of VUV radiation and radicals, while the Si roof allowed only radicals (the Si shades all light below 1.5 µm). The joint effects of VUV, radicals and ions was observed in the open area.

The concentration profile of the oxygen radicals inside the gap $\phi(x)$ can be estimated by considering the kinetics of oxygen radical diffusion, reaction and recombination (consumption) inside the gap as [41]:

$$\phi(x) = \phi_0 \frac{\mathrm{Cosh}\left(\frac{l-2x}{2\zeta}\right)}{\mathrm{Cosh}\left(\frac{l}{2\zeta}\right)} \tag{3.2}$$

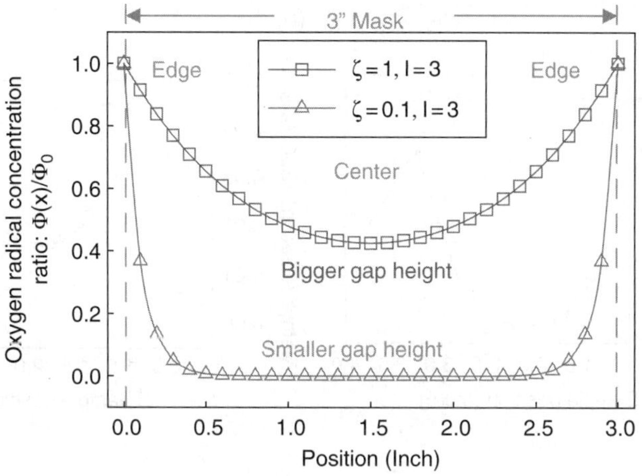


***Figure 3.12*** *Numerical simulations of oxygen radical concentration inside the gap* [41]

$$\zeta = \sqrt{\frac{hD}{\gamma_{reac} + \gamma^{*}_{recom}}} \tag{3.3}$$

where $l$ is the mask length, $h$ the gap height, $D$ the Knudsen diffusivity for oxygen, $\gamma_{reac}$ the reactivity between the oxygen radical and the OSG's methyl groups and $\gamma^{*}_{recom} = \gamma^{mask}_{recom} + \gamma^{low-k}_{recom}$ is the sum of the recombination rate of oxygen radicals on the mask surface and low-$k$ surface. Based on this analytical solution, the oxygen radical concentration corresponding to two gap heights and $\zeta$ values can be deduced as shown in Figure 3.12. Underneath the mask, there are more oxygen radicals close to the edge than that at the center. In addition, by increasing the gap height, the oxygen radical concentration can be increased.

### 3.5.2 Effect of Radical Density

Using the gap structure formed with various masks, experiments were conducted to study the effect of radical density, ion energy, photon energy and intensity on damage formation. Figure 3.13 summarizes the effect of radical density on damage formation in porous OSG low-$k$ dielectrics using a $CO_2$ plasma operated in an ICP chamber, at 5 mtorr, 150 W source power, 100 W bias for 300 seconds. Underneath the mask, more film shrinkage, methyl depletion and moisture uptake were observed close to the edge of the mask than those underneath the center of the mask. Recalling Figure 3.12, there is a good correlation between the observed methyl depletion/moisture uptake observed and the simulated radical density, indicating that the damage induced by the $CO_2$ plasma is proportional to the oxygen radical concentration.

Additional studies on the effect of oxygen radicals were performed as a function of chamber pressure, gap height and plasma power in a similar ICP chamber. As shown in

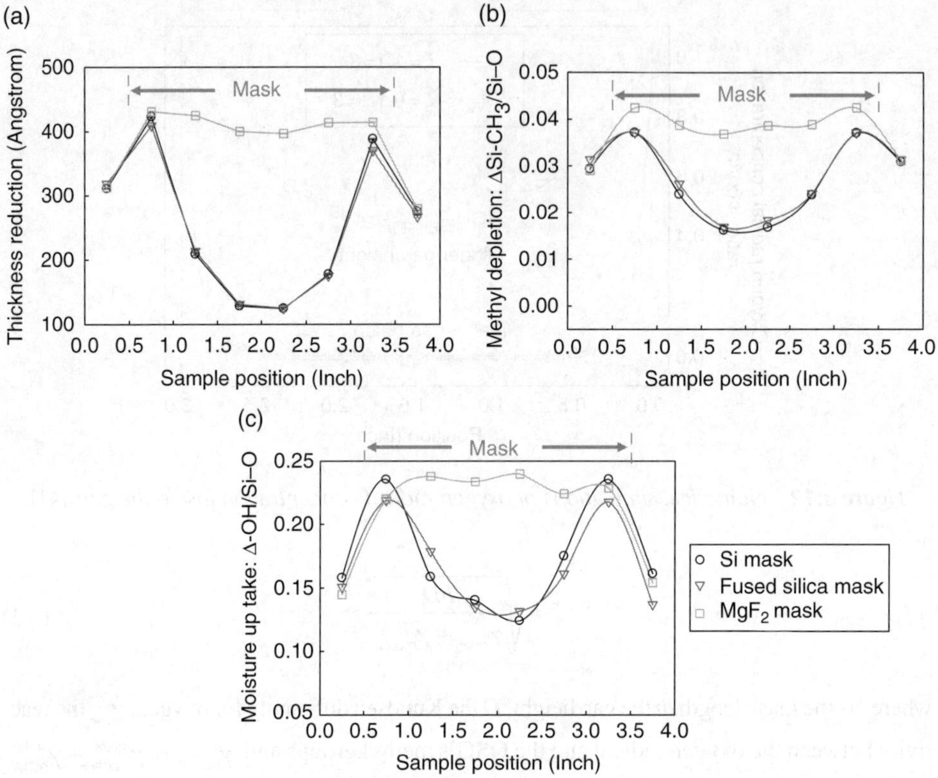


***Figure 3.13*** *(a) Film thickness reduction, (b) methyl depletion and (c) moisture uptake for a k ~2.2 OSG low-k dielectric treated by $CO_2$ plasma in an ICP plasma etch system*

Figure 3.14, with an increase of the gap height from 0.7 mm to 1.4 mm, the film shrinkage and methyl depletion increases. These observations are in agreement with the results presented in Figure 3.12. Similar power and pressure dependences were reported for a source power varying from 200 to 300 W and for chamber pressure ranging between 5 and 60 mtorr by Nagai *et al.* [43] and Worsley *et al.* [44].

### 3.5.3 Effect of Ion Energy

As shown in Figure 3.13, at positions close to the edge of the mask, inside the gap structure, more film modifications were observed than in nonprotected areas outside the gap. These results indicate that ion bombardment can reduce the plasma-induced damage. This is consistent with many other studies [45–47]. Physical sputtering and chemical surface reactions act synergetically and lead to densification of the low-*k* dielectric. This densification, with an appropriate choice of plasma bias power, can be confined to a few nm of the top layers of microporous dielectric films [48]. In this way, the *k* value of the films is not significantly affected; this newly formed densified 'skin' layer acts as a protection or barrier against

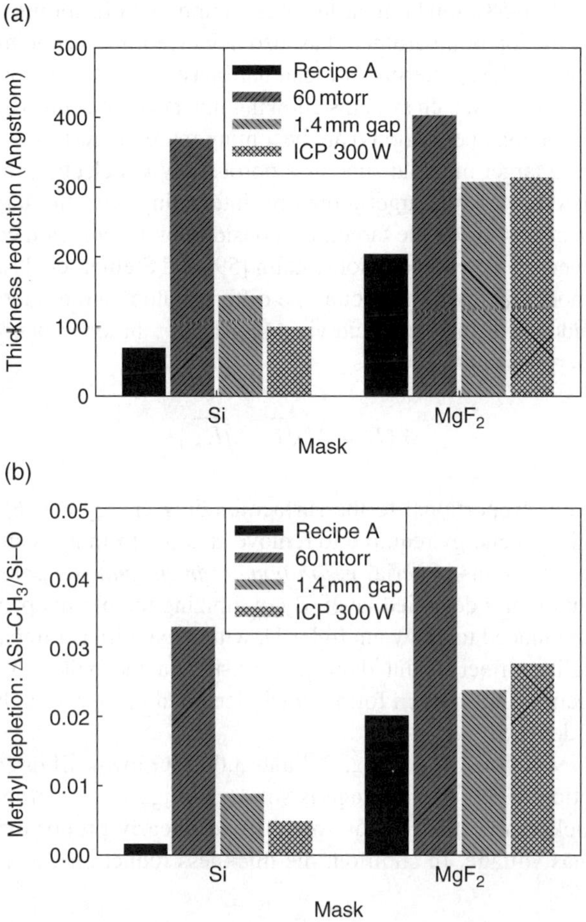


***Figure 3.14*** *(a) Film thickness reduction and (b) methyl depletion for k ~2.2 OSG low-*k *dielectrics treated by $O_2$ plasma in an ICP plasma etcher. Recipe A: source power 200 W, bias 100 W, 5 mtorr, 20 sccm, 0.7 mm gap height, 150 s and 32 °C. All other recipes are based on recipe A, but with a single parameter changed as indicated in the legend. All data were compared at the same position underneath the masks*

oxygen radical diffusion, thus reducing the damage [45, 49]. Ion bombardment induces both pore collapse and the densification of wall materials between the pores [46, 47]. In some cases simultaneous or consecutive effects of ion bombardment and chemistry are more efficient [50–52]. However, the ion-induced densification (sealing) is challenging for ultra-low-*k* (ULK) films because of the larger pore size and, therefore, its technological application during integration might be limited. Nevertheless, damage induced by radical diffusion through the vertical sidewalls cannot be easily reduced by sputtering-driven surface densification, due to obvious geometrical aspects (ion bombardment parallel to the sidewalls).

The effect of ion bombardment is not limited to surface densification, but is an essential component for ion-assisted etching, due to the creation of more active sites, as well as for physical sputtering [53–58]. Ion bombardment is an important factor providing anisotropic etching. However, the ion bombardment has also negative impacts due to the reduction of the etch selectivity and the sputtering of metal hardmasks used for the patterning of low-*k* materials. For those reasons, etch systems allowing independent control of the plasma density (ion flux) and the ion energy (bias) are becoming more interesting.

Once ions enter a target material such as a porous low-*k* dielectric, their momenta and energies are transferred to the target atoms by interacting with the target nuclei or the bound and free electrons inside the target. By considering the collision cascade process of target atoms induced by the energetic ions, Zalm [59] and Steinbrüchel [60] developed the sputtering yield model taking into account the different atom–atom interaction potentials. The energy dependence of the etch yield via a physical or an ion-enhanced chemical process at low ion energy is expressed as

$$Y(E) \approx A\left(\sqrt{E} - \sqrt{E_{th}}\right) \tag{3.4}$$

where $A$ is inversely proportional to the surface binding energy and $E_{th}$ is the threshold energy or the minimum energy required to remove an atom from its position in the lattice. A specific feature of porous material is *ion-bombardment-induced pore collapse*, which leads to the formation of a densified layer. By optimizing the plasma power and bias, this dense layer can be reduced to a few nm [48, 61], without significant impact on the k-value but important sealing effect againt damage diffusion in the bulk. This is not true for mesoporous dielectric films, which form a thick densified layer with a dielectric constant exceeding that of dense $SiO_2$.

The thickness reductions of *k* ~ 2.2, 2.7 and 3.0 OSG low-*k* films induced by an Ar plasma as a function of DC bias voltage is shown in Figure 3.15. Above a certain bias threshold, the thickness reduction was found to be linearly proportional to the square root of the DC bias voltage. In addition, the thickness reduction was found to increase

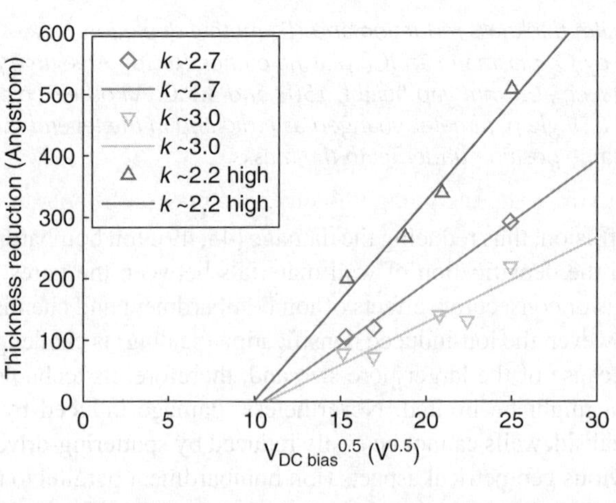


***Figure 3.15*** *Thickness reduction of* k *~2.2, 2.7 and 3.0 OSG low-*k *films induced by Ar plasma treatment in an Oxford plasma etcher as a function of DC bias voltage (ICP 200 W, 5 mtorr, 20 sccm, 150 s, 32 °C)* [62]

with decreasing $k$, probably caused by a reduction in film density and bond strength. Taking into account that both sputtering and densification are initiated by ion bombardment, this linear dependence suggests that both above-mentioned processes are ruled by a similar mechanism.

### 3.5.4 Effect of Photon Energy and Intensity

Processing plasmas are emitting over a very wide spectral range, with wavelengths running from a few tens of nanometers (as an example, the He first line is located at 30.4 nm) up to several micrometers. Jinnai *et al.* also developed a special on-wafer monitoring technique that was calibrated by a UV spectrometer [63]. The technique measures the UV spectra emitted by the plasma and then predicts the type of induced damage. The damage to organosilicate glass (OSG) low-$k$ materials caused by a $CF_4$-based plasma was simulated and good agreement with the experimental results was found. The maximum intensity of VUV–UV emission of typical fluorocarbon-based plasmas used for the etch purpose is located in the range 150–350 nm. The most intense part of the spectrum relates the atomic and molecular electronic transitions, between a few tens of nanometers and ~350 nm. This section describes how VUV (~60–200 nm radiation) can damage exposed low-$k$ surfaces, and what kind of synergy develops as radicals are coupled to VUV. Impinging radiation can have two different effects: chemical and mechanical modifications of the low-$k$ dielectric, caused by chemical bond dissociation, or accumulation of trapped charges, caused by electronic excitations, that all will adversely affect device characteristics (increased capacitance, increased leakage current, decreased breakdown voltage). The depth of VUV damage correlates with the penetration depth of this light into low-$k$ materials [64].

In general, any dielectric exhibits electronic collective excitations that are described by bands. In semiconductors, the bandgap separates the upper fully occupied *valence* band from the lowest fully unoccupied *conduction* band. Depending on the gap energy, at room temperature the material can exhibit semiconducting behaviour (Si, with a gap energy of 1.12 eV) or be insulating ($SiO_2$, with a gap energy around 8.5 eV). VUV irradiation of dielectrics with energies in the range 7–21 eV can generate numerous competing phenomena: (i) photoemission, (ii) photoconduction, (iii) population or depopulation of electrons from trap states and (iv) photoinjection [65]. The photon energy, dielectric thickness and electric properties of the dielectric layer (defect states, valence band structure, bandgap) are critical to determine which of these processes will dominate at a given wavelength. Electron photoemission may occur as the incoming photons have energies higher than the photoemission threshold, which can be either the bandgap + electron affinity[1] for valence electrons, or lower energy for electrons coming from defect states in the bandgap (causing depopulation of defect states). Until they are emitted, recombine or become trapped, created electron–hole pairs can travel through the dielectric, giving rise to photoconduction. Holes have a much shorter mean free path than electrons and thus holes become trapped within the dielectric while electrons might be photoemitted. If the VUV photon energy is less than the photoemission threshold, electrons from defect states can be excited in higher energy defect states and then possibly be photoemitted, but also VUV photons can excite electrons from the valence band into defect states. Finally, photons can be absorbed in the conductive

[1] In $SiO_2$ the bandgap is about 8.5 eV and the electron affinity is 0.9 eV

or semiconductor layer underlying the dielectric material, creating charge that can be injected into the dielectric layer (photoinjection), repopulate the defect states and reduce the amount of trapped positive charges. Therefore, there is a wide range of phenomena that may occur, depending on the band structure of the material, adjacent layers and photon energy. By exposing samples to VUV radiation and then measuring their photoemission, surface potential and trapped charge, Lauer *et al.* [65] were able to map the bandgap structure of porous SiOCH materials. They found that SiOCH has a similar structure to that of $SiO_2$, with a gap of about 9.0 eV and similar defect centers. Irradiation with 8.0 eV photons causes the defect states (~1 eV above the top of the valence band edge) to be depopulated. However, subsequent UV irradiation allows the repopulation of these defect states by photoinjection of electrons from the underlying Si substrate into the SiOCH layer, thus neutralizing the trapped positive charges accumulated during VUV irradiation. UV and VUV irradiation are therefore competing factors. They showed also that UV-cured SiOCH films have more defect states in the bandgap than noncured SiOCH, leading to more charge trapping under VUV exposure [66]. By exposing SiOCH to an ECR plasma covered by a capillary-array window, they could separate the effect of particle bombardment and plasma radiation. Plasma-induced charge accumulation has two contributions: surface ions sticking, due to ion bombardment, and charge accumulation trapped in the bulk, due to photon irradiation [67].

VUV photons also enhance the plasma damage caused by the radicals. As shown in Figures 3.13 and 3.14, the use of an $MgF_2$ mask revealed more film shrinkage, methyl depletion and moisture uptake than those with an Si mask or a fused silica mask. As shown in Figure 3.14, by further increasing the ICP source power or the chamber pressure, the plasma damage was enhanced. This can be attributed to the increase of VUV photon intensity.

The effect of VUV photons was also studied in Takeda's experiment [42]. Using various windows and different locations of the top covers (see Figure 3.16), the authors showed that damage by pure VUV and UV photons remains small if the low-*k* surface is not exposed to $O_2$ and $N_2/H_2$ plasma. Simultaneous exposure of the low-*k* surface to VUV/UV photons and oxygen atoms leads to remarkable modifications, while the effect of oxygen atoms without UV radiation is also small. Finally, the completely 'open' low-*k* surface does not suffer very much, despite of presence of VUV/UV photons and oxygen atoms. This last result is a clear suggestion that ion bombardment plays a positive role by densifying the low-*k* surface.

Using vacuum beam equipment with a calibrated VUV lamp, Lee and Graves [68] showed that 147 nm VUV photons cause scission of Si–$CH_3$ bonds in porous SiOCH, leading to the formation of Si–OH groups. By the addition of a flux of molecular $O_2$, this effect is significantly increased and a more detailed analysis indicate that $O_2$ reacts with a methyl intermediate, desorbing due to scission of Si–C bonds. This generates $CO_2$ and $H_2O$ within the pores, which in their turn react with Si dangling bonds, forming Si–OH. This has been validated by VUV irradiation experiments of $SiO_2$ hardmask-covered SiOCH low-*k* film by Matsunaga *et al.* [69].

These observations are in good agreement with the results reported in a paper by Goldman *et al.* [36]. In this work, using a specially designed plasma chamber and He plasma, it was shown that the degree of low-*k* modification by both VUV photons and He metastable atoms is not significant while the surface densification needs

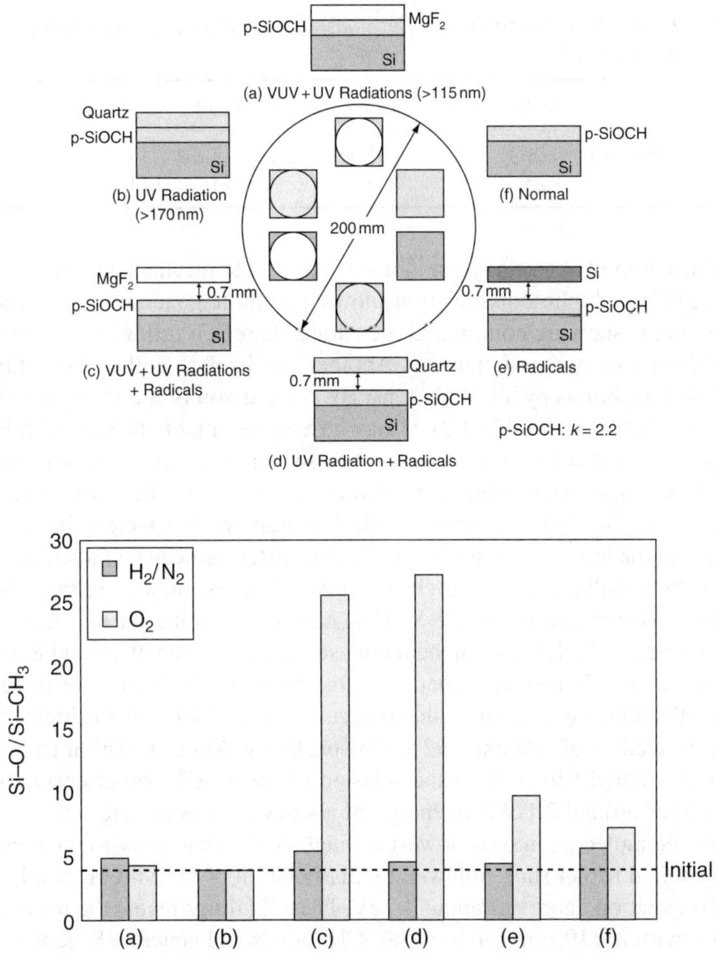


***Figure 3.16*** *Top: schematic diagram of Takeda's experiment. Bottom: Si–O/Si–$CH_3$ ratios of various samples processed in $H_2/N_2$ and $O_2$ CCPs, measured by FTIR spectroscopy; (a) VUV and UV radiation, (b) UV radiation, (c) VUV and UV radiation and radicals, (d) UV radiation and radicals, (e) radicals only and (f) plasma exposure. Reprinted with permission from* [42]. *Copyright 2011, American Institute of Physics*

simultaneous exposure of a low-*k* surface to VUV/UV photons, low-energy $He^+$ ions and He metastable atoms. The strong absorption of VUV photons by Si–$CH_3$ groups and porogen residues lead to their decomposition into a dense hydrocarbon structure consisting of carbon and $CH_2$ chains with cross-linking carbon bonds. The bombardment by low-energy $He^+$ ions stimulates the hydrogen extraction from the hydrocarbon network and induces a modification of this network into a denser carbon network, similar to graphite-like C:H structures.

Several mechanisms can be considered to explain the damage enhancement caused by the VUV photons. Gas-phase VUV-induced photodissociation of $O_2$ and $CO_2$ molecules

***Table 3.2*** *Bond dissociation energy values and associated wavelengths in simple molecules* [71]

| Bond | Si–O | Si–C | Si–H | Si–Si | C–H | C–O | C–C | O–H |
|---|---|---|---|---|---|---|---|---|
| Bond energy (eV) | 8.3 | 4.7 | <3.1 | 3.4 | 3.50 | 11.2 | 6.3 | 4.4 |
| Wavelength (nm) | 150 | 265 | 400 | 366 | 354 | 111 | 197 | 280 |

[70] with formation of active oxygen atoms is a possible mechanism. Early experiments [68] indicate that such photodissociation shows negligible generation of oxygen radicals in plasma systems showing common $O_2$ densities, therefore ruling out this phenomenon as a plausible source of low-*k* damage. Another mechanism is the possibility of direct scission of Si–$CH_3$ bonds by VUV photons. By comparison of the strength of the critical bonds in low-*k* dielectrics (Table 3.2), it may be assumed that both VUV ($MgF_2$ window) and UV (quartz window) photons have energy sufficient to break these bonds of low-*k* materials. However, a clear difference between these two conditions is observed: VUV photons are damaging low-*k* material while UV light with wavelengths above 185 nm does not change the low-*k* composition at all. This difference can be understood by taking into account the results of quantum-chemical calculations, indicating that photons with wavelengths < 190 nm can break the Si–$CH_3$ bonds via excitation into the first excited singlet state. These calculations, on the model substances octamethyl- and tetramethylcyclotetrasiloxane as well as on hexamethyl- and tetramethyldisiloxane, resulted in threshold wavelengths for the excitation of the molecule into the first excited singlet state of 190 and 198 nm as well as of 189 and 192 nm, respectively. After excitation and intersystem crossing in an excited triplet state, the scission of the Si–$CH_3$ bond occurs, gaining an energy benefit of around 2.1 eV. An energy of about 6.2 eV is therefore necessary for the excitation of the molecule into the lowest excited singlet state prior to intersystem crossing into an excited triplet state followed by dissociation of the Si–$CH_3$ bond, despite the low bond dissociation energy of about 4.7 eV. These findings reveal the presumption that only photons with $\lambda < 190$ nm can break Si–$CH_3$ bonds and generate Si centered radicals, which subsequently attract protons from neighboring methyl groups. Therefore, the scission of Si–$CH_3$ bonds need VUV photons with an energy significantly exceeding the Si–C bond energy and has a threshold near 190 nm, which is close to the transmittance threshold of a quartz-based optical window.

The negligible $CH_3$ reduction by VUV photons only, observed in Figure 3.16, is probably related to two phenomena. First, the absorption coefficients of OSG low-*k* material drastically increases below 200 nm and therefore photochemical modification is only localized in a thin surface layer (for He plasma, the penetration depth into $SiO_2$ at 58.4 nm is estimated to be around 15 nm [75]). For the FTIR measurements used in this study, the contribution of the surface layer is too small to be detected. A second possible reason is a difficulty in removing the reaction products from the reaction zone. For this reason, the probability of an inverse reaction of methyl radical with Si dangling bonds can be significant. Increasing the gap height allows a deeper penetration of oxygen radicals, which removes $CH_3$ radicals by the formation of CO, $CO_2$ and $H_2O$. At the same time, the $CH_3$ removal decreases the absorption coefficient of the surface layer and damage is extended into the film.

From a structural perspective, UV-based hardening of low-*k* materials is a key element for lowering the dielectric constant. In order to create porosity, porogen molecules, usually cyclic hydrocarbons, are introduced into an SiOCH film by codeposition with the matrix material. The porogen is removed from the films using UV-assisted thermal curing. The porogen molecules are photodissociated by UV light with the formation of volatile hydrocarbons and nonvolatile carbon-rich porogen residues (PRs). This process is usually performed with broadband UV light, with emission lines in the range 200–600 nm. Simultaneously to porogen removal, a rearrangement of the silica matrix occurs, together with an increased cross-linking (growth of Si–O–Si bonds) and some minor loss of methyl bonds, all leading to enhanced mechanical properties [73]. However, this process was proven to leave porogen residues (cross-linked amorphous $sp^2$ C-like), strongly absorbing in the 200–400 nm range, that hinder the SiOCH matrix cross-linking, therefore deteriorating the mechanical properties of the final film [74]. In addition, it was shown that for porous low-*k* materials, UV treatment led to improvement of the mechanical properties, but at the expense of the hydrophobicity of the material, due to the depletion of Si–$CH_3$ groups. As recently proposed [75], a better strategy is to combine first the full removal of porogen by high-temperature $H_2$-afterglow treatment, followed by UV cure with 172 nm radiation, which cause direct scission of Si–$CH_3$ bonds in contrast to less energetic >200 nm broadband UV light, without impacting Si–O bonds (which photodissociation requires to be shorter than 150 nm UV wavelengths or with significantly higher light intensity).

### 3.5.5 Plasma Damage by Oxidative Radicals

The most popular systems used for the resist strip from low-*k* films are based on downstream plasmas with a high concentration of oxidative or reductive radicals. Such processing conditions, based on chemical etching, allow the achievement of high selectivity because of the absence of ion bombardment. Some systems are also equipped with special grids that shade VUV radiation. In this case, the major reason for plasma damage is the interaction of active radicals with hydrophobic Si–$CH_3$ groups, leading to their removal, and subsequent hydrophilization of the low-*k* material. Such damage can be extended into the low-*k* materials because of the diffusion of active radicals into the porous structure. Therefore a quantitative description of the radical diffusion into the pores is extremely important.

This problem has been extensively studied in heterogeneous catalysis starting from the work by Thiele [76], who first described the effect of in-pore diffusion on chemical reaction rates, using a simple geometric model with isotropic properties. Although this model is based on geometric shapes, it gave a quite good approximation to practical porous materials [77]. A similar analysis has been used to describe the diffusion of radicals into the pores of low-*k* materials. Cho *et al.* [78] described the diffusion of precursors of fluorocarbon polymers into pores. The obtained result had a good agreement with experimental observations and predicted the time required to reach a steady-state concentration of the polymer precursor in a pore ($\approx 10^{-7}$ s) and the time required to seal off pore necks at the surface of the dielectric ($\approx$70 s). Both the model and experimental results show a greater depth of penetration and an enhanced deposition of polymer at higher porosities.

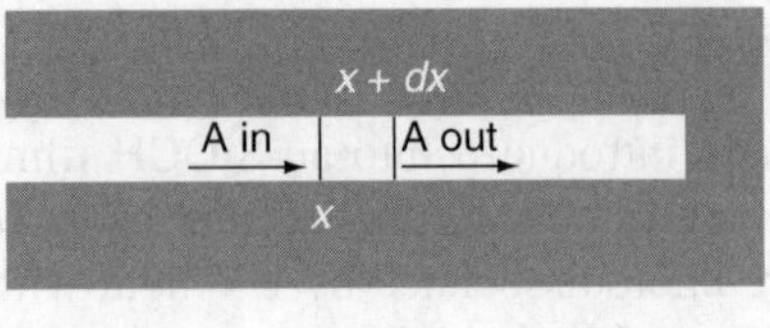


*Figure 3.17* Schematic representation of component A diffusing in an idealized cylindrical pore

Baklanov *et al.* [79] and Shi *et al.* [41] evaluated the penetration depth of oxygen plasma species into porous low-*k* dielectrics. Their analysis was based on the radical diffusion, the first kinetic order reaction and recombination inside the pore structure.

Let us consider radical diffusion into single pores with a diameter $d_p$. It is reasonable to assume that the radical concentration ($C_A$) is uniform in the plasma reactor because of the low pressure. Let us assume that the depth of damage extends up to $x = L$, where $C_A = 0$. The diffusion flux inside the pore is $N_A = -D_A \, \mathrm{d}C_A/\mathrm{d}x$. For an idealized cylindrical pore (Figure 3.17) and an isothermal, isobaric and first-order reaction, the material balance equation along a differential length d$x$ of the pore can be written as

$$\frac{\pi d_p^{\,2}}{4} D_A \left(\frac{dC_A}{dx}\right)_x - \frac{\pi d_p^{\,2}}{4} D_A \left(\frac{dC_A}{dx}\right)_{x+dx} - \pi d_p \, dx \, \gamma_R C_A \ (1) = 0 \qquad (3.5)$$

The first term of Equation (3.5) reflects the radicals flux into the region d$x$, while the second term is the flux out of this region. The third term represents the radical consumption inside the region due to the reaction with Si–$CH_3$ groups ($\gamma_{react}$) and the radical surface recombination ($\gamma_{recom}$). Therefore, $\gamma_R = \gamma_{react} + \gamma_{recom}$. An analytical solution of Equation (3.5) shows that the depth of radical penetration depends on the ratio of $\gamma_R$ to $D_A(\lambda)$, which is called the Thiele modulus $\lambda$ (TM):

$$\frac{d^2 C_A}{dx^2} - \lambda^2 C_A = 0, \quad \text{where } \lambda = \left(\frac{4\gamma_R}{d_p D_A}\right)^{1/2} \qquad (3.6)$$

Assuming the Knudsen diffusivity, the diffusion coefficient can be represented as [80]

$$D_A = 97 \frac{d_p}{2} \sqrt{\frac{T}{M}} \qquad (3.7)$$

where $M$ is the molecular mass of plasma radicals, $d_p$ the pore diameter and $T$ is the absolute temperature of the radical under consideration. The chemical reactions between the oxygen radical and the Si–$CH_3$ groups start with detachment of H, because it is the weakest bond [81]:

$$\text{Si–CH}_3 + \text{O} \rightarrow \text{Si–CH}_2^* + \text{OH} \qquad (3.8)$$

where $\gamma_{react}$ represents the probability of this reaction. The surface recombination of oxygen radicals in conditions of plasma etching (low pressure and small pore size) follows a Langmuir–Hinshelwood mechanism:

$$\text{O phys} + \text{O chem} \rightarrow \text{O}_2 \tag{3.9}$$

where $\gamma_{recom}$ represents the recombination probability of oxygen radicals. The oxygen atom loss probabilities ($\gamma_{react}$ + $\gamma_{recom}$) in three different low-*k* films have been experimentally measured by means of plasma-induced actinometry [81]. The major channel of O and H loss is surface recombination. The consumption of these atoms through reactions with the C-containing hydrophobic groups has a minimal contribution. Thus, the damage depth in low-*k* films is defined by surface recombination of radicals diffusing through it. This is supported by a Monte Carlo simulation [82] of the evolution of the depth profile of Si–$CH_3$ after exposure to O atoms. Coming back to Equation (3.6), it can be deduced that the strongest decay of radical concentration will happen for a large Thiele modulus $\lambda = \sqrt{4\gamma_R/(d_p D_A)}$.

The TM can grow by increasing the recombination rate $\gamma_R$ or decreasing the pore diameter $d_p$. The recombination rate increases with process temperature [79], with a more pronounced growth rate for H than for O and N, which explains the reduction of plasma damage at high temperatures in Ar/$H_2$- or He/$H_2$-based discharges. Also, $\gamma_R$ can be raised by increasing the carbon concentration in the low-*k* dielectric and the surface recombination rate of the oxygen radical on the pore wall ($\gamma_{recom}$) [77]. The effects of pore size and porosity on plasma-induced damage as predicted by this model are consistent with the results of plasma damage studies reported by several groups [44, 83–86]. Namely, as shown in Figure 3.18, the effect of exposing OSG low-*k* films with various *k* values (2.2, 2.5, 2.7) to pure oxygen radicals demonstrated a direct inverse correlation between the decreasing *k* and the level of plasma damage (film thickness reduction and normalized methyl depletion).

In conclusion, further progress in low-*k* materials development includes both the reduction of the *k* value and a decrease of the mean pore size, because it improves their integrability. Starting from certain pore sizes approaching 1–1.5 nm, the classic diffusion approach for calculating O atom concentrations in nanoporous material seems to be irrelevant. Using Monte Carlo simulation and experimental values of the loss probability [82], it was shown that the frequency of oxygen atoms entering into pores is three orders of magnitude smaller than the residence time of oxygen radicals inside the pores. Therefore, only solitary oxygen atoms are instantly present in any porous channels of the SiOCH film and the stationary concentration of oxygen atoms in the porous channels is extremely limited. The problem can therefore no longer be considered in terms of atomic concentration and its gradient. Another issue is that the reduction of the pore diameter to values approaching molecular dimensions results in a dramatic decrease in diffusivity due to *single file diffusion* [87–89]. Recently, several models of mass transport and microporous film modification have been proposed in the literature [82, 90, 91].

### 3.5.6 Hydrogen-Based Plasma

An interesting option for stripping of low-*k* materials is $H_2$-based plasma. Since hydrogen is generally less reactive than oxygen one could hope to find conditions in which hydrogen plasma removes photoresist (breaks up organic polymer) without damaging

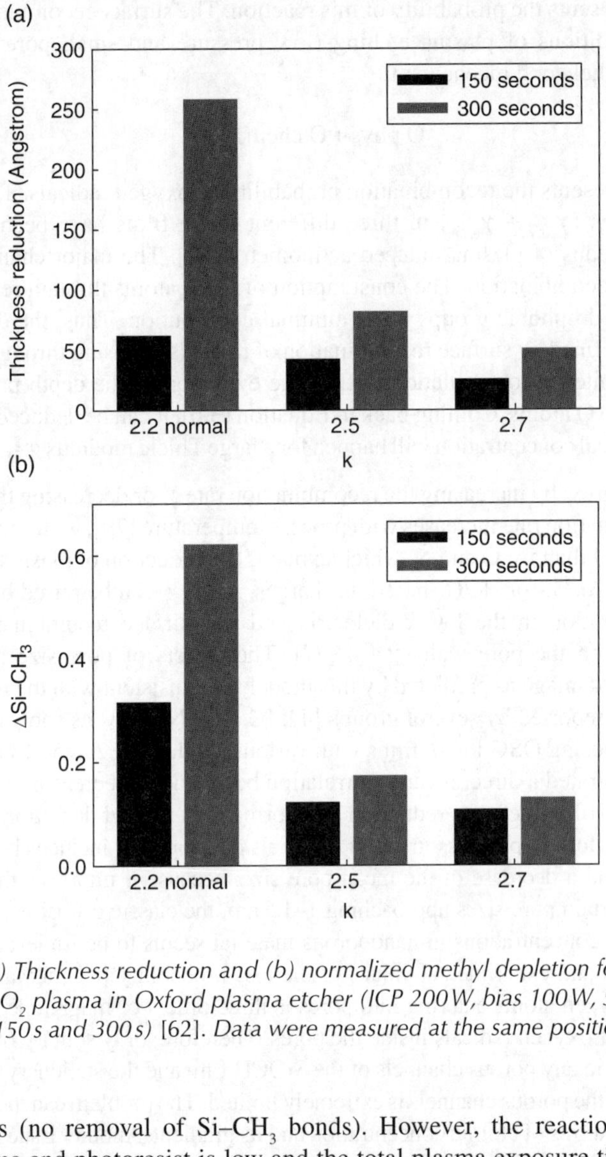


***Figure 3.18*** *(a) Thickness reduction and (b) normalized methyl depletion for OSG low-*k *films treated by $O_2$ plasma in Oxford plasma etcher (ICP 200 W, bias 100 W, 5 mtorr, 20 sccm, Si mask, 32 °C, 150 s and 300 s)* [62]. *Data were measured at the same position underneath the Si mask*

low-*k* materials (no removal of $Si–CH_3$ bonds). However, the reaction rate between hydrogen plasma and photoresist is low and the total plasma exposure time will be significantly longer. In order to compare various stripping plasma from a low-*k* damage point of view, it is necessary to use a normalization technique, where the plasma damage is normalized to the resist removal rate [92]. In other words, the low-*k* material should be exposed to different plasmas for the necessary time needed to remove a fixed amount of resist (e.g. 400 nm). Even though hydrogen plasma is expected to be less damaging, conventional hydrogen plasma processing generated in CCP and ICP chambers (around room temperature and in the presence of VUV light and ion bombardment) has an adverse effect on the low-*k* value, removing $Si–CH_3$ bonds, densifying the film and increasing its dielectric constant [93].

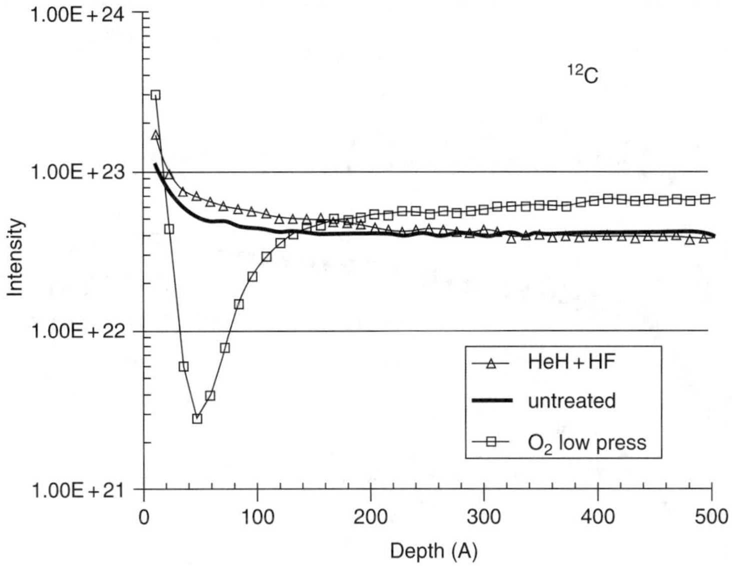


***Figure 3.19*** *SIMS analysis of a porous MSQ subject to the He:$H_2$ sublimation process compared to a low-pressure oxygen strip. Reprinted with the authors' permission* [94]

The hydrogen-based plasma strip has been significantly improved (in low-*k* damage terms) by raising the substrate temperature (increasing the resist removal rate) and removing the energetic species – ions and photons (use of a remote downstream plasma). It was found [94] that He/$H_2$ remote plasma at elevated substrate temperatures (200–400 °C) could remove resist without causing carbon depletion (see Figure 3.19) and a *k* value increase.

The mechanism leading to a damage-free strip by hydrogen radicals is the following. When the hydrogen atoms react with an Si–$CH_3$ group, the first step is the detachment of an hydrogen atom [81]:

$$\text{Si–CH}_3 + \text{H} \rightarrow \text{Si–CH}_2{}^* + \text{H}_2 \tag{3.10}$$

This reaction is similar to the reaction (3.8). However, the subsequent reaction step is very different. Further reaction of the Si–$CH_2$* dangling bond with oxygen atoms leads to its destruction, while the reaction with hydrogen atoms restores the Si–$CH_3$ group (Si–$CH_2$* + H → Si–$CH_3$).

This mechanism was evidenced by Lazzeri *et al.* [95], by exposing low-*k* films to a deuterium plasma. The deuterium replaces hydrogen in the low-*k* matrix, as shown in Figure 3.20, while the total concentration (H + D) remains constant.

In certain PECVD low-*k* films, the $H_2$-based strip at elevated temperatures showed some carbon loss although the Si–$CH_3$ concentration was not changed [96]. The lost carbon was attributed to *carbon porogen residues*, which are left inside the porous low-*k* structure by the manufacturing process. The porogen residues can be observed by UV ellipsometry since they substantially change the optical properties of low-*k* materials in the wavelength

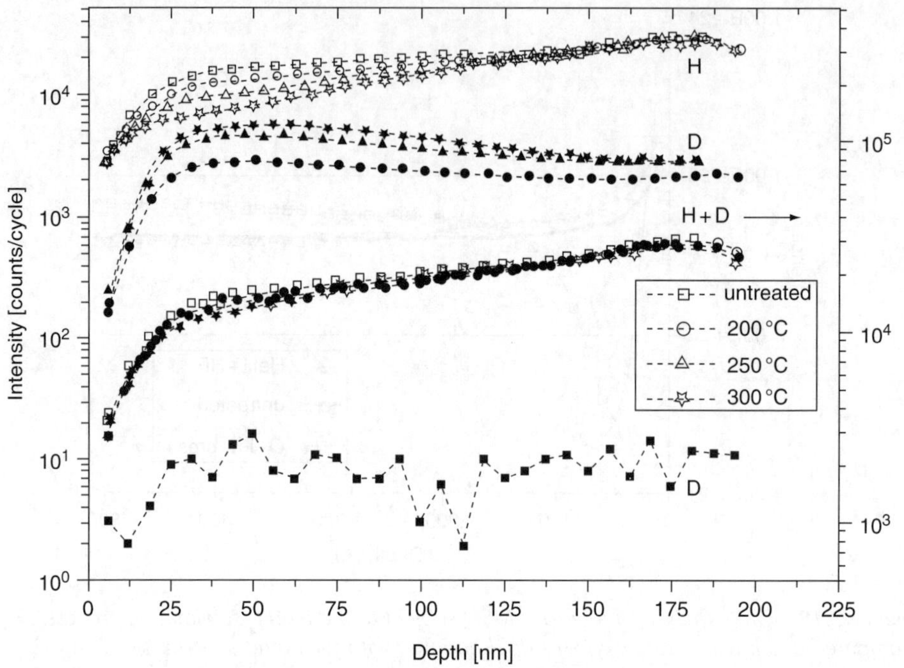


***Figure 3.20*** *H, D and the sum of H + D depth profiles from untreated nanoporous silica (NPS) (squared symbols) and NPS material after $D_2$ plasma exposures at 200 °C (circles), 250 °C (triangles) and 300 °C (stars), negative SIMS. The sum of H + D is plotted on the right axis. Reprinted with permission from* [95]. *Copyright 2006, American Institute of Physics*

range of 150–250 nm [97]. After high-temperature He/$H_2$ remote plasma treatment, the specific signature of porogen residues disappears. It should be noted that the penetration depth of hydrogen atoms (and, therefore, porogen residue removal) is limited, most likely by diffusion-recombination mechanisms. The selective removal of porogen, and porogen residues, by hydrogen atoms without damaging the OSG low-*k* matrix allowed the synthesis of a new low-*k* material with the so-far lowest achieved *k* value among all PECVD materials [98].

### 3.5.7 Minimization of Plasma Damage

*Optimization of Porous Low-*k *Dielectrics*

Several studies have demonstrated that plasma damage can be reduced by engineering the chemical properties of the OSG low-*k* dielectrics. Ryan *et al*. compared the plasma damage of two OSG low-*k* dielectrics with $k \sim 2.2$, one with a carbon concentration of ~35 % and a pore size of 2.1 nm, the other with a lower carbon concentration of ~19 % but a larger pore size of 2.7 nm [99]. The OSG low-*k* dielectrics with the high carbon content suffered less plasma damage than that with the low carbon content. Shi *et al*. reported that increasing the carbon concentration in OSG porous low-*k* dielectrics improved the material resistance to oxidative plasma damage [41]. After treatment with a $CO_2$ plasma, the measured depth of the carbon depleted layer was 23.0 nm for a high carbon concentration (C/Si = 1.66), while it was 60.9 nm for material showing a low carbon concentration (C/Si = 1.22).

*Choice of the Plasma Reactor*

In order to reduce the ashing damage, extensive studies have been carried out by optimizing the configuration of the existing plasma chambers since it determines the plasma density (ions, radicals), the energy distribution of ions, the photon energies and intensities. It is already well known that a CCP $O_2$ discharge generates less plasma damage in comparison with an afterglow $O_2$ plasma or an ICP $O_2$ discharge without sufficient bottom power [62, 100, 101]. A CCP generally has a wider ion energy distribution than an ICP. Without bias power, in an ICP the wafer surface is at the floating potential, yielding typical ion energy from 10 to 30 eV while the ion bombardment energy for the CCP is significantly higher (from 50 up to several hundreds of eV). The CCP discharge induces more surface densification, blocking oxygen radical diffusion into the porous low-*k* dielectric (see Figure 3.21).

However, as already mentioned, almost damage-free strip processing has been achieved by using high-temperature $H_2$/He downstream plasma. A high temperature and microwave source provide a high-resist strip rate and hydrogen atoms without VUV photons are not able to break Si–$CH_3$ bonds [96].

*Choice of the Plasma Species*

Plasma damage can be minimized by changing the plasma ash chemistry, by the addition of carbon-containing species. It has been reported that a $CO_2$-based strip process gave 17 % lower RC delay compared to an $O_2$-based strip process at the same RF power and pressure on a comb test structure [102]. $CO_2$-based ashing helps to minimize the amount of damage on the ULK materials and reduces the hardmask undercut [103].

The CO and the $CO_2$ plasmas contain polymerizing species such as C and $C_2$. Due to the difference in atomic mass between C and O, the Knudsen diffusivity is slightly higher for the C atoms: $D_C = 1.2D_O$. A faster C diffusion leads to the formation of a C-rich passivation layer on the pore surface prior to the exposure to oxygen radicals. This increases the carbon concentration and reduces the pore radius inside the low-*k* dielectric. The additional carbon species can be used as a sacrificial material to consume the oxygen radical and protect the Si–$(CH_3)_x$ groups. This is consistent with the observation by Kong *et al.* that a C-rich passivation layer reduces the $O_2$-induced trench sidewall damage [104]. Shi *et al.* also found that under a similar plasma condition, $CO_2$ plasma induced less damage than $O_2$ plasma, as evidenced by a smaller thickness reduction, a lower methyl depletion and less moisture uptake [41, 62, 105]. The results shown in Figure 3.22 indicate that increasing the CO content in a CO/$O_2$ plasma results in a decrease of the radical induced damage, such as thickness reduction and methyl depletion.

*Optimizing Process Integration: Porogen Removal and He Pre-treatment*

Optimization of process integration is a way to reduce the plasma damage. Calvert and Gallagher first proposed a 'Post Integration Porogen Removal' approach [106]. In this method, the low-*k* dielectric was synthesized by deposition of an MSQ-based matrix polymer

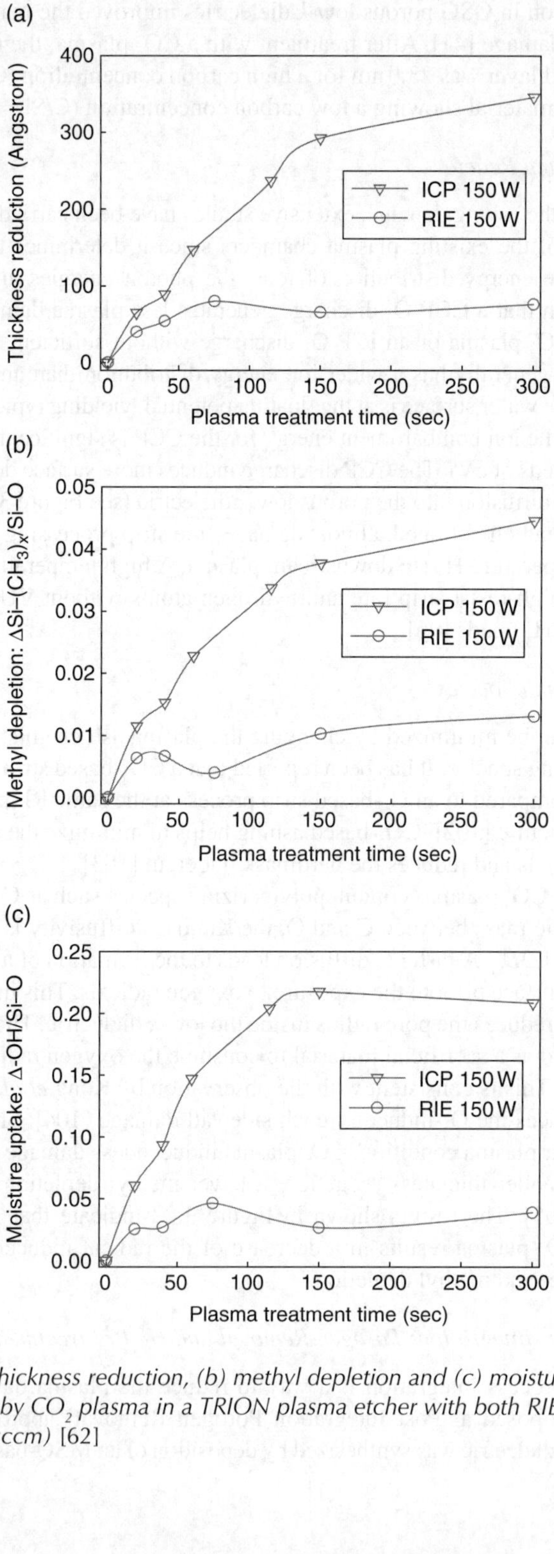


***Figure 3.21*** *(a) Thickness reduction, (b) methyl depletion and (c) moisture uptake for OSG low-k films treated by $CO_2$ plasma in a TRION plasma etcher with both RIE (CCP) and ICP mode (5 mtorr, 20 sccm)* [62]

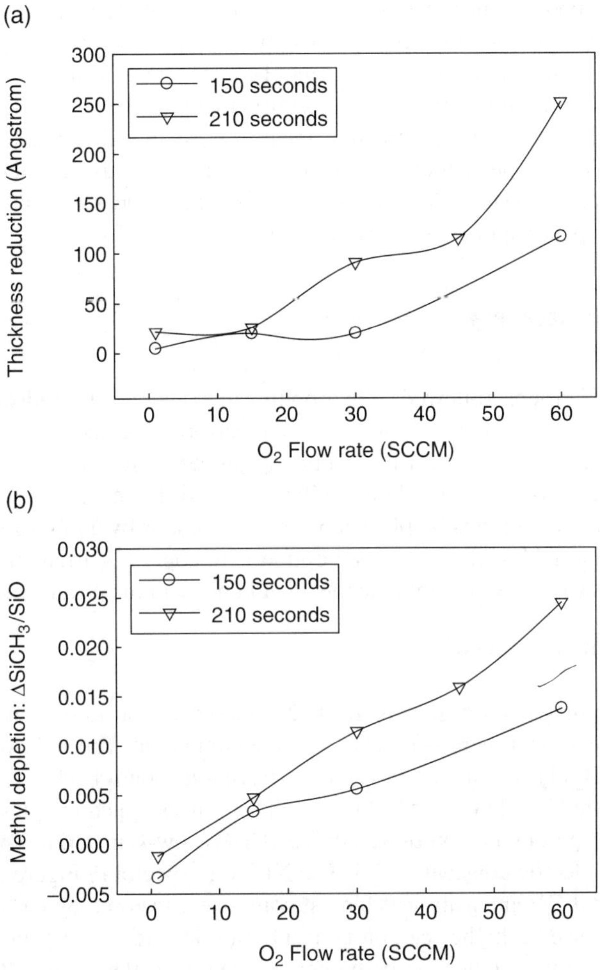


***Figure 3.22*** *(a) Thickness reduction and (b) methyl depletion for a porous OSG low-*k *film induced by a $CO/O_2$ plasma in an Oxford plasma etcher. Data were measured at the same position underneath the Si mask: ICP 200 W, bias 100 W, 20 mtorr, total $[O_2 + CO]$ flow 60 sccm* [62]

together with acrylic polymer-based nanoparticles as the porogen material. The dielectric was covered with a porous mask permeable to the desorption of reaction products. Instead of removing the porogen before the plasma etching and ashing processes, the porosity was created after patterning (or copper filling) by thermal curing. Due to the presence of polymeric nanoparticles (porogen) during the plasma ash process, the pore size and porosity were both reduced and, as a consequence, the plasma damage was reduced. Several other authors used the same approach to improve the integration yield [107, 108]. However, a small shrinkage of low-*k* films was observed during the post-processing porogen removal, leading to the delamination of the barrier layers and a drastic decrease of the reliability of the integrated structures.

An He plasma pre-treatment can be used to reduce the plasma damage through surface densification and reduction of the pore radius [109–111]. However, the application of this approach is also limited. For instance, it cannot be used after micropatterning, because the feature sidewalls are covered by post-etch residues that must be removed before densification. A good application of this treatment is the post-CMP cleaning of Cu/low-*k* structures, before dielectric barrier deposition. Consecutive or simultaneous treatment of this surface by He and reductive plasma allows removal of Cu oxide and seals the low-*k* dielectric, facilitating the deposition of a conformal dielectric barrier [112].

## 3.6 Dielectric Recovery

The loss of methyl groups, followed by hydrophilization and moisture adsorption, increases the dielectric constant and degrades the electrical performance and reliability of the low-*k* materials. There is a large interest in developing processes for dielectric recovery of the porous low-*k* materials, which until now followed two different approaches. The first one is to restore the methyl groups by plasma beam treatment or by applying silylation agents such as TMCS and DMDCS. The second option is to apply UV irradiation together with thermal treatments in order to eliminate the –OH bonds and the physisorbed water.

### 3.6.1 $CH_4$ Beam Treatment

This technique is simple to integrate in the dual damascene process since it does not require any special chamber and there is no chemical contamination. The $CH_x$ ions and radicals, generated in a $CH_4$ plasma, are used to convert the Si–OH bonds back to Si–$CH_3$ bonds for restoring the methyl loss [113–115]. Bao *et al.* [114] investigated the plasma damage and $CH_4$ recovery of porous carbon-doped oxide (CDO) low-*k* materials, with about 25% porosity and a dielectric constant of 2.5. The XPS C1s spectra in Figure 3.23 (a) demonstrate that the total C1s intensity was almost completely restored by $CH_4$ plasma. Angle-resolved XPS showed a higher carbon concentration at a low detection angle of 30° as compared to that at 90°, indicating carbon enrichment near the surface. The thickness of this layer was estimated to be about 7 nm, with its density gradually decreasing from 1.9 to 1.7 g/cm$^3$ at the low-*k* surface. The surface hydrophobicity was also restored, as indicated by the recovery of the water contact angle to a value close to that of the pristine film.

After damage and subsequent $CH_4$ plasma treatment, the low-*k* samples were electrically tested and both C-V and I-V characteristics were improved. Figure 3.24 (a) shows the C-V hysteresis of an $O_2$-ashed low-*k* film before and after the $CH_4$ treatment. The almost complete supression of hysteresis indicates a significant removal of Si dangling bonds and polar silanol groups. The leakage current was also reduced, as shown in Figure 3.24 (b). However, the leakage current reduction in the high electric field range was rather limited, indicating an ineffective recovery of the charge build-up induced by defects in the bulk of the dielectric generated by the $O_2$ ashing process. Aimadeddine *et al.* [113] compared the plasma effects of He/$H_2$, $NH_3$ and $CH_4$, and found that the $CH_4$ plasma treatment is the most effective in improving the dielectric reliability by surface passivation of porous low-*k* sidewalls.

Despite the improved electric performance and surface hydrophobicity, the recovery effectiveness of the $CH_4$ plasma treatment was rather limited. First, the dielectric recovery

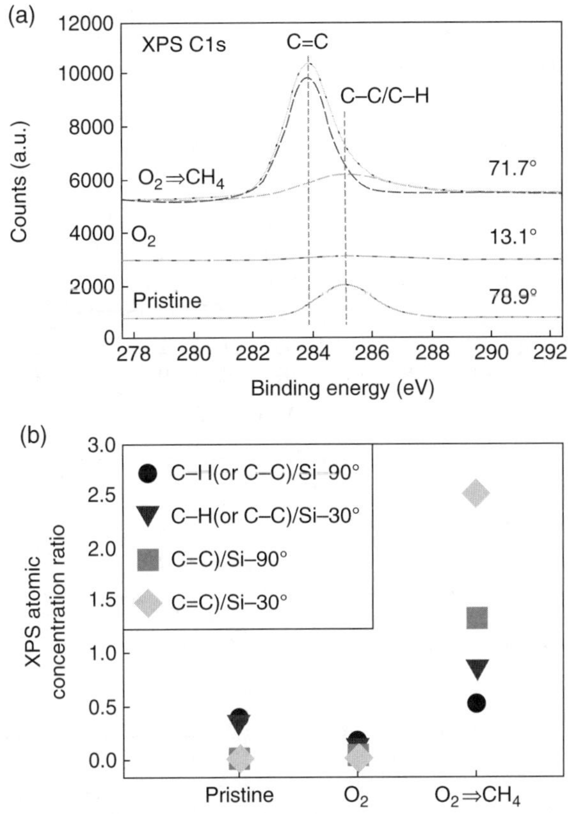


***Figure 3.23*** *(a) XPS C1s spectra and (b) AR–XPS peak area ratio for pristine CDO low-*k*, $O_2$ plasma damaged low-*k* and post-ash $CH_4$ plasma-treated low-*k* materials* [115] *(see color Plate 2)*

by $CH_4$ plasma treatment cannot reach the bulk of the low-*k* film, as shown by the XPS depth profile in Figure 3.25. Also, the top layer after treatment still contains some high dielectric constant components. The deconvolution of the XPS C1s peak in Figure 3.23 (a) revealed the presence of a C=C contribution, i.e. the existence of $sp^2$ C=C bonds in the top polymer layer. This was further confirmed by Raman spectroscopy, as shown in Figure 3.26. The high polarizability of the $sp^2$ C=C bond (1.643 Å$^{-3}$) as compared with that of the C–C bond (0.531 Å$^{-3}$) or C–H bond (0.652 Å$^{-3}$) causes the dielectric constant of the top layer to exceed that of the bulk low-*k* material. This is verified by the high average density (1.85 g/cm$^3$) of the polymer layer as compared with the density of 1.6 g/cm$^3$ for the bulk.

### 3.6.2 Dielectric Recovery by Silylation

The silylation approach has generated a lot of interests [116–125]. In a silylation process, the chemicals are usually chosen to reduce or eliminate the silanol bonds and replace them with the hydrophobic silyl groups. In certain cases, the mechanical properties of the porous

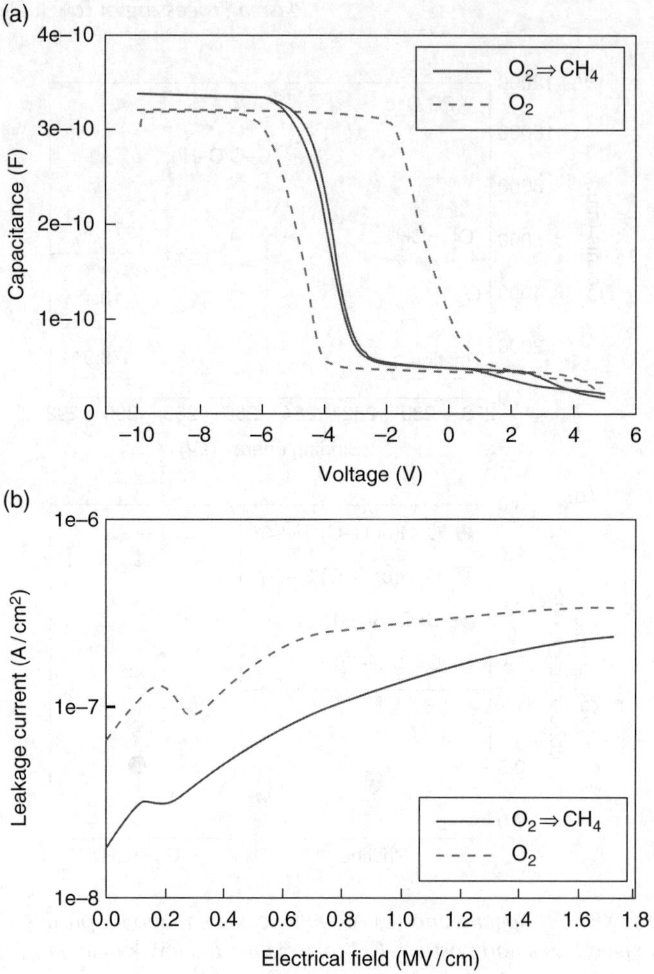


***Figure 3.24*** *(a) C-V and (b) I-V characteristics of the oxygen ashed CDO low-k material before and after plasma $CH_4$ repair* [115]

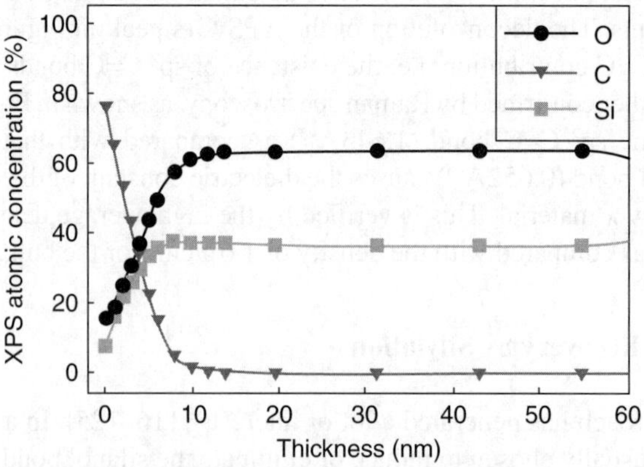


***Figure 3.25*** *XPS depth profiling result for a low-k film that was first damaged by an $O_2$ plasma and then modified by a post-ash $CH_4$ plasma* [115]

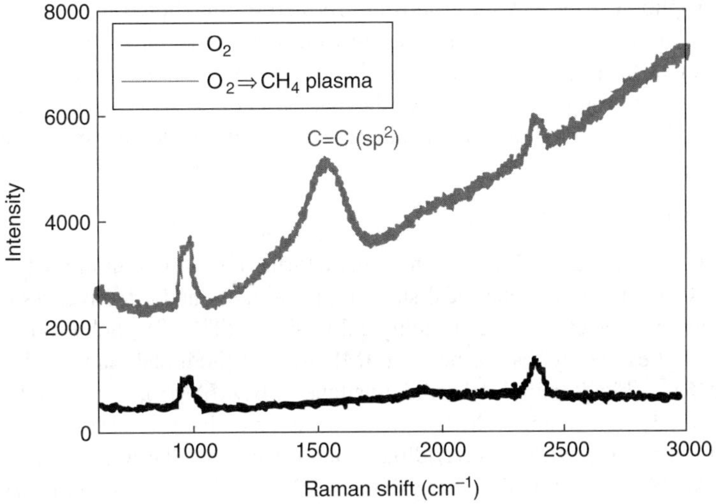


***Figure 3.26*** *Raman spectra for an $O_2$ plasma-damaged CDO low-*k *and post-ash $CH_4$ plasma-treated CDO low-k materials* [115]

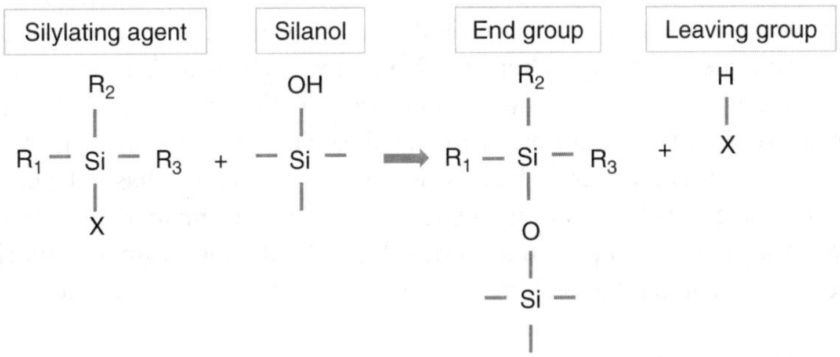


***Figure 3.27*** *Schematic diagram of the silylation process applied to silanol bond of a damaged low-*k *surface*

low-*k* films can also be improved by altering the cage to network ratio of the porous organosilicate structure. In a silylation process, a substituted silyl group ($R_3$Si–) is introduced to interact with the silanol molecule to restore the dielectric loss. The mechanism is described in Figure 3.27.

The silylating agent is formed by connecting a silicon atom to R and X moieties, where R is any hydrogen, alkyl, aryl, allyl, phenyl or vinyl moiety. X is the functional group and could be a silazane, chloro, amino or alkoxy moiety. There can be one, two or three functional groups in one silylation agent molecule, and the agents are called monofunctional, bifunctional and trifunctional accordingly. During the recovery process, silylating agents react with silanol groups on the damaged low-*k* surface, forming end groups and leaving groups. The end groups are bonded to the low-*k* surface and contain R moieties, turning the surface hydrophobic. The leaving groups are formed by functional moieties bonded with the hydrogen atoms, which are carried away from the surface at the end of the process. The efficiency of the silylation depends on the number of functional moiety, the effectiveness of the leaving group, the pore size of the dielectric, the size of the silylating molecule, the medium in which silylation occurs and process conditions (temperature, pressure).

*Choice of Silylation Agents*

A large pool of candidate silylation agents can be formed by various combinations of R and X moieties. Considering the chemical stability, hazardness and effectiveness of the compounds, popular choices include trimethylchlorosilane (TMCS), dichlorodimethylsilane (DMDCS), hexamethyldisilazane (HMDS), (bis)dimethylaminodimethylsilane (BDMADMS), (bis)dimethylaminomethylsilane (BDMAMS), trimethylbromosilane (TMBS), trimethyliodosilane (TMIS), phenyltrimethoxysilane (PTMOS), diphenyltetramethyldisilasane (DPTMDS), phenyldimethylchlorosilane (PDMCS), trimethylmethoxysilane (TMMS), dimethyldimethoxysilane (DMDMS) and methyltrimethoxysilane (MTMS).

The size, the polarizability of the end group and the reactivity of the functional group are the major factors controlling the effectiveness of the silylating agent. In general, a small molecular size induces faster diffusion and less steric hindrance, thus yielding a better recovery. A lower polarizability R moiety corresponds to a lower polarizability of the resulting end group, which directly contributes to lowering the dielectric constant of the recovered low-*k* material. A highly reactive X moiety silylates the surface and pore walls of the low-*k* material with a high reactivity. For example, the chloro-based silylation agents are more reactive than the amino-based or the silazane-based silylation agents, and thus they can provide a better recovery. The choice of the X moiety also has to be considered in order to reduce the corrosiveness of the leaving group. For example, the by-product of the chlorine-based silylation agent, such as DMDCS and TMCS, is hydrogen chloride, which is corrosive and not compatible with the damascene process used to fabricate Cu interconnect structures.

Another important factor to be considered is the number of X moieties per silylating molecule. Difunctional silylation agents are more effective than the monofunctional silylation agents since the latter can react only with a single isolated silanol group on the surface or the pore wall of the low-*k* material. Once an end group is formed, the subsequent reaction of the adjacent silanols with another monofunctional silylation agent will be blocked due to the steric hindrance effect, as illustrated in Figure 3.28 (top). In contrast, a difunctional silylation agent can react with two neighboring silanols, as shown in Figure 3.28 (bottom). Trifunctional silylation agents are unlikely to react simultaneously with three silanols, as this will leave behind an unreacted end group on the low-*k* surface that will induce moisture from the air, leading to an additional silanol formation. Also trifunctional silylation agents are more likely to initiate oligomerization and form a monolayer, blocking the silylation reaction; they can also cross-link and form contaminants in the reaction chamber.

Silylation experiments have been performed by Nitta *et al.* to evaluate the effect of mono-, di- and trifunctional silylation agents with the same leaving group using a liquid medium [126]. The results are shown in Figure 3.29, indicating that the trifunctional silylation agent recovers the most methyl bonds. The difunctional silylation agent is more

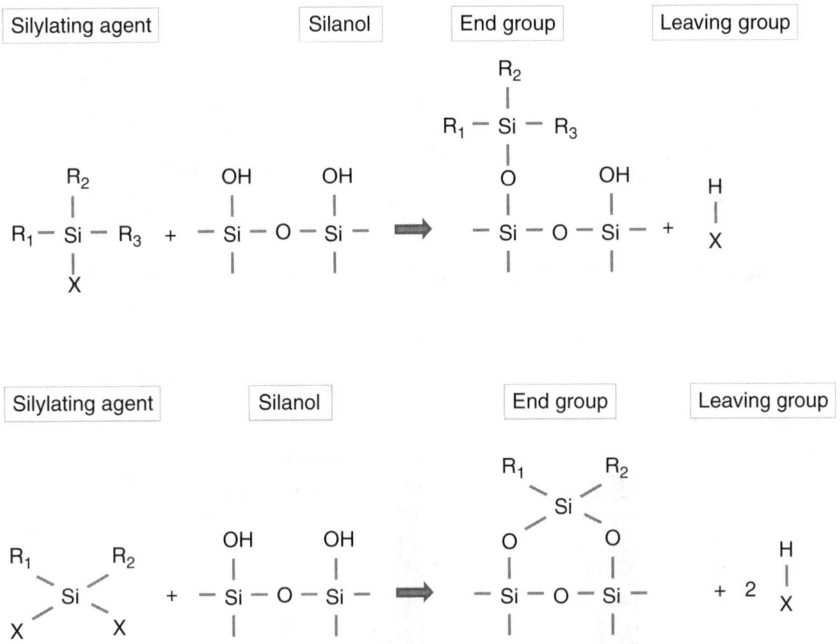


*Figure 3.28* The schematic diagram of the silylation process on a silanol bond of a damaged low-k surface: (top) the steric hindrance effect prevents the simultaneous capture of more than one silanol with a monofunctional silylation agent; (bottom) difunctional silylation agents react with two neighboring silanols simultaneously

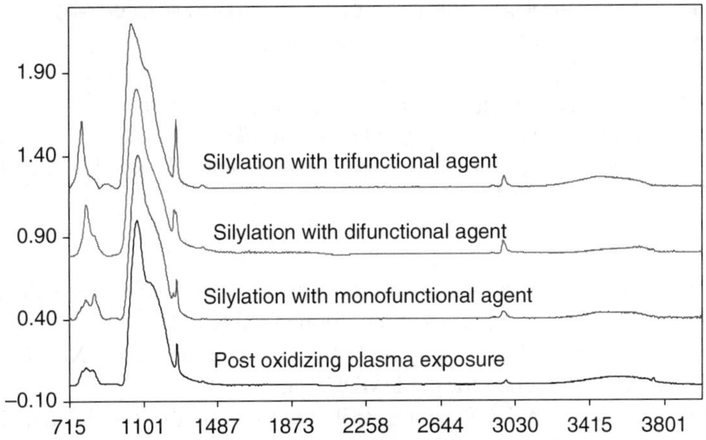


*Figure 3.29* FTIR spectra of post-plasma damage and silylated porous MSSG films with mono-, di- and trifunctional agents [126] (see color Plate 3)

effective in eliminating the silanol bonds and recovering a relatively large number of methyl bonds, thus providing the best recovery effect.

In a separate study, Shi *et al.* chose the silylation agents of TMCS, DMDCS and PTMOS deposited by CVD to evaluate the effects of different functional and leaving groups and the

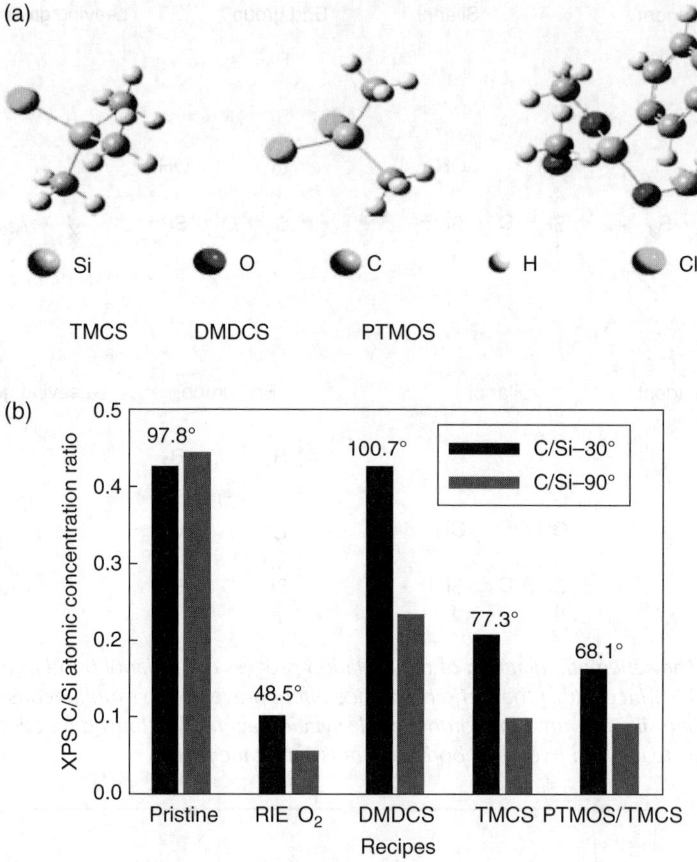


***Figure 3.30*** *(a) Molecular structures of trimethylchlorosilane (TMCS), dimethyldichlorosilane (DMDCS) and phenyltrimethoxysilane (PTMOS). (b) Comparison of C/Si and water contact angle for vapor TMCS, DMDCS and PTMOS/TMCS silylation processes* [62] *(see color Plate 4)*

molecule size. They found that under similar conditions, the DMDCS was the most effective for dielectric recovery of a porous carbon-doped oxide (CDO) low-*k* surface damaged by $O_2$ plasma [62]. The results (see Figure 3.30) show that the largest XPS C/Si ratio and water contact angle are obtained for the DMDCS agent.

The high efficiency of DMDCS can be attributed to its smallest molecular size as compared with TMCS and PTMOS, and also provides the most rapid diffusion and the least steric hindrance effect. On a CDO film, isolated silanol and disilanol (also named germinal silanol) are formed after $O_2$ plasma damage. The TMCS with a single Cl moiety can react with the isolated silanol easily. However, once reacted, the three $CH_3$ groups can cause steric hindrance, blocking any further reaction between other TMCS molecules and a second hydroxyl. This steric hindrance effect has been reported by other groups, including the study by Gun'ko *et al.* [127], who combined quantum chemistry calculations and analytical techniques to investigate the kinetics of the interaction between HMDS and fumed silica. They found that once one of the hydroxyl group in a germinal silanol is silylated by HMDS, the subsequent absorption of another hydroxyl group by HMDS would be hindered, leading to a slowdown of the silylation process. In addition, the DMDCS agent has two chlorine leaving groups while TMCS has only one chlorine departing group, so DMDCS has a better chance of reacting with silanol groups. The PTMOS agent has the least effective leaving group $OCH_3$, slowing down the silylation process compared with DMDCS.

*Silylation Process Condition and Medium*

In order to develop an effective silylation process for repairing the bulk plasma damage, it is essential to use the appropriate silylating medium and an optimum condition for the silylation process. First, for the silylation reaction to work, it is necessary to treat the organosilicate film to have an abundance of silanols, which will ensure that the silylation reaction will occur to a full extent. In most cases, the plasma-damaged low-*k* surface contains sufficient silanols, but in some rare cases a plasma- or UV-assisted oxidative treatment has to be used before silylation. Second, moisture should be removed before the silylation process since the silylation agents are susceptible to water molecules. In the presence of $H_2O$, the di- and trifunctional agents tend to oligomerize to form a monolayer on the surface, which will reduce the reactivity, block the pore, and thus prohibit the dielectric recovery and complicate the reaction mechanism. Third, the silylation process is normally more efficient at an elevated temperature, due to thermal activation of the silylation agent. A high temperature promotes the diffusion of silylation agents into the low-*k* films and the chemical reaction with the silanol group. Therefore, the silylation process is normally conducted at an elevated temperature, high enough to enhance the diffusion and reaction without breaking down the chemical. After silylation, annealing is usually performed as a last step to condense unsilylated silanols in the film to form additional siloxane bonds. The annealing occurs in the range of 350 to 450 °C, high enough to initiate the silanol condensation without thermal breakdown of the low-*k* material.

The medium used in the silylation process is also an important factor, which can be categorized as liquid, supercritical $CO_2$ and vapor-phase silylation processes. There are significant differences in the process setup and the extent of the dielectric recovery. The discussion of the choices of the silylation medium is beyond the scope of this review and will not be presented here.

Silylation is still not widely used in process integration. The main issue is the limited penetration depth of the silylation agents into the porous low-*k* material. Damage recovery is efficient only if the depth of damage is limited to a few nanometers.

### 3.6.3 UV Radiation

The results from the post-ash $CH_4$ plasma treatment and silylation are mostly confined to the dielectric surface, limiting the recovery of the plasma damage, which is normally throughout the bulk of the film. This has led to the development of the UV process for low-*k* dielectric recovery. Post-deposition UV curing is commonly used to improve the

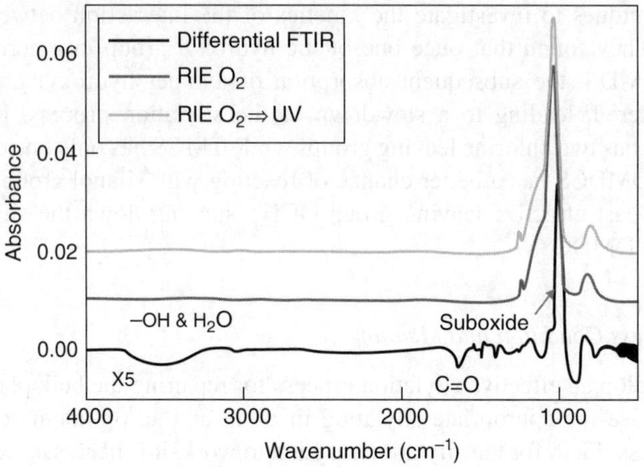


*Figure 3.31* FTIR spectra for $O_2$ plasma-damaged and UV-repaired OSG [105] (see color Plate 5)

mechanical properties of as-deposited low-*k* dielectrics and to remove porogens. After plasma damage, UV radiation is expected to eliminate the silanol and physisorbed water from the bulk of the low-*k* dielectric, reducing the dielectric constant and leakage current.

In a recent study [105], *k* ~2.5 CDO films were first damaged by an oxidative RIE plasma and then UV treated at 350 °C for about 300 seconds. After UV treatment, the film thickness was reduced from 83.1 to 78.5 nm, while the refractive index was reduced from 1.333 to 1.323 and the dielectric constant reduced from 3.26 to 2.55. In Figure 3.31, differential FTIR was used to understand the mechanism behind the dielectric recovery. After UV treatment, –OH, physisorbed $H_2O$ molecules and C=O bonds were almost completely removed; at the same time, there was an increase of the Si–O–Si network and suboxide bonds. The following reaction can be considered:

$$\equiv Si{-}OH + HO{-}Si\equiv \;\rightarrow\; \equiv Si{-}O{-}Si\equiv + H_2O$$
$$\equiv Si{-}OH + H_3C{-}Si\equiv \;\rightarrow \equiv Si{-}O{-}Si\equiv + CH_4$$
$$\equiv Si{-}OH + H{-}Si\equiv \;\rightarrow\; \equiv Si{-}O{-}Si\equiv + H_2$$

As shown in Figure 3.31, there was no methyl depletion or reduction of the Si–H bond after UV curing. Hence, the first equation dominates the UV curing process on $O_2$ plasma-damaged low-*k* dielectrics. In addition, based on the reduction of C=O and –OH bonds, the following reaction is proposed:

$$\equiv Si{-}OH + HOC{-}Si\equiv \;\rightarrow\; \equiv Si{-}Si\equiv + H_2 + CO_2$$

The UV curing effect of low-*k* dielectrics recovery can be optimized with the process parameters, including UV wavelength and intensity, substrate temperature, UV curing time, chamber pressure and ambient gas. Although the UV treatment on plasma-damaged OSG low-*k* dielectrics can reduce the bulk –OH bond and moisture, the surface hydrophobicity was not recovered. On the other hand, the silylation process excels in surface recovery. A combination of UV treatment and silylation can lead to an even better recovery, as was demonstrated by Huang *et al.* [128].

## 3.7 Conclusions

In the last ten years, driven by the ITRS scaling roadmap, low dielectric constant materials have been introduced. Up till now, the most successful materials are porous organosilica glasses (OSG), where methyl bonds are created, together with porosity. Due to their porous and interconnected microstructure, these materials exhibit tremendous sensitivity when exposed to plasma. While previously limited to 1–2 nm for dense $SiO_2$, plasma damage for low-*k* materials can extend up to tens of nanometers. Plasma damage can be separated into two components: vertical damage, i.e. parallel to the ion flux (through horizontal surfaces), and transversal damage, i.e. normal to the ion flux (through the sidewalls of the etched structures). Plasma damage has been shown to be caused mainly by oxygen and fluorine radicals, diffusing into the pores and reacting with the hydrophobic Si–$CH_3$ bonds to form hydrophilic Si–OH bonds. Plasma damage is enhanced by VUV irradiation $< 190$ nm, through photodissociation of the Si–$CH_3$ bonds. Damage can be minimized by ion bombardment, through the formation of a densified surface layer, but this solution applies only to horizontal surfaces. On vertical surfaces, plasma etch damage can be impeded by carefully balancing the C/F ratio of the etch discharge, in order to seal the surface with C-rich fluoropolymers. Another way to decrease plasma damage is to introduce new integration schemes such as the use of hard-mask materials, protecting the underlying low-*k* from radicals, ions and part of the radiation. Bulk chemical modifications (carbon enrichment) and optimized synthesis (porogen residue removal) can also help to reduce the pristine *k* value while keeping acceptable mechanical properties. Optimized strip plasma can also prevent damage by O radicals, for instance by reducing their plasma density and depositing a C-rich passivation layer on the pore surfaces or using $H_2$-based plasmas at high temperatures. Finally, dielectric recovery has been attempted by $CH_4$ plasma, sylilation and UV treatments. $CH_4$ plasma and sylilation suffer from being limited to a depth of a few nanometers. For severely damaged low-*k* materials, repair may be possible only by a combination of $CH_4$ plasma or sylilation with UV cure.

In the coming years, tremendous efforts will still be necessary to understand and optimize low-*k* plasma processing and integration, due to the need to decrease further the integrated *k* value, down to 2.0 and beyond. With interconnect wiring dimensions of about 10–20 nm, plasma damage can be tolerated only over 1–2 nm on each side of the dielectric, i.e. of the order of a single pore dimension. This put a lot of emphasis not only on cleaning plasmas but also on etch plasmas. This will also push the SiOCH-based low-*k* values to their limits, leaving the door open for new materials, for instance fully organic polymer materials.

## References


[1] Maex, K., Baklanov, M.R., Shamiryan, D., Iacopi, F., Brongersma, H. and Yanovitskaya, Z.S. (2003) Low dielectric constant materials for microelectronics. *J. Appl. Phys.*, **93**, 8793.
[2] Tatsumi, T. (2007) Control of reactive plasmas for low-k/Cu integration. *Appl. Surf. Sci.*, **253**, 6716.
[3] Hoofman, R.J.O.M., Verheijden, G.J.A.M., Michelon, J., Iacopi, F., Travaly, Y., Baklanov, M.R., Tőkei, Zs. and Beyer, G.P. (2005) Challenges in the implementation of low-k dielectrics in the back-end of line. *Microelectron. Engng*, **80**, 337–344.
[4] McGahay, V. (2010) Porous dielectrics in microelectronics: wiring applications. *Materials*, **3**, 536.
[5] Tsai, H.-C., Chang, Y.-S. and Chang, S.-Y. (2008) Effect of plasma treatments on interface adhesion between SiOCH ultra-low-*k* film and SiCN etch stop layer, *Microelectron. Engng*, **85**, 1658.
[6] Volksen, W., Miller, R.D. and Dubois, G. (2010) Low dielectric constant materials. *Chem. Rev.*, **110**, 56–110.
[7] Shul, R.J. and Pearton, S.J. (2000) *Handbook of Advanced Plasma Processing Techniques*, Springer, p. 10.
[8] Amanatides, E. and Mataras, D. (2001) Frequency variation under constant power conditions in hydrogen radio frequency discharges. *J. Appl. Phys.*, **89**, 1556; Perret, A., Chabert, P., Jolly, J. and Booth, J.-P. (2005) Ion energy uniformity in high-frequency capacitive discharges. *Appl. Phys. Lett.*, **86**, 021501.
[9] Petit-Etienne, C., Darnon, M., Vallier, L., Pargon, E., Cunge, G., Boulard, F. and Joubert, O. (2010) Reducing damage to Si substrates during gate etching processes by synchronous plasma pulsing. *J. Vac. Sci. Technol. B*, **28** (5), 926.
[10] Ukraintsev, V.A., Kulkarni, M., Baum, C.C., Kirmse, K., Guevremont, M., Lakkapradaga, S., Bhatia, K.N., Herrera, P.P. and Whitney, U.K. (2002) Spectroscopic ellipsometry and reflectometry from gratings (scatterometry) for critical dimension measurements and in-situ, real-time process monitoring. *Proc. SPIE*, **4689**, 189.
[11] Bao, J., Shi, H., Liu, J., Huang, H., Ho, P.S., Goodner, M.D., Moinpour, M. and Kloster, G.M. (2008) Mechanistic study of plasma damage of low *k* dielectric surfaces. *J. Vac. Sci. Technol. B*, **26**, 219.
[12] Baklanov, M.R., Mogilnikov, K.P., Polovinkin, V.G. and Dultsev, F.N. (2000) Determination of pore size distribution in thin films by ellipsometric porosimetry. *J. Vac. Sci. Technol. B*, **18**, 1385.
[13] Baklanov, M.R., Mogilnikov, K.P. and Le, Q.T. (2006) Quantification of processing damage in porous low dielectric constant films. *Microelectron. Engng*, **83**, 2287.
[14] Shamiryan, D., Baklanov, M.R. and Maex, K. (2003) Diffusion barrier integrity evaluation by ellipsometric porosimetry. *J. Vac. Sci. Technol. B*, **21**, 220.
[15] Yonekura, K., Goto, K., Matsuura, M., Fujiwara, N. and Tsujimoto, K. (2005) Low-damage damascene patterning using porous inorganic low-dielectric-constant materials. *Jpn. J. Appl. Phys.*, **44**, 2976.
[16] Marsik, P., Urbanowicz, A.M., Vinokur, K., Cohen, Y. and Baklanov, M.R. (2008) Changes of UV optical properties of plasma damaged low-k dielectrics for sidewall damage scatterometry. Materials and processes for advanced interconnects for microelectronics. *Mater. Res. Soc. Symp. Proc.*, **1079E**, N07-04.
[17] Bouyssou, R., Kodaji, M.El., Licitra, C., Chevolleau, T., Besacier, M., Posseme, N., Joubert, O. and Schiavone, P. (2010) Scatterometric porosimetry: a new characterization technique for porous material patterned structures. *J. Vac. Sci. Technol. B*, **28** (4), L31.
[18] Talanov, V.V., Scherz, A., Moreland, R.L. and Schwartz, A.R. (2006) Noncontact dielectric constant metrology of low-*k* interconnect films using a near-field scanned microwave probe. *Appl. Phys. Lett.*, **88**, 192906.
[19] Seki, H., Inoue, K., Nagai, N., Shimada, M., Inukai, K., Hashimoto, H. and Ogawa, S. (2005) Proceedings of the Advanced Metallization Conference 2004, p. 375.

[20] Baklanov, M.R., Zhao, L., Van Besien, E. and Pantouvaki, M. (2011) Effect of porogen residue on electrical characteristics of ultra low-*k* materials. *Microelectron. Engng*, **88** (5), 990.
[21] Nagai, H., Takashima, S., Hiramatsu, M., Hori, M. and Goto, T. (2002) Behavior of atomic radicals and their effects on organic low dielectric constant film etching in high density $N_2/H_2$ and $N_2/NH_3$ plasmas. *J. Appl. Phys.*, **91**, 2615.
[22] Fuard, D., Joubert, O., Vallier, L. and Bonvalot, M. (2001) High density plasma etching of low *k* dielectric polymers in oxygen-based chemistries. *J. Vac. Sci. Technol. B*, **19**, 447.
[23] Baklanov, M.R., Le, Q.T., Kesters, E., Iacopi, F., Van Aelst, J., Struyf, H., Boullart, W., Vanhaelemeersch, S. and Maex, K. (2004) Proceedings of the 7th International Interconnect Technology Conference, San Francisco, California.
[24] Bailly, F., David, T., Chevolleau, T., Darnon, M., Posseme, N., Bouyssou, R., Ducote, J., Joubert, O. and Cardinaud, C. (2010) Roughening of porous SiCOH materials in fluorocarbon plasmas. *J. Appl. Phys.*, **108**, 014906.
[25] Darnon, M., Chevolleau, T., Joubert, O., Maitrejean, S., Barbe, J.C. and Torres, J. (2007) Undulation of sub-100 nm porous dielectric structures: a mechanical analysis. *Appl. Phys. Lett.*, **91**, 194103.
[26] Xia, L.-Q., Cui, D., Balseanu, M., Nguyen, V., Zhou, K., Pender, J. and Naik, M. (2011) Novel hardmask for sub-20 nm copper/low-*k* backend dual damascene integration. *ECS Trans.*, **35** (4), 651.
[27] Kakuchi, M., Hikita, M. and Tamamura, T. (1986) Amorphous carbon films as resist masks with high reactive ion etching resistance for nanometer lithography. *Appl. Phys. Lett.*, **48**, 835.
[28] Kragler, K., Gunther, E., Leuschner, R., Falk, G., Hammerschmidt, A., von Seggern, H. and Saemann-Ischenko, G. (1995) Scanning tunneling microscopy based lithography employing amorphous hydrogenated carbon as a high resolution resist mask. *Appl. Phys. Lett.*, **67**, 1163.
[29] Cardinaud, C. and Turban, G. (1990) Mechanistic studies of the initial stages of etching of Si and $SiO_2$ in a $CHF_3$ plasma. *Appl. Surf. Sci.*, **45**, 109.
[30] Standaert, T.E.F.M., Matsuo, P.J., Allen, S.D., Orehlein, G.S. and Dalton, T.J. (1999) Patterning of fluorine-, hydrogen-, and carbon-containing $SiO_2$-like low dielectric constant materials in high-density fluorocarbon plasmas: Comparison with $SiO_2$. *J. Vac. Sci. Technol. A*, **17**, 741.
[31] Posseme, N., Chevolleau, T., Joubert, O., Vallier, L. and Mangiagalli, P. (2003) Etching mechanisms of low-*k* SiOCH and selectivity to SiCH and $SiO_2$ in fluorocarbon based plasmas. *J. Vac. Sci. Technol. B*, 21, 2432.
[32] Standaert, T.E.F.M., Joseph, E.A., Oehrlein, G.S., Jain, A., Gill, W.N., Wayner Jr, P.C. and J. L. Plawsky, J.L. (2000) Etching of xerogel in high-density fluorocarbon plasmas. *J. Vac. Sci. Technol. A*, **18**, 2742.
[33] Posseme, N., Chevolleau, T., Joubert, O., Vallier, V. and Rochat, N. (2004) Etching of porous SiOCH materials in fluorocarbon-based plasmas, *J. Vac. Sci. Technol. B*, **22**, 2772.
[34] Iba, Y., Kirimura, T., Sasaki, M., Kobayashi, Y., Nakata, Y. and Nakaishi, M. (2008) Effects of Etch Rate on Plasma-Induced Damage to Porous Low-k Films, *Jpn. J. Appl. Phys*. **47**, 6923.
[35] Bao, J., Shi, H., Huang, H., Ho, P.S., McSwiney, M.L., Goodner, M.D., Moinpour, M. and Kloster, G.M. (2010) Oxygen plasma damage to blanket and patterned ultralow-kappa surfaces. *J. Vac. Sci. Technol. A*, **28** (2), 207.
[36] Goldman, M.A., Graves, D.B., Antonelli, G.A., Behera, S.P. and Kelber, J.A. (2009) Oxygen radical and plasma damage of low-*k* organosilicate glass materials: diffusion-controlled mechanism for carbon depletion, *J. Appl. Phys.*, **106** (1), 01311.
[37] Braginsky, O.V., Kovalev, A.S., Lopaev, D.V., Malykhin, E.M., Mankelevich, Yu.A., Proshina, O.V., Rakhimova, T.V., Rakhimov, A.T., Voloshin, D.G., Vasilieva, A.N., Zyryanov, S.M., Smirnov, E.A. and Baklanov, M.R. (2011) The effect of He plasma treatment on properties of organosilicate glass low-*k* films. *J. App. Phys.*, **109**, 043303.
[38] Zheng, L., Ling, L., Hua, X., Oerhlein, G.S. and Hudson, E.A. (2005) Studies of film deposition in fluorocarbon plasmas employing a small gap structure. *J. Vac. Sci. Technol. A*, **23** (4), 634.
[39] Uchida, S., Takashima, S., Hori, M., Fukasawa, M., Ohshima, K., Nagahata, K. and Tatsumi, T. (2008) Plasma damage mechanisms for low-*k* porous SiOCH films due to radiation, radicals, and ions in the plasma etching process. *J. Appl. Phys.*, **103**, 073303.

[40] Hua, X., Kuo, M., Oehrlein, G.S., Lazzeri, P., Iacob, E., Anderle, M., Inoki, C.K., Kuan, T.S., Jiang, P. and Wu, W. (2006) Damage of ultralow *k* materials during photoresist mask stripping process. *J. Vac. Sci. Technol. B*, **24** (3), 1238.

[41] Shi, H., Huang, H., Bao, J., Im, J., Ho, P.S., Zhou, Y., Pender, J.T., Armacost, M. and Kyser D. (2009) Plasma altered layer model for plasma damage characterization of porous OSG films, Proceedings of the IEEE International Interconnect Technology Conference, p. 78.

[42] Takeda, K., Miyawaki, Y., Takashima, S., Fukasawa, M., Oshima, K., Nagahata, K., Tatsumi, T. and Hori, M. (2011) Mechanism of plasma-induced damage to low-*k* SiOCH films during plasma ashing of organic resists. *J. Appl. Phys.*, **109**, 033303.

[43] Nagai, H., Hiramatsu, M., Hori, M. and Goto, T. (2003) Measurement of oxygen atom density employing vacuum ultraviolet absorption spectroscopy with microdischarge hollow cathode lamp. *Rev. Sci. Instrum.*, **74** (7), 3453.

[44] Worsley, M.A., Bent, S.F., Gates, S.M., Fuller, N.C.M., Volksen, W., Steen, M. and Dalton, T. (2005) Effect of plasma interactions with low-*k* films as a function of porosity, plasma, chemistry, and temperature, *J. Vac. Sci. Technol. B*, **23** (2), 395.

[45] Kunnen, E., Baklanov, M.R., Franquet, A., Shamiryan, D., Rakhimova, T.V., Urbanowicz, A.M., Struyf, H. and Boullart, W. (2010) Effect of energetic ions on plasma damage of porous SiCOH low-*k* materials. *J. Vac. Sci. Technol. B*, **28** (3), 450.

[46] Lee, H.J., Soles, C.L., Lin, E.K. and Wu, W. (2007) Nonuniform structural degradation in porous organosilicate films exposed to plasma, etching, and ashing as characterized by X-ray porosimetry. *Appl. Phys. Lett.*, **91** (17), 172908.

[47] Sun, J.N., Gidley, D.W., Hu, Y., Frieze, W.E. and Ryan, E.T. (2002) Depth-profiling plasma-induced densification of porous low-*k* thin films using positronium annihilation lifetime spectroscopy. *Appl. Phys. Lett.*, **81** (8), 1447.

[48] Lee, H.J., Lin, E.K., Wu, W.L., Fanconi, B.M., Lan, J.K., Cheng, Y.L., Liou, H.C., Wang, Y.L., Feng, M.S. and Chao, C.G. (2001) X-ray reflectivity and FTIR measurements of $N_2$ plasma effects on the density profile of hydrogen silsesquioxane thin films. *J. Electrochem. Soc.*, **148**, F195.

[49] Ryan, E.T., Martin, J., Junker, K., Lee, J.J., Guenther, T., Wetzel, J., Lin, S., Gidley, D.W. and Sun, J. (2002) Integration damage in organosilicate glass films. Proceedings of the 5th IEEE International Interconnect Technology Conference, p. 27.

[50] Urbanowicz, A.M., Baklanov, M.R., Heijlen, J., Travaly, Y. and Cockburn, A. (2007) Damage reduction and sealing of low-*k* films by combined He and $NH_3$ plasma treatment. *Electrochem. Solid-State Lett.*, **10**, G76.

[51] Urbanowicz, A.M., Shamiryan, D., Zaka, A., Verdonck, P., De Gendt, S. and Baklanov, M.R. (2010) Effects of He plasma pretreatment on low-*k* damage during Cu surface cleaning with $NH_3$ plasma. *J. Electrochem. Soc.*, **157**, H565.

[52] Peng, H.-G., Chi, D.-Z., Wang, W.-D., Li, J.-H., Zeng, K.-Y., Vallery, R.S., Frieze, W.E., Skalsey, M.A., Gidley, D.W. and Yee, A.F. (2007) Pore sealing by $NH_3$ plasma treatment of porous low dielectric constant films. *J. Electrochem. Soc.*, **154** (4) G85.

[53] Morikawa, Y., Yasunami, S., Chen, W., Hayashi, T. and Uchida, T. (2001) Low-*k* materials etching in magnetic neutral loop discharge plasma. *J. Vac. Sci. Technol. A*, **19**, 1747.

[54] Shibano, T., Fujiwara, N., Hirayama, M., Nagata, H. and Demizu, K. (1993) Etching yields of $SiO_2$ by low-energy $CF_x^+$ and $F^+$ ions. *Appl. Phys. Lett.*, **63** (17), 2336–2338.

[55] Todorov, S.S. and Fossum, E.R. (1988) Sputtering of silicon dioxide near threshold. *Appl. Phys. Lett.*, **52** (5), 365.

[56] Oehrlein, G.S., Zhang, Y., Vender, D. and Haverlag, M. (1994) Fluorocarbon high-density plasmas. I. Fluorocarbon film deposition and etching using $CF_4$ and $CHF_3$. *J. Vac. Sci. Technol. A*, **12**, 323.

[57] Oehrlein, G.S., Zhang, Y., Vender, D. and Joubert, O. (1994) Fluorocarbon high-density plasmas. II. Silicon dioxide and silicon etching using $CF_4$ and $CHF_3$. *J. Vac. Sci. Technol. A*, **12**, 333.

[58] Rueger, N., Beulens, J.J., Schaepkens, M., Doemling, M.F., Mirza, J.M., Standaert, T.E.F.M. and Oehrlein, G.S. (1997) Role of steady state fluorocarbon films in the etching of silicon dioxide using $CHF_3$ in an inductively coupled plasma reactor. *J. Vac. Sci. Technol. A*, **15** (4), 1881.

[59] Zalm, P.C. (1984) Some useful yield estimates for ion-beam sputtering and ion plating at low bombarding energies. *J. Vac. Sci. Technol. B*, **2** (2), 151.

[60] Steinbruchel, C. (1989) Universal energy-dependence of physical and ion-enhanced chemical etch yields at low ion energy. *Appl. Phys. Lett.*, **55** (19), 1960.

[61] Cui, H., Carter, R.J., Moore, D.L., Peng, H.-G., Gidley, D.W. and Burke, P.A. (2005) Impact of reductive $N_2/H_2$ plasma on porous low-dielectric constant SiCOH thin films. *J. Appl. Phys.*, **97**, 113302.

[62] Shi, H. (2010) Mechanistic study of plasma damage to porous low-*k*: process developement and dielectric recovery. PhD thesis, The University of Texas at Austin.

[63] Jinnai, B., Fukuda, S., Ohtake, H. and Samukawa, S. (2010) Prediction of UV spectra and UV-radiation damage in actual plasma etching processes using on-wafer monitoring technique. *J. Appl. Phys.*, **107**, 043302.

[64] Eslava, S., Eymery, G., Marsik, P., Iacopi, F., Kirschhock, C.E.A., Maex, K., Martens, J.A. and Baklanov, M.R. (2008) Optical property changes in low-*k* films upon ultraviolet-assisted curing. *J. Electrochem. Soc.*, **155**, G115.

[65] Lauer, J.L., Sinha, H., Nichols, M.T., Antonelli, G.A., Nishi, Y. and Shohet, J.L. (2010) Charge trapping within UV and vacuum UV irradiated low-*k* porous organosilicate dielectrics. *J. Electrochem. Soc.*, **157** (8), G177.

[66] Sinha, H., Antonelli, G.A., Jiang, G., Nishi, Y. and Shohet, J.L. (2011) Effect of vacuum ultraviolet radiation on deposited and ultraviolet-cured low-*k* porous organosilicate glass. *J. Vac. Sci. Technol. A*, **29** (3), 030602.

[67] Ren, H., Antonelli, G.A., Nishi, Y. and Shohet, J.L. (2010) Plasma damage effects on low-*k* porous organosilicate glass. *J. Appl. Phys.*, **108**, 094110.

[68] Lee, J. and Graves, D.B. (2010) Synergistic damage effects of vacuum ultraviolet photons and $O_2$ in SiOCH ultra-low *k* dielectric films. *J. Phys. D: Appl. Phys.*, **43**, 425201.

[69] Matsunaga, N., Okumura, H., Jinnai, B. and Samukawa, S. (2010) Hard-mask-through UV-light-induced damage to low-*k* film during plasma process for dual damascene. *Jpn. J. Appl. Phys.*, **49**, 04DB06.

[70] Okabe, H. (1978) *Photochemistry of Small Molecules*, John Wiley & Sons, Inc., New York.

[71] Lide, D.R. (1993–1994) *Handbook of Chemistry and Physics*, Chemical Rubber Publishing Company.

[72] Tatsumi, T., Fukuda, S. and Kadomura, S. (1994) Radiation damage of $SiO_2$ surface induced by vacuum ultraviolet photons of high-density plasma. *Jpn. J. Appl. Phys.*, **33**, 215.

[73] Iacopi, F., Travaly, Y., Eykens, B., Waldfried, C., Abell, T., Guyer, E.P., Gage, D.M., Dauskardt, R.H., Sajavaara, T., Houthoofd, K., Grobet, P., Jacobs, P. and Maex, K. (2006) Short-ranged structural rearrangement and enhancement of mechanical properties of organosilicate glasses induced by ultraviolet radiation. *J. Appl. Phys.*, **99**, 053511.

[74] Urbanowicz, A.M., Vanstreels, K., Verdonck, P., Shamiryan, D., De Gendt, S. and Baklanov, M.R. (2010) Improving mechanical robustness of ultralow-*k* SiOCH plasma enhanced vapor deposition glasses by controlled porogen decomposition prior to UV-hardening, *J. Appl. Phys.*, **107**, 104122.

[75] Urbanowicz, A.M., Vanstreels, K., Verdonck, P., Van Besien, E., Christos, T., Shamiryan, D., De Gendt, S. and Baklanov, M.R. (2011) Effect of UV wavelength on the hardening process of porogen-containing and porogen-free ultralow-*k* plasma-enhanced chemical vapor deposition dielectrics. *J. Vac. Sci. Technol. B*, **29** (3), 032201.

[76] Thiele, E.W. (1939) Relation between catalytic activity and size of particle. *Ind. Engng Chem.*, **31**, 916.

[77] Marschall, J. and Boulter, J.E. (2006) An analytic model for atom diffusion and heterogeneous recombination in a porous medium. *J. Phys. D: Appl. Phys.*, **39** (17), 3849.

[78] Cho, W., Saxena, R., Rodriguez, O., Ojha, M., Achanta, R., Plawsky, J.L. and Gill, W.N. (2005) Polymer penetration and pore sealing in nanoporous silica by $CHF_3$ plasma exposure. *J. Electrochem. Soc.*, **152** (6), F61.

[79] Baklanov, M.R., Urbanowicz, A., Mannaert, G. and Vanhaelemeersch, S. (2006) Low-*k* dielectric materials: challenges of plasma damage. Proceedings of the 8th International Conference on Solid-State Integrated Circuit Technology, Shanghai, China, p. 291.

[80] Chapman, B.N. (1980) *Glow Discharge Processes: Sputtering and Plasma Etching*, John Wiley & Sons, Inc., New York.

[81] Rakhimova, T.V., Braginsky, O.V., Kovalev, A.S., Lopaev, D.V., Mankelevich, Y.A., Malykhin, E.M., Rakhimov, A.T., Vasilieva, A.N., Zyryanov, S.M. and Baklanov, M.R. (2009) Recombination of O and H atoms on the surface of nanoporous dielectrics. *IEEE Trans. Plasma Sci.*, **37** (9), 1697.

[82] Braginsky, O.V., Kovalev, A.S., Lopaev, D.V., Malykhin, E.M., Mankelevich, Yu.A., Rakhimova, T.V., Rakhimov, A.T., Vasilieva, A.N., Zyryanov, S.M. and Baklanov, M.R. (2010) The mechanism of low-*k* SiOCH film modification by oxygen atoms. *J. Appl. Phys.*, **108**, 073303.

[83] Ryan, E.T., Martin, J., Junker, K., Wetzel, J., Gidley, D.W. and Sun, J. (2001) Effect of material properties on integration damage in organosilicate glass films. *J. Mater. Res.*, **16** (12), 3335.

[84] Shamiryan, D., Baklanov, M.R., Vanhaelemeersch, S. and Maex, K. (2002) Comparative study of SiOCH low-*k* films with varied porosity interacting with etching and cleaning plasma. *J. Vac. Sci. Technol. B*, **20** (5), 1923.

[85] Moore, D., Carter, R., Cui, H., Burke, P., McGrath, P., Gu, S.Q., Gidley, D. and Peng, H. (2005) Process integration compatibility of low-*k* and ultra-low-*k* dielectrics. *J. Vac. Sci. Technol. B*, **23** (1), 332.

[86] Moore, D.L., Carter, R.J., Cui, H., Burke, P., Gu, S.Q., Peng, H., Valley, R.S., Gidley, D.W., Waldfried, C. and Escorciac, O. (2005) Damage of low-*k* and ultralow-*k* dielectrics from reductive plasma discharges used for photoresist removal. *J. Electrochem. Soc.*, **152** (7), G528.

[87] Levitt, D.G. (1973) Dynamics of a single-file pore – non-Fickian behaviour. *Phys. Rev. A*, **8** (6), 3050.

[88] Weisz, P.B. (1973) *Chemtech.*, **3**, 498.

[89] Kollmann, M. (2003) Single-file diffusion of atomic and colloidal systems: asymptotic laws. *Phys. Rev. Lett.*, **90** (18), 180602.

[90] Arkhincheev, V.E., Kunnen, E. and Baklanov, M.R. (2011) Active species in porous media: random walk and capture in traps. *Microelectron. Engng*, **88** (5), 694.

[91] Safaverdi, S., Barkema, G., Kunnen, E., Urbanowicz, A. and Maes, C. (2011) Saturation of front propagation in a reaction–diffusion process. *Phys. Rev. B*, **83** (24), 245320 (2011)

[92] Worsley, M.A., Bent, S.F., Gates, S.M., Fuller, N.C.M., Volksen, W., Steen, M. and Dalton, T. (2005) Effect of plasma interactions with low-films as a function of porosity, plasma chemistry, and temperature, *J. Vac. Sci. Technol. B*, **23**, 395.

[93] Grill, A. and Patel, V. (2004) Interaction of hydrogen plasma with extreme low-*k* SiCOH dielectrics. *J. Electrochem. Soc.*, **151**, F133.

[94] Han, Q.Y., White, B., Berry, I.L., Waldfried, C. and Escorcia, O. (2005) *Solid State Phenomena*, **103–104**, 341.

[95] Lazzeri, P., Stueber, G.J., Oehrlein, G.S., McGowan, R., Busch, E., Pederzoli, S., Bersani, M. and Anderle, M. (2006) Time of flight secondary ion mass spectroscopy investigation of ultralow-*k* dielectric modifications in hydrogen and deuterium plasmas. *J. Vac. Sci. Technol. B*, **24**, 2695.

[96] Urbanowicz, A.M., Vanstreels, K., Shamiryan, D., De Gendt, S. and Baklanov, M.R. (2009) Effect of porogen residue on chemical, optical, and mechanical properties of CVD SiCOH low-*k* materials. *Electrochem. Solid-State Lett.*, **12**, H292.

[97] Marsik, P., Verdonck, P., De Roest, D. and Baklanov, M.R. (2010) Porogen residues detection in optical properties of low-*k* dielectrics cured by ultraviolet radiation. *Thin Solid Films*, **518**, 4266.

[98] Urbanowicz, A.M., Vanstreels, K., Verdonck, P., Shamiryan, D., De Gendt, S. and Baklanov, M.R. (2010) Improving mechanical robustness of ultralow-*k* SiOCH plasma enhanced chemical vapor deposition glasses by controlled porogen decomposition prior to UV-hardening. *J. Appl. Phys.*, **107**, 104122.

[99] Ryan, E.T., Gates, S.M., Grill, A., Molis, S., Flaitz, P., Arnold, J., Sankarapandian, M., Cohen, S.A., Ostrovski, Y. and Dimitrakopoulos, C. (2008) Property modifications of nanoporous pSiCOH dielectrics to enhance resistance to plasma-induced damage. *J. Appl. Phys.*, **104** (9), 094109.

[100] Posseme, N., Chevolleau, T., David, T., Darnon, M., Louveau, O. and Joubert, O. (2007) Mechanisms of porous dielectric film modification induced by reducing and oxidizing ash plasmas. *J. Vac. Sci. Technol. B*, **25** (6), 1928.

[101] Hwang, S.W., Lee, G.-R., Min, J. and Moon, S. (2003) Effect of ion energy and incident angle on the modification of a porous low dielectric-constant material in an $O_2$ plasma ashing process. *Surf. Coat. Technol.*, **174**, 835.

[102] Shi, H., Huang, H., Bao, J., Sun, Y., Ho, P.S., Zhou, Y., Pender, J.T.P., Armacost, M. and Ryser, D. (2008) Impact of $CO_2$ plasma on porous organosilicate (OSG) low-*k* dielectrics. *Future Fab. Int.*, **27**, 88.

[103] Liu, H., Widodo, J., Liew, S.L., Wang, Z.H., Wang, Y.H., Lin, B.F. Z. Wu, L.Z., Seet, C.S., Lu, W., Low, C.H., Liu, W.P., Zhou, M.S. and Hsia, L.C. (2009) Challenges of ultra low-*k* integration in BEOL interconnect for 45 nm and beyond. Proceedings of the IEEE International Interconnect Technology Conference 2009, p. 258.

[104] Kong, B., Choi, T., Sirard, S., Kim, D.J. and Lee, N.-E. (2007) Etch induced sidewall damage evaluation in porous low-*k* methyl silsesquioxane films. *J. Vac. Sci. Technol. A*, **25** (4), 986.

[105] Shi, H.L., Bao, J.J., Liu, J.J., Huang, H., Smith, R.S., Zha, Q., Ho, P.S., Goodner, M.D., Moinpour, M. and Moster, G.M. (2008) Dielectric recoveries on $O_2$ plasma damaged organosilicate low-*k* dielectrics. Proceedings of the Advanced Metallization Conference 2007, vol. 23, p. 447.

[106] Calvert, J.M. and Gallagher, M.K. (2003) A new approach to ultralow-*k* dielectrics. *Semiconductor Int.*, 11/1.

[107] Fayolle, M., Jousseaume, V., Assous, M., Tabouret, E., le Cornec, C., Haumesser, P.H., Leduc, P., Feldis, H., Louveau, O., Passemard, G. and Fusalba, F. (2004) Cu/ULK integration using a post integration porogen removal approach. Proceedings of the IEEE International Interconnect Technology Conference 2004, p. 208.

[108] Caluwaerts, R., Van Hove, M., Beyer, G., Hoofman, R.J.O.M., Struyf, H., Verheyden, G.J.A.M., Waeterloos, J., Tokei, Z., Iacopi, F., Carbonell, L., Le, Q.T., Das, A., Vos, I., Demuynck, S. and Maex, K. (2003) Post patterning meso porosity creation: a potential solution for pore sealing. Proceedings of the International Interconnect Technology Conference (IITC 2003), pp. 242–244.

[109] Takahashi, H., Takimoto, Y., Masuda, M. and Ando, Y. (2008) Evaluation of plasma damage in ultra-low-*k* materials with cap film using 'extracted *k*-value' method. *Microelectron. Engng*, **85** (10), 2107.

[110] Wang, Z.L., Li, J.J., Sun, Z.H., Li, Y.L., Luo, Q., Gu, C.Z. and Cui, Z. (2007) Effect of grain size and pores on the dielectric constant of nanocrystalline diamond films. *Appl. Phys. Lett.*, **90** (13), 133118.

[111] Beynet, J., De Roest, D., Rochat, N., Kellens, K., Verdonck, P. and Sprey, H. (2009) Use of MIR-FTIR and *k*-value measurements to assess potential solutions to reduce damage during porous low-*k* integration. *ECS Trans.*, **18** (1), 275.

[112] Urbanowicz, A.M., Shamiryan, D., Zaka, A., Verdonck, P., De Gendt, S. and Baklanov, M.R. (2010) Effects of He plasma pretreatment on low-*k* damage during Cu surface cleaning with $NH_3$ plasma. *J. Electrochem. Soc.*, **157** (5), H565.

[113] Aimadeddine, M., Arnal, V., Roy, D., Farcy, A., David, T., Chevolleau, T., Posséme, N., Vitiello, J., Chapelon, L.L., Guedj, C., Brechet, Y., Volpi, F. and Torres, J. (2006) Effect of $CH_4$ plasma on porous dielectric modification and pore sealing for advanced interconnect technology nodes. Proceedings of the IEEE International Interconnect Technology Conference 2006, p. 81.

[114] Bao, J.J., Shi, H.L., Liu, J.J., Huang, H., Ho, P.S., Goodner, M.D., Moinpour, M. and Kloster, G.M. (2007) Mechanistic study of plasma damage and $CH_4$ recovery of low *k* dielectric surface. Proceedings of the IEEE 2007 International Interconnect Technology Conference, p. 147.

[115] Shi, H., Bao, J., Liu, J.J., Huang, H., Ho, P.S., Goodner, M.D., Moinpour, M. and Kloster, G.M. (2007) Effect of $CH_4$ plasma treatment on $O_2$ plasma ashed organosilicate low-*k* dielectrics. *Mater. Res. Soc. Proc. B*, **990**, B03–12.

[116] Chang, T.C., Mor, Y.S., Liu, P.T., Tsai, T.M., Chen, C.W., Chu, C.J., Pan, F.M., Lur, W. and Sze, S.M. (2002) Trimethylchlorosilane treatment of ultralow dielectric constant material after photoresist removal processing. *J. Electrochem. Soc.*, **149** (10), F145.

[117] Chang, T.C., Mor, Y.S., Liu, P.T., Tsai, T.M., Chen, C.W., Mei, Y.J. and Sze, S.M. (2002) Recovering dielectric loss of low dielectric constant organic siloxane during the photoresist removal process. *J. Electrochem. Soc.*, **149** (8), F81.

[118] Mor, Y.S., Chang, T.C., Liu, P.T., Tsai, T.M., Chen, C.W., Yan, S.T., Chu, C.J., Wu, W.F., Pan, F.M., Lur, W. and Sze, S.M. (2002) Effective repair to ultra-low-*k* dielectric material (*k*-2.0) by hexamethyidisilazane treatment. *J. Vac. Sci. Technol. B*, **20** (4), 1334.

[119] Chang, T.C., Liu, P.T., Mor, Y.S., Tsai, T.M., Chen, C.W., Mei, Y.J., Pan, F.M., Wu, W.F. and Sze, S.M. (2002) Eliminating dielectric degradation of low-*k* organosilicate glass by trimethylchlorosilane treatment. *J. Vac. Sci. Technol. B*, **20** (4), 1561.

[120] Hu, J.C., Wu, C.W., Gau, W.C., Chen, C.P., Chen, L.J., Li, C.H., Chang, T.C. and Chu, C.J. (2003) Self-organized nanomolecular films on low-dielectric constant porous methyl silsesquioxane at room temperature. *J. Electrochem. Soc.*, **150** (4), F61.

[121] Liu, J., Kim, W., Bao, J., Shi, H., Baek, W. and Ho, P.S. (2007) Restoration and pore sealing of plasma damaged porous organosilicate low *k* dielectrics with phenyl containing agents. *J. Vac. Sci. Technol. B*, **25** (3), 906.

[122] Gorman, B.P., Orozco-Teran, R.A., Zhang, Z., Matz, P.D., Mueller, D.W. and Reidy, R.F. (2004) Rapid repair of plasma ash damage in low-*k* dielectrics using supercritical $CO_2$. *J. Vac. Sci. Technol. B*, **22** (3), 1210.

[123] Xie, B. and Muscat, A.J. (2005) The restoration of porous methylsilsesquioxane (p-MSQ) films using trimethylhalosilanes dissolved in supercritical carbon dioxide. *Microelectron. Engng*, **82**, 434.

[124] Smith, C., Mueller, D., Matz, P. and Reidy, R. (2006) Topographical and chemical surface modification of porous MSQ using silylating agents with different numbers of methoxy groups. *Mater. Res. Soc. Symp. Proc.*, **914**, F04–04.

[125] Chakrapani, N., Colburn, M.E., Dimitrakopoulos, C.D., Pfeiffer, D., Purushothaman, S. and Nitta, S.V. (2011) *Recovery of hydrophobicity of low-k and ultra low-k organosilicate films used as inter metal dielectrics*, US Patent 20,110,003,402.

[126] Nitta, S.V., Putushothaman, S., Chakrapani, N., Rodriguez, O., Klymko, N., Ryan, E.T., Bonilla, G., Cohen, S., Molis, and McCullough, K. (2006) Use of difunctional silylation agents for enhanced repair of post plasma damaged porous low *k* dielectrics, Proceedings of the Advanced Metallization Conference 2005, p. 325.

[127] Gun'ko, V.M., Vedamuthu, M.S., Henderson, G.L. and Blitz, J.P. (2000) Mechanism and kinetics of hexamethyldisilazane reaction with a fumed silica surface. *J. Coll. Interf. Sci.*, **228** (1), 157.

[128] Huang, H., Bao, J.J., Huang, H., Liu, J.J., Smith, R.S., Sun, Y., Ho, P.S., McSwiney, M.L., Moinpour, M. and Kloster, G.M. (2008) Dielectric recovery of plasma damaged organosilicate low-*k* films by combing UV and silylation treatments. *Mater. Res. Soc. Symp. Proc. D*, **1079**, N02–10.

# 4

# Wet Clean Applications in Porous Low-*k* Patterning Processes

*Quoc Toan Le, Guy Vereecke, Herbert Struyf, Els Kesters and Mikhail R. Baklanov*

*IMEC, Leuven, Belgium*

## 4.1 Introduction

In back-end-of-line (BEOL) interconnect, the migration from aluminum/silicon dioxide to copper/low-*k* dielectrics represents at the same time opportunities and critical challenges to the semiconductor industry. The surface preparation, cleaning and removal of undesired substances or layers are essential steps in manufacturing semiconductor devices. However, the development of many cleaning steps required in the semiconductor industry has always been dependent on several other factors and materials presented in the stack. These factors can be related to process integration, type of hardmask employed, low-*k* dielectrics properties, patterning scheme, as well as the equipment platform for the clean itself. In addition, with respect to the older aluminum/silicon dioxide technologies, which employed more robust materials with larger CDs, the current copper/low-*k* integration with technology nodes below 65 nm has more complex and difficult cleaning requirements.

One of the most important requirements for a wet clean solution is achieving efficient cleaning of photoresist, post-etch residues or Cu surface without modifying or degrading the properties of the layers presented in the stack, especially the porous dielectric. Each low-*k* material is unique and, as a matter of fact, a given chemistry that is compatible with a low-*k* film is not necessarily compatible with another film in the same family or of similar

*Advanced Interconnects for ULSI Technology*, First Edition. Edited by Mikhail R. Baklanov, Paul S. Ho and Ehrenfried Zschech.

porosity. Therefore, the use of a cleaning chemistry in integration needs to be carefully investigated on a case-by-case basis.

Many process steps in BEOL consist of plasma processing for patterning, where chemical reactive species directly react with the layers present in the structure, including the dielectric material and the photoresist (PR) layer. PR removal refers to the complete cleaning of the post-etch PR layer of the surface without attacking or modifying the underlying layers. In BEOL cleaning, one of these layers is porous dielectrics. Traditionally the remaining PR layer after a plasma etch is removed before copper deposition using an oxidizing plasma process. With the introduction of new porous dielectric materials, this approach is no longer acceptable since they are susceptible to physical and chemical damage upon plasma processing, which may irreversibly affect the dielectric properties during the PR removal process. An alternative approach for removal of the PR layer on porous low-*k* dielectrics by means of wet processes has been considered to minimize damage to the low-*k* material.

Of interest in this chapter are the effect of wet clean process on the change in porous dielectric properties and recent developments of the wet clean applied in BEOL processing. In particular, the removal of PR residues occurring after (in the case of a PR mask) or before the patterning of the dielectrics layer (in the case of a dielectric stack with a metal hardmask), the removal of post-etch sidewall residues formed during the patterning of the low-*k* dielectric stack and cleaning of the copper surface will be described and discussed in this chapter. In addition, this chapter will exclusively deal with the removal of 193-nm PR and post-etch residues generated during the patterning of a porous low-*k* dielectric in a stack containing a PR mask and a metal hardmask.

Ellipsometry, Fourier transform infrared spectroscopy (FTIR) and X-ray photoelectron spectroscopy (XPS) have been the techniques of choice for the analysis of blanket low-*k*, PR and post-etch residues. These methods provide information on the optical properties such as thickness and the refractive index, chemical bonds and elemental composition of the films analyzed, respectively. Metrology techniques for characterization of physical properties and chemical structure are lacking for sidewall polymer residues. To date, top-down inspection using scanning electron microscopy (SEM) has been widely adopted for identification of sidewall residues and assessing the cleanliness of a sidewall after a wet clean process. Other techniques such as angle-resolved X-ray photoelectron spectroscopy and microbeam infrared spectroscopy have emerged as potential methods for quantitative sidewall characterization, but are also time-consuming.

## 4.2 Silica and Porous Hybrid Dielectric Materials

Considering the similarity between porous silica-based low-*k* material and silica, the chemical stability of the former is mainly governed by the silica backbone, Si–O–Si [1]. The following is meant as a brief introduction and comparison between silica and porous silica-based low-*k* dielectrics.

Silica, the well-known and major component of the earth's crust, has been the subject for numerous studies. In its framework structure, cross-linking and sharing of oxygen atoms

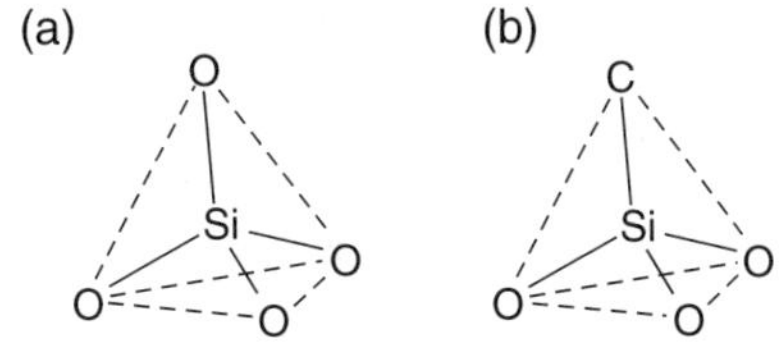


***Figure 4.1*** *Structure of (a) silica unit, $SiO_4$ tetrahedron, and (b) silica-based low-k dielectrics*

by silicon occur. A complete sharing of all four oxygen atoms per $SiO_4$ tetrahedron results in a net chemical formula of $(SiO_2)_n$ or silicon dioxide (Figure 4.1). Silicon dioxide can exist in several forms, ranging from highly ordering of the silicate tetrahedra, such as quartz, to structures with disordering of the tetrahedra where no long-range order exists, as in the case of glass. Depending on the ordering of the tetrahedra units in their structure, the density of silica films can vary between 2 and 3 g/cm$^3$.

In microelectronics, silica films used as dielectrics have a density of 2.1–2.3 g/cm$^3$ and a refractive index of ~1.46 [2], which corresponds to a typical dielectric constant of ~3.9–4.0. Substitution of some of the oxygen atoms in the $(SiO_2)_n$ network by organic groups such as alkyl, aryl or alkoxy gives rise to hybrid dielectrics or silica-based dielectrics with a lower dielectric constant. Note that this class of low-*k* material is also referred to as carbon-doped oxide (CDO) or organosilicate glass (OSG) or SiOCH material in the literature. Methyl represents the most commonly used substituent among the organic substitution groups. As a consequence, the presence of methyl groups in the silica backbone significantly lowers the dielectric constant. The lower dielectric constant of these materials can be attributed to a lower polarizability of the Si–$CH_3$ bond in comparison with the Si–O bond in the $SiO_2$ network. In addition, due to the larger size of the methyl group versus oxygen, the intrinsic porosity (free volume) of the material also increases compared to silicon dioxide, which in turn provides additional decrease in the dielectric constant. For instance, 'dense' CDO films containing ~25–30 % of carbon, as measured using XPS, typically have a density of ~1.5 g/cm$^3$, a refractive index of ~1.40–1.42 and a dielectric constant of 3.0–3.1. Clearly, doping the silicate glasses with carbon by introducing methyl groups significantly lowers the density, refractive index and *k* value of the dielectric films. Characterization using ellipsometric porosimetry revealed that these films are not completely dense but have ~6–8 % porosity, which corresponds to the magnitude of the free volume presented in this type of material.

Other types of low-*k* dielectrics are silsesquioxane (SSQ)-based and organic polymer materials. Silsesquioxanes, with the general formula of $(R\text{–}SiO_{3/2})_n$, are also referred to as T-resin owing to the presence of three oxygen substituents per silicon atom. This type of material typically has two main configurations: a ladder-type structure and a cage-like structure where each unit contains eight silicon atoms situated at the vertices of a cube (T8 cube). Again, among various substituents, R = methyl (methylsilsesquioxane (MSQ)-based low-*k* material) represents the most popular. In terms of composition and bond structure, characterization using FTIR represents a simple and informative method. A typical FTIR spectrum of silica-based or MSQ-based dielectrics shows dominant stretching vibrations in the 1000–1200 cm$^{-1}$ region (Si–O–Si backbone), a smaller absorption band at ~1275 cm$^{-1}$

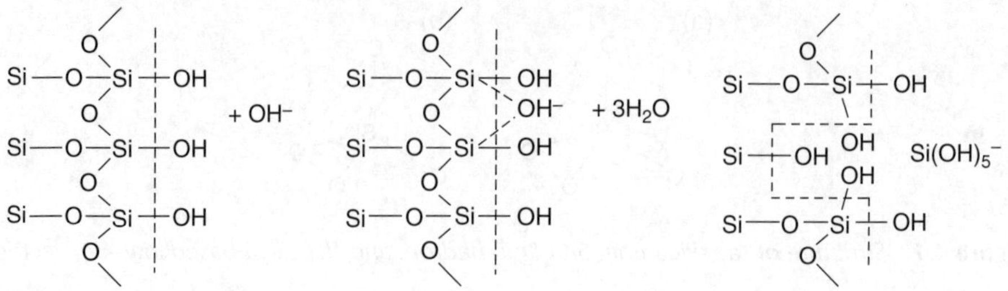


***Figure 4.2*** *Mechanism of silica dissolution in water in the presence of hydroxyl ions as proposed by Iler* [3]. *The dotted line represents the silica–water interface. Reprinted from R.K. Iler,* The Chemistry of Silica, *page 63, Copyright John Wiley & Sons, Ltd 1979, with permission from John Wiley & Sons, Ltd*

($Si–CH_3$) and a relatively large band at 2800–3200 cm$^{-1}$ (C–H). A complete review of the different classes of low-*k* dielectrics, their characteristics and properties are discussed in detail in Reference [2]. This chapter will mainly be restricted to silica-containing low-*k* dielectrics, i.e. silica-based and MSQ-based dielectrics.

Iler [3] stated in his reviews and summaries of the silica surface that the active surface of silica gel was covered by OH groups that directly bind to the $SiO_2$ skeleton. The surface silanol (Si–OH) groups formed by adding water to the oxide surface were also the reactive sites where water molecules are adsorbed through hydrogen bonding. Organic molecules with polar groups also adsorbed at the same sites at the silica surface. The nature of the underlying silica (bulk) did not significantly affect the concentration of the adsorbate. The concentration of SiOH groups at the silica surface was believed to be ~6.6 OH groups/nm$^2$. Prigogine and co-workers [4, 5] suggested that there are two types of adsorption sites at the silica surface: acceptors and donors of protons, where their activities are both related to the electronic density around the silicon atom at the surface.

The soluble form of silicon dioxide is monomeric and is generally formulated as $Si(OH)_4$:

$$(SiO_2)_x + 2H_2O = (SiO_2)_{x-1} + Si(OH)_4$$

The dissolution rate of silica in water is influenced by several factors, among which the structure (amorphous or crystalline) [6], temperature and pH [3] play a major role. Interestingly, regardless of the type of silica involved, the dissolution process always requires the presence of a catalyst. Hydroxyl and fluoride ions are the catalysts responsible for silica dissolution in alkaline and acid solutions, respectively. In alkaline solution, the first step of dissolution consists of adsorption of hydroxyl ions on the silica surface followed by the formation of silicate ions, as shown in Figure 4.2.

In general, in alkaline solution the neutral monomer is in equilibrium with silicate ions via the following reaction:

$$Si(OH)_4 + OH^- = Si(OH)_5^-$$

At pH < 11, the silicate ions in turn hydrolyze to soluble silica, $Si(OH)_4$ and $OH^-$ ions [3], in which the hydroxyl ion is only the catalyst that controls the dissolution rate. For solutions

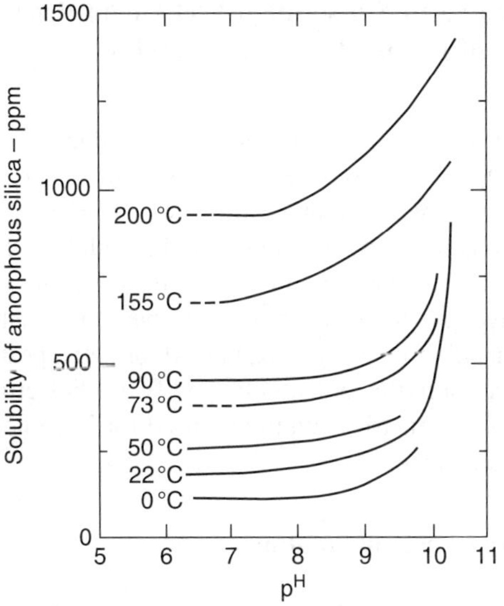


***Figure 4.3*** *Solubility of amorphous silica as a function of pH at different temperatures* [3]. *Reprinted from R.K. Iler,* The Chemistry of Silica, *page 48, Copyright John Wiley & Sons, Ltd 1979, with permission from John Wiley & Sons, Ltd*

with pH > 11, the hydroxyl ions convert $Si(OH)_4$ to silicate ions, thus keeping the solution unsaturated, which makes the dissolution process continue more quickly.

The effect of pH and temperature on solubility of amorphous silica is shown in Figure 4.3. A significant increase in solubility was recorded for pH > 9, which is explained by the formation of silicate ions in addition to the monomer. Also, the solubility was found to increase substantially with temperature within the range of 0–200 °C.

Silicon dioxide reacts with hydrofluoric acid (HF) to produce hexafluorosilicic acid:

$$SiO_2 + 6\,HF \rightarrow H_2SiF_6 + 2\,H_2O$$

As in alkaline medium, Si–OH groups are first formed in acidic solutions by the chemical reaction of the silica surface with water molecules. The first step of the dissolution process consists of the substitution of a surface Si–OH group by a Si–F group [4, 5, 7]. This step is pH dependent and determines the rate of the dissolution process [7]. Prigogine and co-workers observed that there is a correlation between the concentration in Si–F groups at the surface, as well as that of the $HF_2^-$ ion, and the dissolution rate of silica [4, 5]. Using a quartz microbalance to monitor in situ the etch rate, Knotter [7] reported that at low pH (< 1.5) water is eliminated from the protonated silanol group and an $HF_2^-$ ion or an $H_2F_2$ molecule supplies an $F^-$ that binds to the positively charged silicon atom. At higher pH values (> 1.5), the elimination of an $OH^-$ group from the $SiO_2$ surface becomes the major reaction route. Once the $OH^-$ group is eliminated, an $HF_2^-$ molecule supplies an $F^-$ ion. All subsequent reactions to remove the Si–F unit from the $SiO_2$ matrix are fast reaction steps (18–20 times faster) involving $HF_2^-$ addition reactions on $F_xSi$–O bonds.

## 4.3 Impact of Plasma and Subsequent Wet Clean Processes on the Stability of Porous Low-*k* Dielectrics

As mentioned earlier, the chemical stability of silica-based low-*k* films is mainly governed by the silica backbone, Si–O–Si. However, as low-*k* films are porous and contain organic groups, their stability also depends on other intrinsic properties of the film, such as porosity, structure of the porous network, wettability and the ability of moisture uptake of the film. In comparison with dense $SiO_2$, porous low-*k* materials are mechanically weak, less stable thermally and have low chemical stability. In particular, pattern etching and subsequent processes for removal of resist layers and/or fluorinated post-etch residues, by a dry or wet process, are critical steps that potentially modify the dielectric properties. The nature of post-etch residues and chemistries commonly used for its removal will be discussed in more detail in the next section 4.4.

### 4.3.1 Stability in Pure Chemical Solutions

Similarly to silica, the dissolution of low-*k* dielectrics is possible in the presence of a catalyst such as hydroxyl or fluoride ions. Therefore, the use of solutions that contain these ions should be limited at low concentrations. For instance, little or no dissolution was measured when using low-concentration solutions. Using different chemicals with pH ranging from 2.7 (HF 0.025 %) to 5.9 and 5.73 (ammonium fluoride 0.2 and 1 %, respectively) and 12.4 and 13.1 (tetramethylammonium hydroxide (TMAH) 0.125 and 1 %, respectively), it was shown that both as-deposited silica-based and MSQ-based low-*k* materials were not etched upon 5 min of immersion at 40 °C [8]. These treatments only induced a slight modification of the surface, as evidenced by a decrease in the water contact angle as a function of increased pH. A wet treatment using dilute TMAH caused the most hydrophilic surface. From the same study [8], it was also shown that for a given chemistry the degree of modification strongly depends on the nature and properties of the low-*k* dielectrics, i.e. high-porosity and large pore size material (MSQ) is more prone to be degraded by a plasma and/or wet clean process than a lower-porosity one (silica-based low-*k*). Another study comparing the effect of HF 0.05 % on porous spin-on (MSQ) and porous CVD (silica-based) low-*k* dielectrics also demonstrated that the material degradation strongly depends on the nature of its porosity and surface properties [9]. While the porous CVD films were fully compatible with the cleaning solution, the spin-on materials were significantly affected, resulting in film etching, a decrease in porosity and *k*-value degradation. The presence of organic additives such as acetic, glycolic, oxalic, or malic acids in general induced further degradation [9].

With regard to the thickness of porous low-*k* that is removed by a diluted HF solution, Figure 4.4 compares the etch rate for different low-*k* materials using an HF 0.1 % solution at RT [10]. The etch rate was calculated based on the difference in thickness of etch plasma-treated low-*k* films after exposure to an HF 0.1 % solution for 2 min. Since these low-*k* substrates were subjected to an etch plasma prior to the HF etch, this plot also shows the extent of damage induced by the etch plasma process. As demonstrated in Figure 4.4, the higher the open porosity of the low-*k* material, the thicker the layer removed by the HF solution. In other words, for the same plasma process the degree of

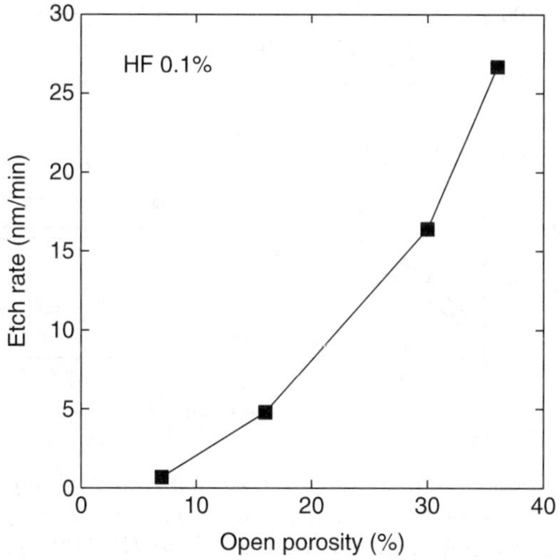


***Figure 4.4*** *Estimated etch rate for various plasma-treated porous low-k materials in HF 0.1 % solution. Data calculated on the basis of the decrease in thickness after a 2 min immersion* [10]. *Reprinted from Q.T. Le, M.R. Baklanov, E. Kesters, A. Azioune, H. Struyf, W. Boullart, J.-J. Pireaux and S. Vanhaelemeersch,* Electrochemical Solid-State Letters, ***8****, F21, Copyright 2005, by permission of ECS – The Electrochemical Society*

damage is higher for higher-porosity films. The high etch rate in HF solution in fact reflects the depth of plasma damage.

Typically, with respect to $SiO_2$, dissolution of porous low-*k* in diluted HF is not linear as a function of immersion time. This is particularly true in the case of a low-*k* film after plasma treatment, where the composition is gradually changed from the surface to the bulk of the film. The plasma-damaged areas, $SiO_2$-like, are etched rapidly in HF media. The thickness reduction measured for a plasma-treated 25 % porosity low-*k* film after immersion in HF 0.1 % solution is shown in Figure 4.5 [11]. Various plasmas were considered for the study including He, $NH_3/N_2$, $C_4H_8/Ar/N_2$ and $CO/O_2$. A thickness loss between 3.5 and 4.5 nm was measured after immersion in HF 0.1 % for 4 min for the low-*k* film treated by one of these plasmas. For patterned structures, taking into account CD variations and cleaning efficiency criteria, it was shown that diluted HF solutions remained compatible with porous ULK integration within a narrow process window [11]. For instance, using diluted HF a lateral etch of ~1.4 nm was recorded for M1 structures (45 nm technology node), in relatively good correlation with the results obtained on blanket wafers.

### 4.3.2 Stability in Commercial Chemistries

Because of their porous structure, apart from a higher etch rate versus their dense counterparts, hybrid dielectrics are prone to absorb the chemicals used during the ash or wet clean steps. For instance, nitrogen or amine-containing species absorbed from an ash or wet clean using formulated chemistries may be released in the next litho step, thus creating a poisoning of the photoresist (193 nm), which will result in development defects [12, 13].

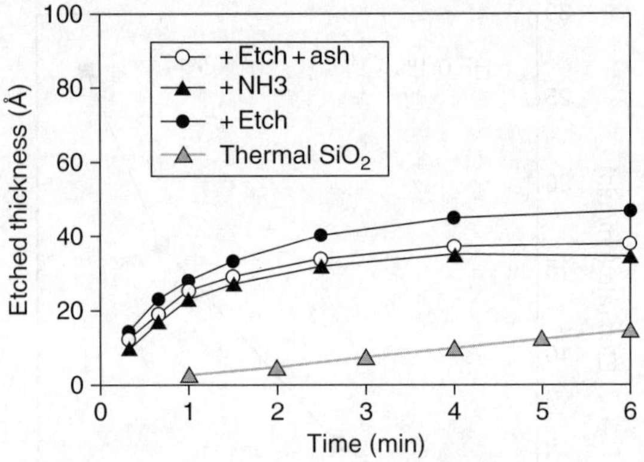


***Figure 4.5*** *Thickness reduction measured for a plasma-treated 25 %-porosity low-k film after immersion in HF 0.1 % solution* [11]. *Reprinted from L. Broussous, W. Puyrenier, D. Rebiscoul, V. Rouessac and A. Ayral, Proceedings of the IITC 2008, p. 87, Copyright 2008, with permission from IEEE*

In addition, the chemical absorption ability of porous materials makes them more exposed to the chemicals used in different processing steps that in turn may potentially lead to modification of material properties. The impact of etch and ash plasmas followed by a wet clean using commercial chemistry on porous low-*k* is described in the following paragraph. This can be considered as a case study, typical for a low-porosity (7 %)/microporous silica-based low-*k* film [14]. The treatments applied were $Ar/CH_2F_2/CF_4/O_2$ (etch) and $O_2/CF_4$ (ash) plasmas. The wet clean was done using a proprietary aqueous, hydroxylamine-based solution containing diluted organic acid [14].

The surface composition and water contact angle measured after each process step are summarized in Table 4.1. The etch plasma results in the formation of fluorocarbon polymer, as evidenced by a large increase in carbon and fluorine concentrations. Both the etch and ash plasmas made the surface more hydrophilic versus the pristine surface, due to the formation of a plasma-modified layer [10]. The low-*k* surface, inherently hydrophobic due to its large concentration in methyl groups, becomes highly hydrophilic after having been subjected to an oxygen-containing plasma, where methyl groups at the low-*k* dielectric surface are replaced by hydroxyl groups. Immersing the plasma-treated samples into the cleaning solution results in a complete wettable surface due to removal of fluorine-containing species from the surface. However, about 2 at% of fluorine still remained after a wet clean of 10 min. This amount of fluorine could correspond to the etch residues incorporated into the pores. Indeed, the cleaning solution appears to be efficient for removal of fluorinated species at the surface, but is not able to 'extract' this species from the pores.

Figure 4.6 shows the FTIR spectra of the silica-based material recorded before and after being subjected to etch and ash plasmas, and the subsequent wet treatment. For the as-deposited low-*k* substrate, besides the presence of a large absorption band between 1200 and 1000 $cm^{-1}$ (Si–O–Si) and a smaller peak at ~1275 $cm^{-1}$ ($Si–CH_3$), two other main absorption bands are also observed at 900–730 $cm^{-1}$ (Si–C and Si–O) and 2980 $cm^{-1}$ (C–H bonds, not shown here). The slight change in the spectral shape recorded at ~1100 $cm^{-1}$ for the plasma-treated sample (curve b) may be reflected by the presence of a small amount of

***Table 4.1*** *Atomic concentration recorded at the surface of the pristine silica-based film before and after different process steps as measured by XPS*

| Treatment | Si2p (atomic %) | C1s (atomic %) | O1s (atomic %) | F1s (atomic %) | Water contact angle (°) |
|---|---|---|---|---|---|
| As-deposited | 30.9 | 29.1 | 39.3 | 0.7 | 90 |
| Etch | 12.2 | 55.7 | 24.7 | 7.4 | 39.2 |
| Etch + ash | 29.4 | 11.7 | 55.4 | 3.5 | 34.5 |
| Etch + ash + wet clean | 29.3 | 15.3 | 53.3 | 2.1 | 0 |

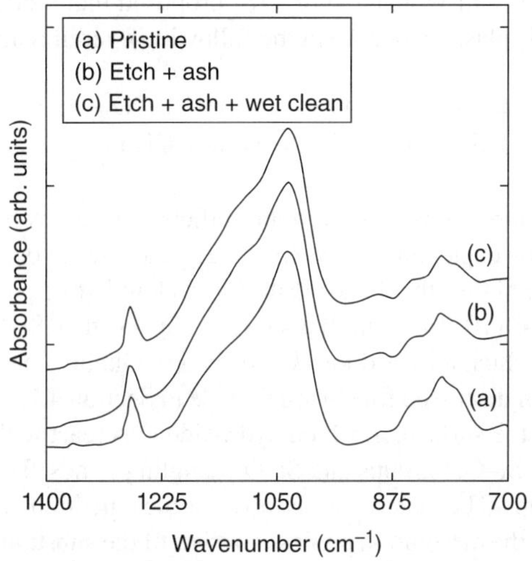


***Figure 4.6*** *FTIR spectra for the low-k film before and after plasma treatments and subsequent wet clean* [14]. *Reprinted from Q.T. Le, J. Van Olmen, R. Vanderheyden, E. Kesters, K. Kenis, T. Conard, W. Boullart, M.R. Baklanov and S. Vanhaelemeersch,* ECS Transactions, ***1*** *(3), 164, Copyright 2005, by permission of ECS – The Electrochemical Society*

fluorinated polymer in the bulk. The intensity of the absorption bands attributed to Si–$CH_3$ bonds remained unchanged, indicating that little or no bulk modification occurred during etch and ash plasma treatment and subsequent wet clean. Ellipsometry results also indicated similar thicknesses and refractive indices after wet treatment.

A wide variety of commercial aqueous and solvent-based chemistries containing organic acid, TMAH, ammonium fluoride ($NH_4F$) and choline was evaluated for fluoride silicon glass (FSG) and SiOCH dielectrics integration [15]. In terms of material compatibility and residue removal efficiency, several conclusions resulted from this study: (a) solvent-based chemistries containing $NH_4F$ are prone to photoresist poisoning and have apparent non-compatibility with the low-*k* films, (b) TMAH and choline-containing chemistries have certain specific problems with FSG and SiOCH dielectrics and (c) aqueous chemistry containing organic acid and diluted HF solution seem to be promising solutions. It should be noted that residue removal using HF is at the expense of dielectric etching, since polymer residues are most likely not dissolved in this solution. The main mechanism of removal is thus substrate undercutting. Therefore, this approach should be carried out under well-controlled conditions to avoid excessive CD loss due to a dielectric etch by HF solution.

Another study by Broussous *et al.* [16] reported an efficient and damage-free cleaning process using a proprietary aqueous solution containing diluted organic acids (Microstrip® 8500 from FFEM). This chemistry was found to be compatible, at 50 °C, with ultra-porous low-*k* films used for 45 and 32 nm technology nodes and achieved a good cleaning efficiency without CD loss and without increasing the integrated *k* value.

In contrast to the plasma process, where low-*k* damage is attributed to preferential loss of $CH_3$ groups caused by dissociation of Si–$CH_3$ bonds in reactions involving O and H radicals [17–20], low-*k* film modification and etching in a wet clean solution mainly involves Si–O bonds. Chang *et al.* [17] and Worsley *et al.* [18] proposed that modifications of the hybrid low-*k* film in $O_2$ and $H_2$ plasmas occur via the following overall reactions, respectively:

$$\equiv \text{Si–CH}_3 + 4\text{O} \rightarrow \equiv \text{Si–OH} + \text{CO}_2 + \text{H}_2\text{O}$$
$$\equiv \text{Si–CH}_3 + 2\text{H} \rightarrow \equiv \text{Si–H} + \text{CH}_4$$

Oxygen plasma is more damaging, i.e. more efficient at removing alkyl groups, than is hydrogen plasma. Carbon depletion can be explained by the low dissociation energy of the Si–C bond compared to those of the other bonds in the hybrid low-*k* structure. The average bond dissociation energy for Si–C (107.9 kcal/mol) is substantially less than Si–O (193.5 kcal/mol) [21].

The mechanism by which a hybrid low-*k* is etched in alkaline, or HF solution (acidic), is similar to what has been discussed for silicon dioxide in Section 4.1. The first step consists of hydrolysis reactions at the surface, in which hydroxide ions react with Si–O–Si bonds, leading to the formation of Si–OH groups and Si–$O^-$ dangling bonds. The latter are in turn converted into Si–OH groups. The material is dissolved as the repetition of the hydrolysis reaction occurs. In HF solution, the dissolution process consists of the substitution of a surface Si–OH group by an Si–F group. As for $SiO_2$, subsequent reactions to remove the Si–F unit from the O–Si–$CH_3$ matrix involve $HF_2^-$ addition reactions on $F_xSi(CH_3)$–O bonds. A particular characteristic of hybrid low-*k* material is the presence of a large amount of hydrophobic methyl groups. The high hydrophobicity of the film partially prevents the reactive ions ($OH^-$ in alkaline and $F^-$ and $HF_2^-$ in HF solutions) from attacking the Si–O bonds at the liquid–low-*k* film interface. As a consequence, the hydrolysis rate is reduced as compared to the silica.

In general, there is no universal cleaning chemistry that at the same time fully preserves the low-*k* integrity and achieves good cleaning performance for all types of post-etch residues generated on low-*k* and Cu surfaces, and particles. The fact that porous low-*k* materials are easily damaged by etch and ash plasmas means that the choice of a wet clean needs to be made accordingly. Among the commercial chemistries available for post-etch cleaning, fluoride-containing chemistries, hydroxylamine-based and basic chemistries [22] may present significant limitations, especially for high-porosity low-*k* films.

### 4.3.3 Hydrophobicity of Hybrid Low-*k* Materials

In contrast to hydrophilic silica surface, hybrid low-*k* films are hydrophobic due to the presence of organic groups (alkyl, aryl, etc.) in their structures. The presence of these groups at a low-*k* surface significantly delays the dissolution process. At a given concentration in

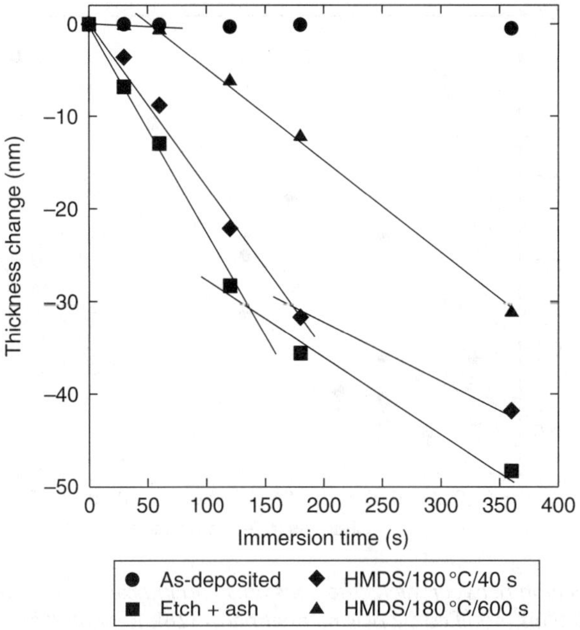


***Figure 4.7*** *Change in thickness of as-deposited and treated silica-based films (30 % porosity) after immersion in HF 0.1 % as measured using spectroscopic ellipsometry* [23]. *Reprinted from Q.T. Le, M.R. Baklanov, E. Kesters, Q. Han, S. Luo, C. Waldfried and S. Vanhaelemeersch, Proceedings of the International Semiconductor Technology Conference, p. 36, Copyright 2006, by permission of ECS – The Electrochemical Society*

active species, e.g. hydroxyl or fluoride ions, the film only started to be etched once its surface becomes wettable by the solution. For instance, as-deposited silica-based low-*k* films with porosity up to 30 % were not etched by HF 0.1 % solution, while the etch rate for silicon dioxide was in the order of several nm/min. The stability of the as-deposited dielectric film in HF 0.1 % solution is mainly ascribed to the hydrophobicity of the film ($\theta_{H_2O}$ ~90°). Upon subjection to a combination of etch ($Ar/CF_4/CH_2F_2/O_2$) and ash ($O_2/CF_4$) plasmas, the low-*k* surface becomes much more hydrophilic ($\theta_{H_2O}$ ~35°) and was readily etched by the same solution (Figure 4.7) [23]. In the present case, of course, the relatively high etch rate measured for the plasma-treated film with respect to the as-deposited one is not solely attributed to the change in surface wettability, but it is also explained by the formation of a damaged layer caused by the plasma. Exposing the plasma-treated film to the vapor of a silylating agent such as hexamethyldisilazane (HMDS), trimethylchlorosilane [24–26] or diphenyltetramethyldisilazane [27] allowed the replacement of hydroxyl termination groups by trimethylsilyl groups ($-Si-(CH_3)_3$) or by phenyldimethyl groups, which greatly recovered the initial hydrophobicity of the surface ($\theta_{H_2O}$ ~76–79°) [23, 27]. In fact, the hydrophobic surface of the HMDS-treated sample, as a consequence of the presence of $CH_3$-terminated siloxane groups, resulted in a delay of etching and in a lower etch rate by HF 0.1 % solution compared to that of the sample without HMDS treatment, as shown in Figure 4.7. For this low-*k* dielectric (porosity ~30 %, mean pore diameter ~2.2 nm), the exposure to HMDS vapor only modified the surface, without significant modification of the inside of the

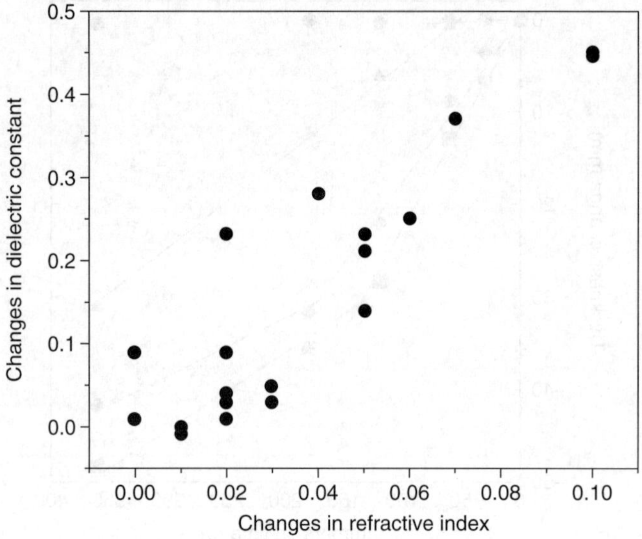


***Figure 4.8*** *Correlation between dielectric constants and indexes of refraction measured from porous low-*k *films modified by plasma treatments* [28]. *Reprinted from A. Grill and V. Patel,* Journal of the Electrochemical Society, ***153****, F169, Copyright 2006, by permission of ECS – The Electrochemical Society*

film [23]. The transmission FTIR spectra recorded for the samples before and after HMDS exposure showed no significant difference between plasma-treated and HMDS-treated samples (figure not shown), testifying to the surface character of the silylation using HMDS. Indeed, considering the size of the $CH_3$-terminated siloxane groups grafted at the surface during the silylation reaction make the pores at the surface smaller, such that the reactive sites available in the bulk are not easily accessible for HMDS molecules.

The ability of moisture uptake of a low-*k* film can be associated with the 'degree' of hydrophobicity/hydrophilicity of the bulk (pore walls) that determines the internal wettability by chemical solution during a wet process. The moisture absorption into dielectric films has a significant effect on the change in refractive index of low-*k* films. Considering the very high dielectric constant of water, ~ 80, a small quantity of absorbed water present in the porous film can result in a significant increase in the *k*-value of the dielectric film. For instance, a small increase in the refractive index of 0.1 already led to a substantial increase of >0.4 in the dielectric constant (Figure 4.8). Therefore, in order to prevent loss of hydrophobicity and *k*-value increase water must be removed before the next processing steps.

Unlike pure silica, silica-based low-*k* materials contain Si–$CH_3$ groups. The organic substitutes not only provide low density and low dielectric constant matrix materials, but also represent a certain weakness of the molecule. For low-*k* dielectric films containing Si–$CH_3$ groups, it was shown that, in the presence of a catalyst such as hydroxyl ion (by autocatalytic reaction), at elevated temperature water molecules were the cause of decomposition of methyl-containing low-*k* films. Note that the presence of a catalyst is indispensable for such a reaction:

$$\text{Si–CH}_3 + \text{H}_2\text{O} \rightarrow \text{Si–OH} + \text{CH}_4$$

As a result, $CH_4$ molecules were found to start desorbing at a temperature close to 400 °C if water was initially present in the low-*k* film. In contrast, $CH_4$ desorption was only detected at 550 °C for the film that contains very low concentrations of water [29]. The hydrophobic low-*k* film became much more hydrophilic due to the formation of Si–OH groups. An improvement of thermal stability, up to 550 °C, could be achieved by a dehydration of the low-*k* film using $XeF_2$ at elevated temperature [29].

Rogojevic *et al.* presented a method for evaluation of moisture absorption in the case of porous silica xerogels using a quartz crystal microbalance [30]. The materials to be investigated were first directly coated on to the quartz crystal prior to exposure to the environment containing water. Under well-controlled experimental conditions, the amount of water absorbed can be quantified. This method, however, requires a rather long exposure time to reach the absorption equilibrium, i.e. between 10 and 15 h.

An efficient method for evaluation of hydrophobicity of the porous network, or measurement of the internal contact angle, was introduced by Baklanov [31]. This method is based on the same principle as that of the ellipsometric porosimetry that is widely used for evaluation of porosity and pore size of porous low-*k* materials. The assessment of internal hydrophobicity of porous films is done using water as the adsorbate instead of an organic solvent. This method allows for determination of the proportion of the water adsorbed into the pores of the low-*k* film, thereby assessing the degree of damage caused by a process step [31]. A linear relation was found between carbon depletion, caused by the plasma process, and hydrophilicity, although the depth of analysis determined by EP is significantly larger than that of TOF–SIMS. This observation also suggests that the hydrophilicity of low-*k* films depends not only on the carbon concentration but also on the nature of chemical bonds presented in the film.

## 4.4 Removal of Post-Etch Residues and Copper Surface Cleaning

In back-end-of-line processing, to enable anisotropic etching, a thin fluorocarbon layer is deposited on the sidewalls of features, thereby inhibiting lateral etching. The presence of this layer also minimizes low-*k* modification such as carbon depletion and densification during etching using fluorocarbon chemistry. For instance, etching of a PR-mask/MSQ-based low-*k* stack using $Ar/CF_4/CH_2F_2/O_2$ plasma created a thin fluorocarbon ($CF_x$) polymer layer on the sidewall, where the concentration in fluorine was measured as ~37 at% by angle-resolved XPS [32]. This fluorocarbon polymer, being formed by recombination of fluorocarbon radicals present in the plasma, must be selectively removed prior to subsequent processing steps to ensure high adhesion and good coverage of materials deposited in the etched features [33, 34]. Also, trapped fluorine species and moisture present in porous low-*k* dielectrics, as well as copper oxidation and a sidewall polymer, have a negative impact on the global yield [35].

Typically, for PR-mask dielectric structures, the fluorocarbon polymer ($CF_x$) is mainly removed by $O_2$, $N_2$ or $H_2$-based plasmas during the photoresist strip. However, incorporation of metallic or other nonvolatile impurities due to the reactive sputtering of via/trench bottom materials makes the polymer removal much more difficult. A subsequent wet clean is thus required to completely remove the post-etch residues remaining on the top surface, dielectric sidewall and Cu surface. Therefore, the performance of a wet clean process and the extent of low-*k* etching are intimately dependent on the ashing process. Oxidizing plasmas are well known to induce damaging to porous low-*k* materials by carbon depletion, film densification, silanol formation and water adsorption [24, 32, 36–40], which result in significant increases in the dielectric constant. Reductive plasmas have received later attention as low-damage ashing processes. However, several studies reported in the literature showed similar damaging behavior of these plasmas toward porous low-*k* dielectrics [40–42]. An alternative ashing process using a mixture of He and $H_2$ was reported to induce less damage to porous low-*k* dielectrics [23, 36].

For a wet clean, a diluted HF solution with concentrations up to ~0.5 % has been the simplest choice for this purpose. Note that the diluted HF solution does not dissolve post-ash fluorocarbon residues; the mechanism of removal here is substrate undercutting. The kinetics of low-*k* etching in diluted HF greatly depends on the extent of plasma modification. Undercutting the residues or particles is an effective method for wafer cleaning [43]. However, the formation of a damaged layer during etch and ash processes leads to significant CD loss, making it inappropriate, especially for small structures (45 nm and below). Assessing a wide range of commercial aqueous and solvent-based chemistries, it was suggested that for porous SiOCH dielectrics (90 nm technology node), organic acid, $NH_4F$, choline and TMAH-containing chemistries were efficient for polymer removal, but they had certain issues regarding low-*k* and Cu compatibility [15]. Nevertheless, for a given chemistry, significant differences in cleaning efficiency and low-*k* modification were also observed, depending on experimental conditions such as temperature and time. In the search for chemistries breaking fluoropolymers into soluble fragments, a more recent post-etch residue removal approach based on reductive chemistry using radical anions has been proposed by Hess's group [44–47]. This method consists of two steps: the formation of naphthalene radical anions by the reaction of metallic sodium and naphthalene in tetrahydrofuran followed by a selective oxidation using aqueous ozone. The treatment using radical anions resulted in defluorination of the polymer (residues) backbone, giving a polymer product with a carbon–carbon double bond in its structure. A selective cleavage of the double bonds was subsequently performed using aqueous ozone to complete the removal of the defluorinated residues. In order to avoid the use of sodium in the process, naphthalene radical anions were also able to be generated electrochemically in *N,N*-dimethylformamide. The authors claimed that the removal mechanism is a synergistic combination of solvent swelling and radical anion defluorination. The method was demonstrated to remove fluorocarbon polymers efficiently for both model system (polymer deposition on a blanket wafer) and patterned structures. Evaluation using an Hg probe showed that this approach was compatible with silica-based (Coral) and MSQ low-*k* materials ($k$ = 2.3–2.4). However, considering the presence of radical anions and the use of ozone, process conditions need to be carefully taken into account in order to minimize damage to porous materials.

In summary, chemistries containing aqueous or semi-aqueous fluoride, solvent/amines and hydroxylamine-based substances represent the major classes of wet clean chemicals that have been used for removal of post-etch residues and copper oxides. The main issues with the amine and hydroxylamine-containing chemistries are the poisoning of 193 nm photoresist, poor compatibility with copper, elevated temperature processing (>55 °C) and poor performance on advanced low-*k* materials [22]. For diluted HF aqueous solutions, a well controlled concentration in oxygen is required to minimize a copper attack. Acidic

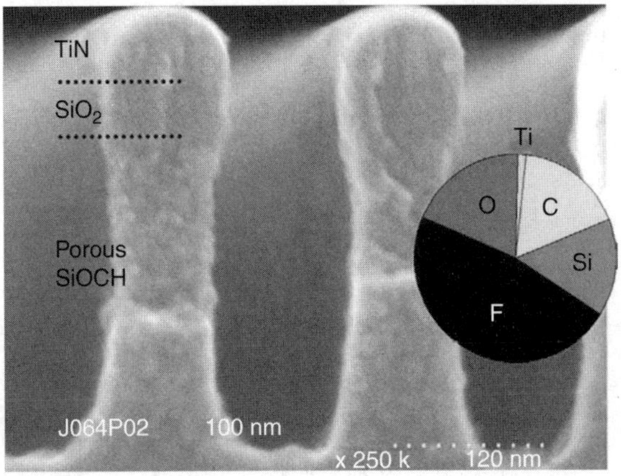


***Figure 4.9*** *SEM micrograph and XPS composition of a porous SiOCH (*k *value = 2.3, porosity ~45 %) trench sidewall after patterning using Ar/$CF_4$ plasma [48]. Reprinted from M. Darnon, T. Chevolleau, D. Eon, R. Bouyssou, B. Pelissier, L. Vallier, O. Joubert, N. Posseme, T. David, F. Bailly and J. Torres,* Microelectronic Engineering, ***85****, 2226, Copyright 2008, with permission from Elsevier*

aqueous (inorganic and organic acids) chemistries and semi-aqueous basic chemistries have also been used for BEOL cleaning. The use of semi-aqueous chemistries at high pH is to be minimized since they have high viscosity and promote copper attack, damage to porous low-*k* and low-*k* delamination [22].

As mentioned above, $CF_x$ polymer residues are mainly removed during the photoresist strip but the strip plasmas based on oxygen, fluoro-oxygen and hydrogen induced the film damage. A possible solution is to use a metal hardmask for the patterning of low-*k* dielectrics thanks to its high selectivity toward low-*k* materials. In this case, the photoresist is stripped just after the hardmask 'opening' thereby minimizing damage of the low-*k* material. Tuning an RIE condition can also minimize the damage during the dielectric etching. In Cu/low-*k* processing, using a metal hardmask – or, more precisely, a TaN or TiN hardmask – for patterning the dielectric layer, there are typically three types of residues after etch and ash: polymer residues (typically on the sidewall), organometallic residues deposited on the metal hardmask surface and metal fluoride crystal residues. The latter type was reported to evolve as a function of storage time [43, 48]. With regard to sidewall composition using fluorocarbon plasma (Ar/$CF_4$) for patterning a TiN hardmask/porous low-*k* stack, the sidewall surface of the porous low-*k* was found to be composed of silicon (15 %), oxygen (19 %), carbon (17 %), Ti (1 %) and fluorine (49 %) (Figure 4.9) [48]. This result indicated (a) the presence of a fluorocarbon passivation layer on the trench sidewalls and (b) a very small amount of TiN-based etch by-products (within the limit of detection by XPS) were deposited on the trench sidewalls during the plasma etch.

Another study performed on a lower porosity film (porosity ~25 % and *k* value ~2.5) showed similar results. Figure 4.10 shows the XPS C1s core-level spectra recorded from the sidewalls of the patterned porous low-*k* structure. The C1s spectrum corresponding to the post-etch sample (spectrum (a) in Figure 4.10) indicates that the residue was composed mainly of CF, $CF_2$ and a lower concentration of $CF_3$ groups. Diluted HF, a

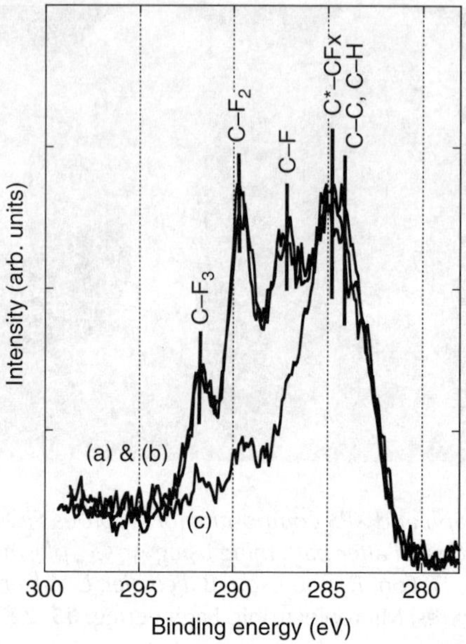


***Figure 4.10*** *XPS C1s spectra measured on the low-k sidewall: (a) after low-k patterning, (b) subsequent 254 nm UV treatment (total dose of 3 J/cm²) and (c) after 254 nm UV/3 J/cm² + wet clean using DMSO/10 % MEA mixture* [49]. *Reprinted from Q.T. Le, F. Drieskens, J.F. de Marneffe, T. Conard, M. Lux, H. Struyf and G. Vereecke, UCPSS 2010 Abstract Book, vol. 50, Copyright 2010, with permission from Trans Tech Publications*

solvent mixture based on DMSO or formulated chemistries commercially available were found not to be efficient in removing this type of residue. A two-step cleaning process consisting of a UV treatment at $\lambda = 254$ nm with total energies $\geq 3$ J/cm$^2$ combined with the use of DMSO/monoethanolamine (MEA) mixtures significantly enhanced the removal efficiency [49]. The positive effect of the UV treatment could be explained by a direct bond scission by UV radiation. UV treatment alone with a total energy of 3 J/cm$^2$ did not lead to any noticeable change in the C1s spectrum (Figure 4.10 (b)). In contrast, a large amount of fluorinated groups was removed from the sidewall after a similar UV exposure followed by a wet clean using DMSO/10 vol % MEA at 50 °C for 5 min (Figure 4.10 (c)). The fluorine concentration on the low-*k* sidewall decreased from ~45 to ~13 at% [49]. In another but similar study, it was found that the cleaning efficiency was further improved when replacing the DMSO–MEA mixture by a proprietary DMSO-additives formulated cleaner [50].

Surface wetting is one of the key requirements in the development of an efficient wet cleaning formulation. Three different approaches have been considered for this purpose. For aqueous cleaning chemistry, an improvement of surface wettability can be achieved using alcohol additives such as isopropyl alcohol (IPA). In this case, one can keep the advantages of aqueous media and the IPA allows the contact angle to be decreased in a controllable way. The low flash points, a characteristic of most monoalcohols, render them less compatible with fab processing, in particular at high concentration. The second way is the use of surfactants. However, evaluation of a number of standard surfactants showed that they have limited compatibility with porous materials. Most of them have large adsorption energy on pore walls, making them difficult to be removed after the wet clean [1]. As a result, the film porosity decreases, resulting in an increase of dielectric constant. Therefore, the use of surfactant as an additive in cleaning solutions could be considered for improving wetting of a low-*k* stack only if the choice of surfactant is made properly. The molecular size of the surfactant must be larger than the pore diameter to avoid absorption into the pores [51]. Cleaning using solvents in supercritical $CO_2$ medium represents another means to improve wettability of a surface. Critical $CO_2$ medium has a zero wetting angle and was shown to be efficient in the presence of appropriate additives for removal of residues on surface and internal pores [52, 53].

It has been widely recognized that via cleaning after dual damascene etching is required for good electrical contact yields and reliabilities. Degradation of the copper interconnect is attributed to the formation of oxides at the via bottom: cupric oxide (CuO) that nucleates on cuprous oxide ($Cu_2O$), which typically forms in contact with metallic copper [54]. In addition, exposure to ambient conditions resulted in the formation of copper hydroxide ($Cu(OH)_2$) and copper carbonate ($CuCO_3$). These contamination species must be removed before the copper is subjected to subsequent processing steps. To this end, the Cu surface can be cleaned using both dry and wet chemistries, including ammonia or hydrogen plasma (atomic hydrogen) at a temperature between 200 and 350 °C [55], alcohol vapor [54] and diluted solutions of HF and organic acids such as citric and oxalic acids. In addition, several industrial suppliers have developed standard recipes, which are able to clean the Cu surface and are compatible with low-*k* dielectrics. Copper cleaning using hydrogen plasma was proven efficient, but has a major concern regarding porous low-*k* compatibility. On the contrary, exposure to a reducing agent like ethyl alcohol is a milder process and has potentially high compatibility toward porous low-*k* materials. In this process, cuprous oxide is first converted to cupric oxide before being reduced to metallic copper. While oxide removal was able to be achieved at a lower temperature (~130 °C), it occurred more efficiently at 200 °C.

A different approach was proposed using HFAc (hexafluoroacetylacetone), which forms volatile products with Cu and gives a principal possibility to remove Cu residues from a via sidewall [56, 57]. HFAc can also be dissolved in $SCCO_2$, which offers a possibility to clean the Cu surface and remove the Cu residue from feature sidewalls [58]. Via cleaning can also be done using the vapor phase of organic acids [59, 60]. The vapor treatment using formic and acetic acid vapor at 200 °C was shown to reduce copper oxides to metallic copper [59]. One interesting characteristic of this approach is the possibility of controlling the temperature and choice of organic acids, thus enabling a control in the reduction rate, copper surface and extent of low-*k* compatibility.

Figure 4.11 compares via contact resistance distribution of via chains for the intermediate level consisting of a $k=2.25$ dielectrics (nanoclustering silica, NCS). The comparison is shown for wet chemicals, organic acids and control samples. The gas-phase organic acid was very effective in reducing CuO. For the sample cleaned with gas-phase organic acids, a low contact resistance was measured together with a small variability of the contact resistance, which was found to perform better than the wet chemical. Also, the yield of the cleaned samples was shown to be enhanced significantly compared to that of the control sample (~80 %).

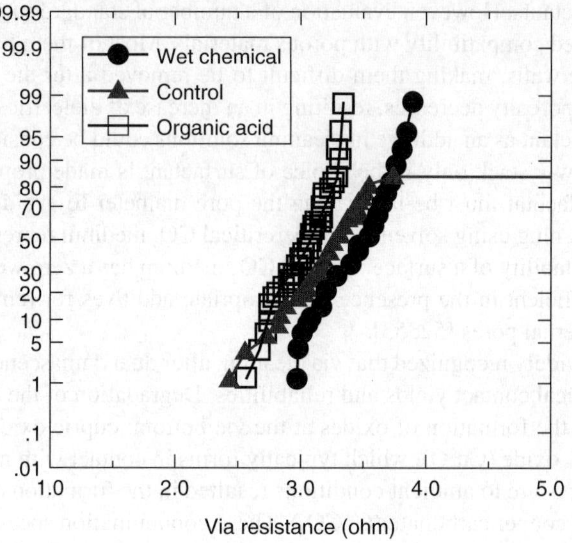


*Figure 4.11* *Via contact resistance distribution of via chains for intermediate level* [60]. *Reprinted from H. Kudo, K. Ishikawa, M. Nakaishi, A. Tsukune, S. Ozaki, Y. Nakata, S. Akiyama, Y. Mizushima, M. Hayashi, Ade. A. Akbar, T. Kouno, H. Iwata, Y. Iba, T. Ohba, T. Futatsugi, T. Nakamura and T. Sugii, Proceedings of the IITC 2008, p. 93, Copyright 2008, with permission from IEEE*

## 4.5 Plasma Modification and Removal of Post-Etch 193 nm Photoresist

### 4.5.1 Modification of 193 nm Photoresist by Plasma Etch

The degradation of PR in plasmas is a complex phenomenon. The intensity and extent of the PR modification is defined by different factors, including the type of plasma applied, the power density, which in turn affects the reactivity of the chemical species generated in the plasma, and other experimental parameters such as pressure and time of exposure. During plasma exposure, the material is exposed to a broad spectrum of electrons, ions, radicals and electromagnetic radiation (vacuum ultraviolet radiation). These species have different energies and penetration depths.

The polymers used for 248 nm lithography are not applicable for 193 nm lithography due to the strong absorbance at 193 nm of conjugated C=C bonds (aromatic ring) usually present in the 248 nm PR structure. Acrylic-based polymers represent one of the polymer classes that satisfy the transparency requirement. However, these polymers have rather poor plasma etch resistance. It was found that the PR etch rate is dependent on the number of oxygen atoms in the monomer unit [61–63]. In other terms, for etch processes with ion bombardment such as in reactive ion plasma etching, the etch rate has a linear dependence on the '$N/(N_C - N_O)$' factor of the etched polymer [61], where $N$, $N_C$ and $N_O$ denote the total number of atoms in a monomer unit, the number of carbon atoms and the number of oxygen atoms in the same monomer unit, respectively. Thus, the dry etch resistance under ion bombardment of a polymer is inversely proportional to the effective carbon content of the

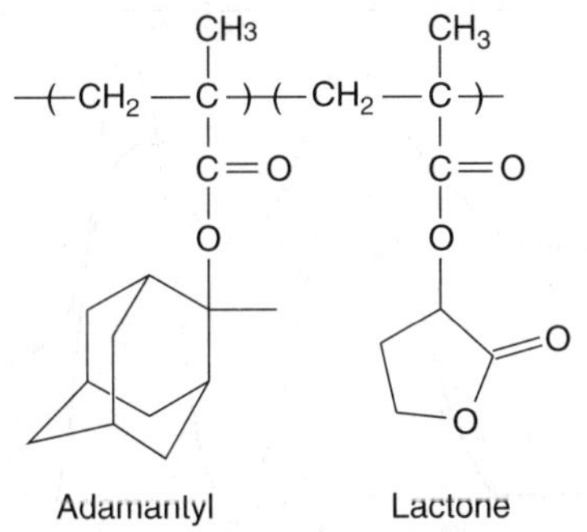


***Figure 4.12*** *Schematic view of typical methacrylate-based monomer units*

material. Adding alicyclic units such as norbornane and adamantane as side-chain groups to the acrylate backbone not only increases the effective number of carbon atoms but also maintains the material transparency at the 193 nm wavelength.

As an example, the methacrylate-based monomer shown in Figure 4.12 is characterized by relatively bulky pendant side groups: an adamantyl for improving the etch resistance property and a lactone for enhancement of its solubility in an organic solvent. Both groups are cleavable upon DUV exposure in the photolithographic process. In terms of plasma etch resistance, following the $N/(N_C - N_O)$ rule described above, it is reasonable to assume that the ester groups in the polymer molecule should be preferentially removed when exposed to plasma.

Modification of 193 nm PR by plasmas has been studied in relation to plasma hardening processes (curing), reduction of line edge roughness (LER) and cleaning. This section deals with the effect of the plasma process carried out after the lithography process, aiming to 'stabilize' the PR layer, i.e. before etching of the BARC layer and metal hardmask and/or before the dielectrics etch. It was reported that the plasmas utilized for pre-treatment of 193 nm PR, usually referred to as 'plasma cure' in the literature, such as HBr, Ar, $H_2$ and $Cl_2$, induce surface roughening, increased line width roughness and substantial modification of the chemical composition of PR [64]. From the FTIR analyses, as shown in Figure 4.13, treatments by one of these plasmas for 60 s led to the removal of the lactone (~1795 $cm^{-1}$) and ester groups (~1730 $cm^{-1}$) in the side chains of the polymer. With respect to the initial PR after coating, the reduction in intensity of carbonyl-containing species was found to be as high as 60–80 % for the HBr and $H_2$ plasma pre-treatments. The Ar-plasma pre-treated PR resulted in about 40 % loss of the C=O containing groups. Among the pre-treatment plasmas described in Reference [64], the HBr or $H_2$ plasmas, which removed a substantial amount of the C=O groups, were the most effective on suppressing the LWR and LER increases caused by the plasma process in the subsequent step to etch the BARC and hardmask.

Using surface-sensitive analytical techniques such as XPS and atomic force microscopy (AFM), it was shown that the PR layer undergoes substantial modification during the exposure to fluorocarbon plasma [65, 66]. A modified, graphitic layer with a higher refractive index as compared to the bulk PR material was formed on the PR surface within a few seconds of plasma exposure. Preferential cleavage of the polymer side chain groups occurred via the removal of oxygen-containing groups. This selective removal of C–O

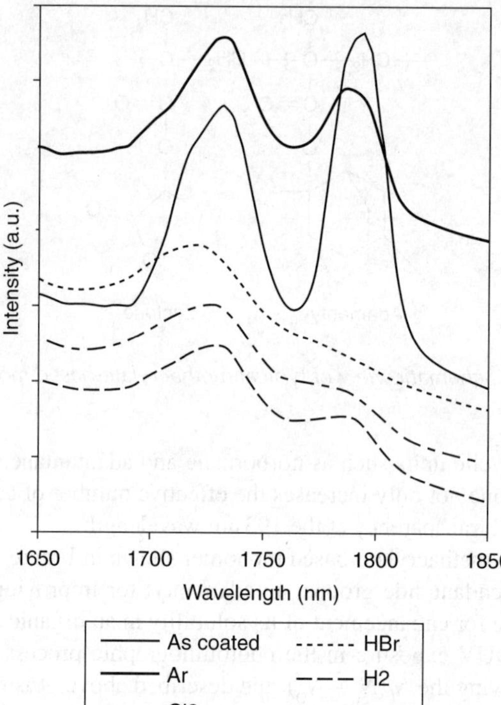


***Figure 4.13*** *IR absorption spectra within the region between 1650 and 1850 cm*$^{-1}$ *wavenumbers (not wavelengths as the authors wrongly indicated in the paper) after various plasma treatments* [64]. *Reprinted from M.C. Kim, D. Shamiryan, Y. Jung, W. Boullart, C.J. Kang and H.K. Cho,* Journal of Vacuum Science Technology B, ***24****, 2645, Copyright 2006, with permission from American Vacuum Society*

groups seen for 193 nm PR indicates that oxygen of the bulk PR enhances the PR etch rate until a fully fluorinated surface has developed. Indeed, the XPS data revealed that cleavage at ester groups of the side chain in the polymer and dangling bond formation took place, leading to chain cross-linking and other intramolecular rearrangements, prior to a fluorinated surface being developed. The evolution of XPS C1s core-level spectra recorded for untreated 193 nm PR and after $C_4F_8$/Ar plasma exposure for different times is shown in Figure 4.14 [65]. Clearly, the intensities of the components corresponding to fluorine-containing species increased rapidly as a function of exposure time. This can also be seen in Figure 4.15, where the change in F/C ratio is shown for 193 nm PR exposed to $C_4F_8$/Ar plasma for various times as a function of the probing depth measured using the angle-resolved XPS analysis. First, the F/C ratio significantly increased for a longer exposure and, second, a slight F/C gradient as a function of depth was also observed at the beginning of the plasma exposure. This gradient decreased for a longer etching time as the F/C ratio increased and started to disappear after ~3 s after the beginning of the exposure, which indicates that a steady-state modified layer had been achieved. Sumiya *et al.* [66] claimed that in order to prevent selective etching of the ester groups in the polymer, thereby reducing

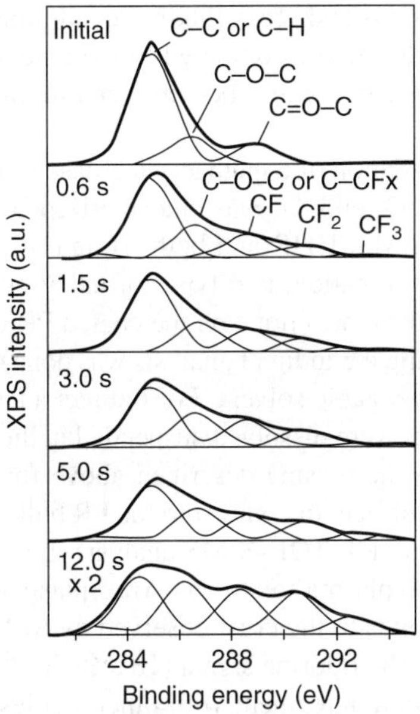


***Figure 4.14*** *Evolution of XPS C1s core-level spectra obtained with untreated 193 nm PR and after a $C_4F_8$/Ar plasma etch for different times [65]. Reprinted from M. Sumiya, R. Bruce, S. Engelmann, F. Weilnboeck and G.S. Oehrlein,* Journal of Vacuum Science Technology B, ***26****, 1637, Copyright 2008, with permission from American Vacuum Society*

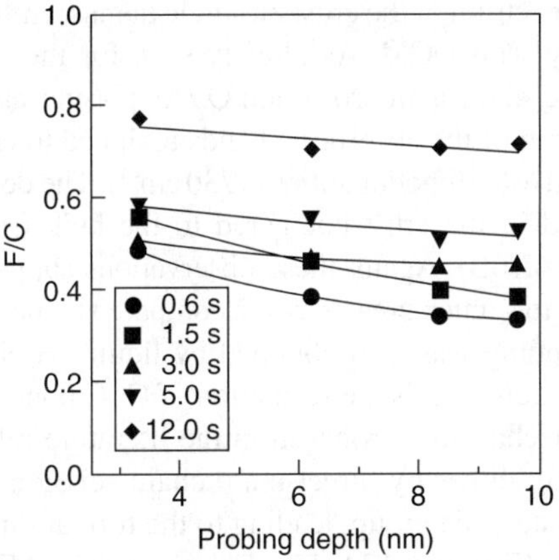


***Figure 4.15*** *Change in the F/C ratio of 193 nm PR exposed to $C_4F_8$/Ar plasma for various times as a function of probing depth measured using the angle-resolved XPS analysis [65]. Reprinted from M. Sumiya, R. Bruce, S. Engelmann, F. Weilnboeck and G.S. Oehrlein,* Journal of Vacuum Science Technology B, ***26****, 1637, Copyright 2008, with permission from American Vacuum Society*

the formation of dangling bonds and cross-linking, conditions that have a relatively large amount of fluorine on the PR surface compared to the amount of oxygen in the PR bulk are required. Fluorine-rich plasmas could be beneficial in minimizing the removal of the ester groups.

In another study, the degradation of a methacrylate-based PR by RIE plasma, a gas mixture containing $CH_2F_2$, $CF_4$, $O_2$ and Ar, was characterized by several spectroscopic techniques such as XPS, TOF–SIMS, FTIR and both $^1$H and $^{13}$C nuclear magnetic resonance [67]. To characterize PR modification, two types of analyses were performed: direct and indirect analyses. Direct analysis was done on the coated PR or remained after dissolution of the bulk on a Si wafer, while the indirect analysis was done on the fraction of the PR dissolved, or undissolved, in an organic solvent. For indirect analysis, blanket PR layers with and without plasma treatment were dissolved/suspended in dichloromethane (DCM) at RT.

After the treatment using the plasma described above for 35–50 s, a modified (crust) layer was formed at the PR surface. In contrast to the PR bulk, the crust layer is not soluble in common organic solvents at RT. TOF–SIMS analysis showed that the surface of the PR is enriched in fluorine by the plasma treatment. After plasma treatment, the top 80 nm of the PR layer, which corresponds to the crust observed by SEM, contains a higher fluorine concentration than the bulk. The fluorine signal (18 at%, by XPS) decreases exponentially as a function of depth. In good agreement with other studies on polymethacrylate-based resins reported in the literature [65, 66, 68, 69], liquid $^1$H NMR results obtained on hot-DMSO-soluble crust samples indicated that the lactone side group was most probably removed together with the ester that links to the polymer backbone. However, contrary to those studies, the loss of side groups occurred preferentially for lactone groups and not for adamantyl groups.

The $^1$H NMR results presented above were in fact confirmed by FTIR. The FTIR spectra in Figure 4.16 show the FTIR spectra recorded on the pristine PR together with DCM-isolated crust and bulk samples from the plasma-etched PR, and an attenuated total reflection-FTIR (ATR-FTIR) spectrum of the crust on single damascene dielectrics structures after dissolving the bulk away with DCM. As already seen for the case of single-component plasmas shown in Figure 4.16, a fluorocarbon/$O_2$/Ar plasma also resulted in significant change in the intensity ratio of the absorption bands assigned to carbonyl stretch vibrations of lactone (1795 cm$^{-1}$) and ester functionalities (1730 cm$^{-1}$). The decrease of the lactone/ester ratio is more pronounced in the crust compared to the bulk (Figure 4.16A) and with a longer etch time (Figure 4.16B). Again, these observations support the statement that lactone–ester groups are lost to a much greater extent compared to the adamantane–ester groups.

Another interesting finding was also obtained by liquid $^1$H NMR. The analysis of the DMSO-soluble crust fraction reveals the formation of C=C bonds after RIE. This is in fact in agreement with the mechanisms proposed in the literature related to chain scission of polymethacrylate [68, 70], either by direct main chain scission or by disproportionation after loss of a complete ester side group, leading to the formation of end-chain C=C bonds and a propagating radical (Figure 4.17A(1)). The observed NMR resonance pattern indicates additional types of unsaturations next to the end-chain C=C bonds. This also points to the formation of C=C bonds in and/or along the backbone due to termination reactions following ester cleavage (Figure 4.17A(2)).

Based on the fact that ester groups were cleaved during plasma exposure, a cross-linking scheme can also be proposed, explaining the insolubility of the crust in organic solvents

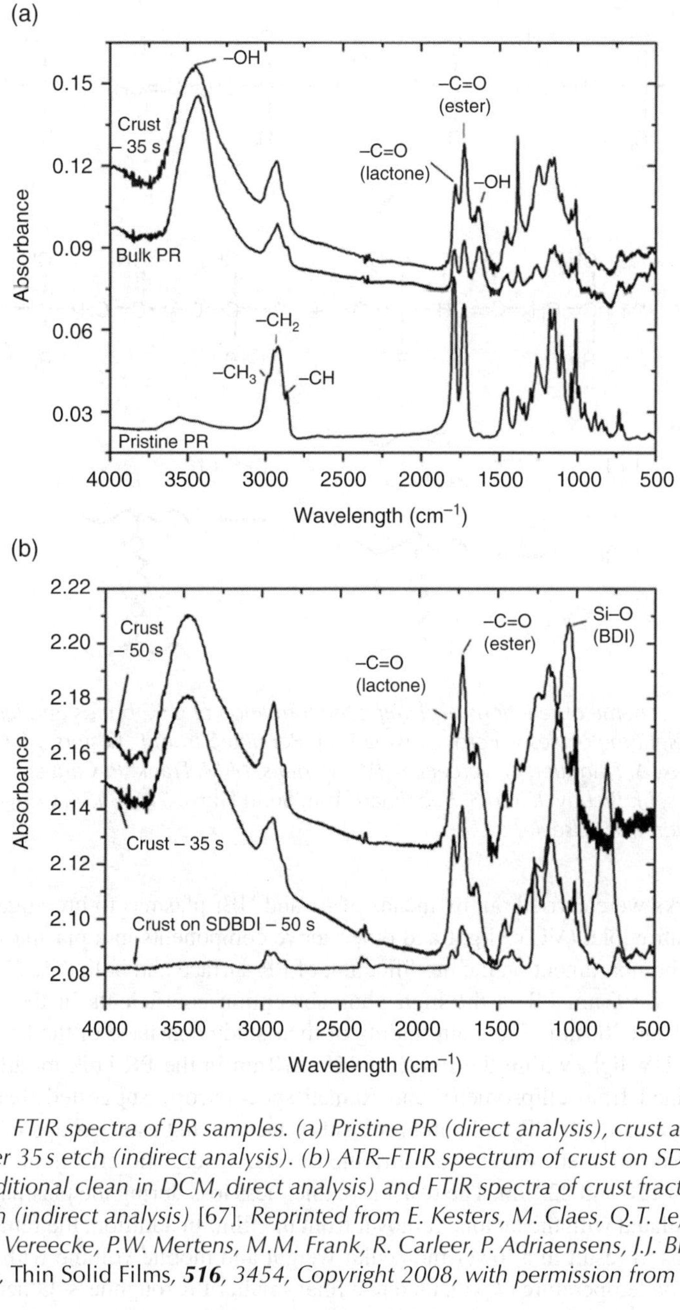


***Figure 4.16*** *FTIR spectra of PR samples. (a) Pristine PR (direct analysis), crust and bulk fractions after 35 s etch (indirect analysis). (b) ATR–FTIR spectrum of crust on SD BDI (50 s etch and additional clean in DCM, direct analysis) and FTIR spectra of crust fractions after 35 and 50 s etch (indirect analysis)* [67]. *Reprinted from E. Kesters, M. Claes, Q.T. Le, M. Lux, A. Franquet, G. Vereecke, P.W. Mertens, M.M. Frank, R. Carleer, P. Adriaensens, J.J. Biebuyck and S. Bebelman,* Thin Solid Films, ***516****, 3454, Copyright 2008, with permission from Elsevier*

(Figure 4.17B). Cross-linking can result from the recombination of two radicals (I + I), leading to the direct coupling of two PR chains by a C–C bond, or by I–II–I coupling with diradical II fragments as the cross-linker. Besides cross-linking, branching can be introduced by the recombination of an intrachain radical with a propagating radical (I + II).

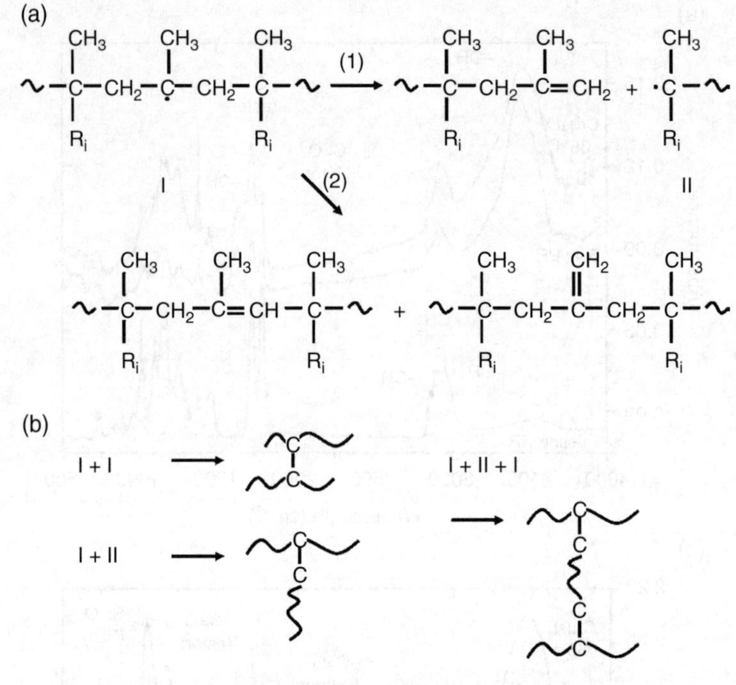


***Figure 4.17*** *Scheme of reactions explaining the formation of C=C bonds and larger molecular weight molecules via cross-linking* [67]. *Reprinted from E. Kesters, M. Claes, Q.T. Le, M. Lux, A. Franquet, G. Vereecke, P.W. Mertens, M.M. Frank, R. Carleer, P. Adriaensens, J.J. Biebuyck and S. Bebelman,* Thin Solid Films, ***516****, 3454, Copyright 2008, with permission from Elsevier*

Other works were carried out by means of Ar and HBr plasmas to investigate the effect of vacuum ultraviolet (VUV) light and other active components in a plasma such as radicals and ion bombardment on the modification of PR surface and bulk [71, 72]. An important feature of 193 nm PR is the high photoabsorption coefficients in the VUV region between 110 and 300 nm. The main finding of their study consisted of the key role played by plasma VUV light within the range of 110–210 nm in the PR bulk modification. The results obtained from ellipsometry and Raman spectroscopy suggested that $sp^2$ carbon atoms are formed in the PR bulk, and removal of lactone and ester groups in the bulk PR layer was also observed, which was attributed to the synergistic effects of VUV plasma light between 170 and 220 nm. The resist film thickness loss during the plasma treatment is directly correlated with the lactone removal from the film. In addition, the removal of C=O species seems to result in a lower molecular weight and mobile volume with a decreased glass transition temperature ($T_g$) that led to a final smaller PR roughness. Other studies carried out on 193 nm PR by Nest *et al.* [73] and Titus *et al.* [74] demonstrated that PR roughening and degradation are mainly caused by simultaneous effects of VUV radiation, ion bombardment and substrate heating.

In summary, the formation of a modified, cross-linked layer occurred at the PR surface after a plasma process using inert gas (Ar, He) or reactive plasmas such as HBr, $Cl_2$ and fluorocarbon mixture. In both cases, preferential removal of the ester function in the side groups of the polymer is the cause of the PR degradation, C=C bond formation and cross-linking.

A similar scenario was also observed for UV-treated polymethyl methacrylate (PMMA), a polymer that can be considered as a model material for methacrylate-based PR. The FTIR spectra of the PMMA films recorded after irradiation with different doses, both in air and in vacuum, are shown in Figure 4.18 [75]. The absorption band corresponding to C=O in the pendant group ($COOCH_3$) of PMMA at 1735–1750 cm$^{-1}$ decreases in intensity and shifts toward higher absorption frequencies as a function of increasing dose, indicating a change of the chemical environment. In addition, the C–O stretching bands at 1270–1150 cm$^{-1}$ also decreased substantially. This behavior is consistent with literature and was explained by the removal of the ester side group followed by a main-chain scission or the generation of unsaturated bonds in the polymer chain [76–78].

### 4.5.2 Wet Removal of 193 nm Photoresist

Traditionally, the remaining PR layer after plasma etch is removed before copper deposition using oxidizing (e.g. $O_2$, $O_2/CF_4$) or reductive ($N_2/H_2$, $NH_3$, $NF_3$) plasmas, followed by a liquid processing step. However, with the introduction of new porous dielectric materials, this approach is no longer acceptable since the new materials are susceptible to physical and chemical damage upon plasma processing [2, 10, 28, 36, 37, 79, 80], which may irreversibly affect the dielectric properties during the PR removal process.

This section describes, first, PR removal methods using organic solvents and organic strippers (a mixture of chemicals in organic solvent). Then the feasibility of wet strip using ozonated aqueous chemistries, which represent an environmentally acceptable alternative method to minimize damage to low-*k* material, will be considered. Finally PR removal using supercritical $CO_2$ will be shortly discussed.

*Photoresist Removal by Organic Solvents*

Polymer dissolution in organic solvents plays an important role in many applications such as microlithography, plastic recycle [81] and the coating industry, in which certain characteristics of the cleaning fluid such as surface tension of solutions and solubility parameters have been used to assist in solvent selection.

Levitin *et al.* [82] investigated the cleaning efficiency of solvent-based cleaning solutions and correlated to solvatochromic parameters by means of the Owens–Wendt analysis for evaluation of the interactions between fluorocarbon-based residues (PR residues), substrates and solvent properties. The Owens–Wendt theory is derived from the combination of Young's equation (relation between the work of adhesion, contact angle and surface tension of the liquid) and Fowkes equation (relation between the interface tension and the polar and dispersive components of the surface energy). This theory divides the surface energy into two components: surface energy due to dispersive interactions and polar interactions. One of the advantages of the Owens–Wendt analysis is that it allows an estimation of the 'dispersive' and 'polar' character of the solutions and surfaces by measurement of contact angles, assuming that the liquid surface tension is known (according to Spelt and Neumann [83], the 'dispersive' and 'polar' components are not directly determined by the

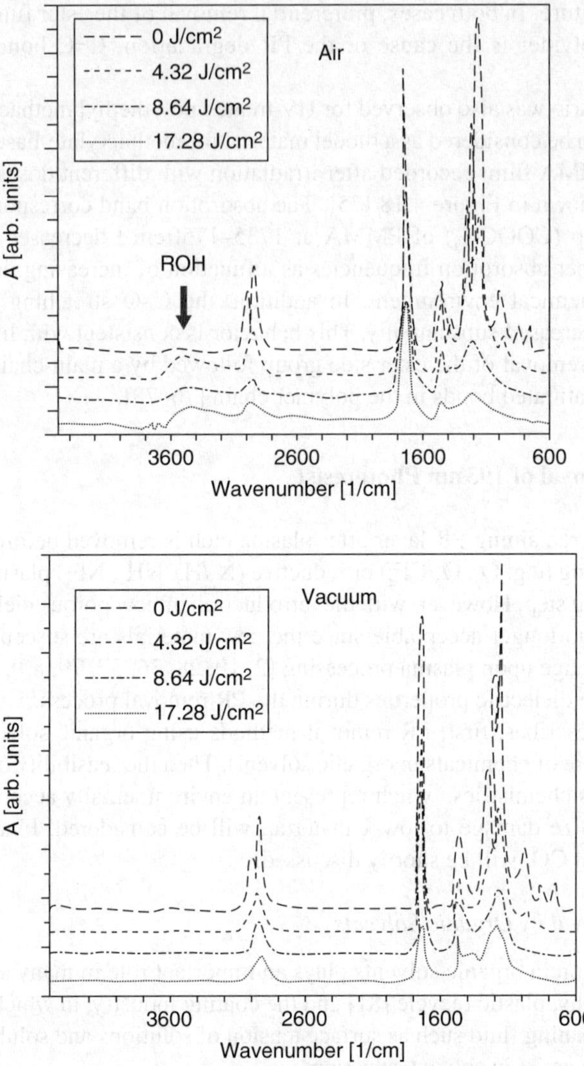


***Figure 4.18*** *IR absorption spectra of a 3.2 μm thick PMMA film after different irradiation doses in air and vacuum [75]. Reprinted from P. Henzi, K. Bade, D.G. Rabus and J. Mohr, Modification of polymethylmethacrylate by deep ultraviolet radiation and bromination for photonic applications,* Journal of Vacuum Science Technology B, **24***, 1755, Copyright 2006, with permission from American Vacuum Society*

relative magnitude of dispersion and polar forces in a given solid, but are only functions of total surface tension; however, this discussion is out of scope of this chapter).

The approach of Levitin *et al.* [82] involved the characterization of both the solid surfaces (among them, the polymer or residues to be studied) and the liquids (solvents used to remove residues). Optimization of the liquid–solid interaction can assist in the selection of appropriate cleaning solutions. The Owens–Wendt analysis predicts that the interaction energy will be maximized when the dispersive and polar contributions of the surface match those of the liquid. The solid surfaces were characterized by measurement of the contact angles formed with the same probe liquids as described in Reference [83]: methylene iodide, which participates only in dispersive interactions, and water, which participates in both dispersive and polar interactions on the surface. As to the liquid characterization, the contact angles of different solvents were measured on a PTFE surface, which participates in only dispersive interactions. The polar contribution of the surface tension is calculated by subtracting the dispersive contribution from the total surface tension of the liquid. The liquids selected for their study were almost exclusively organic solvents, including the two probe liquids (water and methylene iodide), dimethylformamide (DMF), *N*-methyl-2-pyrrolidone (NMP), DMSO, tetrahydrofuran (THF) and acetonitrile. Both fluorocarbon model samples and post-etch PR residue samples were exposed to the above solvents. DMF, NMP and DMSO caused delamination of the film from the silicon substrate within 2 min of immersion. However, the films did not dissolve in these solvents, which were explained by the presence of a highly cross-linked layer. Note that the fluorine concentration of the fluorocarbon model samples used in this study (~50 at%) was three times higher than that measured on the PR surface described in Reference [67] (18 at%, see Section 4.5.1). It reflects the variability in film composition and bonding structure of fluorocarbon crust generated at the PR surface by different plasma etching processes [82, 84, 85]. For several cases, in particular for post-etch PR residue, a good matching between the dispersive characters of solvents and that of the residues did not result in an effective removal. This could be partially explained by the fact that the dispersive character of the solid was determined from the crust, whereas the solid (residue)–substrate interface was formed between the PR bulk and the substrate. Their study also demonstrated that addition of ionic salts to conventional solvents such as water or NMP significantly alters the overall surface tension and alters the distribution of polar and dispersive components describing the interactions.

The Hildebrand solubility parameter was first introduced by Hildebrand and Scott in 1950, and was then developed by Hansen [86, 87]. As for the Owens–Wendt method, the general principle of use of the Hildebrand theory can be summarized as 'like dissolves like', or, in other terms, substances will dissolve in each other if the magnitude of the intermolecular attractions is similar, a simplification that was found to be invalid for many polar substances. Hansen realized that all the cohesive bonds holding a liquid together were broken when it evaporated, including dispersion, polar and hydrogen bonding forces. Hansen proposed to divide the Hildebrand solubility parameter into three specific molecular interactions including polar cohesive forces ($\delta_P$), (nonpolar) dispersive interaction ($\delta_D$) and hydrogen bonding ($\delta_H$):

$$\delta^2 = \delta_P^2 + \delta_D^2 + \delta_H^2$$

All substances exhibit 'nonpolar' dispersive attraction forces. When two molecules are in proximity, the random polarities in each molecule tend to induce corresponding polarities in one another, causing the molecules to fluctuate together. These induced interactions are referred to as London dispersion forces, or induced dipole–induced dipole interactions. Polar cohesive forces are present in most substances. They are produced by permanent dipole–dipole interactions (Keesom interactions) or dipole-induced dipole interactions (Debye interactions). The third component reflects the hydrogen bonding capability, a

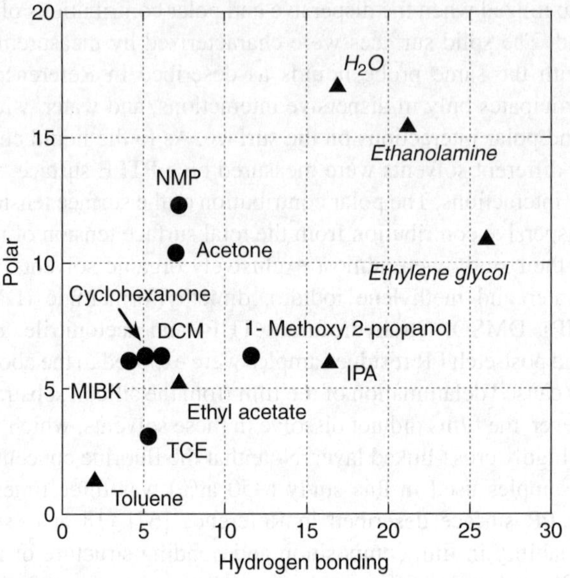


***Figure 4.19*** *Dissolution of (meth)acrylate-based 193 nm PR by different solvents characterized by Hansen solubility parameters: (●) good solvent; (▲) bad solvent* [80]. *Reprinted from Q.T. Le, J. Keldermans, N. Chiodarelli, E. Kesters, M. Lux, M. Claes and G. Vereecke,* Japanese Journal of Applied Physics, ***47****, 6870. Copyright 2008, with permission from The Japan Society of Applied Physics*

particularly strong type of polar (intermolecular) interaction. This interaction occurs in molecules where a hydrogen atom is attached to a strongly electronegative heteroatom, such as oxygen, nitrogen or fluorine.

Good and bad solvents for a (meth)acrylate-based PR are given in Figure 4.19 as a function of their Hansen solubility parameters [80]. Although the Hansen solubility parameters for a solvent is related to three main specific molecular interactions, the dispersive parameters for many common solvents are very similar. Hence a three-dimensional graph could be reduced to a two-dimensional one. For a solvent classification in this case, a solvent was considered as good if the as-deposited PR was completely dissolved within 1 min, whereas a bad one required more than 10 min for PR dissolution or did not dissolve PR at all. It has been found that good solvents for as-deposited PR films have their Hansen polar and hydrogen bonding parameters situated within the range of 4–12 and 3–12, respectively. However, some exceptions could be found for certain solvents. For instance, ethyl acetate had its polar component and hydrogen bonding very close to those of dichloromethane (DCM) and situated inside of the defined solubility window, but it was not a good solvent for the PR used in this work. This could be explained by the fact that the 193 nm PR used in this work was not a pure polymer. It was a copolymer and contained different types of additive for improving certain properties of the PR mixture. In addition, other intrinsic properties of a solvent such as the shape of molecule, molar volume and viscosity also affect solubility behavior.

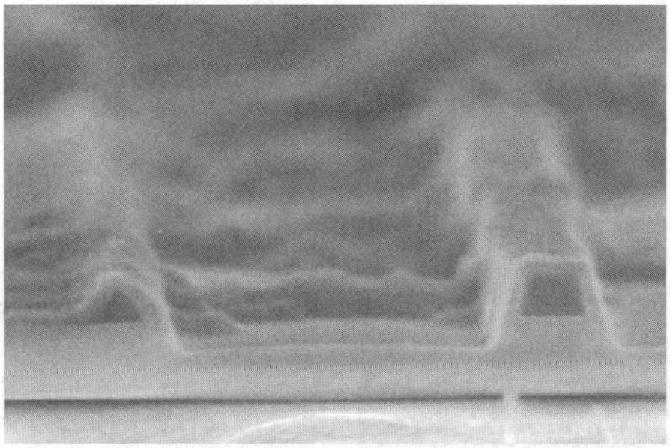

***Figure 4.20*** *SEM micrograph of post-etch (plasma-treated) PR on a dielectrics after immersion in NMP for 15 min, showing bulk dissolution (narrow line) and collapse of the PR crust (large field)* [80]. *Reprinted from Q.T. Le, J. Keldermans, N. Chiodarelli, E. Kesters, M. Lux, M. Claes and G. Vereecke,* Japanese Journal of Applied Physics, ***47****, 6870. Copyright 2008, with permission from The Japan Society of Applied Physics*

In the case of post-etch PR, as discussed in Section 4.5.1, a modified, cross-linked layer is present at the surface. This layer is chemically 'inert' in many organic solvents and aqueous solutions. As a result, immersion of the post-etch PR sample in various solvents usually only leads to dissolution of the less modified part, bulk, underneath the crust layer. As shown in Figure 4.20, the bulk PR was readily dissolved by simple immersion at RT in various types of solvent ranging from halogenated solvents (dichloromethane, trichloroethylene) to ketones (methyl isobutyl ketone, cyclohexanone) or NMP [88, 89]. On the contrary, none of them was able to dissolve the crust layer, even for an extended immersion time of up to 15 min (Figure 4.20). Collapse of the crust was also observed due to dissolution of the bulk PR underneath the crust that makes the crust structure weaker, especially in large areas (Figure 4.20).

The wet process combining the action of organic solvents (bulk dissolution) with physical forces provided by sonic cavitation (for mechanical removal of the plasma-induced crust) can be applied to remove PR. For example, complete removal was achieved for a 150 nm half-pitch structure using a megasonic clean process with NMP (1–2 min/megasonic power of 100 W/850 kHz) [80]. In order to investigate the mechanism of PR removal using solvent combined with megasonics, two separate experiments decoupling the actions of solvent and physical forces were carried out. The first one consisted of a cleaning experiment using a megasonic clean in deionized (DI) water only, for 1 min, under identical conditions for which the entire PR layer was completely removed by the solvent with the megasonic clean. The second experiment was done in two steps. The sample was first immersed in DCM for 15 min followed by a megasonic clean in DI water under identical conditions as the first experiment. A cross-section SEM of the sample cleaned by the megasonic process in DI water indicates that both the crust and bulk PR still remained on top of the dielectric lines (figure not shown). In contrast, immersion in organic solvent prior to a megasonic clean in water resulted in the removal of PR in many areas (Figure 4.21 (b)). These results first suggest that the PR crust is permeable to these solvents and that

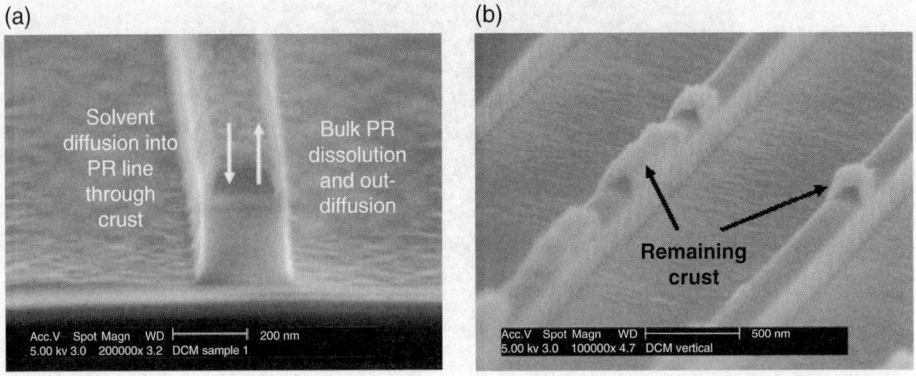


***Figure 4.21*** *SEM images describing the mechanism of PR removal using solvent combined with megasonics: (a) simple immersion in organic solvent (DCM) and (b) solvent immersion prior to a megasonic clean in water [80]. Reprinted from Q.T. Le, J. Keldermans, N. Chiodarelli, E. Kesters, M. Lux, M. Claes and G. Vereecke,* Japan Journal Applied Physics, ***47****, 6870, Copyright 2008, with permission from The Japan Society of Applied Physics*

out-diffusion of dissolved bulk PR also occurred through the crust. Furthermore, these results clearly indicate that bulk dissolution (due to organic solvent) or the cavitation effect alone (due to the megasonic process) is not sufficient for a complete removal of PR lines. Dissolution of bulk PR in organic solvents makes the PR structure more fragile, whereas the sonic energy assists in removing the remaining crust mechanically without dissolving it.

It is noteworthy that a similar approach was actually applied for stripping ion-implanted photoresist in the FEOL. The process was performed at RT using a mixture of IPA and a fluorinated salt such as KF in megasonics [90]. The PR was not decomposed during the stripping process but was in fact physically peeled off from the PR–$SiO_2$ substrate interface. The stripping mechanism of this technology is first due to high wettability of IPA/KF on the photoresist surface, allowing F and IPA molecules to be able to penetrate deeply into the photoresist. The PR is swollen by IPA and silyl groups at the interface are removed by F, and finally the photoresist is physically stripped off a substitute by means of physical forces brought by megasonics. The IPA/KF/megasonic technology was found to be highly effective in stripping hardened photoresist such as ion-implanted photoresist and photoresist with CF polymer accumulation. For a BEOL application, however, the presence of the porous low-*k* layer in the stack somewhat limits the use of high concentrated fluorinated components.

Removal of photoresist by wet cleans combining solely dissolution by organic solvents and physical forces does not seem scalable to advanced technology generations with smaller structures. For example, the process mentioned above, capable of removing PR from 150 nm half-pitch structures, was not efficient for 90 nm half-pitch structures under identical conditions. This can be attributed to a larger proportion of crust in the etched PR pattern, which requires a stronger force to break. Besides, there are concerns about breaking or fracturing fragile structures and materials (e.g. porous dielectrics) by physical forces. Chemical additives are likely to be needed to enhance removal efficiency in more advanced generations to break cross-links in the crust at lower megasonic power. The development of new generations of organic strippers tackling the various challenges of advanced BEOL integrating Cu and porous dielectrics is actively being pursued [91–96].

*Aqueous Photoresist Removal by a Combination of UV Exposure and Ozonated Strip*

New environmental regulations, putting stronger constraints on the ESH impact of organic solvents and chemistries, together with considerations about their cost of ownership (COO), makes an aqueous ozone strip particularly attractive. It was reported that ozone in the gas phase or ozone in water ($O_3/H_2O$) can be used for removal of organic contamination and a Novolak-type PR layer [97, 98]. Kashkoush [99] showed that PR was effectively removed from bare silicon and patterned wafers using ozonated water. De Gendt *et al.* [100] reported that a method to enhance the PR stripping rate is to minimize the thickness of the boundary layer of stagnant ozone solution at the wafer surface. The success of removal of organic layers by aqueous ozone depends on the type of bonds present in the material. This type of process seems particularly attractive for the removal of etched 193 nm PR, which present C=C bonds that are particularly reactive toward cleavage by ozone [101]. The reaction between ozone and C=C bond can be described as follows [102]:

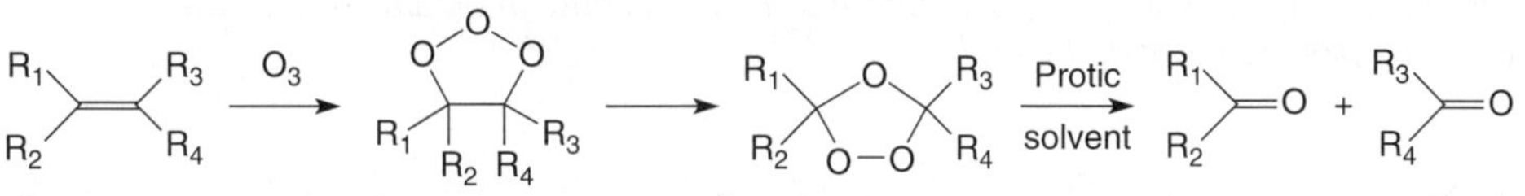


As a consequence of this reaction, smaller fragments result from long polymer chains. This opens a way toward breaking the etched crust network and possibly dissolution of the whole PR layer.

Figure 4.22 shows the FTIR spectra of blanket plasma-treated PR samples, before and after exposure to $O_3/H_2O$ for 5 and 10 min [103]. All samples received a water rinse of 2 min. The absorption band attributed to the lactone group (~1795 $cm^{-1}$) nearly completely vanished after 5 min of exposure. The spectrum remained similar for 10 min of exposure. Furthermore, the absorption band at ~1600–1650 $cm^{-1}$, attributed to the O–H group from residual $H_2O$ and C=C bonds, completely disappeared after the wet process. This strongly suggests that $O_3$ broke C=C bonds, which had been generated by the etch plasma, but that there were no sufficient reaction sites to make all PR reaction products (crust fragments) soluble in water. A final rinse in organic solvent (propylene carbonate, PC) was needed for a complete removal of the PR layer.

The similarity of the degradation of 193 nm PR by etch plasma with PMMA modification by VUV or UV radiations [75–78], which induces the formation of C=C bonds and polymer chain scission, as shown in Section 4.4, has opened a possible route toward a complete aqueous PR removal by using UV pre-treatments.

UV light absorption is a necessary condition for possible reactions to occur. Figure 4.23 shows the UV absorption spectra for pristine and plasma-etch 193 nm PR (treatment in a dual-frequency dielectric etch chamber at RT, HBr, followed by $Cl_2/O_2$ and $Cl_2/HBr$) [104]. Clearly, the absorption of the film substantially increased after being subjected to the plasma treatments, especially within the range of ~170–300 nm.

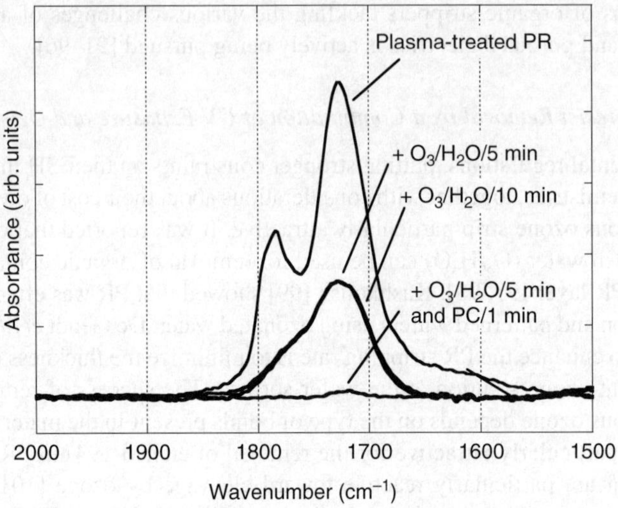


***Figure 4.22*** *Comparison between FTIR spectra of blanket plasma-treated PR film and after subsequent exposure to $O_3/H_2O$ vapor for 5–10 min. A rinse in organic solvent was required for a complete removal of the PR layer*

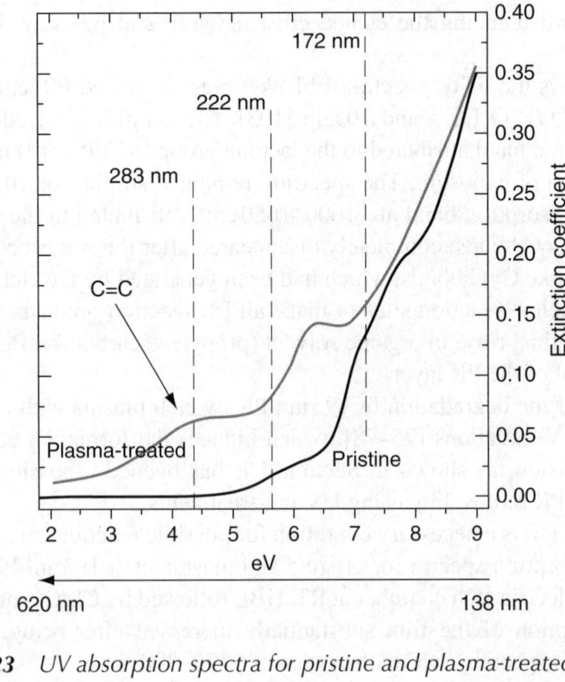


***Figure 4.23*** *UV absorption spectra for pristine and plasma-treated 193 nm PR*

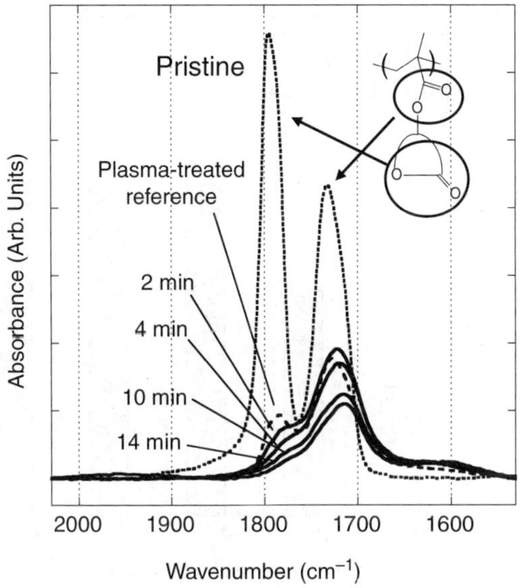


***Figure 4.24*** *FTIR spectra of plasma-treated photoresist film as measured before and after UV irradiation (dual wavelength) under $N_2$ for various exposure times*

Experiments were carried out using UV sources with different wavelengths [104–107]: a mercury low-pressure lamp (dual wavelength source; $\lambda_1 = 185$ nm, $\lambda_2 = 254$ nm; 28 mW/cm$^2$ at 254 nm), an $Xe_2$* excimer lamp ($\lambda = 172$ nm; 8 mW/cm$^2$), a KrCl excimer lamp ($\lambda = 222$; 25 mW/cm$^2$) or an XeBr* excimer lamp ($\lambda = 283$ nm; 20 mW/cm$^2$). The UV treatments using single-wavelength sources were done at $10^{-4}$ mbar, while the treatments using the dual-wavelength lamp were done under an $N_2$ atmosphere.

The formation of C=C bonds was evaluated by FTIR. Experimental results showed that irradiations by photons with wavelengths within the range of 172–283 nm led to substantial changes in FTIR spectra. For instance, for plasma-treated PR, the C=O absorption band attributed to the lactone group (~1795 cm$^{-1}$) and ester (1730 cm$^{-1}$) already drastically decreased after the plasma treatment, and further decreased after the subsequent UV process (Figure 4.24). The loss in C=O groups is very similar to that observed for different plasmas described in Section 4.5.1 and also for the etch plasma used for patterning the low-*k* film [89]. In addition, the intensity of the absorption band centered at ~1620 cm$^{-1}$, assigned to the presence of C=C bonds and the OH from $H_2O$, increased upon 2 min of treatment and appeared to remain constant for longer treatment times.

Figure 4.25 summarizes the evolution of the ratio between the absorption band of OH + C=C (measured at ~1620 cm$^{-1}$) and OH (~3200–3600 cm$^{-1}$) as a function of the treatment time using the dual-wavelength UV source, for both pristine and plasma-treated photoresist films. In general, within the range of treatment time studied, the concentration of C=C bonds clearly increased as a function of treatment time. The increase in C=C concentration was quite limited for pristine PR substrate and much more pronounced for the plasma-treated sample. A comparison of the same ratio (OH + C=C/OH absorption bands) for different wavelengths is shown in Figure 4.26. In comparison to 172 nm, treatments at 222 and

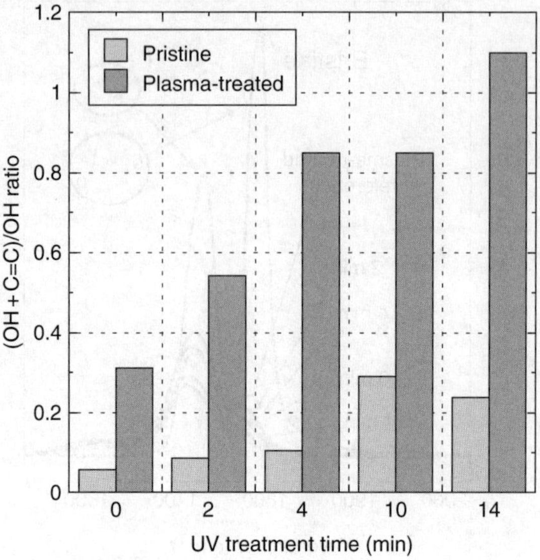


*Figure 4.25* *Evolution of (OH + C=C)/OH peak ratio measured for PR samples after various UV exposure times (dual wavelength) under* $N_2$ [105]. *Reprinted from Q.T. Le, E. Kesters, L. Prager, M. Claes, M. Lux and G. Vereecke,* Solid State Phenomena, ***145–146***, *323, Copyright 2009, with permission from Trans Tech Publications*

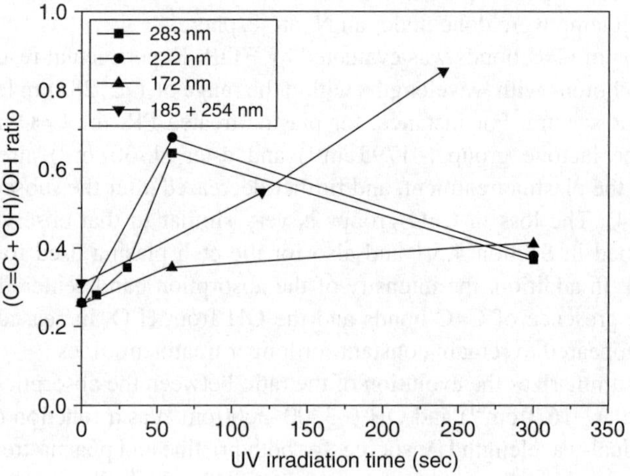


*Figure 4.26* *Evolution of (OH + C=C)/OH peak ratio measured for PR samples after various UV exposure times* [106]. *Reprinted from E. Kesters, Q.T. Le, M. Lux, L. Prager and G. Vereecke,* Microelectronic Engineering, ***87***, *1674, Copyright 2010, with permission from Elsevier*

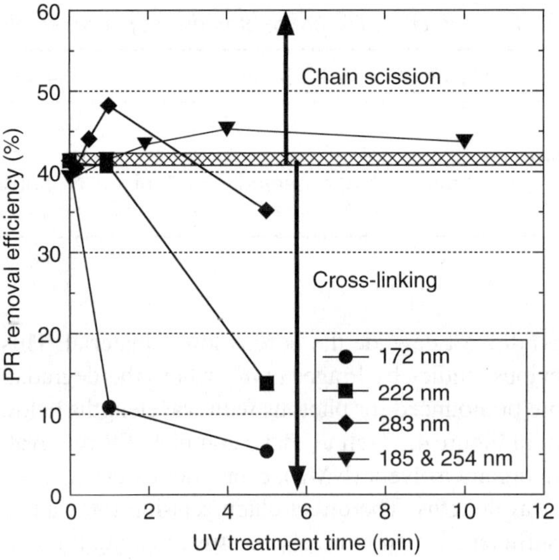


***Figure 4.27*** *Effect of UV irradiation on removal of post-etch PR by subsequent NMP immersion (2 min at 60 °C)*

283 nm were more efficient in terms of generation of C=C bonds. Upon UV irradiations for 60 s, the highest concentration of C=C bonds was reached for 222 and 283 nm. For irradiation times >60 s a continued increase in C=C bond formation was observed for the dual-wavelength UV lamp, while a decrease is observed for $\lambda = 222$ and 283 nm.

Main-chain scission was evaluated by dissolution tests performed in a good organic solvent. Figure 4.27 shows the changes induced by different UV treatments on the removal efficiency of a PR film by subsequent immersion in NMP at 60 °C for 2 min. Treatments by 172 and 222 nm UV sources did not result in a positive effect on PR removal by organic solvent, implying the formation of more cross-linked material, even for a short treatment time of 1 min. In contrast, UV irradiation at 283 nm for 30 and 60 s resulted in a slight improvement in PR removal efficiency, suggesting that direct C–C chain scission also took place, which in turn rendered it more soluble in the solvent [70]. In general, for wavelengths 172, 222 and 283 nm, the PR removal efficiency by dissolution decreased for long UV exposure times. This means that cross-linking was the dominant process under these conditions (exposure times >60 s) [108]. For all UV exposure times with the dual wavelength source, the PR removal efficiency was only slightly improved with respect to the non-UV-treated sample. There was an apparent competition between chain scission and cross-linking. Cross-linking seemed to dominate at shorter wavelengths, i.e. 172 nm and, to some extent, 222 nm. This observation might also explain the nearly neutral effect of the Hg vapor lamp, which generates both a longer (254 nm) and a shorter (185 nm) wavelength band [106]. Actually, tests performed with a 254 nm lamp showed mainly improvement in dissolution from chain scission and little C=C bond formation [107].

A summary of the effects of UV pre-treatment on plasma-treated PR with respect to chain scission, C=C formation, cross-linking and low-*k* compatibility is given in Table 4.2. The compatibility study showed that UV pre-treatments at longer wavelengths (222, 254

***Table 4.2*** *Summary of the effects of UV pre-treatments on plasma-treated PR* [104–107]

| Wavelength | 172 nm | 185 + 254 nm | 222 nm | 254 nm | 283 nm |
|---|---|---|---|---|---|
| Chain scission | – | – | – | ++ | + |
| C=C formation (1 min) | + | ++ | +++ | + | +++ |
| Cross-linking | < 1 min | Not observed | ≥ 1 min | Not observed | ≥ 1 min |
| Low-*k* compatibility* | – | – | + | + | + |

* $\Delta k \leq 0.1$ by Hg probe.

and 283 nm photons) did not degrade the porous low-*k* material. This result was in good agreement with previous studies by Prager *et al*., where the degradation of porous low-*k* was found to be more pronounced for photons with wavelengths below 200 nm [109].

The results shown in Figure 4.27 tell us that a complete PR removal was not possible by simple immersion in organic solvent (NMP), even after extended UV treatments for 5 and 10 min at different wavelengths. The direct chain scission rate caused by UV irradiation was obviously not sufficiently high to break the crust, making it more soluble in a solvent like NMP. Also, we have shown in Figure 4.22 that $O_3$/water alone was not able to completely remove PR and BARC on patterned SD structures, even not after extended exposure of 10 min. However, complete removal was achieved using different UV pre-treatments followed by different $O_3/H_2O$ strip processes, and a final solvent rinse. This is illustrated in Figure 4.28 for UV irradiation at 222 nm and an $O_3$/water vapor wet strip (thin boundary layer process as in Reference [100]). The SEM images in Figure 4.28 show a single damascene low-*k* structure after different process steps. An exposure to $O_3/H_2O$ vapor at 90 °C for 5 min followed by a rinse in organic solvent apparently removed most of the PR but not the BARC layer on top of the TiN hard mask (Figure 4.28 (b)). A pre-treatment using 222 nm UV for 1 min prior to exposure to $O_3$/water vapor for 5 min at 90 °C followed by a rinse in water led to a similar situation where at least the BARC layer still remained on the surface (Figure 4.28 (c)). In contrast, a complete removal of both PR and BARC layers was obtained if a rinse in an organic solvent (PC) was added before the water rinse (Figure 4.28 (d)). The observed complete removal of the PR layer when combining the UV treatment of 222 nm photons and exposure to an oxidizer such as $O_3$ was in turn supportive of the C=C formation due to UV treatment and consistent with the FTIR results shown in Figure 4.26. The effect of UV treatment on BARC is believed to be due to a direct scission of the BARC molecule backbone [110]. Similar results were obtained for UV irradiation at 254 nm and an $O_3$/water immersion wet strip [107].

While the used UV treatments showed good compatibility with this $k = 2.5$ low-*k* material ($\Delta k \leq 0.05$), the $O_3/H_2O$ processes were not fully compatible. The *k*-value of the low-*k* film showed an increase of about 0.3 after being processed for 5 min with the sole $O_3/H_2O$ vapor and liquid immersion wet strips. However, values obtained for the combined UV at 254 nm and liquid immersion strip showed better compatibility, with $\Delta k = 0.04$ and 0.14 for 2 and 5 min $O_3/H_2O$, respectively. This indicated some protective effect of this UV treatment towards low-*k* damage by wet ozone, which would deserve further study. Finally, a separate study with acetic acid as a radical scavenger during $O_3/H_2O$ showed no adverse effect on wet strip efficiency, as the cleavage of C=C bonds by $O_3$ is not a radical process,

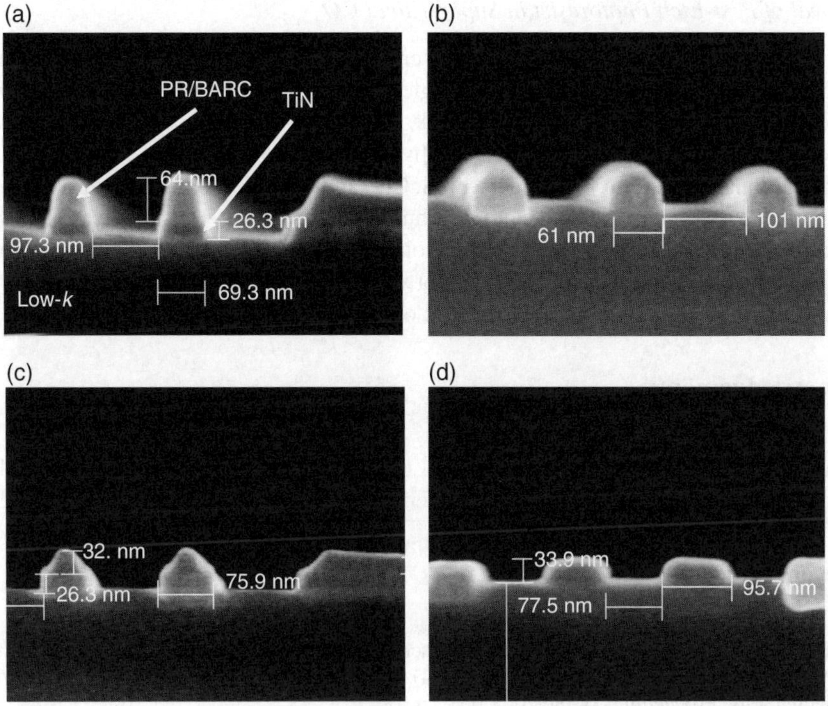


***Figure 4.28*** *SEM images of single damascene low-k structure: (a) after BARC and TiN etch, (b) with subsequent exposure to $O_3/H_2O$ vapor for 5 min and a rinse in organic solvent (PC), (c) 222 nm UV/1 min + $O_3/H_2O$/5 min, (d) 222 nm UV/1 min + $O_3/H_2O$/5 min + PC rinse/1 min* [104]. *Reprinted from Q.T. Le, E. Kesters, L. Prager, M. Lux, P. Marsik and G. Vereecke, MRS Proceedings 1156, pp. D02–09, Copyright 2009, with permission from Cambridge University Press*

while allowing decreasing low-*k* damage [103]. It opens further possibilities to improve the compatibility of the $O_3/H_2O$ strip.

In summary, UV irradiation at certain wavelengths can be applied as a pre-treatment to enhance PR and BARC removal capability using selective oxidizing chemistries. The major benefits of UV treatments for the PR wet strip were found to be the formation of C=C bonds, which can be used as reaction centers toward an oxidizer, and direct chain scission. Experimental results showed that, after a UV treatment (e.g. 1 min) followed by $O_3/H_2O$ immersion (e.g. 5 min), a complete removal of post-etch 193 nm PR and BARC for a damascene low-*k* stack was only achieved with a rinse using an organic solvent, indicating that reaction fragments were still too large to be soluble in water. Recent efforts have led to the development of a complete aqueous process for removal of PR and BARC layers on TiN-containing stacks, by performing UV irradiations under $O_2$. Preliminary results show good compatibility with porous SiOC low-*k* dielectrics and complete aqueous removal on SD structures [111].

*Removal of Post-Etch Photoresist in Supercritical $CO_2$*

Cleaning using various co-solvents in near-critical and supercritical $CO_2$ ($ScCO_2$) media [52, 53, 84, 85] that have a zero wetting angle was demonstrated to be efficient to remove post-etch PR and residues in a low-*k* stack. The cosolvents selected for use in $ScCO_2$ include dimethyl sulfoxide (DMSO), *N*-methyl-2-pyrrolidone (NMP), acetyl acetone, propylene carbonate (PC), etc. Cleaning with $ScCO_2$ and cosolvents appeared promising and was thought to become a major cleaning technology since it was considered as an environmentally friendly and stand-alone application that would allow a substantial increase in throughputs. However, it turned out to be not very cost effective and required a rather long processing time for removal of post-etch PR on patterned structures.

## Acknowledgements

The authors wish to thank Martine Claes and Lutz Prager for valuable discussion, and Marcel Lux for assistance in FTIR and SEM characterization.

## References

[1] Baklanov, M.R., Le, Q.T., Kesters, E., Iacopi, F., Van Aelst, J., Struyf, H., Boullart, W., Vanhaelemeersch, S. and K. Maex, K. (2004) Challenges of clean/strip processing for Cu/low-*k* technology. Proceedings of the 2004 IITC, p. 187.

[2] Maex, K., Baklanov, M.R., Shamiryan, D., Iacopi, F., Brongersma, S.H. and Yanovitskaya, Y.S. (2003) Low dielectric constant materials for microelectronics. *J. Appl. Phys.*, **93**, 8793.

[3] Iler, R.K. (1979) *The Chemistry of Silica*, John Wiley & Sons, Ltd.

[4] M. Prigogine, M. and J. J. Fripiat, J.J. (1979) La réactivité des surfaces de silice. *J. Chim. Phys.*, **76**, 26.

[5] Born, H.H. and Prigogine, M. (1979) The rate equation for the dissolution of silica in hydrochloric–hydrofluoric acid mixture. *J. Chim. Phys.*, **76**, 538.

[6] Van Muylader, J. *et al.* (1974) Silicon, in *Atlas of Electrochemical Equilibria in Aqueous Solutions* (ed. M. Pourbaix), National Association of Corrosion Engineers, Houston, Texas.

[7] Knotter, D.M. (2000) Etching mechanism of vitreous silicon dioxide in HF-based solutions. *J. Am. Chem. Soc.*, **122**, 4345.

[8] Kesters, E.,Le, Q.T., Baklanov, M.R., Boullart, W. and Mertens, P.W. (2005) Effect of chemical solution on the stability of low-*k* films. *Solid State Phenom.*, **103–104**, 349.

[9] Rebiscoul, D., Puyrenier, B., Broussous, L., Louis, D., and G. Passemard, G. (2006) Study of the post-etch cleaning compatibility with dense and porous ULK materials – characterization of the process impact. *Microelectron. Eng.*, **83**, 2319.

[10] Le, Q.T., Baklanov, M.R., Kesters, E., Azioune, A., Struyf, H., Boullart, W., Pireaux, J.-J. and Vanhaelemeersch, S. (2005) Removal of plasma-modified low-*k* layer using dilute HF: influence of concentration. *Electrochem. Solid-State Lett.*, 8, F21.

[11] Broussous, L., Puyrenier, W., Rebiscoul, D., Rouessac, V. and Ayral, A. (2008) Post-etch cleaning for porous low-*k* integration: impact of HF wet etch on 'pore-sealing' and '*k* recovery'. Proceedings of the IITC 2008, p. 87.

[12] Louveau, O., Louis, D., Assous, M., Blanc, R., Brun, P., Lamy, S. and Lajoinie, E. (2002) Challenge of ashing and cleaning on SiOC–H dielectric: characterization and main issues. *Microelectron. Eng.*, **61**, 867.

[13] Louis, D., Assous, M., Blanc, R., Brun, P., Arvet, C., Lajoinie, E. and Holmes, D. (2001) Wet or plasma clean? Competing or complementary? A dual damascene study with SiOCH dielectric on copper wiring. Proceedings of the 2001 IITC, p. 289.

[14] Le, Q.T., Van Olmen, J., Vanderheyden, R., Kesters, E., Kenis, K., Conard, T., Boullart, W., Baklanov, M.R. and Vanhaelemeersch, S. (2005) Efficient cleaning of low-*k* materials using single-wafer cleaning solution: a compatibility study and electrical characterization. *ECS Trans.*, **1** (3), 164.

[15] Broussous, L., Hinsinger, O., Favier, S. and Besson, P. (2003) Post-etch cleaning chemistries evaluation for low *k*-copper integration. *Solid State Phenom.*, **92**, 263.

[16] Broussous, L., Courouble, K., Richard, E., Pernel, C., Loup, V. and Lévy, D. (2010) 'Damage free' cleaning for advanced BEOL interconnections. UCPSS 2010 Abstract Book, p. 52.

[17] Chang, T.C., Liu, P.T., Mor, Y.S., Tsai, T.M., Chen, C.W., Mei, Y.J., Pan, F.M., Wu, W.F. and Sze, S.M. (2002) Eliminating dielectric degradation of low-*k* organosilicate glass by trimethylchlorosilane treatment. *J. Vac. Sci. Technol. B*, **20**, 1561.

[18] Worsley, M.A., Bent, S. F., Gates, S.M., Fuller, N.C.M., Volksen, W., Steen, M. and Dalton, T. (2005) Effect of plasma interactions with low-*k* films as a function of porosity, plasma chemistry, and temperature. *J. Vac. Sci. Technol. B*, **23**, 395.

[19] Darnon, M., Chevolleau, T., Posseme, N., David, T., Licitra, C., Torres, J. and Joubert, O. (2007) Porous SiOCH modifications by $O_2$, $NH_3$ and $CH_4$ plasmas used as ashing and pore sealing steps. Presentation at PESM 2007, Leuven, Belgium, 10–11 September 2007.

[20] Abe, H., Yoneda, M. and Fijiwara, N. (2008) Developments of plasma etching technology for fabricating semiconductor devices. *Jpn. J. Appl. Phys.*, **47**, 1435.

[21] Kerr, J.A. (1984) Strengths of chemical bonds, in *Handbook of Chemistry and Physics* (eds R.C. Weast, M.J. Astle and W.H. Beyer), CRC Press, Inc., pp. F176–F181.

[22] R. Small, R. (2010) Challenges of BEOL wafer cleaning. 2010 SPCC Proceedings.

[23] Le, Q.T., Baklanov, M.R., Kesters, E., Han, Q., Luo, S., Waldfried, C.,and Vanhaelemeersch, S. (2006) Damage of porous low-*k* material: evaluation methodology and means to minimize damage. Proceedings of the International Semiconductor Technology Conference, p. 36.

[24] Mor, Y.S., Chang, T.C., Liu, P.T., Tsai, T.M., Chen, C.W., Yan, S.T., Chu, C.J., Wu, W.F., Pan, F.M., Lur, W. and Sze, S.M. (2002) Effective repair to ultra-low-*k* dielectric material ($k \sim 2.0$) by hexamethyldisilazane treatment. *J. Vac. Sci. Technol. B*, **20**, 1334.

[25] B. Xie, B. and Muscat, A.J. (2004) Repair of porous methylsilsesquioxane films using supercritical carbon dioxide. *MRS Proc.*, **812**, F1.4.

[26] Chang, T.C., Mor, Y.S., Liu, P.T., Tsai, T.M., Chen, C.W., Mei, Y.J. and Sze, S.M. (2002) Recovering dielectric loss of low dielectric constant organic siloxane during the photoresist removal process. *J. Electrochem. Soc.*, **149**, F81.

[27] Liu, J., Kim, W., Bao, J., Shi, H., Baek, W. and Ho, P.S. (2007) Restoration and pore sealing of plasma damaged porous organosilicate low-*k* dielectrics with phenyl containing agents. *J. Vac. Sci. Technol. B*, **25**, 906.

[28] Grill, A. and Patel, V. (2006) The effect of plasma chemistry on the damage induced to porous SiCOH dielectrics. *J. Electrochem. Soc.*, **153**, F169.

[29] Sugahara, S., Kadoya, T., Usami, K.-I., Hattori, T. and Matsumura, M. (2001) Preparation and characterization of low-*k* silica film incorporated with methylene groups. *J. Electrochem. Soc.*, **148**, F120.

[30] Rogojevic, S., Jain, A., Gill, W.N. and Plawsky, J. (2002) Moisture adsorption in nanoporous silica xerogels. *Electrochem. Solid-State Lett.*, **5**, F22.

[31] Baklanov, M.R., Mogilnikov, K.P. and Le, Q.T. (2006) Quantification of processing damage in porous low dielectric constant films. *Microelectron. Eng.*, **83**, 2287.

[32] Furukawa, Y., Patz, M., Kokubo, T. and Snijders, J.H.M. (2003) Material modification of the patterned wafer during dry etching and strip determined by XPS. *Microelectron. Eng.*, **70**, 267.

[33] Demuynck, S., Kim, H., Huffman, C., Darnon, M., Struyf, H., Versluijs, J., Claes, M., Vereecke, G., Verdonck, P., Volders, H., Heylen, N., Kellens, K., De Roest, D., Sprey, H.,and Beyer, G. (2009) Dielectric reliability of 50 nm half pitch structures in Aurora® LK. *Jpn. J. Appl. Phys.*, **48**, 04C018.

[34] Le, Q.T., Claes, M., Conard, T., Kesters, E., Lux, M. and Vereecke, G. (2009) Removal of post-etch photoresist and sidewall residues using organic solvent and additive combined with physical forces. *Microelectron. Eng.*, **86**, 181.

[35] ITRS 2009.
[36] Waldfried, C., Escorcia, O., Han, Q. and Smith, P.B. (2003) Cleaning solutions for p-MSQ low-*k* device applications. *Electrochem. Solid-State Lett.*, **6**, G137.
[37] Kondoh, E., Asano, T., Nakashima, A. and Komatu, M. (2000) Effect of oxygen plasma exposure of porous spin-on-glass films. *J. Vac. Sci. Technol. B*, **18**, 1276.
[38] Liu, P.T., Chang, T.C., Su, H., Mor, Y.S., Yang, Y.L., Chung, H., Hou, J. and Sze, S.M. (2001) Improvement in integration issues for organic low-*k* hybrid-organic-siloxane-polymer. *J. Electrochem. Soc.*, **148**, F30.
[39] Gorman, B.P., Orozco-Teran, R.A., Zhang, Z., Matz, P.D., Mueller, D.W. and Reidy, R.F. (2004) Rapid repair of plasma ash damage in low-*k* dielectrics using supercritical $CO_2$. *J. Vac. Sci. Technol. B*, **22**, 1210.
[40] Moore, D.L., Carter, R.J., Cui, H., Burke, P., Gu, S.Q., Peng, H., Valley, R.S., Gidley, D.W., Waldfried, C. and Escorsia, O. (2005) Damage of low-*k* and ultralow-*k* dielectrics from reductive plasma discharges used for photoresist removal. *J. Electrochem. Soc.*, **152**, G528.
[41] Dalton, T.J., Fuller, N., Tweedie, C., Dunn, D., Labelle, C., Gates, S., Colburn, M., Chen, S.T., Tai, L., Dellaguardia, R., Petrarca, K., Dziobkowski, C., Kumar, K. and Siddiqui, S. (2004) Ash-induced modification of porous and dense SiCOH inter-level-dielectric (ILD) materials during damascene plasma processing. Proceedings of the IITC, p. 154.
[42] Yonekura, K., Sakamori, S., Goto, K., Matsuura, M., Fujiwara, N. and Yoneda, M. (2004) Investigation of ash damage to ultralow-*k* inorganic materials. J. Vac. Sci. Technol. B, **22**, 548.
[43] Lin, M.C., Wang, M.Q., Lai, J., Huang, R., Weng, C.M., Liao, J.H., Tang, J., Weng, C.H., Lu, W., Chen, H.W. and Lee, J.T.C. (2008) Metal hard mask employed Cu/low-*k* film post ash and wet clean process optimization and integration into 65 nm manufacturing flow. *Solid State Phenom.*, **134**, 359.
[44] Timmons, C.L. and Hess, D.W. (2004) Electrochemical cleaning of post-plasma etch fluorocarbon residues using reductive radical anion chemistry. *Electrochem. Solid-State Lett.*, **7**, G302.
[45] Timmons, C.L. and Hess, D.W. (2008) Photoresist and fluorocarbon postplasma etch residue removal using electrochemically generated radical anions. *J. Electrochem. Soc.*, **155**, H771.
[46] Song, I., Timmons, C.L., Levitin, G. and Hess, D.W. (2009) Photoresist removal using alternative chemistries and pressures. *Solid State Phenom.*, **145–146**, 303.
[47] Mertens, P.W., Kim, T.-G., Claes, M., Le, Q.T., Vereecke, G., Kesters, E., Suhard, S., Pacco, A., Lux, M., Kenis, K., Urbanowicz, A., Tokei, Z., Beyer, G. (2009) Challenges and novel approaches for photo resist removal and post-etch residue removal for 22 nm interconnects. IITC 2009 Proceedings, p. 237.
[48] Darnon, M., Chevolleau, T., Eon, D., Bouyssou, R., Pelissier, B., Vallier, L., Joubert, O., Posseme, N., David, T., Bailly, F. and Torres, J. (2008) Patterning of narrow porous SiOCH trenches using a TiN hard mask. *Microelectron. Eng.*, **85**, 2226.
[49] Le, Q.T., Drieskens, F., de Marneffe, J.F., Conard, T., Lux, M., Struyf, H. and Vereecke, G. (2010) Modification of post-etch residues by UV for wet removal. UCPSS 2010 Abstract Book, p. 50.
[50] Le, Q.T., de Marneffe, J.-F., Conard, T., Vaesen, I., Struyf, H. and Vereecke, G. (2011) Effect of UV Irradiation on Modification and Subsequent Wet Removal of Model and Post-Etch Fluorocarbon Residues. *J. Electrochem. Soc.*, submitted.
[51] Le, Q.T., Jeannot, V., Baklanov, M.R., Vanderheyden, R., Boullart, W. and Vanhaelemeersch, S. (2006) Effect of surfactant as additive in wet clean solutions on properties of low-*k* materials. *Electrochem. Solid-State Lett.*, **9**, F17.
[52] Biberger, M.A., Schilling, P., Frye, D. and Mills, M.E. (2000) *Semiconductor Fabtech*, 12th edn, Henley Publishing Ltd, London, p. 239.
[53] Mullee, W.H., Biberger, M.A. and Schilling, P. (2001) *Removal of photoresist and residue from substrate using supercritical carbon dioxide process related applications*. WO Patent 2001/033613.
[54] Satta, A., Shamiryan, D., Baklanov, M.R., Whelan, C.M., Le, Q.T., Beyer, G.P., Vantomme, A. and Maex, K. (2003) The removal of copper oxides by ethyl alcohol monitored *in situ* by spectroscopic ellipsometry. *J. Electrochem. Soc.*, **150**, G300.

[55] Baklanov, M.R., Shamiryan, D.G., Tökei, Zs., Beyer, G.P., Conard, T., Vanhaelemeersch, S. and Maex, K. (2001) Characterization of Cu surface cleaning by hydrogen plasma. *J. Vac. Sci. Technol. B*, **19**, 1201.

[56] Sekiguchi, A., Kobayashi, A., Koide, T., Okada, O. and Hosokawa, N. (2000) Reaction of copper oxide and $\beta$-diketone for *in situ* cleaning of metal copper in a copper chemical vapor deposition reactor. *Jpn. J. Appl. Phys.*, **39**, 6478.

[57] Ueno, K., Donnelly, V.M. and Tsuchiya, Y. (1998) Cleaning of $CHF_3$ plasma-etched $SiO_2$/SiN/Cu via structures using a hydrogen plasma, an oxygen plasma, and hexafluoroacetylacetone vapors. *J. Vac. Sci. Technol. B*, **16**, 2986.

[58] Danel, A., Millet, C., Perrut, V., Daviot, J., Jousseaume, V., Louveau, O. and Louis, D. (2003) Supercritical $CO_2$ for ULK/Cu integration. Proceedings of the 2003 IITC, p. 248.

[59] Ishikawa, K., Yagishita, T. and Nakamura, M. (2003) Vapor treatment of copper surface using organic acids. *MRS Proc.*, **766**, E3.28.

[60] Kudo, H., Ishikawa, K., Nakaishi, M., Tsukune, A., Ozaki, S., Nakata, Y., Akiyama, S., Mizushima, Y., Hayashi, M., Akbar, A.A., Kouno, T., Iwata, H., Iba, Y., Ohba, T., Futatsugi, T., Nakamura, T. and Sugii, T. (2008) Enhancing yield and reliability by applying dry organic acid vapor cleaning to copper contact via-bottom for 32-nm nodes and beyond. Proceedings of the IITC 2008, p. 93.

[61] Gokan, H., Esho, S. and Ohnishi, Y. (1983) Dry etch resistance of organic materials. *J. Electrochem. Soc.*, **130**, 143.

[62] Sarrubi, T.R., Ross, M.F., Neisser, M., Kocab, T., Beauchemin Jr, B.T., Livesay, W.R., Wong, S.S. and Ng, W. (2001) Mechanism studies of scanning electron microscope measurement effects on 193-nm photoresists and the development of improved line-width measurement methods. *Proc. SPIE*, **4345**, 211.

[63] Kim, M.S., Park, J.W., Kim, H.J., Jun, B.J., Gil, M.G., Kim, B.H., Ross, M.F. and Livesay, W.R. (2001) Investigation of electron beam stabilization of 193-nm photoresists. *Proc. SPIE*, **4345**, 737.

[64] Kim, M.C., Shamiryan, D., Jung, Y., Boullart, W., Kang, C.J. and Cho, H.K. (2006) Effects of various plasma pretreatments on 193 nm photoresist and linewidth roughness after etching. *J. Vac. Sci. Technol. B*, **24**, 2645.

[65] Sumiya, M., Bruce, R., Engelmann, S., Weilnboeck, F. and Oehrlein, G.S. (2008) Study of 193 nm photoresist degradation during short time fluorocarbon plasma exposure. I. Studies of modified layer formation. *J. Vac. Sci. Technol. B*, **26**, 1637.

[66] Sumiya, M., Bruce, R., Engelmann, S., Weilnboeck, F. and Oehrlein, G.S. (2008) Study of 193 nm photoresist degradation during short time fluorocarbon plasma exposures. II. Plasma parameter trends for photoresist degradation. *J. Vac. Sci. Technol. B*, **26**, 1647.

[67] Kesters, E., Claes, M., Le, Q.T., Lux, M., Franquet, A., Vereecke, G., Mertens, P.W., Frank, M.M., Carleer, R., Adriaensens, P., Biebuyck, J.J. and Bebelman, S. (2008) Chemical and structural modifications in a 193-nm photoresist after low-*k* dry etch. *Thin Solid Films*, **516**, 3454.

[68] Seki, S., Kawachi, S., Tagawa, S., Egami, A., Kurihara, K., Nakamura, M. and Itoh, H. (2005) Etching efficiency of polymethacrylates for monochromatic low energy ion beams. *J. Photopolym. Sci. Technol.*, **18**, 255.

[69] Dong, L., Hill, D.J.T., O'Donnell, J.H., Pomery, P.J. and Hatada, K. (1996) Effects of various ester groups in radiation degradation of syndiotactic poly(methacrylate)s. *J. Appl. Polym. Sci.*, **59**, 589.

[70] Torikai, A. and Mitsuoka, T. (1995) Electron spin resonance studies of poly(methyl methacrylate) irrdiated with monochromatic light. *J. Appl. Polym. Sci.*, **55**, 1703.

[71] Pargon, E., Menguelti, K., Martin, M., Bazin, A., Chaix-Pluchery, O., Sourd, C., Derrough, S., Lill, T. and Joubert, O. (2009) Mechanisms involved in HBr and Ar cure plasma treatments applied to 193 nm photoresists. *J. Appl. Phys.*, **105**, 094902.

[72] Azarnouche, L., Pargon, E., Menguelti, K., Fouchier, M., Tiron, R., Gouraud, P., Verove, C. and Joubert, O. (2009) Impact of plasma and annealing treatments on 193 nm photoresist line width roughness and profile. 56th AVS Presentation.

[73] Nest, D., Graves, D.B., Engelmann, S., Bruce, R.L., Weilnboeck, F., Oehrlein, G.S., Andes, C. and Hudson, E.A. (2008) Synergistic effects of vacuum ultraviolet radiation, ion bombardment, and heating in 193 nm photoresist roughening and degradation. *Appl. Phys. Lett.*, **92**, 153113.

[74] Titus, M.J., Nest, D.G., Chung, T.-Y. and Graves, D.B. (2009) Comparing 193 nm photoresist roughening in an inductively coupled plasma system and vacuum beam system. *J. Phys. D: Appl. Phys.*, **42**, 245205.

[75] Henzi, P., Bade, K., Rabus, D.G. and Mohr, J. (2006) Modification of polymethylmethacrylate by deep ultraviolet radiation and bromination for photonic applications. *J. Vac. Sci. Technol. B*, **24**, 1755.

[76] Choi, J.O., Moore, J.A., Corelli, J.C., Silverman, J.P. and Bakhru, H. (1988) Degradation of poly(methylmethacrylate) by deep ultraviolet, X-ray, electron beam, and proton beam irradiations. *J. Vac. Sci. Technol. B*, **6**, 2286.

[77] Ruck, D.M., Schulz, J. and Deusch, N. (1997) Ion irradiation induced chemical changes of polymers used for optical applications. *Nucl. Instr. Methods Phys. Res. B*, **131**, 149.

[78] Waali, E.E., Scott, J.D., Klopf, J.M., Vladimirsky, Y. and Vladimirsky, O. (1997) One- and two-dimensional nuclear magnetic resonance spectra of X-ray-degraded poly(methyl methacrylate). *Macromolecules*, **30**, 2386.

[79] Shamiryan, D., Baklanov, M.R., Vanhaelemeersch, S. and Maex, K. (2002) Comparative study of SiOCH low-*k* films with varied porosity interacting with etching and cleaning plasma. *J. Vac. Sci. Technol. B*, **20**, 1923.

[80] Le, Q.T., Keldermans, J., Chiodarelli, N., Kesters, E., Lux, M., Claes, M. and Vereecke, G. (2008) Alternative photoresist removal process to minimize damage of low-*k* material induced by ash plasma. *Jpn. J. Appl. Phys.*, **47**, 6870.

[81] Miller-Chou, B.A. and Koenig, J.L. (2003) A review of polymer dissolution. *Prog. Polym. Sci.*, **28**, 1223.

[82] Levitin, G., Timmons, C. and Hess, D.W. (2006) Photoresist and etch residue removal. *J. Electrochem. Soc.*, **153**, G712.

[83] Spelt, J.K. and Neumann, A.W. (1987) Solid surface tension: the equation of state approach and the theory of surface tension components. Theoretical and conceptual considerations. *Langmuir*, **3**, 588.

[84] Myneni, S. and Hess, D.W. (2003) Post-plasma-etch residue removal using $CO_2$-based fluids. *J. Electrochem. Soc.*, **150**, G744.

[85] Levitin, G., Myneni, S. and Hess, D.W. (2004) Post plasma etch residue removal using $CO_2$–$TMAHCO_3$ mixtures: comparison of single-phase and two-phase mixtures. *J. Electrochem. Soc.*, **151**, G380.

[86] Hansen, C.M. (2000) *Hansen Solubility Parameters – A User's Handbook*, CRC Press, Boca Raton, Florida.

[87] Hansen, C.M. (2004) 50 Years with solubility parameters—past and future. *Prog. Org. Coat.*, **51**, 77.

[88] Le, Q.T., Chiodarelli, N., Blum, I., Kesters, E., Lux, M., Claes, M., Vereecke, G. and Mertens, P.W. (2007) Cleaning of post-etch photoresist layer on patterned surface using organic solvent combined with physical forces. Proceedings of Surface Preparation and Cleaning Conference, Austin, Texas, 25–26 April 2007.

[89] Claes, M., Le, Q.T., Kesters, E., Lux, M., Urionabarrenetxea, A., Vereecke, G., Mertens, P.W., Carleer, R. and Adriaensens, P. (2007) All-wet strip approaches for post-etch photoresist layers after low-*k* patterning. *ECS Trans.*, **11**, 177.

[90] S. Ojima, S., Jizaimaru, T., Omae, S. and Ohmi, T. (1997) Room temperature photoresist stripper. *J. Electrochem. Soc.*, **144**, 4005.

[91] Klipp, A., Vereecke, G., Le, Q.T. and Claes, M. (2009) New cleaning solutions for all-wet-post etch residue removal on Cu BEOL for sub 45nm nodes. Proceedings of Sematech Surface Preparation and Cleaning Conference, Austin, Texas, 23–25 March 2009.

[92] Wu, A., Baryschpolec, E., Rao, M., Schaller, M., Bartsch, C., Leppack, S. and Ott, A. (2009) Development of compatible wet clean stripper for integration of CoWP metal cap in Cu/low-*k* interconnects. Proceedings of Sematech Surface Preparation and Cleaning Conference, Austin, Texas, 23–25 March 2009.

[93] Rector, A., Pfettscher, D., Pollard, K.D., Scheele, D. and Franklin, C. (2009) Preserving the integrity of low-*k* materials with all wet. Proceedings of Sematech Surface Preparation and Cleaning Conference, Austin, Texas, 23–25 March 2009.
[94] Le, Q.T., Klipp, A., Lux, M., Li, Y., Zhao, L. and Vereecke, G. (2009) Removal of photoresist and BARC in Cu BEOL using an all-wet process. *ECS Trans.*, **25** (5), 173.
[95] Franklin, C., Rector, A., Pfettscher, D., Pollard, K.D. and Scheele, D. (2009) Using a novel solvent-based technology to preserve low-*k* material when removing gap fill material. *ECS Trans.*, **25** (5), 117.
[96] Le, Q.T., Demuynck, S., Suhard, S., Klipp, Vereecke, B. and Vereecke, G. (2010) All-wet removal of post-etch photoresist and sidewall residues: electrical characterization of 90 nm and 30 nm ½ pitch structure. Proceedings of Sematech Surface Preparation and Cleaning Conference, Austin, Texas, 22–24 March 2010.
[97] Ohmi, T., Isagawa, T., Kogure, M. and Imaoka, T. (1993) Native oxide growth and organic impurity removal on Si surface with ozone-injected ultrapure water. *J. Electrochem. Soc.*, **140**, 804.
[98] Noda, S., Kawase, K., Horibe, H., Kuzumoto, M. and Kataoka, T. (2005) Development of a method for resist removal by ozone with acetic acid vapor. *J. Electrochem. Soc.*, **152**, G73.
[99] Kashkoush, I. (1997) A novel method for photoresist stripping using ozone/de-ionized water chemistry. 1997 IEEE International Symposium on Semiconductor Manufacturing Conference Proceedings, San Francisco, California, p. 81.
[100] De Gendt, S., Wauters, J. and Heyns, M. (1998) A novel resist and post-etch residue removal process using ozonated chemistry. *Solid State Technol.*, **41**, 57.
[101] Vankerckhoven, H. (2004) PhD thesis, Katholieke Universiteit Leuven, Belgium.
[102] Muraoka, H., Muraoka, R., Sato, A. and Endo, M. (2003) *Method and apparatus for removing organic films*, US Patent 2003/0108823 A1.
[103] Le, Q.T., Lux, M., Kesters, E., Prager, L. and Vereecke, G. (2009) Effect of radical scavenger and UV irradiation on removal of photoresist and BARC using water/ozone in Cu/low-*k* interconnect. *ECS Trans.*, **25** (5), 71.
[104] Le, Q.T., Kesters, E., Prager, L., Lux, M., Marsik, P. and Vereecke, G. (2009) Application of UV irradiation in removal of post-etch 193 nm photoresist. *MRS Proc.*, **1156**, D02–09.
[105] Le, Q.T., Kesters, E., Prager, L., Claes, M., Lux, M. and Vereecke, G. (2009) Modification of photoresist by UV for post-etch wet strip applications. *Solid State Phenom.*, **145–146**, 323.
[106] Kesters, E., Le, Q.T., Lux, M., Prager, L. and Vereecke, G. (2010) Removal of post-etch 193 nm photoresist in porous low-*k* dielectric patterning using UV irradiation and ozonated water. *Microelectron. Eng.*, **87**, 1674.
[107] Kesters, E., Le, Q.T., Lux, M., Onandia, L., Baerts, C. and Vereecke, G. (2009) Influence of UV irradiation on the removal of post-etch photoresist in porous low-*k* dielectric patterning. *ECS Trans.*, **25** (5), 63.
[108] Kishimura, S., Kimura, Y., Sakai, J., Tsujita, K. and Matsui, Y. (1999) Improvement of dry etching resistance of resists by deep UV cure. *Jpn. J. Appl. Phys.*, **38**, 250.
[109] Prager, L., Marsik, P. Gerlach, J.W., Baklanov, M.R., Naumov, S., Pistol, L., Schneider, D., Wennrich, L., Verdonck, P. and Buchmeiser, M.R. (2008) Effect of pressure on efficiency of UV curing of CVD-derived low-*k* material at different wavelengths. *Microelectron. Eng.*, **85**, 2094.
[110] Kesters, E., Le, Q.T., Lux, M., Pittevils, J., Vereecke, G., Struyf, H. and De Gendt, S. (2011) Towards fully aqueous ozone wet strip of 193 nm photoresist and BARC using UV pre-treatments in BEOL low-*k* patterning applications. *J. Electrochem. Soc.*, submitted.
[111] Kesters, E., Vereecke, G., Lux, M., Pittevils, J. and Struyf, H. (2011) Aqueous photoresist and BARC wet strip in BEOL MHM patterning using UV and ozonated chemistry. 4th International Workshop on Plasma Etch and Strip in Microelectronics, PESM 2011, Mechelen, Belgium, 5–6 May 2011.

[93] Zweber, A., [illegible] (2009) [illegible] Cleaning Conference [illegible] 23–25 March 2009 [illegible]
[94] [illegible] (2009) Removal of photoresist [illegible] and BEOL [illegible] process. [illegible]
[95] Franklin, C., [illegible] (2009) [illegible] material. [illegible]
[96] [illegible] (2010) [illegible] and [illegible] Proceedings of [illegible] 22–24 March 2010 [illegible]
[97] [illegible]
[98] [illegible] (2007) [illegible]
[99] [illegible] (1999) [illegible] Semiconductor [illegible]
[100] [illegible] Solid State Technol. 41, [illegible]
[101] [illegible]
[102] [illegible] (2001) [illegible]
[103] [illegible] (2005) [illegible]
[104] [illegible]
[105] [illegible] (2007) [illegible]
[106] [illegible]
[107] [illegible] (2009) [illegible]
[108] [illegible] (1999) [illegible]
[109] [illegible] (2009) [illegible] 204.
[110] [illegible] (2010) [illegible]
[111] [illegible] (2011) [illegible] Workshop on [illegible] May 2011.

# Section II

## Conductive Layers and Barriers

# 5

# Copper Electroplating for On-Chip Metallization

*Valery M. Dubin*

*NANO3D SYSTEMS LLC, Portland, Oregon, USA*

## 5.1 Introduction

Copper damascene metallization was introduced by IBM in 1997 to increase the electromigration resistance and decrease the resistance of interconnects [1]. Electroplated copper was recently extended by IBM to contact level, replacing tungsten at 22 nm technology nodes [2]. A novel Cu metallization scheme enabled low-resistance contacts by successfully filling the high aspect ratio contacts without voids.

Conventional dual damascene metallization consists of additives-assisted Cu electroplating on to a PVD copper seed. On account of the PVD seed deposition process, the seed layer exhibits a considerable overhang at the feature opening. Quantitatively, the seed overhang can be as high as 25 % of the seed thickness in the field of the wafer, thereby considerably reducing the opening available for the electroplating operation. A scaling estimate of the aspect ratios incoming to the electroplating step is shown in Figure 5.1, which indicates that for sub-50 nm features the aspect ratios that electroplating must fill can be as high as 20:1. Besides, the time to fill the sub-20 nm features becomes very short (< 1 s) and is comparable to the wafer immersion time into the plating solution. Therefore, plating additives are required to have high bottom-up fill and fast adsorption rates [3, 4].

*Advanced Interconnects for ULSI Technology*, First Edition. Edited by Mikhail R. Baklanov, Paul S. Ho and Ehrenfried Zschech.

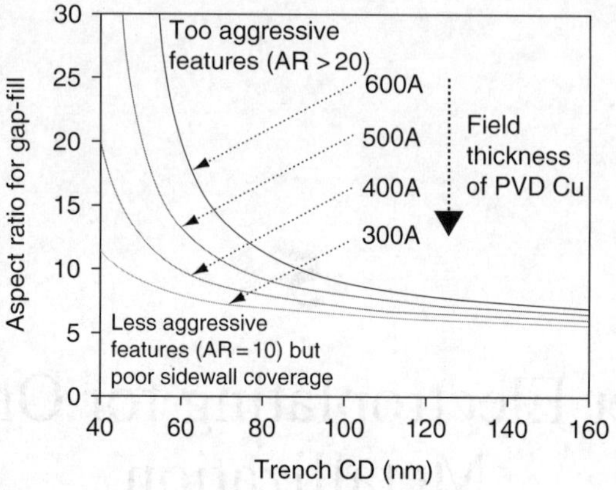


***Figure 5.1*** *Scaling estimate of the via/trench aspect ratios after seed layer deposition*

## 5.2 Copper Electroplating Techniques

Copper electroplating solutions contain Cu salt (such as copper sulfate with $Cu^{2+} > 10$ g/l), acid (like $H_2SO_4 < 100$ g/l), halide (e.g. $Cl^- > 10$ ppm) and proprietary organic additives. The use of organic additives and chloride ions in acid copper plating solution enables the void-free copper deposition. Based on their function, the additives can be broadly classified into two types: accelerator and suppressors. The accelerator additives typically belong to the mercapto family, while the suppressor type additives typically belong to the polyglycol family. The glycols behave as suppressors only in the presence of chloride ions. Chloride ions on their own act as the mild suppressants for Cu deposition. Some commercially available chemistries have a third additive, called leveler, that is also a suppressor but often has a nitrogen-containing functional group.

D.C. deposition in a sulfuric acid-based electrolyte is widely used for fabricating damascene Cu interconnections in a three-component additive plating system containing accelerator, suppressor and leveler. The improvement of gap filling performance and film properties such as the reduction of the film roughness, the formation of uniform grains with twin boundaries and the increase of the mechanical strength without decreasing the ductility were reported in work [5–12] using pulse deposition. Under optimum conditions, pulse deposition led to a 50 % enhancement in Cu (111) intensity and a 30 % reduction in resistivity compared to the constant potential deposition [13]. Moreover, pulsating the current (or voltage) permits the plating process to be operated at a higher average current density than d.c. plating without the formation of dendrites as well as defect-free gap filling of damascene structures in leveler-free electrolyte [7]. The use of pulse plating also allows control of the topography of the Cu deposits [14].

Organic additives are consumed in the plating bath. Therefore, the additives need to be analyzed and replenished. Methods that have been used or considered for automated plating-bath analysis include titration (potentiometric, colorimetric or pH), HPLC (high-performance liquid chromatography), X-ray fluorescence (XRF), photometry, pH monitoring, conductivity monitoring, oxidation–reduction potential monitoring (ORP) and mass spectrometry [15]. The replenishment systems typically utilize a parameter-based model to set the replenishment rates or volumes and will adjust the parameters algorithmically based on analysis results.

## 5.3 Copper Electroplating Superfill

Numerous groups have tried to explain the superfill phenomena. An early model proposed the diffusion-limited consumption of suppressor, e.g. polyethylene glycol (PEG) as the primary requirement for the bottom-up fill [16]. Only a large PEG whose MW exceeds 2000 g/mol can effectively polarize the cathode, in turn inducing the catalytic effect of the bis(3-sulfopropyl) disulfide (SPS) accelerator on copper deposition, resulting in a synergistic interaction between the suppressor and accelerator on the microvia filling [17]. Another model proposed the local accumulation of accelerator at the feature bottom, where the surface area reduction during deposition induced the preferential enrichment of the accelerator additive. This mechanism provided the driving force for superfill [18]. Bottom-up fill was also explained based on the role played by the accelerator and the corresponding breakdown products formed during via filling. The breakdown of accelerator resulted in very strong accelerating species formation, such as sodium 3-mercaptopropylsulfonate (MPS), whose presence during the deposition helps in the formation of Cu(I)–thiolate that could lead to reduction of Cu ions [19].

### 5.3.1 The Role of Accelerator

Proposed surface functions for chloride, thiol and sulfonate groups in an electrodeposition bath are fourfold [20]: (1) the thiol group causes MPS to adhere to or stay in the vicinity of the metal surface, (2) the sulfonate group attracts copper ions and facilitates ion transfer to the surface, (3) $Cl^-$ attenuates MPS adhesion and attraction of copper ions to MPS and (4) $Cl^-$ reduces the surface potential. Proton and C-13 NMR spectroscopic studies of possible interactions of the copper electroplating accelerant (SPS) and its reduced form (MPS) with both cuprous chloride and cupric chloride confirmed that MPS reacted with CuCl to yield Cu(I)–thiolate products. No evidence for the formation of a solution Cu(I)–SPS complex from CuCl and SPS was found [21].

In situ scanning tunneling microscopy showed that accelerating species are responsible for disrupting and preventing the formation of the inhibiting PEG–Cl layer. When as-received SPS is added, individual SPS and MPS molecules are imaged as a mobile two-dimensional gas diffusing on the $Cl^-$ adlattice. SPS-derived accelerating species of different mobility and tunneling contrast appear with exposure time. The accelerator–$Cl^-$ surfactant accounts for many aspects of the additive function previously observed and stipulated by the curvature-enhanced accelerator model of superconformal film growth [22].

The accelerator-assisted copper deposition rate is very similar to deposition rates observed in additives-free electrolytes, indicating that the accelerator is a depolarizer. LSV of the acid copper bath (VMS) containing accelerator and suppressor additives indicates polarized kinetics at short times (forward scan in Figure 5.2) followed by 'accelerator-like'

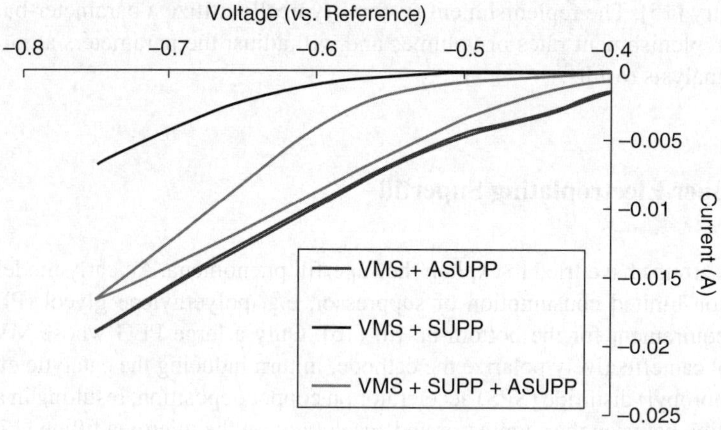


***Figure 5.2*** *Linear-sweep voltammetry in Cu electrolyte (VMS) with accelerator (ASUPP), VMS with SUPP and VMS with SUPP and ASUPP. Hysteresis in the presence of SUPP and ASUPP is indicative of the transient additives interactions*

kinetics at longer times (reverse scan in Figure 5.2). This marked hysteresis and depolarization seen from the anodic and cathodic directional sweeps indicates the interaction between the suppressor and the accelerator on the surface during the deposition reaction. LSV experimental results also confirm that transient effects involving kinetics of additive reactions (adsorption, desorption and decomposition), as was previously described [23], must be comprehended in explaining the bottom-up fill process and obtaining an accurate model.

### 5.3.2 The Role of Suppressor

It is generally accepted that the inhibition proceeds through the formation of an adsorbed complex of suppressor with Cu(I) and chloride [24]. Recently, it was shown by some of the authors that also cupric ions and protons affect inhibition by suppressor [25]. $Cl^-$ plays a central role in the binding of PEG to the surface, although the nature of interaction remains an open question as to whether or not this is mediated by $Cu^+$ or some $Cl^-$-induced alteration of interface hydration [26–33]. The halide component also serves to hinder sulfide formation kinetically with the underlying Cu metal that otherwise occurs rapidly if direct contact between Cu and the disulfide or thiol compounds is allowed. This accounts for the well-known observation that deposition in the absence of $Cl^-$ leads to brittle Cu deposits due to sulfide formation derived from thiol or disulfide additives [34].

Besides PEG, other polyether additives were compared to PEG in plating solutions of varying acidity, such as polypropylene glycol (PPG) and triblock copolymers of PEG and PPG, those with ethylene oxide terminal blocks termed EPE and propylene oxide terminals termed PEP [35]. The extent of suppression observed on a rotating disk electrode varied by over 50 mV, with PEG suppressing the least and PPG the most. When both homopolymers were present, results were very near those of PEG alone. Also, a variation

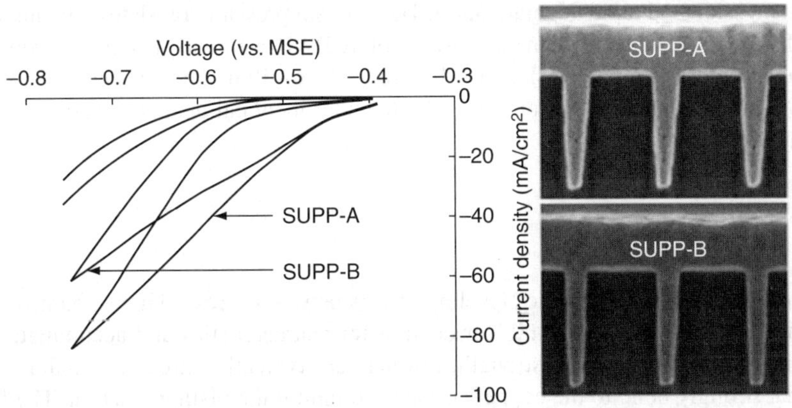


***Figure 5.3*** *LSV response indicates that SUPP-B causes stronger inhibition and slower depolarization of the copper deposition rate than SUPP-A. The suppressors were tested for their gap-fill capability in the presence of an accelerator and a leveler in the plating bath. Cross-sectional SEM indicates good gap-fill in the presence of SUPP-B and poor gap-fill (centerline void near the via bottom) in the presence of the weaker SUPP-A*

in Tafel slope was observed, with PPG exhibiting a more activated reaction. These results were interpreted by a significant difference in surface coverage between PEG and PPG. Blanket copper films plated in the presence of each of the suppressors showed significant differences in surface luster and roughness, with the copolymer EPE resulting in extremely bright and smooth deposits [36]. The superior performance of EPE was attributed to its high suppression strength, which is greater than either PEG or PPG at the relevant plating potentials. EPE performance is also improved as SPS concentration decreases [37]. Several other suppressing compounds such as polyoxyethylated naphthol and polyoxyethylene lauryl ether also exhibit superior inhibition, thus enabling the bottom-up fill of high aspect ratio features, which cannot be achieved by PEG. Plating experiments of high aspect ratio features confirm the advantage provided by this class of additives [38].

The transient effects during superfill are associated with the following process steps: (a) the instantaneous initial depletion of the PEG within the via by fast adsorption, followed by (b) the slow diffusion of the PEG from the via exterior bulk to the via bottom, the slow transport of the PEG being due to its fast adsorptive consumption on the via sidewalls, (c) adsorption of the SPS, which is not diffusion limited, on the additives-free via bottom in the absence of the diffusion-limited PEG and (d) slow displacement from the copper surface of the adsorbed PEG by the stronger bonding accelerator species. This acceleration, primarily provided by the transport–adsorption process, is further amplified by the local surface area reduction in the via bottom [38, 39].

The fast 'bottom-up' fill plating by employing strongly suppressing additives was studied [40]. Figure 5.3 shows linear sweep voltammograms (LSV) of two suppressing agents, SUPP-A and SUPP-B. The LSV response for SUPP-A indicates strong polarization at short times (forward potential scan) followed by slow depolarization by the accelerator (reverse scan). In contrast, SUPP-B was easily displaced by the accelerator, thereby causing poor suppression and rapid depolarization. The poor suppression translates into uninhibited sidewall growth during gap-fill in the presence of SUPP-B eventually leading to 'bottom-up' fill-related centerline voids, as shown in Figure 5.3. Gap-fill in the presence of the stronger SUPP-B shows a superior 'bottom-up' fill rate under identical plating conditions, thereby enabling void-free gap-fill.

### 5.3.3 The Role of Leveler

A leveler, when present, provides leveling effects between recessed areas, bumps and the field surface. These levelers exhibit mass transfer characteristics and accumulate on the areas under high electric field strength. Levelers are typically nitrogen-containing molecules that strongly bond to the copper surface and inhibit the plating reaction. The LSV in Figure 5.4 shows the reduced hysteresis and inhibited plating rate of copper in the presence of all three additives components, e.g. suppressor, accelerator and leveler, as compared to the enhanced deposition rate observed when the plating bath contains only the suppressor and the accelerator.

The fast 'bottom-up' fill plating by employing strongly suppressing additives was studied [3]. It was shown that levelers with stronger suppression (high suppression strength levelers (HSSL), based on LSV) enable a higher superfill rate according to a partial gap fill study by SEM (Figure 5.5). The stronger suppressing strength additives also provide complete planarization of several micrometer-wide features and superfill of sub-50 nm trenches (Figure 5.6). The films deposited from the plating bath with HSS additives have a strong (111) texture, smooth surface (RMS <4 nm) and low trace impurities (ppm).

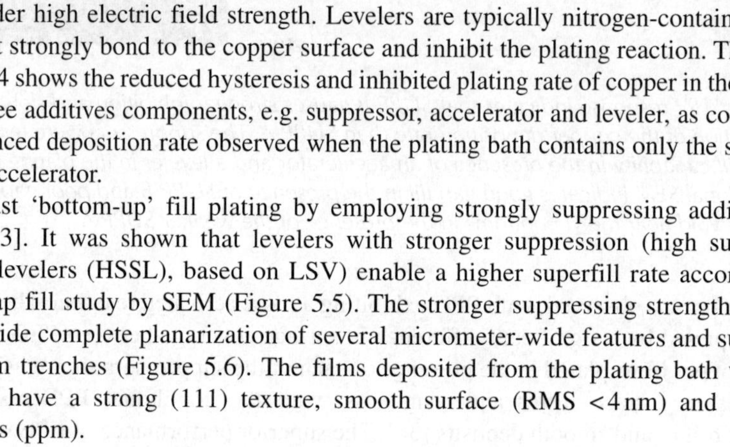


*Figure 5.4 Linear-sweep voltammetry in VMS with SUPP, ASUPP and LEV showing greater suppression and diminished hysteresis than VMS with SUPP and ASUPP only. The greater suppression is indicative of the stronger bonding capability of the leveler, which prevents it from being displaced by the ASUPP, thereby maintaining prolonged inhibition of the plating rate*

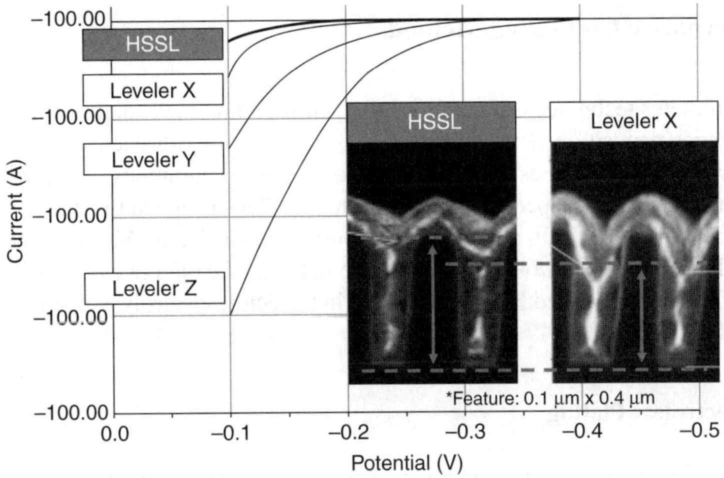


***Figure 5.5*** *Partial fill of trenches with various suppression strength levelers. Notice that the higher suppression strength (HSSL) results in a higher bottom-up fill rate*

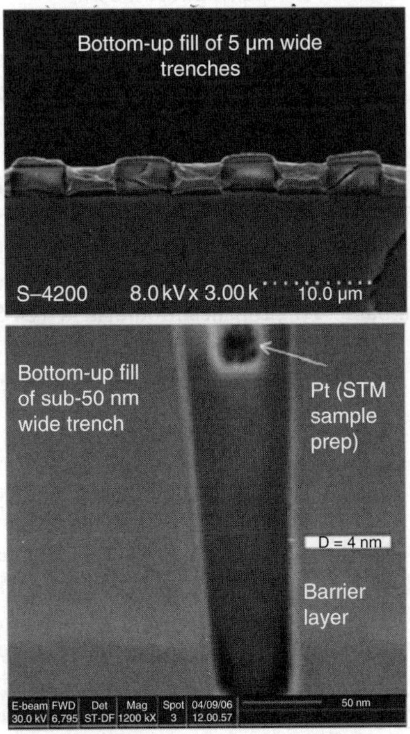


***Figure 5.6*** *Cu superfill in suppression-based chemistry for the trenches ranges from sub-50 nm to over 5 μm wide*

## 5.4 Alternative Cu Plating Methods

In submicrometer technology generations, the Cu seed layer is usually deposited using a physical vapor deposition (PVD) method. As the trench width decreases to the tens of nanometer range, it becomes more difficult to obtain a continuous and conformal seed layer on the trench wall, especially at the trench base. This is due to the directional deposition of PVD, which results in a 'bottom void' during trench filling. Moreover, the coverage of the seed layer becomes worse by the damage in the acidic plating electrolyte. The native Cu oxide formed on the seed layer is dissolved in the acidic electrolyte, resulting in defects in the seed layer [41, 42].

### 5.4.1 Electroless Plating

Before adopting methods such as atomic layer deposition and electroless plating in the seed layer fabrication, seed layer repair has been suggested as a bridge between PVD and the methods mentioned. Seed layer repair is a technique that improves the continuity of the seed layer by connecting sparse parts of Cu seed through the addition of a few Cu layers. Electroless plating [43], alkaline-based electroplating [44] and SnPd activation [45] have been reported as the PVD seed layer repairing methods.

Electroless copper deposition can be used to form the seed layer and to fill the damascene feature, therefore replacing PVD seed and Cu electroplating steps. Excellent step coverage capability is expected on high aspect ratio and nanometer-scaled structures with a good copper thickness uniformity achievable on large substrates. Void-free filling of high aspect ratio via-holes was also reported [46–48] with the proper choice of plating additives such as SPS and mercapto alkyl carboxylic acids as accelerators and PEG–PPG triblock copolymers as inhibitors [49–51]. Void-free electroless copper deposition in submicrometer trenches was obtained in the presence of the single poly(ethylene glycol) (PEG) as inhibitor [52] as well as accelerating species such as SPS [53, 54] and its derivatives, 3-*N*,*N*-dimethylamino-dithiocarbamoyl-1-propanesulfonic acid (DPS) [55] and 2-mercapto-5-benzimidazole-sulfonic acid (MBIS) [56].

### 5.4.2 Direct Plating

Direct plating of Cu on to a relatively conformal liner is critical to enable extension of electroplated Cu interconnects. Although Ru is not a diffusion barrier [57], Ru is of interest because it has a higher electrical conductivity than Ta, is insoluble in acid Cu plating baths, the bulk phases of Cu and Ru are immiscible and, in addition, thin conformal (sub-10 nm) film can be deposited by CVD or ALD methods.

One of the problems with ruthenium is that the metal is susceptible to oxidation in air and aerated water [58–60] so that copper may be plated on to an oxide rather than the metal barrier film. If left untreated, subsequent Cu electroplating on such Ru substrates leads to pronounced three-dimensional (3D) nucleation/growth kinetics with incomplete Cu surface coverage. It was also pointed out that residual oxide results in weak Cu/Ru adhesion, low nucleation density and poor trench filling [58].

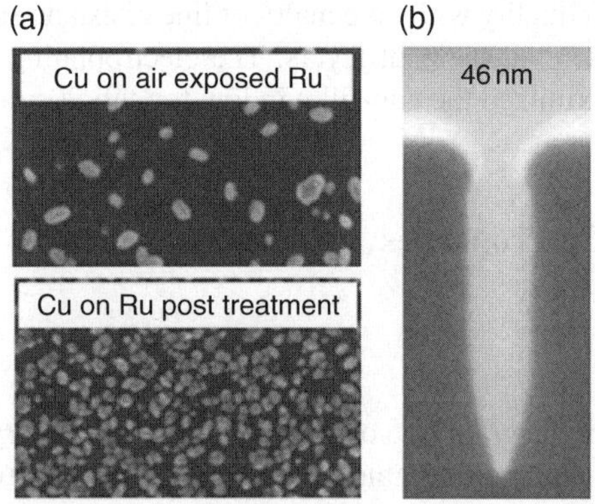


***Figure 5.7*** *(a) Improvement in nucleation density of Cu on Ru by treatment of the Ru surface to remove the surface oxide. (b) Treatment of the Ru surface on patterned wafers followed by electroplating using a conventional electroplating additives chemistry results in void-free fill of sub-50 nm features*

The Ru surface oxide can be reduced if the Ru substrate is subjected to potential in an acidic environment or it has been chemically treated in an acidic solution containing additives. Treatment of the Ru surface on patterned wafers followed by electroplating using a conventional electroplating additives chemistry results in void-free fill of sub-50 nm features, as shown in Figure 5.7. However, electroplating on thin (<5 nm) Ru is challenging due to the terminal effect, leading to nonuniform Cu nucleation and gap-fill. This issue needs to be addressed by better design of the plating hardware to enable plating on highly resistive substrates.

It was reported [61] that after the modifications to the Ru deposition and Cu plating processes, samples easily passed adhesion tape tests. Using the more quantitative four-point bend test, multiple samples consisting of 1 µm electroplated Cu on PVD Ru were analyzed. In each case, when the optimized Cu plating process was used and the Cu/Ru/oxide interfaces were engineered to enhance adhesion, the film stack survived up to the cohesive failure of the epoxy used to join two samples together. A critical strain energy release rate of 29.27 $J/m^2$ was observed, indicating excellent film stack adhesion. Given the published correlations between Cu adhesion and stress/electromigration performance, the excellent Cu/Ru/oxide interfacial bond strength indicates that this new material system may deliver good device reliability performance. X-ray diffraction data collected for Cu directly electroplated on to 200 Å PVD Ru showed the same predominant Cu(111) film texture that is characteristic of Cu electroplated on a PVD Cu seed.

Besides Ru and its alloys, other direct plateable materials with low copper solubility were studied, such as PVD RuTa [62], CVD Co [63, 64], W [65] and Ta [66]. Copper electrodeposited directly on to a tungsten and tantalum diffusion barrier layer in an acidic copper–ammonium–citrate (Cu–$NH_3$–Cit) bath exhibit dense growth with sufficient wetting and adhesion. PVD $Ru_{0.9}Ta_{0.1}$/TaN and Ru/TaN film stacks showed compatible oxidation and Cu diffusion barrier properties to the Ta/TaN bilayer liner. CVD Co demonstrated diffusion barrier performance to Cu (by triangular voltage sweep), but not to $O_2$. The use of enhancement layers of Ru or Co improved copper wetting on barrier and good gap-fill.

A better Cu interconnect quality with no penalty of line resistance and improved reliability is obtained by use of the enhancement layers. The electroplating on enhancement layers without copper seed maximized the gap-fill window for sub-20 nm technology nodes.

## 5.5 Electroplated Cu Properties

### 5.5.1 Resistivity

In the as-deposited form, the grain size of plated copper film is very small. To ensure good electrical properties, reliability and stable CMP removal rate, in situ anneal with $N_2$ and $H_2$ was carried out after copper plating, accompanied by copper grain size grown and about 20% resistance drop on blanket wafers. In fact, recrystallization dynamics within a high aspect ratio or a very small feature size trench could be very different from that in blanket copper films due to the physical constraints of the trench sidewalls. Proper temperature of the anneal should be chosen based on the actual process condition. The overburden thickness above the damascene features is also a critical factor in terms of the overburden grain extension inside the trenches [67].

This increase in Cu line resistivity arises from the increased electron scattering from grain boundaries and interfaces. Damascene line resistivities were measured with line widths ranging from 75 nm to 500 nm, excluding the barrier, and varying thicknesses as a function of temperatures to determine the contribution of different electron scattering mechanisms (Table 5.1 and Figure 5.8) on Cu line resistivity [68].

### 5.5.2 Impurities

Additives in the plating bath largely control the material properties of copper deposits [69–71]. To design electroplating chemistry better for next-generation ULSI devices, an understanding of additive effects on copper material property is important.

The authors in work [72] used mass spectrum analysis to quantify accelerator by-products and found that the SPS by-products correlated with the formation of Cu post-CMP defects because by-products influenced the properties of electroplated Cu films (such as tensile stress) and the chemical corrosion rate.

***Table 5.1*** *Relative electron scattering contribution from interface, grains, impurities and bulk at 20 K and 300 K for selected line widths: 75 nm, 250 nm and 520 nm*

| Line width | Interface | | Grains | | Impurities | | Bulk | |
|---|---|---|---|---|---|---|---|---|
| | 20 K | 300 K | 20 K | 300 K | 20 K | 300 K | 20 K | 300 K |
| 75 nm | 23% | 13% | 40% | 19% | 21% | 5% | 16% | 63% |
| 250 nm | 28% | 7% | 30% | 8% | ≤1% | ≤1% | 42% | 84% |
| 520 nm | 17% | 3% | 22% | 4% | ≤1% | ≤1% | 61% | 92% |

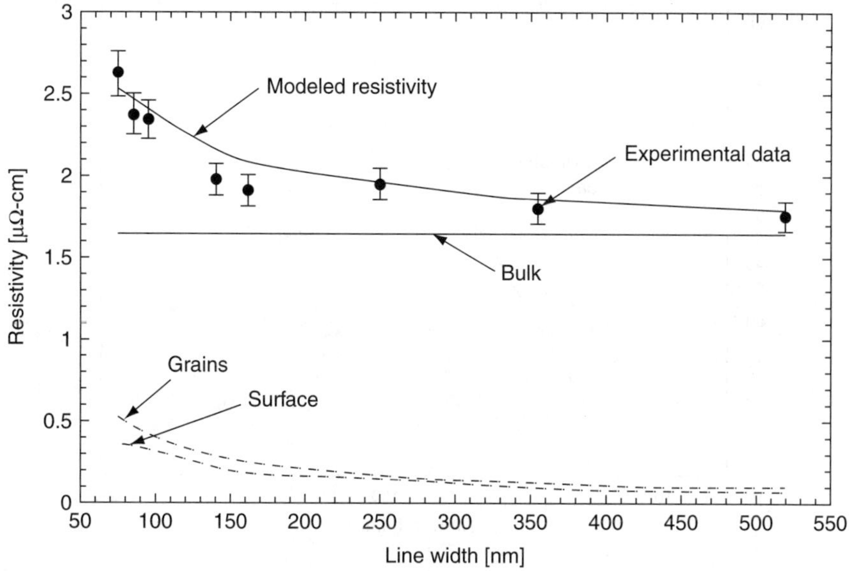


***Figure 5.8*** *Resistivity versus line width at 300 K with analytic model fit parameters: mean free path = 39 nm, p = 0.2, C = 1, R = 0.3 and the grain size is equal to the line width*

The nature of a leveler dominates the impurity concentration in electroplated Cu films according to research [73], which in turn may have impacts on grain recrystallization, texture, stress and corrosion property. The XRD patterns of copper films deposited on a PVD copper surface that had a predominant (111) texture. The trend of decreasing (200) with leveler dosage is observed. Higher stress was observed with a high dosage of leveler.

In the case of a lower impurity content (like Cl, S and C) incorporated in Cu during the electrochemical deposition, a higher resistance against electromigration was verified. The electromigration activation energy for metallizations with small contaminations was found to be (1.00 ± 0.06) eV, whereas Cu interconnect lines with high nonmetallic impurities revealed an activation energy of (0.65 ± 0.03) eV. The electromigration-induced degradation by void formation starts at the top interface between Cu and the dielectric cap layer. Probably, this results from high mechanical stresses due to differences in material properties or due to an interface weakening by the segregation of S and C impurities [74]. To control the incorporation of the impurities into copper film, the additives need to be evaluated toward the degree of their incorporation and breakdown products also need to be controlled by using a proper feed-bleed method for a Cu plating bath in a wafer processing tool.

### 5.5.3 Electromigration

A simple method used to improve the electromigration lifetime is to dope the Cu with impurities, such as Al, Ag, Ti or Mn, using an alloy seed layer or co-plating. During subsequent anneals, the impurities segregate at grain boundaries and interfaces, including the critical interface between the Cu and the capping layer. The presence of the impurities at

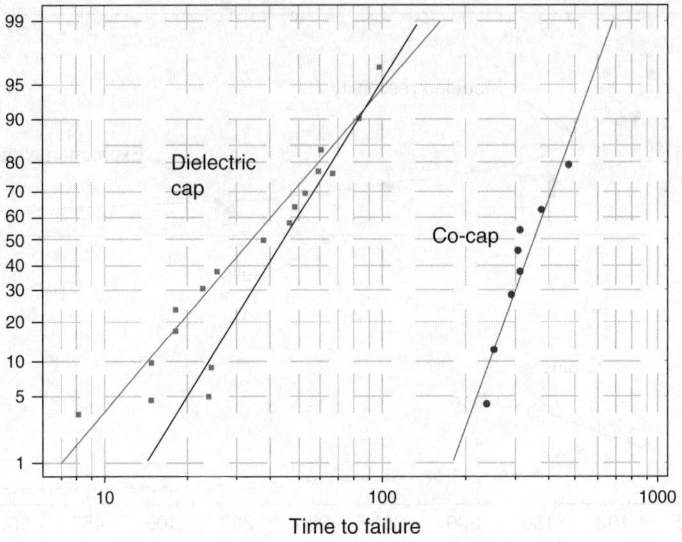


***Figure 5.9*** *Electromigration lifetime of non-Co-capped and Co-capped interconnect structures*

the interfaces reduces Cu diffusion, resulting in an enhancement of electromigration lifetimes of up to 10 times. The main challenge with the use of alloy seed layers is the increase in resistivity caused by doping. With alloy seed layers, there is always a trade-off between increased reliability and increased resistivity [75–78].

Electroless diffusion barrier (such as CoWP, CoWB, CoWBP, NiWP/NiReP, etc.) capping processes have been proposed, resulting in higher interfacial bonding strengths relative to that of the Cu/dielectric interface [43, 79, 80]. The electroless metal cap deposition process is particularly appealing, because of its low cost, intrinsic selectivity and superior film properties (excellent adhesion to copper, good corrosion and diffusion properties, etc.). The addition of an electroless barrier increased (×10) the electromigration resistance (Figure 5.9) and, should there be a need for the dielectric etch stop, resulted in the reduction of the effective ILD capacitance.

A 5 % reduction in RC delay was demonstrated with an electroless cap replacing the dielectric cap layer [81]. It also achieved a 100 % yield of line to line leakages, via chain and metal line resistances.

## 5.6 Conclusions

In copper plating, 32 nm and beyond technology nodes require a high bottom-up fill rate of damascene structures of high pre-plating aspect ratios (>20) while narrow features are filling at high speed (<1 s). High suppression strength and high adsorption rate plating additives are required to enable void-free filling of narrow features with high aspect ratios.

Extension of copper damascene plating to sub-20 nm technology generation will require alternative liners (such as Ru, Co, etc.) that have better conformality and wetting, higher corrosion resistance and a larger gap-filling window. Electroless Cu seed-fill is also being investigated.

Cu film properties such as impurities, resistivity and electromigration are significantly influenced by plating bath compositions that require further optimization of additives. Electromigration resistance of damascene Cu metallization can be improved >×10 (compare to pure copper with a dielectric cap) by alloying (Al, Mn, etc.), which results in a resistivity increase or selective electroless barrier capping with an achievable RC reduction of over 5 %.

## References

[1] Andricacos, P.C., Uzoh, C., Dukovic, J.O., Horkans, J. and Deligianni, H. (1998) Damascene copper electroplating for chip interconnections. *IBM J. Res. Dev.*, **42**, 567.
[2] Haran, B.S., Kumara, A., Adam, L., Changa, J., Basker, V., Kanakasabapathy, S., Horak, D., Fan, S., Chen, J., Faltermeier, J., Seo, S., Burkhardt, M., Burns, S., Halle, S., Holmes, S., Johnson, R., McLellan, E., Levin, T.M., Zhu, Y., Kuss, J., Ebert, A., Cummings, J., Canaperi, D., Paparao, S., Arnold, J., Sparks, T., Koay, C.S., Kanarsky, T., Schmitz, S., Petrillo, K., Kim, R.H., Demarest, J., Edge, L.F., Jagannathan, H., Smalley, M., Berliner, N., Cheng, K., LaTulipe, D., Koburger, C., Mehta, S., Raymond, M., Colburn, M., Spooner, T., Paruchuri, V., Haenscha, W., McHerron, D. and Doris, B. (2008) 22 nm technology compatible fully functional 0.1 $\mu m^2$ 6T-SRAM cell. Proceedings of the Electron Device Meeting, IEEE International, San Francisco, California, 15–17 December 2008.
[3] Dubin, V.M. (2008) Suppression-based Cu electroplating chemistry. Extended Abstracts of 214th ECS Meeting, Honolulu, Hawaii, 12–17 October, 2008.
[4] Dubin, V.M. (2003) Electrochemical aspects of new materials and technologies in microelectronics. *Microelectron. Engng*, **70**, 461.
[5] Seah, C.H., Mridha, S. and Chan, L.H. (2001) DC/pulse plating of copper for trench/via filling. *J. Mater. Proc. Technol.*, **114**, 233.
[6] Tantavichet, N. and Pritzker, M.D. (2002) Low and high frequency pulse current plating of copper onto a rotating disk electrode. *J. Electrochem. Soc.*, **149**, C289.
[7] Chang, S.-C., Shieh, J.-M., Dai, B.-T. and Feng, M.S. (2002) Investigations of pulse current electrodeposition for damascene copper metals. *J. Vac. Sci. Technol. B*, **20**, 2295.
[8] Chandrasekar, M.S. and Pushpavanam, M. (2008) Pulse and pulse reverse plating–conceptual, advantages and applications. *Electrochim. Acta*, **53**, 331.
[9] Emekli, U. and West, A.C. (2009) Effect of additives and pulse plating on copper nucleation onto Ru. *Electrochim. Acta*, **54**, 1177.
[10] Lu, L., Shen, Y., Chen, X., Qian, L. and Lu, K. (2004) Ultrahigh strength and high electrical conductivity in copper. *Science*, **304**, 422.
[11] Dao, M., Lu, L., Shen, Y.F. and Suresh, S. (2006) Strength, strain-rate sensitivity and ductility of copper with nanoscale twins. *Acta Materialia*, **54**, 5421.
[12] Lu, L., Chen, X., Huang, X. and Lu, K. (2009) Revealing the maximum strength in nanotwinned copper. *Science*, **323**, 607.
[13] Kim, M.J., Cho, S.K., Koo, H.-C., Lim, T., Park, K.J. and Kim, J.J. (2010) Pulse electrodeposition for improving electrical properties of Cu thin film. *J. Electrochem. Soc.*, **157**, D564.
[14] Stöckgen, U., Wehner, S., Heinrich, J., Kiesel, A. and Liske, R. (2010) Integration challenges for copper damascene electroplating. *ECS Trans.*, **25**, 3.
[15] Taylor, T., Ritzdorf, T., Lindberg, F., Carpenter, B. and LeFebvre, M. (1998) Electrolyte composition monitoring for copper interconnect applications. Proceedings of 195th ECS Meeting, San Diego, California, 4–5 May 1998.

[16] West, A.C. (2000) Theory of filling of high-aspect ratio trenches and vias in presence of additives. *J. Electrochem. Soc.*, **147**, 227.

[17] Dow, W.-P., Huang, H.-S., Yen, M.-Y., and Chen, H.-H. (2005) Influence of molecular weight of polyethylene glycol on microvia filling by copper electroplating. *J. Electrochem. Soc.*, **152**, C77.

[18] Josell, D., Wheeler, D., Huber, W.H. and Moffat, T.P. (2001) Superconformal electrodeposition in submicron features. *Phys. Rev. Lett.*, **87**, 016102.

[19] Schultz, Z.D. Feng, Z,V., Biggin, M.E. and Gewirth, A. (2006) Vibrational spectroscopic and mass spectrometric studies of the interaction of bis(3-sulfopropyl)-disulfide with Cu surfaces. *J. Electrochem. Soc.*, **153**, C97.

[20] Guymon, C.G., Harb, J.N., Rowley, R.L. and Wheeler, D.R. (2008) MPSA effects on copper electrodeposition investigated by molecular dynamics simulations. *J. Chem. Phys.*, **128**, 044717.

[21] Garcia-Cardona, E., Wong, E.H. and Barkey, D.P. (2011) NMR spectral studies of interactions between the accelerants SPS and MPS and copper chlorides. *J. Electrochem. Soc.*, **158**, D143.

[22] Moffat, T.P. and Ou Yang, L.-Y. (2010) Accelerator surface phase associated with superconformal Cu electrodeposition. *J. Electrochem. Soc.*, **157**, D228.

[23] Dubin, V.M., Brewer, R., Simka, H. and Shankar, S. (2002) Characterization and control of Cu electroplating chemistries for on-chip metallization. *Future Fab. Int.*, **13**, 244.

[24] Kelly, J.J. and West, A.C. (1998) Copper deposition in the presence of polyethylene glycol. *J. Electrochem. Soc.*, **145**, 3472.

[25] Long, J.G., Searson, P.C. and Vereecken, P.M. (2006) Electrochemical characterization of adsorption-desorption of the cuprous-suppressor–chloride complex during electrodeposition of copper. *J. Electrochem. Soc.*, **153**, C258.

[26] Moffat, T.P., Wheeler, D. and Josell, D. (2004) Electrodeposition of copper in the SPS–PEG–Cl additive system. *J. Electrochem. Soc.*, **151**, C262.

[27] Huerta Garrido, M.E. and Pritzker, M.D. (2009) Inhibition of copper deposition by polyethylene glycol and chloride. *J. Electrochem. Soc.*, **156**, D36.

[28] Huerta Garrido, M.E. and Pritzker, M.D. (2009) Inhibition of copper deposition by polyethylene glycol and chloride. *J. Electrochem. Soc.*, **156**, D175.

[29] Chen, H.-M., Parulekar, S.J. and Zdunek, A. (2008) Interactions of chloride and polyethylene glycol in acidic copper sulfate electrolyte. *J. Electrochem. Soc.*, **155**, D341.

[30] Hebert, K.R. (2005) Role of chloride ions in suppression of copper electrodeposition by polyethylene glycol. *J. Electrochem. Soc.*, **152**, C283.

[31] Walker, M.L., Richter, L.J. and Moffat, T.P. (2005) In situ ellipsometric study of PEG/$Cl^-$ co-adsorption on Cu, Ag, and Au. *J. Electrochem. Soc.*, **152**, C403.

[32] Feng, Z.V., Li, X. and Gewirth, A.A. (2003) Inhibition due to the interaction of polyethylene glycol, chloride, and copper in plating baths: a surface-enhanced Raman study. *J. Phys. Chem. B*, **107**, 9415.

[33] Chai, L., Goldberg, R., Kamp, N. and Klein, J. (2008) Selective adsorption of poly(ethylene oxide) onto a charged surface mediated by alkali metal ions. *Langmuir*, **24**, 1570.

[34] Anderson, D., Haak, R., Ogden, C., Tench, D. and White, J. (1985) Tensile properties of acid copper electrodeposits. *J. Appl. Electrochem.*, **15**, 631.

[35] Hung, C.-C., Wang, Y.-L., Lee, W.-H. and Chang, S.-C. (2008) Competitive adsorption between bis(3-sodiumsulfopropyl disulfide) and polyalkylene glycols on copper electroplating. *J. Electrochem. Soc.*, **155**, H669.

[36] Gallaway, J.W. and West, A.C. (2008) PEG, PPG, and their triblock copolymers as suppressors in copper electroplating. *J. Electrochem. Soc.*, **155**, D632.

[37] Gallaway, J.W., Willey, M.J. and West, A.C. (2009) Copper filling of 100 nm trenches using PEG, PPG, and a triblock copolymer as plating suppressors. *J. Electrochem. Soc.*, **156**, D287.

[38] Mendez, J., Akolkar, R. and Landau, U. (2009) Polyether suppressors enabling copper metallization of high aspect ratio interconnects. *J. Electrochem. Soc.*, **156**, D474.

[39] Akolkar, R. and Landau, U. (2009) Mechanistic analysis of the 'bottom-up' fill in copper interconnect metallization. *J. Electrochem. Soc.*, **156**, D351.

[40] Dubin, V.M., Akolkar, R., Cheng, C.C., Chebiam, R., Fajardo, A. and Gstrein, F. (2007) Electrochemical materials and processes in Si integrated circuit technology. *Electrochim. Acta*, **52**, 2891.

[41] Weiss, K., Riedel, S., Schulz, S.E., Schwerd, M., Helneder, H., Wendt, H. and Gessner, T. (2000) Development of different copper seed layers with respect to the copper electroplating process. *Microelectron. Engng*, **50**, 433.

[42] Reid, J., Mayer, S., Broadbent, E., Klawuhn, E. and Ashtiani, K. (2000) Factors influencing damascene feature fill using copper PVD and electroplating. *Solid State Technology*, **43**, 86.

[43] Dubin, V.M., Lopatin, S., Kohn, A., Petrov, N., Eizenberg, M. and Shacham-Diamand, Y. (2005) *Electroless Barrier and Seed Layers for On-Chip Metallization*, (eds M. Datta, T. Osaka and J.W. Schultze), Ch. 3, CRC Press.

[44] Lee, S.-W., Shi, F.G. and Lopatin, S.D. (2002) Dual damascene advanced interconnects: new copper seed layer enhancement process metrology using ion chromatography. *Microelectron. J.*, **33**, 945.

[45] Cho, S.K., Lim, T., Lee, H.-K. and Kim, J.J. (2010) A study on seed damage in plating electrolyte and its repairing in Cu damascene metallization. *J. Electrochem. Soc.*, **157**, D187.

[46] Shingubara, S., Wang, Z., Yaegashi, O., Obata, R., Sakaue, H. and Takahagi, T. (2004) Bottom-up fill of copper in deep submicrometer holes by electroless plating. *Electrochem. Solid-State Lett.*, **7**, C78.

[47] Wang, Z., Yaegashi, O., Sakaue, H., Takahagi, T. and Shingubara, S. (2004) Bottom-up fill for submicrometer copper via holes of ULSIs by electroless plating. *J. Electrochem. Soc.*, **151**, C781.

[48] Shingubara, S., Wang, Z., Yaegashi, O., Obata, R., Sakaue, H. and Takahagi, T. (2003) Bottom-up fill of copper in high aspect ratio via holes by electroless plating. International Electron Devices Meeting Technical Digest, Washington, DC, 8–10 December 2003.

[49] Wang, Z., Obata, R., Sakaue, H., Takahagi, T. and Shingubara, S. (2006) Bottom-up copper fill with addition of mercapto alkyl carboxylic acid in electroless plating. *Electrochim. Acta*, **51**, 2442.

[50] Wang, Z., Liu, Z., Jiang, H. and Wang, X. (2006) Bottom-up fill mechanisms of electroless plating with addition of mercapto alkyl carboxylic acid. *J. Vac. Sci. Technol. B*, **24**, 803.

[51] Yang, Z., Li, N., Wang, X., Wang, Zh. and Wang, Z. (2010) Bottom-up filling in electroless plating with an addition of PEG–PPG triblock copolymers. *Electrochem. Solid-State Lett.*, **13**, D47.

[52] Hasegawa, M., Yamachika, N., Shacham-Diamand, Y., Okinaka, Y. and Osaka, T. (2007) Evidence for 'superfilling' of submicrometer trenches with electroless copper deposit. *Appl. Phys. Lett.*, **90**, 101916.

[53] Lee, C.H., Lee, S.C. and Kim, J.J. (2005) Bottom-up filling in Cu electroless deposition using bis-(3-sulfopropyl)-disulfide (SPS). *Electrochim. Acta*, **50**, 3563.

[54] Lee, C.H., Lee, S.C. and Kim, J.J. (2005) Improvement of electrolessly gap-filled Cu using 2,2'dipyridyl and bis-(3-sulfopropyl)-disulfide (SPS). *Electrochem. Solid-State Lett.*, **8**, C110.

[55] Lee, C.H., Kim, A.R., Kim, S.-K., Koo, H.-C., Cho, S.K. and Kim, J.J. (2005) Electroless Cu bottom-up filling using 3-*N*,*N*-dimethylaminodithiocarbamoyl-1-propanesulfonic acid. *Electrochem. Solid-State Lett.*, **8**, J27.

[56] Lee, C.H., Kim, A.R., Koo, H.-C. and Kim, J.J. (2009) Effect of 2-mercapto-5-benzimidazolesulfonic acid in superconformal Cu electroless deposition. *J. Electrochem. Soc.*, **156**, D207.

[57] Kim, H., Koseki, T., Ohba, T., Ohta, T., Kojima, Y., Sato, H. and Shimogaki, Y. (2005) Cu wettability and diffusion barrier property of Ru thin film for Cu metallization. *J. Electrochem. Soc.*, **152**, G594.

[58] Moffat, T.P., Walker, M., Chen, P.J., Bonevich, J.E., Egelhoff, W.F., Richter, L., Witt, C., Aaltonen, T., Ritala, M., Leskela, M. and Josell, D. (2006) Electrodeposition of Cu on Ru barrier layers for damascene processing. *J. Electrochem. Soc.*, **153**, C37.

[59] Liu, J., Lei, J., Magtoto, N., Rudenja, S., Garza, M. and Kelber, J.A. (2005) The effects of an iodine surface layer on Ru reactivity in air and during Cu electrodeposition. *J. Electrochem. Soc.*, **152**, G115.

[60] Guo, L., Radisic, A. and Searson, P.C. (2006) Electrodeposition of copper on oxidized ruthenium. *J. Electrochem. Soc.*, **153**, C840.
[61] Johnston, S., Feller, D., Dubin, V. and Moon, P. (2004) Copper electrodeposition on Ru liner. Proceedeedings of the Advanced Metallization Conference 1 September–1 October 2003, Materials Research Society, Pittsburgh, Pennsylvania.
[62] Yang, C.-C., Cohen, S., Shaw, T., Wang, P.-C., Nogami, T. and Edelstein, D. (2010) Characterization of 'ultrathin-Cu'/Ru(Ta)/TaN liner stack for copper interconnects. *IEEE Electron Dev. Lett.*, **31**, 722.
[63] Nogami, T., Maniscalco, J., Madan, A., Flaitz, P., DeHaven, P., Parks, C., Tai, L.,B Lawrence, B.S., Davis, R., Murphy, R., Shaw, T., Cohen, S., Hu, C.-K., Cabral, C., Chiang, S., Kelly, J., Zaitz, M., Schmatz, J., Choi, S., Tsumura, K., Penny, C., Chen, H.-C., Canaperi, D., Vo, T., Ito, F., Straten, O., Simon, A., Rhee, S.-H., Kim, B.-Y., Bolom, T., Ryan, V., Ma, P., Ren, J., Aubuchon, J., Fine, J., Kozlowski, P., Spooner, T. and Edelstein, D. (2010) CVD Co and its application to Cu damascene interconnects. Extended Abstracts of International Interconnect Technology Conference, Burlingame, California, 6–9 June 2010.
[64] Huang, H.Y., Hsieh, C.H., Jeng, S.M., Tao, H.J., Cao, M. and Mii, Y.J. (2010) A new enhancement layer to improve copper interconnect performance. Extended Abstracts of International Interconnect Technology Conference, Burlingame, California, 6–9 June 2010.
[65] Park, K.-S. and S. Kim, S. (2010) Seedless copper electrodeposition onto tungsten diffusion barrier. *J. Electrochem. Soc.*, **157**, D609.
[66] Kim, S. (2010) Seedless copper electrodeposition onto tantalum diffusion barrier by two-step deposition process. *Electrochem. Solid-State Lett.*, **13**, D83.
[67] Carreau, V., Maitrejean, S., Brechet, Y., Verdier, M., Bouchu, D. and Passemard, G. (2008) Cu grain growth in interconnects trenches – experimental characterization of the overburden effect. *Microelectron. Engng*, **85**, 2133.
[68] Plombon, J.J., Andideh, E., Dubin, V. and Maiz, J. (2006) Influence of phonon, geometry, impurity and grain size on Cu line resistivity. *Appl. Phys. Lett.*, **89**, 113124.
[69] Vas'ko, V.A., Tabakovic, I., Riemer, S.C. and Kief, M.T. (2004) Effect of organic additives on structure, resistivity, and room-temperature recrystallization of electrodeposited copper. *Microelectron. Engng*, **75**, 71.
[70] Ahmed, S., Ahmed, T.T., O'Grady, M., Nakahara, S. and Buckley, D.N. (2008) Stress and microstructure in electrodeposited copper nanofilms by substrate curvature and in-situ electrochemical afm measurements. *ECS Trans.*, **11**, 1.
[71] Lim, G., Kim, J. and Kim, H. (2008) Effects of organic additives on residual stress and surface roughness of electroplated copper for flexible PCB. Extended Abstracts of the 213th ECS Meeting, Phoenix, Arizona, 18–22 May 2008.
[72] Hung, C.-C., Lee, W.-H., Hu, S.-Y., Chang, S.-C., Chen, K.-W. and Wang, Y.-L. (2008) Effect of bis-(3-sodiumsulfopropyl disulfide) byproducts on copper defects after chemical mechanical polishing. *J. Vac. Sci. Technol. B*, **26**, 255.
[73] Chen, Q.,X. Lin, X.,V. Paneccasio, V.,P. Ye, P. and R. Hurtubise, R. (2009) Leveler effect on electroplated copper material properties. Extended Abstracts of the 215th ECS Meeting, San Francisco, California, 24–29 May 2009.
[74] Stangl, M., Lipták, M., Acker, J., Hoffmann, V., Baunack, S. and Wetzig, K. (2009) Influence of incorporated non-metallic impurities on electromigration in copper damascene interconnect lines. *Thin Solid Films*, **517**, 2687.
[75] Hu, C.K., Gignac, L., Rosenberg, R., Liniger, E., Rubino, J., Sambucetti, C., Stamper, A., Domenicucci, A. and Chen, X. (2003) Reduced Cu interface diffusion by CoWP surface coating. *Microelectron. Engng*, **70**, 406.
[76] Yokogawa, S. and Tsuchiya, H. (2007) Effects of Al doping on the electromigration performance of damascene Cu interconnects. *J. Appl. Phys.*, **101**, 013513.
[77] Isobayashi, A., Enomoto, Y., Yamada, H., Takahashi, S. and Kadomura, S. (2005) Thermally robust Cu interconnects with Cu–Ag alloy for sub 45 nm node. IEIC Technical Report 104, p. 33.
[78] Koike, J., Haneda, M., Iijima, J. and Wada, M. (2010) Cu alloy metallization for self-forming barrier process. Extended Abstracts of International Interconnect Technology Conference, Burlingame, California, 5–7 June 2010.

[79] Osaka, T., Aramaki, H., Yoshino, M., Ueno, K., Matsuda, I. and Shacham-Diamand, Y. (2009) Fabrication of electroless CoWP/NiB diffusion barrier layer on $SiO_2$ for ULSI devices. *J. Electrochem. Soc.*, **156**, H707.

[80] Moon, P., Dubin, V., Johnston, S., Leu, J., Raol, K. and Wu, C. (2003) Process roadmap and challenges for metal barriers. International Electron Device Meeting Technical Digest, Washington, DC, 8–10 December 2003.

[81] Ko, T., Chang, C.L., Chou, S.W., Lin, M.W., Lin, C.J., Shih, C.H., Su, H.W., Tsai, M.H., Shue, W.S. and Liang, M.S. (2003) High performance/reliability Cu interconnects with selective CoWP cap. Symposium on VLSI Technology Digest of Technical Papers, Kyoto, Japan, 10–12 June 2003.

# 6

# Diffusion Barriers

*Michael Hecker*[1] *and René Hübner*[2]

[1]*Center for Complex Analysis, GLOBALFOUNDRIES Dresden Module One LLC & Co. KG, Dresden, Germany*
[2]*Fraunhofer Institute for Non-Destructive Testing IZFP, Dresden Branch, Dresden, Germany*

## 6.1 Introduction

### 6.1.1 Cu Metallization, Barrier Requirements and Materials

Due to the replacement of Al by Cu metallization in 180 nm CMOS technology nodes and below, yielding many gains as the reduction of the interconnect resistivity and the enhancement of electromigration resistance, a price had to be paid too: so-called liners or barrier films became mandatory to prevent potential diffusion of Cu atoms into the surrounding dielectrics and into active silicon. In contrast to other three-dimensional (3D) transition metals, Cu has a high diffusivity in silicon (diffusion coefficient $D = 3\times10^{-4}$ $cm^2/s$ [1]), where it can impact the function of active elements by creating deep trap states or yielding shortcut paths in case of agglomeration. Further, Cu forms silicides as $Cu_3Si$ at temperatures as low as 200 °C [2], modifying the electronic structure and yielding long-range lattice expansion (cf. Figure 6.1).

The Cu interconnects (cf. Figure 6.2 for a typical structure) need to be completely embedded into barrier films, which apart from their direct barrier function have to meet several other requirements, such as to ensure sufficient mechanical stability and adhesion between

*Advanced Interconnects for ULSI Technology*, First Edition. Edited by Mikhail R. Baklanov, Paul S. Ho and Ehrenfried Zschech.

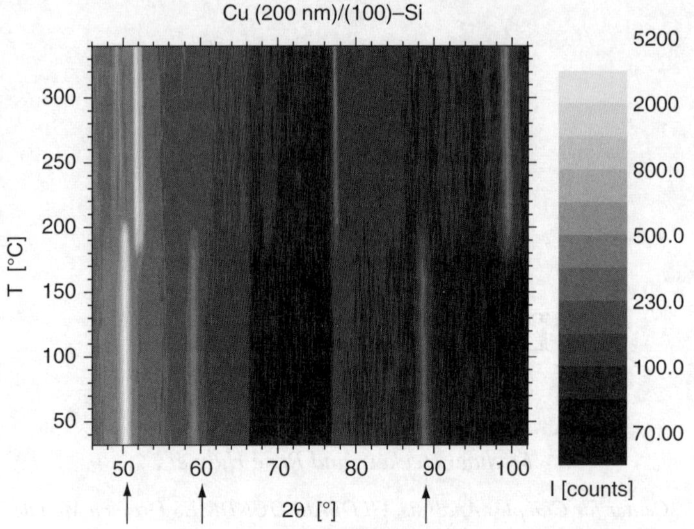


***Figure 6.1*** *In situ X-ray scattering intensity measurement during annealing of a Cu film deposited onto Si (heating rate 4 K/min; measurement in vacuum). Arrows mark peak positions of the Cu reflections, indicating $Cu_3Si$ phase formation at ~200 °C* [3]

***Figure 6.2*** *TEM bright-field image of a Cu/dielectric interconnect stack with horizontally arranged metal levels and vertical interconnect access (VIA) structures embedded into Ta-based diffusion barriers best seen as bright lines in the high-angle annular dark-field STEM image (inset) showing atomic number contrast* [4]

dielectric and interconnects. One of the basic requirements for the barrier/interconnect stack is to show a resistivity as low as possible and to be surrounded by materials with low dielectric permittivity. Typically, dielectric diffusion barriers used as etch stop layers on top of the interconnects have the highest dielectric constant in the metallization stack, whereas the metallic diffusion barrier covering the Cu line sidewalls and its bottom have much higher resistivity than Cu. Therefore, it is important to improve and scale down the barrier films, even more since the dimensions of the Cu lines and the spacing between them continue to shrink. The bonding properties between copper and diffusion barriers are also critical for the reliability of the Cu lines, in particular with respect to their electromigration behavior [5].

Introduction of selectively grown metal barrier films on top of the Cu interconnects, replacing the former copper–dielectric interface, has been shown to improve the electromigration lifetime significantly. Research is still ongoing to enhance the selectivity in order to prevent other reliability issues such as time-dependent dielectric breakdown and to minimize the impact of the barriers on an interconnect resistivity increase.

Apart from the choice of the right material, the microstructure of the diffusion barriers can be tuned to improve their function [6], e.g. by:

(i) blocking grain boundaries by impurity atoms (e.g. O, N, C, Zr) or oxides,
(ii) fabrication of a nanocrystalline microstructure without elongated grain boundaries perpendicular to the film surface,
(iii) deposition of amorphous barriers, e.g. by introduction of Si atoms into the barrier material, or
(iv) utilization of single-crystalline barriers without grain boundaries acting as fast diffusion paths.

Regarding the last two options, amorphous barriers are usually preferred due to the more difficult process integration of single-crystalline films. Various further requirements on barrier systems have been compared for different barrier systems [7]. Due to the low diffusion coefficient for Cu in Ta-based systems, they are of particular interest [8]. TaN/Ta-based films are widely used in industry and will be discussed in Section 6.2.1. The subject of present research is new materials with the potential of an improved trade-off between low resistivity (lower than for existing barrier materials), good wettability and high barrier function, as well as with technological advantages, such as the option of direct Cu electroplating without seed layers and a reduced number of process steps. Such materials include, for example, Pd, Ru and Rh among others [9] and will be discussed in more detail in Section 6.3.1.

### 6.1.2 Barrier Deposition Techniques

Present barriers consist of films with a few nanometer thickness and interface properties critical for performance and reliability of the 3D-interconnect system, which has to be fully embedded into dielectric materials. Correspondingly, barrier deposition processes are challenging. Furthermore, various barrier types are needed in interconnect systems, such as those for contacts, for covering interconnect bottom and sidewalls, and for interconnect caps. The deposition processes have to be optimized for the different barriers separately.

Since the introduction of the Cu metallization, PVD (physical vapor deposition)-based approaches have been widely used to obtain good barrier properties. In particular, for covering the bottom and sidewalls of Cu interconnects, ionized PVD techniques utilizing magnetron sputtering are applied and have the potential to be further optimized for depositing present liner materials down to the 22 nm CMOS technology node [10]. To deposit conformal films in high-aspect ratio structures with reduced thickness, ALD (atomic layer deposition) is under evaluation. In particular, for alternative barriers such as Ru, TaN/Ru or Cu(Mn) (cf. Section 6.3.1) ALD solutions are promising.

For barrier films on the contact level, where tungsten fill is used, CVD (chemical vapor deposition) processes are also suitable. However, also for these films (e.g. Ti/Ti$N_x$ or W$N_x$ in case of W contacts) ALD processes have the potential to improve the contact properties.

For covering the top of interconnects, modified barriers with special requirements, e.g. to be functional as etch-stop layers, are used. While PECVD-based dielectric films such as $SiC_xN_y$ are applied in CMOS technology nodes down to 32 nm, selective metal deposition of $CoW_xP_y$, $NiMo_xP_y$, Co or Ru is considered to be promising for additional capping layers or complete barriers in following technology generations (cf. Reference [11], Section 6.3.2).

In addition, advanced techniques based on supercritical fluid deposition [12], growth of self-assembled monolayers or barrier-free metallization [13] are evaluated. An important criterion for utilizing these techniques in volume manufacturing is the number of necessary process steps. Therefore, barrier systems avoiding the deposition of separate films and enabling the reduction of subsequent steps, such as seedless Cu fill, are of high interest (cf. Sections 6.3.1 and 6.3.3).

### 6.1.3 Characterization of Barrier Performance

The evaluation of the effectiveness of barrier layers against Cu interdiffusion can be done by analytical as well as electrical characterization and involves systematic annealing experiments. While heat treatment at several temperatures for a fixed period of time allows for the determination of the critical temperatures for particular degradation processes, annealing at constant temperature and varying periods of time is suitable for analyzing diffusion processes in more detail. Based on this approach, structure and performance alterations of diffusion barriers as they might occur during microprocessor fabrication and operation can be monitored.

To analyze changes in microstructure, layer setup and particular functional properties, analytical characterization of diffusion barriers in the as-deposited state and after annealing can be performed using X-ray scattering, spectroscopic and microscopic methods. In particular, these complementary techniques allow the determination of the barrier stability and a detailed analysis of the barrier failure modes. Characterization of the layer structure can be done using *X-ray reflectometry* (XRR) as well as *X-ray diffraction* (XRD) at a small incidence angle or in Bragg–Brentano geometry ($\theta$–$2\theta$ scans). For layer stacks, XRR experiments provide information about film thickness, interface roughness and the depth profile of the electron density. In addition, XRD investigations are suitable for determining the degree of crystallinity and analyzing phase composition, crystallite size, texture and lattice strain. Besides *Auger electron spectroscopy* (AES) and *X-ray photoelectron spectroscopy* (XPS), *glow discharge optical emission spectroscopy* (GDOES) is an appropriate technique for element depth profile analysis [14]. *Secondary ion mass spectroscopy* (SIMS) is applied for Cu trace analysis in diffusion barrier layers [15]. With a detection limit of $2 \times 10^{16}$ Cu atoms in 1 $cm^3$ silicon, SIMS depth profiles can be recorded for small sample areas in the order of $10^4$ $\mu m^2$. Another analytical approach for determining the Cu

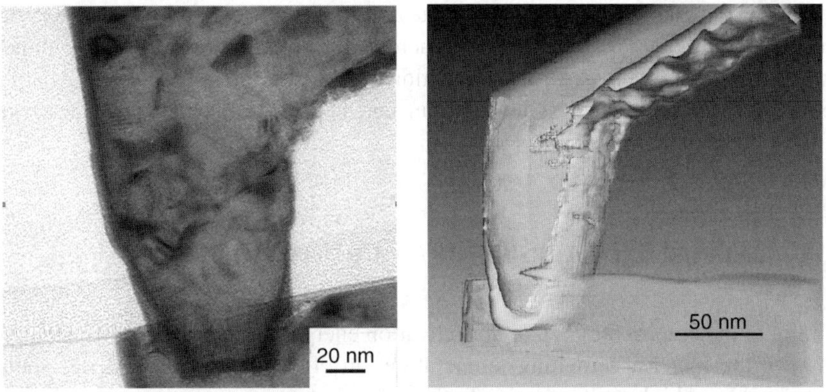


***Figure 6.3*** *TEM bright-field image (left) of a Cu contact manufactured in 45 nm CMOS technology node with a TaN/Ta barrier system. Conformal barrier deposition on the right via sidewall is questionable. Clarification requires advanced techniques, such as electron tomography, showing in 3D clearly the sites of a missing barrier for the same contact (right)* [17]

contents in the barrier layer, the dielectric and the Si substrate after thermal treatment is given by selective wet-chemical etching of the different materials followed by Cu concentration measurement in each etch solution using *atomic absorption spectrometry* (AAS). For the latter method, detection limits for copper in silicon and in $SiO_2$ of $2.5 \times 10^{14}$ atoms/cm$^3$ and $5.0 \times 10^{16}$ atoms/cm$^3$, respectively, have been obtained [16]. Besides the application of trace-analytical techniques, a sensitive proof of Cu diffusion into the substrate is also possible by detecting $Cu_3Si$ formation using XRD for Cu/barrier/Si samples. *Transmission electron microscopy* (TEM) possesses the advantage to carry out chemical and microstructure investigations at cross-sections of layer stacks with high lateral resolution, including TEM bright-field imaging for film thickness determination and electron diffraction for phase analysis. Barrier characterization is challenging due to the needed high sensitivity for trace elements simultaneously with atomic resolution. Thus, advanced TEM techniques are required, such as TEM combined with energy dispersive X-ray spectroscopy (EDXS) and electron energy loss spectroscopy (EELS), allowing determination of the chemical composition in small sample volumes, or 3D imaging techniques (Figure 6.3).

Electrical characterization of device structures offers the potential for highly sensitive Cu trace analysis [18]. There are, however, significant differences in the sensitivity for Cu detection between p- and n-type silicon, as well as between several devices, like a p–n junction, Schottky diode or MOS capacitor. Since for reverse-biased shallow p–n junctions the increase of leakage current caused by annealing occurs only after Cu precipitate formation in the junction area, this method is not very sensitive. For reverse-biased Schottky diodes formed by a conductive, nonreacting diffusion barrier on p-type Si, the presence of Cu at the contact interface has a strong influence on the current–voltage characteristics. The method is highly sensitive to Cu penetration in the form of barrier height, leakage current and ideality factor. The most functional characterization of the diffusion barrier stability is given by analyzing MOS capacitors, i.e. two-terminal devices composed of a thin dielectric layer between a metallic Cu/barrier field plate and an Si substrate. Subjecting MOS devices to bias–temperature stress (BTS) leads to increasing Cu ion mobility and to commencing thermal diffusion as well as to a degradation of the dielectric properties. All of these processes have an influence on leakage current, capacitance–voltage characteristics or generation lifetime measurements [18].

## 6.2 Metal-Based Barriers as Liners for Cu Seed Deposition

According to the empirical rule that the activation energy for a particular diffusion process scales with the material's melting temperature, an effective diffusion barrier should be characterized by a high melting point $T_m$. For a given temperature, the effect of a specific diffusion phenomenon (e.g. grain boundary diffusion) is thus reduced compared to materials with low $T_m$. As a consequence, refractory metals with melting points $T_m > 2000\,°C$ are suitable candidates as diffusion barriers for Cu interconnects. They can be found among the following metals which are currently investigated for this purpose, e.g. Cr [19], Ti [20], Mo [21], W [22], Ta [23] and Nb [24], and are generally characterized by a low electrical resistivity. Thermodynamically stable barrier materials can be also achieved by forming compounds such as:

(1) polycrystalline and amorphous Me–N, Me–C, Me–O and Me–B compounds, such as $TiN_x$ [25], $VN_x$ [26], $ZrN_x$ [27], $NbN_x$ [28], $MoN_x$ [21], $HfN_x$ [29], $WN_x$ [30], $TaN_x$ [31], $WC_x$ [32], $TaC_x$ [33], $MoO_x$ [34], $TaO_x$ [35] and $TiB_2$ [36],
(2) polycrystalline and amorphous Me–Si compounds, such as $MoSi_x$ [37], $WSi_x$ [38] and $TaSi_x$ [39],
(3) polycrystalline and amorphous Me alloys, such as $TiW_x$ [40], $TaCo_x$ and $TaFe_x$ [41], $TaW_x$ [42], $NiNb_x$ [43] and $CuZr_x$ [44].

Due to their importance for applications, in the following Ta-based, W-based and Ti-based diffusion barriers will be discussed in more detail.

### 6.2.1 Ta-Based Barriers

*Pure Ta Diffusion Barriers*

The refractory metal tantalum ($T_m = 3020\,°C$ [45]) fulfills important requirements for an application as diffusion barrier. Intermetallic Cu–Ta compounds do not exist [45, 46] and the solid solubility of tantalum in copper and vice versa is very small [45]. The growth of Cu clusters on a Ta surface during annealing can be described by Ostwald ripening mediated by surface diffusion. Corresponding surface diffusion coefficients at $T = 550\,°C$ were determined to $D_S^{\beta-Ta} = 2.6_{1.0}^{3.7} \times 10^{-11}\,cm^2/s$ and $D_S^{\alpha-Ta} = 2.0_{1.5}^{3.1} \times 10^{-12}\,cm^2/s$ [47]. According to angular resolved X-ray photoelectron spectroscopy (ARXPS) investigations, the Ta/Si and the Ta/$SiO_2$ interfaces have a layered internal structure with $TaSi_2$ formation observed for the Ta/Si interface and Ta suboxide formation due to $SiO_2$ reduction and eventually $Ta_5Si_3$ growth occurring for the Ta/$SiO_2$ interface [48]. Both interfaces are thermally very stable. As shown by Hübner *et al.* [49], annealing a Ta layer at $T = 500\,°C$ leads only to a further increase of the intrinsic compressive stress, which is in agreement to results from Cabral *et al.* [50]. While an amorphous Ta–Si intermixing layer is formed between tantalum and silicon at $T = 550\,°C$, the growth of $TaSi_2$ crystallites at the interface to the substrate is detected for the first time at $T = 620\,°C$ [49]. These findings are in good agreement with experimental results by Cheng and Chen [51] and Nguyen Tan *et al.* [52] as well as theoretical considerations by Walser and Bené [53]. A significant reaction between Ta and silicon oxide to orthorhombic $Ta_2O_5$ and tetragonal $Ta_5Si_3$ occurs at $T = 1000\,°C$ [49], which is in accordance with theoretical considerations reported by Pretorius *et al.* [54] and Beyers [55]. Further engineering of the Ta/Si interface is possible by implanting Si ions [56]. At a temperature of $T \approx 150\,°C$, the formation of an amorphous Ta(Si) interface layer is observed. The width of the mixed layer increases linearly with implanted dose and temperature. Substrate amorphization results in an increase of the contact resistance, which can be improved by implantation with $5 \times 10^{16}$ Si/cm$^2$ at $T = 400\,°C$. In the latter case, $TaSi_2$ formation is observed at the interface [56]. The adhesion of tantalum to silicon oxide is worse when compared with the Ta nitride adhesion to the same dielectric [57]. Regarding the interface with the metallization material, tantalum shows strong adhesion to copper [58]. Calculations by Simka *et al.* [59] predict increasing adhesion strength in the order of TiN <TaN <Ta and are consistent with wetting experiments done using 10 nm thick Cu layers. According to Kim and Duquette [60], the delamination energy in the steady state varies between $G_{ss} = 5$ and 23 J/m$^2$ for Cu film thicknesses between 180 and 570 nm. Cu adhesion to tantalum can be further improved by inserting an ultrathin layer of poly(acrylic acid) [61] as well as an Ru or Co film [62]. In the latter case, however, oxygen and carbon are incorporated into CVD Co films, native Co oxide grows rapidly within a few hours, oxygen diffuses through cobalt and cobalt diffuses through copper at normal processing temperatures [63].

So far, three different Ta modifications have been mainly observed: (i) thermodynamically stable body-centered cubic (bcc) α-Ta, (ii) thermodynamically metastable tetragonal β-Ta and (iii) amorphous Ta (cf. Table 6.1). Depending on the deposition method and the deposition parameters, a Ta layer contains one or a mixture of the three phases. At room temperature, metastable β-Ta usually grows on silicon and $SiO_2$ [51, 64, 65], while α-Ta is deposited at elevated substrate temperatures ($T > 400\,°C$) [66]. The thermodynamically stable modification will also be observed if tantalum is prepared onto thin (>0.3 nm) Nb [67] or Ti films [68]. Annealing experiments can be used to transform metastable β-Ta into stable α-Ta, whereby the transformation temperature is significantly determined by the

***Table 6.1*** *Summary of various properties for the three different Ta modifications*

| | α-Ta | β-Ta | Amorphous Ta |
|---|---|---|---|
| Space group | $Im\bar{3}m$ (229) | $P\bar{4}2_1m$ (113) | – |
| Lattice parameter (single crystal) | $a = 5.43010(5)$ Å [73, 74] | $a = 10.211(3)$ Å, $c = 5.3064(10)$ Å [75] | – |
| Resistivity $\rho$ (layer) | 15–50 μΩ cm [76] | 150–220 μΩ cm [76] | 150–180 μΩ cm [72] |
| Coefficient of thermal expansion $\alpha$ | $(6.9 \pm 0.9)\ 10^{-6}\,°C^{-1}$ [77] | $(6.0 \pm 0.8)\ 10^{-6}\,°C^{-1}$ [77] | |
| Biaxial modulus $M$ | (270 ± 30) GPa [77] | (175 ± 20) GPa [77] | |

deposition conditions. For a Ta layer with intrinsic compressive stress, the transformation occurs abruptly at $T = 750\,°C$ and is accompanied by a stress relief [49, 69, 70]. According to Hoogeveen *et al.* [71], the transformation temperature decreases with increasing Ta layer thickness. Ultrathin amorphous Ta films can be prepared by atomic layer deposition [72].

Due to the excellent, above-mentioned properties, thin Ta films have been intensively investigated as diffusion barriers between copper and silicon in the last few years [6, 78]. Polycrystalline Ta layers are characterized by a high thermal stability, which seems to be independent of the Ta crystal structure [6, 79]. Holloway and Fryer [65] detected significant Cu diffusion through a 50 nm thick Ta barrier together with $TaSi_2$ formation after a $T = 600\,°C$ / $t = ½$ h anneal. In a later study [80], they concluded that Cu diffusion into the Si substrate promotes and, thus, precedes the nucleation of Ta silicide. The order in which Cu and Ta silicide formation occurs depends on the thickness of the diffusion barrier (cf. Laurila *et al.* [23]). Thick Ta layers (~ 100 nm) are able to prevent Cu diffusion into silicon, even at the $TaSi_2$ formation temperature. In the case of thin Ta films (10–50 nm), however, Cu atoms can penetrate the barrier before $TaSi_2$ nucleation and react with silicon to form $Cu_3Si$ [23]. This finding is in good agreement to the results from Hecker *et al.* [31] and Hübner [81], who detected the commencement of Cu diffusion through a 10 nm thick Ta barrier by $Cu_3Si$ formation at $T = 550\,°C$ (Figure 6.4). Due to the nanocrystalline microstructure of tetragonal β-Ta with grain sizes in the order of the film thickness, there are grain boundaries that act as fast paths for Cu diffusion. According to XRD results, $Cu_3Si$ formation is detected prior to $TaSi_2$ nucleation, which occurs at $T = 575\,°C$ [81]. This Ta silicide formation temperature is at least 45 K lower compared to the corresponding temperature for a sample without Cu capping layer [49]. Thus, it is concluded that Cu diffusion into silicon promotes $TaSi_2$ grain growth [81]. The Ta barrier stability depends not only on the vacuum base pressure during deposition [82] but also on the residual gas pressure during annealing. Yin *et al.* [83] demonstrated that heat treatment in vacuum at pressure $p_0 \approx 1$ Pa leads to O diffusion along Cu grain boundaries into the Cu/Ta interface and, consequently, to the formation of a thin amorphous Ta oxide layer. Such an interface oxide layer acts as an additional barrier against Cu diffusion, as has been proven by Laurila *et al.* [84]. Recently, partial Ta barrier oxidation was found to be the root cause of an unusually large resistance increase during stress migration testing at $T = 275\,°C$ [85]. Various attempts have been undertaken to further improve the thermal stability of Ta diffusion barriers. Kang *et al.* [86] deposited Ta layers under Ar ion bombardment, which results in a higher packing density of the grain boundaries and, thus, in an increased barrier stability. Further possibilities include the incorporation of $CeO_2$ into Ta grain boundaries [87] as well as the insertion of an additional Zr [88] or V film [89] into the Ta barrier. Using ALD, the deposition of amorphous Ta films is possible [72]. In comparison to PVD–Ta films deposited between Cu and silicon, ALD–Ta layers of similar thickness are characterized by a $Cu_3Si$ formation temperature that is at least 70 K higher [72].

Ta diffusion barriers possess a significantly higher thermal stability when deposited onto $SiO_2$ [90–92]. Wang *et al.* [92] have shown, that there is no Ta compound formation for a 25 nm thick Ta layer between copper and $SiO_2$ even after annealing at $T = 800\,°C$. In contrast, for a Ta barrier prepared onto silicon, Ta silicide formation occurs at $T = 650\,°C$ [92]. However, also for Ta films deposited onto $SiO_2$ thermally induced processes are observed. For example, Jang *et al.* [93] noticed a diffusion of Ta atoms through the Cu layer to the sample surface after heat treatment at $T = 800\,°C$. According to Hübner *et al.* [94],

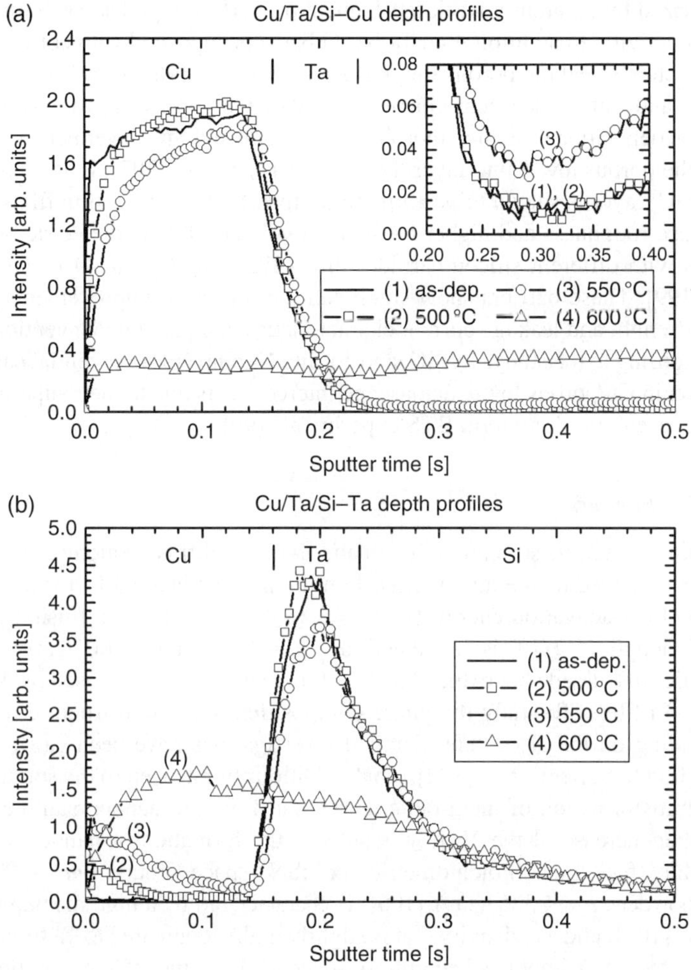


***Figure 6.4*** *GDOES depth profiles of the Cu distribution (a) and of the Ta distribution (b) for a Cu/Ta/Si sample in the as-deposited state and after annealing at several temperatures for t = 1 h. The profiles indicate the onset of Ta diffusion towards the surface at 500 °C and of Cu diffusion across the barrier at 550 °C. Reprinted from Reference* [81], *with permission from Nova Science Publishers, Inc.*

tetragonal β-Ta transforms into bcc α-Ta at $T = 700\,°C$. TEM analyses show that the α-Ta crystallites grow on the $SiO_2$ substrate and within the original Ta layer [94]. Due to its relatively high electrical permittivity, silicon oxide is being replaced by low-*k* and ultra-low-*k* interlayer dielectrics. Thus, a further decrease of the metallization-induced *RC* delay in signal propagation can be achieved. In the 45 nm CMOS technology node, low-*k* materials such as organosilicate glasses with *k* values between 2.7 and 3.1 have been implemented. However, for further reduction of interconnect *RC* delay for the 22 nm CMOS technology node and beyond, the introduction of porous (ultra) low-*k* dielectrics targeting *k* values between 2.3 and 2.6 is necessary [95]. This poses many challenges, as porous dielectrics are characterized by inherently weak mechanical properties, high leakage current and low thermal conductivity. In addition, thin Ta-based barriers deposited with conventional PVD do not satisfactorily seal the porous low-*k* sidewall surfaces [96]. To achieve better barrier performance and better electrical characteristics of Cu/porous ultra-low-*k* interconnect structures, the deposition of a thin amorphous hydrogenated $SiC_x$ layer between a PVD–Ta barrier and the porous low-*k* interlayer dielectric was proposed [97]. In the case of porous SILK™, an $SiC_x$ layer is shown to perform better than $Si_3N_4$ or $SiO_2$ thin films in terms of smaller surface roughness and higher breakdown voltage [98]. Ta/dielectric bilayer diffusion barriers with different silicon carbide films ($SiC_x$, $SiC_xN_y$, $SiC_xO_y$) were studied by Chen *et al.* [99]. These barriers show significant performance improvements in terms of breakdown strength and leakage current characteristics compared to conventional PVD-Ta barriers. According to the study, Ta/$SiC_x$ and Ta/$SiC_xN_y$ bilayers are potential candidates for an application in Cu/porous low-*k* damascene interconnects due to their superior electrical performance, even after long-term BTS experiments [99].

### *$TaN_x$ Diffusion Barriers*

Due to the polycrystalline structure of α-Ta and β-Ta, Cu diffusion along grain boundaries is observed at elevated temperatures. Based on the pre-exponential factor $D'_{0,Cu/Ta} = 9.0 \times 10^{-4}$ cm²/s and the activation energy $Q'_{Cu/Ta} = 2.3$ eV [100] for this diffusion process, the characteristic length $l = (Dt)^{1/2}$ is calculated to $l_{Cu/Ta} \approx 4$ nm for $T = 600$ °C, $t = 1$ h. Besides the possibilities discussed above, the barrier stability can also be increased by N incorporation into the Ta films. To study the phase composition and microstructure of such $TaN_x$ layers depending on the N content, detailed investigations have been carried out [101–105]. According to Stavrev *et al.* [102], small additions of nitrogen to the sputter gas argon lead to the transformation of metastable tetragonal β-Ta into nanocrystalline bcc Ta(N), while a further increase of the $N_2$ flow results in the formation of nanocrystalline face-centered cubic (fcc) TaN. Stoichiometric fcc TaN (space group: $Fm\bar{3}m$ (225), lattice parameter (powder): $a = 4.3363(1)$ Å [106]) is characterized by a high melting temperature ($T_m = 3090$ °C [107]) and a resistivity that is slightly higher compared to Ta films [102]. The deposition of fcc ALD–TaN has been rarely successful. Using $TaCl_5$ as the precursor and $NH_3$, high-resistivity $Ta_3N_5$ films were obtained and the use of hydrazine as the reducing agent produced similar results [108]. Low-resistivity fcc TaN films with $\rho_{TaN} \approx 1500\,\mu\Omega$ cm but high impurity incorporation were prepared using additional reducing agents like Zn, trimethyl-aluminum or amines [109–111]. However, Kim *et al.* [112] were successful in depositing low-resistivity ($\rho_{TaN} \approx 350\,\mu\Omega$ cm) fcc TaN films by PEALD using $TaCl_5$ and a plasma of both nitrogen and hydrogen. Characterizing the barrier continuity by an HF etching test, ALD–TaN layers deposited onto silicon oxide are continuous and free from pinholes [113]. According to XPS analyses by Zier *et al.* [114], TaN deposition onto Si starts with a 0.8 nm thick $Si_3N_4$ interlayer followed by TaN islands with a height of about 1.5 nm at full coverage. In contrast to that, TaN islands grow directly on top of silicon oxide [114], leading to strong adhesion at the TaN/$SiO_2$ interface [57]. Heat treatment of a PVD–TaN layer up to $T = 800$ °C leads not only to a decrease of the amorphous phase fraction but also to fcc TaN crystallite growth. In particular, the mean TaN grain size for a 50 nm thick layer increases from $d_{TaN} = 3.5$ nm in the as-deposited state to $d_{TaN} = 10$ nm after annealing at $T = 900$ °C. Further heat treatment leads to phase transformations into

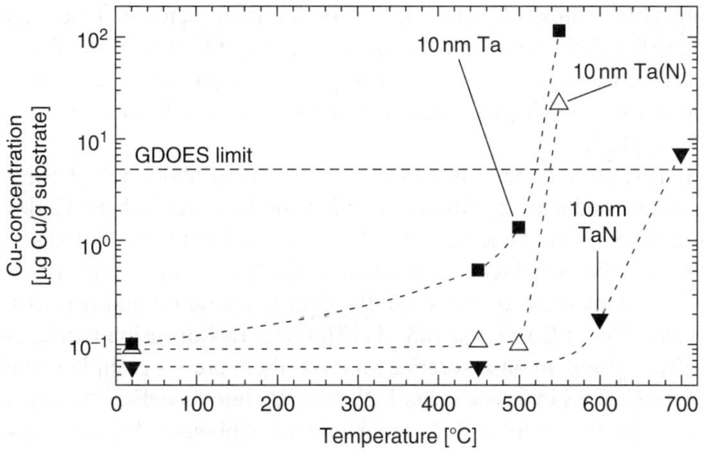


***Figure 6.5*** *Copper concentration in the substrate and barrier versus annealing temperature as measured by AAS. The dashed line shows the limit below which no enhanced Cu signal was detected in the substrate by GDOES. Reprinted from Reference* [31]*, with permission from Elsevier*

thermodynamically stable tetragonal $Ta_4N_5$ and subsequently hexagonal $Ta_5N_6$ [115]. There are, however, no signs for a reaction with $SiO_2$ up to $T = 1100\,°C$ [81]. The thermal stability with a pure Si substrate is also very high. Only annealing at $T = 1050\,°C$ results in a reaction between $Ta_4N_5$ and silicon to $TaSi_2$ [81]. Cu films can be directly electrodeposited on an oxide-free TaN diffusion barrier from specially designed electrolytes that contribute to good adhesion between copper and TaN [116]. The adhesion of the $SiO_2$/TaN/Cu interfaces depends sensitively on film and interface chemistry, e.g. on the barrier nitrogen content, as derived from four-point bending experiments. For example, the TaN/$SiO_2$ interface (N content 50 at%) shows an interface fracture energy $G_c$ of ~ 10 J/m$^2$, more than twice the Ta/$SiO_2$ interface adhesion ($G_c$ ~ 4 J/m$^2$ [57]). Also, solid-state wetting experiments have been used to derive the work of adhesion for pertinent Cu/liner interfaces using contact-angle measurements [117]. These experiments show better adhesion of Cu to PVD–Ta (work of adhesion 2170 mJ/m$^2$) compared to PVD– or ALD–$TaN_x$ (1850 mJ/m$^2$) and a significant enhancement in Cu wetting characteristics for ALD-$TaN_x$ films after surfactant treatment with indium (2280 mJ/m$^2$) [117].

Investigations concerning the influence of the N content on the diffusion barrier properties of $TaN_x$ films between copper and silicon show that the thermal stability increases with an increasing $x_N/x_{Ta}$ ratio [31, 92, 118, 119]. Based on X-ray diffraction experiments, Wang *et al.* [92] detected the formation of Ta silicide and Cu silicide for 25 nm thick Ta and TaN barriers after annealing at $T = 650\,°C$ and 800 °C, respectively. According to AAS studies [31], Cu trace diffusion through 10 nm thick Ta, $Ta_{20}N_{80}$ and TaN barriers can be observed after one hour of heat treatment at $T = 450\,°C$, 550 °C and 600 °C, respectively (Figure 6.5). Kim *et al.* [120] showed that TaN films prepared by PEALD from $TaCl_5$ and a plasma of both hydrogen and nitrogen have diffusion barrier properties that are comparable to those of PVD–TaN layers. A further increase of the thermal stability can be obtained for $TaN_x$ diffusion barriers deposited by PEALD from PDMAT precursor in the presence of hydrogen and/or nitrogen plasma. In particular, a 0.6 nm thick $TaN_{0.8}$ layer with a resistivity of 350 μΩcm and nanocrystalline fcc TaN microstructure was stable up to $T = 700\,°C$ [121].

Investigations regarding the failure mechanisms of amorphous $TaN_x$ barriers with $Ta_2N$ composition were conducted by Min *et al.* [122] and Chen and Chen [123]. Caused by a crystallization process commencing at $T = 450\,°C$ as well as thermally induced cracks, Cu atoms diffuse into silicon and react with it to form $Cu_3Si$. A slightly higher N content in the barrier leads to an increase of the crystallization temperature and a retardation of the subsequent formation of $Cu_3Si$ and $TaSi_2$ [123]. For a TaN diffusion barrier characterized by small fcc TaN grains in an amorphous matrix, there are no grain boundaries running directly from the Cu/TaN interface to the TaN/substrate interface. Heat treatment at elevated temperatures leads, however, to TaN grain growth. Although the mean crystallite size remains smaller than the TaN film thickness, several TaN grain boundaries connecting the Cu metallization layer with the Si substrate and enabling fast Cu diffusion seem to exist. Based on grain boundary diffusion parameters $D'_{0,Cu/TaN} = 2.8 \times 10^{-10}\,cm^2/s$ and $Q'_{Cu/TaN} = 1.3\,eV$ [124], a mean diffusion distance of $l_{Cu/TaN} \approx 9\,nm$ is estimated for annealing at $T = 800\,°C$ and $t = 1\,h$. This result is in good agreement with the experimental findings by Hübner [81]. Commencing $Cu_3Si$ formation and, thus, significant Cu diffusion through a 10 nm thick TaN layer into silicon occur during an equivalent thermal treatment. As in the case of the Cu/Ta/Si sample, the Cu diffusion process leads to the generation of sufficient free Si atoms, which can diffuse along grain boundaries into the TaN layer and finally react with the barrier to form $TaSi_2$. Compared to the sample without a Cu capping layer, this reaction temperature is thus lowered by $\Delta T \approx 200\,K$ [81].

### *Multilayered Ta/TaN$_x$ Diffusion Barriers*

Based on the research results obtained for Ta and $TaN_x$ films, a combination of these single layers promises further optimization of the diffusion barrier properties [125–127]. There are at least two advantages for an application of a bilayer as a diffusion barrier. Firstly, the two different barrier sides allow for an optimization of physical properties at both adjacent interfaces. Secondly, the bilayer should be characterized by a thermal stability that is at least as high as for the more stable single layer. For a successful application of a bilayer diffusion barrier, there is, however, the pre-condition of a chemically stable interface between both single layers. The implementation of a Ta/TaN layer stack as a Cu diffusion barrier was proposed in References [7] and [128]. Choosing appropriate deposition parameters, Edelstein *et al.* [7] were able to deposit hexagonally close-packed (hcp) TaN with only a small fraction of the cubic modification. Compared to fcc TaN, films of hcp Ta nitride are characterized by lower intrinsic stress, lower resistivity and comparably strong adhesion to $SiO_2$. The Ta layer, which adheres well to copper, consists of the electrically more conducting α-Ta modification, which forms during deposition onto TaN [128]. Using blanket substrates, Traving *et al.* [129] demonstrated that the TaN layer can be as thin as 0.7 nm to grow the α-Ta phase onto Ta nitride. In the case of trench structures, however, the nominal TaN thickness has to be increased with decreasing feature sizes to obtain pure α-Ta [129]. According to Ho *et al.* [130], a PVD–Ta/ALD–TaN bilayer was successfully

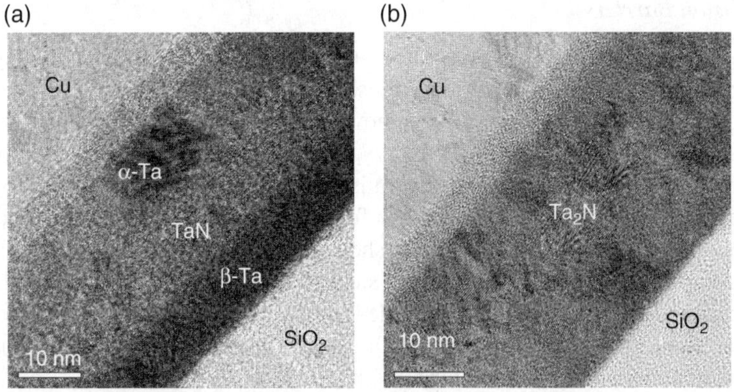


***Figure 6.6*** *Cross-sectional TEM image of a Cu/Ta/TaN/Ta/$SiO_2$/Si layer stack in the as-deposited state (a) and after heat treatment at T = 600 °C for t = 1 h (b). Reprinted from Reference [94], with permission from Elsevier*

integrated into a dual-damascene Cu/low-*k* interconnect structure with via sizes of 110 nm achieving an *RC* reduction of up to 16% relative to the PVD baseline [130]. For a similar bilayered barrier, Moon *et al.* [131] observed a deep penetration of ALD–TaN into the as-deposited porous ultra-low-*k* material ($k = 2.3$; pore size: 2–3 nm) facilitated by the interconnected pore structure. However, etching the ultra-low-*k* dielectric drastically changed the surface due to plasma damage and redeposition of etched species and, hence, pore sealing. Furthermore, the plasma damage resulted in hydrophilic sidewalls, promoting the growth of the ALD–TaN layer without penetration into the etched dielectric [131].

The thermal stability of a graded Ta/TaN/Ta diffusion barrier was characterized by Hübner *et al.* [94]. In the as-deposited state, β-Ta is located in the Ta film directly on the substrate, whereas the Ta film below the Cu metallization consists of α-Ta and the layer in between of fcc TaN (Figure 6.6). Regarding low-temperature annealing, Ta/TaN/Ta diffusion barriers deposited onto silicon and silicon oxide behave similarly. Nitrogen diffusion from the TaN layer into both adjacent Ta films starts at $T = 300\,°C$ and leads to hexagonal $Ta_2N$ formation at $T = 500\,°C$ (Figure 6.6). For the Cu/Ta/TaN/Ta/Si sample, nucleation of hexagonal $TaSi_2$ occurs at $T = 650\,°C$. This reaction temperature is comparable to the value determined for the Ta/Si sample [81]. Thus, $TaSi_2$ nucleation in the case of the Cu-coated sample is not catalyzed by Cu diffusion. However, since $Cu_3Si$ formation does not occur until $T = 700\,°C$, grain boundaries due to $TaSi_2$ formation may slightly accelerate Cu diffusion into silicon. Based on the critical temperatures for $Cu_3Si$ formation determined by glancing angle X-ray diffraction, the thermal stabilities ($S_{barrier}$) of the corresponding diffusion barriers obey the relation: $S_{Ta} < S_{Ta/TaN/Ta} < S_{TaN}$. In particular, a 30 nm thick Ta/TaN/Ta barrier is characterized by a lower thermal stability than a 10 nm thick TaN film, which is part of the trilayer. This result can be explained with the N redistribution in the Ta–TaN layer stack and the subsequent $Ta_2N$ formation, which lead to a degradation of the thermally stable TaN layer. Continuing experiments should clarify whether TaN layers with $x_N/x_{Ta} \leq 1$ can prevent N diffusion into the adjacent Ta films [81, 94].

*$TaSi_x$ Diffusion Barriers*

Although Stavrev *et al.* [102] demonstrated that a higher N content in a $TaN_x$ layer results in the formation of smaller Ta nitride grains, an amorphous diffusion barrier cannot be produced based on this approach. This is, however, possible by incorporating silicon into tantalum [132–134]. According to Lee *et al.* [39], the failure of such amorphous $TaSi_x$ layers deposited between copper and silicon is determined by two mechanisms: (i) barrier crystallization into Ta silicides and (ii) $Cu_3Si$ formation caused by a reaction of copper with silicon from the barrier. Kolawa *et al.* [135] showed that the thermal stability of $TaSi_x$ layers is higher compared to pure Ta films. Si $n^+p$ shallow junction diodes with Cu metallization and an integrated 180 nm thick Ta layer failed during annealing at $T = 550\,°C$ for $t = ½$ h, while those diodes with a 100 nm thick $Ta_{74}Si_{26}$ diffusion barrier remained intact even during heat treatment at $T = 600\,°C$ for $t = ½$ h [135].

*$TaSi_xN_y$ Diffusion Barriers*

With regard to binary $TaN_x$ diffusion barriers, further improvement of the thermal stability of $TaSi_x$ layers seems to be possible by additional incorporation of nitrogen into the barrier. Besides a high resistance against oxidation [136, 137] and low intrinsic film stress [138, 139], PVD–$TaSi_xN_y$ barriers show better adhesion to copper, promote a stronger Cu <111> texture component and result in improved electromigration behavior compared to Ta and TaN films [140, 141]. In combination with a $TaSi_xO_y$ layer, $TaSi_xN_y$ films have been investigated as a capping layer for surface acoustic wave (SAW) structures with embedded finger electrodes [142] or a high-efficiency humidity diffusion barrier for solar cell encapsulation [143]. Regarding their application as a diffusion barrier in Cu metallization systems, $TaSi_xN_y$ barriers are characterized by a high thermal stability between copper and silicon [144–146] as well as between copper and $SiO_2$ [147].

Systematic studies to characterize the failure mechanisms and, in particular, their dependency on the chemical composition of the Ta–Si–N diffusion barrier were carried out by Hübner *et al.* [16, 81, 148–150]. For this purpose, 10 nm thick $Ta_{73}Si_{27}$, $Ta_{56}Si_{19}N_{25}$ and $Ta_{30}Si_{18}N_{52}$ films were prepared onto pure and thermally oxidized Si substrates and partially covered with a 50 nm Cu layer. As in the case of Ta and binary TaN, the thermal stability of ternary $TaSi_xN_y$ films with the interfacing dielectric is mandatory for an application of these Ta-based materials as a Cu diffusion barrier. Since the X-ray amorphous microstructure remains stable for all $TaSi_xN_y$ layers up to $T = 900\,°C$, the required chemical stability with the dielectric is assured. Eventual degradation of the $TaSi_xN_y$ layers is governed by (1) a reaction with silicon oxide to form $Ta_2O_5$ and (2) film crystallization preceding the reaction process for N-containing layers. In the case of an Si substrate, the amorphous $TaSi_xN_y$ microstructure is degraded by crystallization. As for $Ta_{73}Si_{27}$, this process can be significantly accelerated by Si diffusion from the substrate into the Ta-based layer. With increasing N content in the $TaSi_xN_y$ film, the impact of the Si substrate on the crystallization process is reduced. Annealing at higher temperatures leads to $TaSi_2$ nucleation. According to XRD results, the propensity for a reaction with silicon decreases with increasing N content in the $TaSi_xN_y$ layer. However, for a film with a particular composition, the reaction with silicon is always detected at a lower temperature than the reaction with silicon oxide [81, 150].

The detection of $Cu_3Si$ by glancing angle X-ray diffraction can be used to prove Cu diffusion into silicon. Since Cu silicide does not appear during annealing of the

$Cu/TaSi_xN_y/SiO_2/Si$ samples, $Cu_3Si$ formation in the case of the $TaSi_xN_y$ layers deposited directly onto silicon is the result of a reaction between copper and the Si substrate. Furthermore, it can be shown that copper is the diffusing species during this reaction. Additionally, for the $Cu/TaSi_xN_y/Si$ samples, the critical temperatures for $Cu_3Si$ formation are lower than the crystallization temperatures for the corresponding specimens without a Cu cap. Thus, the $Cu_3Si$ formation temperatures can be used to compare the thermal stabilities $S_{barrier}$ of the $TaSi_xN_y$ films. Including the results for Ta and TaN barriers of the same thickness, the relation $S_{Ta} < S_{Ta_{73}Si_{27}} < S_{Ta_{56}Si_{19}N_{25}} < S_{Ta_{30}Si_{18}N_{52}} < S_{TaN}$ is obtained, which means that an amorphous $Ta_{73}Si_{27}$ diffusion barrier is characterized by a higher thermal stability than a polycrystalline Ta layer. Further improvement is achieved by depositing $TaSi_xN_y$ films. For comparable N contents, a $Ta_{30}Si_{18}N_{52}$ layer is, however, less stable against Cu diffusion than a TaN barrier [81, 150].

Besides the determination of the thermal stabilities of the amorphous $TaSi_xN_y$ films, the characterization of the failure mechanisms is of particular interest [150]. For a $Ta_{73}Si_{27}$ barrier deposited between copper and silicon, $Cu_3Si$ formation and, thus, Cu diffusion are detected for the first time under comparable annealing conditions as the commencing crystallization of the same barrier prepared onto silicon oxide. Therefore, it can be assumed that a $Ta_{73}Si_{27}$ diffusion barrier fails by crystallization and associated defect formation. The slightly lower $Cu_3Si$ nucleation temperature for the $Cu/Ta_{73}Si_{27}/Si$ sample ($T$ = 590 °C) compared to the crystallization temperature for the $Cu/Ta_{73}Si_{27}/SiO_2/Si$ sample ($T$ = 600 °C) might be explained by simultaneous Ta diffusion from the barrier to the sample surface and Si diffusion from the Si substrate into the barrier layer, which in combination lead to accelerated barrier crystallization. After diffusing along grain boundaries, Cu atoms react with silicon to form $Cu_3Si$. During this process, free Si atoms are generated. They diffuse into the crystallized barrier and accelerate the formation of hexagonal $TaSi_2$.

For a $Ta_{56}Si_{19}N_{25}$ barrier, $Cu_3Si$ and $TaSi_2$ formation are detected during the same annealing at $T$ = 630 °C for $t$ = 1 h. In particular, the $TaSi_2$ nucleation temperature is thus reduced by at least $\Delta T$ = 370 K in the presence of copper. Furthermore, $Cu_3Si$ formation occurs at the same thermal treatment as crystallization of the $Ta_{56}Si_{19}N_{25}$ layer. Consequently, Cu atoms seem to diffuse along grain boundaries. These considerations are confirmed by trace-analytical SIMS and AAS analyses [16, 81, 150] (Figure 6.7).

In the case of a $Ta_{30}Si_{18}N_{52}$ barrier, $Cu_3Si$ formation and, consequently, Cu diffusion are observed at significantly lower thermal stresses than barrier crystallization, which occurs for the $Cu/Ta_{30}Si_{18}N_{52}/SiO_2/Si$ sample at $T$ = 900 °C. An equivalent annealing temperature is necessary for $TaSi_2$ nucleation in the case of the $Cu/Ta_{30}Si_{18}N_{52}/Si$ sample. If there were already massive barrier defects at commencing $Cu_3Si$ formation, Si atoms should be able to diffuse along them and to react with the barrier, as in the case of a $Ta_{56}Si_{19}N_{25}$ layer. Since this behavior is not observed, Cu atoms seem to diffuse through an amorphous $Ta_{30}Si_{18}N_{52}$ barrier [81]. These considerations are in good accordance with experimental results by Reid [151] for a $Ta_{36}Si_{14}N_{50}$ barrier.

As in the case of Ta and TaN layers, $Ta_{73}Si_{27}$ and $Ta_{30}Si_{18}N_{52}$ films can be combined to form layer stacks. Although the critical temperature for N diffusion from the $TaSi_xN_y$ film into the adjacent Ta–Si layers is increased to $T$ = 500 °C, there is no stability improvement for the graded $Ta_{73}Si_{27}$–$Ta_{30}Si_{18}N_{52}$ barriers compared to a $Ta_{30}Si_{18}N_{52}$ single layer [152].

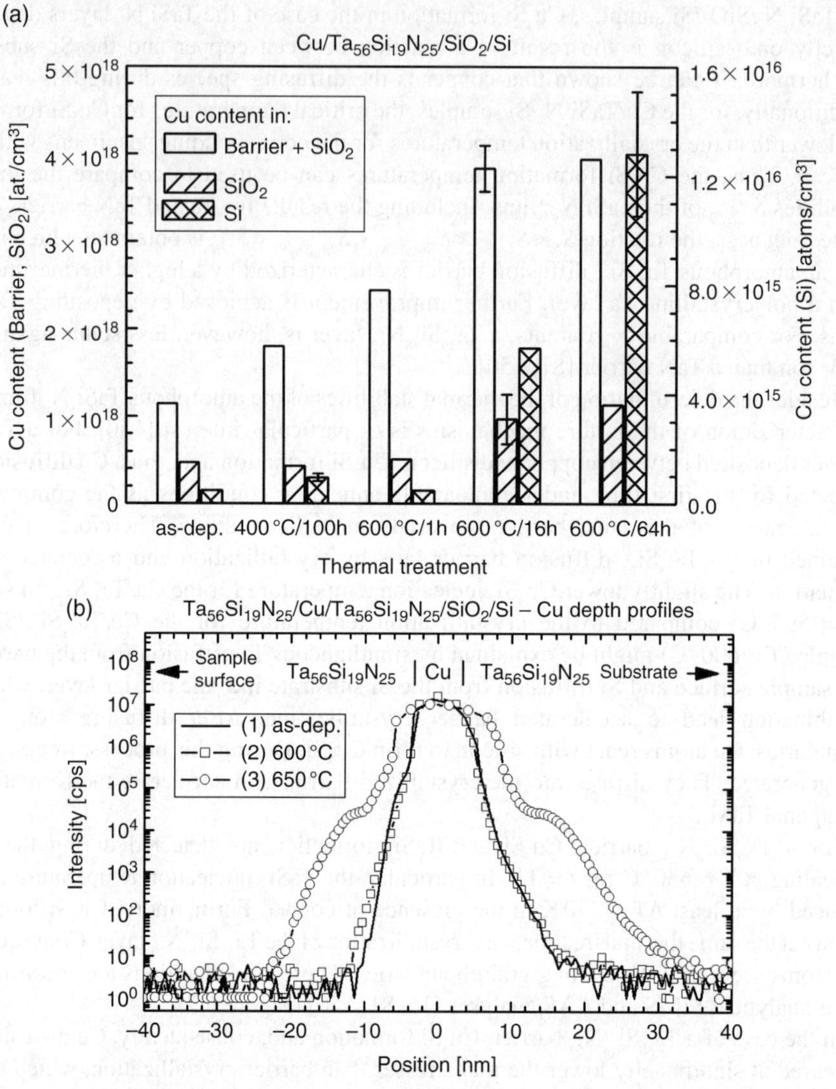


***Figure 6.7*** *(a) Cu concentrations determined by AAS for the Cu/$Ta_{56}Si_{19}N_{25}$/$SiO_2$/Si sample in the as-deposited state and after several heat treatments. Reprinted from Reference* [81] *with permission from Nova Science Publishers, Inc. (b) SIMS depth profiles of the Cu distribution for the $Ta_{56}Si_{19}N_{25}$/Cu/$Ta_{56}Si_{19}N_{25}$/$SiO_2$/Si samples in the as-deposited state and heat-treated at several temperatures for t = 1 h. For comparison, the zero point of the length scale was shifted to the position of maximum Cu intensity. Reprinted from Reference* [150], *with permission from Elsevier*

### 6.2.2 W-Based Barriers

Tungsten-based films show promise when applied as barrier layers for various reasons, such as:

(i) From a thermodynamical point of view, tungsten as a refractory material with a very high melting point ($T_m$ = 3407 °C) should have at least as good diffusion barrier properties as tantalum. However, the microstructure of pure tungsten with a strong tendency for columnar grain growth and related grain boundaries acting as fast diffusion paths impedes a direct application of pure W films.

(ii) Tungsten does not have to be introduced as new extra material into present microprocessor fabrication, since it is used anyway, e.g. as a contact material.

(iii) For W compounds such as $WN_x$ or $WC_x$, precursors for CVD and ALD are available, enabling very conformal film deposition. The processes and compositions can be tuned in a way to prevent fast diffusion paths in grain boundaries or to avoid grain boundaries at all by setting up an amorphous microstructure with sufficient stability.

As a consequence, various investigations on W-based barrier films (cf. Reference [153]) have been performed. In the case of N contents below 50 %, an electrical resistivity of about 200 μΩ cm can be achieved for 10 nm thick barrier films using a plasma-enhanced CVD process (PECVD) from WF–N–H chemistry [154]. Such films are typically amorphous after deposition. Firstly, one challenge here is to obtain sufficient thermal stability of the amorphous microstructure to prevent fast diffusion paths after annealing, since there is a tendency for crystalline phase formation ($W_2N$). Secondly, high adhesion of the barrier to the Cu needs to be achieved, which is also challenging, e.g. due to the presence of trace atoms such as carbon with the potential to yield weak bonding.

In contrast to Ta films showing a tendency of tantalum silicide formation during annealing at elevated temperatures and subsequently avalanche-like barrier failure [31], for $WN_x$ films no corresponding tungsten silicide formation occurs [30]. However, phase formation (occurrence of crystalline W and $W_2N$ at 600 °C anneals; see Figure 6.8) indicates a risk of Cu trace diffusion. It seems to be promising to optimize these barriers further by stuffing diffusion paths and stabilizing the amorphous barrier structure, which can be achieved by incorporating silicon.

An amorphous microstructure of $WN_x$ films can also be obtained by ALD techniques, which bring in the additional advantage of very conformal deposition at reduced film thicknesses. Furthermore, for W contacts such barrier films would be a matching material with the potential of reducing the number of process steps. The replacement of the conventional Ti/TiN liner stack used for W contacts by a thin $WN_x$ barrier has also the potential of further significant reduction of the overall contact resistance [155–157].

By addition of silane in the CVD of $WN_x$ films, fully amorphous $WSi_xN_y$ films can be obtained [158, 159]. However, the Si introduction typically yields an increase in electrical resistivity. This can be counteracted by reducing the N content, but with the risk of reduced thermal stability of the barrier structure. Thus, optimization of the deposition parameters, probably in the range of low silane and nitrogen flow, is necessary and a compromise has to be found. A further challenge is to obtain homogeneous ternary films without significant compositional depth gradients.

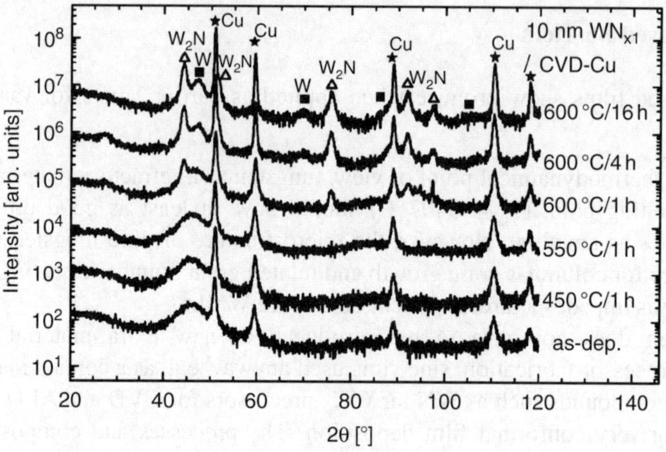


***Figure 6.8*** *XRD diagrams of 10 nm thick $WN_x$ barrier films deposited onto $SiO_2$ and capped by Cu. Curves measured after different anneal steps are shifted vertically for clarity. Reprinted from Reference* [30], *with permission from Elsevier*

Annealing of $WSi_xN_y$ films results in the formation of the same crystalline phases as in the case of binary $WN_x$ films (cf. Figure 6.8), but at higher temperatures. This enhanced thermal stability depends sensitively on the process parameters. Process optimization to achieve a stable amorphous film structure (e.g. by Ar flow modification [158] or pre-anneal before Cu deposition [160]) can improve the applicability of these films as Cu diffusion barriers in CMOS technology. However, due to their low roughness and good contact properties, amorphous $WSi_xN_y$ films are deposited in related applications, e.g. for heterostructure high electron-mobility transistor structures [161–163].

### 6.2.3 Ti-Based Barriers

Titanium was investigated as a diffusion barrier several years ago [7], but not implemented into Cu metallization systems due to some integration issues: (1) fast Ti diffusion into copper leading to a significant resistance increase, (2) voiding/delamination caused by volatile $TiF_4$ as the reaction product of Ti with residual fluorine from dielectric fluorinated silica glass, (3) significant overhang formation during sputtering, resulting in insufficient via step coverage, and (4) corrosion during chemical–mechanical polishing of $TiN_x$. Nowadays, cost reduction for memory products is the main driving force for reconsidering Ti as barrier material. In addition, Ti-based layers are considered as diffusion barriers for so-called through-silicon vias, i.e. vertical electrical connections passing completely through a silicon wafer or a die to enable 3D system integration. Most of the aforementioned concerns are no longer an integration issue. Low-*k* and ultra-low-*k* dielectrics do not contain fluorine, high plasma density PVD sputtering significantly reduces overhang formation due to high ionization rates and optimization of the polishing chemistry prevents galvanic corrosion during $TiN_x$ CMP [164]. Compared to Ta-based systems, Ti-containing diffusion barriers allow better control of moisture penetration from porous ultra-low-*k* dielectrics [165]. A Ti/TiN stack with a stoichiometric TiN film would be able to address the issue of Ti diffusion into Cu, but lacks Cu adhesion on the stoichiometric Ti nitride. The problem can be solved by depositing a thin Ti layer on top of the Ti/TiN bilayer. This Ti flash layer needs to be thin enough not to cause significant resistance increase but thick enough for sufficient Ti segregation at the Cu/$SiC_x$ interface, leading to enhanced interface adhesion and, hence, improvement in electromigration performance. The first Ti layer of the Ti/TiN/Ti sandwich structure is needed to avoid TiN at the via bottom directly contacting the underling copper [164]. The effect of density on the diffusion barrier properties of $TiN_x$ films between Cu and Si was investigated by Lee *et al.* [166]. Using rf sputtering from a TiN target, nanocrystalline $TiN_x$ films were obtained showing a higher density and a smaller grain size for increasing substrate bias. Characterizing the thermal stability, a 25 nm $TiN_{0.81}$ film ($\rho = 4.99\,g/cm^3$) fails during annealing at $T = 600\,°C$ for $t = 1\,h$, while the integrity of the Cu/$TiN_{0.86}$ ($\rho = 5.12\,g/cm^3$)/Si stack is not lost before $T = 800\,°C$. Hence, Lee *et al.* [166] concluded that a high density $TiN_x$ film is of primary importance in achieving good diffusion barrier stability. However, TiN is a substantially inferior barrier compared to $WN_x$ and TaN showing comparable stability for 10 and 20 nm thick barriers between Cu and $SiO_2$ [167]. An improvement of the thermal stability of TiN diffusion barriers is possible by adding zirconium [168]. Considering the impact of various porous low-*k* materials on TiN diffusion barrier integrity, Bonitz *et al.* [169] found that a smaller pore size facilitates a better $SiO_2$ liner and TiN barrier growth. For $SiO_2$ aerogel dielectrics with 7 nm pore size, the CVD liner which is used for pore sealing has no effect on the samples with 10 nm TiN. However, for the ultra-low-*k* material with a 3 nm pore size a slight improvement is observed with an integrated $SiO_2$ liner. In addition, reduced TiN penetration into the dielectric is detected for patterned samples compared to blanket wafers, which is caused by partial pore sealing during ultra-low-*k* etching [169]. According to Guedj *et al.* [170], the best reliability for porous $SiC_xO_yH_z$ ($k = 2.4$) is obtained with a 12 nm thick CVD TiN barrier. With this barrier, the porous ultra-low-*k* dielectric can even outperform dense PECVD $SiC_xO_yH_z$ ($k =$ 2.4) regarding time-dependent dielectric breakdown (TDDB) and breakdown voltages. Suh *et al.* [171] investigated $TiZr_x$ alloy as a new barrier material for low-cost and high-performance Cu/low-*k* interconnects. In the $TiZr_x$-$TiZr_xN_y$ bilayer, the ternary nitride is used as a diffusion barrier, while the binary alloy ensures good adhesion to copper. In combination with titanium, Majumder and Takoudis [172] characterized the thermal stability of sputter-deposited Mo and MoN films. While a Ti (5 nm)/Mo (5 nm) bilayer is found to break down during annealing at $T = 700\,°C$ for $t = 5\,min$ in $N_2$ atmosphere, a Ti (5 nm)/MoN (5 nm) barrier stack fails only at $T = 800\,°C$. An equivalent thermal stability was observed for an Mo (5 nm)/$WN_x$ (5 nm) bilayer stack between Cu and Si [173]. Comparing 50 nm thick reactively sputtered Zr-based films, ZrN outperforms $ZrGe_xN_y$, which fails during heat treatment at $T = 600\,°C$ for $t = 1\,h$ [174]. The growth and diffusion barrier properties of ALD $NbN_x$ thin films were studied by Alén *et al.* [175]. According to their results, addition of Ta or Ti to the barrier layers does not lead to an increase of the breakdown temperature.

### 6.2.4 Further Systems

To meet the challenges of downscaling in the metallization system, modification of presently used Ta/TaN- and Ti/TiN-based barrier films by addition of more and/or application of new elements has been the subject of various investigations. Typically, a compromise has to be found between requirements such as stabilization of an amorphous microstructure with a high resistivity against diffusion processes and retaining a small electrical resistivity. Among the various approaches, the following main routes can be distinguished:

(i) Addition of further elements into $TaN_x$, $WN_x$ or $TiN_x$ barriers. As discussed in Sections 6.2.1 and 6.2.2, adding Si into $TaN_x$ and $WN_x$ films helps to improve their thermal stability and barrier function. The same behavior has been shown for $TiSi_xN_y$ barriers, where the Si content helps to block Cu diffusion paths via amorphization of the microstructure [176]. Further elements supporting the barrier function are oxygen, leading, however, to an unavoidable resistance increase, carbon or boron, as shown for $TaC_xN_y$ [177], $TaB_xN_y$ [178], $TaSi_xC_y$ [179] and $WN_xC_y$ films [180], germanium in the case of $TaGe_x(O)N_y$ [181] and $WGe_xN_y$ films [182], or transition metals as, for example, Co in the case of $TaCo_xN_y$ films [183].

(ii) Establishment of ternary barriers on the basis of other elements than those discussed above. Apart from the Ta-, W- and Ti-based barrier systems, very effective diffusion barriers can be established on the basis of Mo (such as $MoSi_xN_y$ [37]) and Zr (such as $ZrC_xN_y$, $ZrTi_xN_y$ or $ZrAl_xN_y$ [184]). For such systems the achievement of sufficiently low electrical resistivity is a challenge. However, with scaling down of the barrier dimensions, such films are of interest for application as diffusion barriers. Further systems based on Ni or Co, e.g. $CoMo_xP_y$, $CoW_xP_y$ or $CoW_xB_y$, are specially suited for an application as a capping layer of the Cu metallization. They will be discussed in more detail in Section 6.3.2.

(iii) Stacking of different binary systems. A further degree of freedom in tuning barrier systems is obtained by stacking. Examples are Mo/$WN_x$ bilayer stacks [173], Ge/$HfN_x$ stacks [185], MgO/TiN stacks [186] and TiN/Al/TiN stacks [187]. As in the case of the widely used TaN/Ta stack (Section 6.2.1), the different interfaces can be separately tuned to match an optimum connection to both sides, the Cu metallization and the interlayer dielectric.

(iv) Development of barriers with more than three constituent elements. Structural barrier improvements, such as stuffing of potential grain boundaries, can be further obtained by introducing more than three elements constituting the barrier films. One option is given by moderate addition of O, as investigated for the $TiAl_xN_yO_z$ system [188]. A special class showing promise for barrier applications is that of high-entropy alloys (HEAs) consisting of five or more principle elements and possessing properties such as high thermal stability, low diffusion rates of atoms and an amorphous microstructure. Due to their high mixing entropy, the simple solid-solution structure is stabilized. As an example, sputter-deposited Al–Cr–Ta–Ti–Zr–N and Al–Mo–Nb–Si–Ta–Ti–V–Zr films have been evaluated for barrier applications [189, 190]. Future challenges for such systems are optimization of the composition and proof of sufficiently large stable process windows for industrial application.

## 6.3 Advanced Barrier Approaches

### 6.3.1 Barriers for Direct Cu Plating

Attractive materials for potential application as a diffusion barrier are those enabling direct electroplating of Cu onto the barrier without the need for an additional Cu seed layer. For this purpose, metals different from conventionally used refractory metals, which prevent

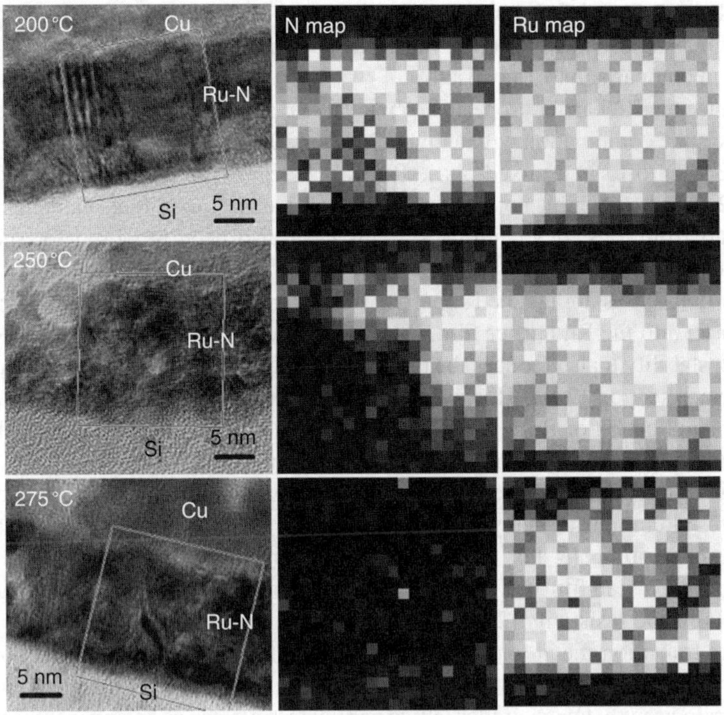


***Figure 6.9*** *TEM bright-field images (left) as well as N and Ru distributions (middle and right) for an $RuN_x$ diffusion barrier after several heat treatments, indicating N redistribution at elevated temperatures. Reprinted from Reference [200], with permission from ECS*

direct electroplating due to surface oxidization, are needed. Lane *et al.* identified various metals, such as Pt, Pd, Ru, Rh, Ir, Ag, Te and Tc, with sufficiently small enthalpy of formation of metal oxides, that allow direct Cu plating [9]. Among those, Ru attracted particular attention due to several properties facilitating its producibility within the CMOS process [191–193]. For the 22 nm CMOS technology node, a performance gain of 10% and narrowing of the operating-speed distribution were predicted using an Ru barrier compared to the Ta barrier [194]. Ru forms a highly adhesive interface to Cu with good electromigration performance, has low solubility in Cu, and thus low impact on Cu resistivity, and is compatible with CVD and ALD technologies [195]. Further, it has almost half of Ta resistivity and (even very thin) Ru films can act as a diffusion barrier enabling direct plating of copper [196]. For magnetic recording applications, even sub-nm Ru films have been shown to be an effective barrier against C diffusion on top of $CoCr_xTa_y$ alloys [197].

However, thin Ru films are not very effective as a Cu diffusion barrier. Above 300 °C, 5 nm thick Ru films can lose their barrier property, depending sensitively on surface plasma or sputter treatments [198, 199]. The weak thermal stability is also influenced by the columnar grain structure of crystalline Ru which is unfavorable to block Cu diffusion. Although N addition to Ru leads to a mixed nanocrystalline-amorphous microstructure, N out-diffusion is observed already at low annealing temperatures (Figure 6.9) [200].

It should be mentioned that further problems associated with pure Ru films need to be solved, such as their bad adhesion to the oxide and their tendency to enhance Cu oxidization [201] in comparison to Ta barriers.

To enhance the Ru barrier properties, obviously an analogous solution to that applied for Ta barriers is the combination with a thin TaN film [202, 203]. Both subfilms can be deposited by PEALD to form a mixed-phase Ru–TaN barrier, with a composition of an Ru:Ta ratio of 12 found to exhibit both a good diffusion barrier and direct-plate characteristics [204]. Apart from TaN, other nitride systems have also been proposed for such mixed-phase systems, e.g. Ru–$WC_xN_y$ [193].

Another option to stabilize the Ru barrier is to include additions of other elements, such as Ta, P or C, into the Ru barrier film. Whereas C addition does not only help to improve the thermal stability it also reduces the tendency of enhanced Cu oxidization resulting from Ru barrier application [201]. However, high C levels have a negative impact on the Cu resistivity. Application of $RuP_x$ films has been shown even for high P contents between 5 and 23 at% acceptable barrier properties after annealing [205, 206].

Since direct integration of such new barrier films into the industrial manufacturing process requires alterations of many process steps, stepwise modification of present diffusion barriers is more probable. For example, promising results have been achieved by ALD deposition of a Cu seed film onto Ru [207], shown to outperform PVD Cu seed with respect to morphology and resistivity [208].

A further transition metal under evaluation as a diffusion barrier with the potential for direct Cu plating is Co, forming also a highly adhesive interface to Cu and enabling an even coherent grain growth due to similar lattice parameters. Its use as an enhancement layer for Cu deposition can improve the adhesion and surface nucleation of subsequent Cu metallization considerably [62, 63, 209]. Though Co is sensitive to oxidization, this oxide can be removed in a galvanic bath. However, the resulting sensitivity of the Co/Cu system for galvanic corrosion will still require further process optimization. This question needs to be solved mandatorily for industrial application [63, 210]. However, in combination with further elements such as $CoW_xP_y$, these films have shown very promising results as capping layers for Cu metallization (cf. Section 6.3.2).

### 6.3.2 Metal Capping Layers

In conventional Cu technology, the interconnect lines are embedded into metal-based barrier films (such as TaN/Ta) at the bottom and at the sidewalls, and are capped by a dielectric film (based, for example, on PECVD $SiN_x$). These capping layers have various functions, such as to act as an etch stop, Cu oxidation protection and a Cu diffusion barrier. Efforts have been undertaken to reduce the dielectric constant of these capping layers and to find what can be achieved by addition of further elements such as carbon [211], but with the risk of reducing the adhesion strength to Cu [212]. Finally, the presence of dielectric capping layers increases the effective *k* value of the dielectric in between the Cu lines by an unwanted amount, resulting in the need for further downscaling of these insulating layers. However, this downscaling is limited to a thickness of about 20 nm, since below this limit the barrier properties begin to degrade significantly [213]. A further critical issue is the high process sensitivity of the Cu/cap interface with respect to necessary clean procedures. As a result, there is a strong impact on reliability-limiting degradation mechanisms. For

$SiC_xN_y$-based films, preferred paths for electromigration have been proven to be concentrated at the Cu/cap interface [214]. Related issues with respect to this interface are void formation and also sensitivity of time-dependent dielectric breakdown.

Other reliability concerns such as stress migration are also known to appear for metallization schemes with dielectric capping at Cu/barrier interface sites, e.g. at via bottoms. Therefore, alternative metal-based materials with the potential to replace a dielectric/Cu interface by a metal/Cu interface are evaluated. In particular, the transition metals Co and Ni are the subject of many promising investigations, related to their ability to form a strongly adhesive interface to copper. Co has the additional advantage of not forming a mixture with Cu, but rather segregating and keeping a coherently bonded interface, also during anneals up to 500 °C [215]. By adding small amounts of refractory metals, e.g. W or Mo, the hardness and melting point can be increased, thus yielding an improvement in the diffusion barrier function [11]. Further addition of light elements such as B or P can help to tune the microstructure by grain boundary stuffing and supporting film amorphization, thus making these films effective diffusion barriers [216–219].

A special advantage of utilizing metal-based Cu capping layers is their ability for electroless plating. This process offers low cost and simplicity of operation, excellent step-coverage capability for filling high-aspect-ratio structures, low deposition temperatures and high selectivity [220]. Tuning of composition is needed, since there are limits in electroless plating, such as for films with P contents above 12% [221], and since an increase of Cu resistivity must be prevented. Furthermore, tuning of the microstructure is needed to optimize the barrier performance and can be done by adjusting electroless plating parameters. Accordingly, $CoW_xP_y$ films can be deposited, for example, in the amorphous state, polycrystalline or crystalline with strong preferred orientation [222].

A large number of electromigration studies clearly proved that the introduction of metal-based capping layers improves the reliability significantly. Also, smaller amounts of stress and smaller stress gradients are introduced by metal capping layers, such as $CoW_xP_y$ [223], being important for reliability. In particular, the option of at least partially coherent growth of, for example, $CoW_xP_y$ enables interfaces to be obtained with improved properties (Figure 6.10). If challenges, such as the necessary highly selective deposition of metal capping

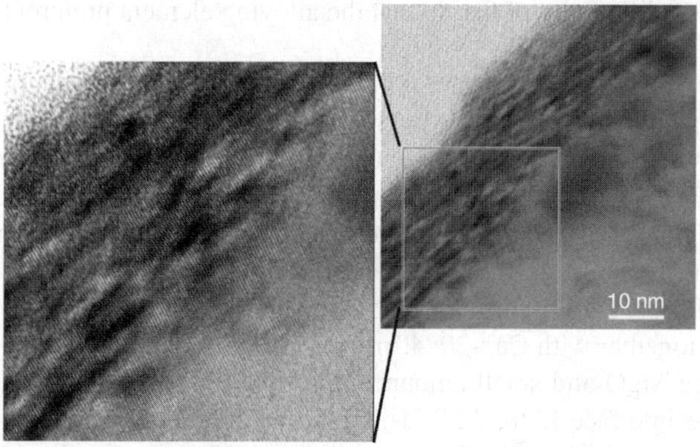


***Figure 6.10*** *TEM bright-field image of a Cu/$CoW_xP_y$ interface [224]. Lattice planes visible in both the $CoW_xP_y$ (dark region) and the Cu underneath indicate coherent film growth without disordered regions of intermixing. © Carl Hanser Verlag München*

layers on the Cu top, the stability of the process and complete process integration, are overcome, metal-based capping layers have a high potential to enable further technology scaling and to improve the reliability properties, such as resistivity against electromigration, stress migration and time-dependent dielectric breakdown for the 32 nm technology node and beyond.

### 6.3.3 Self-Forming Diffusion Barriers

The concept of self-forming diffusion barriers is based on the general idea of adding an alloying element to Cu, letting it migrate to and subsequently react with the Cu/dielectric interface to form a stable phase that can act as a reliable barrier. For proper selection of the alloying element, thermodynamic driving forces and kinetic factors have to be considered. As described by Koike [225], there are two driving forces associated with the growth of self-forming barriers. The first one is related to oxide formation at the Cu/dielectric interface. In this regard, metals with a standard free energy of oxide formation slightly larger than $SiO_2$ are preferred alloying elements, since atom migration to the interface and subsequent oxidation will occur, but strong reduction of $SiO_2$ associated with releasing free Si atoms is not expected. The second driving force is related to the expulsion of the alloying element and, hence, to its activity coefficient $\gamma$ in copper, which characterizes the deviation from the state of an ideal solid solution by the presence of a chemical reaction. For $\gamma > 1$, the alloying element shows negligible solid solubility in Cu and tends to move out by migrating to the film surface or the Cu/dielectric interface followed by oxide formation. This behavior is preferred for self-forming barrier layers. However, if grain boundary segregation occurs, film resistivity will increase, which is unfavorable for interconnect applications. For $\gamma < 1$ (i.e. for Al, Mg, Ti, Zr), the alloying element tends to remain in solid solution (leading to a resistivity increase) and to form intermetallic compounds (unless the first driving force is particularly strong). Hence, alloying elements of the latter type are disadvantageous for self-forming barriers. Besides the existence of a driving force, kinetic factors influence the growth of self-forming diffusion barrier layers. Since the growth rate should not be controlled by the diffusivity of the alloying element, the latter one has to be faster than the self-diffusivity of Cu, so that the alloying element preferentially migrates to the Cu/dielectric interface [225].

Cu(Mg) alloys belong to one material system that has been intensively investigated for an application as a self-forming diffusion barrier layer. In the as-deposited state, Cu(5–12 at% Mg) layers consist primarily of a Cu–Mg solid solution with trace amounts of orthorhombic $CuMg_2$ [226]. Upon annealing to $T = 400\,°C$, Mg segregates to the surface and the Cu(Mg) grains grow from an average size of 20 nm to 60 nm, resulting in a ~25–40 % decrease in film resistivity. In the same temperature regime, the $CuMg_2$ phase dissolves and fcc $Cu_2Mg$ forms. During further heat treatment, Mg segregates to the film/silica interface, penetrates together with Cu ~20–40 nm into silica, and reduces $SiO_2$ to form a 35 nm thick layer of fcc MgO and small amounts (≤1 vol%) of monoclinic $CuMgSi_2O_6$ on the silica side of the interface [226, 227]. This interfacial reaction of the Cu(Mg) film with $SiO_2$ is kinetically limited by the Mg reduction of $SiO_2$, characterized by an activation energy $E_a = (2.7 \pm 0.1)$ eV [228]. Si atoms released during this reaction diffuse into the metal film with $E_a = (2.2 \pm 0.1)$ eV, which is consistent with the lattice diffusion of Si in

Cu [228]. As a result of the Si diffusion process, the electrical resistivity of Cu(8–12 at% Mg) films increases between ~40% and 190% [226]. In combination with the growth of Cu–Mg intermetallic compounds and the formation of a barrier layer with relatively large thickness and low thermal stability, Cu(Mg) alloys are rather unattractive for an application as a self-forming diffusion barrier. Comparable to Cu(Mg), Ti-containing Cu alloy films are characterized by similar properties. Heat treatment at a sufficient temperature results in the formation of intermetallic $Cu_4Ti$. Additionally, Si diffusion into Cu(Ti) as a consequence of $SiO_2$ reduction leads to increasing electrical resistivity, which is unsuitable for interconnect applications [229].

Cu(Mn) alloys have a high potential for implementation as a seed layer to grow a self-forming diffusion barrier at the Cu/dielectric interface [230]. Manganese has a large solid solubility of about 20 at% at room temperature and is completely soluble in copper at high temperatures. Consequently, grain-boundary segregation or precipitation does not occur. The standard free energy of Mn oxide formation is slightly larger than that for Si oxide. In addition, the diffusivity of Mn atoms in Cu is faster than the Cu self-diffusivity. The activity coefficient of Mn in Cu is larger than one, indicating that Mn atoms can be easily expelled from Cu if another chemical reaction is energetically more favorable. Especially in Cu(Mn) metallization layers, Mn tends to diffuse to the surface and the Cu/dielectric interface, where oxidation reactions take place [230]. For Cu(7.9 at% Mn) films deposited by simultaneous sputtering of Cu and Mn targets onto TEOS-based $SiO_2$, annealing at $T = 450\,°C$ for $t = 30$ min leads to the formation of a 3–4 nm thick, thermally stable Mn-containing amorphous oxide layer at the Cu/$SiO_2$ interface. Residual Mn atoms migrate to the sample surface to form an oxide film, leading to a drastic resistivity decrease within the Cu layer [231]. The implementation of a Cu(2 at% Mn) seed layer into the dual damascene process results in the formation of an about 2 nm thick continuous $MnSi_xO_y$ diffusion barrier at the Cu/$SiO_2$ interfaces of the via sidewalls and the interconnect trenches. Due to missing oxygen, there is no barrier at the via bottom resulting in a 50% reduction of the via resistance compared to samples with a 15 nm Ta barrier as well as excellent performance regarding electromigration and stress-induced voiding [232, 233]. The growth of the $MnSi_xO_y$ barrier follows a logarithmic rate law at $T = 350\,°C$ and $450\,°C$, which represents a field-enhanced growth mechanism in the early stage and self-limiting growth behavior in the late stage. Besides annealing temperature and time, the barrier layer thickness is influenced by the Mn concentration and ranges between 2 nm and 8 nm (Figure 6.11). The self-formed $MnSi_xO_y$ layer is thermally stable during annealing at $T = 450\,°C$ for $t = 100$ h and at $T = 600\,°C$ for $t = 10$ h [234–236]. For a Cu(4 at% Mn) film deposited onto TEOS-based $SiO_2$, the dielectric constant of the barrier is determined after annealing at $T = 450\,°C$ for $t = 0.5$ h and 5 h to $k = 11.4$ and 5.1, respectively [237]. Based on X-ray absorption fine structure analysis, Ablett *et al.* [238] showed that for Cu(8 at% Mn) alloy films on porous low-*k* material the barrier layer is composed of $MnSiO_3$ and MnO with almost complete Mn segregation from the alloy film, while the self-forming process will not be fully terminated on thermally grown $SiO_2$ substrates. Optimum annealing conditions to reduce the resistivity of Cu(Mn) alloy films in the self-forming barrier process were investigated by Iijima *et al.* [239]. While a resistivity decrease to 2.0 μΩ cm by annealing in Ar gas at $T = 350\,°C$ and $p_{O2} = 1 \times 10^{-3}$ Pa is achieved by external oxidation of Mn, unchanged resistivity by annealing in vacuum or in argon at $p_{O2} > 1$ Pa can be explained by an insufficient O supply leaving Mn in the alloy film and by internal oxidation of the Cu(Mn) alloy, respectively [239]. Over the

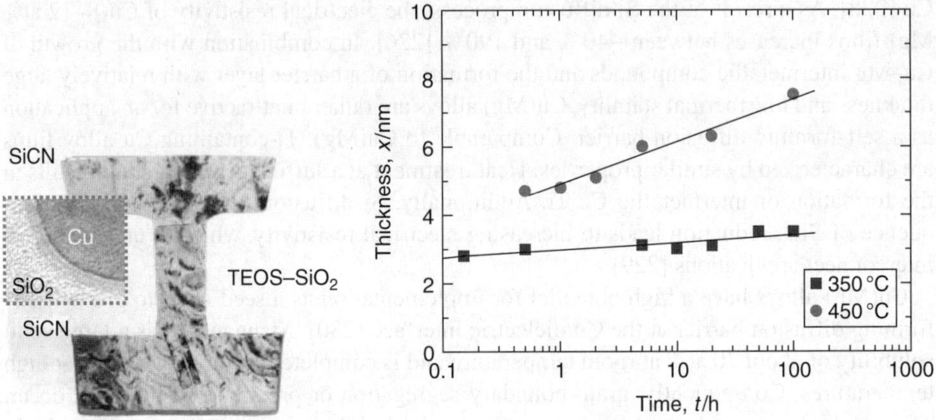


***Figure 6.11*** *Cross-sectional TEM image of a dual-damascene line structure with a Cu(Mn) self-formed barrier layer at the Cu/SiO$_2$ interface and variation of barrier thickness during annealing at 350 °C and 450 °C [230]. © 2006 IEEE*

past few years, Cu(Mn)-based self-formed barriers have been intensively investigated for Cu/ultra-low-*k* dual-damascene interconnects. Implementing this barrier technology in combination with a porous $SiO_xC_y$ ($k = 2.3$)/porous polyarylene ether ($k = 2.3$) hybrid dielectric, more than 90 % yield for wiring and via-chain and 70 % reduction in via resistance compared to the conventional Ta-based process were reported [240]. Further decrease of the dielectric constant of the porous $SiO_xC_y$ layer to $k = 2.0$ resulted in a 5 % interlayer capacitance reduction [241]. According to Pan *et al.* [242], implementing Cu(Mn)-based self-formed barrier layers in combination with ultra-low-*k* dielectrics can be a successful approach for future technology nodes. However, it should be taken into account that the requirements for dielectric material selection, patterning approaches and integration processes might be higher than those for conventional Ta-based barriers [242]. Furthermore, sputtering Cu(Mn) seed layers might lead to conformality issues in small via and trench structures. For that reason, Gordon *et al.* [243] proposed a CVD process where vapor of an Mn compound reacts with the silica surface to form a thin amorphous $MnSi_xO_y$ layer, which shows good barrier properties against diffusion of Cu, $O_2$ and $H_2O$ and is completely conformal inside the surfaces of holes and trenches [243]. According to Neishi *et al.* [244], Mn oxide layers with a uniform thickness between 2.6 nm and 10 nm can be formed by CVD for deposition temperatures ranging between $T = 100$ °C and 400 °C. Heat treatment of the Cu/CVD–Mn/$SiO_2$ samples at $T = 400$ °C up to $t = 100$ h does not lead to interdiffusion processes, indicating good barrier stability [244, 245].

### 6.3.4 Self-Assembled Molecular Nanolayers and Polymer-Based Barriers

The application of self-assembled molecular nanolayers (MNLs) is another novel approach to obtain conformal diffusion barriers with thicknesses significantly below 5 nm. MNLs are thin organic films that form spontaneously on solid surfaces after being adsorbed from either the vapor or liquid phases. Self-assembly is driven by bond formation of an appropriately functionalized head group onto the substrate surface, such as silanes on silica, thiols on metals or acids on metal oxides. Coordinated van der Waals interactions between the hydrocarbon tails of the molecules, generally alkane chains and/or aromatic rings, provide stability and order within the monolayer, while the terminal group comprising, for example, a methyl, phenyl, amine, carboxylic acid or alcohol unit is ideal for interface engineering. Since suitable adjustment of the chain length or the molecular termini can be used to tune their attractive properties, MNLs have been implemented into a number of applications, such as molecular devices, lithography and micromachines. Furthermore, MNLs provide the flexibility to be compatible with the common semiconductor manufacturing technology. MNLs are likely to have good step coverage in high-aspect ratio structures due to the high probability of one end of the molecules sticking to the substrate and the low tendency for multilayer formation. For an application such as a diffusion barrier, the near-zero thickness of the molecular layers will maximize the fraction of the total via volume available for low-resistivity Cu filling.

In one of the first studies employing MNLs as a barrier layer against Cu diffusion, Krishnamoorthy *et al.* [246] showed that Cu/$SiO_2$/Si(001) MOS capacitors with integrated 3-[2-(trimethoxysilyl)ethyl]pyridine (TMSEP) MNL at the metal/dielectric interface and annealed at 200 °C in a 2 MV/cm electrical field are characterized by more than $10^4$ times lower leakage current and a four times higher time to failure when compared to samples without MNLs. Molecular nanolayers with short tail lengths (phenyl-trimethoxy-silane) or aliphatic terminal groups (*n*-propyl-trimethoxy-silane) are ineffective in hindering Cu diffusion, indicating that the molecular length and chemical configuration are key factors determining the efficacy of MNLs as barriers [246]. According to Ramanath *et al.* [247], integrating a ~0.7 nm thick 3-mercapto-propyl-trimethoxy-silane (MPrTMS) MNL between Cu and $SiO_2$ results in enhanced interface adhesion (three times higher interface debond energy) and inhibition of Cu diffusion at the Cu/$SiO_2$ interface (four times higher time to failure), which is due to a strong chemical interaction between Cu and the sulfur-containing MNL termini [247]. Further improvement of the MPrTMS diffusion barrier property can be achieved by irradiation with UV light, leading to a conversion of the mercaptan termini to sulfonates [248]. Annealing Cu/MPrTMS/$SiO_2$ structures at temperatures between 400 and 700 °C results in interfaces that are five times tougher ($G_{MPrTMS} > 20\ J/m^2$) than pristine Cu/$SiO_2$ structures. This toughening effect arises from thermally activated interfacial siloxane bridging that enables the MNL to be strongly linked to both adjacent layers at the interface and suppresses MNL desorption [249]. Depositing MPrTMS onto the Cu surface results in oxide reduction and subsequent formation of a protective molecular bilayer with Si–O–Si linkages [250]. As shown by Ganesan *et al.* [251], amine-terminated MNLs (3-amino-propyl-trimethoxy-silane, APrTMS) at the Cu/$SiO_2$ interface increase the Cu diffusion-induced device failure time by a factor of 3 compared to interfaces without a barrier. Carboxyl-terminated MNLs obtained by grafting succinyl chloride onto APrTMS show more than an additional factor of 4 increase in failure time. In the latter case, Cu immobilization seems to be caused by coordination complex formation through strong interactions between COOH and $Cu^+$ [251]. The effect of solution concentration of organosilane MNLs with two different N-containing termini (TMSEP and *p*-amino-phenyl-trimethoxy-silane (APhTMS)) on the Cu/$SiO_2$ interface adhesion was investigated by Ganesan *et al.* [252]. MNLs from a 5 mM solution form monolayers and enhance adhesion, while 50 mM solutions form multilayers via unregulated self-polymerization, leading to inferior adhesion. Since the Cu/MNL interface is the weakest one, MNLs with termini that bond strongly with Cu are expected to show improvements in interface adhesion [252]. Besides preventing Cu diffusion across Cu/dielectric interfaces, MNLs can also inhibit in-plane interfacial Cu transport [253]. Testing three organosilane MNLs with mercaptan, amino-phenyl and amino-propyl termini, APhTMS was shown to be the most effective in hindering in-plane leakage, possibly due to Cu–N complex formation [253]. As shown above, MNLs are good coupling agents and diffusion barriers at Cu/$SiO_2$ and Cu/low-*k* interfaces and are therefore considered as important elements of future all-wet ULSI metallization with Cu deposited by electroless plating (ELD). To find conditions allowing for ELD of low-resistivity Cu, $SiO_2$/Si substrates were coated with APrTMS, subsequently activated with Au nanoparticles, and finally covered with 30–100 nm thick Cu films. As a result, the lowest Cu resistivity of $\rho \approx (4.0 \pm 0.8)\ \mu\Omega$ cm was achieved after activation with 5 nm Au particles [254, 255]. According to a comprehensive screening test, APrTMS emerged as the most promising MNL barrier against Cu diffusion [256].

A further potential class of organic-based barrier layers is given by polyelectrolytes comprising appropriate functional groups to immobilize metal ions. Forming ultrathin polymeric films through electrostatic adsorption is a simple, versatile technique that can be integrated with current fabrication methods. Furthermore, the use of polymeric structures as barriers could open up possibilities of integrating them, or their constituents, with low-*k* dielectric materials or their precursors, thereby avoiding the need for dedicated barrier layers for Cu/low-*k* metallization technology [257]. Ganesan *et al.* [257] investigated 3.5 nm thick polyelectrolyte layers comprising polyethyleneimine (PEI) and polyacrylic acid (PAA) in a Cu/PAA/PEI/$SiO_2$/Si layer stack and showed a five times longer failure time compared to Cu/$SiO_2$/Si structures. Increasing the number of PAA/PEI bilayers did not result in further improvement of the barrier properties [257]. According to Gandhi *et al.* [258], implementing polyallylamine hydrochloride (PAH)–polystyrene sulfonate (PSS) nanolayers leads to a fourfold enhancement in device failure times during BTS at $T = 200\,°C$, $E = 2$ MV/cm compared to pristine Cu/$SiO_2$ interfaces. Although the bonding at both Cu/PSS and PAH/$SiO_2$ interfaces is strong, the interfacial toughness is only ~2 J/m$^2$, which is due to weak electrostatic bonding at the PSS/PAH interface. Similar behavior is observed for Cu/$SiO_2$ interfaces modified with other polyelectrolyte bilayers that inhibit Cu diffusion [258].

Thin films of poly(*p*-xylylene) (parylene-N or PPX) have been shown to act as effective pore sealants and prevent precursor penetration during subsequent MOCVD for metal deposition. Controlled PPX penetration into porous dielectrics may result in a substantial fracture toughness increase of the porous dielectric. In addition, PPX deposition is selective; i.e. it will deposit on the sidewalls of the porous materials but not on the metal surface at the bottom of via or trench structures. PPX does not contain oxygen and is a barrier for metal ion penetration, in particular Cu diffusion. As shown by Senkevich *et al.* [259], BTS of a MOS capacitor with a 5 nm PPX layer at $T = 150\,°C$, $E = 1$ MV/cm and previously annealed at $T = 250\,°C$ in Ar–3%$H_2$ does not result in a flatband voltage shift, which indicates a stable polymer-capped dielectric. Tantalum deposited onto PPX-caulked porous MSQ is characterized by the low-resistivity $\alpha$ phase [260]. Mallikarjunan *et al.* [261] investigated polyarylether (PAE) polymer films with regard to their stability and figured out that a combination of BTS and TVS techniques is necessary for detailed Cu drift characterization, since it helps to resolve the intrinsic polymer effect from that of metal drift-related instability. Furthermore, the high mobility and ion concentration detected in PAE necessitate the implementation of a robust diffusion barrier to prevent Cu penetration into the polymer [261].

## 6.4 Conclusions

For materials research, the further development of diffusion barrier systems is a challenging process. Many and partially contradictory requirements have to be fulfilled. The development of systems having various properties, such as lower resistivity, stronger bonding, better barrier function against Cu diffusion and reduced thickness, are likely to be based on a trade-off between the different requirements. Furthermore, application in industrial processes requires compatibility with Cu/porous ultra-low-*k* systems at low material and process costs.

Thus, it has to be expected that the number of material systems considered for barrier applications will further increase. Improved Ta-, Ti- and W-based barriers, hybrid barriers containing, for example, Ru, metal capping layers such as $CoW_xP_y$ and zero thickness barriers based on reactions at the Cu surface are fields of ongoing developments. Inclusion of systems needed for three-dimensional integrated circuits with requirements, for example, for high-aspect ratios is one requirement for next-generation technology nodes. There is a close interrelation between the choice of appropriate barrier systems and the surrounding materials, i.e. Cu interconnects and dielectrics, resulting in developments, for example, for seedless Cu deposition directly onto the diffusion barrier or pore sealing of ultra-low-*k* dielectrics.

A variety of techniques for barrier deposition will be required for future technology nodes. Whereas improvements in ionized PVD technology will enable PVD application still at the 22 nm CMOS technology node, ALD-based barriers are developed simultaneously. The latter systems are expected to become the predominant future solution for copper [10]. However, improvements of current PVD- and CVD-based processes as well as the introduction of advanced barrier deposition techniques into the manufacturing process flow are expected to happen stepwise. As an example, the conformity of barrier deposition has been significantly enhanced by the various techniques during the last few years and will continue to be improved in the future.

Key for development of optimized processes will be a detailed understanding of defect mechanisms of barrier systems. This requirement is closely connected with the development of corresponding analytical techniques, especially those with high spatial resolution, such as advanced TEM (cf. Figure 6.3) or 3D atom probe techniques, which will enable analysis of morphology, composition and microstructure on an atomic scale.

## References

[1] Istratov, A.A., Flink, C., Hieslmair, H., Weber, E.R. and Heiser, T. (1998) Intrinsic diffusion coefficient of interstitial copper in silicon. *Phys. Rev. Lett.*, **81**, 1243–1246.
[2] Cros, A., Aboelfotoh, M.O. and Tu, K.N. (1990) Formation, oxidation, electronic, and electrical properties of copper silicides. *J. Appl. Phys.*, **67**, 3328–3336.

[3] Image source: Hecker, M. (2001) IFW, Dresden.
[4] Image source: Engelmann, H.J. (2009) GLOBALFOUNDRIES, Dresden.
[5] Meyer, A.M., Aubel, O., Feustel, F., Engelmann, H.J., Zienert, I., Poppe, J., Gehre, D. and Witt, C. (2009) The evolution of barrier properties during reliability testing of Cu interconnects. *AIP Conf. Proc.*, **1143**, 97–103.
[6] Stavrev, M., Fischer, D., Praessler, F., Wenzel, C. and Drescher, D. (1999) Behavior of thin Ta-based films in the Cu/barrier/Si system. *J. Vac. Sci. Technol. A*, **17**, 993–1001.
[7] Edelstein, D., Uzoh, C., Cabral, C., DeHaven, P., Buchwalter, P., Simon, A., Cooney, E., Malhotra, S., Klaus, D., Rathore, H., Agarwala, B. and Nguyen, D. (2002) An optimal liner for copper damascene interconnects. MRS Proceedings of the Advanced Metallization Conference, AMC 2001 (ULSI XVII), pp. 541–547.
[8] Wenzel, C. and Engelmann, H.-J. (2001) Moderne Barrieresysteme für die Kupfermetallisierung höchstintegrierter Halbleiterbauelemente. *Vakuum in Forschung und Praxis*, **1**, 20–28.
[9] Lane, M.W., Murray, C.E., McFeely, F.R., Vereecken, P.M. and Rosenberg, R. (2003) Liner materials for direct electrodeposition of Cu. *Appl. Phys. Lett.*, **83**, 2330–2332.
[10] Semiconductor Industry Association, (2009). *The International Technology Roadmap for Semiconductors (ITRS)*.
[11] Shacham-Diamand, Y. and Lopatin, S. (1999) Integrated electroless metallization for ULSI. *Electrochim. Acta*, **44**, 3639–3649.
[12] Ohde, H., Kramer, S., Moore, S. and Wai, C.M. (2004) Low-temperature deposition of conformal copper films in supercritical $CO_2$ by catalytic hydrogen reduction of copper hexafluoroacetylacetonate. *Chem. Mater.*, **16**, 4028–4031.
[13] Chu, J.P., Lin, C.H. and John, V.S. (2009) Barrier-free Cu metallization with a novel copper seed layer containing various insoluble substances. *Vacuum*, **83**, 668–671.
[14] Hoffmann, V., Hecker, M. and Hübner, R. (2005) Focussing and defocussing effects at radio frequency glow discharge optical emission spectroscopy analyses of thin films with partly nonconductive components. *Z. Metallkd.*, **96**, 983–987.
[15] Gu, C., Pivovarov, A., Garcia, R., Stevie, F., Griffis, D., Moran, J., Kulig, L. and Richards, J.F. (2004) Secondary ion mass spectrometry backside analysis of barrier layers for copper diffusion. *J. Vac. Sci. Technol. B*, **22**, 350–354.
[16] Hübner, R., Hecker, M., Mattern, N., Voss, A., Acker, J., Hoffmann, V., Wetzig, K., Engelmann, H.-J., Zschech, E., Heuer, H. and Wenzel, Ch. (2004) Influence of nitrogen content on the crystallization behavior of thin Ta–Si–N diffusion barriers. *Thin Solid Films*, **468**, 183–192.
[17] Image source: Utess, D. (2009) GLOBALFOUNDRIES, Dresden.
[18] Kohn, A. and Eizenberg, M. (2009) Diffusion barriers for ultra-large-scale integrated copper metallization, in *Advanced Nanoscale ULSI Interconnects: Fundamentals and Applications* (eds Shacham-Diamand, Y., Osaka, Y., Datta M. and Ohba T.), Springer Science+Business Media, New York, Dordrecht, Heidelberg, London.
[19] Ezer, Y., Härkönen, J., Sokolov, V., Saarilahti, J., Kaitila, J. and Kuivalainen, P. (1998) Diffusion barrier performance of thin Cr films in the Cu/Cr/Si structure. *Mater. Res. Bull.*, **33**, 1331–1337.
[20] Braud, F., Torres, J., Palleau, J., Mermet, J.L. and Mouche, M.J. (1995) Ti-diffusion barrier in Cu-based metallization. *Appl. Surf. Sci.*, **91**, 251–256.
[21] Chuang, J.-C., Tu, S.-L., and Chen, M.-C. (1999) Sputter-deposited Mo and reactively sputter-deposited Mo–N films as barrier layers against Cu diffusion. *Thin Solid Films*, **346**, 299–306.
[22] Mercier, M., Weber, S., Jacques, A., Hirabayashi, H., Ohkawa, H. and Kinoshita, M. (1997) Evaluation of sputtered tungsten as diffusion barrier materials for copper. *Defect Diffusion Forum*, **143–147**, 1285–1290.
[23] Laurila, T., Zeng, K., Kivilahti, J.K., Molarius, J. and Suni, I. (2000) Failure mechanism of Ta diffusion barrier between Cu and Si. *J. Appl. Phys.*, **88**, 3377–3384.
[24] Ono, H., Nakano, T. and Ohta, T. (1994) Diffusion barrier effects of transition metals for Cu/M/Si multilayers (M = Cr, Ti, Nb, Mo, Ta, W). *Appl. Phys. Lett.*, **64**, 1511–1513.
[25] Riedel, S., Schulz, S.E., Baumann, J., Rennau, M. and Gessner, T. (2001) Influence of different treatment techniques on the barrier properties of MOCVD TiN against copper diffusion. *Microelectron. Engng*, **55**, 213–218.

[26] Takeyama, M.B., Itoi, T., Satoh, K., Sakagami, M. and Noya, A. (2004) Application of thin nanocrystalline VN film as a high-performance diffusion barrier between Cu and $SiO_2$. *J. Vac. Sci. Technol. B*, **22**, 2542–2547.

[27] Chen, C.-S., Liu C.-P., Yang, H.-G. and Tsao, C.Y.A. (2004) Influence of the preferred orientation and thickness of zirconium nitride films on the diffusion property in copper. *J. Vac. Sci. Technol. B*, **22**, 1075–1083.

[28] Alén, P., Ritala, M., Arstila, K., Keinonen, J. and Leskelä, M. (2005) The growth and diffusion barrier properties of atomic layer deposited $NbN_x$ thin films. *Thin Solid Films*, **491**, 235–241.

[29] Ou, K.L. (2006) Integrity of copper–hafnium, hafnium nitride and multilayered amorphous-like hafnium nitride metallization under various thicknesses. *Microelectron. Engng*, **83**, 312–318.

[30] Hecker, M., Hubner, R., Ecke, R., Schulz, S., Engelmann, H.-J., Stegmann, H., Hoffmann, V., Mattern, N., Gessner, T. and Zschech, E. (2002) Effect of annealing on the microstructure of ultrathin tungsten nitride diffusion barriers for copper metallization. *Microelectron. Engng*, **64**, 269–277.

[31] Hecker, M., Fischer, D., Hoffmann, V., Engelmann, H.-J., Voss, A., Mattern, N., Wenzel, C., Vogt, C. and Zschech, E. (2002) Influence of N content on microstructure and thermal stability of Ta–N thin films for Cu interconnection. *Thin Solid Films*, **414**, 184–191.

[32] Wang, S.J., Tsai, H.Y., Sun, S.C. and Shiao, M.H. (2001) Thermal stability of sputtered tungsten carbide as diffusion barrier for copper metallization. *J. Electrochem. Soc.*, **148**, G500–G506.

[33] Tsai, H.Y., Sun, S.C. and Wang, S.J. (2000) Characterization of sputtered tantalum carbide layer for copper metallization. *J. Electrochem. Soc.*, **147**, 2766–2772.

[34] Charai, A., Hörnström, S.E., Thomas, O., Fryer, P.M. and Harper, J.M.E. (1989) A comparison between aluminum and copper interactions with high-temperature oxide and nitride diffusion barriers. *J. Vac. Sci. Technol. A*, **7**, 784–789.

[35] Hartman, J.W., Yeh, H., Atwater, H.A. and Hashim, I. (1999) Agglomeration of Cu electroplating seed layers on ultra-thin Ta, $Ta_{1-x}N_x$, $Ta_{1-x}O_x$, contaminated Ta, and composite Ta/$Ta_{1-x}N_x$ diffusion barriers. *Mater. Res. Soc. Symp. Proc.*, **564**, 257–262.

[36] Pelleg, J. and Sade, G. (2002) Diffusion barrier properties of amorphous $TiB_2$ for application in Cu metallization. *J. Appl. Phys.*, **91**, 6099–6104.

[37] Reid, J.S., Kolawa, E., Ruiz, R.P. and Nicolet, M.-A. (1993) Evaluation of amorphous (Mo, Ta, W)–Si–N diffusion barriers for <Si>|Cu metallizations. *Thin Solid Films*, **236**, 319–324.

[38] Wang, M.T., Lin, Y.C., Lee, J.Y., Wang, C.C. and Chen, M.C. (1999) Effective improvement on barrier capability of chemical vapor deposited $WSi_x$ using $N_2$ plasma treatment. *J. Electrochem. Soc.*, **146**, 1583–1592.

[39] Lee, Y.-J., Suh, B.-S., and Park, C.-O. (1999) Co-sputter deposited Ta–Si diffusion barrier between Si and Cu: the effects of Si content on the barrier property. *Thin Solid Films*, **357**, 237–241.

[40] Wang, S.-Q., Suthar, S., Hoeflich, C. and Burrow, B.J. (1993) Diffusion barrier properties of TiW between Si and Cu. *J. Appl. Phys.*, **73**, 2301–2320.

[41] Fang, J.S., Hsu, T.P. and Chen, G.S. (2006) Crystallization and failure behavior of Ta–TM (TM = Fe, Co) nanostructured/amorphous diffusion barriers for copper metallization. *J. Electron. Mater.*, **35**, 15–21.

[42] Noya, A., Takeyama, M.B. and Sase, T. (2005) Diffusion-barrier properties of $Ta_{1-x}W_x$ alloy films and silicidation-induced Cu penetration in Cu/Si contacts. *J. Vac. Sci. Technol. B*, **23**, 280–287.

[43] Thomas, R.E., Guo, K.J., Aaron, D.B., Dobisz, E.A., Perepezko, J.H. and Wiley, J.D. (1987) Investigation of amorphous $Ni_{0.60}Nb_{0.40}$ diffusion barriers. *Thin Solid Films*, **150**, 245–252.

[44] Takeyama, M., Kagomi, S., Noya, A., Sakanishi, K. and Sasaki, K. (1996) Application of amorphous Cu–Zr binary alloy as a diffusion barrier in Cu/Si contact systems. *J. Appl. Phys.*, **80**, 569–573.

[45] Subramanian, P.R. and Laughlin, D.E. (1989) The Cu–Ta (copper–tantalum) system. *Bull. Alloy Phase Diagrams*, **10**, 652–655.

[46] Kaufman, L. (1991) Coupled thermochemical and phase diagram data for tantalum based binary alloys. *Calphad*, **15**, 243–259.

[47] Fillot, F., Tökei, Zs. and Beyer, G.P. (2007) Surface diffusion of copper on tantalum substrates by Ostwald ripening. *Surf. Sci.*, **601**, 986–993.
[48] Zier, M., Oswald, S., Reiche, R., Kozlowska, M. and Wetzig, K. (2004) Interface formation and reactions at Ta–Si and Ta–$SiO_2$ interfaces studied by XPS and ARXPS. *J. Elec. Spec. Relat. Phenom.*, **137-140**, 229–233.
[49] Hübner, R., Hecker, M., Mattern, N., Hoffmann, V., Wetzig, K., Engelmann, H.-J. and Zschech, E. (2004) Comparison of the annealing behavior of thin Ta films deposited onto Si and $SiO_2$ substrates. *Anal. Bioanal. Chem.*, **379**, 568–575.
[50] Cabral Jr, C., Clevenger, L.A. and Schad, R.G. (1994) Repeated compressive stress increase with 400 °C thermal cycling in tantalum thin films due to increases in the oxygen content. *J. Vac. Sci. Technol. B*, **12**, 2818–2821.
[51] Cheng, J.Y. and Chen, L.J. (1991) Growth kinetics of amorphous interlayers by solid-state diffusion in ultrahigh vacuum deposited polycrystalline Nb and Ta thin films on (111) Si. *J. Appl. Phys.*, **69**, 2161–2168.
[52] Nguyen Tan, T.A., Azizan, M. and Derrien, J. (1987) Interfacial reaction between Ta ultrathin films and Si(111) substrate. *Surf. Sci.*, **189/190**, 339–345.
[53] Walser, R.M. and Bené, R.W. (1976) First phase nucleation in silicon–transition-metal planar interfaces. *Appl. Phys. Lett.*, **28**, 624–625.
[54] Pretorius, R., Harris, J.M. and Nicolet, M.-A. (1978) Reaction of thin metal films with $SiO_2$ substrates. *Solid State Electron.*, **21**, 667–675.
[55] Beyers, R. (1984) Thermodynamic considerations in refractory metal–silicon–oxygen systems. *J. Appl. Phys.*, **56**, 147–152.
[56] Peikert, M., Bhandari, R., Wieser, E., Wenzel, C., Lipp, D., Reuther, H. and Mücklich, A. (2004) The effect of silicon ion implantation on the structure of tantalum–silicon contacts. *Thin Solid Films*, **449**, 187–191.
[57] Lane, M., Dauskardt, R.H., Krishna, N. and Hashim, I. (2000) Adhesion and reliability of copper interconnects with Ta and TaN barrier layers. *J. Mater. Res.*, **15**, 203–211.
[58] Chen, L., Ekstrom, B. and Kelber, J. (1999) The effect of surface oxides on Cu/Ta interfacial interactions. *Mater. Res. Soc. Symp. Proc.*, **564**, 287–292.
[59] Simka, H., Shankar, S., Duran, C. and Haverty, M. (2005) Fundamentals of Cu/barrier-layer adhesion in microelectronic processing. *Mater. Res. Soc. Symp. Proc.*, **863**, 283–288.
[60] Kim, S. and Duquette, D.J. (2007) Quantitative measurement of interfacial adhesion between seedless electrodeposited copper and tantalum-based diffusion barriers for microelectronics. *Electrochem. Solid State Lett.*, **10**, D6–D9.
[61] Zong, Y., Shan, X. and Watkins, J.J. (2004) Sacrificial adhesion promotion layers for copper metallization of device structures. *Langmuir*, **20**, 9210–9216.
[62] Huang, H.Y., Hsieh, C.H., Jeng, S.M., Tao, H.J., Cao, M. and Mii, Y.J. (2010) A new enhancement layer to improve copper interconnect performance. Proceedings of the IEEE International Interconnect Technology Conference, p. 4.2.
[63] Nogami, T., Maniscalco, J., Madan, A., Flaitz, P., DeHaven, P., Parks, C., Tai, L., Lawrence, B.S., Davis, R., Murphy, R., Shaw, T., Cohen, S., Hu, C.-K., Cabral, C., Chiang, S., Kelly, J., Zaitz, M., Schmatz, J., Choi, S., Tsumura, K., Penny, C., Chen, H.-C., Canaperi, D., Vo, T., Ito, F., Straten, O., Simon, A., Rhee, S.-H., Kim, B.-Y., Bolom, T., Ryan, V., Ma, P., Ren, J., Aubuchon, J., Fine, J., Kozlowski, P., Spooner, T. and Edelstein, D. (2010) CVD Co and its application to Cu damascene interconnections. Proceedings of the IEEE International Interconnect Technology Conference, p. 3.1.
[64] Liu, L., Wang, Y. and Gong, H. (2001) Annealing effects of tantalum films on Si and $SiO_2$/Si substrates in various vacuums. *J. Appl. Phys.*, **90**, 416–420.
[65] Holloway, K. and Fryer, P.M. (1990) Tantalum as a diffusion barrier between copper and silicon. *Appl. Phys. Lett.*, **57**, 1736–1738.
[66] Fischer, D., Meissner, O., Bendjus, B., Schreiber, J., Stavrev, M. and Wenzel, C. (1997) AFM characterization of Ta-based diffusion barriers for use in future semiconductor metallization. *Surf. Interface Anal.*, **25**, 522–528.
[67] Face, D.W. and Prober, D.E. (1987) Nucleation of body-centered-cubic tantalum films with a thin niobium underlayer. *J. Vac. Sci. Technol. A*, **5**, 3408–3411.

[68] Chen, G.S., Lee, P.Y. and Chen, S.T. (1999) Phase formation behavior and diffusion barrier property of reactively sputtered tantalum-based thin films used in semiconductor metallization. *Thin Solid Films*, **353**, 264–273.
[69] Mutscheller, A., Clevenger, L.A., Harper, J.M.E., Cabral Jr, C. and Barmak, K. (1992) Effect of deposition conditions on intrinsic stress, phase transformation and stress relaxation in tantalum thin films. *Mater. Res. Soc. Symp. Proc.*, **239**, 51–56.
[70] Clevenger, L.A., Mutscheller, A., Harper, J.M.E., Cabral Jr, C. and Barmak, K. (1992) The relationship between deposition conditions, the beta to alpha phase transformation, and stress relaxation in tantalum thin films. *J. Appl. Phys.*, **72**, 4918–4924.
[71] Hoogeveen, R., Moske, M., Geisler, H. and Samwer, K. (1996) Texture and phase transformation of sputter-deposited metastable Ta films and Ta/Cu multilayers *Thin Solid Films*, **275**, 203–206.
[72] Kim, H., Cabral Jr, C., Lavoie, C. and Rossnagel, S.M. (2002) Diffusion barrier properties of transition metal thin films grown by plasma-enhanced atomic-layer deposition. *J. Vac. Sci. Technol. B*, **20**, 1321–1326.
[73] Mueller, M.H. (1997) The lattice parameter of tantalum. *Scr. Metall.*, **11**, 693.
[74] Smirnov, Yu.M. and Finkel, V.A. (1966) Crystal structure of tantalum, niobium, and vanadium at 110–400 K. *Soviet Physics JETP*, **22**, 750–753.
[75] Arakcheeva, A., Chapuis, G. and Grinevitch, V. (2002) The self-hosting structure of β-Ta. *Acta Cryst. B*, **58**, 1–7.
[76] Baker, P.N. (1972) Preparation and properties of tantalum thin films. *Thin Solid Films*, **14**, 3–25.
[77] Knepper, R. and Baker, S.P. (2007) Coefficient of thermal expansion and biaxial elastic modulus of β phase tantalum thin films. *Appl. Phys. Lett.*, **90**, 181908.
[78] Moshfegh, A.Z. and Akhavan, O. (2000) Bias sputtered Ta modified diffusion barriers in Cu/Ta($V_b$)/Si(111) structures. *Thin Solid Films*, **370**, 10–17.
[79] Catania, P., Doyle, J.P. and Cuomo, J.J. (1992) Low resistivity body-centered cubic tantalum thin films as diffusion barriers between copper and silicon. *J. Vac. Sci. Technol. A*, **10**, 3318–3321.
[80] Holloway, K., Fryer, P.M., Cabral Jr, C., Harper, J.M.E., Bailey, P.J. and Kelleher, K.H. (1992) Tantalum as a diffusion barrier between copper and silicon: failure mechanism and effect of nitrogen additions. *J. Appl. Phys.*, **71**, 5433–5444.
[81] Hübner, R. (2007) Advanced Ta-based diffusion barriers for Cu interconnects, in *New Research on Thin Solid Films* (ed. Benjamin M.G.), Nova Science Publishers Inc., New York.
[82] Clevenger, L.A., Bojarczuk, N.A., Holloway, K., Harper, J.M.E., Cabral Jr, C., Schad, R.G., Cardone, F. and Stolt, L. (1993) Comparison of high vacuum and ultra-high-vacuum tantalum diffusion barrier performance against copper penetration. *J. Appl. Phys.*, **73**, 300–308.
[83] Yin, K.-M., Chang, L., Chen, F.-R., Kai, J.-J., Chiang, C.-C., Chuang, G., Ding, P., Chin, B., Zhang, H. and Chen, F. (2001) Oxidation of Ta diffusion barrier layer for Cu metallization in thermal annealing. *Thin Solid Films*, **388**, 27–33.
[84] Laurila, T., Zeng, K., Kivilahti, J.K., Molarius, J. and Suni, I. (2001) Effect of oxygen on the reactions in the Si/Ta/Cu metallization system. *J. Mater. Res.*, **16**, 2939–2946.
[85] Aubel, O., Meyer, M.A., Feustel, F., Engelmann, H.J., Zienert, I., Poppe, J., Schmidt, H., Gehre, D., Geisler, H., Langer, E., Limbecker, P., Foltyn, T., Witt, C., Yao, W., Thierbach, S., Koschinsky, F. and Zistl, C. (2008) Extensive investigations of temperature influence on barrier integrity during reliability testing. *Microelectron. Engng*, **85**, 2042–2046.
[86] Kang, B.-S., Lee, S.-M., Kwak, J.S., Yoon, D.-S. and Baik, H.K. (1997) The effectiveness of Ta prepared by ion-assisted deposition as a diffusion barrier between copper and silicon. *J. Electrochem. Soc.*, **144**, 1807–1812.
[87] Yoon, D.-S., Baik, H.K. and Lee, S.-M. (1998) Tantalum–microcrystalline $CeO_2$ diffusion barrier for copper metallization. *J. Appl. Phys.*, **83**, 1333–1336.
[88] Kwak, J.S., Baik, H.K., Kim, J.-H. and Lee, S.-M. (1998) Improvement of Ta diffusion barrier performance in Cu metallization by insertion of a thin Zr layer into Ta film. *Appl. Phys. Lett.*, **72**, 2832–2834.
[89] Kwak, J.S., Baik, H.K., Kim, J.H., Lee, S.-M., Ryu, H.J. and Je, J.H. (1999) Effect of thin V insertion layer into Ta film on the performance of Ta diffusion barrier in Cu metallization. *J. Appl. Phys.*, **85**, 6898–6903.

[90] Hu, C.-K., Chang, S., Small, M.B. and Lewis, J.E. (1986) Diffusion barrier studies for Cu. Proceedings of the Third International VLSI Multilevel Interconnection Conference, pp. 181–187.
[91] Olowolafe, J.O., Mogab, C.J. and Gregory, R.B. (1993) Evaluation of Ta, Ti and TiW encapsulations for corrosion and diffusion protection of Cu interconnects. *Thin Solid Films*, **227**, 37–43.
[92] Wang, M.T., Lin, Y.C. and Chen, M.C. (1998) Barrier properties of very thin Ta and TaN layers against copper diffusion. *J. Electrochem. Soc.*, **145**, 2538–2545.
[93] Jang, S.-Y., Lee, S.-M. and Baik, H.K. (1996) Tantalum and niobium as a diffusion barrier between copper and silicon. *J. Mater. Sci.: Mater. Electron.*, **7**, 271–278.
[94] Hübner, R., Hecker, M., Mattern, N., Hoffmann, V., Wetzig, K., Wenger, Ch., Engelmann, H.-J., Wenzel, Ch., Zschech, E. and Bartha, J.W. (2003) Structure and thermal stability of graded Ta–TaN diffusion barriers between Cu and $SiO_2$. *Thin Solid Films*, **437**, 248–256.
[95] Mayer, U. (2010) Metrology challenges in high volume ULK production. Invited talk, Semicon Europe 2010, 20 October 2010, Dresden.
[96] Maex, K., Baklanov, M.R., Shamiryan, D., Iacopi, F., Brongersma, S.H. and Yanovitskaya, Z.S. (2003) Low dielectric constant materials for microelectronics. *J. Appl. Phys.*, **93**, 8793–8841.
[97] Chen, Z., Prasad, K., Li, C.Y., Lu, P.W., Su, S.S., Tang, L.J., Gui, D., Balakumar, S., Shu, R. and Kumar, R. (2004) Dielectric/metal sidewall diffusion barrier for Cu/porous ultralow-*k* interconnect technology. *Appl. Phys. Lett.*, **84**, 2442–2444.
[98] Lau, W.-S., Tan, H.-J., Chen, Z. and Li, C.-Y. (2007) A comparison of various dielectric/metal sidewall diffusion barriers for Cu/porous ultra-low-*k* interconnect technology in terms of leakage current and breakdown voltage. *Vacuum*, **81**, 1040–1046.
[99] Chen, Z., Prasad, K., Jiang, N., Tang, L.J., Lu, P.W. and Li, C.Y. (2005) Silicon carbide based dielectric composites in bilayer sidewall barrier for Cu/porous ultralow-*k* interconnects. *J. Vac. Sci. Technol. B*, **23**, 1866–1872.
[100] Gupta, D. (1995) Diffusion in several materials relevant to Cu interconnection technology. *Mater. Chem. Phys.*, **41**, 199–205.
[101] Sun, X., Kolawa, E., Chen, J.-S., Reid, J.S. and Nicolet, M.-A. (1993) Properties of reactively sputter-deposited Ta–N thin films. *Thin Solid Films*, **236**, 347–351.
[102] Stavrev, M., Fischer, D., Wenzel, C., Drescher, K. and Mattern, N. (1997) Crystallographic and morphological characterization of reactively sputtered Ta, Ta–N and Ta–N–O thin films. *Thin Solid Films*, **307**, 79–88.
[103] Baumann, J., Schwalbe, G., Zimmermann, S., Kaufmann, C. and Gessner, T. (2002) Phase, structure and properties of sputtered Ta and $TaN_x$ films. MRS Proceedings of the Advanced Metallization Conference, AMC 2001 (ULSI XVII), pp. 613–618.
[104] Shin, C.-S., Kim, Y.-W., Gall, D., Greene, J.E. and Petrov, I. (2002) Phase composition and microstructure of polycrystalline and epitaxial $TaN_x$ layers grown on oxidized Si(001) and MgO(001) by reactive magnetron sputter deposition. *Thin Solid Films*, **402**, 172–182.
[105] Bae, J.W., Lim, J.-W., Mimura, K. and Isshiki, M. (2007) Ion beam deposition of α-Ta films by nitrogen addition and improvement of diffusion barrier property. *Thin Solid Films*, **515**, 4768–4773.
[106] Mashimo, T., Tashiro, S., Toya, T., Nishida, M., Yamazaki, H., Yamaya, S., Oh-Ishi, K. and Syono, Y. (1993) Synthesis of the B1-type tantalum nitride by shock compression. *J. Mater. Sci.*, **28**, 3439–3443.
[107] Greenwood, N.N. and Earnshaw, A. (1997) *Chemistry of the Elements*, 2nd edn, Butterworth-Heinemann, Oxford.
[108] Ritala, M., Kalsi, P., Riihelä, D., Kukli, K., Leskelä, M. and Jokinen, J. (1999) Controlled growth of TaN, $Ta_3N_5$, and $TaO_xN_y$ thin films by atomic layer deposition. *J. Chem. Mater.*, **11**, 1712–1718.
[109] Alén, P., Juppo, M., Ritala, M., Sajavaara, T., Keinonen, J. and Leskelä, M. (2001) Atomic layer deposition of Ta(Al)N(C) thin films using trimethylaluminum as a reducing agent. *J. Electrochem. Soc.*, **148**, G566–G571.
[110] Juppo, M., Ritala, M. and Leskelä, M. (2000) Use of 1,1-dimethylhydrazine in the atomic layer deposition of transition metal nitride thin films. *J. Electrochem. Soc.*, **147**, 3377–3381.

[111] Alén, P., Juppo, M., Ritala, M., Leskelä, M., Sajavaara, T. and Keinonen, J. (2002) *tert*-Butylamine and allylamine as reductive nitrogen sources in atomic layer deposition of TaN thin films. *J. Mater. Res.*, **17**, 107–114.

[112] Kim, H., Kellock, A.J. and Rossnagel, S.M. (2002) Growth of cubic-TaN thin films by plasma-enhanced atomic layer deposition. *J. Appl. Phys.*, **92**, 7080–7085.

[113] Besling, W.F.A., Ignacimouttou, M.-L., Humbert, A., Mellier, M. and Torres, J. (2004) Continuity and morphology of TaN barriers deposited by atomic layer deposition and comparison with physical vapor deposition. *Microelectron. Engng*, **76**, 60–69.

[114] Zier, M., Oswald, S., Reiche, R. and Wetzig, K. (2005) XPS and ARXPS investigations of ultra thin TaN films deposited on $SiO_2$ and Si. *Appl. Surf. Sci.*, **252**, 234–239.

[115] Violet, P., Blanquet, E. and Le Bacq, O. (2006) Density functional study of the stability and electronic properties of $Ta_xN_y$ compounds used as copper diffusion barriers. *Microelectron. Engng*, **83**, 2077–2081.

[116] Kim, S. and Duquette, D.J. (2006) Multiple bath usage for adhesion enhancement of directly electrodeposited copper on TaN. *Electrochem. Solid State Lett.*, **9**, C38–C40.

[117] van der Straten, O., Zhu, Y., Rullan, J., Dunn, K. and Kaloyeros, A.E. (2006) Study of copper-refractory metal interfaces via solid-state wetting for emerging nanoscale interconnect applications. *J. Mater. Res.*, **21**, 255–262.

[118] Burgess, S.R., Donohue, H., Buchanan, K., Rimmer, N. and Rich, P. (2002) Evaluation of Ta and TaN-based Cu diffusion barriers deposited by advanced hi-fill (AHF) sputtering onto blanket wafers and high aspect ratio structures. *Microelectron. Engng*, **64**, 307–313.

[119] Zimmermann, S., Baumann, J., Kaufmann, C. and Gessner, T. (2003) Thermal stability of thin Ta and $TaN_x$ films as diffusion barriers for copper metallization. MRS Proceedings of the Advanced Metallization Conference, AMC 2002 (ULSI XVIII), pp. 859–864.

[120] Kim, H., Lavoie, C., Copel, M., Narayanan, V., Park, D.-G. and Rossnagel, S.M. (2004) The physical properties of cubic plasma-enhanced atomic layer deposition TaN films. *J. Appl. Phys.*, **95**, 5848–5855.

[121] Kim, H., Detavenier, C., van der Straten, O., Rossnagel, S.M., Kellock, A.J. and Park, D.-G. (2005) Robust $TaN_x$ diffusion barrier for Cu-interconnect technology with subnanometer thickness by metal–organic plasma-enhanced atomic layer deposition. *J. Appl. Phys.*, **98**, 014308.

[122] Min, K.-H., Chun, K.-C. and Kim, K.-B. (1996) Comparative study of tantalum and tantalum nitrides ($Ta_2N$ and TaN) as a diffusion barrier for Cu metallization. *J. Vac. Sci. Technol. B*, **14**, 3263–3269.

[123] Chen, G.S. and Chen, S.T. (2000) Diffusion barrier properties of single- and multilayered quasi-amorphous tantalum nitride thin films against copper penetration. *J. Appl. Phys.*, **87**, 8473–8482.

[124] Oku, T., Kawakami, E., Uekubo, M., Takahiro, K., Yamaguchi, S. and Murakami, M. (1996) Diffusion barrier property of TaN between Si and Cu. *Appl. Surf. Sci.*, **99**, 265–272.

[125] Nogami, T., Joo, Y.-C., Lopatin, S., Romero, J., Bernard, J., Blum, W., Engelmann, H.-J., Gray, J., Tracy, B., Chen, B., Lukanc, T., Brown, D., Besser, P., Morales, G. and Cheung, R. (1999) Graded Ta/TaN/Ta barrier for copper interconnects for high electromigration resistance. MRS Proceedings of the Advanced Metallization Conference, AMC 1998 (ULSI XIV), pp. 313–319.

[126] Chen, G.S., Huang, S.C., Chen, S.T., Yang, T.J., Lee, P.Y., Jou, J.H. and Lin, T.C. (2000) An optimal quasisuperlattice design to further improve thermal stability of tantalum nitride diffusion barriers. *Appl. Phys. Lett.*, **76**, 2895–2897.

[127] Fréty, N., Bernard, F., Nazon, J., Sarradin, J. and Tedenac, J.C. (2006) Copper diffusion into silicon substrates through TaN and Ta/TaN multilayer barriers. *J. Phase Equilib. Diff.*, **27**, 590–597.

[128] Ishikawa, K., Miyauchi, M., Ashihara, H., Saitoh, T., Tanaka, U., Ohshima, T., Torii, K., Ishihara, S., Aoki, H., Yamaguchi, H., Hinode, K. and Saito, T. (2002) Electromigration resistance improvement of dual-damascene copper interconnection using TaN/Ta barrier formed by ionized bias sputtering. MRS Proceedings of the Advanced Metallization Conference, AMC 2001 (ULSI XVII), pp. 487–492.

[129] Traving, M., Zienert, I., Zschech, E., Schindler, G., Steinhögl, W. and Engelhardt, M. (2005) Phase analysis of TaN/Ta barrier layers in sub-micrometer trench structures for Cu interconnects. *Appl. Surf. Sci.*, **252**, 11–17.
[130] Ho, P., Rajagopalan, R., Chong, H., Chung, H. and Yu, J. (2004) ALD/PVD barrier reduces RC and improves reliability. *Semiconductor Int.*, **27** (6), 61–62, 64, 67.
[131] Moon, B.K., Iijima, T., Malhotra, S., Simon, A., Shaw, T., Ryan, E.T., Labelle, C., Fuller, N., Bolom, T., Dunn, D., Flaitz, P., Mehta, S., Inoue, K. and McGahay, V. (2005) Integration of ALD-TaN liners on nanoporous dielectrics. *Mater. Res. Soc. Symp. Proc.*, **863**, 221–226.
[132] Murarka, S.P. and Fraser, D.B. (1980) Silicide formation in thin cosputtered (tantalum + silicon) films on polycrystalline silicon and $SiO_2$. *J. Appl. Phys.*, **51**, 1593–1598.
[133] Baiocchi, F.A., Lifshitz, N., Sheng, T.T. and Murarka, S.P. (1988) Equilibration of nonstoichiometric Ta–Si deposits on polycrystalline silicon at high temperatures. *J. Appl. Phys.*, **64**, 6490–6495.
[134] Fischer, D., Stavrev, M., Urbansky, N., Wenzel, C., Neubauer, E., Bauer, J.G., Scherg, T. and Schulze, H.-J. (1999) Barrier and contact behavior of tantalum-based thin films for use in copper metallization scheme. MRS Proceedings of the Advanced Metallization Conference, AMC 1998 (ULSI XIV), pp. 337–344.
[135] Kolawa, E., Chen, J.S., Reid, J.S., Pokela, P.J. and Nicolet, M.-A. (1991) Tantalum-based diffusion barriers in Si/Cu VLSI metallizations. *J. Appl. Phys.*, **70**, 1369–1373.
[136] Pokela, P.J., Reid, J.S., Kwok, C.-K., Kolawa, E. and Nicolet, M.-A. (1991) Thermal oxidation of amorphous ternary $Ta_{36}Si_{14}N_{50}$ thin films. *J. Appl. Phys.*, **70**, 2828–2832.
[137] Cabral Jr, C., Saenger, K.L., Kotecki, D.E. and Harper, J.M.E. (2000) Optimization of Ta–Si–N thin films for use as oxidation-resistant diffusion barrier. *J. Mater. Res.*, **15**, 194–198.
[138] Kwok, C.-K., Kolawa, E. Nicolet, M.-A. and Lee, R.C. (1991) Stress and resistivity in reactively sputtered amorphous metallic Ta–Si–N films. *Mater. Res. Soc. Symp. Proc.*, **226**, 261–266.
[139] Venkatraman, R., Mendonca, J., Ong, T.P., Hamilton, G., Rogers, B., Frisa, L., Simpson, C., Kaushik, V., Gregory, R., Apen, E., Coronell, D., Angyal, M., Jain, A., Herrick, M., Capasso, C., Fiordalice, R., Klein, J. and Weitzman, E. (1998) Process development and integration of PVD Ta–Si–N films for copper interconnect applications. MRS Proceedings of the Advanced Metallization Conference, AMC 1997 (ULSI XIII), pp. 63–70.
[140] Hara, T., Yoshida, Y. and Toida, H. (2002) Improved barrier and adhesion properties in sputtered TaSiN layer for copper interconnects. *Electrochem. Solid State Lett.*, **5**, G36–G39.
[141] Lee, H. and Lopatin, S.D. (2005) The influence of barrier types on the microstructure and electromigration characteristics of electroplated copper. *Thin Solid Films*, **492**, 279–284.
[142] Reitz, D., Heuer, H., Baunack, S., Hübner, R., Hoffmann, V., Menzel, S., Wenzel, C. and Wetzig, K. (2005) Investigation of a Ta–Si–O/Ta–Si–N bilayer system for embedded SAW finger structures. *Microelectron. Engng*, **82**, 301–306.
[143] Heuer, H., Wenzel, C., Herrmann, D., Hübner, R., Zhang, Z.L. and Bartha, J.W. (2006) Thin tantalum–silicon–oxygen/tantalum–silicon–nitrogen films as high efficiency humidity diffusion barriers for solar cell encapsulation. *Thin Solid Films*, **515**, 1612–1617.
[144] Lee, Y.-J., Suh, B.-S., Kwon, M.S. and Park, C.-O. (1999) Barrier properties and failure mechanism of Ta–Si–N thin films for Cu interconnection. *J. Appl. Phys.*, **85**, 1927–1934.
[145] Fischer, D., Scherg, T., Bauer, J.G., Schulze, H.-J. and Wenzel, C. (2000) Study of Ta–Si–N thin films for use as barrier layer in copper metallization. *Microelectron. Engng*, **50**, 459–464.
[146] Aouadi, S.M., Zhang, Y., Bohnhoff, A., Lee, J. and Williams, M. (2005) Real-time spectroscopic ellipsometry study of Ta–Si–N ultrathin diffusion barriers. *J. Vac. Sci. Technol. A*, **23**, 1359–1363.
[147] Angyal, M.S., Shacham-Diamand, Y., Reid, J.S. and Nicolet, M.-A. (1995) Performance of tantalum–silicon–nitride diffusion barriers between copper and silicon dioxide. *Appl. Phys. Lett.*, **67**, 2152–2154.
[148] Hübner, R., Hecker, M., Mattern, N., Hoffmann, V., Wetzig, K., Wenger, Ch., Engelmann, H.-J., Wenzel, Ch. and Zschech, E. (2004) Degradation mechanisms of Ta and Ta–Si diffusion barriers during thermal stressing. *Thin Solid Films*, **458**, 237–245.
[149] Hübner, R., Reiche, R., Hecker, M., Mattern, N., Hoffmann, V., Wetzig, K., Heuer, H., Wenzel, Ch., Engelmann, H.-J. and Zschech, E. (2005) Void formation in the Cu layer during thermal treatment of $SiN_x/Cu/Ta_{73}Si_{27}/SiO_2/Si$ systems. *Cryst. Res. Technol.*, **40**, 135–142.

[150] Hübner, R., Hecker, M., Mattern, N., Hoffmann, V., Wetzig, K., Heuer, H., Wenzel, Ch., Engelmann, H.-J., Gehre, D. and Zschech, E. (2006) Effect of nitrogen content on the degradation mechanisms of thin Ta–Si–N diffusion barriers for Cu metallization. *Thin Solid Films*, **500**, 259–267.

[151] Reid, J.S. (1995) Amorphous ternary diffusion barriers for silicon metallizations. PhD thesis, California Institute of Technology, Pasadena, California.

[152] Hübner, R., Hecker, M., Wetzig, K., Heuer, H., Wenzel, Ch., Engelmann, H.-J. and Zschech, E. (2006) Thermal behavior of graded Ta–Si/Ta–Si–N diffusion barriers for Cu metallization. *AIP Conf. Proc.*, **817**, 59–64.

[153] Becker, J.S. and Gordon, R.G. (2003) Diffusion barrier properties of tungsten nitride films grown by atomic layer deposition from bis(*tert*-butylimido)bis(dimethylamido) tungsten and ammonia. *Appl. Phys. Lett.*, **82**, 2239–2241.

[154] Ecke, R., Schulz, S.E., Hecker, M. and Gessner, T. (2002) Development of PECVD $WN_x$ ultrathin films as barrier layer for copper metallization, *Microelectron. Engng*, **64**, 261–268.

[155] Abell, T., Shamiryan, T., Schuhmacher, J., Besling, W., Sutcliffe, V. and Maex, K. (2003) Precursor penetration and sealing of porous CVD SiCOH low k dielectric for atomic layer deposition of $WC_xN_y$ barrier. MRS Proceedings of the Advanced Metallization Conference, AMC 2002 (ULSI XVIII), pp. 717–723.

[156] Chen, Y.-C., Hung, T.-Y., Chang, Y.-L., Shieh, K., Hsu, C.-L., Huang, C., Yan, W.H., Ashtiani, K., Pisharoty, D., Lei, W., Chang, S., Huang, F., Collins, J. and Tzou, S.F. (2007) Optimizing ALD WN process for 65 nm node CMOS contact application. Proceedings of the IEEE International Interconnect Technology Conference, pp. 105–107.

[157] Yamamoto, N., Hanaoka, Y. and Yoshida, T. (2005) Thermal stability of nitrogen in $WN_x$ barriers applied to polymetal gates. *J. Vac. Sci. Technol. B*, **23**, 1664–1673.

[158] Ecke, R., Schulz, S.E., Hecker, M., Mattern, N. and Gessner, T. (2003) Influence of $SiH_4$ on the WN-PECVD process. *Microelectron. Engng*, **70**, 346–351.

[159] Qu, X., Lu, H., Peng, T., Ru, G. and Li, B. (2004) Effects of preannealing on the diffusion barrier properties for ultrathin W–Si–N thin film. *Thin Solid Films*, **462–463**, 67–71.

[160] Vomiero, A., Marchi, E.B., Frabboni, S., Quaranta, A., Mea, G.D., Mariotto, G., Felisari, L. and Butturi, M. (2004) Effects of thermal annealing on the structural properties of sputtered W–Si–N diffusion barriers. *Mater. Sci. Semicon. Proc.*, **7**, 325–330.

[161] Shiojima, K., McInturff, D.T., Woodall, J.M., Grudowski, P.A., Eiting, C.J. and Dupuis, R.D. (1999) Electrical characteristics and thermal stability of W, WSiN, and Nb contacts to p- and n-type GaN. *J. Electron. Mater.*, **28**, 228–233.

[162] Würfl, J., Hilsenbeck, J., Nebauer, E., Tränkle, G., Obloh, H. and Österle, W. (2000) Reliability of AlGaN/GaN HFETs comprising refractory ohmic and Schottky contacts. *Microelectron. Reliab.*, **40**, 1689–1693.

[163] Lau, W.S., Singh, B.P., Gunawan, S., Tan, J.B.H., Jiang, Y.Y., Wynn, R.T., Lim, V.S.W. and Trigg, A.D. (2005) WSiN cap for ohmic contacts to n-GaN with better morphology. *Phys. Status Solidi (c)*, **2**, 2548–2550.

[164] Wu, W., Wu, H.-J., Dixit, G., Shaviv, R., Gao, M., Mountsier, T., Harm, G., Dulkin, A., Fuchigami, N., Kailasam, S.K., Klawuhn, E. and Havemann, R.H. (2008) Ti-based barrier for Cu interconnect applications. Proceedings of IEEE International Interconnect Technology Conference, pp. 202–204.

[165] Sakata, A., Yamashita, S., Omoto, S., Hatano, M., Wada, J., Higashi, K., Yamaguchi, H., Yosho, T., Imamizu, K., Yamada, M., Hasunuma, M., Takahashi, S., Yamada, A., Hasegawa, T. and Kaneko, H. (2006) Reliability improvement by adopting Ti-barrier metal for porous low-*k* ILD structure. Proceedings of the IEEE International Interconnect Technology Conference, pp. 101–103.

[166] Lee, W.-H., Kuo, Y.-L., Huang, H.-J. and Lee, C. (2004) Effect of density on the diffusion barrier property of $TiN_x$ films between Cu and Si. *Mater. Chem. Phys.*, **85**, 444–449.

[167] Kizil, H. and Steinbrüchel, C. (2004) TiN, TaN and $W_xN$ as diffusion barriers for Cu on $SiO_2$: capacitance-voltage, leakage current, and triangular-voltage-sweep tests after bias temperature stress. *Thin Solid Films*, **449**, 158–165.

[168] Lee, C. and Kuo, Y.-L. (2007) The evolution of diffusion barriers in copper metallization. *JOM J. Min. Met. Mater. Soc.*, **59**, 44–49.

[169] Bonitz, J., Schulz, S.E. and Gessner, T. (2004) Ultra thin CVD TiN layers as diffusion barrier films on porous low-*k* materials. *Microelectron. Engng*, **76**, 82–88.

[170] Guedj, C., Arnal, V., Guillaumond, J.F., Arnaud, L., Barnes, J.P., Toffoli, A., Jousseaume, V., Roule, A., Maitrejean, S., Chapelon, L.L., Reimbold, G., Torres, G. and Passemard, G. (2005) Influence of the diffusion barriers on the dielectric reliability of ULK/Cu advanced interconnects. Proceedings of the IEEE International Interconnect Technology Conference, pp. 57–59.

[171] Suh, B.-S., Choi, S.-M., Wee, Y., Lee, J.-E., Lee, J., Lee, S.-J., Lee, S.-G., Shin, H., Lee, N.-I., Kang, H.-K. and Suh, K. (2005) Integration and reliability of a noble TiZr/TiZrN alloy barrier for Cu/low-*k* interconnects. Proceedings of the IEEE International Interconnect Technology Conference, pp. 138–140.

[172] Majumder, P. and Takoudis, C. (2008) Thermal stability of Ti/Mo and Ti/MoN nanostructures for barrier applications in Cu interconnects. *Nanotechnology*, **19**, 205202.

[173] Majumder, P. and Takoudis, C.G. (2007) Investigation on the diffusion barrier properties of sputtered Mo/W–N thin films in Cu interconnects. *Appl. Phys. Lett.*, **91**, 162108.

[174] Leu, L.C., Sadik, P., Norton, D.P., McElwee-White, L. and Anderson, T.J. (2008) Comparative study of ZrN and Zr–Ge–N thin films as diffusion barriers for Cu metallization on Si. *J. Vac. Sci. Technol. B*, **26**, 1723–1727.

[175] Alén, P., Ritala, M., Arstila, K., Keinonen, J. and Leskelä, M. (2005) The growth and diffusion barrier properties of atomic layer deposited $NbN_x$ thin films. *Thin Solid Films*, **491**, 235–241.

[176] Park, J.-S. and Kang, S.-W. (2004) Plasma-enhanced ALD of titanium–silicon nitride using $TiCl_4$, $SiH_4$, and $N_2/H_2/Ar$ plasma. *Electrochem. Solid State Lett.*, **7**, C87–C89.

[177] Aouadi, S.M., Zhang, Y., Basnyat, P., Stadler, S., Filip, P., Williams, M., Hilfiker, J.N., Singh, N. and Woollam, J.A. (2006) Physical and chemical properties of sputter-deposited $TaC_xN_y$ films. *J. Phys.: Condens. Matter*, **18**, 1977–1986.

[178] Lin, S. and Lee, C. (2006) Characteristics of sputtered Ta–B–N thin films as diffusion barriers between copper and silicon. *Appl. Surf. Sci.*, **253**, 1215–1221.

[179] Lin, T., Cheng, H., Chin, T., Chiu, C. and Fang, J. (2007) 5-nm-thick TaSiC amorphous films stable up to 750 °C as a diffusion barrier for copper metallization. *Appl. Phys. Lett.*, **91**, 152908.

[180] Koh, W., Kumar, D., Li, W.-M., Sprey, H. and Raaijmakers, I.J. (2005) Meeting the Cu diffusion barrier challenge using ALD tungsten nitride carbide. *Solid State Technol.*, **6**, 54–58.

[181] Rawal, S. Norton, D.P. Ajmera, H. Anderson, T.J. and McElwee-White, L. Properties of Ta-Ge-(O)N as a diffusion barrier for Cu on Si. *Appl. Phys. Lett.,* **90,** 051913 (2007).

[182] Rawal, S., Norton, D.P., Anderson, T.J. and McElwee-White, L. (2005) Properties of W–Ge–N as a diffusion barrier material for Cu. *Appl. Phys. Lett.*, **87**, 111902.

[183] Fang, J.S., Ke, M.L. and Chen, H. C. (2007) Evaluation of DC-sputtered glassy TaCoN thin film for copper metallization. *J. Electron Mater.*, **36**, 1462–1468.

[184] Ruan, J.-L., Huang, J.-L., Chen, J.S. and Lii, D.F. (2005) Effects of substrate bias on the reactive sputtered Zr–Al–N diffusion barrier films. *Surf. Coat. Technol.*, **200**, 1652–1658.

[185] Rawal, S., Norton, D.P., Kim, K., Anderson, T.J. and McElwee-White, L. (2006) Ge/$HfN_x$ diffusion barrier for Cu metallization on Si. *Appl. Phys. Lett.*, **89**, 231914.

[186] Kim, K.H., Norton, D.P., Christen, D.K., Cantoni, C., Paranthaman, M. and Aytug, T. (2009) Epitaxial growth of MgO/TiN multilayers on Cu. *Vacuum*, **83**, 897–901.

[187] Shin, Y.-H. and Shimogaki, Y. (2004) Diffusion barrier property of TiN and TiN/Al/TiN films deposited with FMCVD for Cu interconnection in ULSI. *Sci. Technol. Adv. Mater.*, **5**, 399–405.

[188] Kim, H.C. and Alford, T.L. (2004) Investigation on diffusion barrier properties of reactive sputter deposited $TiAl_xN_yO_z$ thin films for Cu metallization. *Thin Solid Films*, **449**, 6–11.

[189] Chang, S.-Y. and Chen, D.-S. (2009) 10-nm-thick quinary (AlCrTaTiZr)N film as effective diffusion barrier for Cu interconnects at 900 °C. *Appl. Phys. Lett.*, **94**, 231909.

[190] Tsai, M.-H., Yeh, J.-W. and Gan, J.-Y. (2008) Diffusion barrier properties of AlMoNbSiTaTiVZr high-entropy alloy layer between copper and silicon. *Thin Solid Films*, **516**, 5527–5530.

[191] Chyan, O., Arunagiri, T.N. and Ponnuswamy, T. (2003) Electrodeposition of copper thin film on ruthenium: a potential diffusion barrier for Cu interconnects. *J. Electrochem. Soc.*, **150**, C347–350.

[192] Sari, W., Eom, T.-K., Jeon, C.-W., Sohn, H. and Kim, S.-H. (2009) Improvement of the diffusion barrier performance of Ru by incorporating a $WN_x$ thin film for direct-plateable Cu interconnects. *Electrochem. Solid State Lett.*, **12**, H248–H251.

[193] Greenslit, D., Chakraborty, T. and Eisenbraun, E. (2009) Development of plasma-enhanced atomic layer deposition grown Ru–WCN mixed phase films for nanoscale diffusion barrier and copper direct-plate applications. *J. Vac. Sci. Technol. B*, **27**, 631–636.

[194] Nakamura, N., Takigawa, Y., Soda, E., Hosoi, N., Tarumi, Y., Aoyama, H., Tanaka, Y., Kawamura, D., Ogawa, S., Oda, N., Kondo, S., Mori, I. and Saito, S. (2009) Design impact study of wiring size and barrier metal on device performance toward 22 nm-node featuring EUV lithography. Proceedings of the IEEE International Interconnect Technology Conference, pp. 14–16.

[195] Rossnagel, S. (2005) Ru–Cu interconnect technology. *Solid State Technol.*, **2**, 1–4.

[196] Arunagiri, T.N., Zhang, Y., Chyan, O., El-Bouanani, M., Kim, M.J., Chen, K.H., Wu, C.T. and Chen, L.C. (2005) 5 nm ruthenium thin film as a directly plateable copper diffusion barrier. *Appl. Phys. Lett.*, **86**, 083104.

[197] Wang, J., Zhang, S., Wang, H., Xu, P., Iha, D. and Sawasaki, S. (2008) Ruthenium interlayer as diffusion barrier under carbon overcoat in magnetic recording media. *J. Nanosci. Nanotechnol.*, **8**, 2613–2617.

[198] Lim, J., Park, H. and Lee, C. (2005) Enhancement of Ru nucleation by pretreatments of the underlying TaSiN film surface in Ru MOCVD. *Thin Solid Films*, **475**, 194–197.

[199] Musschoot, J., Xie, Q., Deduytsche, D., De Keyser, K., Longrie, D., Haemers, J., Van den Berghe, S., Van Meirhaeghe, R.L., D'Haen, J. and Detavernier, C. (2010) Texture of atomic layer deposited ruthenium. *Microelectron. Engng*, **87**, 1879–1883.

[200] Damayanti, M., Sritharan, T., Mhaisalkar, S.G., Engelmann, H.J., Zschech, E., Vairagar, A.V. and Chan, L. (2007) Microstructural evolution of annealed ruthenium–nitrogen films. *Electrochem. Solid State Lett.*, **10**, P15–P17.

[201] Ding, S.-F., Xie, Q., Waechtler, T., Lu, H.-S., Schulz, S.E. and Qu, X.-P. (2010) Inhibition of enhanced Cu oxidation on ruthenium. Proc. of IEEE International Interconnect Technology Conference, p. 8.9.

[202] Qu, X.-P., Tan, J.J., Zhou, M., Chen, T., Xie, Q., Ru, G.-P. and Li, B.-Z. (2006) Improved barrier properties of ultrathin Ru film with TaN interlayer for copper metallization. *Appl. Phys. Lett.*, **88**, 151912.

[203] Choi, B.H., Lim, Y.H., Lee, J.H., Kim, Y.B., Lee, H.-N. and Lee, H.K. (2010) Preparation of Ru thin film layer on Si and TaN/Si as diffusion barrier by plasma enhanced atomic layer deposition. *Microelectron. Engng*, **87**, 1391–1395.

[204] Kumar, S., Greenslit, D., Chakraborty, T. and Eisenbraun, E.T. (2009) Atomic layer deposition growth of a novel mixed-phase barrier for seedless copper electroplating applications. *J. Vac. Sci. Technol. A*, **27**, 572–576.

[205] Perng, D.-C., Hsu, K.-C., Tsai, S.-W. and Yeh, J.-B. (2010) Thermal and electrical properties of PVD Ru(P) film as Cu diffusion barrier. *Microelectron. Engng*, **87**, 365–369.

[206] Henderson, L.B. and Ekerdt, J.G. (2009) Time-to-failure analysis of 5 nm amorphous Ru(P) as a copper diffusion barrier. *Thin Solid Films*, **517**, 1645–1649.

[207] Gebregziabiher, D.K., Kim, Y.-G., Thambidurai, C., Ivanova, V., Haumesser, P.-H. and Stickney, J.L. (2010) Electrochemical atomic layer deposition of copper nanofilms on ruthenium. *J. Cryst. Growth*, **312**, 1271–1276.

[208] Waechtler, T., Ding, S.F., Hofmann, L., Mothes, R., Xie, Q., Oswald, S., Detavernier, C., Schulz, S.E., Qu, X.-Q., Lang, H. and Gessner, T. (2011) ALD-grown seed layers for electrochemical copper deposition integrated with different diffusion barrier systems. *Microelectron. Engng*, **88**, 684–689.

[209] Lu, J., Ha, H.C., Aubuchon, J., Ma, P., Yu, S.H. and Narasimhan, M. (2008) Conformal CVD Co deposition for enhancement of Cu gapfill application. Advanced Metallization Conference (AMC), pp. 21–22.

[210] Heylen, N., Yunlong, L., Kellens, K., Carbonell, L., Volders, H., Santoro, G., Gravey, V., Cockburn, A., Wang, Y., Shah, K., Leunissen, L., Beyer, G.P. and Tokei, Z. (2010) CMP process optimization for improved compatibility with advanced metal liners. Proceedings of the IEEE International Interconnect Technology Conference, p. 5.6.

[211] Prasad, K., Yuan, X.L., Li, C.Y. and Kumar, R. (2002) Evaluation of diffusion barrier layers in Cu interconnects. Proceedings of the IEEE International Interconnect Technology Conference, pp. 373–376.

[212] Wang, T.C., Cheng, Y.L., Wang, Y.L., Hsieh, T.E., Hwang, G.J. and Chen, C.F. (2006) Comparison of characteristics and integration of copper diffusion-barrier dielectrics. *Thin Solid Films*, **498**, 36–42.

[213] Nguyen, S., Haigh Jr, T., Shaw, T., Molis, S., Dziobkowski, C., Zahakos, C., Cohen, S., Shobha, H., Grill, A., Bonilla, G. and Klymko, N. (2010) Ultrathin (5–35 nm) SiCNH dielectrics for damascene Cu cap application: thickness scaling and oxidation barrier performance limitation. Proceedings of the 218th ECS Meeting, Abstract 1685.

[214] Vairagar, A.V., Mhaisalkar, S.G., Meyer, A.M., Zschech, E., Krishnamoorthy, A., Tu, K.N. and Gusak, A.M. (2005) Direct evidence of electromigration failure mechanism in dual-damascene Cu interconnect tree structures. *Appl. Phys. Lett.*, **87**, 081909.

[215] Hecker, M., Pitschke, W., Tietjen, D. and Schneider, C.M. (2002) X-ray diffraction investigations of structural changes in Co/Cu multilayers at elevated temperatures. *Thin Solid Films*, **411**, 234–239.

[216] Koo, H.-C., Cho, S.K., Kwon, O.J., Suh, M.-W., Im, Y. and Kim, J.J. (2009) Improvement in the oxidation resistance of Cu films by an electroless Co–alloy capping process. *J. Electrochem. Soc.*, **156**, D236–D241.

[217] Hamid, Z.A., Aal, A.A., Shaaban, A. and Hassan, H.B. (2009) Electrodeposition of CoMoP thin film as diffusion barrier layer for ULSI applications. *Surf. Coat. Technol.*, **203**, 3692–3700.

[218] Zylberman, A., Shacham-Diamand, Y., Swerdlov, Y. and Petrov, N. (2003) Comparative study of electroless Co(W,P) and Co(Mo,P) thin films for capping and barrier layer for Cu metallization. MRS Proceedings of the Advanced Metallization Conference, AMC 2002 (ULSI XVIII), pp. 767–773.

[219] Einati, H., Bogush, V., Sverdlov, Y., Rosenberg, Y. and Shacham-Diamand, Y. (2005) The effect of tungsten and boron on the Cu barrier and oxidation properties of thin electroless cobalt–tungsten–boron films. *Microelectron. Engng*, **82**, 623–628.

[220] Osaka, T., Takano, N., Kurokawa, T., Kaneko, T. and Ueno, K. (2002) Electroless nickel ternary alloy deposition on $SiO_2$ for application to diffusion barrier layer in copper interconnect technology. *J. Electrochem. Soc.*, **149**, C573–C578.

[221] Kohn, A., Eizenberg, M., Shacham-Diamand, Y., Israel, B. and Sverdlov, Y. (2001) Evaluation of electroless deposited Co(W,P) thin films as diffusion barriers for copper metallization. *Microelectron. Engng*, **55**, 297–303.

[222] Dulal, S.M.S.I., Yun, H.J., Shin, C.B. and Kim, C.-K. (2009) Electrodeposition of CoWP film V. Structural and morphological characterisations. *Appl. Surf. Sci.*, **255**, 5795–5801.

[223] Murray, C.E., Besser, P.R., Witt, C. and Toney, M. (2010) *In situ* evolution of stress gradients in Cu films induced by capping layers. *Appl. Phys. Lett.*, **96**, 261903.

[224] Zschech, E., Engelmann, H.-J., Meyer, M.A., Kahlert, V., Vairagar, A.V., Mhaisalkar, S.G., Krishnamoorthy, A., Yan, M., Tu, K.N. and Sukharev, V. (2005) Effect of interface strength on electromigration-induced inlaid copper interconnect degradation: experiment and simulation. *Int. J. Mater. Res.*, **96**, 966–971.

[225] Koike, J. (2009) Physical vapor deposition barriers for Cu metallization – PVD Barriers, in *Advanced Nanoscale ULSI Interconnects: Fundamentals and Applications* (eds Shacham-Diamand, Y., Osaka, T., Datta, M. and Ohba, T.), Springer Science+Business Media, New York, Dordrecht, Heidelberg, London.

[226] Frederick, M.J., Goswami, R. and Ramanath, G. (2003) Sequence of Mg segregation, grain growth, and interfacial MgO formation in Cu–Mg alloy films on $SiO_2$ during vacuum annealing. *J. Appl. Phys.*, **93**, 5966–5972.

[227] Frederick, M.J. and Ramanath, G. (2004) Interfacial phase formation in Cu–Mg alloy films on $SiO_2$. *J. Appl. Phys.*, **95**, 3202–3205.

[228] Frederick, M.J. and Ramanath, G. (2004) Kinetics of interfacial reaction in Cu–Mg alloy films on $SiO_2$. *J. Appl. Phys.*, **95**, 363–366.

[229] Kohama, K., Ito, K., Tsukimoto, S., Mori, K., Maekawa, K. and Murakami, M. (2008) Characterization of self-formed Ti-rich interface layers in Cu(Ti)/low-*k* samples. *J. Electron. Mater.*, **37**, 1148–1157.

[230] Koike, J., Haneda, M., Iijima, J. and Wada, M. (2006) Cu alloy metallization for self-forming barrier process. Proceedings of the IEEE International Interconnect Technology Conference, pp. 161–163.

[231] Koike, J. and Wada, M. (2005) Self-forming diffusion barrier layer in Cu–Mn alloy metallization. *Appl. Phys. Lett.*, **87**, 041911.

[232] Usui, T., Nasu, H., Koike, J., Wada, M., Takahashi, S., Shimizu, N., Nishikawa, T., Yoshimam, M. and Shibata, H. (2005) Low resistive and highly reliable Cu dual-damascene interconnect technology using self-formed $MnSi_xO_y$ barrier layer. Proceedings of the IEEE International Interconnect Technology Conference, pp. 188–190.

[233] Usui, T., Nasu, H., Takahashi, S., Shimizu, N., Nishikawa, T., Yoshimaru, M., Shibata, H., Wada, M. and Koike, J. (2006) Highly reliable copper dual-damascene interconnects with self-formed $MnSi_xO_y$ barrier layer. *IEEE Trans. Electron Devices*, **53**, 2492–2499.

[234] Iijima, J., Haneda, M. and Koike, J. (2006) Growth behavior of self-formed barrier using Cu–Mn alloys at 350 to 600 °C. Proceedings of the IEEE International Interconnect Technology Conference, pp. 155–157.

[235] Haneda, M., Iijima, J. and Koike, J. (2007) Growth behavior of self-formed barrier at Cu-Mn/$SiO_2$ interface at 250–450 °C. *Appl. Phys. Lett.*, **90**, 252107.

[236] Koike, J., Haneda, M., Iijima, J., Otsuka, Y., Sako, H. and Neishi, K. (2007) Growth kinetics and thermal stability of a self-formed barrier layer at Cu–Mn/$SiO_2$ interface. *J. Appl. Phys.*, **102**, 043527.

[237] Chung, S.-M. and Koike, J. (2009) Analysis of dielectric constant of a self-forming barrier layer with Cu–Mn alloy on TEOS–$SiO_2$. *J. Vac. Sci. Technol. B*, **27**, L28–L31.

[238] Ablett, J.M., Woicik, J.C., Tökei, Zs., List, S. and Dimasi, E. (2009) Phase identification of self-forming Cu–Mn based diffusion barriers on *p*-SiOC:H and $SiO_2$ dielectrics using X-ray absorption fine structure. *Appl. Phys. Lett.*, **94**, 042112.

[239] Iijima, J., Fujii, Y., Neishi, K. and Koike, J. (2009) Resistivity reduction by external oxidation of Cu–Mn alloy films for semiconductor interconnect application. *J. Vac. Sci. Technol. B*, **27**, 1963–1968.

[240] Watanabe, T., Nasu, H., Usui, T., Minamihaba, G., Kurashima, N., Gawase, A., Shimada, M., Yoshimizu, Y., Uozumi, Y. and Shibata, H. (2007) Self-formed barrier technology using CuMn alloy seed for copper dual-damascene interconnect with porous-SiOC/porous-PAr hybrid dielectric. Proceedings of the IEEE International Interconnect Technology Conference, pp. 7–9.

[241] Watanabe, T., Hayashi, Y., Tomizawa, H., Usui, T., Gawase, A., Shimada, M., Watanabe, K. and Shibata, H. (2008) Robust BEOL process integration with ultra low-*k* ($k = 2.0$) dielectric and self-formed $MnO_x$ barrier technology for 32 nm-node and beyond. Proceedings of the IEEE International Interconnect Technology Conference, pp. 208–210.

[242] Pan, S.C., Chi, C.C., Ko, C.C., Kuo, H.H., Huang, F.M., Lee, S.C., Lin, M.H., Huang, H.Y., Hsieh, C.H., Jeng, S.M., Tao, H.J., Cao, M. and Mii, Y.J. (2010) Interface effect on Mn-containing self-formed barrier formation with extreme low-*k* dielectric integration. Proceedings of the IEEE International Interconnect Technology Conference, pp. 1–3.

[243] Gordon, R.J., Kim, H., Au, Y., Wang, H., Bhandari, H.B., Liu, Y., Lee, D.K. and Lin, Y. (2009) Chemical vapor deposition (CVD) of manganese self-aligned diffusion barriers for Cu interconnections in microelectronics. MRS Proceedings of the Advanced Metallization Conference, AMC 2008 (ULSI XXIV).

[244] Neishi, K., Aki, S., Matsumoto, K., Sato, H., Itoh, H., Hosaka, S. and Koike, J. (2008) Formation of a manganese oxide barrier layer with thermal chemical vapor deposition for advanced large-scale integrated interconnect structure. *Appl. Phys. Lett.*, **93**, 032106.

[245] Matsumoto, K., Neishi, K., Itoh, H., Sato, H., Hosaka, S. and Koike, J. (2009) Deposition behavior and diffusion barrier property of CVD $MnO_x$. Proceedings of the IEEE International Interconnect Technology Conference, pp. 197–199.

[246] Krishnamoorthy, A., Chanda, K., Murarka, S.P., Ramanath, G. and Ryan, J.G. (2001) Self-assembled near-zero-thickness molecular layers as diffusion barriers for Cu metallization. *Appl. Phys. Lett.*, **78**, 2467–2469.
[247] Ramanath, G., Cui, G., Ganesan, P.G., Guo, X., Ellis, A.V., Stukowski, M., Vijayamohanan, K., Doppelt, P. and Lane, M. (2003) Self-assembled subnanolayers as interfacial adhesion enhancers and diffusion barriers for integrated circuits. *Appl. Phys. Lett.*, **83**, 383–385.
[248] Gandhi, D.D., Tisch, U., Singh, B., Eizenberg, M. and Ramanath, G. (2007) Ultraviolet-oxidized mercaptan-terminated organosilane nanolayers as diffusion barriers at Cu–silica interfaces. *Appl. Phys. Lett.*, **91**, 143503.
[249] Gandhi, D.D., Lane, M., Zhou, Y., Singh, A.P., Nayak, S., Tisch, U., Eizenberg, M. and Ramanath, G. (2007) Annealing-induced interfacial toughening using a molecular nanolayer. *Nature*, **447**, 299–302.
[250] Ganesan, P.G., Kumar, A. and Ramanath, G. (2005) Surface oxide reduction and bilayer molecular assembly of a thiol-terminated organosilane on Cu. *Appl. Phys. Lett.*, **87**, 011905.
[251] Ganesan, P.G., Singh, A.P. and Ramanath, G. (2004) Diffusion barrier properties of carboxyl- and amine-terminated molecular nanolayers. *Appl. Phys. Lett.*, **85**, 579–581.
[252] Ganesan, P.G., Cui, G., Vijayamohanan, K., Lane, M. and Ramanath, G. (2005) Effects of amine- and pyridine-terminated molecular nanolayers on adhesion at Cu–$SiO_2$ interfaces. *J. Vac. Sci. Technol. B*, **23**, 327–331.
[253] Gandhi, D.D., Ganesan, P.G., Chandrasekar, V., Gan, Z., Mhaisalkar, S.G., Li, H. and Ramanath, G. (2007) Molecular-nanolayer-induced suppression of in-plane Cu transport at Cu–silica interfaces. *Appl. Phys. Lett.*, **90**, 163507.
[254] Inberg, A., Glickman, E., Asher, T., Fishelson, N. and Shacham-Diamand, Y. (2009) Electrical properties of sub-100 nm Cu films deposited by electroless plating on amino-terminated silicon oxide activated with Au nano-particles. *Surf. Coat. Technol.*, **204**, 520–524.
[255] Asher, T., Inberg, A., Glickman, E., Fishelson, N. and Shacham-Diamand, Y. (2009) Formation and characterization of low resistivity sub-100 nm copper films deposited by electroless on SAM. *Electrochim. Acta*, **54**, 6053–6057.
[256] Maestre Caro, A., Maes, G., Borghs, G. and Whelan, C.M. (2008) Screening self-assembled monolayers as Cu diffusion barriers. *Microelectron. Engng*, **85**, 2047–2050.
[257] Ganesan, P.G., Gamba, J., Ellis, A., Kane, R.S. and Ramanath, G. (2003) Polyelectrolyte nanolayers as diffusion barriers for Cu metallization. *Appl. Phys. Lett.*, **83**, 3302–3304.
[258] Gandhi, D.D., Singh, A.P., Lane, M., Eizenberg, M. and Ramanath, G. (2007) Copper diffusion and mechanical toughness at Cu–silica interfaces glued with polyelectrolyte nanolayers. *J. Appl. Phys.*, **101**, 084505.
[259] Senkevich, J.J., Wang, P.I., Wiegand, C.J. and Lu, T.-M. (2004) Bias-temperature stability of ultrathin parylene-capped dielectrics: influence of surface oxygen on copper ion diffusion. *Appl. Phys. Lett.*, **84**, 2617–2619.
[260] Senkevich, J.J., Karabacak, T., Bae, D.-L. and Cale, T.S. (2006) Formation of body-centered-cubic tantalum via sputtering on low-*k* dielectrics at low temperatures. *J. Vac. Sci. Technol. B*, **24**, 534–538.
[261] Mallikarjunan, A., Murarka, S.P. and Lu, T.-M. (2004) Separation of copper ion-induced and intrinsic polymer instabilities in polyarylether using triangular voltage sweep. *J. Appl. Phys.*, **95**, 1216–1221.

# Section III

## Integration and Reliability

# 7

# Process Integration of Interconnects

*Sridhar Balakrishnan, Ruth Brain and Larry Zhao*

*Intel Corp., Hillsboro, Oregon, USA*

## 7.1 Introduction

This chapter aims to introduce the reader to the challenges of on-die interconnect integration as the semiconductor industry continues to march down the scaling path with leading companies exercising high-volume production at 45 nm and lower nodes as of 2010. Before we dive into the state-of-the-art technology nodes, a brief review of the progression of complexity and incorporation of novel materials into the on-die interconnect architecture during the last twenty years should prove beneficial.

## 7.2 On-Die Interconnects in the Submicrometer Era

The early 1990s saw memory and logic technologies pushing into the submicrometer arena. Multiple levels of metallization at the 0.8 μm node became the norm [1] and W-filled plugs were introduced for contacts and vias in conjunction with subtractive patterned Al wires in silicon dioxide ILD. The integration challenges of good dielectric fill between metal lines and achieving the level of planarization required to pattern submicrometer features accurately after stacking multiple layers of interconnects resulted in the implementation of improved PECVD $SiO_2$ multistep deposition techniques [2], spin-on glass

*Advanced Interconnects for ULSI Technology*, First Edition. Edited by Mikhail R. Baklanov, Paul S. Ho and Ehrenfried Zschech.

and etch-back schemes [3] and dielectric chemical mechanical polish (CMP) [4]. Plug fill with CVD W proved to be very scalable, but the industry moved towards using CMP to remove the bulk tungsten and adhesion layer over the dielectric rather than etch back processes [5]. Electromigration (EM) improvements required the use of dopants in the Al and the use of underlayer and overlayer shunts, and antireflective layers further enhanced EM performance [1, 5].

During the mid 1990s, the introduction of 0.25 μm technologies saw an increase in the number of metal layers including investigation of W- and Ti-based local interconnects [6]. Already the move to replace W filled plug and subtractive Al conductors was afoot due to the realization that at the narrower pitches expected for the sub-180 nm nodes they would require extremely high aspect ratio lines, which made their patterning and subsequent ILD fill extremely challenging. This led the industry to consider the use of Cu as a replacement given its lower resistivity and improved electromigration performance, but the challenges associated with subtractive patterning of Cu were even more daunting than that of Al, and hence the focus moved to the damascene or inlay approach to integrating Cu [7]. In this approach, rather than patterning the conductor and then filling the dielectric between the lines, the ILD is first patterned and then a barrier layer, which also serves as an adhesion layer, and copper are deposited with a combination of PVD and electroplating, so as to not only fill the patterned features but then overfill so as to produce a blanket of copper across the wafer. Through the use of CMP, the excess copper and barrier films between the lines are removed, resulting in a flat surface that is ready for the deposition of the next layer ILD stack. While the contact layer remained W filled, technologies moved to use of copper to fill the vias and lines in a single step, which was termed dual damascene. In order to keep the copper from diffusing into the ILD, in addition to the previously mentioned PVD barrier film, the post-CMP top surface needs to be capped with a dielectric diffusion barrier. In the dual damascene architecture, this layer also served as an etch-stop film for the next layer via. The first manufacturing technology to make use of copper interconnects produced in the manner described above was reported in the late 1990s [8].

On the ILD front, one of the first steps to reduce the dielectric constant versus $SiO_2$ ($k = 4.0$) came from the adoption of SiOF, whose dielectric constant was close to 3.5. Technologies targeted around 180 nm and 130 nm nodes saw the integration of SiOF with conventional Al/W-based conductors and DUV- and i-line-based patterning [9], but also with dual damascene copper interconnects [10]. The presence of the etch-stop layers detracts some of the benefits of moving to lower $k$ bulk ILD, because they typically have a substantially higher dielectric constant, thus increasing the effective dielectric constant ($k_{eff}$) of the ILD stack. The International Technology Roadmap for Semiconductors (ITRS), the roadmap for interconnects from 2000 [11], predicted a rapid march down the $k_{eff}$ roadmap to sub-100 nm nodes, and an excerpt of this roadmap is shown in Table 7.1. Note the prediction that, by 2003, the bulk dielectric constant of the ILD was expected to be well below 2.5. In reality, integration of the next generation of ILDs, which ended up being carbon-doped $SiO_2$ (CDO), was more challenging than anticipated. Toward the end of the 1990s, two types of low-$k$ ILD candidates were being investigated – so-called spin-on films, which relied on spinning a film, much as a photoresist is spun, followed by post-treatments to remove solvents and set the final film versus CVD low-$k$ films [12].

*Table 7.1* *ITRS 2000 interconnect roadmap*

| Year | 1999 | 2000 | 2001 | 2002 | 2003 | 2004 | 2005 |
|---|---|---|---|---|---|---|---|
| Technology Node | 180nm | | | 130nm | | | 100nm |
| MPU ½ pitch | 230 | 210 | 180 | 160 | 145 | 130 | 115 |
| MPU gate length (nm) | 140 | 120 | 100 | 90 | 80 | 70 | 65 |
| Minimum global wiring pitch (nm) | 1050 | 945 | 850 | 765 | 690 | 620 | 560 |
| Global wiring A/R (Al) | 2 | 2.1 | 2.2 | 2.3 | 2.4 | ** | ** |
| Global wiring dual damascene A/R (Cu wire/via) | 2.2/2.4 | 2.3/2.6 | 2.4/2.7 | 2.5/2.7 | 2.6/2.8 | 2.7/2.8 | 2.7/2.8 |
| Cu thinning global wiring due to dishing and erosion (nm), 10% × height, 80% areal density, 15 micron wide wire | 116 | 109 | 102 | 95 | 90 | 84 | 76 |
| Cu global wiring dishing (nm), 15 micron wide wire, 10%× height | 116 | 109 | 102 | 95 | 90 | 84 | 76 |
| Cu thinning global wiring due to dishing (nm), 100 micron wide feature | 80 | 72 | 65 | 59 | 53 | 48 | 43 |
| Conductor effective resistivity (μΩ-cm) Al wiring | 3.3 | 3.3 | 3.3 | 3.3 | 3.3 | ** | ** |
| Conductor effective resistivity resistivity, (μΩ-cm) Cu wiring* | 2.2 | 2.2 | 2.2 | 2.2 | 2.2 | 2.2 | 2.2 |
| Barrier/cladding thickness (for Cu wiring) (nm)*** | 17 | 16 | 14 | 13 | 12 | 11 | 10 |
| Interlevel metal insulator—effective dielectric constant (κ) | 3.5–4.0 | 3.5–4.0 | 2.7–3.5 | 2.7–3.5 | 2.2–2.7 | 2.2–2.7 | 1.6–2.2 |
| Interlevel metal insulator — effective dielectric constant (κ) | 3.5–4.0 | 3.5–4.0 | 2.9–3.5 | 2.9–3.5 | 2.2–2.9 | 2.2–2.9 | 1.6–2.2 |
| Interlevel metal insulator (minimum expected), —bulk dielectric constant (κ) | 2.9 | 2.9 | 2.7 | 2.7 | 2.0 | 2.0 | 1.3 |

* *Assumes a conformal barrier/nucleation layer*
** *This technology is not expected to extend to this node*
*** *Calculated for a conformal layer in local wiring to meet minimum effective conductor resistivity*
*Solutions Exist* *Solutions Being Pursued* *No Known Solutions*

One of the first manufacturable technologies employing CDO was at the 90 nm node [13] with as-deposited dielectric constants below 3.0. One of the key challenges that made the integration of low-*k* ILD difficult was the significant reduction in modulus (*E*) and hardness (*H*) versus FSG and oxide. Poorer mechanical properties made the CMP of these

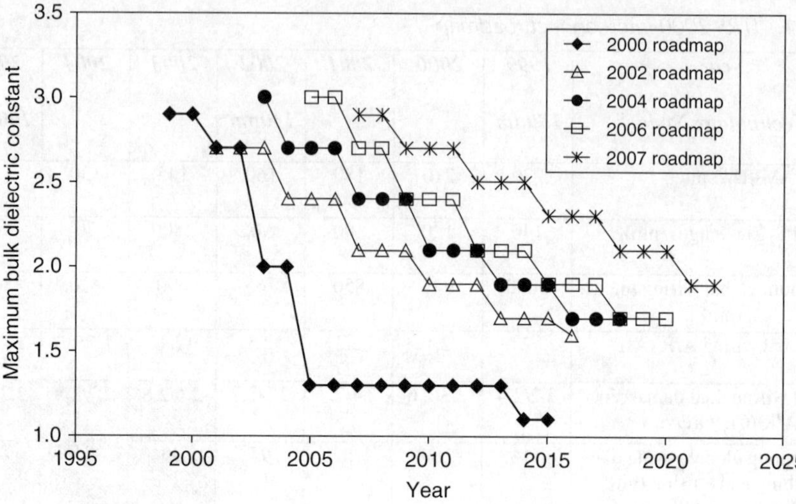


*Figure 7.1* Predicted bulk dielectric constant from ITRS roadmaps

materials more difficult and also posed a thermomechanical risk for packaging and reliability [13]. Additional difficulties with integrating low-*k* dielectrics surfaced as the attempt to drive down *k* by including porosity in the film raised a whole new host of issues such as *k* damage, poorer patterning fidelity and even further degradation in mechanical properties. A later section in this chapter will provide an in-depth review of these challenges, especially the class of films referred to as extreme low-*k* (ELK) or ultra-low-*k* (ULK). This level of integration difficulties was not anticipated, and is best exemplified in a review of the ITRS roadmaps from 2000 to 2007, which show a progressive slip in the anticipated implementation dates for low-*k* films (Figure 7.1) approaching and lower than 2.5.

## 7.3 On-Die Interconnects at Sub-100 nm Nodes

The number of metal layers at the sub-100 nm nodes has grown rapidly, consistent with Rent's rule, to enable interconnection between the ever-increasing densities of transistors. The move from the 90 nm node to the 65 nm node saw the number of metal layers reach eight [14] to ten [15]. The interconnect layers became increasingly stratified, with the lowest layers with the tightest pitches being used in a local or short distance interconnect use, while the mid and upper layers, which had intermediate to loose pitches, were used for interconnects over distances of hundreds of micrometers to millimeters for signals and power distribution. On the patterning front, the move from 248 nm exposure tools to 193 nm at the sub-100 nm node helped with the ability to pattern the critical lower layers, but the steady beat of Moore's law quickly saw the minimum pitch needing to be supported outpace the ability to move to smaller wavelengths. In fact, as of this writing, the exposure wavelength in high-volume manufacturing remains tied at 193 nm for the last decade. Publications on the 65 nm node [16] speak of difficulties in finding an adequate process window in terms of the depth of focus and exposure latitude across a wide range of pattern widths, densities and configurations. Continuous improvements in photoresist and antireflective materials, as well as source design in conjunction with design rule restrictions and sophisticated optical proximity correction and mask-making techniques, enabled the continued scaling from a lithographic perspective. Process integration approaches such as the use of hardmasks [17] and self-aligned schemes have been used to retain or improve upon the lithographic pattern fidelity during the subsequent pattern transfer into the dielectric films or allowed for less stringent registration requirements, while avoiding yield and reliability issues such as inadvertent shorting or high field conditions. At these nodes, given the difficulties in integrating low-*k* dielectrics in the 2.5 range, some $k_{eff}$ reduction can be obtained by reducing the dielectric constant and thickness of etch-stop layers. Reducing the thickness of etch-stop layers impact on the selectivity needed for conventional full via etch processes, which rely on adequate stopping power of the etch stop to comprehend nonuniformities in ILD thickness and etch rates across wafers and across structures. Thus, scaling etch-stop thickness requires changes in the patterning loop such as the use of partial via flows [18]. Changes to the dielectric constant of the etch stop film has also been achieved through changing from silicon nitride-based films, which have a dielectric constant as high as 7, to SiC and SiCN films, which have *k* values of around 5.

With this review of architectural and materials changes, we are ready to delve deeper into process integration efforts in more recent sub-65 nm nodes in the areas of integration of low-*k* dielectrics, integration of conductors, patterning approaches and a review of the current state-of-the-art in novel interconnect integration such as incorporation of air-gaps.

## 7.4 Integration of Low-*k* Dielectrics in Sub-65 nm Nodes

We will now take a closer look at the challenges and solution spaces identified in integrating low-*k* dielectrics in the sub-65 nm nodes. While there is no prescribed partitioning of dielectrics used in on-die interconnects in terms of dielectric constant, a typical approach is to refer to those with *k* of around 3 as low-*k* while those with *k* of around 2.5 are referred to as extreme low-*k* (ELK), though the term ultra-low-*k* (ULK) has started to be used to refer to dielectrics in this and even lower *k* ranges. We will focus on the dielectrics that are around $k = 2.5$, though many of the concerns and solution paths do carry forward into the ULK space. It should be noted that, quite often, papers refer to the as-deposited *k* in defining the class of film, without necessarily considering the integrated end-of-line $k_{eff}$. The drive for integrating dielectrics with ever lower *k* comes primarily from the parasitic delay associated with the RC product of the interconnect network as well as the ever-increasing impact that interconnect dynamic switching capacitance ($C_{dyn}$) has on power consumption by the device. The proportionality between this parasitic power and $C_{dyn}$, the operating voltage ($V$) and operating frequency ($f$) of the device is indicated below:

$$\text{Switching power } \alpha = C_{\text{dyn}} V^2 f$$

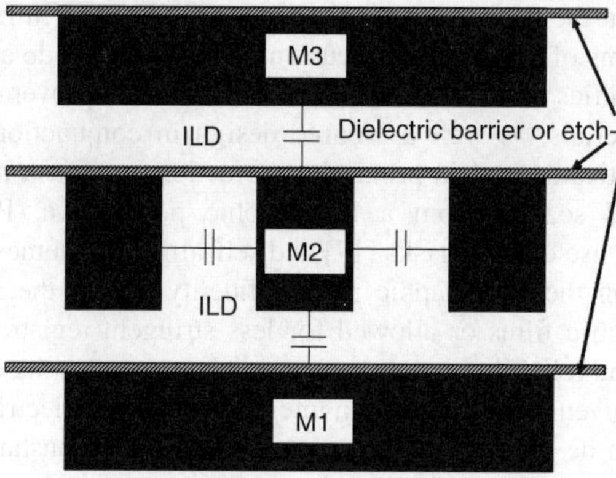


*Figure 7.2* Capacitance associated with typical multilevel interconnects

In the world of shrinking power envelopes and ever-increasing transistor density, which necessitates tighter pitch interconnects and additional levels of metallization, this parasitic power consumption is becoming increasingly important to reduce. The capacitance of a given interconnect layer depends upon the geometric components within the layer as well as those of adjacent layers and the choice of materials. Figure 7.2 is a representational drawing of the capacitances associated with a typical multilevel metallization scheme as used in modern on-die interconnects. In this case, we can see that the capacitance associated with a wire at the second metal layer (M2) consists of the intralayer capacitance associated with the adjacent wires within the same level, but is also due to the interlayer capacitance associated with coupling to wires running in adjacent M3 and M1 layers.

The capacitance per unit length of conductor can then be clearly seen to depend on material parameters, such as the dielectric constant of the ILD and the dielectric barrier or etch-stop layers as well as dimensions such as the M2 line-to-line space, M2 trench height and dielectric thickness between the adjacent metal layers. With the ILD material, it is the integrated $k$ value at the end-of-line that will dictate the observed capacitance, not just the starting as-deposited $k$, and therein lies one of the key challenges with effective deployment of low-$k$ and ELK dielectrics. The presence of even thin damage layers introduced while integrating these films through a full dual-damascene patterned and metalized layer can be highly detrimental when delivering lower capacitance at the end-of-line. Figure 7.3 shows the results of simulating a damage layer of $k = 4.0$ on the integrated $k_{eff}$ at a 56 nm pitch. We can see that even a 2 nm thick damage layer present for a $k = 2.5$ ILD can make it equivalent to a $k = 2.7$ ILD without any damage.

A similar scenario can be envisioned for any integrated 'bulk' damage to the dielectric that may occur during processing. It is therefore critical that claims of ELK integration should include the resistance versus total capacitance on a multilayer structure as in Figure 7.3 to have merit. Since many of the process steps to ensure reliability of ELK-based interconnects can be damaging, true interconnect benchmarking should also include reliability parameter performance for any new dielectrics such as electromigration, bias temperature stress and of course die yield.

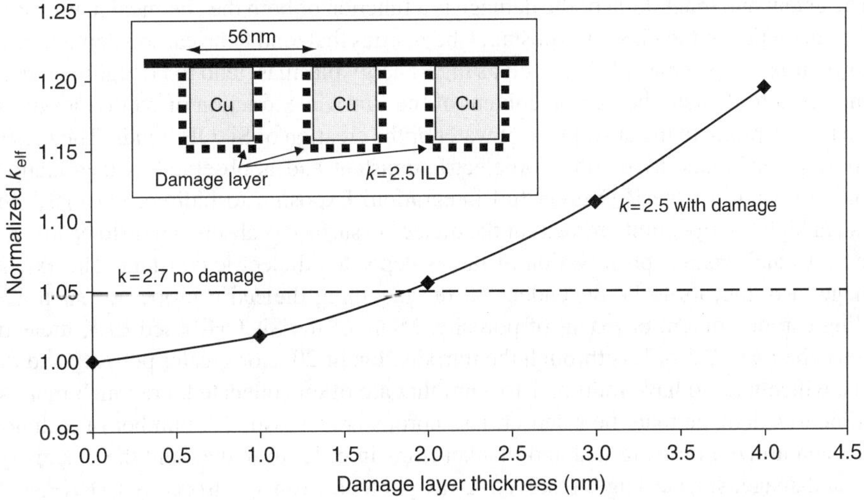


***Figure 7.3*** *Electrostatic simulation of* k *impact from the damage layer at a 28 nm half-pitch*

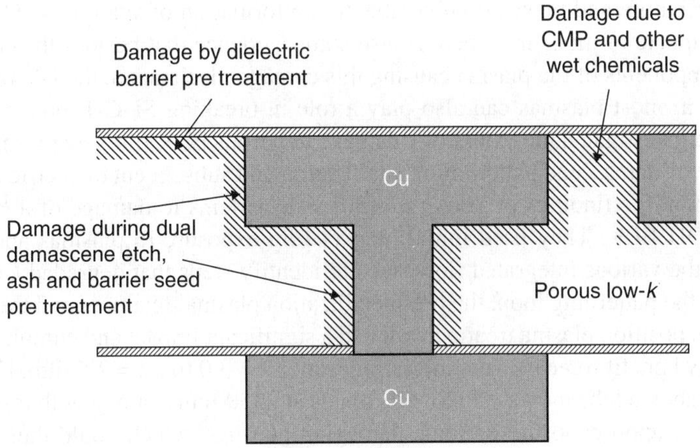


***Figure 7.4*** *Regions prone to* k *damage and their potential sources*

### 7.4.1 Degradation of Dielectric Constant during Integration

The propensity of low-*k* and ELK dielectrics to suffer such *k* damage has been widely reported in the literature [19–21]. Such damage is seen to occur during exposure of these dielectrics to plasmas such as those used during dry etch and ash, as well as plasma pre-treatments that are commonly used prior to metallization or next-layer dielectric deposition. Figure 7.4 shows the interfaces that can suffer damage and result in *k* degradation in a fully integrated dual-damascene stack.

The extent and magnitude of the damage is a function of both the chemical nature as well as the energetics of the plasma exposure. Chemistries that tend to be carbon depleting, such as oxygen-based plasmas [22], as well as high-energy plasmas, tend to be highly damaging as they tend to deplete the carbon content of the film via extraction of methyl groups and make it susceptible to the absorption of water or the creation of Si–OH bonds. The presence of these polar bonds drives up the dielectric constant and is observed as degradation in effective $k$ value of the ILD upon full integration. Exposure to indirect plasmas, where physical high-energy bombardment of the dielectric surface is absent, tends to be less damaging [20] and enables preservation of the as-deposited dielectric constant. The extent of damage also appears to be dependent on the incoming dielectric properties such as the starting carbon content or extent of porosity. Many of the SiCOH-based ELK dielectrics achieve their $k$ of 2.5 or less through the introduction of 20 % or greater porosity and these pores, which tend to have radii of 1 to 2 nm that are often connected, are much more susceptible to $k$ damage than their denser, less porous predecessors. A number of techniques have been utilized on patterned and blanket films in order to understand the chemistry of plasma damage, such as angle-resolved X-ray photoelectron spectroscopy (ARXPS) [19], Fourier transformed infrared spectroscopy (FTIR) [20], contact angle measurements [21] and TEM-enabled chemical profiling [22] such as electron energy loss spectroscopy (EELS). The dominant damage mechanisms appear to involve stripping of carbon (which is typically in the form of Si–$CH_3$ bonds), which causes the film to then lose its initial hydrophobic nature and makes it vulnerable to the formation of silanol (Si–OH bonds) as well as the uptake of moisture. There is also some evidence that besides the chemical and physical components of the plasma causing this damage mechanism, the UV radiation that is prevalent in most plasmas can also play a role in breaking Si–$CH_3$ bonds [23]. Many authors have investigated the extent of $k$ damage to porous ELK films by plasma processes utilized in dual-damascene patterning, metallization and subsequent dielectric depositions. We will look at the findings of Yoo *et al.* [20] as it pertains to damage of a $k = 2.5$ class porous SiCOH ELK. They used initial blanket film exposure to plasmas and chemicals involved in the various integrated flow steps to identify steps that degraded $k$ and showed that steps in the patterning loop, the pre-metallization plasma treatment and the pre-dielectric barrier deposition plasma treatment all had a significant impact and cumulatively could eliminate any benefit forecast from migrating from a $k = 3.0$ to a $k = 2.5$ film. FTIR spectra (Figure 7.5) showed the increase in Si–OH bonds at ~3500 $cm^{-1}$, along with loss of Si–$CH_3$ at ~1000 $cm^{-1}$, upon exposure to some damaging plasmas, which could then be reduced through optimization of plasma parameters.

While some of the $k$ increase resulting from carbon stripping making the film hydrophilic can be eliminated by simple thermal anneals, changes to the film network resulting from replacement of Si–$CH_3$ bonds with Si–OH are harder to remove, though some success has been reported with using $CH_4$ or silylating agent post-treatments. Yoo *et al.* [20] showed that the full expected benefit for an integrated $k = 2.5$ film over a $k = 3.0$ film could be observed through careful choice of plasma conditions coupled with the use of silylation repair techniques as seen in the plot of end-of-line capacitance of a meander-fork structure versus the inverse of the meander resistance, as seen in Figure 7.6.

The use of a hardmask on top of the porous ELK during plasma-based patterning of the trenches and vias to the underlying layer can also help with reducing $k$ damage. Hence the

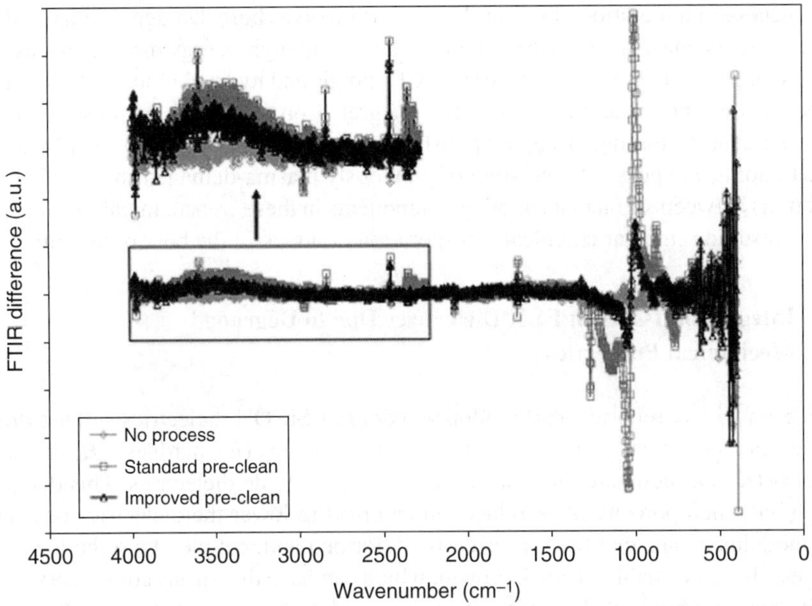


***Figure 7.5*** *FTIR spectra showing the impact of the pre-clean process on porous SiCOH ILD*

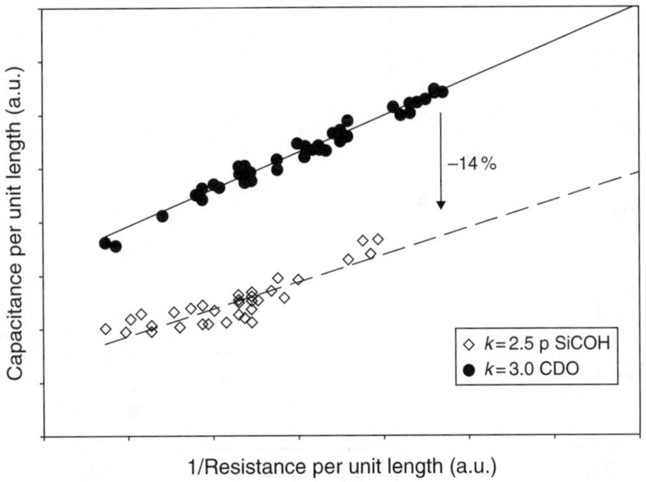


***Figure 7.6*** *Full* k *benefit realized by optimal process integration for a* k = *2.5 porous SiCOH*

key to successful integration of porous ELK films involves both damage avoidance through careful choice of plasma chemistries, hardware to avoid high-energy direct plasmas as well as the use of repair techniques to restore Si–$CH_3$ bonds and hydrophobicity. Besides plasma treatments, wet chemicals involved in the integrated process, such as those used during dual-damascene patterning or copper polish, can also result in $k$ degradation [24] and may be due to not just exposure to moisture of previously plasma-damaged surfaces but also to interactions between surfactants or other components in these wet chemicals and the porous network resulting in polar chemical incorporation or attack in the bulk of the film.

### 7.4.2 Integration Issues in ELK Dielectrics Due to Degraded Mechanical Properties

As has been widely reported, carbon-doped oxides or SiCOH dielectrics suffer a dramatic reduction in mechanical properties such as elastic modulus ($E$), hardness ($H$) and fracture toughness ($K_{IC}$) as compared to conventional silicon dioxide dielectrics. This degradation is amplified when porosity is introduced in an effort to lower the dielectric constant. The lower modulus means that the film is easily deformed and, coupled with the low fracture toughness, becomes liable to crack growth, which can have disastrous consequences on the integrity and reliability of the interconnect stack. McGahay *et al.* [15] show the dramatic increase in crack driving force ($G$) for films they studied as $k$ reduced from 3.0 to 2.7, as shown in Figure 7.7.

The low modulus makes these films susceptible to deformation during subsequent processing steps such as CMP, which is typically used to remove the excess electroplated copper and barrier material from the surface of the ILD. Since CMP involves application of a force between the wafer and polish pad in the presence of abrasive slurry particles,

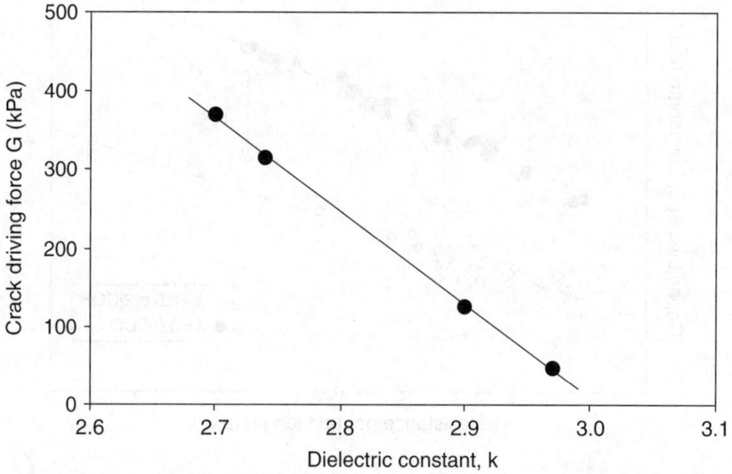


***Figure 7.7*** *Crack driving force increase for SiCOH films versus as-deposited* k

low-modulus materials can be deformed and manifest themselves as bent or wavy lines that can impact electrical connectivity. This, along with other CMP defects such as watermarks [25], can be mitigated through careful selection of slurry, CMP down-force and hardware, along with recipe conditions. One approach taken is to add a higher modulus cap whose mechanical and wettability properties are more favorable to Cu CMP, but this usually comes at the penalty of leaving behind a higher *k* layer between the Cu lines. Another area where low modulus of these ELK materials is making integration difficult is during dual-damascene patterning, especially with the use of hardmask films. The film stress of the hardmask layers, such as sputtered Ti, TiN or TaN, can be large enough to cause trench patterns to buckle or deform at narrow spaces. Even if the wafer processing is tailored to enable wafer-level integration of these relatively fragile porous low-*k* dielectrics, a substantial challenge needs to be overcome in terms of their thermomechanical reliability upon packaging using either wirebond or flip-chip package technologies. In flip-chip technologies, where the die to package connection is made using solder balls on the package and bumps on the wafer, the joint is made by raising the temperature above the melting point of the solder in the presence of flux to cause the solder to wet and make contact with the bumps on the die. During the cool-down cycle, the substantial difference in the coefficient of thermal expansion (CTE) between the package puts substantial stress on the ILD, especially under the bumps near the die corners. The extent of stress transmission into the ILD stack on the die depends on a number of parameters, such as the extent of CTE mismatch, the melting temperature of the solder, geometrical considerations and the compliance of the die-to-package connection. If the solder connection is able to deform plastically during the cool-down cycle or if the modulus is low, then less stress is coupled into the dielectric stack on the die. A number of publications can be found that describe efforts to model the stress associated with flip-chip packaging on ILD [26, 27]. If the stress at a given ILD layer exceeds a certain threshold, then cracks can propagate in either the bulk of the film or potentially along weak interfaces in the backend stack, with disastrous results. The mechanical properties of the dielectric, such as its fracture toughness, are therefore a key consideration when integrating porous low-*k* dielectrics. Many literature citations are available that speak to the challenges in this regard [28, 29], including the added complexity resulting from Pb elimination from the die-to-package connection. Similar high stress conditions can also occur during state-of-the-art wirebond packaging. For example, the mold compound that is used as an encapsulant after wirebonding can induce stress post-cure, with key modulators being CTE mismatch and the mechanical properties of the package, mold compound and die thickness [30]. In addition, the dicing and wirebonding process itself may also need to be optimized to handle these more fragile ELK dielectrics, since these processes can apply stresses that can exceed the fracture toughness and result in crack propagation. Matsuura *et al.* [31] mention the use of laser-based dicing to avoid the chipping and cracking risk associated with conventional mechanical sawing when dealing with ELK films on die. Similarly, Tseng *et al.* [32] point out that the presence of ELK under the bond pad can significantly tighten the available bonding process window between not having adequate force for good bond versus causing cracking and peeling in the ELK layer. Quite often, the impact of assembly stresses is not immediately apparent after the assembly process and hence completion of a robust set of reliability tests such as temperature cycling and thermal shock, as well as exposure to moisture tests, are highly recommended to completely characterize and ensure robust integration of ELK dielectrics.

## 7.5 Patterning Integration at Sub-65 nm Nodes

Let us now consider the integration challenges and potential solutions in patterning the interconnects at the small geometries of today's state-of-the-art technology nodes. Integrated circuit companies have been very successful at improving circuit density and performance through the use of transistor scaling. Transistor scaling and performance optimization have resulted in dense and high-speed devices. The two-year cadence of 0.7 times the pitch scaling from node to node has resulted in circuits scaling by 0.5 times the area every two years. At the same time, new device features have been added to on-die circuitry for improved feature integration and overall cost reduction. The result of these scaling trends has been that transistors have become smaller and faster with a given circuit area reduced by 2 times each generation, but die sizes have shrunk less than this due to added features. The interconnect system metal cross-section continues to reduce by approximately 2 times each generation to keep pace with circuit scaling and avoid being signal line limited, but the interconnect length is scaling by only 0.7 times to 1 time. This scaling phenomena results in interconnect delays that are increasingly large portions of the overall circuit delay. To maintain the density improvements while simultaneously managing the RC delay, additional interconnect layers have been added at newer technology nodes. Different companies have had somewhat different layers at a given node due to differences in circuit topology, as well as routing and power delivery concerns, but this is a consistent theme across the industry. As an example of the interconnect layer trend, we reference Intel's high-performance logic technology process evolution: the 0.35 μm technology node had four layers [5], the 0.18 μm technology node had six layers [9], the 65 nm technology node had eight layers [14] and the 32 nm technology node had nine layers [33]. Tighter pitch and higher density interconnects are used at the lower metal layers to provide the necessary routing and connections, and progressively looser pitch layers are used at the upper metal layers to provide longer length, global routing with a better performance. The total number of layers is kept to a minimum within these routing and performance constraints to control die cost. For recent technologies, a range of pitches needs to be fully integrated to optimize the interconnect system across the interconnect layers.

There are several other strategies to improve interconnect performance delays that all impact the ability to pattern and integrate the system to the levels of die yield and reliability required for high-performance interconnect systems. There are geometrical options including: (1) increase the metal line aspect ratio (AR) or the height/width of the metal line, to reduce the RC delay somewhat, (2) increase the height of the ILD between metal layers to reduce the layer-to-layer capacitance and (3) use the possibility of material changes that decrease the bulk resistivity of the metal and/or the dielectric constant of the interlayer dielectric (ILD) to improve R and/or C. The trend towards increased number of metal layers, and hence increased die cost, was slowed by making geometrical as well as significant material changes to interconnect systems. Again referencing Intel's high-performance logic technology, there were ILD material changes made at the 180 nm technology node as fluorine-doped $SiO_2$ systems were introduced to improve the dielectric constant [9]. Additional changes were made at the 130 nm technology node to improve further the dielectric constant and to introduce copper as a replacement for higher resistance aluminum and tungsten [34], along with a change in the patterning approach from subtractive for aluminum to dual damascene for copper. Even lower *k* carbon-doped oxide was introduced at the 90 nm technology node [35] with corresponding new challenges in patterning them. In parallel with these material changes, the geometries and pitch at a given layer was optimized for best density and RC performance, and via heights were increased to reduce layer-to-layer capacitance. Similar trends can be seen at the 45 nm and 32 nm technology nodes, showing interconnect systems with improvements in materials/pitches and geometry optimization to have the best possible system for density, performance and cost [33, 36].

One of the critical outcomes of the patterning process is the resistance and capacitance of the metal and ILD system after the patterning process is completed. This performance requirement needs to be evaluated in the context of a fully integrated process that meets the manufacturability and reliability requirements as well. The difficulty of achieving this optimized system can be understated by either doing this RC evaluation at a loose pitch, where via-to-metal shorting and ILD damage concerns are minimized, or by avoiding several of the constraints that are ultimately required of the interconnect system, such as supporting high current densities in the metal for power delivery reliability. As described in an earlier section of this chapter, low-*k* ILD films are susceptible to damage during processing that will increase the capacitance of the film. The patterning process must ultimately address all these constraints while delivering a tight pitch/highly design-friendly system. Low-resistance metals are susceptible to oxidation as well. Ultimately the patterning process must deliver the results on a low-*k* dielectric with the best possible metal conductor system for optimized RC, while still maintaining the high density required by designs and the manufacturability required to keep the die cost low.

### 7.5.1 Patterning Challenges

There are many challenges associated with dual-damascene interconnect patterning, which has been coupled with copper metallization processes to become the industry standard since the late 1990s to meet the best possible metal line resistance. The choice of metal material determines many of the other materials needed to support fabrication. There needs to be a diffusion barrier material between the Cu and the ILD to avoid Cu diffusion into the ILD and line-to-line leakage issues; this has been most typically accomplished with a conductive Ta diffusion barrier layer around the bottom and sides of the Cu and an $SiC_xN_y$ on top of the Cu. Patterning dry and wet etches must be compatible with the Cu, Cu diffusion barrier, ILD and the dielectric diffusion barrier (e.g. $SiC_xN_y$). The wet and dry etch processes are further constrained by the lithography capability at a given pitch.

In addition to the basic dry and wet etch challenges caused by the metal and ILD choices, there has been and continues to be significant development of sacrificial materials to enable the lithography process. These materials must be developed to support the antireflective needs of the optical lithography as well as having good interface interactions with the photoresist to enable good profile resolution prior to etch. In addition, these materials need to be compatible with the dry and wet etch chemistries needed to etch the ILD and often directly support a particular integration scheme, and thus are likely to have additional requirements. As an example, in standard dual-damascene processing the ILD is deposited and then via lithography, via dry etch and via cleans are completed. Next the wafer needs to have the trench features patterned without any change to the already imprinted via

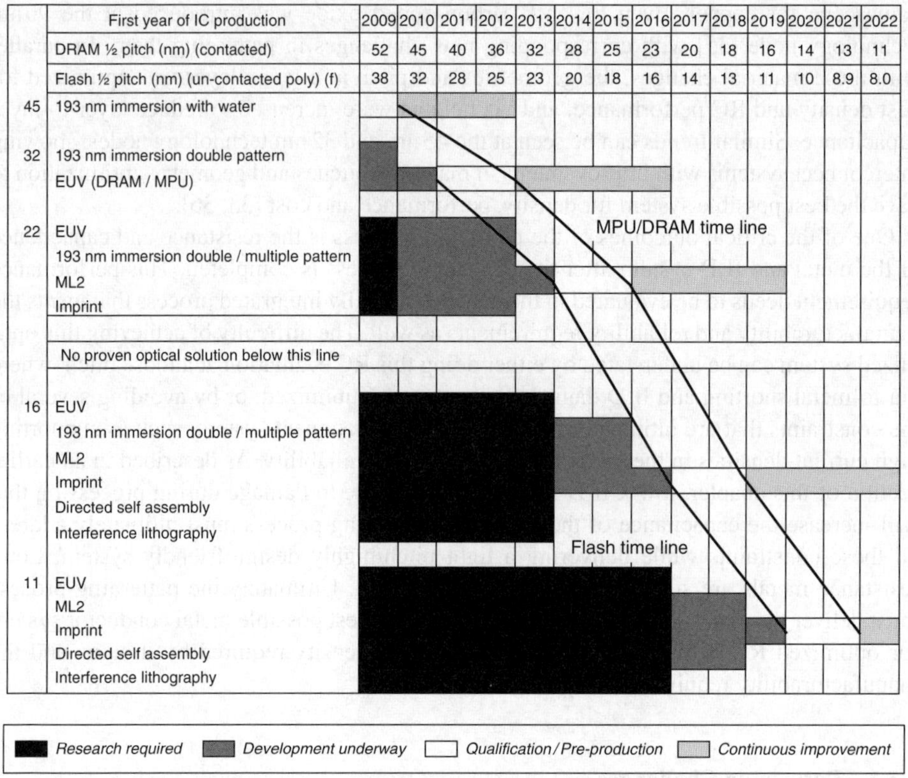


***Figure 7.8*** *Lithography exposure tool potential solutions*

pattern, and this metal lithography needs to be completed on a nonplanar surface covered with vias. The sacrificial material for this example needs to have good planarity after deposition, good via gapfill capability, sufficient antireflective properties at the exposure wavelength and acceptable chemical compatibility with the photoresist and developer. In addition, the dry and wet etch process must also be developed to handle these materials and still provide excellent profiles and tight process control for a given ILD; thus the relative material composition of the sacrificial material and the ILD, and hence the required etch chemistries, are important. A variety of sacrificial materials have been developed to assist in planarizing this surface prior to metal lithography to assist in good depth-of-focus to enable high fidelity and tight pitch patterns. There are many references in published literature to different integration schemes, sacrificial materials and the trade-offs between these; an example for a fluorine-doped $SiO_2$ ILD [37] and for a polymer ILD [38] are good case studies of the differences required. To enable a high-yield, high-performance interconnect patterning process it is critical to optimize for the combination of ILD materials, photoresist, dry and wet etch chemistries and the underlying metal materials simultaneously.

As the critical interconnect pitches have shrunk over the last 15–20 years, the lithography wavelength used to pattern at the required pitches has not kept pace. The tightest metal interconnects transitioned to using 248 nm lithography at around the 350 nm technology node, but did not transition to the 193 nm lithography until around the 90 nm technology node. As can be seen in Figure 7.8, adapted from the ITRS 2009 roadmap for lithography, 193 nm-based patterning is forecast to remain a mainstay for some time.

Due to the fact that pitches have continued to tighten without a corresponding decrease in the lithography wavelength over the past decade, there has been increasing complexity in the designs caused by increasingly restrictive line and space configurations to enable resolution of the critical features. Design rules are critical to ensure that a manufacturable process can be created that enables high die yields without being limited by line-to-line shorting and line openings [40]. Advanced lithography techniques, including optical proximity correction (OPC), can also be used greatly to enhance the patterning capability of two-dimensional features such as line-end shape and the line width control when surrounded by anything other than a grid of similarly drawn metal [41]. These developments have been critical to enable continued interconnect scaling at high die yield as pitches continue to shrink.

In addition to tightly controlled layout design rules for metals, careful via placement within metal lines are critical to enable good functionality for layer-to-layer connections. Small vias are desirable from the aspect of being able to support tight pitch, high-density connections between layers, whereas larger vias are sometimes needed to support a power grid that generally needs to have high current densities through the metal line and via, and needs to have a limited voltage drop to the transistor. The lithography for these small vias has become more difficult and via sizes are also scaling faster than lithography wavelengths. Via-to-metal shorting issues have become more difficult to solve as the metal pitch has reduced due to several factors. Lower-*k* ILDs typically are softer and there is additional loss of ILD material around the via during the metal patterning process, due to additional exposure to sacrificial materials, dry etch chemistries and wet cleans. In addition to degradation of the low-*k* ILD for minimum capacitance, the ILD damage created during dry etches can lead to material loss during the subsequent cleans processes. This leads to a size increase of the via that must be constrained to avoid shorting to a nearby, independent metal line. One solution to this is to require ever-tighter alignment of the printed metal layer to the already etched via layer. This path quickly becomes unmanufacturable as the requirements for alignment of a full die to the previously printed pattern is less than half the pitch and as the via size increases lead to further reduction of alignment margin.

Patterning integration architecture choices can significantly impact the ability to integrate low-*k* interlayer dielectrics into a high-density, high-performance and reliable interconnect stack. Significantly, it can also increase the options to enable the desired line/space optimization within a constrained lithography condition. The use of a hardmask and lithography wavelength-specific sacrificial materials, which suppress reflections from previous layers, can allow the lithography process more latitude in the line-to-space ratio at a given pitch for the optimum resolution. It can also enable subsequent etch processes to adjust the line-to-space ratio by varying the etch taper to meet the desired linewidth for a given design requirement.

Another opportunity to enable lower capacitance interconnect stacks, without improving the ILD dielectric constant, includes significant scaling of the relatively higher *k* 'etchstop' layer that is normally used to protect the previously completed Cu layer from the subsequent layers of noncompatible chemistries. Aggressive scaling of the etch-stop layer puts additional pressure on the via dry and wet etch processes either to stop prior to the underling Cu layer with excellent process control or to identify acceptable chemistries that are compatible with the Cu layer and still meet the profile and resist thickness requirements.

Altering the via and trench patterning order is another significant step to enable excellent metal and via dimension control, while reducing the potential damage to a low-*k* ILD by minimizing the exposure to dry and wet etch chemistries. During standard dual-damascene processing the via etch is completed first so the ILD surrounding the via area is exposed to potentially damaging etches for both the via and trench dry and wet etch processes. In the self-aligned via architecture, the metal line is patterned first but only to a depth that clears the hardmask material with minimal etch into the low-*k* ILD. The metal lines at this point are only a fraction of their final depth. After the metal line pattern transfer is complete into the hardmask material, the via patterning is overlaid on to the hardmask lines and etched into the ILD. Except for the areas where vias are being etched, the low-*k* ILD is covered and protected during this etch process. This reduced the low-*k* exposure to potential damage. Only during the final steps of the process is the trench depth fully etched and cleaned. Details of this self-aligned via flow have been described elsewhere [42].

In addition to limiting the potential ILD damage, in a self-aligned via process the via etch is designed to only remove the ILD and leave the hardmask intact, hence forming vias only where metal lines were previously patterned. This results in vias that are well aligned with the metal line patterning and have only a small encroachment into the space between the metal lines. This is another opportunity to enable good via pitch scaling without stressing the photolithography resolution requirements for the via patterning.

Future options to enable interconnect pitch scaling without a corresponding lithography wavelength reduction include pitch splitting (PS) of a metal layer or a spacer patterning (SP) technique [39]. Pitch splitting is simply printing the tighter pitch in two separate exposures. In this scenario every other line is printed in the first exposure and the remaining lines are printed in the second exposure. This enables a looser pitch to be printed, although with small resist dimensions. In the spacer patterning technique, every other line is printed in resist on a single mask. After these resist features are printed, a spacer is deposited around this resist. The neighboring lines are formed by the areas left around the spacers. The spacer is etched and the resist removed, creating the full set of interconnect lines to be patterned into the hardmask or ILD.

## 7.6 Integration of Conductors in Sub-65 nm Nodes

As stated in the introduction to this chapter, current state of the art on-die interconnects utilize dual-damascene copper-filled interconnects at most layers, while W-filled features are still the primary means to make contact to the active transistors. In terms of electrical performance, the line and via resistance of the conductor system are vital parameters in determining RC delay, voltage droop induced across the die and also due to the vertical stack of interconnects between the transistor and the interconnect to the package. In addition, these interconnects have to be reliable and meet lifetime requirements in the face of reliability challenges such as electromigration (EM), bias temperature stress (BTS) and stress-induced voiding (SIV) – to name a few. As stated earlier, the interconnect stack utilizes varying linewidths and minimum pitches based on the electrical needs, so typically the tightest pitch can be found at the lower metal layers since they need to access the discrete contacts to the source, drain and gate connections to the transistor, which are laid out as densely as feasible. These lower layers are also used for local interconnections and hence tend to be relatively short in wire length. Commensurate with the narrow critical dimensions (CD) of these wires is higher resistance due not only to the smaller cross-sectional area but also higher resistivity due to the size effect, which we will discuss shortly. The smaller cross-sectional area can also impact the reliable current-carrying capability while still meeting electromigration requirements. Intermediate layers feature wider CD and pitch and are used to route connections across circuit blocks and between adjacent blocks, while the upper layers are used as global interconnects and power distribution networks across the die and are designed to maximize the cross-sectional area, thereby minimizing the resistance penalty.

### 7.6.1 Narrow Line Copper Resistivity

The so-called 'size effect' is where the resistivity of copper wires starts to increase over and above the material bulk resistivity as dimensions start to approach the mean free path for electron scattering, which at room temperature is around 39 nm [43]. The increase in resistivity is caused by increased scattering of electrons at these narrow dimensions, with grain boundaries and surfaces such as the barrier/liner or dielectric barrier capping the Cu lines post-Cu CMP, as well as the roughness associated with patterning of the trench. The fundamentals of scattering mechanisms in metals has been well studied, and most current day attempts to model and predict effective resistivity of sub-50 nm wide Cu interconnects revolve around solving equations proposed by Fuchs [44] and Sondheimer [45], addressing surface scattering, and Mayadas and Schatzkes [46], who deals with the impact of polycrystallinity and grain boundary-induced scattering. Figure 7.9 shows the measured data from Steinhogl *et al.* [47] for Cu narrow line resistivity versus models that apply Mathiesen's rule to combine the grain boundary and surface scattering mechanisms proposed by the Fuchs–Sondheimer (FS) and Mayadas–Schatzkes (MS) models. There are a number of parameters in these models, such as the surface specularity $p$ and grain boundary reflectivity coefficient $R$, and there is still uncertainty on precise values to use for typical dual-damascene interconnect architectures.

A number of papers have also been published that use sophisticated Monte Carlo simulations and attempt to account for a variety of electron–phonon interactions, which result in a range of elastic, quasi-elastic and inelastic scattering [48, 49]. Since this size effect is intrinsic to scaling, ways to mitigate this include minimizing the cross-section of the high-resistance barrier or liner, as well as attempts to increase grain size [50, 51] and modulate surface scattering by minimizing line edge roughness and understanding the impact of the choice of liner/barrier [52]. Other factors that can affect the end-of-line resistivity include the presence of impurities, which may be due to the use of additives to enable bottom-up fill as well as dopants in the Cu lines that may be introduced to improve reliability characteristics of the interconnect.

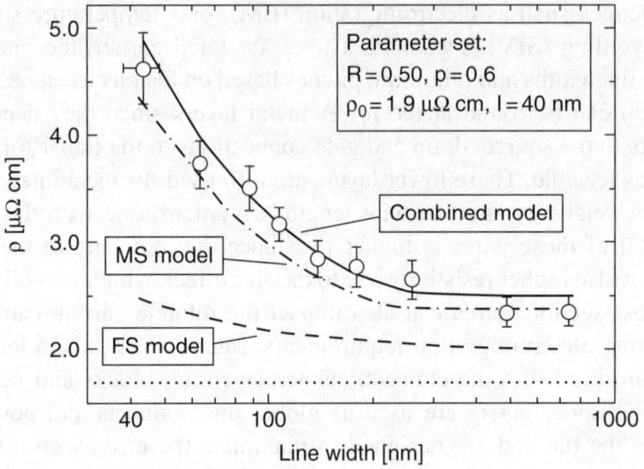


***Figure 7.9*** *Measured narrow Cu line resistivity (circles) versus a combination of FS and MS models. Reproduced with permission from Reference* [47]

### 7.6.2 Integrating Novel Barrier/Liner Materials and Deposition Techniques for Cu Interconnects

While the mainstay for dual-damascene copper interconnects remains the use of PVD deposited Ta-based barrier/liner followed by PVD Cu seed and electroplated Cu utilizing bottom-up fill chemistries, continued scaling is challenging the extendibility of the PVD barrier and seed process steps and necessitating evaluating novel deposition methods like ALD and CVD as well as novel liner materials like ruthenium in conjunction with improved Cu gapfill electroplating chemistries. PVD films tend to have difficulty maintaining step coverage when going into narrow and high aspect ratio features. They also exhibit 'overhang', where the film deposition is thicker at the very top of the feature and results in a high degree of nonuniformity in thickness at the top of the via or trench versus further down the sidewall. This effect is particularly problematic for the Cu seed. As feature sizes go down, the ability to get enough Cu on the sidewall of high aspect ratio vias while leaving an adequate opening at the top to allow void-free bottom-up fill by electroplating is becoming very challenging. This requires us to look at alternate materials and deposition techniques. Integration challenges exist with both types of changes. In going from the PVD process to a ALD or CVD process for barrier deposition, such as ALD TaN, one has to be concerned with precursor incorporation into ILD, especially porous ELK [53], as seen in Figure 7.10.

This necessitates some sort of pore sealing approach to prevent this. People have proposed plasma-based pore sealing [54] as well as depositing material to fill pores at the surface [55], but great care must be taken to not deteriorate the $k_{eff}$ of the ILD stack. In addition, these processes can typically have some residual impurities from the precursor or co-reactant that can negatively impact film adhesion, barrier properties or reliability performance. In the case of ALD TaN, due to poor adhesion between it and Cu, a Ta film is required, which currently requires use of PVD to give adequate purity;

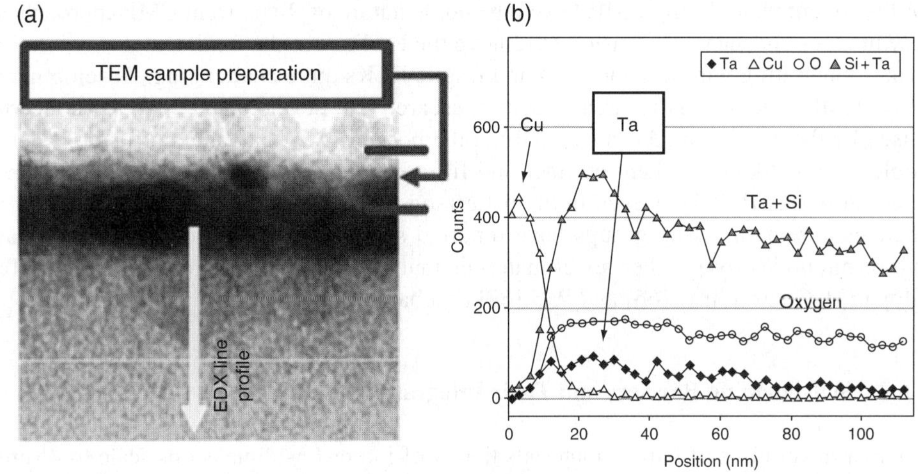


***Figure 7.10*** *(a) TEM showing Ta incorporation and (b) EDX line scans showing a Ta signal well into porous ILD. Reproduced with permission from Reference* [53]

therefore insertion of ALD TaN does not necessarily help with a wider opening post-barrier-seed deposition. The PVD Ta process and equipment continue to be the first choice of most IC fabrication companies and high-bias processes have managed to keep pace with the feature size reduction of Moore's law. The use of high bias for barrier deposition does have some concerns with ILD damage for fragile ELK dielectrics [56]. As stated previously, the PVD Cu seed deposition process is more prone to poor step coverage and is therefore a prime candidate for replacement. Approaches include using a CVD or ALD process to deposit the Cu seed, and many papers have been published on this over the last fifteen years [57, 58]. If the technology relies on seed doping as a means to deliver required EM performance [59], then this complicates the elimination of PVD deposited seed. These conformal Cu deposition processes face challenges in delivering a high-purity film required for low resistivity and good adhesion to the barrier/liner to avoid poor reliability. Deposition also needs to occur at temperatures where the thin, highly conformal copper films are not prone to agglomeration and produce continuous films of an adequate thickness to enable good gapfill during the following electroplate step. Another approach is to rely on a noble material liner, like ruthenium (Ru) and plate directly on it without requiring a Cu seed. While Ru, by virtue of it being a noble metal, has excellent resistance to ingress of oxygen or moisture from the dielectric into the conductor, there is evidence that thin-film Ru is not an adequate diffusion barrier [60] to copper and also may not have very good adhesion to the ILD; hence it is often used with a more conventional Ta-based barrier underneath. Still, it can be of interest since the ability to deposit high-quality CVD or ALD Ru is widely reported, as is the ability to plate on it directly [61] or with a discontinuous Cu seed. Alloying the Ru with phosphorus is reported to improve the barrier properties to Cu and can provide a path for an integrated liner/barrier process [62]. While Ru looks very promising on a number of fronts, it is not without integration challenges, foremost among which is the ability to remove it during CMP. Given the noble nature of Ru, current CMP approaches rely more on mechanical abrasion to remove the Ru lines in the field, and this naturally makes integrating with low-modulus and fragile ELKs quite challenging. Development of Ru CMP processes is an area of active research and development. Another concern raised by the corrosion resistance of Ru is the possibility of galvanic corrosion of less noble metals, like Cu, when the liner and fill are exposed to wet chemistries encountered during CMP [63] or even in the patterning loop. In summary, while a number of challenges remain, many groups have reported successful integration of Ru liners at sub-65 nm nodes [64]. Other novel materials that have been investigated include RuTa alloy [65], Co as a liner [66] and WN [67] as a barrier.

### 7.6.3 Self-Forming Barriers and Their Integration

Another novel class of barrier materials that is of interest as dimensions scale to 45 nm nodes and beyond is the self-forming barrier approach, wherein an alloy of copper is deposited and this alloying element reacts with the ILD to form the barrier. One extensively studied system is Cu–Mn [68–70], but there has also been investigations of other systems. For an Mn-based self-forming barrier, it is widely believed that the Mn in the CuMn seed that is deposited on the ILD reacts to produce an Mn silicate, which then acts as a Cu diffusion barrier. The elimination of a separate barrier has many advantages, namely the potential for improved line and via resistance, as well as providing for a physically larger opening for the Cu electroplate step. The latter point cannot be overstressed as even a few nanometers are a meaningful fraction of the total available real estate at advanced IC nodes. There are, however, some additional concerns with the replacement of Ta barrier with such a novel material, including compatibility with subsequent processing, such as the absence of a meaningful polish stop layer making careful control of dielectric loss while ensuring complete removal of Cu more challenging. Additionally, exposure of this Mn silicate to next-layer cleans chemistry due to presence of unlanded vias needs to be considered while integrating such materials. Since the barrier functions to not only contain the copper and keep it from entering the ILD, which would result in leakage or premature dielectric breakdown, the barrier should also have appropriate resistance to the ingress of oxygen or moisture that might be present in the ILD into the copper line itself; hence reliability testing of such novel barriers in terms of TDDB, BDV and EM performance [70] is highly recommended. The absence of a via bottom barrier means that Cu atoms are free to migrate from layer to layer under EM stress, unlike in the case of conventional Ta barriers where the presence of the Ta blocking boundary at the bottom of the via produces a region of Cu atom flux divergence. While this may be a beneficial feature when there is a large 'source' of Cu atoms feeding a small 'sink', one can also imagine a condition, as shown in Figure 7.11, where the electron flow from the W contact into the lowest SFB dual damascene layer, designated as M1, of relatively small interconnect length and immediately into a much longer upper SFB layer, designated as V1M2, would cause a rapid depletion of copper in the M1 layer due to absence of the blocking boundary at the M1 to V1 interface. Therefore, careful consideration of the inherent EM capabilities and design rules may be relevant to an integrated process.

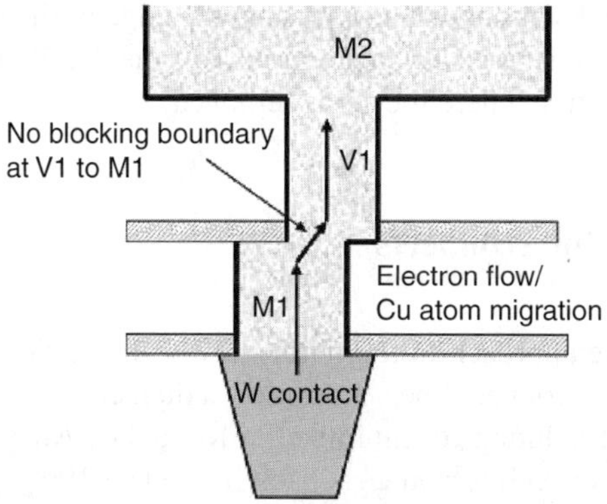


*Figure 7.11* *EM concern due to the absence of a blocking boundary from the self-forming barrier process*

### 7.6.4 Integration to Enable Reliable Copper Interconnects

Let us take a look at some of the integration issues encountered while meeting reliability requirements for copper interconnects at sub-65 nm nodes. One of the major issues is that, with scaling, EM lifetimes continue to drop even if the same current density is applied [71], while the actual demand for higher $J$ rises. This requires improved EM performance from the interconnect system. The EM performance is dictated by the ability of the electron wind to cause the migration of Cu atoms and produce a void that causes either an unacceptable increase in resistance or even a complete failure. Hence, an understanding of fast diffusion paths can prove beneficial in designing interconnect architectures with improved EM performance. It is widely believed that the Cu to dielectric barrier interface is a prime location for such a fast diffusion path, and hence attempts to modulate this are expected to impact EM performance. Alloying the Cu seed with Al has been shown to improve EM [72], albeit at a small increase in resistivity. Yokogawa *et al.* [59] show that the Al segregates to the post-CMP Cu to SiCN dielectric barrier interface and they hypothesize that this elevated level of Al reduces the EM induced diffusion of Cu at this interface. Another way in which dramatic EM improvements have been obtained is through the use of selective self-aligned metal capping layers post-Cu CMP. Candidate capping films include electroless CoWP [73] and CoWB [74] as well as films like Ni or W. A large amount of literature is available for the first two, including electromigration evaluations [75, 76], which show the remarkable improvement in EM performance by more than an order of magnitude in times-to-fail [75]. There are, however, some unique integration challenges associated with the use of such capping layers, including the interaction of these electroless chemistries with porous ELK ILD, wherein dielectric leakage and premature electrical breakdown with commensurate poor BTS performance can occur due to incorporation of components from the chemistry. This issue can be resolved through the use of a pore sealing strategy. Another issue that has been reported is the need to carefully engineer the compatibility of next layer wet cleans that may be used as part of the patterning process [77]. Incompatibility can result in removal of the Co cap underneath the dielectric barrier for a substantial distance beyond the edge of the next layer vias, and this can have disastrous impacts on electrical continuity of the via fill and also poor reliability due to the narrow separation where the Co is removed.

## 7.7 Novel Air-Gap Interconnects

Air-gap is an alternative approach to the porous low-*k* dielectrics to achieve low capacitance between metal interconnect lines. As air has a dielectric constant of 1.0, it could be potentially the ultimate solution to minimize an RC delay. Air-gap-based interconnects were first investigated for Al interconnects back in the late 1990s [78, 79] and have been widely studied for Cu interconnects since then. Many IC companies have demonstrated integration processes for air-gap Cu interconnects [80–85]. However, air-gap fabrication typically requires more process steps, which means higher manufacturing costs. In addition, there are remaining concerns regarding their reliability, though some data showing viability is available [85]. As a result, air-gap-based interconnects have not yet become a mainstream process for high-volume manufacturing as of this writing in 2010. On the other hand, ultra-low-*k* integration is becoming more and more challenging as the low-*k* dielectric continues to incorporate a higher porosity and more organic components in order to achieve lower dielectric constant values. Porosity can be as high as 45–50 % for ultra-low-*k* materials with a dielectric constant of 2.0, which can lead to poor mechanical integrity and less resistance to damage from plasma etching, wet cleans, barrier deposition and Cu CMP. Therefore, integration of porous low-*k* materials with $k \leq 2.0$ is deemed to be extremely difficult. This keeps air-gap-based interconnects as a viable option for future technologies.

Air-gaps refer to the air cavities integrated into dielectric spacings between metal interconnect lines. Several integration schemes and structures for air-gap formation have been reported. Those schemes in general can be divided into two categories. The first one involves a nonconformal deposition of dielectric inside patterned trenches to form air-gaps. The second category is to use a sacrificial material that can be removed at late process steps to create air-gaps. Both of those schemes will be discussed in this section.

### 7.7.1 Unlanded Via Integration with Air-Gap Interconnects

Before we start to discuss air-gap integration schemes, it is necessary to explain one of the key process concerns for air-gap Cu interconnects, namely the presence of so-called unlanded vias, where the via is partially landed on the metal below the via and partially landed on the dielectric. Unlanded vias are caused by misalignment of the via lithography to the metal layer below, which is allowed in design rules in order to give enough process window to the via patterning process. Figure 7.12 shows the potential issue caused by an unlanded via in air-gap Cu interconnects. For illustration purposes, the air-gap is present in the M1 layer. As can be seen, an unlanded via at the next layer causes that unlanded V1 via to connect to the air-gap underneath. During the subsequent metal depositions to fill V1 and M2, metal intrusion into the air-gap could occur, causing shorts or other reliability issues. This issue has to be addressed for any air-gap integration schemes.

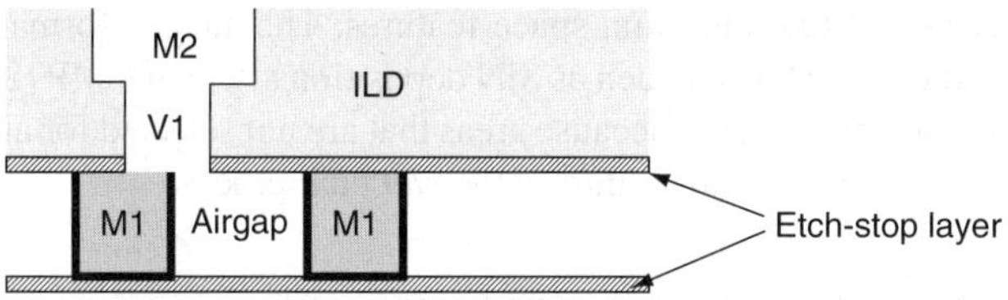


***Figure 7.12*** *Schematic of process issues caused by an unlanded via in air-gap Cu interconnects*

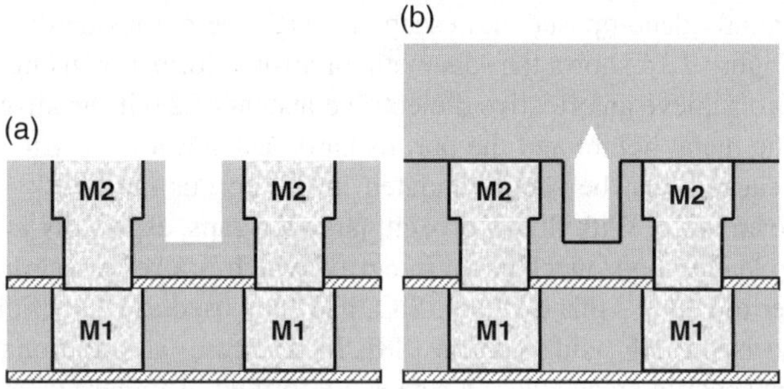


***Figure 7.13*** *A process sequence used to incorporate air-gaps in a specific metal level by using the nonconformal CVD technique. (a) A specific lithography step to define air-gap trenches in between copper lines. (b) Air-gap formation by depositing a conformal dielectric barrier and nonconformal dielectric layer*

### 7.7.2 Air-Gap Formation Using Nonconformal Dielectric Deposition

Air-gap formation using non-conformal CVD $SiO_2$ deposition was reported by Arnal *et al.* [80] as early as 2001. As shown in Figure 7.13, the basic idea of this approach is first to remove dielectric material between metal interconnect lines (see Figure 7.13 (a)) and then use a nonconformal dielectric deposition to form air-gaps by pinch-off of the dielectric layer (see Figure 7.13 (b)). The shape of the air-gap (i.e. closure-point height, width, volume of the cavity) depends very much on the aspect ratio, defined as the ratio between the height and the width of the trenches. The pinch-off profile is more difficult to achieve as the trench width increases or the aspect ratio decreases. For practical applications, a specific lithographic step is needed to define the exact areas for air-gap formation. This step provides the opportunity to place air-gaps in the regions where they will be most useful, namely the narrow space regions, while avoiding wide spaces where the low aspect ratio makes it difficult for the nonconformal next-layer dielectric to pinch-off. The masking step can also prevent air-gaps from forming under potential next-layer unlanded vias.

Recently, nonconformal low-*k* deposition has also been demonstrated to reduce the capacitance further by Yoo *et al.* [85]. By using a low-*k* dielectric, 23 % and 28 % capacitance reductions were achieved, respectively, on 32 nm and 22 nm node test vehicles when air-gaps were inserted at the minimum space features. The air-gap formation can also be introduced after a dielectric barrier such as SiN deposition after Cu CMP [87]. This scheme enables trench creation by wet etch because areas that are not selected for air-gap formation are protected by the dielectric barrier during the wet etch process.

### 7.7.3 Air-Gap Formation Using a Sacrificial Material

In this approach, a sacrificial material is placed in between the Cu lines where air-gaps are further required. Daamen *et al.* [81] have demonstrated a process with a stack consisting of a sacrificial material with a porous hardmask on top. The sacrificial material is a polymer that can be thermally decomposed and escape through the porous hardmask after the Cu CMP process. Figure 7.14 shows the schematic of air-gap formation using this technique.

It is possible to achieve an effective dielectric constant of 2.0 if the air-gap height takes up to 75 % of the metal height and the porous hardmask has a dielectric constant of 2.4. The unlanded via risk can be accommodated in the porous hardmask, but the process window may be an issue. With this approach, large air-gaps, especially at wide dielectric spaces, can also lead to poor mechanical integrity, which is a key concern for this type of scheme. In order to address this concern, $SiO_2$ has been used as a sacrificial layer, which can be removed by an HF acid solution [88]. In this case, an additional photomask is needed to define specific areas where air-gaps are formed. Another techniquc to improve the mechanical integrity has been proposed by Nakamura *et al.* [83]. This stack consists of a PAr film applied as a trench layer material and an SiOC film applied as a via layer material. After making several intermediate/semiglobal layers using the conventional hybrid BEOL process and the monolithic ILD process, through-holes are formed at the region where the PAr materials should be removed. The PAr layer is then removed by an RIE process. Moisture uptake through the gas outlet has been found to be a major issue for this technique.

The unlanded via issue can be addressed by two different options in this scheme. One option is to make the via etch stop at the porous hardmask layer. The other option is to do the sacrificial layer removal after fabrication of the upper layer interconnects. In this case, an etch-stop layer is not desirable because it will block the removal of the sacrificial layer. Furthermore, a self-aligned barrier on top of the Cu is needed to

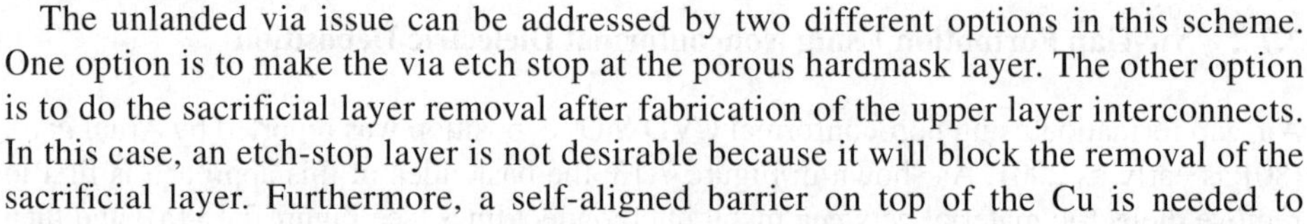


***Figure 7.14*** *Schematic of air-gap formation through sacrificial layer removal: (a) after deposition of the sacrificial material and porous hardmask; (b) after patterning, metallization and Cu CMP; (c) sacrificial material removed to form air-gaps*

prevent Cu diffusion and improve electromigration. Regarding the porous low-*k* hardmask, a high porosity is more desirable for sacrificial material removal but less desirable for good mechanical integrity. Therefore, an appropriate trade-off between sacrificial removal and mechanical integrity has to be maintained. Other approaches leverage the susceptibility of plasma-damaged ELK to be rapidly etched to produced air-gaps next to lines.

In summary, the nonconformal dielectric deposition technique to form air-gaps is a relatively more mature process compared with the technique of using sacrificial material removal. It also offers more mechanical integrity at higher manufacturing costs. It is clear that scaling challenges will require aggressive integration solution development in all the fronts reviewed in this chapter, namely ELK dielectrics, their patterning, metallization and also the use of alternatives like air-gap interconnects.

## References

[1] Joshi, R., Klepner, S., Basavaiah, S. *et al.* (1991) Process integration for a 2 ns cycle/4 ns access 512K CMOS SRAM. Symposium on VLSI Technology, 1991. Digest of Technical Papers, pp. 31–32.

[2] Kaanta, C., Cote, W., Cronin, J. *et al.* (1998) Submicron wiring technology with tungsten and planarization. Proceedings of the Fifth International IEEE VLSI Multilevel Interconnection Conference 1988, pp. 21–28.

[3] Doan, T., Bellersen, M., de Bruin, L. *et al.* (1998) A double level metallization system having 2 μm pitch for both levels. Proceedings of the Fifth International IEEE VLSI Multilevel Interconnection Conference 1988, pp. 13–20.

[4] Thomas, M.E., Sekigahama, S., Renteln, P. *et al.* (1990) The mechanical planarization of interlevel dielectrics for multilevel interconnect applications. Proceedings of the Seventh International IEEE VLSI Multilevel Interconnection Conference 1990, pp. 438–440.

[5] Bohr, M., Ahmed, U., Brigham, L. *et al.* (1994) A high performance 0.35 μm logic technology for 3.3 V and 2.5 V operation. Technical Digest, International Electron Devices Meeting 1994, pp. 273–276.

[6] Izawa, T., Katsube, M., Yokoyama, Y. *et al.* (1994) A novel embedded SRAM technology with 10-$\mu m^2$ full-CMOS cells for 0.25-μm logic devices. Technical Digest, International Electron Devices Meeting 1994, pp. 941–943.

[7] Stamper, A.K., McDevitt, T.L. and Luce, S.L. (1998) Sub-0.25-micron interconnection scaling: damascene copper versus subtractive aluminum. IEEE/SEMI Advanced Semiconductor Manufacturing Conference and Workshop 1998, pp. 337–346.

[8] Heidenreich, J., Edelstein, D., Goldblatt R. *et al.* (1998) Copper dual damascene wiring for sub-0.25 μm CMOS technology. Proceedings of the IEEE 1998 International Interconnect Technology Conference, pp. 151–153.

[9] Yang, S., Ahmed, S., Arcot, B. *et al.* (1998) A high performance 180 nm generation logic technology. Technical Digest, International Electron Devices Meeting 1998, pp. 197–200.

[10] Thompson, S., Alavi, M., Arghavani, R. *et al.* (2001) An enhanced 130 nm generation logic technology featuring 60 nm transistors optimized for high performance and low power at 0.7–1.4 V. Technical Digest, International Electron Devices Meeting 2001, pp. 11.6.1–11.6.4.

[11] Semiconductor International Association (2000) *The International Technology Roadmap for Semiconductors*, 2000 edn, International SEMATECH, Austin, Texas.

[12] Chiang, S.K. and Lassen, C.L. (1999) The market for low-*k* interlayer dielectrics. *Solid State Technol.*, **42**, 42.

[13] Jan, C.-H., Bielefeld, J., Buehler, M. *et al.* (2003) 90 nm generation, 300 mm wafer low *k* ILD/Cu interconnect technology. Proceedings of the IEEE 2003 International Interconnect Technology Conference, pp. 15–17.

[14] Bai, P., Auth, C., Balakrishnan, S. *et al.* (2004) A 65 nm logic technology featuring 35 nm gate lengths, enhanced channel strain, 8 Cu interconnect layers, low-*k* ILD and 0.57 $\mu m^2$ SRAM cell. Technical Digest, International Electron Devices Meeting 2004, pp. 28.2.1–28.2.4.
[15] McGahay, V., Bonilla, G., Chen, F. *et al.* (2006) 65 nm Cu integration and interconnect reliability in low stress $k = 2.75$ SiCOH. Proceedings of the IEEE 2006 International Interconnect Technology Conference, pp. 9–11.
[16] DellaGuardia, R., Kwong, R., Li, W. *et al.* (2004) BEOL lithography for early development at the 65 nm node. *Proc. SPIE*, **5377**, 980–987.
[17] Posseme, N., Maurice, C., Brun, Ph. *et al.* (2006) New etch challenges for the 65 nm technology node low-*k* integration using an enhanced trench first hard mask architecture. Proceedings of the IEEE 2006 International Interconnect Technology Conference, pp. 36–38.
[18] Xie, Z., Bodke, A., Fu, J. *et al.* (2007) PVD TiN hardmask for copper metallization. 2007 IEEE International Symposium on Semiconductor Manufacturing, pp. 1–3.
[19] Fuller, N.C.M., Worsley, M.A., Nitta, S. *et al.* (2006) Analysis of plasma-induced modification of ULK and eULK materials: dual damascene processing challenges for 45nm ($k = 2.4$) and beyond BEOL technologies. Proceedings of the IEEE 2006 International Interconnect Technology Conference, pp. 24–26.
[20] Yoo, H.J., Bielefeld, J., Balakrishnan, S. *et al.* (2008) Plasma damage and repair of porous low-*k* carbon-doped oxide films. Advanced Metallization Conference 2007, pp. 705–709.
[21] Bao, J.J., Shi, H.L., Liu, J.J. *et al.* (2007) Mechanistic study of plasma damage and $CH_4$ recovery of low *k* dielectric surface. Proceedings of the IEEE 2007 International Interconnect Technology Conference, pp. 147–149.
[22] Iacopi, F., Richard, O. and Van Aelst, J. *et al.* (2006) Understanding integration damage to low-*k* films: mechanisms and dielectric behavior at 100 kHz and 4 GHz. Proceedings of the IEEE 2006 International Interconnect Technology Conference, pp. 12–14.
[23] Jinnai, B., Nozawa, T. and Samukawa, S. (2008) Damage mechanism in low-dielectric (low-*k*) films during plasma processes. *J. Vac. Sci. Technol. B*, **26** (6), 1926–1932.
[24] Kodera, M., Takahashi, T. and Mimamihaba, G. (2010) Evaluation of dielectric constant through direct chemical mechanical planarization of porous low-*k* film. *Jpn. J. Appl. Phys.*, **49**, 04DB07.
[25] Kondo, S., Fukaya, K., Ohashi, N. *et al.* (2006) Direct CMP on porous low-*k* film for damage-less Cu integration. Proceedings of the IEEE 2006 International Interconnect Technology Conference, pp. 164–166.
[26] Wang, G., Groothuis, S., Merrill, C. *et al.* (2004) Investigation of interfacial delamination for Cu/low-*k* structures during flip-chip packaging Proceedings of the Ninth Intersociety Conference on Thermal and Thermomechanical Phenomena in Electronic Systems 2004, vol. 2, pp. 211–218.
[27] Uchibori, C.J., Zhang, X., Ho, P.S. and Nakamura, T. (2006) Impact of chip-package interaction on reliability of Cu/ultra low-*k* interconnects for 65 nm technology and beyond. Proceedings of the 8th International Conference on Solid State and Integrated Circuit Technology 2006, pp. 314–317.
[28] Yeoh, A., Chang, M., Pelto, C. *et al.* (2006) Copper die bumps (first level interconnect) and low-*k* dielectrics in 65 nm high volume manufacturing. Proceedings of the 56th Electronic Components and Technology Conference 2006, pp. 1611–1615.
[29] Hou, S.Y., Shih, C.W., Wu, W.C. *et al.* (2010) Lead-free flip chip solution for 40 nm extreme low-*k* interconnect system. Proceedings of the IEEE 2010 International Interconnect Technology Conference, pp. 1–3.
[30] Sweeney, F., Kim, J., Morison, T. *et al.* (2005) Reducing the effect of mechanical stress on low-*k* die – case study. Proceedings of the 7th Electronic Packaging Technology Conference 2005, vol. 2, pp. 720–722.
[31] Matsuura, Y., Watanabe, A. and Kawakami, S. (2006) Stacked die SIP technology suitable for devices with low-*k* inter-layer dielectric. Proceedings of the 56th Electronic Components and Technology Conference 2006, pp. 1381–1385.
[32] Tseng, A., Lin, B., Huang, L. *et al.* (2008) BGA assembly process development for 45 nm ELK CUP devices. Proceedings of International Conference on Electronic Packaging Technology and High Density Packaging 2008, pp. 1–5.

[33] Natarajan, S., Armstrong, M., Bost, M. *et al.* (2008) A 32 nm logic technology featuring 2nd-generation high-$k$ + metal-gate transistors, enhanced channel strain and 0.171 $\mu m^2$ SRAM cell size in a 291 Mb array. Technical Digest, International Electron Devices Meeting 2008, pp. 1–4.
[34] Tyagi, S., Alavi, M., Bigwood, R. *et al.* (2000) A 130 nm generation logic technology featuring 70 nm transistors, dual Vt transistors and 6 layers of copper interconnects. Technical Digest, International Electron Devices Meeting 2000, pp. 567–570.
[35] Thompson, S., Anand, N., Armstrong, N. *et al.* (2002) A 90 nm logic technology featuring 50 nm strained silicon channel transistors, 7 layers of Cu interconnects, low $k$ ILD and 1 $\mu m^2$ SRAM cell. Technical Digest, International Electron Devices Meeting 2002, pp. 61–64.
[36] Mistry, K., Allen, C., Auth, C. *et al.* (2007) A 45 nm logic technology with high-$k$ + metal gate transistors, strained silicon, 9 Cu interconnect layers, 193 nm dry patterning and 100 % Pb-free packaging. Technical Digest, International Electron Devices Mecting 2007, pp. 247–250.
[37] Hussein, M., Sivakumar, S., Brain, R. *et al.* (2002) A novel approach to dual damascene patterning. Proceedings of the IEEE 2002 International Interconnect Technology Conference, pp. 18–20.
[38] Hussein, M., Brain, R., Turkot, R. *et al.* (2003) Dual damascene patterning of polymer inter-layer dielectrics. Proceedings of the IEEE 2003 International Interconnect Technology Conference, pp. 33–35.
[39] Semiconductor International Association (2009) *The International Technology Roadmap for Semiconductors – Lithography*, 2009 edn, International SEMATECH, Austin, Texas.
[40] Moon, P. (2008) High volume manufacturing issues for on-die interconnects at the 45nm process node. Proceedings of the IEEE 2008 International Interconnect Technology Conference, pp. 213–215.
[41] Schellenberg, F.M. (2005) A history of resolution enhancement technology. *Opt. Rev.*, **12** (2), 83–89.
[42] Brain, R., Agrawal, S., Becher, D. *et al.* (2009) Low-$k$ interconnect stack with a novel self-aligned via patterning process for 32 nm high volume manufacturing. Proceedings of the IEEE 2009 International Interconnect Technology Conference, pp. 249–251.
[43] Mallikarjunan, A., Sharma, S. and Murarka, S.P. (2000) Resistivity of copper films at thicknesses near the mean free path of electrons in copper. *Electrochem. Solid State Lett.*, **3** (9), 437–438.
[44] Fuchs, K. (1938) The conductivity of thin metallic films according to the electron theory of metals. *Proc. Cambridge Phil. Soc.*, **34**, 100–108.
[45] Sondheimer, E.H. (1952) The mean free path of electrons in metals. *Advances in Physics*, **1** (1), 1–42.
[46] Mayadas, A.F. and Shatzkes, M. (1970) Electrical resistivity model for polycrystalline films: the case of arbitrary reflection at external surfaces. *Phys. Rev. B*, **1** (4), 1382–1389.
[47] Steinhogl, W., Schindler, G., Steinlesberger, G. *et al.* (2002) Size-dependent resistivity of metallic wires in mesoscopic range. *Phys. Rev. B*, **66**, 075414.
[48] Zong, Z., Mohammadzadeh, S., Cao, Y. *et al.* (2010) Electrical resistivity calculations for copper nanointerconnect. *Microelectron. Engng*, **87**, 402–405.
[49] Eiguren, A., Hellsing, B., Chulkov, E.V. *et al.* (2003) Phonon-mediated decay of metal surface states. *Phys. Rev. B*, **67**, 235423.
[50] Zhang, W., Brongersma, S.H., Heylen, N. *et al.* (2005) Geometry effect on impurity incorporation and grain growth in narrow copper lines. *J. Electrochem. Soc.*, **152** (12), C832–C837.
[51] Vanstreels, K., Brongersma, S.H., D'Haen, J. *et al.* (2007) High sputter bias super secondary growth grain initiation (in structures). Advanced Metallization Conference 2006, pp. 97–103.
[52] Kitada, H., Suzuki, T., Akiyama, S. *et al.* (2009) Influence of titanium liner on resistivity of copper interconnects. *Jpn. J. Appl. Phys.*, **48** (4), 04C026-1.
[53] Moon, B.K., Iijima, T., Malhotra, S. *et al.* (2005) Integration of ALD-TaN liners on nanoporous dielectrics. *Mater. Res. Soc. Symp. Proc.*, **863**, 221.
[54] Aimadeddine, M., Arnal, M., Roy, D. *et al.* (2006) Effect of $CH_4$ plasma on porous dielectric modification and pore sealing for advanced interconnect technology nodes. Proceedings of the IEEE 2006 International Interconnect Technology Conference, pp. 81–83.

[55] Guedj, C., Arnaud, L., Fayolle, M. *et al.* (2004) Effect of pore sealing on the reliability of ULK/Cu interconnects. Proceedings of the IEEE 2004 International Interconnect Technology Conference, pp. 148–150.
[56] Zhao, L., Baklanov, M.R., Pantouvaki, M., Tökei, Z. and Beyer, G.P. (2010) Ultra low-*k* materials: challenges of scaling. *Electrochem. Soc. Trans.*, **33**, 12.
[57] Park, K.C., Choi, S.-M., Lee, S.-J. *et al.* (2000) Process integration of CVD Cu as a seed application for Cu electroplating and a plug-fill application. Proceedings of the IEEE 2000 International Interconnect Technology Conference, pp. 43–45.
[58] Moon, D.-Y., Kwon, T.-S., Kang, B.-W. *et al.* (2010) Copper seed layer using atomic layer deposition for Cu interconnect. 2010 IEEE 3rd International Nanoelectronics Conference (INEC 2010), pp. 450–451.
[59] Yokogawa, S., Tsuchiya, H., Kakuhara, Y. *et al.* (2008) Analysis of Al doping effects on resistivity and electromigration of copper interconnects. *IEEE Trans. Device and Mater. Reliab.*, **8** (1), 216–221.
[60] Kim, H., Koseki, T., Ohba, T. *et al.* (2005) Cu wettability and diffusion barrier property of Ru thin film for Cu metallization. *J. Electrochem. Soc.*, **152** (8), G594–G600.
[61] Abe, M., Ueki, M., Tada, M. *et al.* (2007) Highly-oriented PVD ruthenium liner for low-resistance direct-plated Cu interconnects. Proceedings of the IEEE 2007 International Interconnect Technology Conference, pp. 4–6.
[62] Shin, J., Kim, H-W., Agapiou, K. *et al.* (2008) Effects of P on amorphous chemical vapor deposition Ru–P alloy films for Cu interconnect liner applications. *J. Vac. Sci. Technol. A*, **26** (4), 974–979.
[63] Maruyama, K., Shiohara, M., Yamada, K. *et al.* (2009) Galvanic corrosion control in chemical mechanical polishing of Cu interconnects with ruthenium barrier metal film. *Jpn. J. Appl. Phys.*, **48** (4), 04C022.
[64] Yang, C.-C., Spooner, T., Ponoth, S. *et al.* (2006) Physical, electrical and reliability characterization of Ru for Cu interconnects. Proceedings of the IEEE 2006 International Interconnect Technology Conference, pp. 187–189.
[65] Yang, C.-C., Cohen, S., Shaw, T. *et al.* (2010) Characterization of 'ultrathin-Cu'/Ru(Ta)/TaN liner stack for copper interconnects. *IEEE Electron Device Lett.*, **31** (7), 722–724.
[66] Nogami, T., Maniscalco, J., Madan, A. *et al.* (2010) CVD Co and its application to Cu damascene interconnections. Proceedings of the IEEE 2010 International Interconnect Technology Conference, pp. 1–3.
[67] Lee, B.H. and Yong, K. (2004) Diffusion barrier properties of metalorganic chemical vapor deposition of $-WN_x$ compared with other barrier materials. *J. Vac. Sci. Technol. B*, **22** (5), 2375–2379.
[68] Ablett, J.M., Woicik, J.C., Tokei, Zs. *et al.* (2009) Phase identification of self-forming Cu–Mn based diffusion barriers on p-SiOC:H and $SiO_2$ dielectrics using X-ray absorption fine structure. *Appl. Phys. Letters*, **94**, 042112.
[69] Koike, J., Haneda, M., Iijima, J. *et al.* (2006) Cu alloy metallization for self-forming barrier process. Proceedings of the IEEE 2006 International Interconnect Technology Conference, pp. 161–163.
[70] Usui, T., Nasu, H., Takahashi, S. *et al.* (2006) Highly reliable copper dual-damascene interconnects with self-formed $MnSi_xO_y$ barrier layer. *IEEE Trans. on Electron Devices*, **53** (10), 2492–2499.
[71] Gall, M., Hauschildt, M. and Hernandez, R. (2010) Large-scale statistical analysis of early failures in Cu electromigration, Part II: scaling behavior and short-length effects. *J. Appl. Phys.*, **108**, 013524.
[72] Vanypre, T., Mourier, T., Torres, J. *et al.* (2008) CuAl alloy: a robust solution for 45/32 nm integration. Advanced Metallization Conference 2007, pp. 97–98.
[73] Hu, C.-K., Gignac, L.M., Rosenberg, R. *et al.* (2004) Atom motion of Cu and Co in Cu damascene lines with a CoWP cap. *Appl. Phys. Letters*, **84** (24), 4986–4988.
[74] Tagami, M., Furutake, N., Inoue, N. *et al.* (2009) Highly-reliable molecular-pore-stack (MPS)–SiCOH/Cu interconnects with CoWB metal-cap films. Proceedings of the IEEE 2009 International Interconnect Technology Conference, pp. 11–13.

[75] Aubel, O., Hohage, J., Feustel, F. *et al.* (2009) Process options for improving electromigration performance in 32 nm technology and beyond. Proceedings of the IEEE 2009 International Reliability Physics Symposium, pp. 832–836.
[76] Moon, P., Dubin, V., Johnston, S. *et al.* (2003) Process roadmap and challenges for metal barriers. Technical Digest, International Electron Devices Meeting 2003, pp. 35.1.1–35.1.4.
[77] Wu, A., Baryschpolec, E., Rao, M. *et al.* (2010) Development of compatible wet-clean stripper for integration of CoWP metal cap in Cu/low *k* interconnects. *Microelectron. Engng*, **87**, 1685–1688.
[78] Bassman, L.C., Vinci, R.P., Shieh, B.P. *et al.* (1997) Simulation of the effect of dielectric air gaps on interconnect reliability. *Mater. Res. Soc. Symp. Proc.*, **473**, 323–328.
[79] Shieh, B., Saraswat, K.C., McVittie, J.P. *et al.* (1998) Air-gap formation during IMD deposition to lower interconnect capacitance. *IEEE Electron Device Lett.*, **19** (1), 16–18.
[80] Arnal, V., Torres, J., Gayet, P. *et al.* (2001) Integration of a 3 level Cu–$SiO_2$ air-gap interconnect for sub 0.1 micron CMOS technologies. Proceedings of the IEEE 2001 International Interconnect Technology Conference, pp. 298–300.
[81] Daamen, R., Verheijden, G.J.A.M., Bancken, P.H.L. *et al.* (2005) Air gap integration for the 45 nm node and beyond. Proceedings of the IEEE 2005 International Interconnect Technology Conference, pp. 240–242.
[82] Nitta, S., Edelstein, D. and Ponoth, S. (2008) Performance and reliability of airgaps for advanced BEOL interconnects. Proceedings of the IEEE 2008 International Interconnect Technology Conference, pp. 191–192.
[83] Nakamura, N., Matsunaga, N., Kaminatsui, T. *et al.* (2008) Cost-effective air-gap interconnects by all-in-one post-removing process. Proceedings of the IEEE 2008 International Interconnect Technology Conference, pp. 193–195.
[84] Chen, H.-W., Jeng, S.-P., Tsai, H.-Y. *et al.* (2009) A self-aligned air gap interconnect scheme. Proceedings of the IEEE 2009 International Interconnect Technology Conference, pp. 146–148.
[85] Yoo, H.J., Balakrishnan, S., Bielefeld, J. *et al.* (2010) Demonstration of a reliable high-performance and yielding air gap interconnect process. Proceedings of the IEEE 2010 International Interconnect Technology Conference, pp. 1–3.
[86] Gosset, L.G., Arnal, V., Prindle, C. *et al.* (2003) General review of issues and perspectives for advanced copper interconnections using air gap as ultra low-*k* material. Proceedings of the IEEE 2003 International Interconnect Technology Conference, pp. 65–67.
[87] Harada, T., Takahashi, M., Murakami, K. *et al.* (2006) Manufacturable low $k_{eff}$ ($k_{eff}$ <2.5) Cu interconnects by selective/low damage air gap formation. Proceedings of the IEEE 2006 International Interconnect Technology Conference, pp. 15–18.
[88] Gras, R., Gaillard, F., Bouchu, D. *et al.* (2008) 300 mm multi level air gap integration for edge interconnect technologies and specific high performance applications. Proceedings of the IEEE 2008 International Interconnect Technology Conference, pp. 196–198.

# 8

# Chemical Mechanical Planarization for Cu–Low-*k* Integration

*Gautam Banerjee*

*Air Products and Chemicals Inc., Allentown, Pennsylvania, USA*

## 8.1 Introduction

Chemical mechanical planarization or CMP is an enabling process technology for advanced submicrometer integration using copper damascene. The copper CMP process must meet increasingly stringent requirements for metal loss. The ITRS 2005 roadmap [1] suggested a total metal loss budget (at 10 % of total metal) of 120 Å at 65 nm, progressively reducing to 60 Å for the 32 nm node. The ITRS roadmap also suggests simultaneous incorporation of a low-*k* dielectric layer, which is likely to be porous for logic devices in sub-30 nm technology nodes. Porous dielectrics integrated with the existing dielectric barriers and hardmask layers are susceptible to cohesive and adhesive failures, when processed under planarization conditions, which are typically optimized for conventional dielectrics. Consequently, low-*k* process integration must include a planarization technology, which offers low mechanical impact to the target materials. Additionally, a CMP process must offer the ability to planarize wafers with minimal over-polish to achieve metal loss and defectivity targets. While incorporation of porous low-*k* for sub-40 nm technology nodes is very challenging for interconnect integration, planarization of all films used in advanced semiconductors and MEMS technologies requires processing in an optimized process regime to maintain control of wafer removal profiles across a 300 mm wafer.

---

*Advanced Interconnects for ULSI Technology*, First Edition. Edited by Mikhail R. Baklanov, Paul S. Ho and Ehrenfried Zschech.

## 8.2 Back to Basics

So, what is CMP? Chemical mechanical planarization (CMP) is a process used to flatten (or 'planarize') a macroscopically flat silicon wafer at the device level. This process is critical in establishing reliable multilevel interconnects in the fabrication of integrated circuits. It is now being used also for MEMS (microelectromechanical systems) and TSV (through silicon vias) processes. The main objectives of CMP are to smooth surface topography of dielectric deposits to enable multilevel metallization or to remove excess coating material to produce inlaid metal damascene structures and shallow isolation trenches. The application of CMP in electronic device fabrication is significant in both memory and logic (microprocessor) device fabrication. As the device and interconnect wiring dimensions continuously shrink with the technology advancement in device manufacturing, the output specifications of the CMP process have become more stringent. For example, the allowable dish depth has come down from 19 nm at the 65 nm node for a 100 micrometer wide feature to 10 nm for the same feature at the 32 nm node and many such process specifications change as the industry progresses into integrating much weaker low-*k* dielectric materials with copper.

In a CMP process, the wafer being planarized is pressed (facing down) on to a polymer polishing pad and is abraded in the presence of slurry, which contains active chemicals and abrasive particles. Thus, CMP can be described as a process that uses the combination of mechanical energy from the pad and abrasives and chemical energy from the slurry chemicals to polish and remove material from the wafer surface. The process flow can be described by the schematic diagram in Figure 8.1. The schematic of the CMP process is presented in Figure 8.2 [2].

In an industrial manufacturing process, a CMP polisher generally has multiple platens (polishing pad holder) so that different steps can be processed on different wafers in sequence to increase the output. A commercial polisher is automated with dry-in/dry-out logic. Dry wafers are loaded via a FOUP (front opening universal pod) to go through polishing and subsequently cleaning sequences before the clean and dry wafers are loaded back on to another FOUP. Figure 8.3 shows two typical commercial CMP tools from the front side in a clean room [3, 4].

For CMP, therefore, one needs to use some consumables such as a polishing pad, wafer, polishing slurry, pad conditioning disk and post-CMP cleaning solution. There are varieties of pads, slurries, cleans and even pad conditioning disks for different types of films to be planarized. Typical films on wafers normally planarized are Cu, Ta, W, TEOS, various low-*k* dielectric films, $Si_3N_4$, SiON, SiCN, Co, Mn, Ru, poly-Si, etc. The most commonly used CMP tools today for the ULSI integration process have 300 mm diameter wafer holders since the 65 nm technology node was introduced into manufacturing. Of course, 200 mm wafer diameter wafers and tools for the same process are also being used for older technology nodes.

## 8.3 Mechanism of the CMP Process

Although CMP is now extensively used for fabricating integrated circuits, the polishing process itself has been used in optical lens manufacturing for several centuries. Sir Isaac Newton in 1695 observed that the scratch size decreased as the abrasive size is

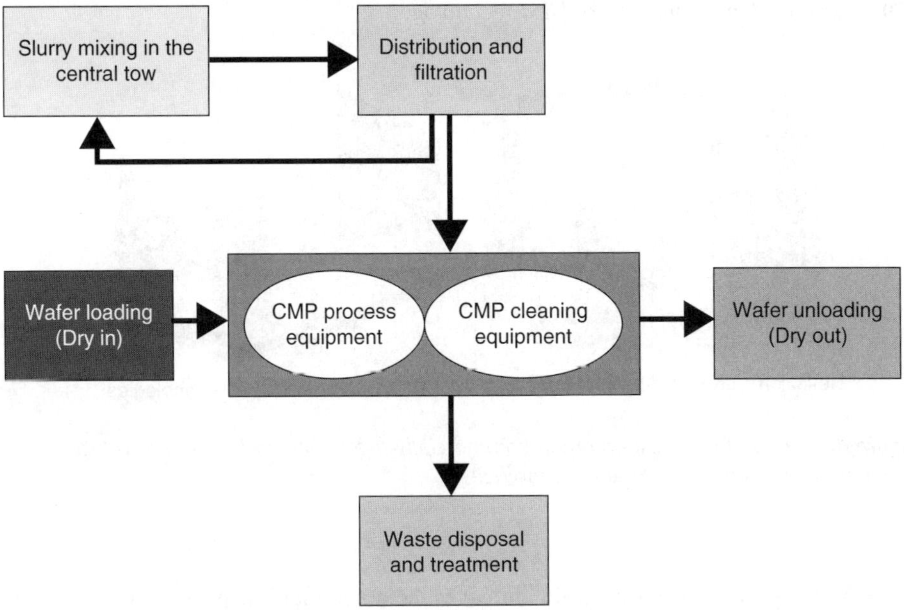


***Figure 8.1*** *Schematic flow diagram of a CMP process* [2]

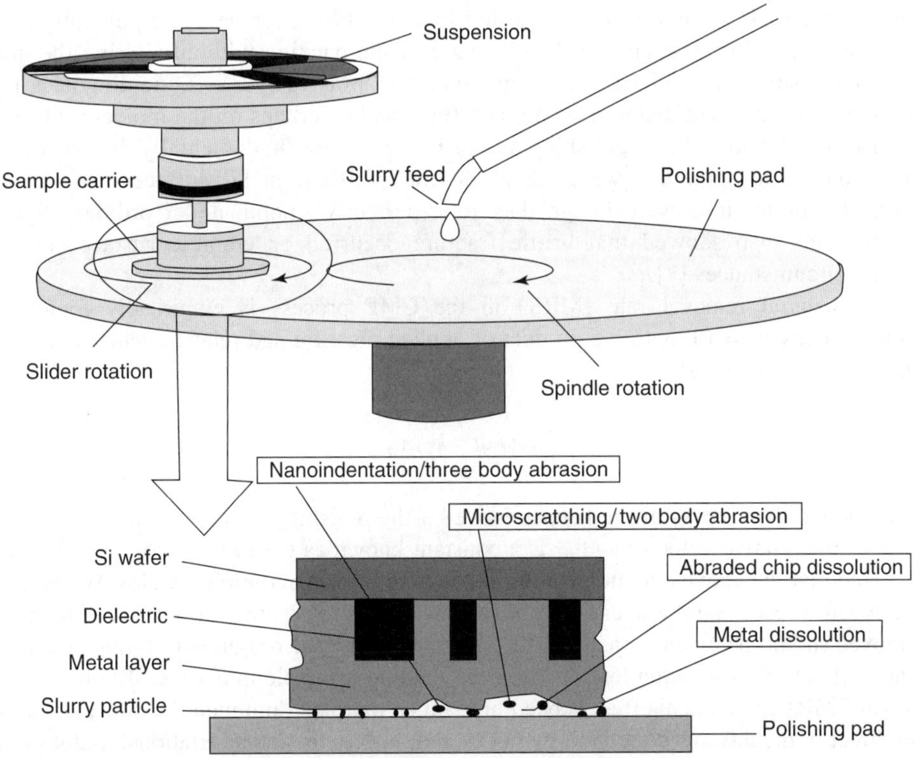


***Figure 8.2*** *Schematic diagram of a typical CMP polisher* [2]

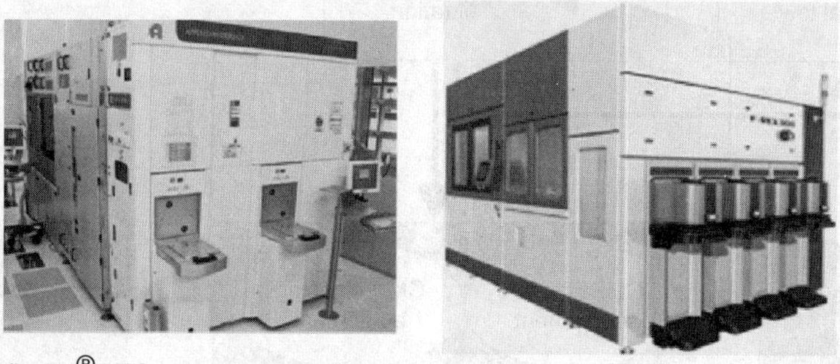


***Figure 8.3*** *Two of the most common commercially used CMP tools (see the website of Applied Materials Inc. and Ebara Technologies)*

decreased. By using very fine grain abrasives, he noted that it is possible to continually fret and wear away the glass and produce a polished surface on which the scratches become too small to be visible. Along the line of this hypothesis, Lord Rayleigh found in 1901 [5] that the polishing process produced highly reflective, structure-less facets in a discontinuous fashion. Further polishing does not improve the quality of the facets but extends their boundary. He suggested that the difference between polishing and grinding can occur by changing the character of the backing without altering the grit size. In polishing, a smaller force is applied on the abrasive by a soft, yielding backing and therefore removes material on a much finer scale, perhaps on the molecular level. Later research showed that polishing and grinding are not fundamentally different [6]. Very fine scale scratches were found on the polished metal surfaces by phase-contrast illumination, even though they look perfectly smooth under ordinary light. Experiments also showed that brittle fracture occurred on ceramic surfaces under certain circumstances [7].

The material removal rate (MRR) in the CMP process is empirically found to proportionally increase with the product of applied pressure and relative velocity, which may be expressed as [8]

$$\mathrm{d}h / \mathrm{d}t = k_p p v_R$$

where $h$ is the thickness of the layer removed, $t$ the polishing time, $p$ the nominal pressure, $v_R$ the relative velocity and $k_p$ is a constant known as the Preston constant. It may be noted that the above equation can be used on both local and global scales. When it is employed to estimate the average removal rate across the wafer surface, the thickness removed should be much larger than the variation in surface roughness. It may be noted that while the Preston equation has been the guiding principle in early CMP processes, for Cu CMP the chemical effect is becoming much more predominant than the mechanical effect in the advanced technology nodes and, hence, the linear relationship does not always hold true.

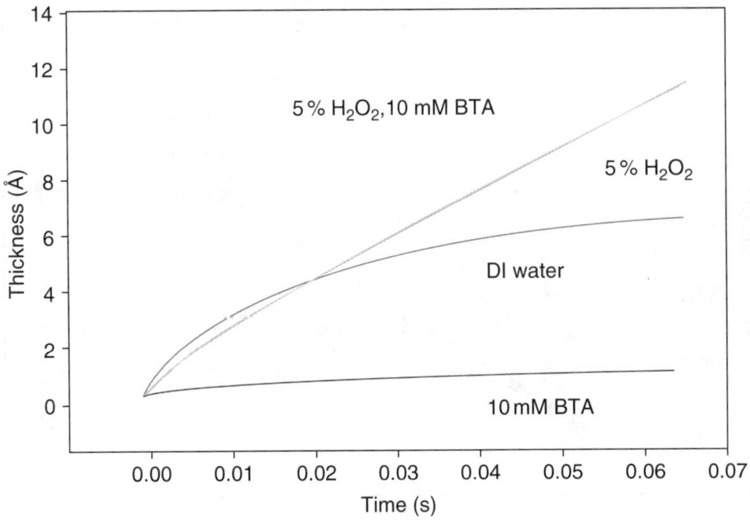


***Figure 8.4*** *Variation of the rate of a surface layer formation in Cu with different slurry chemistries* [2] *(see color Plate 6)*

## 8.4 CMP Consumables

### 8.4.1 Slurry

For the CMP process, a slurry is a specially formulated chemistry combining some chemicals as well as abrasive(s) in water. Typically, slurries are mixed with an oxidizer either at the time of manufacturing or at the point-of-use before loading on to the CMP tool. Today, the most common oxidizer is hydrogen peroxide and is mixed at the point-of-use. Among other ingredients in slurry, usually one finds organic complexing agents, corrosion inhibitors and surfactants of various types.

The slurry dependence of the removal rate of a particular material that is being polished is due to slurry chemistry, i.e. chemical action of the slurry chemicals of the material, the mechanical abrasion of the particles on the polished materials, interplay of the different complexing agents, oxidizers and corrosion inhibitors. Chemicals such as oxidizers and corrosion inhibitors vastly affect the reaction rate of the slurry, with a similar particle nature, size and distributions. Figure 8.4 shows the variation of the reaction rate of the different slurry components on Cu when the reaction kinetics was studied using electrochemical chronoamperometry [9, 10]. It can be seen from Figure 8.4 that the surface rate kinetics reaches about 60 A°/s when Cu is immersed in deionized water. The reaction rate increases to around 120 A°/s when 5 % $H_2O_2$ is added. However, with addition of 10 mM BTA(benzotriazole), the reaction rate came down considerably.

Depending on the end goal, different types of abrasives are used for different types of slurries, such as colloidal silica, fumed silica, alpha alumina, calcined ceria, etc. For the Cu CMP process, the most common abrasive being used today is colloidal silica. Different

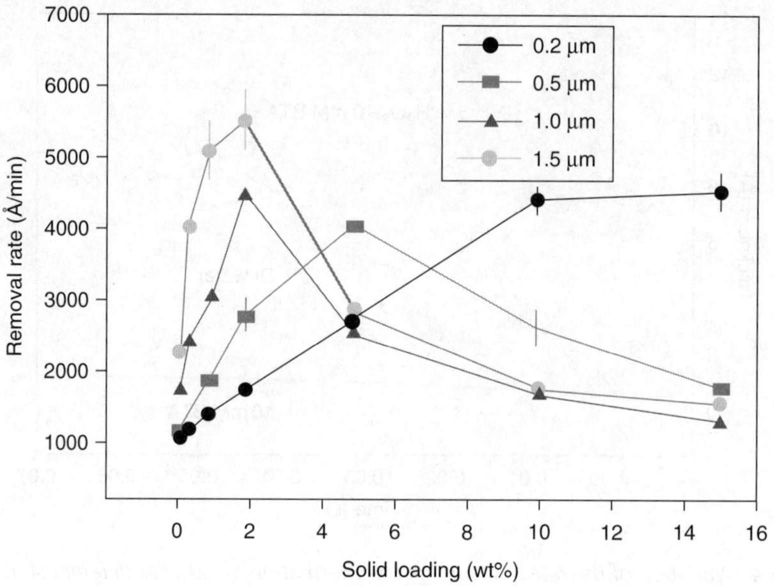


***Figure 8.5*** *Removal rate of silica with different particle sizes and concentrations* [11]

manufacturers of colloidal silica claim to have specific differences in physical properties of their end-product, such as particle size distribution, average particle diameter, purity, manufacturing process difference, etc. Typically, average particle size and concentration in slurry affect the removal rate, as seen in Figure 8.5 [11].

While a single specific type of slurry is used for an Al CMP or W CMP or oxide CMP process, for a Cu CMP process, typically two or even three different types of slurries are used to complete the Cu CMP process. The Cu CMP process requires three distinctive steps – removal of bulk Cu, planarization of the remaining Cu or so-called 'soft landing' to clear Cu and removal/planarization of the barrier layer beneath the Cu layer. As the technology node shifts below 30 nm, the thickness of deposited Cu is decreasing compared to what was the norm prior to that. As a result, the first two steps may be combined into one single step in the future.

Corrosion inhibitors are used in slurries to reduce or prevent corrosion of both copper and barrier layers as appropriate. In a pattern wafer, a galvanic contact between the two films becomes exposed at the end of the CMP process, which is a source of the onset of galvanic corrosion. It is essential to prevent this, which otherwise leads to corrosion defects.

After the barrier layer is planarized, the dielectric layer is exposed to the slurry. Therefore, it is also essential to formulate slurry to ensure that no damage to the dielectric layer occurs and the removal rate of this layer is in accordance with the goal set by device designers.

In general, an ideal CMP slurry should be able to achieve the required removal rates for all the films associated with the integration of a specific process, achieve excellent local and global planarization, should prevent galvanic, pitting or any other form of corrosion to any metal film in the total film stack, provide good surface finish, low defectivity and appropriate selectivity as desired for a specific device pattern wafer.

*Table 8.1* Evolution of a polish down force [12]

| Application | 130 nm | 45 nm |
|---|---|---|
| Oxide | 5–6 psi | |
| STI | 3–4 psi | 2–3 psi |
| Tungsten | 5 psi | 3 psi |
| Copper | 3–5 psi | 2 psi |
| Copper–low-*k* | | <2 psi |

Notable global suppliers of slurries are Cabot Microelectronics Corporation (USA), Dow Electronics Materials (USA), Fuji Film Electronic Materials (USA), DuPont Air Products Nano Materials LLC (USA), Fujimi Corporation (Japan), Hitachi Chemicals Company (Japan), JSR Corporation (Japan), Cheil (South Korea) and BASF (Germany). There are some other smaller local companies in different countries trying to enter the market without a major market share.

### 8.4.2 Pad

Conventional pad technology includes a top-pad layer with or without an under-pad layer to provide a polishing surface. Polishing pads provide local planarization due to inherent material hardness and global uniformity due to the flexing action of the polishing layer. While such pads have served the industry well, inherent coupling of the key output parameter – planarization and WIWNU (with-in-wafer nonuniformity) – limits the overall performance potential (Table 8.1) [12].

Conventional pad systems suffer from two primary limitations. Firstly, the continuous nature of the top polyurethane layer gives rise to an edge rebound effect as it passes under the wafer or, in some cases, the (wafer) retaining ring. While retaining rings help mitigate the edge rebound effect, significant edge-related nonuniformity remains. Secondly, global compliance or WIWNU considerations necessitate that the total stack has limited stiffness, thus impacting within-die planarization. For example, a soft pad stack has a lower flex modulus and compression modulus, which enables good contact at low pressures. A hard pad stack, on the other hand, has a much higher flex modulus and compression modulus; hence, it offers limited contact at lower pressures. This process leads to nonuniform contact at low pressures as well as variation in contact pressure leading to potential contact 'hot spots'.

For new-generation ultra-low-*k* dielectric film integration with a Cu interconnect, the down force requirement is extremely low due to the reduced mechanical strength of porous films and the conventional pad no longer provides the most optimum efficiency. A new generation of pads is being explored by various CMP pad suppliers. Some of these pads overcome to a great extent the global planarity and nonuniformity issues associated with the conventional pads.

The most notable CMP pad supplier in the market today is Dow Electronic Materials (USA) which traces its historical market share due to the early market penetration by

Rodel, Inc. (now part of Dow Electronic Materials). Among other notable pad suppliers are Cabot Microelectronics Corporation (USA), Thomas West Inc. (USA), Fujibo (Japan), JSR Corporation (Japan) and NexPlanar Corporation (USA).

### 8.4.3 Pad Conditioner

The CMP process is very dependent on the surface condition of the pad. During operation the surface of the CMP pad soon becomes loaded with debris from the polishing operation, which leads to glazing on the surface. This means that there are no asperities to hold the abrasive grits, which leads to inefficient polishing and possible microscratching on the surface wafer. To prevent this and extend the life of the CMP pad, it must be dressed regularly with a diamond conditioner. The diamond CMP conditioner has two functions:

(a) Initially to machine a pad to create a flat surface and subsequently to remove glazing from the pad during operation to expose asperities. Even a new pad will have an unacceptably rough surface – this will have to be dressed before it can be used for polishing.
(b) To restore the asperities within the pad's surface to enable the abrasive grits to perform their polishing action efficiently.

A traditional diamond pad conditioner or dresser is made by attaching diamond grits on to a metal disk. Diamond is difficult to bond with other materials due to its natural properties. The diamond industry has developed three distinct methods to hold diamond grits in place: by burying them in electroplated nickel, by retaining them in sintered metal powder or by brazing them with a specialty alloy. The first generation of diamond pad conditioners contained chaotically distributed diamond grits, typically retained in an electroplated nickel layer. The electroplated metal only holds diamonds in place mechanically, so the displacement of diamonds from the matrix is inevitable. The loss of diamond particles during the CMP process affects wafer performance. The second-generation pad conditioners had clusters of diamond grits that were held by braze. In these clusters, the braze layer became thickened. As a consequence, not only does the distribution of diamonds become random, the diamond height also varies considerably. The third-generation pad conditioners have a regular distribution of diamond grits with each diamond strongly bonded by braze. Currently it has become the world standard for most pad conditioners to be constructed, as shown in Figure 8.6 [13].

Instead of attaching individual diamond grits to a metal substrate, the newly designed diamond conditioner employs serrated blades made of sintered polycrystalline diamond, as shown in Figure. 8.7. The cross-sectional SEM images of the pores on the polishing pad after 60 min of dressing formed by a new type of conditioner (BDD) and an old type of conditioner (DG) are shown in Figure 8.8. It is observed that most pores still remain round, like that of the unused polishing pad, when dressed by the BDD, as shown in Figures 8.8 (a) to (c). On the other hand, with the DG, the shapes of the pores materialize in an irregular and rough pattern, as shown in Figure 8.8 (d). It is observed that less residual chips of pad material adhered to the pores dressed by BDD. The residual chips trapped in the pores decreased the area for slurry transport, and this should be taken into account during dressing in the CMP process. Among the several traditional pad conditioning disc manufactures, popular names are 3M (USA), Kinik (Taiwan) and Asahi (Japan).

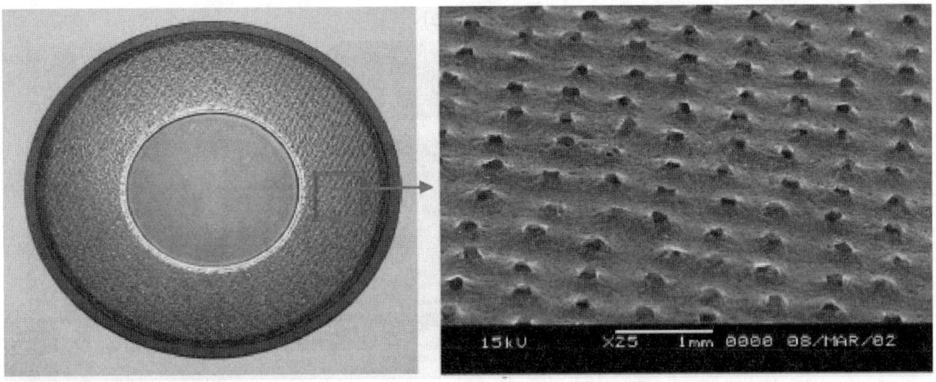


***Figure 8.6*** *Traditional pad conditioner as used (left) and SEM images of the disk surface (right)* [13]

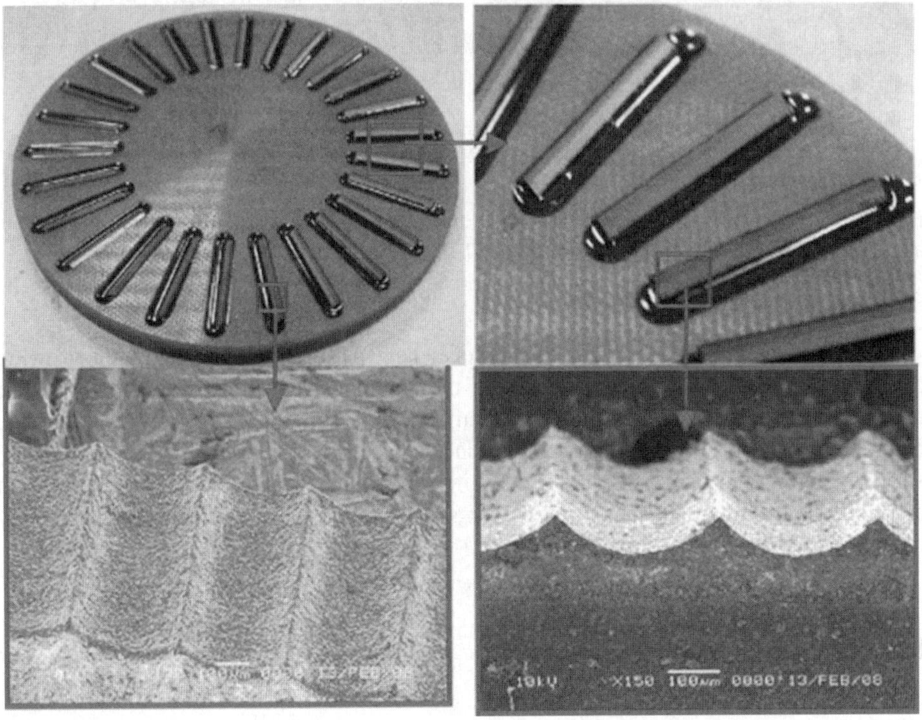


***Figure 8.7*** *Profiles of the blade diamond disk and blade with cutting teeth* [13] *(see color Plate 7)*

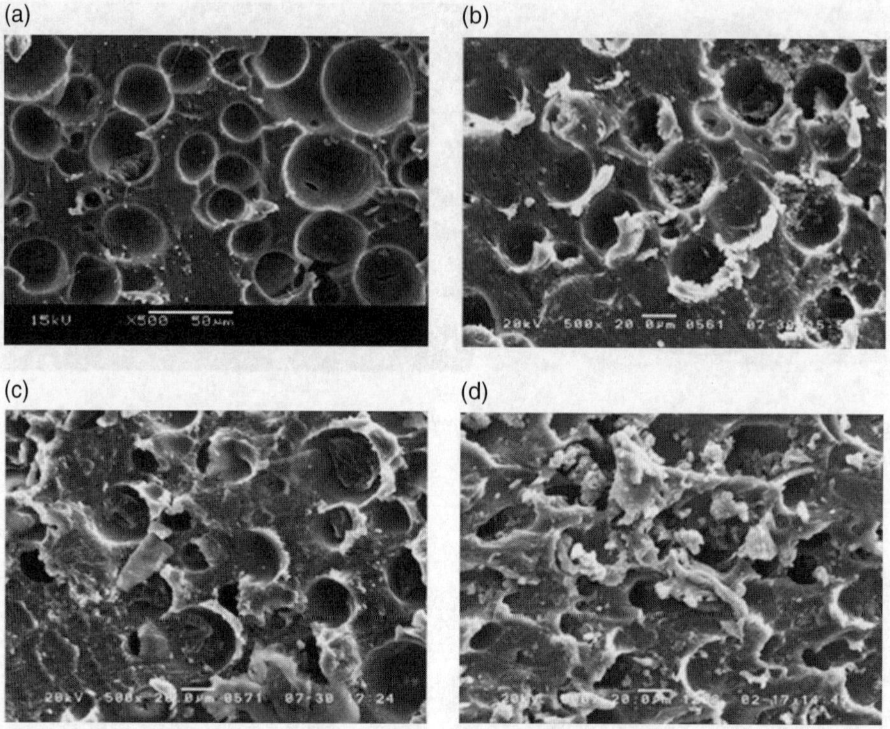


*Figure 8.8* SEM images of an IC1000 pad formed by (a) new pad, (b) BDD (penetration angle of 90°), (c) BDD (penetration angle of 120°) and (d) DG, after 60 min of dressing

Recently, Entegris has brought to market a new pad conditioner based on growing diamond over the entire pad surface using the chemical vapor deposition method. The all-diamond surface provides superior diamond retention by chemically bonding the diamond grit to the substrate. This all-diamond surface is also chemically inert to all CMP slurries, eliminating diamond pullout due to corrosion or wear of the matrix material. In addition, the high-hardness diamond prevents mechanical erosion of the bond matrix by hard CMP slurry particles [14].

## 8.5 CMP Interactions

The targets for dishing and erosion on Cu lines continue to decrease with each successive generation of devices. According to the ITRS roadmap, the maximum allowable Cu line thinning in an array is already less than 15 nm and is projected to be only 6 nm by the 32 nm node. Achieving this target consistently in manufacturing will require slurries formulated to the specific materials system, pads with excellent surface texture control and processes with razor-sharp endpoint and/or high selectivity to a stop layer. Dishing and

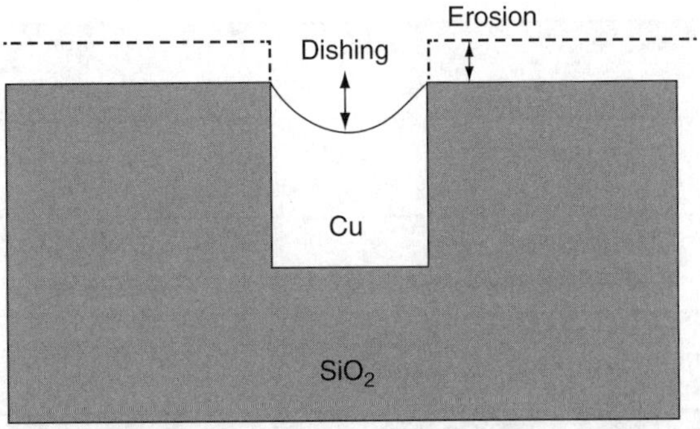


***Figure 8.9*** *Illustration of dishing and erosion* [15]

erosion– which together account for nearly 50 % of yield losses in copper processing – are the most important effects to be minimized in the copper CMP process. Dishing and erosion are forms of local planarization where areas of the wafer polish faster than others. In dishing, the copper is dished out of the lines where as in erosion whole sections of the die are polished faster than others. Figure 8.9 shows a schematic where both dishing and erosion have occurred [15].

The CMP process subjects the surface of the wafer to a rather hostile environment, with the wafer encountering mechanical forces of both pressure and shear while immersed in aqueous chemical slurry, which usually contains tiny suspended particles. As might be expected, the films deposited on the wafer must be capable of withstanding this environment without suffering undesirable irreversible harm.

As mentioned previously, next-generation interconnect structures incorporate lower dielectric constant ILD materials. In the early stages of ULK exploration, it was recognized that several film types, such as carbon-networked dielectrics, would enable reduction in the $k$ value, but in most cases the film's mechanical properties were significantly degraded [16]. This trend in mechanical strength based on film class is demonstrated in Figure 8.10 where materials such as parylene demonstrate $k$ values of 2.0 and below but also have 10 times less mechanical strength than FSG films. The industry has worked to balance the mechanical properties and the need for capacitance reductions due to the considerable concerns for packaging reliability failures [17]. Currently organosilicate (OSG) glass films have been adopted both for dense OSG and now extending down to $k$ values ranging from 2.5 (40 nm for 28 nm nodes) to 2.2 and 2.0. The $k$ values are scaled by addition of porosity (air has a dielectric of 1.0), enabling the industry to maintain the same class of materials but continue to reduce the capacitance impact as structures are scaled. This trend is shown in Figure 8.11 [18]. Although, generally, the addition of porosity will reduce the film's mechanical strength, there have been several improvements in ULK films to optimize the backbone strength, whereby making the ULK film strength comparable to bulk OSG films [19, 20].

Mechanical failure modes include delamination, cracking, and inelastic deformation. For low-$k$ and ULK dielectrics, the materials are characterized in terms of modulus, crack propagation velocity, and yield strength [21–23]. As the industry transitions to

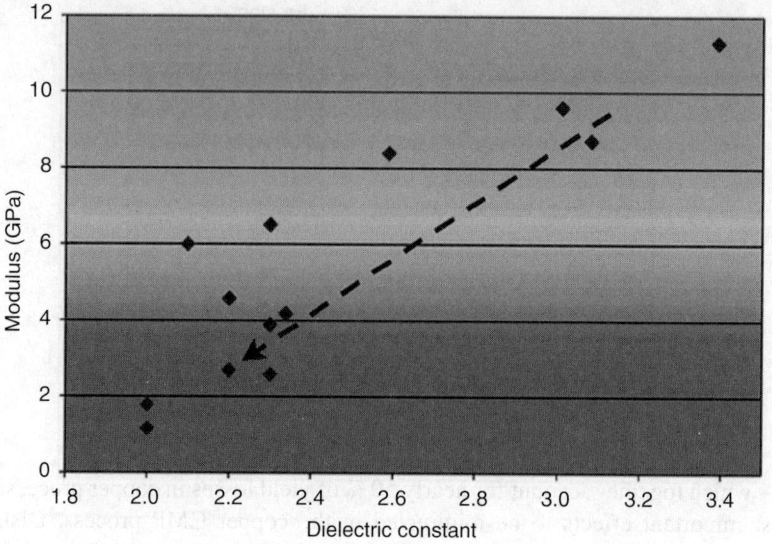


***Figure 8.10*** *Young's modulus as a function of the dielectric constant for a wide range of low-k and ULK films (see color Plate 8)*

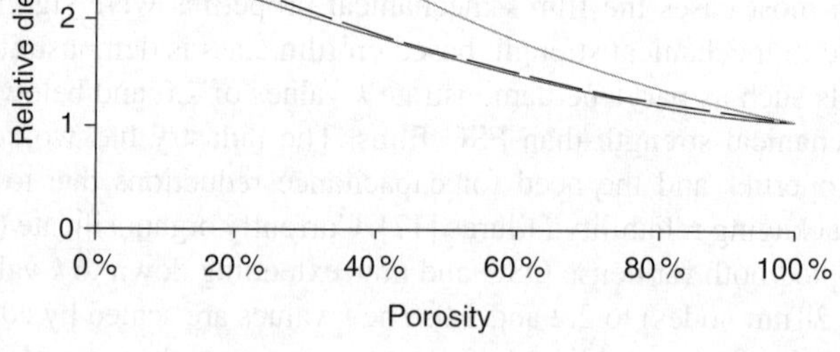


***Figure 8.11*** *Calculated dielectric constant versus porosity*

these materials, the CMP process must accommodate the mechanical fragility of ULK by moving toward lower pressure and lower shear process recipes. These two factors are related through the dynamic coefficient of friction, which is itself related to a large number of factors: platen rotation speed, pad composition, surface texture, pad conditioning

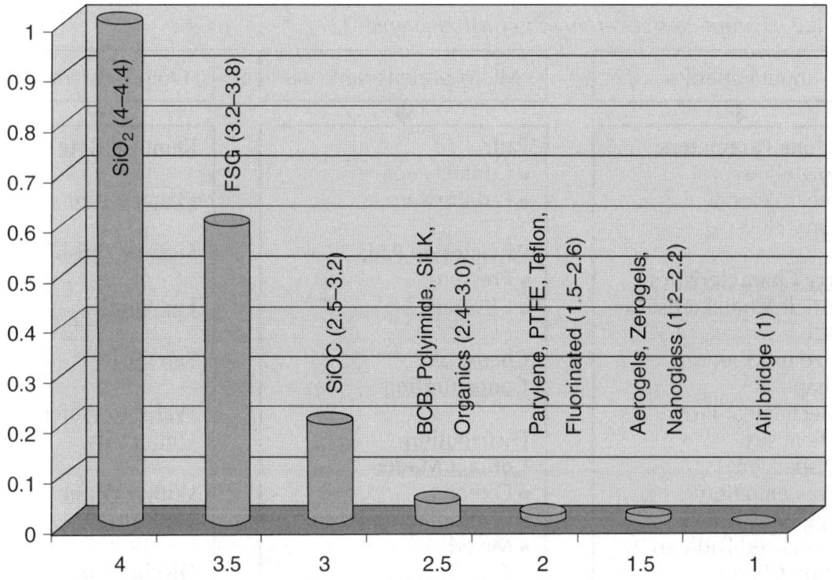


***Figure 8.12*** *Mechanical integrity scale of different dielectric materials normalized against* $SiO_2$

parameters, particle size and morphology, solids concentration, slurry chemistry/composition (especially surfactant choice and concentration), among others. While the relationship among these variables is often not linear, it is generally consistent that lower pressure leads to lower shear; thus most copper and barrier polishing recipes are being developed at 2 psi and below. This maximum pressure is expected to continue, dropping as ULK drives toward lower *k* values, which demand larger fractional pore volumes and the accompanying degradations in mechanical strength [24], as shown in Figure 8.12. With the focus on ULK mechanical integrity, it is sometimes easy to forget that polishing slurries also contain a mixture of chemical ingredients, e.g. oxidizers, complexing agents, passivation additives, selectivity modifiers and surfactants. Slurry developers historically benefited from having dielectric films, such as TEOS or FSG, which were reasonably inert and easy to clean. This is no longer the case with ULK materials, which typically have a strong tendency to absorb water (and other chemicals) from the slurry or at least trap them in any exposed pores [25]. This often results in a higher *k* value than the as-deposited film [26] and/or a lower time-dependent dielectric breakdown (TDDB), either of which is problematic for the interconnect device. Development of a CMP process for any of these advanced integrations must be approached as a multilevel, multidisciplinary project. The properties of the ULK film, the barrier material, the copper deposition technique, annealing cycles and many other factors will influence the CMP process. This difficulty is compounded by the complexity of developing sophisticated integrations with porous ULKs. Low-*k* materials have to be mechanically strong enough in order to resist delamination and mechanical breakdown during the CMP process. The challenge lies not only in identifying and characterizing the candidate materials but also in devising the best methods to integrate those materials.

*Table 8.2* *Parameters governing the CMP dynamics* [2]

**Input Variables** ↓

**Machine Parameters:**
- Down Pressure
- Linear Velocity
- Slurry Flow

**Slurry Characteristics:**
- Particle Characteristics
  - ✓ Size
  - ✓ Size distribution
  - ✓ Shape
  - ✓ Mechanical Properties
  - ✓ Chemistry
  - ✓ Dispersion
  - ✓ Concentration
  - ✓ Agglomeration
  - ✓ Oversized Particles
- Slurry Chemistry
  - ✓ Oxidizers
  - ✓ pH and pH drifts
  - ✓ pH stabilizers
  - ✓ Complexing Agents
  - ✓ Dispersants
  - ✓ Concentration

**Pad Characteristics:**
- Mechanical Properties
- Topography
- Conditioning
- Pad Uniformity

**Substrate Characteristics:**
- Feature Size
- Feature Density

**Micro Scale Parameters** ↓

**Pad:**
- Contact Area
- Pad Pressure

**Particles on Pad:**
- Pressure
- Coverage

**Chemical Concentration**

**Distribution Contact Mode:**
- Direct
- Hydroplaning
- Mixed

**Nanoscale Parameters** ↓

**Chemical Etching**

**Mechanical Removal**

**Chemo-Mechanical:**
- Surface Layer Formation
- Thickness
- Uniformity
- Rate of Formation
- Layer Removal

**Output Parameters** ↓

**Removal Rate**

**Planarization**

**Surface Finish**

**End Point**

**Selectivity**

**Wafer to Wafer Uniformity**

**Within Wafer Uniformity**

**Dishing and Erosion Resistance**

As material dielectric constants get closer to 1, the materials become less robust and harder to integrate. In Table 8.2, Zantye *et al.* [3] have summarized various parameters governing the CMP dynamics.

One of the challenges that has occurred in conjunction with low-*k* adoption is the need to remove the hardmask layer during the CMP process in order to optimize the interconnect capacitance. In the early days, there were concerns that slurries or post-CMP cleaning chemistries might damage the low-*k* film properties and, hence, there was an effort to retain the hardmask layer (typically $SiO_2$, $Si_3N_4$, etc.), as may be seen in Figure 8.13 [3]. While this process ensured no damage to the dielectric layer due to the inherent barrier properties for typical hardmask layers, it also increased the effective *k*-value of the total

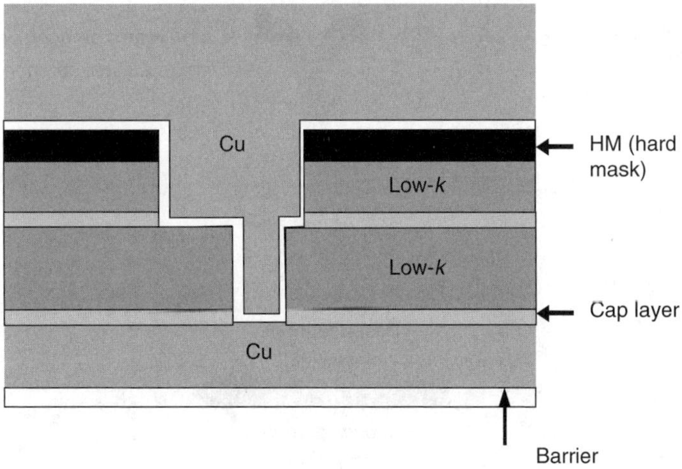


*Figure 8.13* *Dual-damascene structure*

stack. Over time, the slurry and post-CMP cleaning chemistry suppliers developed formulations that were compatible with the low-*k* materials [27]. Therefore, uncapped low-*k* dielectric films are now routinely used in 45 nm and below technology node applications. With the decrease in minimum feature sizes, barrier and hardmask layer dimensions become critical in order to maintain the dielectric performance. Recent studies show that a copper interconnect system may indeed be feasible without a barrier and hardmask layer [28].

At present, the 32 nm technology is ramping into production, implemented mostly in 2010, and market intelligence indicates that porous OSG materials is the dominant choice for the interlayer dielectric (ILD), whereby a *k* value of 2.5 is being adopted by many logic manufacturers. As the industry moves toward 20 nm and 16/14 nm nodes, it is expected that the logic devices will require a porous ultra-low-*k* (ULK) dielectric with a *k* value equal to 2.2 in order to achieve the desired performance targets. For the range of *k* values from 2.5 and below, an example of this film type is porous DEMS® or PDEMS®, which are registered trademarks of Air Products and Chemicals, Inc., respectively, as illustrated by the deposition process diagram in Figure 8.14.

## 8.6 Post-CMP Cleaning

Post-CMP cleaning is an integral part of the CMP process module to clean up the debris and other chemical residues left on wafers during CMP. There are two types of cleaning technology in use: megasonic bath and double-sided scrubber with PVA brushes. Both have advantages and disadvantages and may be employed individually or in combination as needed to achieve effective cleaning of the wafers. The megasonic approach is sometimes useful in removing particles from narrow recesses of a polished wafer if the surface

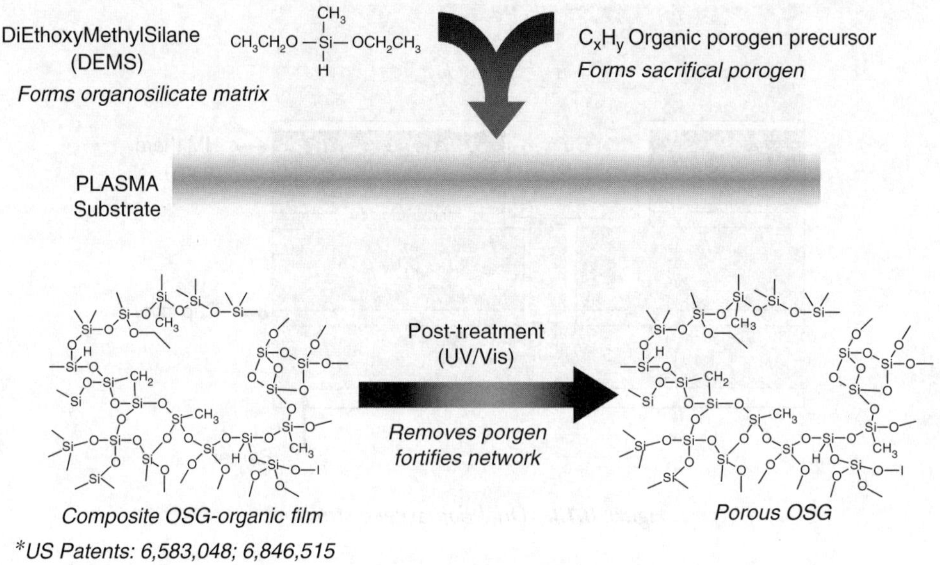


**Figure 8.14** *PDEMS™ 2.2 from Air Products and Chemicals, Inc.*

features are not completely coplanar. Brush scrubbers, on the other hand, are particularly useful at dislodging particles that may be physically attracted to the copper surface. It is relatively common to use the combination of both megasonic cleaning and brush scrubbing to give a more robust cleaning process. There are cases of fabs using only one technique and the more common single method is brush scrubbing. This practice is possible due to a new generation of cleaning chemistries that enables complete removal of particles from the copper surface through chemical and electrochemical means without the need of megasonic energy to physically lift the silica particles out of copper and dielectric film surfaces.

The combinatorial application of using chemistry and electrochemistry to remove particles from a copper surface has become necessary to address the constantly shrinking size of what are considered 'killer' defects. In general, this includes any particle greater than the line width of the smallest feature, which is now smaller than the particle sizes used in most commercial slurries. Silica particles, for example, will stick on the metal surface by electrostatic bonding if the slurry is acidic in pH. In this case, silica acquires a net positive charge while the surface charge of copper becomes negative. The electrostatic bonding is fairly strong and, traditionally, megasonic energy was needed to lift the particles off the copper surface. With the new generation of acidic post-CMP cleans, the particles are lifted off the copper surface by a very slight etching of underlying copper, which is not enough to damage the copper surface topography. In contrast, if alkaline slurry is used for the final step in the CMP process, the silica becomes negatively charged and floats near the copper surface. The images in Figure 8.15 show the ability of a new-generation cleaning formulation to clean the Cu surface without using megasonic cleaning. The Cu wafers were polished with a commercially available barrier CMP slurry and cleaned with (a) deionized water, (b) a previous-generation clean solution that would require megasonic cleaning to

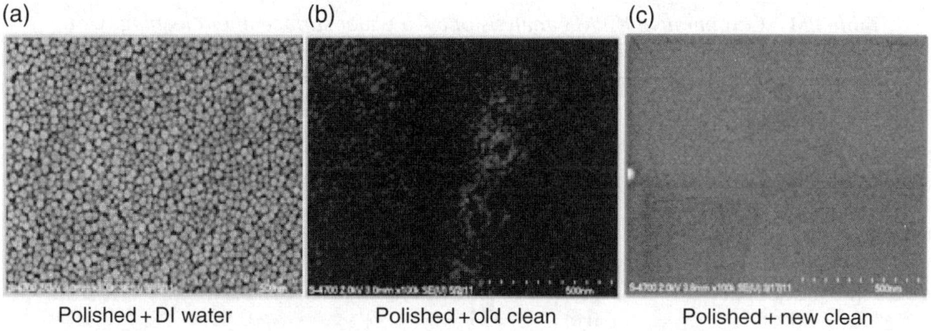


***Figure 8.15*** *Comparison of two different cleans to demonstrate the ability of a new formulation to clean a Cu surface without the additional step of megasonic cleaning*

***Table 8.3*** *Zeta potential values on low-k and silica surfaces in different pH cleaning formulations from Air Products*

| Point-of-use dilution | BDIIX | Silica |
|---|---|---|
| Acidic clean | −24.11 | −30.74 |
| Alkaline clean | −81.48 | −53.69 |

aid in complete removal of particles and (c) a new-generation cleaner that does not require any megasonic cleaning step.

On the dielectric surface, the particle removal depends on the zeta potential depending on the pH of the cleaning formulation, as may be seen in Table 8.3 [28]. A stronger zeta potential value is better at particle removal efficiency. Surfactants can also be useful to form a layer around the particles, which assists in not only removing them from the wafer surface but also preventing their re-deposition.

Post-CMP cleaning processes must address more than just particle removal. Other killer defects like unwanted metal particles or trace metal ions need to be removed as well, especially at 65 nm nodes and below. The new generations of post-CMP cleaning formulations are able to capture such metallic ions more efficiently, leaving a cleaner wafer surface than ever before (Table 8.4).

As mentioned earlier, most of the low-*k* and ULK films are hydrophobic in nature. In order to clean particles efficiently as well as trace metals and organics off the surface of the wafer, it is essential to make the film hydrophilic, at least temporarily, to enable the smooth flow of aqueous cleaning solution followed by rinse water. This is also necessary to avoid leaving water marks on the dielectric film surface. As the images in Figure 8.16 show, one of the more efficient ways to do this is to use an appropriate wetting agent in the cleaning formulation to reduce the contact angle from a typical 50 degrees to below 20 degrees. After a normal cleaning process, no water marks are found. If the contact angle remains high, water marks are very common, and can also leave behind unwanted high levels of trace metals on the wafer surface.

When the dielectric surface is made temporarily hydrophilic with a wetting agent, there is often a concern about damaging the dielectric properties, especially since uncapped dielectric film is now in use. In our early studies, it was observed that certain molecular

*Table 8.4* *Comparative ICPMS analysis of a Cu wafer surface after cleaning with an old cleaner and a new cleaner following the CMP process*

| | | | Method Detection Limits | Old | New |
|---|---|---|---|---|---|
| 1. | Aluminum | (Al) | 0.3 | 49 | 9.5 |
| 2. | Calcium | (Ca) | 0.3 | 4.3 | 1.4 |
| 3. | Chromium | (Cr) | 0.06 | <0.06 | 0.061 |
| 4. | Copper | (Cu) | 0.02 | 0.73 | 1.8 |
| 5. | Iron | (Fe) | 0.1 | 0.93 | 0.18 |
| 6. | Magnesium | (Mg) | 0.1 | 0.37 | 0.23 |
| 7. | Nickel | (Ni) | 0.05 | 0.076 | <0.05 |
| 8. | Potassium | (K) | 0.2 | 160 | 200 |
| 9. | Sodium | (Na) | 0.2 | 3.0 | 1.4 |
| 10. | Titanium | (Ti) | 0.05 | 0.18 | 0.19 |
| 11. | Zinc | (Zn) | 0.05 | 10 | 0.098 |
| 12. | Total# Defects (COMPASS) | | | 229 | 22 |
| | | | SURFACE CONCENTRATION ($\times 10^{10}$ atoms/cm$^2$) | | |

***Figure 8.16*** *Water mark on a nonporous low-k dielectric film (left) with a previous-generation clean formulation and no water mark (right) with a new-generation clean formulation*

structures within CMP and post-CMP formulations can impact the low-*k* electrical properties (Figure 8.17 (a)). Figure 8.17 (b) shows that for nonporous dielectric films, a well-formulated slurry chemistry does not change the film properties before and after cleaning. Even for porous low-*k* film such as PDEMS®, where a cap layer needs to be used, there are concerns about what the impact will be if the cap layer is breached. Experiments performed on uncapped PDEMS® ULK suggest that there is a slight shift in *k* value and modulus when processed with a standard post-CMP cleaning process. However, a subsequent IPA rinse drives out the liquid that penetrates the surface pores of the film. After this step was performed, the PDEMS® returned to its normal properties with no apparent residual effects.

Besides the low-*k* integration issues, both the CMP slurries and post-CMP clean formulations must be electrochemically tuned to prevent any galvanic corrosion of any of the metallic film layers in the integration scheme. Too often it is found that either the

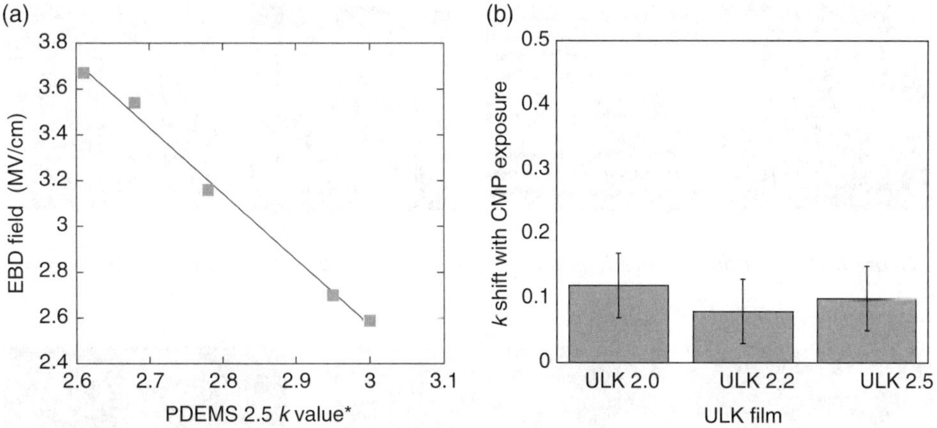


***Figure 8.17*** *(a) ULK 2.5 k value and electrical breakdown field results for a systematic variation in the CMP barrier slurry composition. The figure shows a direct correlation between an increased ULK k value and an increase in the ULK leakage and a decrease in the electrical breakdown field, respectively. *PDEMS k value measured after a p-CMP barrier slurry polish, p-CMP cleans, followed by IPA dry. (b) The k shifts for ULK 2.0, 2.2 and 2.5 films after a CMP polish with CMP68900 followed by p-CMP. Error for k shift measurements ± 0.05*

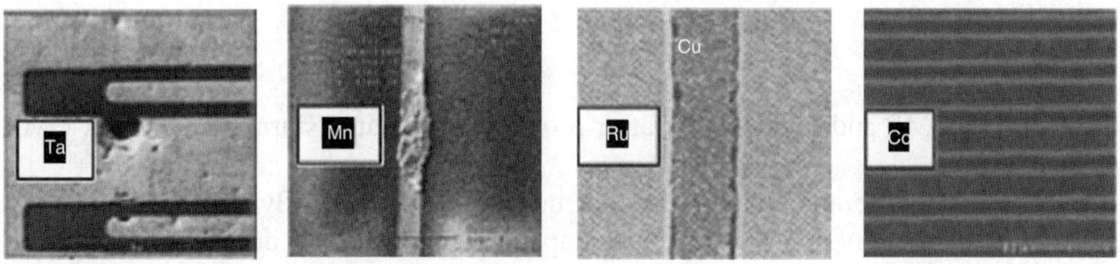


***Figure 8.18*** *Galvanic corrosion of pattern lines with Ta, Mn and Ru as barrier films underneath Cu using improperly formulated slurries or clean and no corrosion of Cu lines with Co as the barrier film under Cu, using properly formulated slurries and post-CMP clean. Reproduced from http://www.businesswire.com/news/appliedmaterials/20040928005768/en with permission of Air Products and Chemical Inc.*

Cu or barrier film suffers from galvanic corrosion, typically found in the form of line edge roughening in high-density lines, after the post-CMP cleaning step, when the defect metrology is usually performed. This is because galvanic corrosion of either the Cu or the underlying barrier metal happens where the two different metals are in contact with each other and exposed to the electrochemically active slurry or post-CMP clean solution. Further investigation by partitioning the defect review after CMP only and after a post-CMP clean can establish the step after which galvanic corrosion might have occurred. Nonetheless, such corrosion is not acceptable as it leads to the electrical test failure. Figure 8.18 shows various galvanic corrosion failures of either Cu or barrier

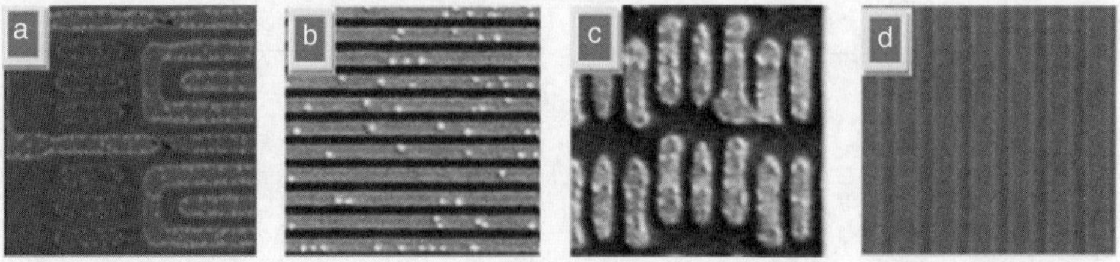


*Figure 8.19* Dendrites on various wafers (a–c) and a dendrite-free wafer (d) using CP98

*Figure 8.20* Other defects (scratch/organic residue/carbon residue)

films during CMP and good wafers after properly formulated slurries and a post-CMP clean was used.

Finally, as the semiconductor process development is clearly focused on 28 nm and 22 nm technology nodes at various companies, for Cu–low-*k* integration one of the critical challenges to implement a post-CMP clean formulation is the occurrence of dendrites or white spots on high-density Cu lines, which is usually thought of as Cu–oxide nodules. Figure 8.19 shows typical images using 28 nm and 22 nm test wafers using a new clean.

### 8.6.1 Other Defects

Typical other common defects on wafers are scratches (which normally come from the agglomerated slurry abrasive particles), organic residues or carbon residues (from slurry components or dielectric film erosion), as can be seen from the images in Figure 8.20.

### 8.6.2 Surface Finish

After the CMP process, a very smooth surface of Cu is expected. Figure 8.21 shows AFM images on Cu surface treated with alkaline and acidic cleans with a root mean square (rms) surface roughness ~0.5 nm.

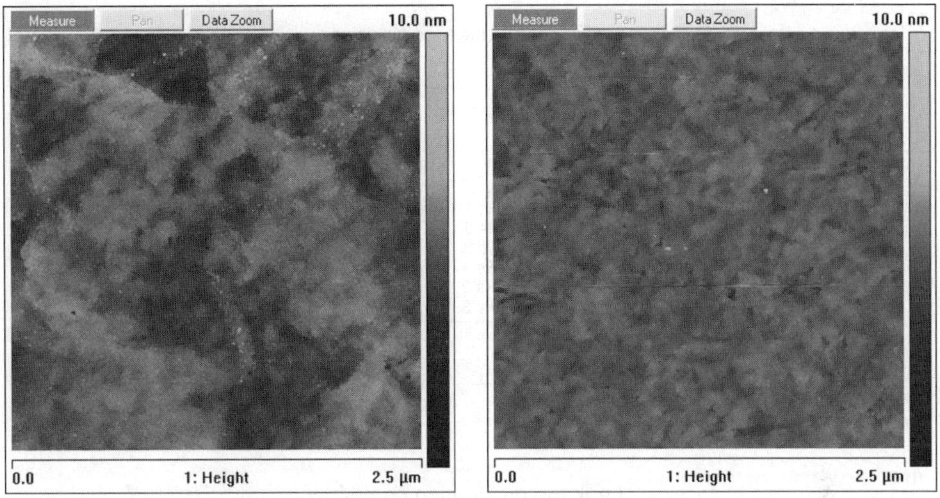


*Figure 8.21* AFM images of Cu wafers after a post-CMP clean: alkaline (left) and acidic (right) (see color Plate 9)

### 8.6.3 E-Test

If the CMP slurry and post-CMP clean step do not introduce any defect at either the metal level or the dielectric level, the electrical test results should show improvement in Kelvin resistance and leakage current performance. Figure 8.22 shows a comparison of these two parameters between a fab baseline process and CMP consumables from Air Products (slurries from DA Nano Materials LLC, DEMS low-$k = 2.5$ film and post-CMP clean from Air Products). As can be seen, the new formulations resulted in improvement over the fab baseline process [26].

## 8.7 Future Direction

The future direction for the interconnect module at 28 nm through 10 nm appears to be a nanoporous ULK dielectric in a capped integration scheme for the logic devices and dense low-*k* for the memory devices for 30 nm technology nodes. There are various liner films such as Ru, Mn, Co, etc., that are being investigated to be used below Cu instead of the traditional Ta/TaN. In most common cases, a Ta/TaN film may be beneath the new liner films as part of the integration. Combinations of these complex film stacks have made the slurry and post-CMP clean formulation development more challenging. However, it is evident from results/images given in this article that it is possible to develop appropriate formulations to accomplish CMP and post-CMP cleaning steps without damaging any film in the stack with a thorough understanding of the complexities, application, film deposition metallurgy, etc.

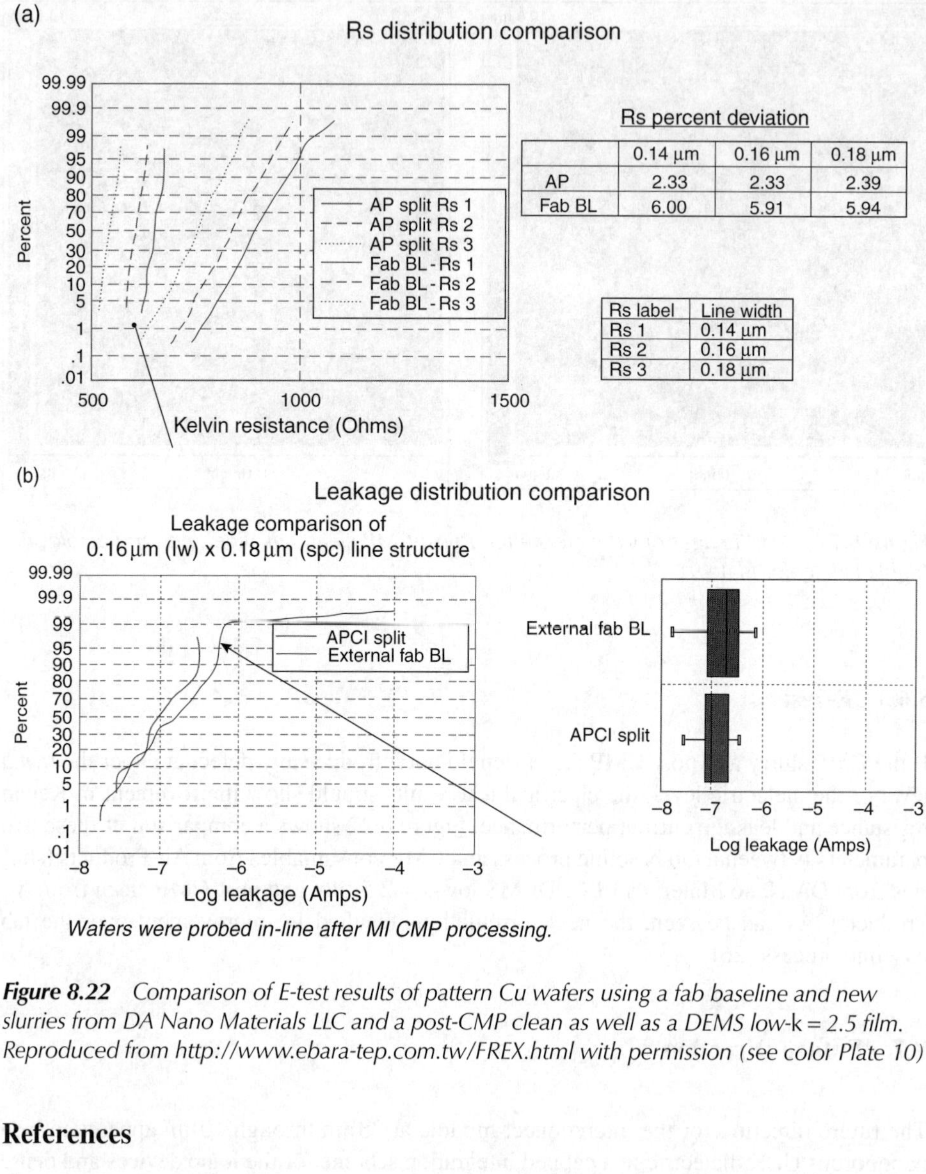


***Figure 8.22*** *Comparison of E-test results of pattern Cu wafers using a fab baseline and new slurries from DA Nano Materials LLC and a post-CMP clean as well as a DEMS low-k = 2.5 film. Reproduced from http://www.ebara-tep.com.tw/FREX.html with permission (see color Plate 10)*

## References

[1] http://www.itrs.net/links/2005itrs/Interconnect2005.pdf.
[2] http://www.itrs.net/Links/2009ITRS/Home2009.htm.
[3] Zantye, P.B., Kumar, A. and Sikder, A.K. (2004) *Mater. Sci. Engng*, **R45**, 89–220, 108.
[4] http://www.businesswire.com/multimedia/home/20040928005768/en/1120068/Applied-Materials-Extends-CopperLow-Leadership-Innovative-CMP.
[5] http://www.ebaratech.com/images/products/FREX300_home.jpg.
[6] Rayleigh, Lord (1901) *Nature*, **64**, 385–388.
[7] Aghan, R.L. and Samuels, L.E. (1970) *Wear*, **16**, 293–301.
[8] Komanduri, R., Lucca, D.A. and Tani, Y. (1997) *Ann. CIRP*, **46**, 545–596.
[9] Preston, F.W. (1927) *J. Soc. Glass Technol.*, **11**, 214–256.

[10] Kaufman, K.B., Thompson, D.B., Broadie, R.E., Jaso, M.A., Guthrie, W.L., Pearson, D.J. and Small, M.B. (1991) *J. Electrochem. Soc.*, **138** (11), 3460.
[11] Luo, J. and Dornfeld, D.A. (2001) *IEEE Trans. Semicond. Manuf.*, **14** (2), 112.
[12] Tsai, T.-H. and Yen, S.-C. (2003) *Appl. Surf. Sci.*, **210** (3–4), 190.
[13] Carpio, R., Pham, J., Tolic, F., Hymes, S. and Bajaj R. (2006) IEEE Conference on VLSI Multilevel Interconnection (VMIC), Fremont, California.
[14] Tsai, M.-Y. (2010) *J. Mater. Proc. Technol.*, **210**, 1095–1102.
[15] http://www.entegris.com/ProductCategory_divEmerging_catCMP_Pad_Cond.aspx.
[16] Merchant, S., Kang, S.H., Sanganeria, M., Schravendijk, B.V. and Mountsier, T. (2001) *J. Minerals*, **52**, 43.
[17] Wang, G., Ho, P.S. and Groothuis, S. (2005) *MicroElectronics Reliability*, **45** (7/8), 1079–1093.
[18] Semiconductor technology, Feature article for Sematech Symposium, San Diego, California, October 2004; http://www.seniconductor-technology.com/features/feature225/.
[19] O'Neill, M.L, Haas, M.K., Peterson, B.K., Vrtis, R.V., Weigel, S.J., Wu, D., Bitner, M.D. and Karwacki, E.G. (2006) *MRS Proc.*, **914**, F01-12.
[20] O'Neill, M.L., Vincent, J.L., Vrtis, R.N., Lukas, A.S., Bitner, M.D., Karwacki, E.J., Peterson, B.K. and McGuigan, S.A. (2005) *MRS Proc.*, **913**.
[21] Guyer, E. P., Patz, M., and Dauskardt, R.H. (2006) Fracture of nanoporous methyl silsesquioxane thin-film glasses. *J. Mater. Res.*, **21** (4), 882.
[22] Liu, X.H., Shaw, T.M., Lane, M.W., Liniger, E.G., Herbst, B.W. and Questad, D.L. (2007) IEEE Conference on Chip-Package-Interaction Modeling of Ultra Low-*k* Copper Back End of Line (IITC), pp. 13–15.
[23] Gage, D.M., Haas, M.K., Matz, L.M., O'Neill, M.L., McDaniel, P.L., Vrtis, R.N., Gao, X., Vratsanos, M.S. and Dauskardt, R.H. (2008) MRS Proceedings.
[24] Bavel, M. V., Iacoi, F., Baklanov, M. and Maex, K. (2007) *Future Fab Int.*, **17**.
[25] Rao, M.R., Tamboli, D., Wieder, T., O'Neill, M.L., Weigel, S.J., Haas, M.K. (2007) MRS Spring Proceedings.
[26] Matz, L.M., Meyers, A.M., Sawayda, R., Rao, M.B., Haas, M.K., Langan, J.G. and O'Neill, M.L. (2008) Advanced Metallization Conference, San Diego, California.
[27] Borst, C.L., Gill, W.N. and Gutmann, R.J. (2002) *Chemical–Mechanical Polishing of Low Dielectric Constant Polymers and Organosilicate Glasses: Fundamental Mechanisms and Application to IC Interconnect Technology*, Kluwer Academic Publishers, Boston, Massachusetts.
[28] Tamboli, D. (2011) Internal result, Air Products, April 2011.

# 9

# Scaling and Microstructure Effects on Electromigration Reliability for Cu Interconnects

*Chao-Kun Hu*[1], *René Hübner*[2], *Lijuan Zhang*[3], *Meike Hauschildt*[4] *and Paul S. Ho*[5]

[1]*IBM Research Division, T.J. Watson Research Center, Yorktown Heights, New York, USA*
[2]*Fraunhofer Institute for Non-Destructive Testing IZFP, Dresden, Germany*
[3]*IBM System and Technology Group, Hopewell Junction, NY 12533*
[4]*GLOBALFOUNDRIES Dresden Module One LLC & Co. KG,Dresden, Germany*
[5]*Microelectronics Research Center, The University of Texas at Austin, Texas, USA*

## 9.1 Introduction

Following the implementation of low-*k* and ultra-low-*k* dielectric materials, future CMOS technology nodes will require further reduction of the Cu on-chip interconnect dimensions. The technology has led to the development of sophisticated back-end-of-line (BEOL) manufacturing processes and integration of advanced thin-film materials. As a result, the microstructure of the Cu lines will change significantly, affecting the size, shape, orientation distributions of the Cu grains, the alloy solute distribution and segregation, and the quality of the interfaces in the interconnect structure [1]. This will increase the contribution of the grain boundaries to the mass transport and alter the characteristics and mechanisms of

*Advanced Interconnects for ULSI Technology*, First Edition. Edited by Mikhail R. Baklanov, Paul S. Ho and Ehrenfried Zschech.

the reliability-limiting degradation processes, such as electromigration (EM) and stress-induced voiding (SIV).

Copper was introduced into the on-chip interconnect to replace aluminum in 1997 by IBM due to its better electrical resistivity and EM performance. With continuous scaling, the electrical resistivity was found to increase as the Cu grain size approaches the mean free path length of the electrons (40 nm at $T = 20\,°C$). This was attributed to the increase in electron scattering by grain boundaries, surface, and interfaces. This so-called size effect for line widths smaller than 100 nm was first reported in 1998 [2] and confirmed by several recent studies [3, 4]. In general, with decreasing line geometries, physical properties, such as electrical resistivity, and reliability performance, such as EM, are increasingly influenced by the Cu microstructure, which has to be carefully controlled and precisely monitored in narrow lines. A comprehensive analysis of the Cu interconnect would require the determination of the general microstructure functions specifying the grain size, texture, stress, and various types of defects [5].

For Cu lines beyond the 65 nm node with line widths reduced below 90 nm, significant changes in Cu microstructure were observed, where small random grains emerged to mix with the bamboo-like grain structure, leading to further lifetime degradation due to the additional grain boundary diffusion. In this case, multiple small grains were usually found to agglomerate in the lower half of the Cu line [6–8]. Two possible mechanisms have been proposed: (1) grain growth inside the line and (2) invasion by grains originating from the overburden in the electroplating process. For lines wider than 90 nm, the extension of the overburden grains into damascene trenches will give rise to a bamboo-like grain structure. For narrower lines, only partial extension of the overburden grains will occur, where the grain structure in the line will depend on the line height and the thermal budget. Consequently, the Cu microstructure is controlled by the grain growth mechanisms inside the line. In this case, impurities trapped in Cu grain boundaries may reduce grain boundary diffusivity in narrow lines. Furthermore, sidewall pinning forces arising from preferential Cu grain orientations may neutralize grain boundary mobility and hence retard grain growth in the lines [9].

EM test results showed that lifetime in on-chip interconnect structures generally decreased for every new interconnect generation, even when tested at comparable current densities [10]. Since Cu interface diffusion controls the mass flow, the EM flux is constrained to the interface areas and its amount relative to other diffusion processes is proportional to the ratio of the interface area to the line area. At the same time, scaling of the via size will reduce the critical void length to cause line failure. These two factors will reduce the EM lifetime and impact the current density design rule for Cu interconnects. This has stimulated great interest from the industry to develop methods to improve Cu EM reliability, including the use of Cu alloys, the incorporation of a metal capping layer at the Cu/dielectric interface, and by improving the Cu microstructure.

Standard EM accelerated testing is performed using Cu damascene line structures containing multiple wiring levels under accelerated conditions at elevated temperatures. The results obtained are used to project the EM lifetime at field (use) conditions. The standard test is limited by the small sample size, typically 20 to 100 samples, which are not adequate to extrapolate the lifetime at the use condition. Therefore, a statistical test methodology for a massive-scale statistical study of EM lifetimes was developed based on the well-known Wheatstone bridge technique [11–13]. For this method, the test structures are arranged into units of parallel and serial interconnects. The advantage of using such statistical test structures is to observe the early time to fail of the lines in a short time and to predict the lifetime due to early failures with a high precision.

This chapter reviews the current understanding of the scaling and microstructure effects on EM reliability for Cu interconnects. It is organized into five parts. First we review in Section 9.2 the basic physics of the EM phenomenon, focusing on EM mass flow, lifetime scaling rule, test structures, and the relationship between Cu EM lifetime and current density. This follows with a discussion in Section 9.3 on the techniques developed recently for Cu microstructure analysis and the correlation of Cu microstructure to EM-induced interconnect damage. The third part in Section 9.4 presents the techniques developed for improving EM lifetime using alloys, upper-level dummy vias, Cu surface treatments, and surface metal coating. The fourth part in Section 9.5 describes the results from a recent study on the effect of grain size on EM lifetime and statistics. Finally, in the last part in Section 9.6, we discuss massive-scale statistical studies using the Wheatstone Bridge method and conclude with a summary in Section 9.7.

## 9.2 Electromigration Fundamentals

### 9.2.1 EM Mass Flow

The mass flow produced by an EM driving force $F_e$ is given by $J_e = n v_d$, where $n$ and $v_d$ are the atomic density and drift velocity, respectively. The mass flow is expressed as $v_d = (D_{eff}/k_B T)F_e$, where $F_e = Z^* e\rho j$; $e$ is the absolute value of the electronic charge, $Z^*$ is the apparent effective charge number, $\rho$ is the metallic resistivity, $D_{eff}$ is the effective diffusivity of atoms diffusing through a metal line, $T$ is the absolute temperature, and $k_B$ is the Boltzmann constant. If mass flow along different paths were independent, then the effective diffusivity in a given line at one cross-section can be written as

$$D_{eff} = \Sigma n_i D_i \tag{9.1}$$

where the subscript $i$ refers to the $i$th diffusion path, and $n_i$ and $D_i$ are the fraction of atoms and diffusivity in the $i$th diffusion path, respectively. Diffusivity is the dominant factor for the mass transport and atoms diffusing along the fast diffusion paths will control the atomic movement. The possible fast diffusion paths in a Cu damascene line are grain boundaries and the Cu/dielectric interface. The slow diffusivities of the Cu/Ta liner interface and dislocation pipe diffusion are not considered [1]. For a Cu bamboo-like or polycrystalline damascene line, the Cu EM drift velocity can be written as

$$v_d = [Z_I^* (D_I / k_B T)\delta_I (1/h)]\, e\rho j \tag{9.2a}$$

or

$$v_d = [Z_{GB}^* (D_{GB} / k_B T)\, f(\delta_{GB} / d)]\, e\rho j \tag{9.2b}$$

respectively, where $h$ and $d$, and the subscripts $I$ and $GB$ refer to the metal height and grain size, and Cu/dielectric cap interface and grain boundary (GB), respectively, and $\delta$ and $D$ denote the effective width and diffusivity, respectively, and $f$ is a geometric factor. For test structures consisting of bamboo–polycrystalline sections, the drift velocity would be

$$v_d = Z^*_{eff}\,(D_{eff}\,/\,k_B T)\,e\rho j \tag{9.3}$$

In this case $v_d$ becomes rather complicated since mass flow along interfaces and grain boundaries are significantly different. These two mass flows are strongly interdependent and mass is moved in parallel or series depending on the microstructure. Here, Equation (9.1) for $D_{eff}$ is no longer appropriate. The typical EM activation energy, $Q_{EM}$, for interface and GB diffusion were about 1.0 and 0.8 eV, respectively [1]. For example, the mass flow in a polycrystalline section between two bamboo sections will change under EM. The two bamboo grain sections act as mass flow blocking boundaries, which are similar to the phenomena of the 'Blech short length effect' [14] in the polycrystalline section. The fast mass flow in the polycrystalline section will be reduced in time by the compressive stress build-up at the front bamboo-like grain section, although the mass flow in the bamboo-like section is increased by the resultant stress gradient from the back and front polycrystalline sections. The net drift velocity is dependent on the fraction and length of the polycrystalline sections and net mass flow at polycrystalline–bamboo boundaries. For small polycrystalline section lengths (<< Blech critical short length) in the Cu lines, the effective mass flow will be controlled by the slow interface mass flow in the bamboo sections. For the case of a long polycrystalline section (>> Blech length) in the Cu line, fast GB diffusion will dominate. For the intermediate case, it becomes an interesting subject to be studied and is not easy to predict since the bamboo grains are not completely blocking boundaries as in the case of the Blech short length effect because of mass leakage at the interface.

### 9.2.2 EM Lifetime and Scaling Rule

In the standard damascene structures, the top surface is covered by a dielectric diffusion barrier layer, e.g. $SiN_x$ or $SiC_xN_yH_z$, and the bottom and sides are covered with a Ta-based liner. The EM flux is constrained to the top interface in an area of $\delta_I w$, where $\delta_I$ is the effective thickness of the interface region and $w$ is the line width. The relative amount of flux, at constant line current density, flowing through the interface region is proportional to the ratio of the interface area to the line area, $\delta_I w/(wh)$, or $\delta_I/h$, where $h$ is the line thickness. The lifetime $\tau$ for near-bamboo-like structure with a Cu mass flow blocking boundary at both ends of the line can then be simplified to

$$\tau = \Delta L_F / v_d = \Delta L_F\, h\, k_B\, T / (\delta_I D_I F_I) \tag{9.4}$$

where $\Delta L_F$ is the critical void length to cause the line to fail, $D_I$ is the Cu/dielectric interface diffusivity, and $F_I$ is the EM driving force at the interface. In the present case, $\Delta L_F$ is closely equal to the via diameter, $w$, which is about the same as the line/via overlap

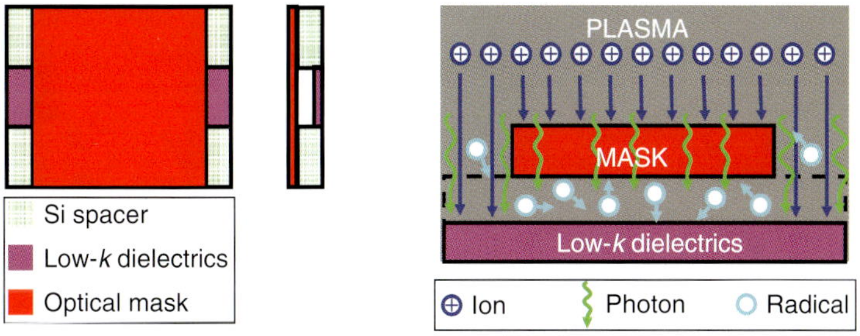


***Plate 1*** *Schematic drawing of gap structure: (left) top view and side view, (right) detailed side view*

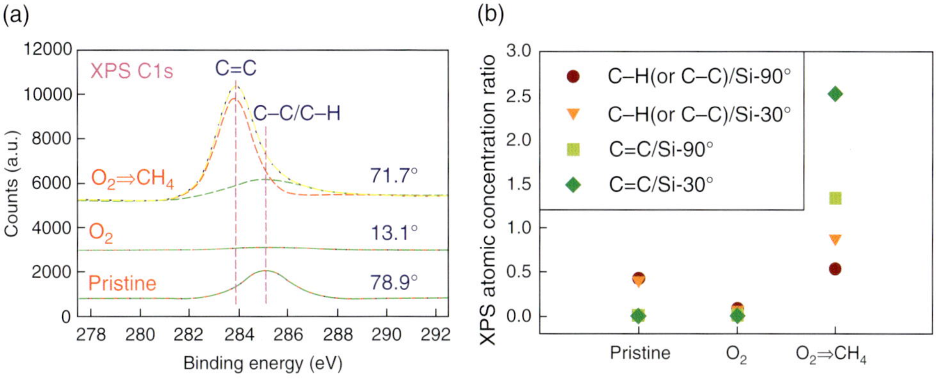


***Plate 2*** *(Top) XPS C1s spectra and (bottom) AR-XPS peak area ratio for pristine CDO low-*k*, $O_2$ plasma damaged low-*k*, and post-ash $CH_4$ plasma treated low-*k

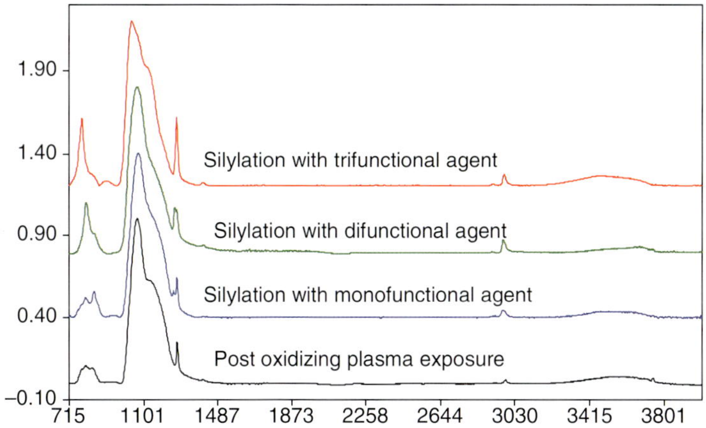


***Plate 3*** *FTIR spectra of post-plasma damage and silylated porous MSSG films with mono-, di- and trifunctional agents*

*Advanced Interconnects for ULSI Technology*, First Edition. Edited by Mikhail R. Baklanov, Paul S. Ho and Ehrenfried Zschech.

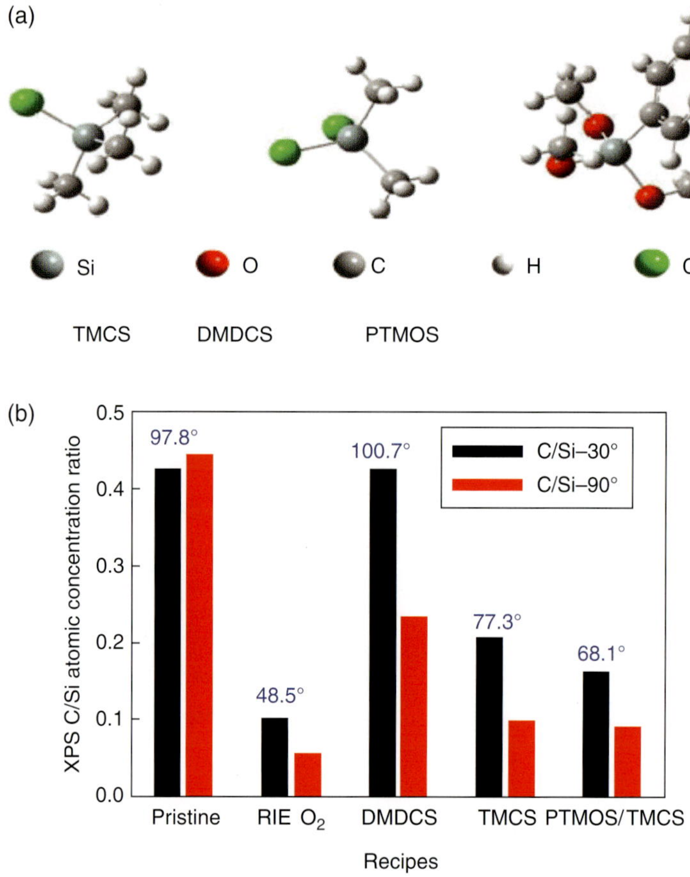


***Plate 4*** *(a) molecular structures of Trimethylchlorosilane (TMCS), Dimethyldichlorosilane (DMDCS), and Phenyltrimethoxysilane (PTMOS). (b) Comparison of C/Si and water contact angle for vapor TMCS, DMDCS, and PTMOS/TMCS silylation processes*

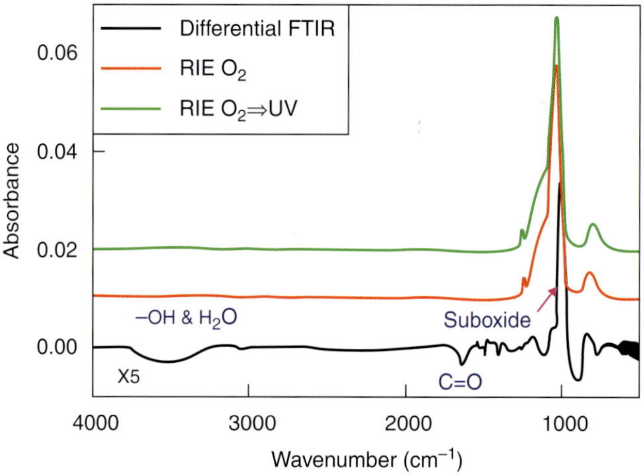


***Plate 5*** *FTIR spectra for $O_2$ plasma damaged and UV repaired OSG*

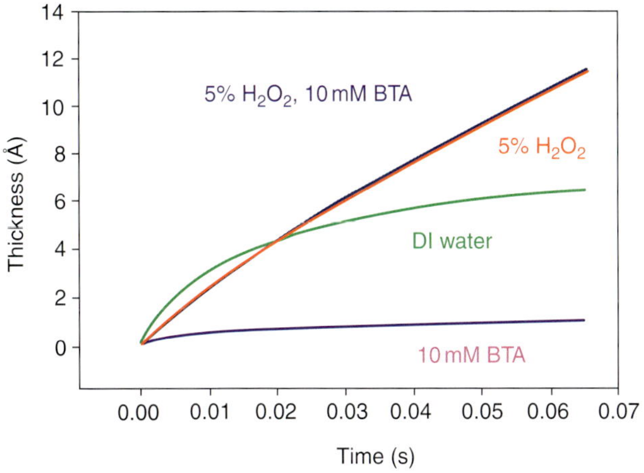


***Plate 6*** *Variation of rate of surface layer formation in Cu with different slurry chemistry*

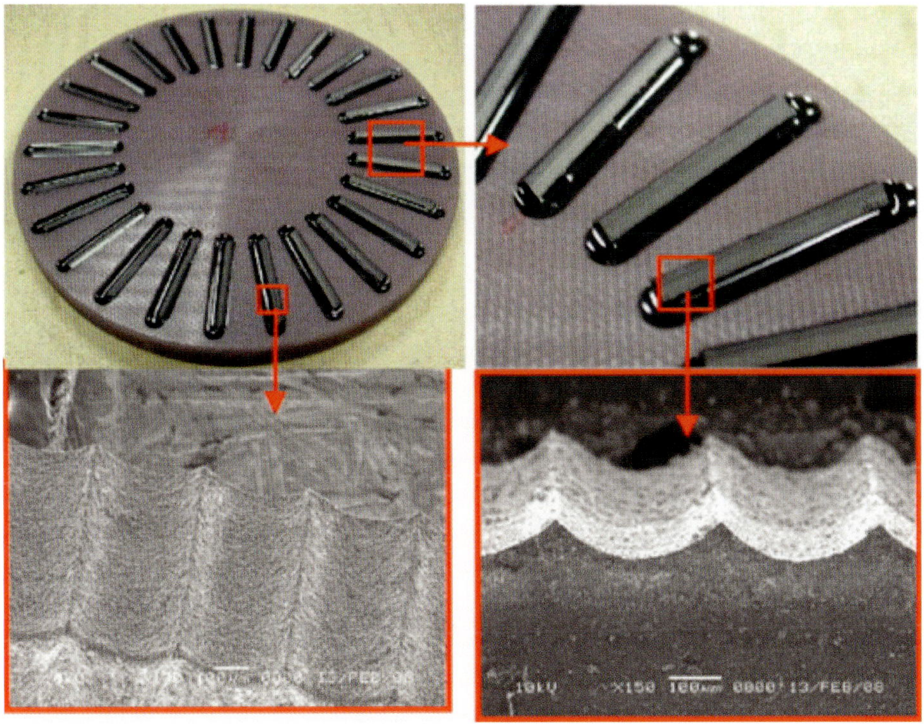

***Plate 7*** *Profiles of the blade diamond disk and blade with cutting teeth*

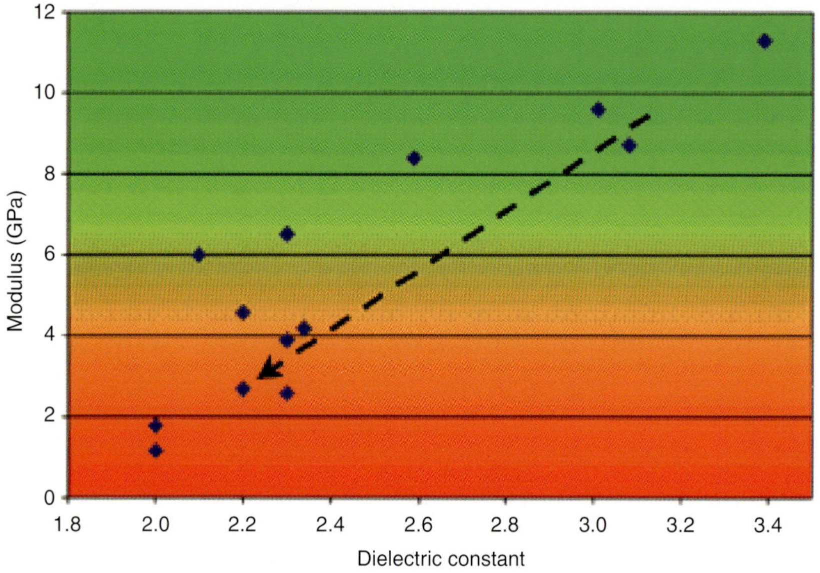


***Plate 8*** *Young's modulus as a function of dielectric constant for a wide range of low-*k *and ULK films*

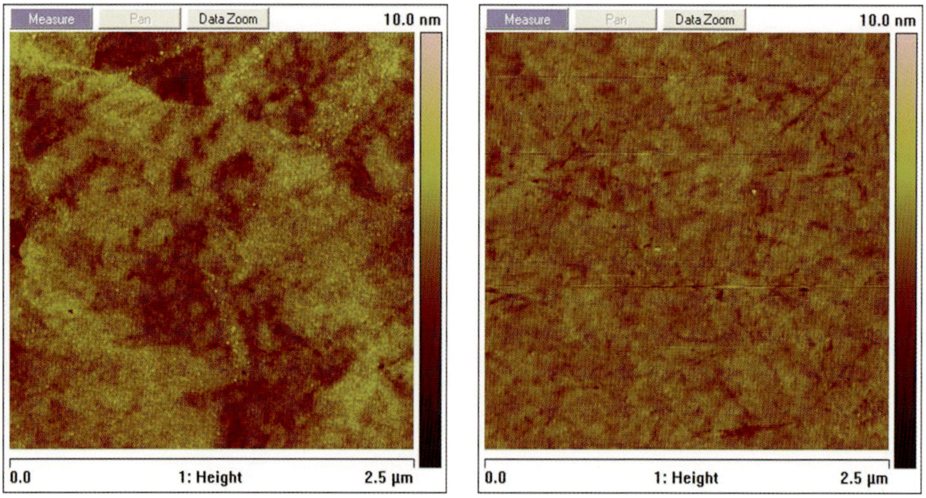


***Plate 9*** *AFM images of Cu wafers after post CMP Clean – alkaline (left) and acidic (right)*

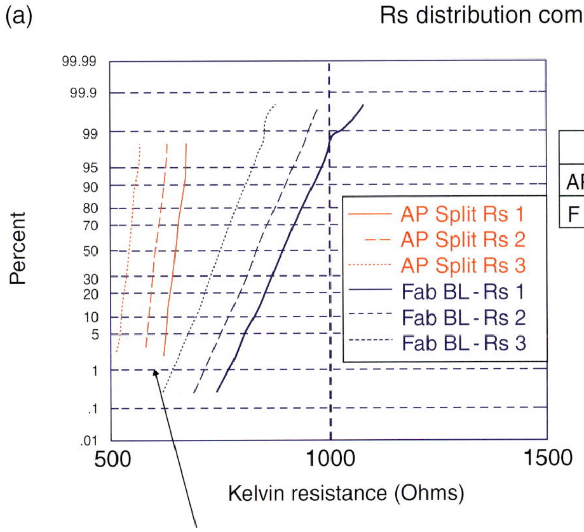


Rs percent deviation

| | 0.14 μm | 0.16 μm | 0.18 μm |
|---|---|---|---|
| AP | 2.33 | 2.33 | 2.39 |
| F ab BL | 6.00 | 5.91 | 5.94 |

| RS label | Line width |
|---|---|
| RS 1 | 0.14 μm |
| RS 2 | 0.16 μm |
| RS 3 | 0.18 μm |

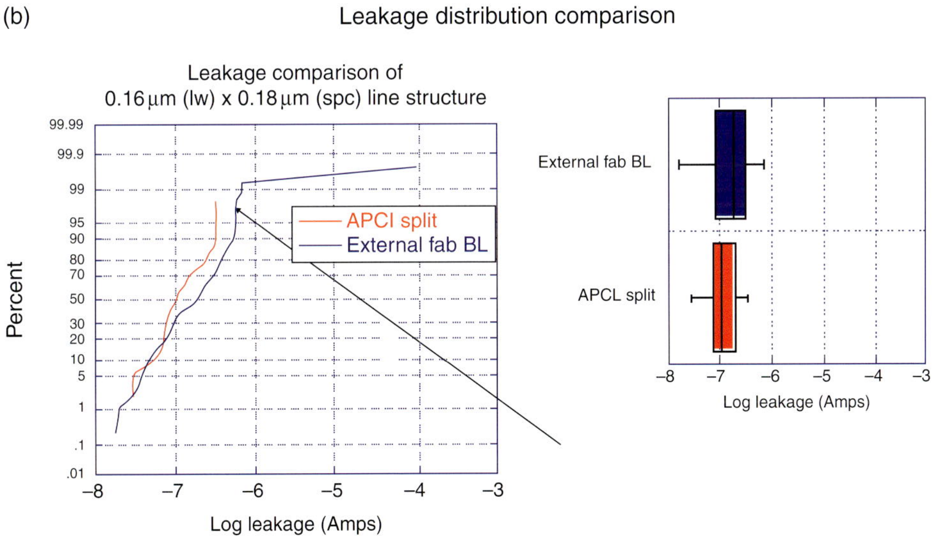


*Wafers were probed in-line after MI CMP processing.*

***Plate 10*** *Comparison of E-test results of pattern Cu wafers using a fab baseline and new slurries from DA Nano Materials LLC and post CMP clean as well as DEMS low-k=2.5 film from Air Products and Chemicals, Inc*

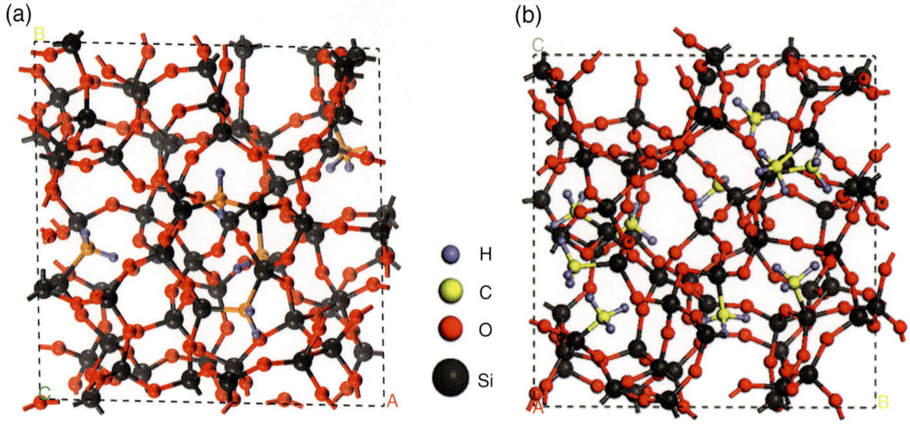


***Plate 11*** *Unit cell of (a) type-I OSG with five methylene cross-links, and (b) type-II OSG with ten methyl terminal groups after relaxation*

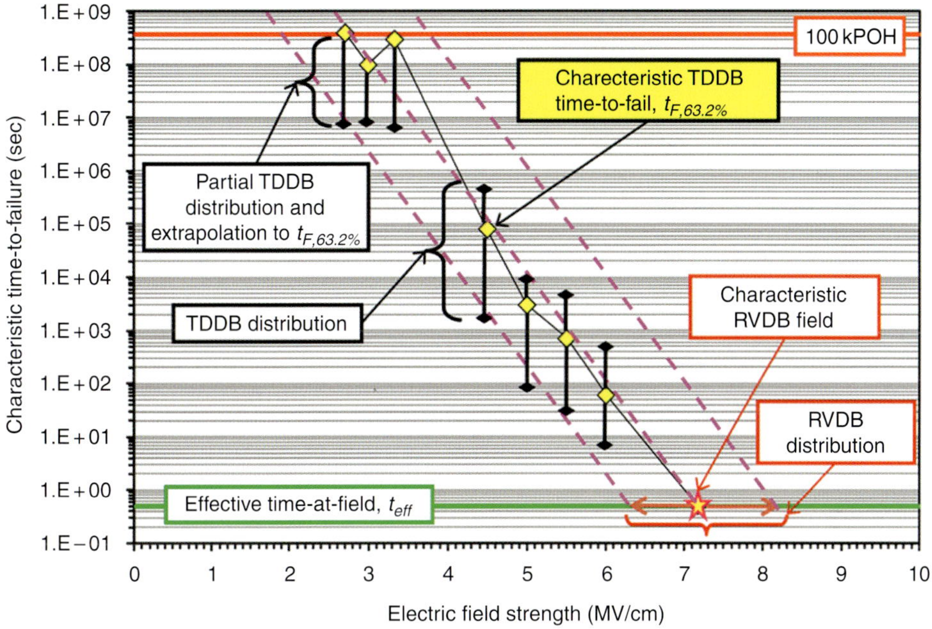


***Plate 12*** *This graphic explains the conjugate relationship between RVDB and TDDB using the Berman model. The effective-time-at-field,* $t_{eff}$*, can be viewed as a fixed time in which a population of capacitors shows variable breakdown voltage. On a TDDB lifetime projection plot (here using an E-model), the RVDB distribution can be plotted on the horizontal axis at the value* $t_{eff}$*. Note that a slower ramp rate will correspond to a higher* $t_{eff}$*. This RVDB distribution will then provide expected time-bounds within which capacitors should fail under TDDB stress at a fixed field. Note that in TDDB, the higher lifetimes do not reach the upper bounds expected from an RVDB projection. A rapid RVDB will cause a capacitor to fail in essentially a 'frozen state' so that it will be sensitive to poorly fabricated dielectrics; however, TDDB would be expected to allow defects to form in places within the capacitor that are advantageous for earlier breakdown. Therefore, it might be expected that TDDB will show a tighter breakdown distribution for higher lifetime devices than would be expected from a rapid RVDB projection. Much slower RVDB may provide better agreement for the entire distribution because some element of TDDB-like damage generation will take place. Of course, the benefit of rapid testing turnaround provided by the RVDB will be lost. This discussion demonstrates that the RVDB distribution is a clear measure of the 'initial quality' of the capacitor population that can fail under TDDB stress, so that it is a necessary condition that RVDB distributions be clean and true. The graphic is adapted from J. Kim* et al.*, International Relative Physics Symposium (2007) (© 2004 IEEE; permission by IEEE)*

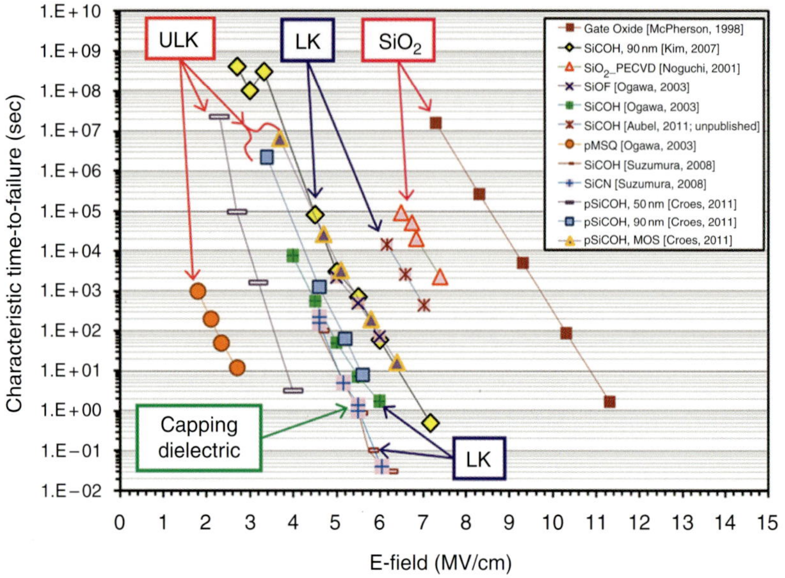


***Plate 13*** *Dielectric Breakdown comparison is shown for different dielectrics. Note that there is inevitable overlap because of differences in test structure area and layout, integration approach, process maturity, etc.; however, clear trend to lower breakdown performance with decreasing k-value is evident. There are also a couple of interesting observations that demonstrate the importance of process/integration quality: (1) For similar LK, more recent process shows superior TDDB performance compared to earlier results; (2) In comparatively integrated structures, both LK with* $SiC_xN_y$ *capping dielectric and* $SiC_xN_y$*-only structures show very similar TDDB, indicating that capping dielectric breakdown is also important; (3) TDDB in MOS structure with ULK only shows better nominal TDDB (needs area-scaling correction, however) than integrated ULK and tighter space integrated-ULK shows worse performance than wider space ULK*

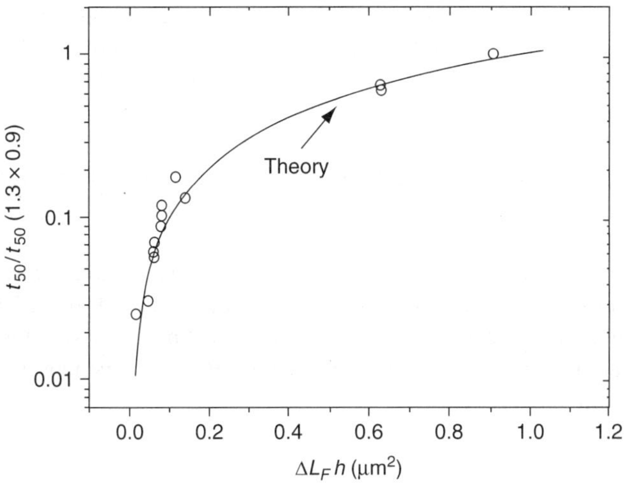


***Figure 9.1*** *Normalized median lifetime versus $\Delta L_F h$ for various Cu interconnect generations. Reprinted from* [10]*, with permission from IEEE*

length $\Delta L_o$. Then the lifetime would be proportional to $w\,h$, the line area from Equation (9.4). For every new generation of CMOS technology, the width and thickness of metal lines are scaled down by a factor of about 0.7. The minimum gate length is also about 0.7 of the minimum metal line width in each generation; for example 90 nm CMOS technology has a minimum metal line width of 0.13 μm. The ratio of the median lifetime $t_{50}$ ($w$) relative to the lifetime for 1 μm technology $t_{50}$ (1.3×0.9 μm²) for each CMOS generation from 90 nm to 1 μm technology is plotted in Figure 9.1 as a function of $\Delta L_F h$. The $t_{50}$ clearly decreases in each CMOS generation, which is related to the decrease in the area $\Delta L_F h$. The solid line in Figure 9.1 is the curve according to Equation (9.4) with no adjustable parameter [10]. The experimental data points follow the theoretical curve closely, which supports the validity of the model. The activation energy for EM, $Q_{EM}$, for electroplated Cu lines from different generations ranging from 1.0 to 0.09 μm CMOS technology appears to be constant at about 1 eV.

As interconnect dimensions continue to scale down, the contribution of interfacial mass transport increases, resulting in a reduction of the Cu EM lifetime and the allowed current density specification for each new generation of interconnects. This trend was found to be independent of the Cu deposition method, i.e. by physical vapor deposition (PVD), chemical vapor deposition (CVD), or electrochemical deposition (ECD) techniques, illustrating the universal nature of the problem. To develop reliable Cu interconnects beyond the 45 nm technology, the present Cu interconnect structures have to be changed, either by implementing Cu alloys, thick liner, large via size, or an improved Cu/dielectric interface [10]. The Cu alloy and thick liner approaches offer reliable Cu interconnects at the expense of Cu line conductivity. Large vias can improve the EM lifetime since a large void is required to fail the line and a large via has a better chance of connecting exposed liners between levels when a void is formed at the via/line interface.

However, the interconnect as well as the device density would be adversely affected by a large via design. A better method for maintaining high conductivity, high performance, and reliability of Cu interconnects is to alter or modify the Cu/dielectric interface to reduce the Cu transport along the interfaces. This can be achieved by using either a selective CoWP cap or surface treatment on top of the Cu line. The former can be performed after Cu CMP by a selective CoWP process [1, 15], whereby a thin layer coats the Cu conductor surface but does not coat the dielectric surface. The CoWP coating should be sufficiently thin but still able to reduce atomic interface transport in the Cu conductor in order to provide maximum reliability without overly complicating the fabrication process. In practice, the manufacturing of CoWP in wafer fabrication is quite challenging; thus the Cu alloy interconnect technique is still the most favored choice in IC manufacturing. The techniques for lifetime improvement will be discussed in detail in Section 9.4.

### 9.2.3 Statistical Test Method

Standard EM testing is performed in Cu line structures containing multiple wiring levels fabricated by the damascene technique. The EM test structures are either a fine line or a large number of multiple parallel and serial fine lines connected to wide and short underlying and overlying lines, as shown in Figures 9.2(a) and (b), respectively. The standard test is limited by the small sample size, typically 20 to 100 samples, which are not adequate for extrapolating the lifetime to the use condition for the on-chip interconnect system. For this reason, it is difficult to test enough parts using the standard single link structure and therefore a statistical testing methodology based on the well-known Wheatstone bridge technique was developed [11–13]. For this method, the test structures are usually arranged into units of parallel and serial interconnects. An example of a unit composed of 10 interconnects in parallel is shown schematically in Figure 9.2(a). The basic unit is repeated multiple (typically 20) times to build a large ensemble of line segments, as shown in Figure 9.2(b). Four of these identical interconnect arrays are arranged in a Wheatstone bridge configuration, as shown in Figure 9.3. In this way, the resistance resolution and therefore the sensitivity to detect void formation is greatly increased by the Wheatstone bridge technique. The advantage of using such statistical test structures is to observe the early time to fail (lifetime) of the lines in a short time and also to predict the lifetime at early failure probability, e.g. $-6\sigma$, with a better precision. Data analysis is based on the assumption that the failure probability distribution for all the lines remained the same during the test time. Some examples of using the Wheatstone bridge method for a massive-scale statistical study of EM early failures for Cu damascene structures are described in Section 9.6.

The failure time distributions are analyzed as lognormal or multilognormal distributions, depending on the test structures used and the failure modes. The tested samples were analyzed for void growth rate, void location, and void morphology to examine the failure mechanism and kinetics controlling the lifetime. Cu line/via microstructures are analyzed by focused ion beam (FIB) microscopy, transmission electron microscopy (TEM), X-ray diffraction (XRD), or electron backscatter diffraction (EBSD) in the scanning electron

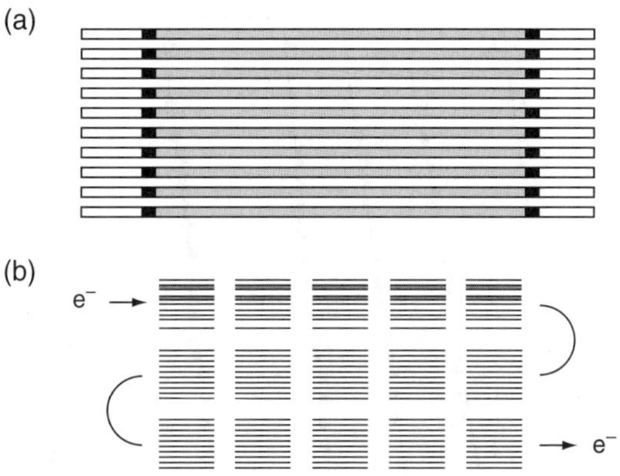


*Figure 9.2* Building blocks for large-scale EM testing: (a) basic unit of 10 interconnects in parallel and (b) interconnect array formed of basic units in series

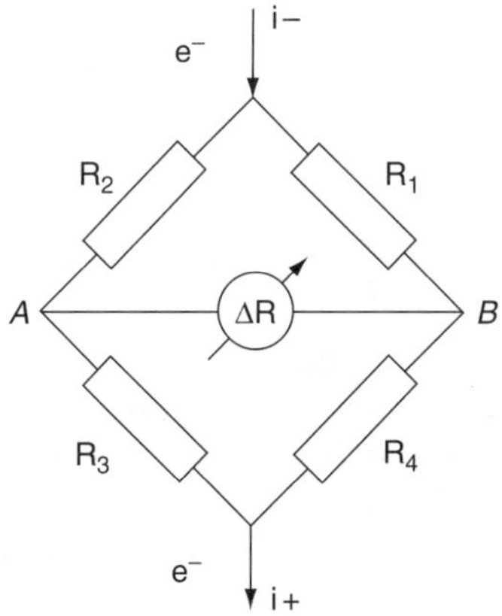


*Figure 9.3* Schematic of the Wheatstone bridge arrangement for large-scale EM testing

microscope (SEM). The crystallographic Ta phases and the Cu grain orientations are determined by XRD, while void growth and extrusion of the tested lines are examined by FIB and SEM on cross-sections.

### 9.2.4 Effect of Current Density on EM Lifetime

For a test structure of a pure Cu line connected to a Cu diffusion blocking boundary, the EM mass flow can be obtained by the void growth rate at or near the cathode end blocking boundary. Void growth can cause the line resistance to increase. Lifetime $\tau$ is defined by $\tau = \Delta L_F / v_d = (\Delta L_F k_B T)/(D_{eff} Z^* e\rho j)$, where $\Delta L_F$ is the critical void length for the lifetime $\tau$.

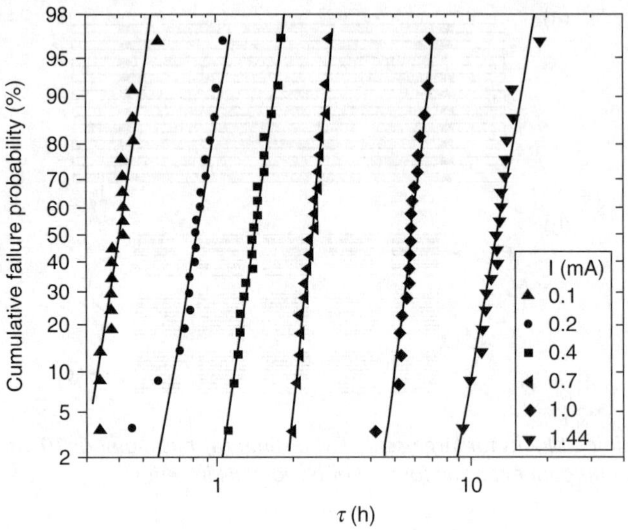


*Figure 9.4* Plot of lifetime versus cumulative lifetime probability for various currents used (lognormal scale). Reprinted from [6] with permission, copyright (2007), American Institute of Physics

It is customary to express $\tau = c_0 \exp(Q_{EM}/k_B T)$, where $c_0$ is assumed to be constant and $Q_{EM}$ is the activation energy for EM.

Figure 9.4 is a lognormal plot of the cumulative failure (lifetime) probability for 80 nm wide Cu M1 lines tested as a function of current at 348 °C [6]. The Cu M1 line is connected to a Cu diffusion blocking boundary W via at both ends of the line. The lifetime data followed a lognormal distribution function, where the straight lines are the least squares fits of the data to a lognormal distribution function. The values of sigma deduced from the slopes are found to be $<0.2$. The low sigma value was due to the electron flow across a blocking boundary (W via) to the Cu line and a physically stable redundant liner in these samples. When a void forms beyond or outside the Cu/W overlap region, the exposed liner layer will be stable enough to serve as a redundant liner to carry the applied current. Liner redundancy allowed the EM-induced void in the Cu line to be continuously grown without a sharp increase in line resistance so that it reduced the void size effect on lifetime distribution. Figure 9.5 shows a plot of the median lifetime $t_{50}$ as a function of current densities at $T = 348$ °C and 272 °C. In Equation (9.4), the lifetime for Cu lines can be expressed as $\tau = \tau_0 j^{-1}$ where $\tau_0 = (\Delta L_F k_B T)/(D_{eff} Z^* e\rho)$ [6]. However, it is customary to express the current density dependence of the lifetime as $\tau = \tau_0 j^{-m}$, where $m$ is called the current exponent and $\tau_0$ is assumed to be a constant at a fixed sample temperature. The values of $m$ in Figure 9.5 determined from the data of 80 nm wide pure Cu lines for currents up to 1 mA were found to be $1.1 \pm 0.2$. The data point at 1.44 mA was excluded. For the 80 nm wide line, a current of 1 mA gives a current density of 149 mA/μm$^2$ for Cu and 338 mA/μm$^2$ for the redundant liner alone without Cu. The large fraction of liner amounts to 30 % in these lines, providing a good

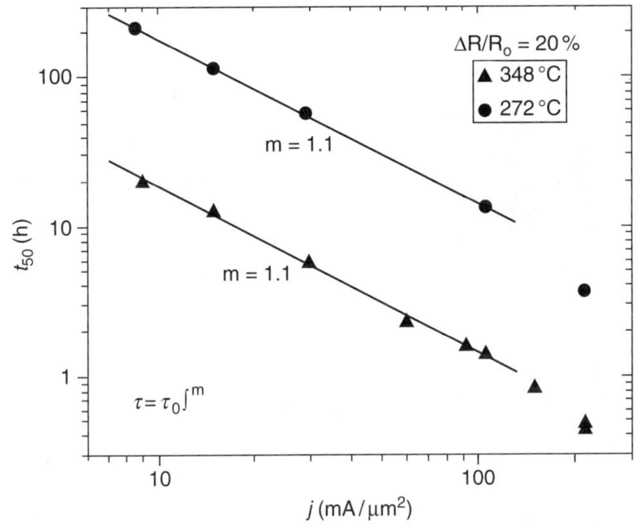


*Figure 9.5* Plot of median lifetime as a function of current applied at two different sample temperatures. Reprinted from [6] with permission, copyright (2007), American Institute of Physics

redundancy for carrying current in the EM-damaged Cu line. In comparison, for 0.28 μm wide lines with a $Cu/SiO_2$ structure, $m$ was found to be $1.1 \pm 0.2$ for $j < 25\,mA/\mu m^2$ and 1.8 for $25\,mA/\mu m^2 \leq j \leq 140\,mA/\mu m$ [2, 16]. For fat Cu damascene lines, 1 μm wide and 1 μm thick, in a poor thermally conductive polyimide dielectric, a value of $m = 2.2$ was observed [17], which greatly deviated from 1. The deviation of $m$ from 1 suggests that its value reflects the effects of local joule heating and liner redundancy. In general, $m$ was found to depend on the test structure, the power used, the liner thickness/resistivity, the contact resistance, the surrounding material, and the thermal conductivity of the insulator.

## 9.3 Cu Microstructure

### 9.3.1 X-ray Diffraction (XRD)

*Laboratory-Based XRD*

In interconnect line stacks, Cu grains are usually not oriented randomly. Rather, at Cu deposition (i.e. during nucleation, agglomeration of the nuclei, and subsequent Cu grain growth), particular grain orientations are preferred, resulting in the formation of specific texture components. Furthermore, the geometric constraints imposed by the dual-damascene process influence the texture evolution during grain growth. In addition, preferred Cu grain orientations may vary during subsequent processing or use, induced by diffusion processes, such as creep, EM, stress-induced voiding, or plastic deformation.

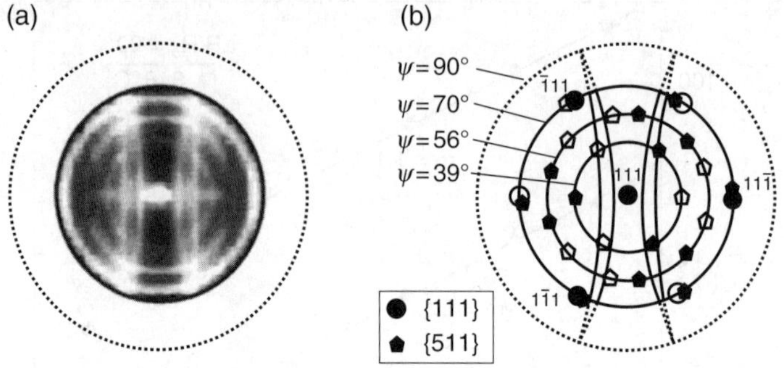


***Figure 9.6*** *{111} pole figure measurement of parallel Cu lines (a) and designation of the main texture components in a schematic {111} pole figure (b). The dotted ring corresponding to ψ = 90° is out of the range of the measurement. Due to the symmetry of the Cu lines running in vertical direction, apart from the regular poles also their mirror images appear, yielding e.g. six 111 maxima instead of three on a regular ψ = 70.5° ring. This property is reflected by open and closed symbols, corresponding to the alignment of crystallite orientations at the left and right line sidewall, respectively* [24]. *Copyright Wiley-VCH Verlag GmbH & Co. KGaA, reproduced with permission)*

Texture characterization for Cu thin films is commonly performed simply by acquiring a $\theta$–$2\theta$ scan, measuring the peak heights of particular Bragg reflections to determine texture coefficients by calculating corresponding peak height ratios [18]. However, this approach is only valid to characterize the strength of a fiber texture with the fiber axis of the specimen lying in the goniometer plane during the diffraction experiment [19]. In general, a more comprehensive texture analysis is necessary for complete microstructure characterization. This can be achieved by X-ray pole figure analysis. In the case of fiber textures with no preferred in-plane orientation of the Cu crystallites, measurement of pole figure cuts at a constant specimen rotation is sufficient. Cu thin films deposited on amorphous substrates are prominent examples [20, 21]. For complete texture characterization, several (*hkl*) pole figures have to be recorded for the same specimen area (for copper at least three pole figures, e.g. (111), (200), and (220) [22, 23]), providing the necessary information to calculate the orientation distribution function (ODF) of the crystallites [5].

Figure 9.6 shows an example for a {111} pole figure measurement of a test specimen with parallel Cu lines [24]. The high intensity in the center is due to a preferred <111> orientation of the Cu grains parallel to the surface normal. Since the outer ring at $\psi$ = 70.5° is structured, these crystallites also show preferred orientations within the film plane. In particular, the six distinct diffraction maxima point to two preferred in-plane orientations with the {112} planes parallel to the sidewalls of the Cu lines. This structure is also reflected by the <511> twins (additional rings at $\psi$ = 39° and 56°). Furthermore, two main vertical stripes in the pole figure are induced by Cu grains with the <111> orientation perpendicular to the sidewall surface. In more detail, the two stripes show a splitting of about 10° into substripes, which are caused by a 5° deviation from the vertical sidewall direction, indicating the strong geometrical influence on Cu microstructure evolution [24, 25].

*Synchrotron-Based XRD*

An essential limitation of X-ray texture analysis is the low spatial resolution resulting in the fact that the location of individual Cu crystallites contributing to the diffraction pattern cannot be determined. Modern laboratory-based X-ray microdiffraction tools with suitable primary-beam optics and a small beam size (down to 10 μm) facilitate the measurement of pole figures for small test patterns of inlaid copper interconnects. However, the beam shape and intensity conditions are still insufficient to resolve single Cu crystallites. For this purpose, the application of high-brightness synchrotron radiation along with improved X-ray focusing optics, such as Fresnel zone plates, tapered capillaries, Kirkpatrick–Baez mirrors, and compound refractive lenses, is needed. Beam diameters of about 500 nm FWHM are now routine and a 90 nm focus has been demonstrated [26].

In the case of Laue microdiffraction, white X-ray beams are focused by a Kirkpatrick–Baez mirror system to a spot size below 1 micrometer [27]. Scanning the specimen under the X-ray beam and recording Laue patterns at each scan position, the orientation of single crystallites in a polycrystalline material can be determined with an accuracy in the order of 0.01°. Furthermore, small shifts of the Bragg reflection positions from their ideal positions can be used to determine the elastic deformation state of the unit cell at a constant volume (the deviatoric component of the strain tensor). The total strain tensor can be obtained by calculating its dilatational component via the measurement of the wavelength of one reflection [28].

Laue microdiffraction has been successfully employed for grain orientation measurements at Cu lines [29, 30]. In particular, in situ EM experiments of Cu interconnects with a bamboo-like line structure showed inhomogeneous stress distributions pointing to flux divergences at particular grain boundaries, which can be explained by anisotropic surface/interface diffusion depending on the crystal orientation [30]. In addition, plastic deformation has been observed in Cu interconnect test structures before the onset of visible microstructure changes [32]. In lines with a strong <111> texture, the extent of plastic deformation is found to be relatively large compared to Cu lines with a weaker texture. This seems to be consistent with the observation that the plastic deformation in the grain strongly correlates with the parallelism of the <112> crystal direction to the Cu line [33].

### 9.3.2 Electron Backscatter Diffraction in the Scanning Electron Microscope

Electron backscatter diffraction (EBSD) is commonly employed as a characterization technique in a scanning electron microscope (SEM) for determining individual grain orientations, local texture, and point-to-point orientation correlations on polycrystalline surfaces of bulk materials [34]. The flat and highly polished specimen is typically tilted to an angle of 70° to allow a maximum number of low energy-loss backscattered electrons to exit the sample surface after they are channeled and subjected to path differences leading to constructive and destructive interferences. The resulting backscatter Kikuchi pattern – a regular arrangement of parallel bright bands on a steep continuous background – is displayed by projection on to a phosphor screen, which converts the diffracted electrons into light patterns suitable for a charge-coupled device (CCD) camera to record. Based on the crystallographic structure of the material studied, the recorded EBSD pattern is

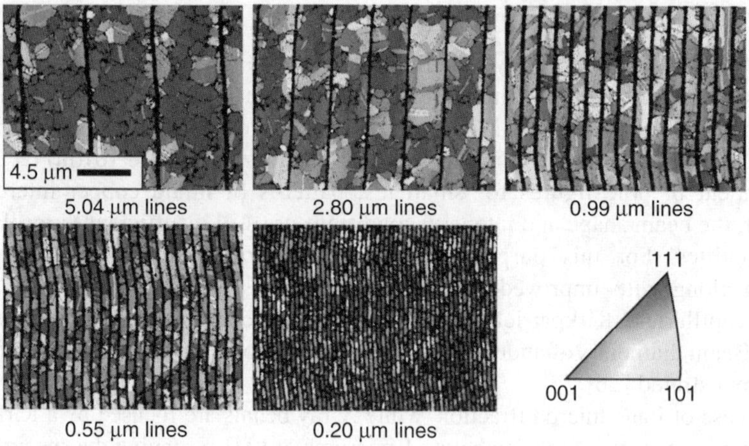


***Figure 9.7*** *IPF maps perpendicular to the wafer surface for Cu lines with different widths. Reprinted from* [43], *with permission from Elsevier*

compared with a simulated one in real time. To obtain a complete map, the electron beam is scanned across the sample collecting Kikuchi patterns at each scan point. After recording the orientation information, the data are analyzed.

The sample surface conditions can significantly impact the quality of the Kikuchi patterns and the accuracy of subsequent indexing. Other than a cover layer on top of the Cu surface, shadowing effects for samples with high surface roughness can cause energy loss and reduce the total number of backscattered electrons. Furthermore, charging effects on insulating regions of the scan area, such as interlayer dielectrics (ILD) between Cu interconnect lines, can cause a drift and degrade the Kikuchi pattern quality. One approach to prevent charging is to deposit a thin and smooth conductive layer on the sample surface, such as an evaporated chromium layer of less than 5 nm [35]. Targeted surface smoothening employing the focused ion beam technique (FIB) is particularly useful for cross-sectioned specimens [36]. Depending on the nature of the cover layer, several techniques can be applied for its removal. Carbon contamination from previous SEM imaging, for instance, is removed by oxygen plasma cleaning. ILD passivation layers can be completely removed by wet-chemical processes or partially thinned by reactive ion etching (RIE) [37].

Due to a smooth surface without perturbing capping layers, EBSD analyses at Cu interconnect structures are ideally performed after chemical–mechanical polishing. For process control and development, two factors influencing the Cu grain size and orientation distribution can be particularly well characterized by EBSD: (1) geometrical aspects [38] and (2) process parameters and new materials [39, 40]. Basic EBSD analyses have been carried out for blanket Cu thin films to investigate the substrate influence on grain evolution at elevated temperatures [41] and the effect of Cu film thickness on the evolution of annealing texture [42]. Muppidi *et al.* [43] studied the effect of barrier materials (sputtered Ta and TaN) and their layering sequence on the microstructure of Cu films and of Cu interconnects with line widths between 0.2 and 5 μm (Figure 9.7). All Cu films were found to have a predominant <111> texture comparable for the different barrier layers. The side-wall constraint in narrow damascene trenches leads to a preferred [110] orientation along

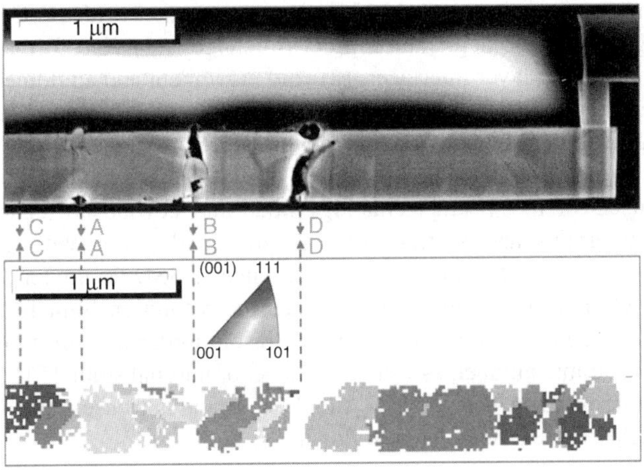


*Figure 9.8* EBSD study of an EM test structure with the CoWP layer at the top Cu interface after an in situ SEM analysis. Reprinted with permission from [50], copyright (2007), American Institute of Physics

the line, which is in good agreement with the results from XRD measurements discussed above [24]. The in-plane texture was found to strengthen with decreasing line width. For an intermediate width of about 1 μm, the Cu grains were orientated most randomly, with the highest fraction of twin boundaries (Figure 9.7) [43, 44]. Geiss and Reid [45] showed that neither the texture nor the grain size of the overburden, as represented by large contact pads, propagated into 100 nm wide lines, although they did propagate into wider lines. A subsequent study by Meyer *et al.* [35] found that EBSD measurements are suitable for Cu microstructure characterization of interconnects with line widths down to 60 nm.

To understand reliability-related failure in Cu interconnects, the analysis of the effect of microstructure on the degradation mechanisms is necessary [46]. Applying in situ transmission X-ray microscopy [47, 48] or in situ scanning electron microscopy [49], mass transport and failure mechanisms in Cu interconnects can be visualized. Subsequent EBSD analysis is suitable to correlate the observed changes to the microstructure of the Cu line. In particular, cross-sections of fully embedded Cu via/line test structures can be continuously imaged during in situ SEM experiments, while EM stressing is performed at elevated temperatures and current densities. For a subsequent EBSD study, the remaining passivation layer is removed and a cross-section is cut through the middle of the line using FIB milling. An example for a postmortem SEM/EBSD analysis is shown in Figure 9.8 [50]. The results obtained from such studies revealed several distinct phases of the EM-induced degradation process for Cu lines > 100 nm with $SiN_x$ capping layer: (1) void formation at the Cu/$SiN_x$ interface, (2) void movement and agglomeration at the line end, (3) void growth into the via, and (4) continued void growth that causes a further resistance increase leading to EM failure [51]. To improve the EM lifetime by increasing the bonding strength of the top interface, several approaches are discussed in the literature, e.g. (1) selective electroless CoWP, CoSnP, or Pd coatings on top of the polished Cu line [52], (2) Ta/TaN layer stack deposition on top of the polished Cu line [53], and (3) local alloying of copper interconnects, e.g. using CuAl [54]. As a consequence of the Cu/capping layer interface strengthening, the degradation process changes: (1) void formation at the Cu/liner and the Cu/capping layer interfaces, (2) void movement and agglomeration, (3) void shape modification, pinning at grain boundaries and triple points, and void growth, and (4) failure due to growth of a pinned void across the entire cross-section or void movement into the via [51].

Cu interconnects of the 45 nm technology node and beyond are characterized by a mixture of large grains and sections of multiple small grains with grain diameters significantly smaller than 50 nm. As a consequence, EBSD is no longer suitable for comprehensive Cu microstructure analysis, since the resolution limit for this technique ranges between 20 and 80 nm depending on the resolution/beam current response of the SEM, the sample atomic number, as well as the level of internal strain [55]. An appropriate alternative for microstructure analysis of small Cu grains is to use transmission electron microscopy (TEM), as described in the following section.

### 9.3.3 Orientation Imaging Microscopy in the Transmission Electron Microscope

The most frequently used TEM technique to characterize nanostructured materials is the application of bright-field (BF) imaging, which is primarily based on transmitted electrons of the direct beam. This method has been employed not only for plan-view and cross-sectional specimens of thin Cu films [56, 57] but also for microstructure characterization of Cu line structures [8, 58]. However, BF imaging with the direct electron beam as well as dark-field (DF) imaging with particular diffracted electron beams suffer from inadequate grain-to-grain contrast necessary for grain shape and grain size determination [59]. In particular, complex contrast features due to defects, strain and thickness contours, or artifacts introduced by sample preparation may not allow for correct grain boundary tracing. A suitable alternative is given by Yao *et al.* [60]. Based on hollow-cone DF imaging with a slightly convergent beam, high-contrast micrographs suitable for distinguishing neighboring grains can be obtained. Although this technique can be employed for grain size quantification of nanocrystalline materials, it does not allow the orientation determination of individual crystallites.

Grain orientation determination in the TEM can be based on electron diffraction (ED). In principle, there are three fundamentally different diffraction techniques: (1) selected area electron diffraction (SAED), (2) convergent beam electron diffraction (CBED) leading to spot patterns, and (3) Kikuchi diffraction (KD) resulting in line patterns. For a detailed description of the individual ED techniques, the reader is referred to pertinent textbooks, e.g. Williams and Carter [61]. The main advantage of nanobeam electron diffraction (NBED), i.e. CBED with a very small convergence angle, compared to SAED is a much finer focused electron beam leading to a higher spatial resolution of a few nanometers. Compared to spot patterns, Kikuchi line patterns have the advantage to provide more precise orientation determination ($<0.1°$). However, the sensitivity to lattice defects and the difficulty of automating the detection of line positions by software algorithms are serious disadvantages of using KD patterns [62].

As discussed above, automated grain orientation analysis of nanocrystalline materials works best with the evaluation of spot diffraction patterns. One possible technique was

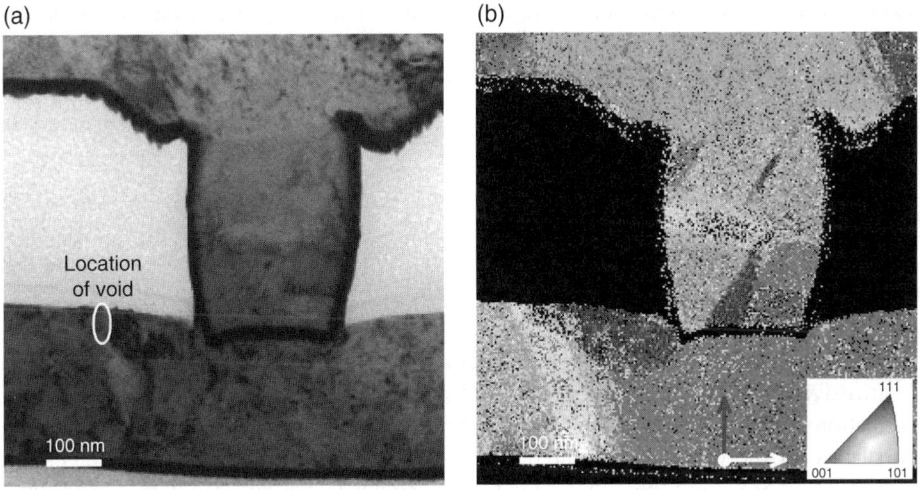


***Figure 9.9*** *Cross-sectional TEM BF image of the structure of interest (stress-induced void and surrounding Cu grains) after (a) a reliability test and (b) the IPF map perpendicular to the sample surface. Reprinted with permission from [69], copyright (2009), American Institute of Physics*

introduced by Rauch *et al.* [63]. Spot patterns are collected with a CCD camera while the specimen area of interest is scanned by the electron beam, which can be controlled by an external source, like the 'spinning star' precession device developed by Nano-MEGAS [64]. Local crystallographic orientation and/or phase information are obtained through an algorithm that compares the recorded spot patterns with pre-calculated (simulated) templates for all possible orientations and/or phases [65, 66]. Identification through template matching can be significantly improved by the application of electron precession, since the number of diffraction spots in the patterns is increased and dynamical effects such as Kikuchi line contrast are reduced [63, 67].

The conical dark-field (CDF) technique is another approach for automated grain orientation determination in the TEM. Contrary to the method previously described, CDF analysis is based on systematic recording of a series of DF images followed by automated reconstruction and indexing of spot patterns for each point in the view of interest [55, 68]. With a precision $< 1°$ [55], Hübner *et al.* [68] showed that the lateral resolution of the CDF technique is better than 5 nm. As depicted in Figure 9.9, CDF analysis has already been successfully employed to study the influence of the Cu microstructure on SIV in multilevel dual-damascene Cu/low-*k* interconnect stacks [69]. In particular, orientation information was obtained for the region next to the Cu via after performing the reliability test. The Cu grains in the lower line are preferentially <111> oriented to the wafer surface. The stress-induced void itself forms at the edge of the via bottom between <111> and high-index Cu grains, where a high local stress gradient exists to drive vacancy agglomeration and void formation [70]. Although characterized by high spatial resolution, the CDF technique is challenging in regard to data acquisition as well as data analysis. Besides a correct alignment of the microscope, specimen drift and beam-induced sample alterations have to be minimized. For data evaluation, the parameter set has to be chosen appropriately. Manual verification of the orientation results is recommended, particularly in the case when reconstructed diffraction patterns from several grains are superimposed [71].

## 9.4 Lifetime Enhancement

### 9.4.1 Effect of a Ta Liner

As stated above, EM mass flow is mainly controlled by the effective Cu diffusivity. The Cu diffusivity in a Cu line is very sensitive to and strongly dependent on the materials and fabrication methods. Cu interconnects are fabricated using a damascene process where the top surface of the Cu line is covered with a thin dielectric diffusion barrier layer, e.g. $a$-$SiC_xN_yH_z$ or $SiN_x$, and the bottom and the two sidewalls are covered with a metal liner, e.g. Ta-based films [72]. The different crystallographic phases of Ta were achieved by changing the substrate surface by optimizing the sputter etching conditions during PVD metal liner deposition. In this section, the effect of Ta liner phases on Cu EM is discussed. Figure 9.10 is a plot of median lifetimes $t_{50}$ of 0.13 μm wide Cu M1 lines using a failure criterion of $\Delta R/R_o = 3\,\%$ for a sample temperature of 300 °C and a current density of 24 mA/μm [2, 73]. The Cu M1 line is connected to a Cu diffusion blocking boundary W via at both ends of the line. The lifetime with a 3 % failure criteria is close to the resistance incubation times, $\tau_i$, thus it is a good metric for comparing void growth rates. Depending on the void growth, $\tau_i$ is the time required to grow a void either beyond the Cu/W overlap region or by depleting Cu out of the grain or the grain cluster in contact with the via. The similar values of $t_{50}$ at 3 % failure criteria among the wafers with different Ta phases suggest that EM mass flow in the Cu lines are all about the same, despite the difference in Ta phases and thicknesses. This behavior was also observed at test temperatures of 255 °C, 320 °C, and 348 °C.

The median lifetime $t_{50}$ for the Cu damascene lines with various Ta phase liners is plotted as a function of $1/T$ in Figure 9.11 [73]. The solid lines are the least squares fits. The median lifetime $t_{50}$ is usually expressed as $t_{50} = \tau_c e^{(Q_{EM}/k_B T)}$, where $\tau_c$ is a constant and $Q_{EM}$ is the activation energy for EM. $Q_{EM}$ in these Cu damascene lines can be extracted from the slopes of the lines in Figure 9.11 and are found to be 0.91 eV to 1.0 eV, with an uncertainty of about 0.04 eV. A similar $Q_{EM}$ was obtained for different wafers, further demonstrating that the mass flow rate is independent of the Ta phase in the Cu lines. $t_{50}$ with a 3 % resistance increase corresponds to the time for a critical void length $\Delta L_F$ of 0.5 μm. The drift velocity can then be estimated by using Equation (9.3) and $v_d = \Delta L_F/\tau_i$. It is customary to express $Z^* = Z^*_{el} - [a/\rho(T)]$, where $Z^*_{el}$ is the effective charge number from the electrostatic force and $a/\rho(T)$ is the effective charge number from the electron wind force [74]. In general, the electron wind force is much larger than the electrostatic force in metals, so it is reasonable to assume that the temperature dependence of $Z^*(T) \approx -a/\rho(T) = -(a/\rho_0)/(1 + bT) = -Z^*_0/(1 + bT)$ and $\rho(T) = \rho_0(1 + bT)$, where $b$ is the temperature coefficient of the resistivity, and $Z^*_0$ and $\rho_0$ refer to 0°C, respectively. Thus the extracted activation energy from $\delta_I Z^*_0 D_I$ versus

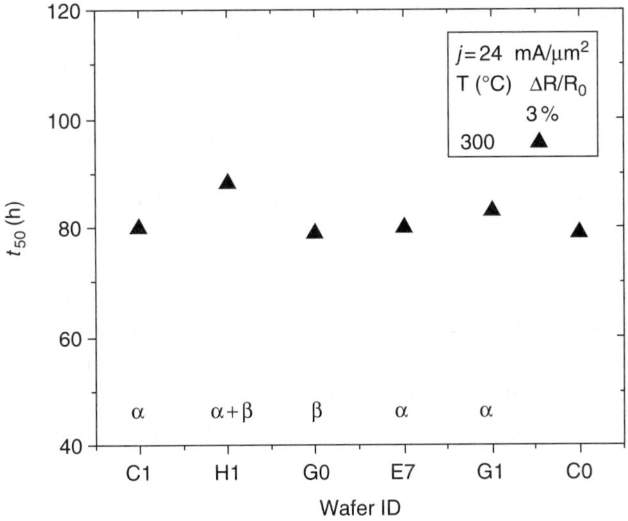


***Figure 9.10*** *Plot of the median lifetimes for the various wafers for 3 % failure criteria with a current density of 24 mA/μm² at 300 °C. The measured Ta liner phases from XRD are included. Reprinted from [73] with permission, copyright (2005), American Institute of Physics*

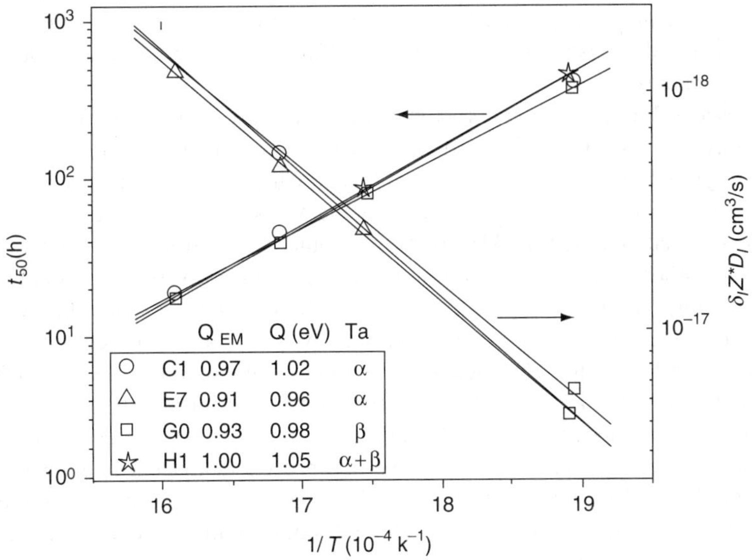


***Figure 9.11*** *Plot of $t_{50}$ and $\delta_I Z^*_0 D_I$ versus (1/T) with extracted values for Cu activation energy of EM and diffusion for various Ta liner phases. Reprinted from [73] with permission, copyright (2005), American Institute of Physics*

$1/T$ would be the activation energy of diffusion. The plot of $(\delta_I Z^*_0 D_I) = \Delta L_F h k_B T/(\tau_i e\rho_0 j)$ as a function of $1/T$ is also plotted in Figure 9.11 [73]. Activation energies for Cu interface diffusivity $Q$ of 0.96 to 1.05 eV with an average value of 1.0 ± 0.04 eV were obtained from the slopes of these plots. The difference between the activation energy for EM and diffusion is about 0.05 eV. Although different $Q_{EM}$ for different Ta liner phases have been reported, 0.70 eV[75] and 0.75 eV[76] for the β-Ta liner and 0.84 eV [76] and 0.98 eV [75] for the α-Ta liner, here the values obtained for the samples with the β-, α- or α+β-Ta liners are about the same and similar to the values obtained for Cu lines capped with $SiN_x$ [77]. This suggests that replacing a $SiN_x$ cap by an *a*-$SiC_xN_yH_z$ cap with a lower dielectric constant did not degrade the Cu lifetime. The result that similar activation energies and void growth rates were extracted from different wafers is consistent with the assumption that the dominant diffusion path in these Cu lines is along the Cu/*a*-$SiC_xN_yH_z$ interface. Results from these samples clearly demonstrated that EM is controlled by diffusion along the Cu/dielectric interface and not affected by the Ta phases in the liner.

### 9.4.2 Upper-Level Dummy Vias

In this section, we describe a new technique for improved EM lifetimes by modifying the interconnect structure with additional upper-level dummy vias on top of a Cu line [7]. The upper dummy via structure can interrupt the Cu mass flow along the top interface to provide a lifetime enhancement without an increase in the line resistance. For the dummy via structures, lines with a near-bamboo microstructure showed a significant increase in EM lifetime compared to lines with more polycrystalline grain sections. The upper level dummy via structure also proved to be a powerful tool to understand the Cu microstructure effect and to distinguish various fast diffusion paths in the line, as described in the following paragraph.

Cu damascene lines, 70 nm and 1.7 μm wide, were used to illustrate the effect of the upper dummy vias on Cu EM. The test structure consisted of a Cu M1 test line connected to the underlying NiSi line and M2 through W and Cu contact vias, respectively, at each end of the Cu M1 line. The damascene dummy via/M2 bar is 40 % wider than the Cu M1 line and was placed on top of the M1 line with no other connections. The custom-designed structure has either a single upper dummy via bar 2 μm away from the contact via or repeating dummy via bars every 5 μm or 10 μm. The dummy via spans the full M1 line width and has M2 on top. Figures 9.12(a) and (b) are plots of the cumulative failure (lifetime) probability of the Cu lines as a function of via size and lifetime. Tests were performed at 324 °C for 1.7 μm and 301 °C for 70 nm wide lines with and without upper level dummy vias, respectively [7]. The applied current density is less than the Blech critical current density, $j_c$ [14], assuming the dummy via distance $x$ is equivalent to the short length ($j_c x < 120$ mA/μm). As compared to the no-dummy via structure, enhancements of 10 times for 1.7 μm and 2–3 times for a 70 nm wide line were observed. The upper dummy vias on the top surface of the Cu lines blocked the Cu mass flow along the top interface, which should yield a significant lifetime improvement. However, the EM improvement would be diminished if the dummy vias were on top of the polycrystalline Cu sections and/or the distance between the dummy vias on bamboo-like Cu grains were longer than the Blech length. The larger improvement in the 1.7 μm wide line samples can

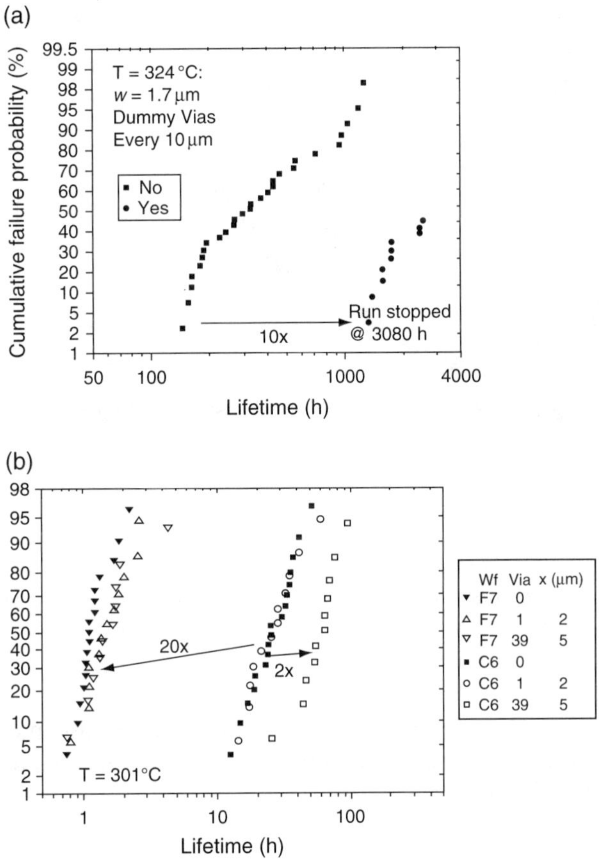


***Figure 9.12*** *Cumulative failure probability versus lifetime for (a) 1.7 μm and (b) 70 nm wide lines, respectively. The total number of dummy vias and the via spacing, x, on a 200 μm M1 test line are also described. Reprinted from [9] with permission, copyright (2007), American Institute of Physics*

be attributed to the presence of more bamboo sections and larger grains compared to the 70 nm wide lines. The data for the one-dummy via structure (open circles in Figure 9.12(b)) showed little or no EM improvement as compared to the lifetime of the no-dummy via samples. This is because the present structure with an embedded via requires a depletion of Cu out of the grain or the Cu grain cluster in contact with the via to cause EM failure [73]. Thus any improvement in the EM lifetime by transferring the EM-induced voids beyond the dummy via would be small. Figure 9.12(b) also includes data from samples with abnormally short lifetimes reduced by 20 times. These wafers with an abnormally short lifetime came from a different experimental lot, which was fabricated with a metal liner/seed deposition that deviated from the standard liner/seed processing steps. These samples were tested with and without dummy vias and the results are plotted as solid and open triangles, respectively. Negligible lifetime improvement was observed. This result suggests that the abnormal samples had a faster Cu/liner interfacial diffusivity than the

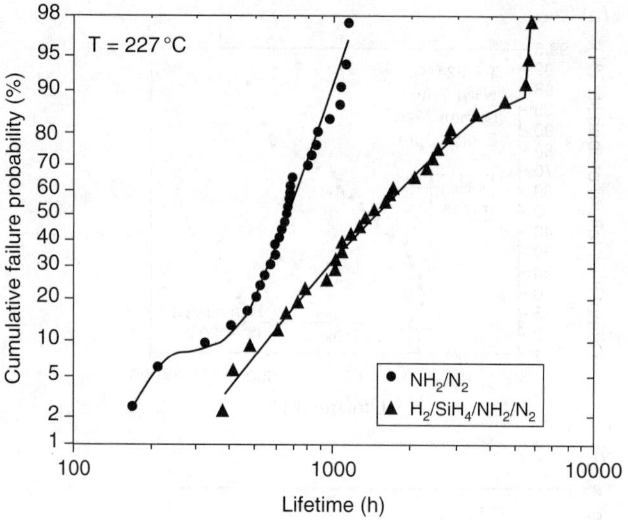


*Figure 9.13* Plot of the cumulative lifetime probability for Cu lines from samples with various Cu surface treatments. The solid curves are the least-squares bimodal fits. Reprinted from [7] with permission, copyright (2009), American Institute of Physics

normal samples. Thus the upper-level dummy via structure also provided additional understanding of the Cu microstructure effect and was useful for identifying the fast diffusion path in these samples.

### 9.4.3 Plasma Pre-clean and $SiH_4$ Soak

The effect of plasma treatment to improve the EM lifetime before chemical vapor deposition (CVD) of the cap layer on top of the Cu line has recently been investigated. The study included the effect of an $H_2$-based plasma pre-clean on a silicon nitride cap layer [78] and a plasma clean for a plasma-enhanced CVD of an *a*-$SiC_xN_yH_z$ cap film [79]. In general, the amount of EM improvement has to be traded off with the increase in the copper resistivity induced by the plasma treatment. In Figure 9.13, we compare the EM lifetime distributions from two groups of samples, one pre-cleaned with $NH_3/N_2$ plasma and the other subjected to $H_2$ plasma/$SiH_4$ soak/$NH_3/N_2$ plasma at the same test temperature of 227 °C [7]. The second group yielded a moderate EM lifetime improvement of 2 times with an ~8 % increase in line resistance, as compared to the first group with an $NH_3/N_2$ plasma clean. In a previous study [79], we reported an improvement of over 10 times in the median EM lifetime after plasma treatment, but that was accompanied with a 2 times increase in initial line resistance after annealing the test samples at 350 °C for 14 hours. The improved Cu lifetime in the case of $H_2$ plasma with $SiH_4$ soak and $NH_3/N_2$ plasma was believed to be due to the formation of a $CuSi_xN_y$ interfacial layer, while the increase in the copper resistivity was attributed to silicon diffusion into the copper grains from the cap/copper interface [79]. The transport of silicon can be reduced by exposure to $NH_3/N_2$ plasma after

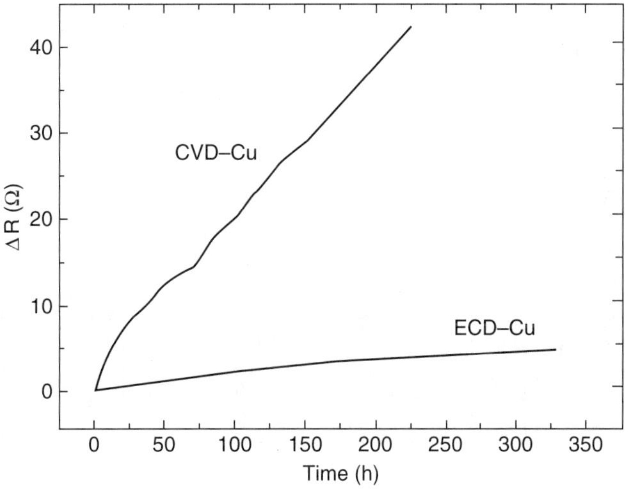


***Figure 9.14*** *Cu line resistance as a function of the EM stress time for CVD Cu and ECD Cu. Reprinted from* [80] *with permission, copyright (2010), American Institute of Physics*

Cu silicide formation or can be prevented by adding nitrogen to the plasma or removing the seasoning layer. A trade-off between improved EM and increased copper resistivity was observed.

### 9.4.4 CVD and ECD Cu and the Effect of Nonmetallic Impurities

Figure 9.14 shows the resistance plots of 2 μm wide Cu damascene lines fabricated by electrochemical deposition (ECD) and chemical vapor deposition (CVD) as a function of the EM stress time. The ECD Cu lines were fabricated using a standard post-Cu deposition annealing step prior to CMP and contained Cu M1 lines connected to W vias at both ends, while the CVD Cu damascene lines were fabricated by filling the trench/via with CVD Cu after PVD liner/Cu seed depositions [80]. The line resistance change observed can be accounted for by Equation (9.3), showing that the EM damage in the CVD Cu line occurred faster than that of the ECD Cu line. In addition, the EM activation energy of CVD Cu was found to be 0.8 eV, lower than the value of 1.1 eV obtained for the ECD Cu damascene lines [81, 82]. The superior EM reliability of ECD Cu over that of the CVD or PVD Cu can be attributed to either the microstructure (bamboo versus polycrystalline) and/or impurities in the ECD Cu [81]. These two possibilities were further evaluated by studying the electroplating impurity effect on Cu EM. Two electroplating solutions, A and B, were used to form a Cu trench and for via fill. The concentrations of Cl, O, C, and S in the electroplated Cu films were measured by SIMS, with the results shown in Figure 9.15. The Cu films using Recipe A had an order of magnitude higher of Cl, O, S, and C impurity concentrations than films formed using Recipe B. Similar large impurity concentration variations in the Cu lines from these two electroplating solutions were also observed [83].

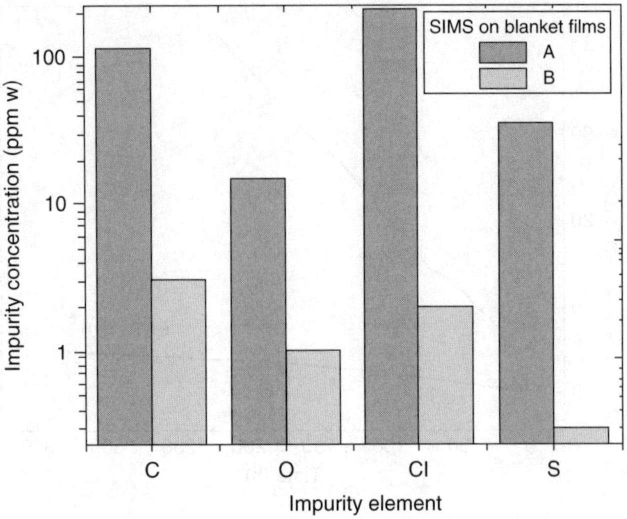


*Figure 9.15* *Impurity concentrations obtained by SIMS on ECD Cu films using Recipes A and B, respectively. Reprinted from* [80] *with permission, copyright (2010), American Institute of Physics*

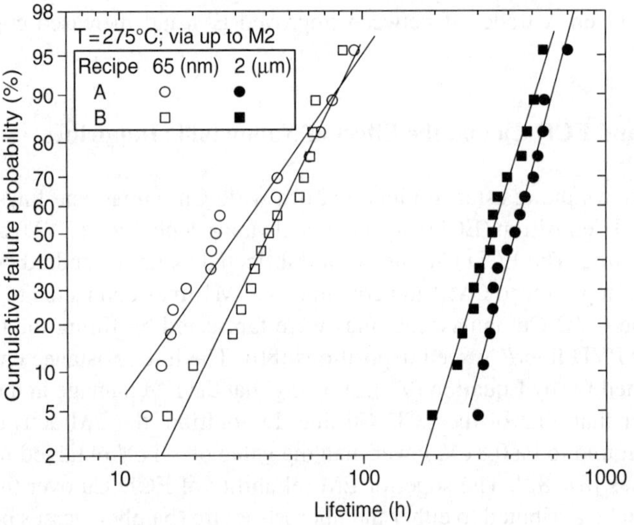


*Figure 9.16* *Cumulative failure probability for 65 nm (O,□) and 2 μm (●,■) wide Cu lines fabricated with ECD Cu Recipes A and B. Reprinted from* [80] *with permission, copyright (2010), American Institute of Physics*

Figure 9.16 plots the cumulative failure probability versus lifetime for the 65 nm and 2 μm wide Cu M2 lines formed using ECD Recipes A and B tested at the same temperature of 275 °C. The test lines had an M1–V1–M2 EM configuration where the V1–M2 dielectric was ULK with $k = 2.4$ [80]. The lifetime data followed a lognormal distribution function

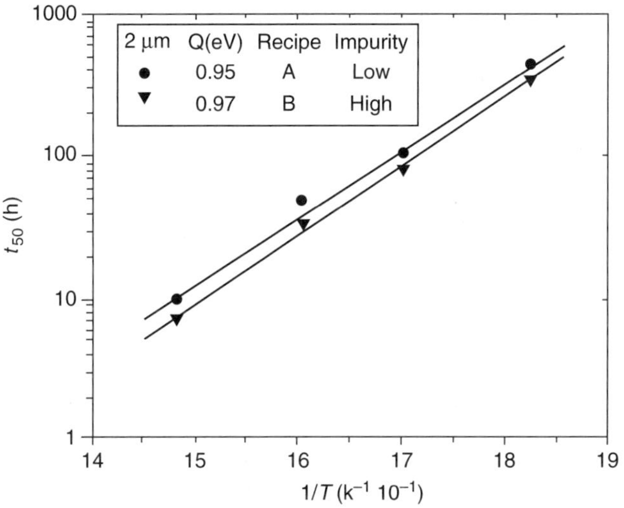


***Figure 9.17*** *Plot of* $t_{50}$ *versus 1/*T *for high and low levels of impurities in 2 µm wide lines. Reprinted from* [80] *with permission, copyright (2010), American Institute of Physics*

where the straight lines represented the least squares fits of the data to a lognormal distribution function. The lifetimes for the 2 µm wide lines were found to be significantly higher than the 65 nm lines, about 10 times higher in the lifetime together with a 0.5 reduction in $\sigma$. In a previous study of Cu damascene lines formed with a different low-$k$ dielectric ($k$ = 2.7), the lifetime difference between 65 nm and 2 µm wide lines was found to be considerably less [6]. This suggests an effect due to defects existing in the 65 nm wide Cu line/ULK samples. Since the lifetime distributions for 2 µm wide damascene lines in ULK were observed to be similar to those of the Cu/LK structure, the 2 µm wide lines were used to investigate the plating impurity effect. Figure 9.17 shows the EM median lifetime and activation energy $Q_{EM}$ obtained for the 2 µm wide lines using ECD Recipes A and B [80]. $Q_{EM}$ values of 0.95 ± 0.05 eV and 0.97 ± 0.05 eV were obtained for the 2 µm lines using Recipes A and B, respectively. The small differences in lifetime and activation energy observed for the two sets of test samples could very well be accounted for by a small variation in the test structures, such as the line area or the line/via overlap. Thus, there seemed to be little or no effect due to ECD impurities, Cl, C, O, and S, on Cu EM. These observations are contradictory to a paper that described a large nonmetallic impurity effect on Cu EM [84]. We believe that the amount of impurities (<0.1 %) in the electroplated Cu reported is too low to make a large contribution to interface and GB mass transport.

The superiority of ECD Cu over CVD or PVD Cu could be attributed to the microstructure effect and not to the presence of nonmetallic impurities in the ECD Cu line. ECD Cu lines with a larger grain size than CVD or PVD Cu lines are the result of an abnormal grain growth behavior of ECD Cu during fabrication [85]. Polycrystalline-only and near-bamboo-like grain structures in 2 µm wide Cu damascene lines were observed for PVD/CVD Cu and the ECD Cu, respectively. Therefore, grain boundary diffusion dominates in PVD/CVD Cu lines and interface diffusion dominates for ECD Cu, although the grain

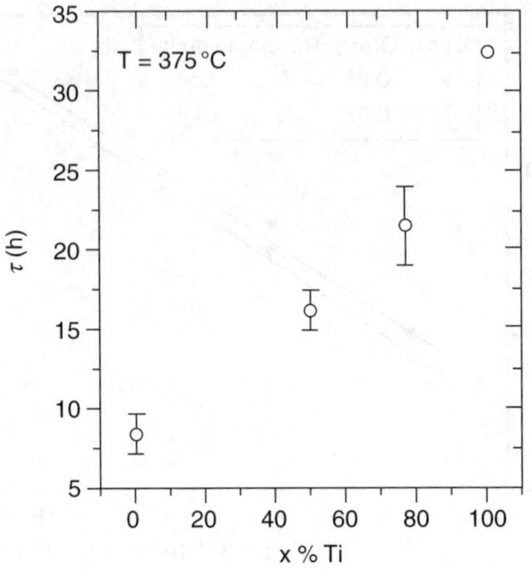


*Figure 9.18* *Plot of mean lifetime τ versus Ti concentration [at%] in Ta in 1 µm wide lines. Reprinted from* [80] *with permission, copyright (2010), American Institute of Physics*

size of CVD or PVD Cu may be limited by the Cu film thickness. The measured EM activation energy of 0.8 eV from CVD Cu is consistent with that obtained by GB diffusion using radioactive isotopes [86] and by EM experiments performed with polycrystalline ECD Cu lines. The results will be discussed further in Section 9.4.6.

### 9.4.5 Cu Alloys

The effect of impurities Co, In, Mg, Sn, Al, or Ti on Cu EM in Cu lines is reviewed in this section. A Cu alloy line can be obtained by using either a thin film layer of impurity or an alloy liner. The redistribution of impurities [80, 87] from Cu alloy seed, electroless selective chemical deposition CoWP on top of the Cu surface, or Ta(Ti) liner into Cu GBs and interfaces was performed by annealing samples at 400 °C for 2 h before the EM test. The diffusion length along GBs was estimated to be > 100 µm from this annealing step.

*In, Sn, Mg Impurities and Ta(Ti) Liner*

Cu lines with an In or Sn impurity have been shown to have an improved Cu EM lifetime and reduced void growth rate measured by in situ [88] and ex situ [89] SEM techniques as compared to pure Cu lines. Figure 9.18 shows a plot of a mean lifetime as a function of Ti concentration in a Ta liner in 1 µm wide polycrystalline lift-off line structures at a sample temperature of 375 °C [80]. The EM enhancement progressed as Ti and Sn concentrations increased. These results suggest that the solutes Sn, In, or Ti decrease Cu diffusivities in grain boundaries and at free surfaces. These observed EM behaviors are similar to Pd in Cu(Pd) [90]. In contrast, the impurity Mg was shown to degrade the Cu lifetime of the polycrystalline lines. The combination of a small Cu grain size in PVD Cu(Mg) and the lack

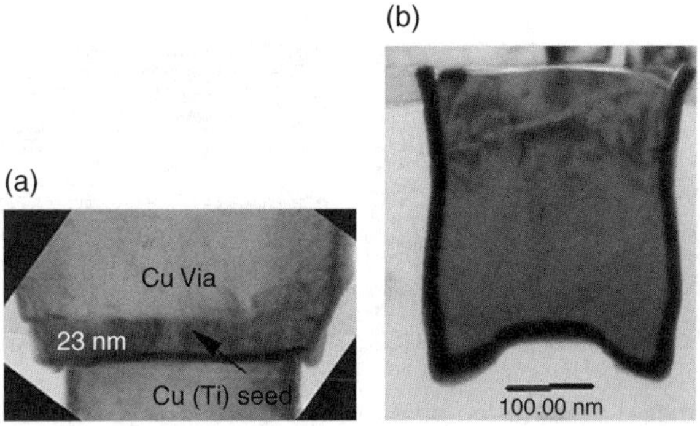


***Figure 9.19*** *Cross-section TEM images of (a) Cu via/Cu line with fine-grain Cu(Ti) alloy seed layer and (b) Cu(Ti) line after 400 °C anneal. Reprinted from* [6] *with permission, copyright (2007), American Institute of Physics*

of a pinning effect of Cu by Mg along the grain boundaries resulted in enhanced Cu polycrystalline line damage [91]. The observation of reduced Cu grain boundary and surface diffusion during EM in Cu(Sn), Cu(In), and Cu(Ti) alloys can be qualitatively interpreted in terms of the solute reducing the grain boundary and/or surface energy [92] and/or acting as a trapping site [93, 94] for Cu. The number of free Cu atoms or vacancies is drastically reduced in the fast paths because of the solute–Cu atom or solute–vacancy interactions, which depend on the diffusion mechanisms in the grain boundaries and interfaces.

*Cu(Ti) Alloy Seed*

The cases of a Cu(0.5 wt% Ti) and a Cu(2.5 wt% Ti) seed layer in 0.22 μm wide dual-damascene lines are also discussed in this section. The initial volume fraction occupied by the Cu alloy seed in the metal line was about 30 %. Comparing to pure Cu lines, the line resistance of Cu(Ti) alloy interconnects, after annealing at 400 °C for a couple hours, was found to increase by 5 % for the Cu(0.5 wt% Ti) and 17 % for the Cu(2.5 wt% Ti) alloy seeds. However, no EM enhancement was observed in the Cu line using Cu(0.5 wt% Ti) seed. Figures 9.19(a) and (b) show TEM cross-sectional images of the Cu via/Cu line and Cu line, respectively, from wafers with the PVD Cu(Ti) alloy seed [6]. The wafers were annealed at 100 °C after Cu electroplating and before CMP. The initial fine-grain structure of the pure PVD Cu seed layer was usually dissolved in the large electroplated Cu grain during abnormal grain growth, which can occur at room temperature or at 100 °C. However, Figure 9.19(a) shows that the Cu(Ti) seed layer covering the bottom and sidewalls of the via did not convert to have large grains. The fine grains of the Cu(Ti) seed in contact with a single-crystal electroplated Cu grain are shown. This indicates that the impurity Ti pins the PVD Cu grain boundary movement and grain growth. The fine Cu(Ti) grains disappeared into the large plated Cu grain after 400 °C annealing, as shown in Figure 9.19(b) [6].

The Cu(2.5 wt% Ti) seed layer was used in this case to replace the pure Cu seed on the line trenches/vias before the ECD Cu processing step. The Ti concentrations in the Cu grains and at the surface were below the detection limit of EDXS, but Ti was detected in

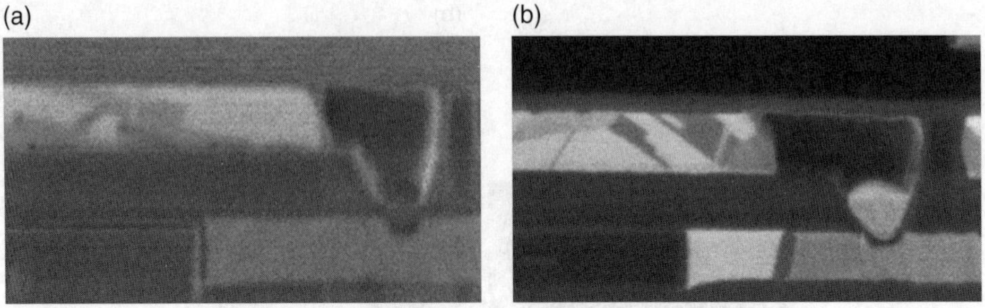


*Figure 9.20* *FIB images of EM tested pure Cu lines cap at 267 °C with total test times of (a) 839 h and (b) 1011 h, respectively. Reprinted from* [6] *with permission, copyright (2007), American Institute of Physics*

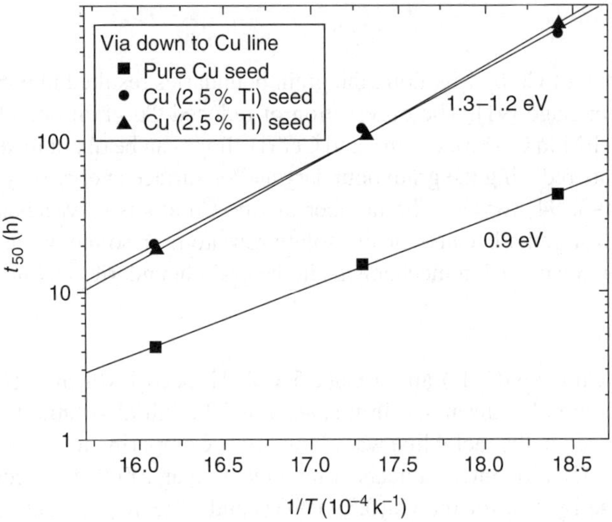


*Figure 9.21* *Plot of* $t_{50}$ *versus (1/T) for pure Cu and Cu(Ti) samples. Reprinted from* [6] *with permission, copyright (2007), American Institute of Physics*

the Ta liner, indicating that the majority of the Ti impurity was observed in the Ta liner and not in Cu. This fact suggested one possible reason why no EM enhancement was observed in the sample using Cu(0.5 wt% Ti) seed. Figures 9.20(a) and (b) are FIB images of pure dual-damascene Cu failed lines [6]. EM-induced void formation was observed in the line and/or at the via, as shown in Figure 9.20, where the direction of electron flow in the samples was from the Cu via upstream to the Cu line. A median lifetime of 870 h and deviation ($\sigma$) of 0.4 were measured. However, no failure was observed in Cu(Ti) alloy lines after 2600 h. Actual Cu(Ti) line resistance decreased during the test. This suggests that an impurity purification process had taken place. The median lifetime $t_{50}$ obtained with a 3% resistance increase failure criterion is plotted in Figure 9.21 for the Cu and Cu(Ti) damascene lines as a function of $1/T$ [6]. The solid lines are the least squares fits. The

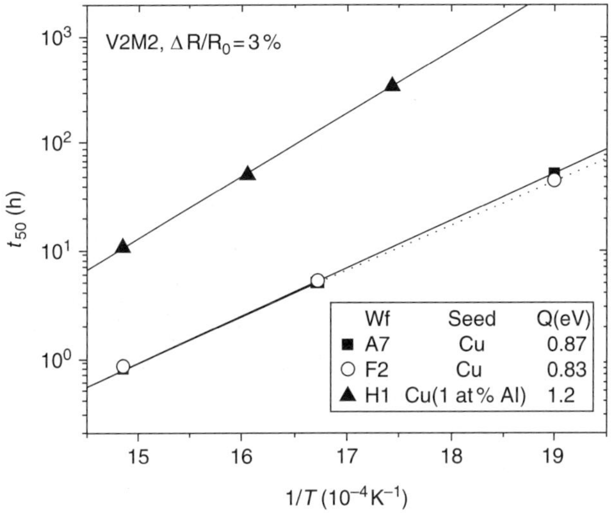


***Figure 9.22*** *Median lifetime versus 1/T for 65 nm wide lines from samples using pure Cu and Cu(Al) seeds. Reprinted from* [80] *with permission, copyright (2010), American Institute of Physics*

values of the activation energy for EM, $Q_{EM}$, in these Cu and Cu(Ti) damascene lines are found to be 0.91 eV and 1.3 eV, respectively, with uncertainties of about 0.05 eV, illustrating an improved chip lifetime for the Cu(2.5 wt% Ti) alloy. These energies are in good agreement with previously reported values [95].

*Near-Bamboo and Polycrystalline Cu(Al) Lines*

To study EM in 65 nm and 1.5 μm wide Cu(Al) lines, a Cu(1 to 2 at% Al) seed layer was used. The Cu(Al) damascene interconnections were produced by replacing the pure Cu seed by a thin Cu(Al) alloy seed deposited on the line trenches or vias before the ECD Cu processing step. A polycrystalline–bamboo microstructure in 65 nm wide lines was obtained using a subsequent post-ECD Cu annealing step at 100 °C. Near-bamboo and polycrystalline-only lines 1.5 μm wide were obtained by processing with and without the 100 °C post-ECD Cu annealing step, respectively, and was followed by CMP right after Cu plating. The median lifetimes $t_{50}$ obtained using a 3 % resistance increase failure criterion are compared for the 65 nm wide Cu damascene lines with a pure Cu seed versus that of a Cu(1 at% Al) seed as a function of *1/T* in Figure 9.22 [80]. Similar Al-enhanced Cu EM lifetime results were also observed in 1.5 μm wide large grain lines with 100 °C annealing after the Cu plating step (data not shown). As compared to pure Cu samples, more than a 10 times enhancement in the EM lifetime was obtained for Cu(Al) at 400 °C, as shown in Figure 9.22. $Q_{EM}$ in these Cu damascene lines can be extracted from the slopes of the lines in Figure 9.22 and are found to be 0.85 ± 0.05 eV and 1.2 ± 0.1 eV for pure Cu and Cu(Al), respectively. These results are in good agreement with the literature [96]. During the course of this investigation, a set of large-grained wafers were exposed to air for a short

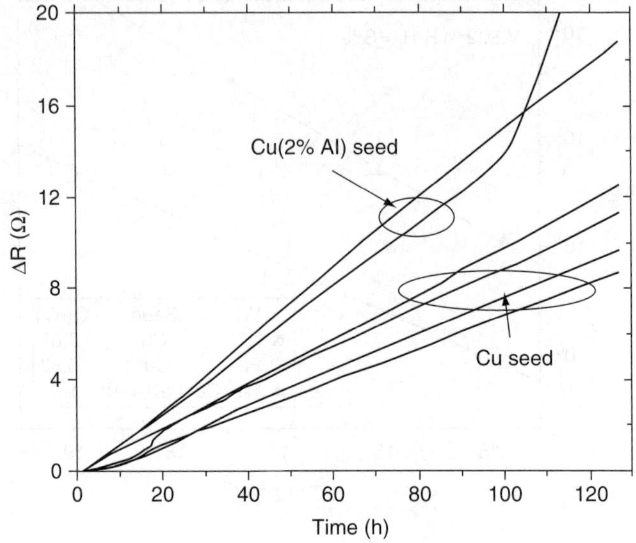


***Figure 9.23*** *Plot of the line resistance change for 1.5 μm wide Cu(Al) and pure Cu lines. Reprinted from* [80] *with permission, copyright (2010), American Institute of Physics*

time while the wafers were still hot. The oxidized Cu(Al) line surfaces were then followed by a plasma pre-clean before depositing the *a*-$SiC_xH_yN_z$ dielectric cap. The Cu lifetimes of these samples were greatly decreased, indicating that a slight Cu(Al) surface oxidation can have a large impact in degrading the Cu/*a*-$SiC_xH_yN_z$ interface.

Figure 9.23 shows the line resistance change versus EM stress time for pure Cu and Cu(2 % Al) seed samples designed for drift velocity measurement [80]. The samples are 1.5 μm wide, 100 μm long, and have a polycrystalline fine-grain microstructure. A faster EM line damage rate was observed in Cu(Al) than in the pure Cu samples. The void lengths in pure Cu and Cu(Al) 100 μm long samples were 15 μm and 23 μm, respectively. A similar EM behavior was also observed in Cu(1 % Al) seed samples. These observations suggest that Al enhanced Cu grain boundary diffusion and increased the EM-induced void growth rate. The observed behavior was consistent with the reported literature [97], suggesting that Al mitigated the Cu migration along the Cu interfaces but not along grain boundaries. The fact that enhanced Cu lifetimes in Cu(Al) were obtained in 65 nm bamboo–polycrystalline and 1.5 μm wide near-bamboo Cu(Al) lines but not 1.5 μm wide polycrystalline-only fine-grained lines indicated that the existence of bamboo grains plays a key role in slowing down the void growth rate. Similar observations were also found in CoWP wafers and will be presented in the next section. We believe, as previously discussed, that the bamboo grains act as Cu diffusion blocking boundaries for GB mass flow and generate a mechanical back-flow stress to reduce the EM GB mass flow. Bamboo grains were observed using cross-sectional TEM within a 1 μm length from the polycrystalline grains along the 65 nm wide lines. A sub-1 μm polycrystalline length section is shorter than the estimated Blech critical length of 10 μm in Cu [1]. This result is also consistent with our observations of no EM line damage in 5 μm or 10 μm long lines. Since future Cu interconnects may have longer polycrystalline line sections because Cu grain growth is pinned by the line size effect, the advantage of using Al in Cu for EM enhancement could be diminished. The EM behavior in Cu alloy structures with varying polycrystalline line lengths and number of bamboo grains remains an important and interesting topic to be studied.

### 9.4.6 CoWP Cap Near-Bamboo and Polycrystalline Cu Lines

An improvement of EM lifetime has been reported for Cu lines with a thin (3 nm to 20 nm) surface layer of electroless CoWP on top of the Cu surface [15]. The measured EM activation energy in a bamboo-like Cu damascene line with CoWP was found to be close to that of Cu bulk diffusion [15]. However, the effect of Co in Cu GB EM has not been extensively studied. Cu movement in GBs is of increasing importance in Cu interconnects since there are more polycrystalline sections in Cu nanowiring. To investigate the effect of Co in Cu GB EM, a 2 μm wide fine-grained polycrystalline Cu line capped with CoWP was used. The polycrystalline damascene lines were fabricated by skipping post-ECD Cu annealing and with the CMP step performed directly after Cu ECD. Co distribution in the Cu grain boundaries was achieved by a high-temperature $a$-$SiC_xN_yH_z$ deposition step and a final 400 °C anneal for 2 h.

Figure 9.24(a) is a plan-view TEM image of a 2 μm wide Cu polycrystalline damascene line [80]. The hand-traced lines on the graph are used to illustrate the location of Cu grain boundaries. Fine-grain sizes and a polycrystalline line structure with a network of GB paths are clearly shown. Figures 9.24(b) and (c) show the plots of the cumulative failure probability versus Cu lifetime and median lifetime versus $1/T$ for the samples with and without a CoWP cap, respectively [80]. Unlike $Q_{EM}$ of 2 eV observed in Cu bamboo-like grains capped with CoWP [15], the EM activation energies were found to be $0.79 \pm 0.05$ eV and $0.72 \pm 0.05$ eV from samples with and without a CoWP cap, respectively. The measured $Q_{EM}$ of 0.79 eV is in good agreement with the activation energy of Cu grain boundary diffusion [86]. Slightly reduced EM lifetimes and $Q_{EM}$ in Cu lines with a CoWP cap suggest that the Co addition in Cu grain boundaries slightly enhanced Cu GB diffusion and hence reduced the EM lifetime.

The selective electroless CoWP capping process can result in a greatly improved Cu lifetime in the bamboo–polycrystalline mixed line structure, although Co did not mitigate Cu GB diffusion, as shown in Figure 9.25 [98]. It shows the 70 nm wide M1 line resistance as a function of time under testing with a 0.6 mA applied current and at 347 °C. The Cu lines with a CoWP cap can still have more than 100 times improvements in the Cu lifetime as compared to the samples without a metal cap. The small line resistance increase for Cu lines with a CoWP cap seen in Figure 9.25 was due to the diffusion and low solubility of Co in Cu [15]. This enhancement was believed to be due to the CoWP capping on Cu lines with mixed bamboo and polycrystalline sections, and the bamboo section completely shut down the Cu fast diffusion. The polycrystalline section is less than 1 μm long in 45 nm node interconnects. As discussed in the case of Cu(Al), short Cu polycrystalline sections with two blocking bamboo sections on each end generated a short length effect similar to the Blech effect. This makes CoWP more effective than Al in slowing down the Cu top surface diffusion to yield a very long lifetime.

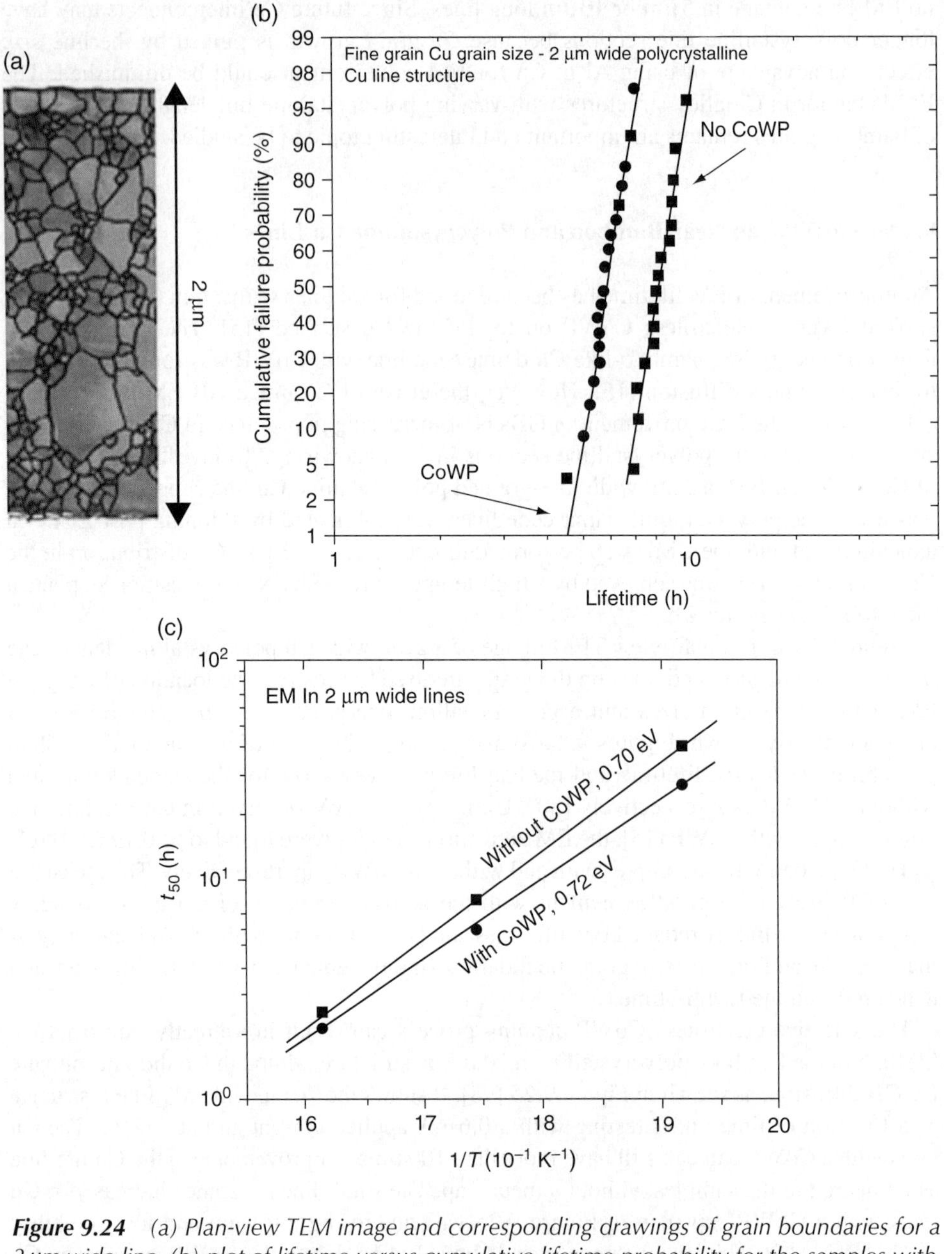


***Figure 9.24*** *(a) Plan-view TEM image and corresponding drawings of grain boundaries for a 2 μm wide line, (b) plot of lifetime versus cumulative lifetime probability for the samples with and without a CoWP cap, and (c) plot of* $t_{50}$ *versus 1/*T *with and without a CoWP cap on the polycrystalline lines. Reprinted from* [80] *with permission, copyright (2010), American Institute of Physics*

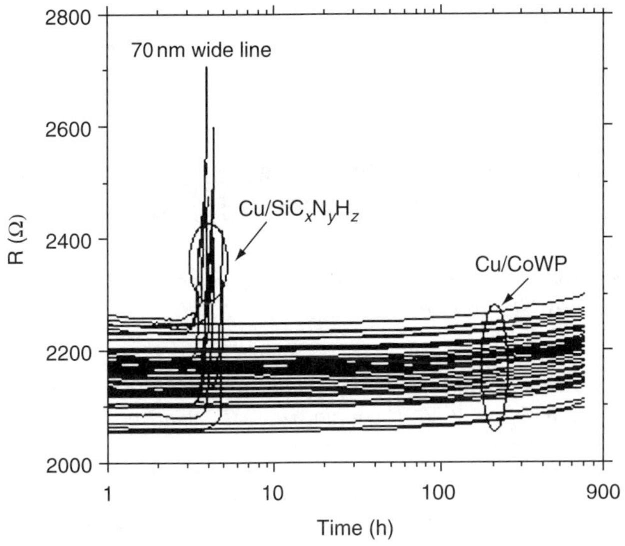


***Figure 9.25*** *Cu test line resistance as a function of time for lines with and without a CoWP cap. Reprinted from* [98] *with permission, copyright (2007), Materials Research Society, Cambridge University Press*

## 9.5 Effect of Grain Size on EM Lifetime and Statistics

To study the microstructure effects, we investigate the effect of grain size on EM lifetime and failure modes for a standard SiCN cap and a CoWP metal cap. The test structure used for this study was a single-linked EM structure consisting of three-level interconnects. The dimensions of the M2 test line were 72 nm wide, 144 nm thick, and 200 μm long. The M1 and M3 feeder lines were made wider to minimize the EM failure probability in these lines. The test structures were fabricated using the standard damascene process. The Cu lines and vias were surrounded by Ta-based liners at the bottom and the sidewalls, and were coated with a low-*k* dielectric SiCN cap on the top. For the M2 test line, two different grain sizes, large grain (LG) and small grain (SG), were obtained by changing the process parameters during electroplating. Two types of cap layers, SiCN with and without a CoWP layer, were applied to the M2 Cu line. In this way, four sets of test structures were obtained and investigated: Cu lines of LG/SG structures with SiCN/CoWP caps.

The TEM images of the M2 line with LG and SG structures are shown in Figure 9.26 and the results of grain size analysis are summarized in Table 9.1. The results show no observation of a bamboo-like grain structure and a clear difference in these two types of grain structures. The LG structure has a larger average grain size at the trench top (~215 nm) than at the trench bottom (~181 nm), which is consistent with the previous results reporting [99] more small-grain agglomerations at the trench bottom in dual-damascene Cu lines narrower than 90 nm. In comparison, the average grain sizes at the

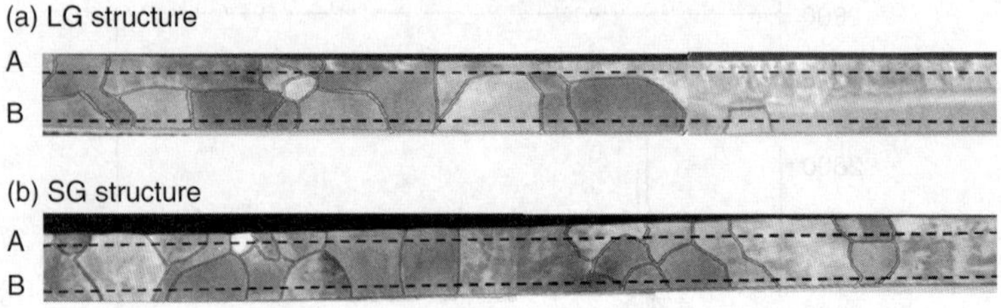


*Figure 9.26* Cross-sectional TEM images of Cu interconnects with two different grain sizes: (a) LG structure and (b) SG structure

*Table 9.1* Average grain size for large and small grain structures along the lines A and B in Figure 9.26. The grain size was calculated by averaging along the dashed line

| Grain structure | Average grain size along line A (nm) | Average grain size along line B (nm) |
|---|---|---|
| Large grain | 215 | 181 |
| Small grain | 123 | 126 |

trench top (~123 nm) and bottom (~126 nm) of the SG structure are comparable, and both are considerably smaller than those of the LG structure. However, a mixture of bamboo grain and polycrystalline sections was still observed in the LG structure.

The EM results obtained for the four sets of test structures at 330 °C with a downstream current flow of 1.03 MA/cm$^2$ are summarized in Figure 9.27. The cap layer and grain size effects are clearly observed from the EM lifetime distributions. Compared with the SiCN cap, the CoWP capped structures show a significantly improved EM lifetime. For the LG and SG structures, the improvement is over 100 times and ~24 times, respectively. This improvement in EM lifetime for the CoWP cap can be attributed to the reduction in Cu/cap interface diffusion, resulting from the highly ordered crystalline interface between Cu and CoWP [50], with a better bonding strength [100] between Cu and Co as compared with Cu and amorphous SiCN. The difference in the interface crystalline quality was investigated using high-resolution TEM (HR-TEM) analysis along the Cu/SiCN and Cu/CoWP interfaces. The results are shown in Figure 9.28, where the Cu line shows a sharp interface with the SiCN cap with no distinct crystalline features. In contrast, for the Cu/CoWP interface, almost perfect crystalline planes of Cu extend all the way through the CoWP metal cap with no distinguishable interface in between. Thus the Cu/CoWP interface is characterized by highly ordered crystalline structures with good bonding strength to suppress the interface diffusion. The TEM results thus reveal a highly disordered Cu/$SiN_x$ interface in contrast to a highly ordered, dominantly metallic bonding Cu/Co interface. The strong bonding of the Cu/Co interface significantly improves the EM lifetime, a result that is consistent with the results reported previously [50, 101, 102].

The EM lifetimes in Figure 9.27 show a clear grain size effect for both cap layers. For the SiCN cap, the EM lifetime degrades by half when changing from an LG to an SG structure. This effect becomes more significant for the CoWP cap, for which the EM

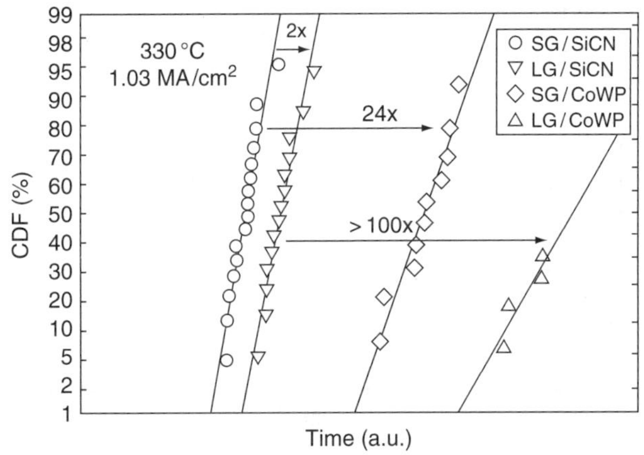


***Figure 9.27*** *Cumulative distribution function (CDF) plots of four sets of Cu interconnects: LG/SG structures with SiCN/CoWP caps. EM tests were performed at 330 °C and 1.03 MA/cm²*

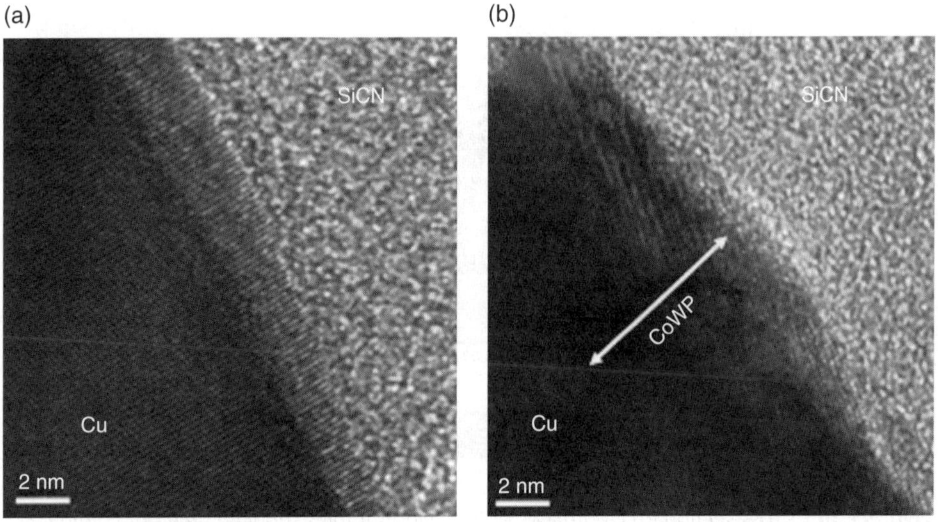


***Figure 9.28*** *HR-TEM images of different Cu/cap interfaces: (a) SiCN cap and (b) CoWP cap*

lifetime degrades to about one-fifth of the LG structure. The degradation in the EM lifetime of the SG structure can be attributed to the additional grain boundary diffusion contribution to the mass transport. Previous studies show that for very weak interfaces, the interface diffusivity is much larger than that of the grain boundary diffusion. In this case, the grain size effect on the EM lifetime can be neglected, even for the polycrystalline structure [103, 104]. However, when the Cu/cap interface strength is improved, the grain size effect becomes more evident. Furthermore, when the cap interface is strengthened by the metal cap, the interface diffusion is markedly suppressed so that it contributes little to the atomic

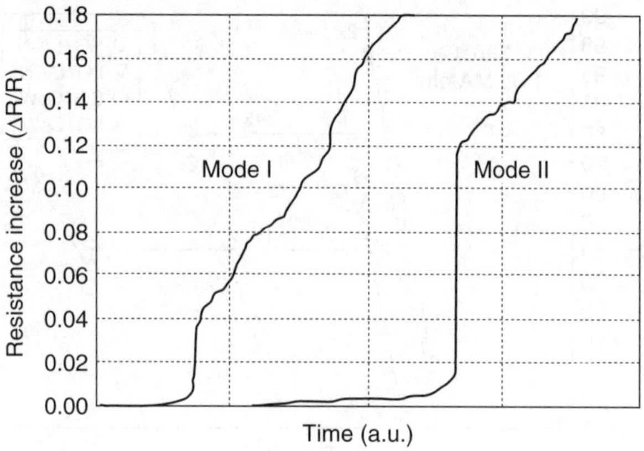


***Figure 9.29*** *Typical resistance traces of mode I and mode II failures in SiCN capped samples: mode I with a small initial resistance step and mode II with a large initial resistance jump*

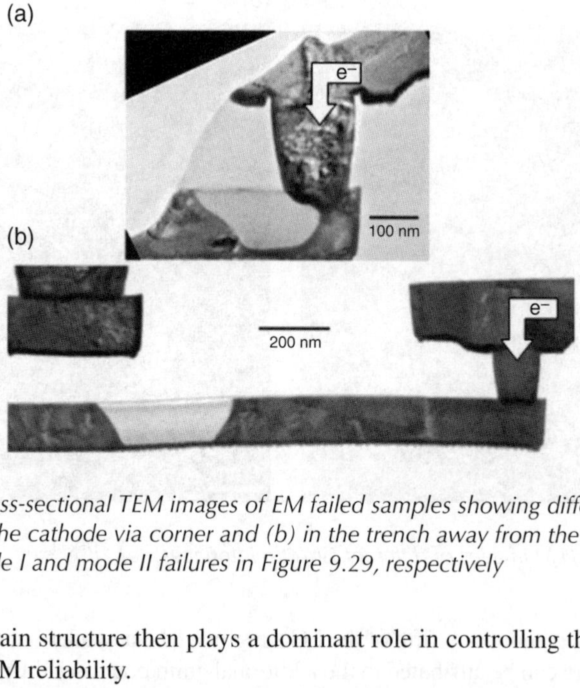


***Figure 9.30*** *Cross-sectional TEM images of EM failed samples showing different voiding locations: (a) at the cathode via corner and (b) in the trench away from the cathode via, representing mode I and mode II failures in Figure 9.29, respectively*

transport. The grain structure then plays a dominant role in controlling the mass transport and in turn the EM reliability.

A postmortem failure analysis was performed using FIB cross-sectioning techniques together with SEM and TEM observations. For SiCN capped samples, two failure modes were observed, as manifested by different forms of initial resistance increase, shown in Figure 9.29. Mode I failure shows a small initial resistance increase followed by a gradual resistance increase. TEM images in Figure 9.30(a) show that mode I failure is due to void formation located at the cathode via corner. This often results from a small critical void size

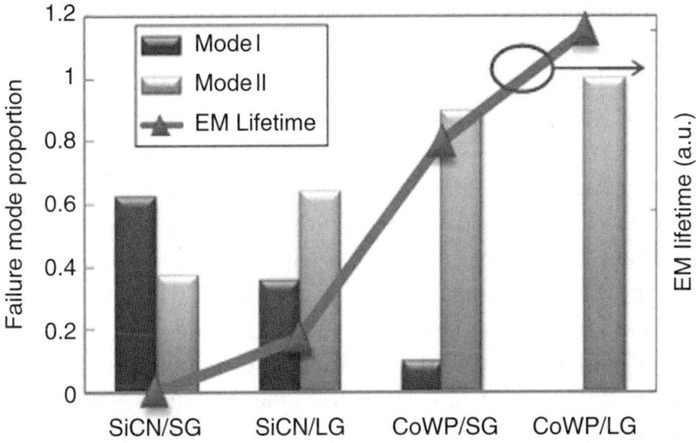


***Figure 9.31*** *Comparison of failure mode proportions and relative EM lifetimes for M2 Cu interconnects with different cap layers and grain sizes*

during EM failure, leading to a small initial resistance step. In comparison, the mode II failure showed a larger initial resistance jump, reflecting a larger critical void size to induce the EM failure. This mode was mostly due to void formation by mass depletion of individual grains in the M2 trench line located several micrometers away from the cathode via, as shown in Figure 9.30(b). Overall, mode II failures are most likely to correspond to a longer EM lifetime as compared with mode I failures.

The different void locations for these two failure modes can be attributed to two factors: one is the flux divergence sites where a vacancy can agglomerate, the other is the effective diffusivity controlling void diffusion in Cu interconnects. During EM, Cu atoms diffuse along the electron flow direction driven by the momentum transfer due to scattering by the electrons. Voids form at flux divergence sites where an unbalanced Cu flow rate exists; i.e. Cu atoms that come from the cathode end of the sites are less than those that diffuse away from the sites. For downstream EM, flux divergence sites can reside at the cathode via corner where the Ta barrier at the via bottom blocks the Cu atoms in the via from diffusing down to the trench line. Flux divergence sites can also occur at interface/grain boundary intersections due to either process defects or the grain texture induced interface diffusivity variations [105]. After voids are nucleated at these flux divergence sites, they will evolve as current stressing continues. In some cases they can diffuse along the Cu line opposite to the electron current flow direction. Voids can change their shape and size during evolution. In other cases, voids will become trapped at these locations and then grow larger until failure occurs. This process of void evolution and growth is significantly affected by the cap interface adhesion strength and Cu grain structures [105].

The failure modes of the four sets of structures are summarized in Figure 9.31 together with their EM lifetimes. The EM lifetime increases from SiCN/SG and SiCN/LG to CoWP/SG and CoWP/LG structures, corresponding to a combination of cap layer and grain structure changing from the weakest case to the strongest case. This is accompanied with a decrease in the proportion of mode I failures and an increase of mode II failures, as shown in Figure 9.31. The latter trend can be understood by considering the effective diffusivity difference in Cu interconnects and by considering contributions from both the cap interface and grain boundary diffusion. These observations are unlike the case of a near-bamboo-like Cu line with a CoWP cap, where bulk diffusion dominates and the EM-induced voids are observed at the cathode end Cu line/Ta liner interface [15]. For the SiCN cap, the cap interface diffusivity is large. Voids can readily move along the interface and eventually accumulate at the cathode via corner to fail the line. For the CoWP cap with significantly suppressed interface diffusion, voids can easily become trapped at locations with either an interface defect or a large grain triple junction, where the voids will grow further, leading to mode II failures. Similarly, compared with the SG structure, the LG structure reduces the grain boundary mass transport and the effective diffusivity for void diffusion. This will lead to void formation due to mass depletion at local grains, consequently making the Cu line more prone to mode II failures. The combination of these two factors, cap material and Cu grain size, shift the failure mode from mode I to mode II, yielding a superior EM performance for the CoWP capped LG structure.

## 9.6 Massive-Scale Statistical Study of EM

An important aspect of the statistical characteristics of the EM phenomena is bimodality found in lifetime distributions. Especially the observation of strong bimodality for the electron up-flow direction in dual-inlaid Cu interconnects has added complexity, but is now widely accepted. More recently, bimodality has also been reported in down-flow EM, leading to very short lifetimes due to small, slit-shaped voids under vias [11, 106, 107]. In addition to shorter lifetimes, these early failure modes can have different activation energy, current exponent, and sigma values compared to the late mode depending on the diffusion and void formation mechanisms. Specific test structures were designed to investigate the early failure phenomena, based on statistical tests using the Wheatstone bridge technique [11, 108]. Using this approach, the resistance change resolution and therefore the sensitivity to detect EM-induced void formation are greatly improved, simultaneously enabling an increase in the number of tested links per device.

In a recent study, statistical test structures were designed with each basic unit containing an array of 230 interconnect segments, providing each Wheatstone bridge circuit with 4 units containing 920 possible failure links. Down-flow EM experiments were performed in a wide temperature range between 150 °C and 325 °C and current densities in the range of 0.15 MA/cm$^2$ and 1.5 MA/cm$^2$. Up-flow tests were performed at 325 °C and 1.5 MA/cm$^2$ [109]. The massive-scale Wheatstone bridge testing encompassed a total number of examined interconnects of 115 920 and 686 400 for up-flow and down-flow directions in 90 nm technology, respectively. In addition, 306 000 down-flow interconnects were tested using 65 nm processing, increasing the total sample size past 1.1 million. Figure 9.32 represents a plot of lifetime distributions for down-flow Wheatstone bridge devices and the corresponding single-link samples tested at 325 °C and 1.5 MA/cm$^2$. The lines through the failure data using a monomodal lognormal fit are for illustrative purposes only. A significant decrease in lifetimes can be discerned for the Wheatstone samples compared to standard single-link devices. This is in agreement with the weakest link

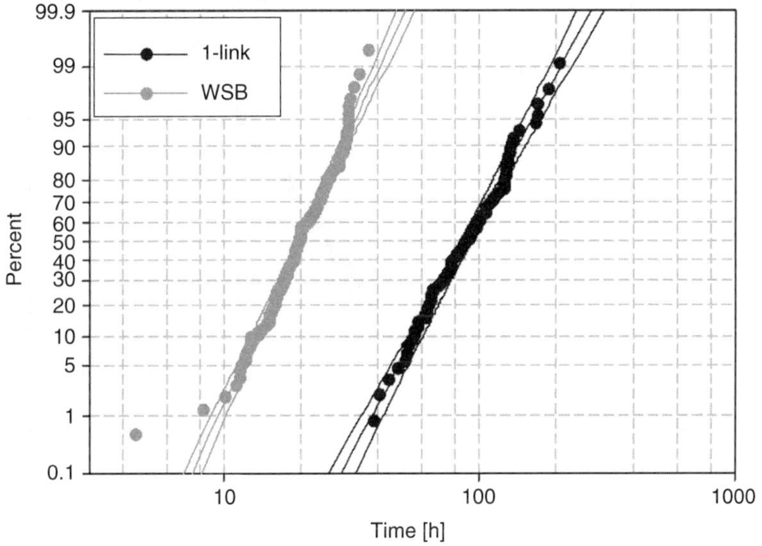


***Figure 9.32*** *Lifetime distributions for down-flow Wheatstone bridge devices and the corresponding single-link samples at 325 °C*

approach, i.e. the failure of the weakest link determines the lifetime of the whole assembly of multiple links.

In order to assess the EM early failure mechanism, the Wheatstone bridge failure data shown in Figure 9.32 need to be deconvoluted with respect to the single interconnect level; i.e. the limited information about the nonfailed links in the EM failure distribution needs to be included to enable a direct comparison between Wheatstone and single-link structures. The statistical deconvolution is accomplished using conditional reliabilities in conjunction with censored data analysis [11]. As a result, each failure time is newly associated with a cumulative failure percentage. Deconvoluted lifetime data for down-flow Wheatstone bridges tested at 325 °C and 1.5 MA/cm$^2$ are shown in Figure 9.33 together with single-link data. Note that the lifetimes are identical to the ones depicted in Figure 9.32, only the probabilities are different. The deconvoluted Wheatstone bridge data and the single link results align well at the transition point. However, it is evident that the majority of the Wheatstone bridge data deviate from the monomodal behavior, which is represented by the straight line fit through the single-link data. This observation clearly indicates the existence of an early failure mechanism for approximately 0.1 % of down-flow samples. A failure analysis using focused ion beam (FIB) cutting and scanning electron microscope (SEM) imaging was performed on single-link samples with short and median lifetimes. Two basic void locations and shapes were identified [11], which are consistent with previous studies [106, 107]. Early failure samples show slit voids under the via in the metal 1 line. In contrast, samples with a median lifetime have faceted voids in the metal 1 line under the via or adjacent to the via. This observation supports the existence of two distinct failure mechanisms. Void formation either starts under the via possibly at a defect location induced

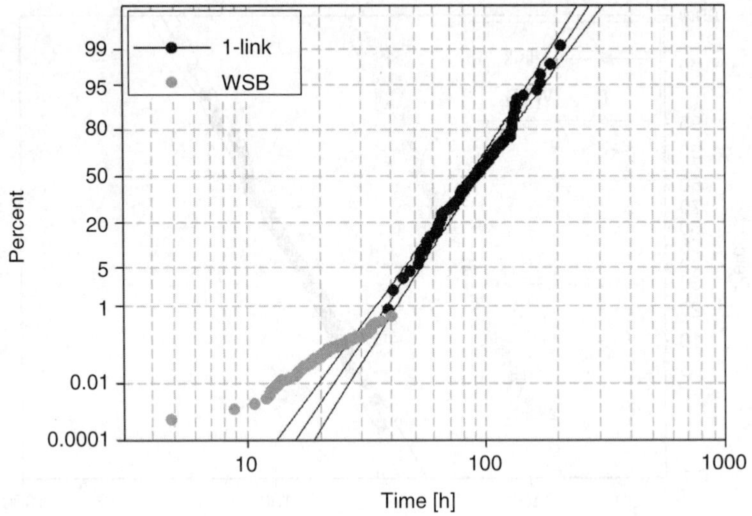


***Figure 9.33*** *Deconvoluted lifetime data for down-flow Wheatstone bridge devices and the corresponding single-link data at 325 °C*

during via processing, such as via etch and cleaning steps, leading to faster failure, or voids form at the interface away from the via and evolve toward the line end. It is important to note that a distinct separation of an early and a late failure mode was not possible using only the single-link data due to the small number of early failures. However, a detailed characterization of these failures is essential, since the early failures determine the reliability of the entire product.

The early failure population can display different characteristics in terms of the lognormal $\sigma$, activation energy, and current density exponent [11]. Lifetime distributions of down-flow Wheatstone bridge devices were measured as a function of temperature in the range of 150 °C to 325 °C and at 1.5 MA/cm$^2$. The lifetime distributions and sigma values were found to be independent of temperature, indicating no change in the failure mechanisms. In general, the early failures have a higher sigma value, and thus EM failures at the device level are induced by the formation of slit-like voids under the via. A possible reason for a void initiating directly under the via is defectivity at the via bottom induced during via processing, such as via etch and cleaning steps.

With the continuing scaling of interconnects, EM lifetimes in single-link tests were found to decrease, even if the current density was kept constant, as mentioned earlier. Both the critical void length and the metal height influence this phenomenon. To investigate the early failure scaling behavior, Wheatstone bridges were designed for 65 nm technology and compared to the 90 nm samples. The scaling factors for the line width and height were about 0.75 and 0.6, respectively. Lifetime distributions from down-flow Wheatstone Bridge tests for both generations are shown in Figure 9.34(a). The test structures used for 65 nm samples were designed differently compared to the 90 nm structures. Each test circuit contains 2448 interconnects compared to 920 in the case of 90 nm technology. Thus, deconvolution of the data is necessary to compare the EM performance adequately. Figure 9.34(b)

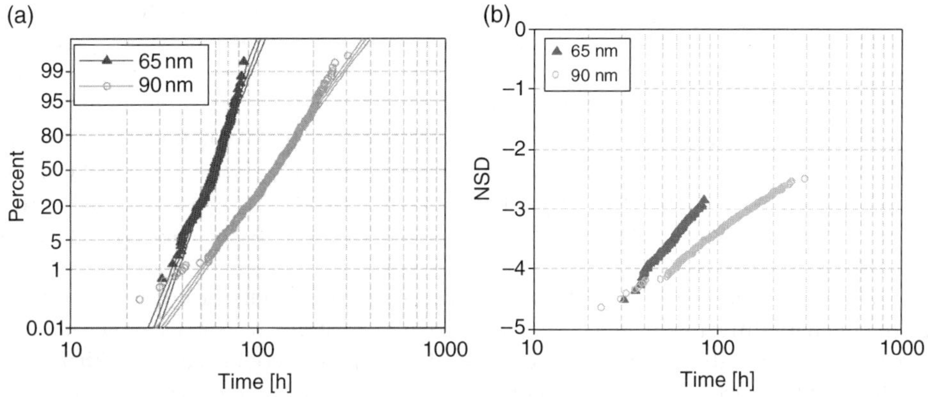


*Figure 9.34* *Lifetime distributions for down-flow Wheatstone bridge devices for 90 nm and 65 nm technologies at 325 °C: (a) raw data and (b) deconvoluted data (NSD = Number of standard Deviation)*

displays the deconvoluted lifetime distributions for both data sets. The ratio between the 65 nm and 90 nm lifetimes is ~0.5, which is in good agreement with the dimensional reductions, considering that the via sizes were scaled by about 0.77 in this particular case. It is interesting to note that the lifetimes of the 65 nm interconnect show a tighter distribution compared to the 90 nm test structures. It is possible that differences in tools and processes between the two technologies led to a cleaner interface between the via and metal interconnect, thus reducing the risk for very early failures. Additional studies are needed for a further assessment of this possibility.

## 9.7 Summary

This paper reviews the effects of scaling on the microstructure and EM performance for Cu damascene interconnects. The grain structure in Cu interconnects varies from near-bamboo-like, to polycrystalline mixing with single-crystal bamboo-like grains, to polycrystalline-only fine grains. For Cu microstructure characterization, several techniques are available, including X-ray diffraction (XRD), electron backscatter diffraction (EBSD) in the scanning electron microscope (SEM), or orientation imaging microscopy (OIM) in the transmission electron microscope (TEM). Using test patterns with Cu interconnect lines, texture analysis can be performed by X-ray pole figure analysis in the laboratory, which has been considerably improved by the introduction of area detectors. However, the resolution of lab-based X-ray analysis is limited to the order of micrometers, making it difficult to determine the location and orientation of individual Cu crystallites. For this purpose, the application of high-brightness synchrotron radiation along with improved X-ray focusing optics is needed. As Cu interconnect line dimensions are reduced to significantly smaller than 100 nm, synchrotron-based microdiffraction may hit its limits in the near future. A breakthrough could be achieved by a combination of third-generation synchrotron sources with novel high-throughput focusing devices and highly sensitive area detectors, which are potentially energy resolving. Electron backscatter diffraction is usually employed in a scanning electron microscope to determine individual grain orientations, local texture, and point-to-point orientation correlations on polycrystalline surfaces of bulk materials. Since Cu interconnects in a 45 nm CMOS technology node and beyond are characterized by a mixture of large grains and sections of multiple small grains with grain diameters significantly smaller than 50 nm, the SEM-based technique is no longer suitable for comprehensive Cu microstructure analysis. An appropriate alternative is given by the application of TEM-based methods, such as the conical dark-field technique.

The mass transport in Cu interconnects occurs mainly by interface diffusion and/or grain boundary diffusion, which depend on the Cu microstructure. The Cu atoms migrate along these fast diffusion paths, whereupon diffusion is accompanied by either edge displacement or surface grain thinning mechanisms to form the voids. However, in some cases, the void is observed to form at a distance away from the cathode end of the line and then moves to the cathode end of the line/via, or becomes trapped at a certain favorable location. After that void movement has mostly ceased, continuous void growth leads to a larger void size and finally results in line failure.

Upper dummy via structure samples are shown to interrupt the Cu mass flow along the top interface and result in lifetime enhancement without increasing Cu resistivity. For dummy via structures, wide lines with a near-bamboo microstructure show increased EM enhancement compared to fine lines with a larger number of polycrystalline grain sections. The upper-level dummy via structure also proved to be a powerful tool for helping to understand the Cu microstructure and to distinguish different fast diffusion paths in the line.

The Cu void growth rate at the cathode end of the line with a redundant liner is almost linearly proportional to the inverse of the current density. No difference is observed between high and low nonmetallic impurities of Cl, C, S, and O on Cu lifetimes, suggesting that the microstructure plays a key role in the superiority of ECD Cu compared to CVD or PVD Cu. The near-bamboo-like ECD Cu is characterized by interface diffusion with $Q_{EM} = 1\,\text{eV}$ and polycrystalline CVD Cu by grain boundary diffusion with $Q_{EM} = 0.8\,\text{eV}$. The fast diffusion in Cu damascene lines can be greatly reduced with the addition of metallic impurities, such as Ti, Al, and Sn. The Ta(Ti) liner also improves polycrystalline PVD Cu line lifetimes with increased $Q_{EM}$ of Cu grain boundaries from 0.8 eV to 1 eV. In and Sn impurities within Cu also lead to increased Cu lifetimes in the polycrystalline line. Using Cu(1 at% Al) or Cu(2.5 wt% Ti) seed layers for the standard 65 nm and 250 nm wide damascene lines enhanced EM lifetimes and increased $Q_{EM}$ from 0.9 eV to 1.15 eV and 0.9 eV to 1.3 eV, respectively. The study of 1.5 μm to 2 μm wide polycrystalline-only lines with Al and Co yielded reduced lifetimes and an increased void growth rate, suggesting that Al and Co mitigate Cu interface diffusion but not grain boundary diffusion. The observed Cu lifetime enhancements for 65 nm wide Cu(Al) lines and pure Cu lines with a CoWP cap is due to the presence of bamboo grains in the bamboo–polycrystalline line structures. According to TEM analyses, the average bamboo grain separation is $< 1\,\mu\text{m}$ along the 65 nm wide lines. It is proposed that Al or Co on top of bamboo grain surfaces act as Cu diffusion blocking boundaries for Cu grain boundary mass flow, resulting in a similar effect to the 'Blech short length'. The alloying process can produce lines with very long EM lifetimes but at the expense of metal conductivity. Hydrogen-based plasma precleans prior to Cu passivation with a silicon carbon nitride is also found to mitigate EM failure in Cu interconnects, but, as in the case of alloys, there is a trade-off between reliability and performance. All these techniques increase the complexity of wafer processing. Further techniques for improving a Cu nanowire microstructure to increase the number of bamboo-like grain sections as well as EM performance could be critical for the development of 14 nm and sub-14 nm technologies.

The study of the grain size effect on EM in a Cu line with the CoWP cap demonstrates that, with the suppression of interfacial diffusion by the CoWP cap, the grain boundary diffusion becomes important in controlling the mass transport. As a result, a small change in the grain structure can yield a large change in the overall Cu diffusivity as well as the EM lifetime. Nevertheless, the benefit of the CoWP cap over the SiCN cap remains significant, even for the 32 nm technology node. The use of the Wheatstone bridge technique was shown to be capable of increasing the tested sample size well past standard single-link test capabilities. The number of tested samples used encompassed a total of 115 920 and 414 000 for up-flow and down-flow directions, respectively, for the 90 nm technology node. Additionally, 144 432 down-flow interconnects were tested using 65 nm processing. The test results clearly identify early failure modes in EM lifetime distributions. This improves the reliability of the extrapolation to the operating temperature using test results from the down-flow early mode. The need for using this methodology for testing EM reliability for the 22 nm technology node and beyond is demonstrated.

## Acknowledgments

Part of the work for this chapter was performed by Research Alliance Teams at various IBM Research and Development Facilities. Two of the authors, L.J. Zhang and P.S. Ho, gratefully acknowledge the research support from the Semiconductor Research Corporation and the collaboration with GLOBALFOUNDRIES.

## References

[1] Hu, C.-K., Gignac, L., Rosenberg, R. (2006) Electromigration of Cu/low dielectric constant interconnects. *Microelectron. Reliab.*, **46**, 213–231.

[2] Chen, F. and Gardner, D. (1998) Influence of line dimensions on the resistance of Cu interconnections. *IEEE Electron Device Lett.*, **19**, 508–510.

[3] Steinhögl, W., Schindler, G., Steinlesberger, G. and Engelhardt, M. (2002) Size-dependent resistivity of metallic wires in the mesoscopic range. *Phys. Rev. B*, **66**, 075414.

[4] Sun, T., Yao, B., Warren, A.P., Barmak, K., Toney, M.F., Peale, R.E. and Coffey, K.R. (2009) Dominant role of grain boundary scattering in the resistivity of nanometric Cu films. *Phys. Rev. B*, **79**, 041402(R).

[5] Bunge, H.J. and Schwarzer, R.A. (2001) Orientation stereology – a new branch in texture research. *Adv. Engng Mater.*, **3**, 25–39.

[6] Hu, C.-K., Gignac, L.M., Baker, B., Liniger, E., Yu, R., Flaitz, P. and Stamper, A.K. (2007) Electromigration reliability of advanced interconnects. *AIP Conf. Proc.*, **945**, 27–41.
[7] Hu, C.-K., Gignac, L.M., Liniger, E., Huang, E., Greco, S., McLaughlin, P., Yang, C.-C. and Demarest, J.J. (2009) Electromigration challenges for nanoscale Cu wiring. *AIP Conf. Proc.*, **1143**, 3–11.
[8] Zhang, L., Im, J. and Ho, P.S. (2009) Line scaling effect on grain structure for Cu interconnects. *AIP Conf. Proc.*, **1143**, 151–155.
[9] Maitrejean, S., Carreau, V., Thomas, O., Labat, S., Kaouache, B., Verdier, M., Lepinoux, J., Bréchet, Y., Legros, M., Douin, J., Brandstetter, S., Cayron, C., Sicardy, O., Weygand, D., Dubreuil, O. and Normandon, P. (2009) Cu grain growth in damascene narrow trenches. *AIP Conf. Proc.*, **1143**, 135–150.
[10] Hu, C.-K., Canaperi, D., Chen, S.T., Gignac, L.M., Herbst, B., Kaldor, S., Krishnan, M., Liniger, E., Rath, D.L., Restaino, D., Rosenberg, R., Rubino, J., Seo, S.-C., Simon, A., Smith, S. and Tseng, W.-T. (2004) Effects of overlayers on electromigration reliability improvement for Cu/low-*k* interconnects. Proceedings of the IEEE International Reliability Physics Symposium, IRPS, pp. 222–228.
[11] Hauschildt, M., Gall, M., Justison, P., Hernandez, R. and Herrick, M. (2007) Large-scale statistical study of electromigration early failure for Cu/low-*k* interconnects. *AIP Conf. Proc.*, **945**, 66–81 (2007).
[12] Gall, M. (1999) Investigation of electromigration reliability in Al(Cu) interconnects. PhD dissertation, The University of Texas at Austin, Texas.
[13] Gall, M., Capasso, C., Jawarani, D., Hernandez, R., Kawasaki, H. and Ho, P.S. (2001) Statistical analysis of early failures in electromigration. *J. Appl. Phys.*, **90**, 732–740.
[14] Blech, A. (1976) Electromigration in thin aluminum films on titanium nitride. *J. Appl. Phys.*, **47**, 1203–1208.
[15] Hu, C.-K., Gignac, L.M., Rosenberg, R., Herbst, B., Smith, S., Rubino, J., Canaperi, D., Chen, S.T., Seo, S.C. and Restaino, D. (2004) Atom motion of Cu and Co in Cu damascene lines with a CoWP cap. *Appl. Phys. Lett.*, **84**, 4986–4988.
[16] Hu, C.-K., Gignac, L. and Rosenberg, R. (2005) Electromigration in Cu thin film, in *Diffusion Processes in Advanced Technical Materials* (ed. D. Gupta), William Andrew, Inc., Norwich, New York.
[17] Hu, C.-K. (1998) Reliability and copper interconnections with low dielectric constant materials. *Mater. Res. Soc. Symp. Proc.*, **511**, 305–316.
[18] Lee, H. and Lopatin, S.D. (2005) The influence of barrier types on the microstructure and electromigration characteristics of electroplated copper. *Thin Solid Films*, **492**, 279–284.
[19] Schwarzer, R.A. (1998) Crystallography and microstructure of thin films studied by X-ray and electron diffraction. *Mater. Sci. Forum*, **287–288**, 23–60.
[20] Hübner, R. (2007) Effect of a Ta–Si–N diffusion barrier on the texture formation in thin Cu films. *J. Appl. Phys.*, **101**, 093512.
[21] Rosenberg, R., Edelstein, D.C., Hu, C.-K. and Rodbell, K.P. (2000) Copper metallization for high performance silicon technology. *Annu. Rev. Mater. Sci.*, **30**, 229–262.
[22] Pantleon, K. and Somers, M.A.J. (2006) In situ investigation of the microstructure evolution in nanocrystalline copper electrodeposits at room temperature. *J. Appl. Phys.*, **100**, 114319.
[23] Pantleon, K. and Somers, M.A.J. (2007) Evolution of the microstructure in nanocrystalline copper electrodeposits during room temperature storage. *Z. Kristallogr. Suppl.*, **26**, 261–266.
[24] Hecker, M. (2003) X-ray scattering techniques, in *Metal Based Thin Films for Electronics* (eds K. Wetzig and C.M. Schneider), Wiley-VCH GmbH & Co. KGaA, Weinheim, Germany.
[25] Zschech, E., Blum, W., Zienert, I. and Besser, P.R. (2001) Effect of Cu line geometry and process parameters on interconnect microstructure and degradation processes. *Z. Metallkd.*, **92**, 803–809.
[26] Budai, J.D., Liu, W., Tischler, J.Z., Pan, Z.W., Norton, D.P., Larson, B.C., Yang, W. and Ice, G.E. (2008) Polychromatic X-ray micro- and nanodiffraction for spatially-resolved structural studies. *Thin Solid Films*, **516**, 8013–8021.

[27] MacDowell, A.A., Celestre, R.S., Tamura, N., Spolenak, R., Valek, B., Brown, W.L., Bravman, J.C., Padmore, H.A., Batterman, B.W. and Patel, J.R. (2001) Submicron X-ray diffraction. *Nucl. Instrum. Meth. Phys. Res. A*, **467–468**, 936–943.

[28] Chung, J.-S. and Ice, G.E. (1999) Automated indexing for texture and strain measurement with broad-bandpass X-ray microbeams. *J. Appl. Phys.*, **86**, 5249–5255.

[29] Tamura, N., MacDowell, A.A., Spolenak, R., Valek, B.C., Bravman, J.C., Brown, W.L., Celestre, R.S., Padmore, H.A., Batterman, B.W. and Patel, J.R. (2003) Scanning X-ray microdiffraction with submicrometer white beam for strain/stress and orientation mapping in thin films. *J. Synchrotron Rad.*, **10**, 137–143.

[30] Spolenak, R., Barr, D.L., Gross, M.E., Evans-Lutterodt, K., Brown, W.L., Tamura, N., MacDowell, A.A., Celestre, R.S., Padmore, H.A., Valek, B.C., Bravman, J.C., Flinn, P., Marieb, T., Keller, R.R., Batterman, B.W. and Patel, J.R. (2001) Microtexture and strain in electroplated copper interconnects. *Mater. Res. Soc. Symp. Proc.*, **612**, 1–7.

[31] Spolenak, R., Tamura, N. and Patel, J.R. (2006) X-ray microdiffraction as a probe to reveal flux divergences in interconnects. *AIP Conf. Proc.*, **816**, 288–295.

[32] Budiman, A.S., Hau-Riege, C.S., Besser, P.R., Marathe, A., Joo, Y.-C., Tamura, N., Patel, J.R. and Nix, W.D. (2007) Electromigration-induced plasticity and texture in Cu interconnects. *AIP Conf. Proc.*, **945**, 56–65.

[33] Budiman, A.S., Tamura, N., Valek, B.C., Gadre, K., Maiz, J., Spolenak, R., Nix, W.D. and Patel, J.R. (2006) Crystal plasticity in Cu damascene interconnect lines undergoing electromigration as revealed by synchrotron X-ray microdiffraction. *Appl. Phys. Lett.*, **88**, 233515.

[34] Schwarzer, R.A., Field, D.P., Adams, B.L., Kumar, M. and Schwartz, A.J. (2009) Present state of electron backscatter diffraction and prospective developments, in *Electron Backscatter Diffraction in Materials Science* (eds A.J. Schwartz, M. Kumar, B.L. Adams and D.P. Field), Springer.

[35] Meyer, M.A., Zienert, I. and Zschech, E. (2005) Electron backscatter diffraction: application to Cu interconnects in top-view and cross section, in *Materials for Information Technology* (eds E. Zschech, C. Whelan and Th. Mikolajick), Springer.

[36] Geisler, H., Zienert, I., Prinz, H., Meyer, M.-A. and Zschech, E. (2003) Potential and limits of texture measurement techniques for inlaid copper process optimization. *AIP Conf. Proc.*, **683**, 469–479.

[37] Nowell, M.M. (2002) Ion beam preparation of passivated copper integrated circuit structures for electron backscatter diffraction/orientation imaging microscopy analysis. *J. Electron. Mater.*, **31**, 23–32.

[38] Mirpuri, K. and Szpunar, J. (2004) Impact of trench aspect ratio on microstructure variation in as-deposited and annealed damascene Cu interconnect lines. *Micron*, **35**, 575–587.

[39] Jiang, Q.-T., Nowell, M., Foran, B., Frank, A., Havemann, R.H., Parihar, V., Augur, R.A. and Luttmer, J.D. (2002) Analysis of copper grains in damascene trenches after rapid thermal processing or furnace anneals. *J. Electron. Mater.*, **31**, 10–15.

[40] Neuner, J., Zienert, I., Peeva, A., Preuße, A., Kücher, P. and Bartha, J.W. (2010) Microstructure in copper interconnects – influence of plating additive concentration. *Microelectron. Engng*, **87**, 254–257.

[41] Mirpuri, K.J. and Szpunar, J.A. (2005) Influence of substrate on high-temperature behavior of copper film studied in situ by electron backscatter diffraction. *J. Electron. Mater.*, **34**, 1509–1520.

[42] Park, N.-J., Field, D.P., Nowell, M.M. and Besser, P.R. (2005) Effect of film thickness on the evolution of annealing texture in sputtered copper films. *J. Electron. Mater.*, **34**, 1500–1508.

[43] Muppidi, T., Field, D.P., Sanchez Jr, J.E. and Woo, C. (2005) Barrier layer, geometry and alloying effects on the microstructure and texture of electroplated copper thin films and damascene lines. *Thin Solid Films*, **471**, 63–70.

[44] Field, D.P., Muppidi, T. and Sanchez Jr, J.E. (2003) Electron backscatter diffraction characterization of inlaid Cu lines for interconnect applications. *Scanning*, **25**, 309–315.

[45] Geiss, R.H. and Read, D.T. (2007) Need for standardization of EBSD measurements for microstructural characterization of thin film structures. *AIP Conf. Proc.*, **931**, 168–172.

[46] Zschech, E., Ho, P.S., Schmeisser, D., Meyer, M.A., Vairagar, A.V., Schneider, G., Hauschildt, M., Kraatz, M. and Sukharev, V. (2009) Geometry and microstructure effect on EM-induced copper interconnect degradation. *IEEE Trans. Device Mater. Reliab.*, **9**, 20–30.
[47] Schneider, G., Denbeaux, G., Anderson, E.H., Bates, B., Pearson, A., Meyer, M.A., Zschech, E., Hambach, D. and Stach, E.A. (2002) Dynamical X-ray microscopy investigation of electromigration in passivated inlaid Cu interconnect structures. *Appl. Phys. Lett.*, **81**, 2535–2537.
[48] Schneider, G., Guttmann, P., Rudolph, S., Heim, S., Rehbein, S., Meyer, M.A. and Zschech, E. (2006) X-ray microscopy studies of electromigration in advanced copper interconnects. *AIP Conf. Proc.*, **817**, 217–222.
[49] Meyer, M.A., Herrmann, M., Langer, E. and Zschech, E. (2002) In situ SEM observation of electromigration phenomena in fully embedded copper interconnect structures. *Microelectron. Engng*, **64**, 375–382.
[50] Meyer, M.A. and Zschech, E. (2007) New microstructure-related EM degradation and failure mechanisms in Cu interconnects with CoWP coating. *AIP Conf. Proc.*, **945**, 107–114.
[51] Zschech, E., Meyer, M.A. and Langer, E. (2004) Effect of mass transport along interfaces and grain boundaries on copper interconnect degradation. *Mater. Res. Soc. Symp. Proc.*, **812**, 361–372 (2004).
[52] Hu, C.-K., Gignac, L., Rosenberg, R., Liniger, E., Rubino, J., Sambucetti, C., Domenicucci, A., Chen, X. and Stamper, A.K. (2002) Reduced electromigration of Cu wires by surface coating. *Appl. Phys. Lett.*, **81**, 1782–1784.
[53] Hu, C,-K., Canaperi, D., Chen, S.T., Gignac, L.M., Kaldor, S., Krishnan, M., Malhotra, S.G., Liniger, E., Lloyd, J.R., Rath, D.L., Restaino, D., Rosenberg, R., Rubino, J., Seo, S.-C., Simon, A., Smith, S. and Tseng, W.-T. (2006) Electromigration Cu mass flow in Cu interconnections. *Thin Solid Films*, **504**, 274–278.
[54] Meyer, M.A., Grafe, M., Engelmann, H.-J., Langer, E. and Zschech, E. (2006) Investigation of the influence of the local microstructure of copper interconnects on void formation and evolution during electromigration testing. *AIP Conf. Proc.*, **817**, 175–184.
[55] Dingley, D.J. (2006) Orientation imaging microscopy for the transmission electron microscope. *Microchim. Acta*, **155**, 19–29.
[56] Anderoglu, O., Misra, A., Wang, H., Ronning, F., Hundley, M.F. and Zhang, X. (2008) Epitaxial nanotwinned Cu films with high strength and high conductivity. *Appl. Phys. Lett.*, **93**, 083108 (2008).
[57] Wang, G., Jiang, Z., Lian, J. and Jiang, Q. (2009) The grain refinement mechanism of electrodeposited copper. *J. Mater. Res.*, **24**, 3226–3236.
[58] An, J.H. and Ferreira, P.J. (2006) In situ transmission electron microscopy observations of 1.8 μm and 180 nm Cu interconnects under thermal stresses. *Appl. Phys. Lett.*, **89**, 151919.
[59] Brandstetter, S., Carreau, V., Maîtrejean, S., Verdier, M. and Legros, M. (2010) Grain morphology of Cu damascene lines. *Microelectron. Engng*, **87**, 383–386.
[60] Yao, B., Sun, T., Warren, A., Heinrich, H., Barmak, K. and Coffey, K.R. (2010) High contrast hollow-cone dark field transmission electron microscopy for nanocrystalline grain size quantification. *Micron*, **41**, 177–182.
[61] Williams, D.B. and Carter, C.B. (1996) *Transmission Electron Microscopy – A Textbook for Materials Science*, Springer Science + Business Media, LLC, New York.
[62] Zaefferer, S. (2002) Computer-aided crystallographic analysis in the TEM. *Advances in Imaging and Electron Physics*, **125**, 355–415.
[63] Rauch, E.F., Véron, M., Portillo, J., Bultreys, D., Maniette, Y. and Nicolopoulos, S. (2008) Automatic crystal orientation and phase mapping in TEM by precession diffraction. *Microscopy and Analysis*, **22**, S5–S8.
[64] NanoMEGAS: www.nanomegas.com.
[65] Rauch, E.F. and Dupuy, L. (2005) Rapid spot diffraction pattern identification through template matching. *Arch. Metall. Mater.*, **50**, 87–99.
[66] Rauch, E.F. and Veron, M. (2005) Coupled microstructural observations and local texture measurements with an automated crystallographic orientation mapping tool attached to a TEM. *Mat.-wiss. u. Werkstofftech.*, **36**, 552–556.

[67] Vincent, R. and Midgley, P.A. (1994) Double conical beam-rocking system for measurement of integrated electron diffraction intensities. *Ultramicroscopy*, **53**, 271–282.
[68] Hübner, R., Engelmann, H.-J. and Zschech, E. (2010) Small grain and twin characterization in sub-100 nm Cu interconnects using the conical dark-field technique in the transmission electron microscope. Proceedings of the IEEE International Interconnect Technology Conference, IITC, pp. 1–3.
[69] Zschech, E., Huebner, R., Chumakov, D., Aubel, O., Friedrich, D., Guttmann, P., Heim, S. and Schneider, G. (2009) Stress-induced phenomena in nanosized copper interconnect structures studied by X-ray and electron microscopy. *J. Appl. Phys.*, **106**, 093711.
[70] Zschech, E., Hübner, R., Aubel, O. and Ho, P.S. (2010) EM and SM induced degradation dynamics in copper interconnects studied using electron microscopy and X-ray microscopy. Proceedings of the IEEE International Reliability Physics Symposium, IRPS, pp. 574–580.
[71] Hübner, R., Engelmann, H.-J. and Zschech, E. (2010) Conical dark-field analysis for small grain characterization in narrow Cu interconnect structures: potential and challenges. *AIP Conf. Proc.*, **1300**, 39–46.
[72] Hu, C.-K., Small, M.B., Kaufman, F. and Pearson, D.J. (1990) Copper–polyimide wiring technology for VLSI circuit. Workshop on CVD W, Cu and Other Advanced Metals for ULSI/VLSI, September 1989, Materials Research Society, Warrendale, Pennsylvania, pp. 369–373.
[73] Hu, C.-K., Gignac, L.M., Liniger, E., Detavernier, C., Malhotra, S.G. and Simon, A. (2005) Effect of metal liner on electromigration in Cu damascene lines. *J. Appl. Phys.*, **98**, 124501-08.
[74] Huntington, H.B. (1974) Electromigration in metals, in *Diffusion in Solids: Recent Developments* (eds A.S. Nowick and J.J. Burton), Ch. 6, Academic, New York.
[75] Demuynck, S., Tokei, Z.S., Bruynseraede, C., Michelon, J. and Max, K. (2004) Alpha-Ta formation and its impact on electromigration. MRS Proceedings of the Advanced Matellization Conference, AMC 2003 (ULSI XIX), pp. 355–359.
[76] Lin, J.C., Park, S.K., Pfeifer, K., Augur, R., Blaschke, V., Shue, S.L. and Liang, M.S. (2003) Electromigration reliability study of self-ionized plasma barriers for dual damascene Cu metallization. Proceedings of the Advanced Metallization Conference, AMC 2002 (ULSI XVIII), pp. 233–237.
[77] Hu, C.-K., Rosenberg, R., Rathore, H.S., Nguyen, D.B. and Agarwala, B. (1999) Scaling effect on electromigration in on-chp Cu wiring. Proceedings of the IEEE International Interconnect Technology Conference, IITC, pp. 267–269.
[78] Stamper, A.K., Baks, H., Cooney, E., Gignac, L., Gill, J., Hu, C.-K., Kane, T., Liniger, E., Wang, Y.-Y. and Wynne, J. (2006) Damascene copper integration impact on electromigration and stress migration. Proceedings of the Advanced Metallization Conference, AMC 2005 (ULSI XXI), pp. 727–733.
[79] Ryan, E.T., Martin, J., Bonilla, G., Molis, S., Spooner, T., Widodo, J., Kim, J.-H., Liniger, E., Grill, A. and Hu, C.-K. (2007) H-base plasma modifications of SiCN/Cu interface to mitigate EM failure. *J. Electrochem. Soc.*, **154**, H604–610.
[80] Hu, C.-K., Angyal, M., Baker, B.C., Bonilla, G., Cabral, C., Canaperi, D.F., Choi, S., Clevenger, L., Edelstein, D., Gignac, L., Huang, E., Kelly, J., Kim, B.Y., Kyei-Fordjour, V., Manikonda, S.L., Maniscalco, J., Mittal, S., Nogami, T., Parks, C., Rosenberg, R., Simon, A., Xu, X., Vo, T.A. and Witt, C. (2010) Effect of impurity on Cu EM. *AIP Conf. Proc.*, **1300**, 57–67.
[81] Hu, C.-K. and Luther, B. (1995) Electromigration in two-level interconnects of Cu and Al alloys. *Mater. Chem. Phys.*, **41**, 1–7.
[82] Hu, C.-K. and S. Reynolds, S. (1997) CVD Cu interconnections and electromigration. *Electrochem. Soc. Proc.*, **97–25**, 1514–1523.
[83] Parks, C. and Kelly, J., private communication.
[84] Stangl, M., Liptak, M., Acker, J., Hoffmann, V., Baunack, S. and Wetzig, K. (2009) Influence of incorporated non-metallic impurities on electromigration in copper damascene interconnect lines. *Thin Solid Films*, **517**, 2687–2690.

[85] Harper, J.M.E., Cabral Jr, C., Andricacos, P.C., Gignac, L., Noyan, I.C.,K.P. Rodbell, K.P. and Hu, C.-K. (1999) Mechanisms for microstructure evolution in electroplated copper thin films. *J. Appl. Phys.*, **86**, 2516–2525.
[86] Surholt, T. and Herzig, C. (1997) Grain boundary self-diffusion in Cu polycrystals of different purity. *Acta Mater.*, **45**, 3817.
[87] Ito, K., Ohmori, K., Kohama, K., Mori, K., Maekawa, K., Asai, K. and Murakami, M. (2010) Application of Ti-based self-formation barrier layers to Cu dual-damascene interconnects. *AIP Conf. Proc.*, **1300**, 91–97.
[88] Lee, K.L., Hu, C.-K. and Tu, K.N. (1995) In situ scanning electron microscope comparison studies on electromigration of Cu and Cu(Sn) alloys for advanced chip interconnects. *J. Appl. Phys.*, **78**, 4428–4437.
[89] Hu, C.-K., Luther, B., Kaufman, F.K., Hummel, J., Uzoh, C. and Pearson, D.J. (1995) Cu interconnection integration and reliability. *Thin Solid Films*, **84**, 262–292.
[90] Park, C.W. and Vook, R.W. (1993) Electromigration-resistant Cu–Pd alloy films. *Thin Solid Films*, **226,** 238–247.
[91] Hu, C.-K., Lee, K.Y., Lee, K.L., Crabal, C., Colgan, E. and Stanis, C. (1996) Electromigration drift velocity in Al-alloy and Cu-alloy lines. *J. Electrochem. Soc.*, **143**, 1001–1006.
[92] Amir, Q.M. and Gupta, D. (1995) Cellular precipitation and precipitate coarsening in a Mg–Al alloy. *Can. Metall. Quarterly*, **34**, 43–50.
[93] Rosenberg, R. (1971) Inhibition of electromigration damage in thin films. *J. Vac. Sci. Technol.*, **9**, 263–270.
[94] Hu, C.-K. and H.B. Huntington, H.B. (1988) Diffusion and electromigration of impurities in lead solders, in *Diffusion Phenomena in Thin Films and Microelectronic Materials* (eds D. Gupta and P.S. Ho), Noyes Data Corporation, Park Ridge, New Jersey, p. 555.
[95] Tonegawa, T., Hiroi, M., Motoyama, K., Fujii, H. and Miyamoto, H. (2003) Suppression of bimodal stress-induced voiding using high-diffusive dopant from Cu-alloy seed layer. Proceedings of the IEEE International Interconnect Technology Conference, IITC, pp. 216–218.
[96] Yokogawa, S. and Tsuchiya, H. (2007) Effects of Al doping on the electromigration performance of damascene Cu interconnects. *J. Appl. Phys.*, **101**, 013513–18.
[97] Michael, N.L. and Kim, C.U. (2001) Electromigration in Cu thin films with Sn and Al cross strips. *J. Appl. Phys.*, **90**, 4370–4377.
[98] Hu, C.-K., Gignac, L.M., Baker-O'Neal, B.C., Bonilla, G., Liniger, E.G. and Flaitz, P.L. (2007) Electromigration reliability in nanoscale Cu interconnects. *Mater. Res. Soc. Symp. Proc.*, **1036E**, M05-02.
[99] Hu, C.-K., Gignac, L., Baker, B., Liniger, E., Yu, R. and Flaitz, P. (2007) Impact of Cu microstructure on electromigration reliability. Proceedings of the IEEE International Interconnect Technology Conference, IITC, pp. 93–95.
[100] Lane, M.W., Liniger, E.G. and Lloyd, J.R. (2003) Relationship between interfacial adhesion and electromigration in Cu metallization. *J. Appl. Phys.*, **93**, 1417–1421.
[101] Gambino, J., Wynne, J., Gill, J., Mongeon, S., Meatyard, D., Lee, B., Bamnolker, H., Hall, L., Li, N. and Hernandez, M. (2006) Self-aligned metal capping layers for copper interconnects using electroless plating. *Microelectron. Engng*, **83**, 2059–2067.
[102] Aubel, O., Thierbach, S., Seidel, R., Freudenberg, B., Meyer, M.A., Feustel, F., Poppe, J., Nopper, M., Preusse, A., Zistl, C. and Weide-Zaage, K. (2008) Comprehensive reliability analysis of CoWP metal cap unit processes for high volume production in sub-μm dimensions. Proceedings of the IEEE International Reliability Physics Symposium, IRPS, pp. 675–676.
[103] Hu, C.-K., Rosenberg, R. and Lee, K.Y. (1999) Electromigration path in Cu thin-film lines. *Appl. Phys. Lett.*, **74**, 2945–2947.
[104] Hau-Riege, C.S. and Thompson, C.V. (2001) Electromigration in Cu interconnects with very different grain structures. *Appl. Phys. Lett.*, **78**, 3451–3453.
[105] Sukharev, V., Zschech, E. and Nix, W.D. (2007) A model for electromigration-induced degradation mechanisms in dual-inlaid copper interconnects: effect of microstructure. *J. Appl. Phys.*, **102**, 053505-1–053505-14.
[106] Lee, S.-C. and Oates, A.S. (2006) Identification and analysis of dominant electromigration failure modes in copper/low-*k* dual damascene interconnects. Proceedings of the IEEE International Reliability Physics Symposium, IRPS, pp. 107–114.

[107] Li, B., Christiansen, C., Gill, J., Filippi, R., Sullivan, T. and Yashchin, E. (2006) Minimum void size and 3-parameter lognormal distribution for EM failures in Cu interconnects. Proceedings of the IEEE International Reliability Physics Symposium, IRPS, pp. 115–122.
[108] Hauschildt, M., Gall, M. and Hernandez, R. (2008) Large-scale statistics for Cu electromigration. *AIP Conf. Proc.*, **1143**, 31–46.
[109] Hauschildt, M., Gall, M. and Hernandez, R. (2009) Large-scale electromigration statistics for Cu interconnects. *Mater. Res. Soc. Symp. Proc.*, **1156**, D06-06.

# 10

# Mechanical Reliability of Low-*k* Dielectrics

*Kris Vanstreels[1], Han Li[2] and Joost J. Vlassak[3]*

[1]*IMEC, Leuven, Belgium*
[2]*IBM T. J. Watson Research Center, Yorktown Heights, New York, USA*
[3]*School of Engineering and Applied Sciences, Harvard University, Cambridge, Massachusetts, USA*

## 10.1 Introduction

Over the last several decades, the semiconductor industry has continuously demanded miniaturization of microelectronic devices and higher transistor package density. This brings along an increase in the interconnection density by reducing the metal pitch and increasing the number of interconnect levels. This quest for continuous improvement in the performance of microelectronic devices has been the driving force for the semiconductor industry to pursue Moore's law, which states that the integrated circuit density per unit area doubles every one and a half to two years [1]. However, in recent years it has become necessary to introduce new technology elements in order to keep up with this trend. The introduction of porous dielectrics with reduced permittivity is one of the significant material changes required to keep performance improvements on pace with the increasing device density [2]. Two different deposition approaches of porous low-*k* dielectrics are used in the microelectronics industry: spin-on (from liquid solutions/gels) and plasma-enhanced chemical vapor deposition (PECVD). Over the years, the spin-on approach has

*Advanced Interconnects for ULSI Technology*, First Edition. Edited by Mikhail R. Baklanov, Paul S. Ho and Ehrenfried Zschech.

been well explored [3]. Spin-on low-*k* dielectrics with a wide range of porosity up to 99 % have been achieved using various methods, including silica particle nanotemplating [4], the sacrificial porogen method or templated sol–gel polymerization of bridged silsesquioxane precursors [5]. In contrast, PECVD low-*k* dielectrics have been less explored even though they are more popular in the microelectronics industry due to their better compatibility with the technological needs [6–9]. Porosity in these materials is mainly created using sacrificial porogen molecules, usually cyclic hydrocarbons [10]. Downscaling of dimensions in the deep submicrometer range implies considerable challenges for the patterning and deposition technologies, as well as for thin-film reliability and physical characterization. One of the major reliability issues for electronic devices is mechanical failure and interfacial debonding driven by stresses in multilayer structures. These stresses can either be created during deposition (residual intrinsic stresses) or be induced by thermal excursions during processing or device operation. Typically, the dual-damascene process for the integration of Cu/low-*k* structures consists of a number of processes such as the deposition of insulating layers, photoresist patterning, photoresist stripping, etching, ashing, metallization, chemical mechanical polishing (CMP), etc. During these various integration processes, porous low-*k* dielectrics are subjected to different thermal stresses, which in turn may lead to a fracture of the low-*k* dielectric and/or interfacial delamination due to low mechanical strength or a lack of adhesion strength. In the microelectronics industry, Young's modulus is often used to assess the mechanical behavior and process compatibility of an interlayer dielectric material. When Young's modulus cannot meet a minimum standard, the material simply cannot be used for dual-damascene integration. Other important properties are fracture toughness and interfacial adhesion.

The goal of this chapter is to give an overview of the mechanical reliability of low-*k* materials with a focus on hybrid low-*k* dielectrics. The content is organized into two parts, each of them focusing on different properties to quantify the mechanical strength of low-*k* materials: mechanical properties (Young's modulus, hardness) and fracture properties (interfacial adhesion, fracture toughness). Each part will first contain a brief overview of the most common measurement techniques to characterize the corresponding properties, followed by several case studies. The first section is dedicated to the mechanical properties of low-*k* dielectrics. The different effects of intrinsic film porosity, organic cross-linking and/or terminal groups (–$CH_3$, –OH, etc.) on the mechanical properties of organosilicate glasses is discussed. Furthermore, the impact of organic porogen content and UV radiation on the hardening process of organosilicate glasses is described. The second section is devoted to fracture properties of porous low-*k* materials, where the focus lies on both interfacial adhesion and fracture toughness. The impact of porosity and network structure, UV radiation and environmental effects on the fracture behavior of organosilicate glass films is discussed in detail.

## 10.2 Mechanical Properties of Porous Low-*k* Materials

### 10.2.1 Techniques to Measure Mechanical Properties of Thin Films

A wide range of techniques is available to characterize the mechanical properties of thin films on substrates [11, 12]. Mechanical properties of thin films can be measured using indirect or direct measurement methods, as schematically shown in Figure 10.1. Indirect

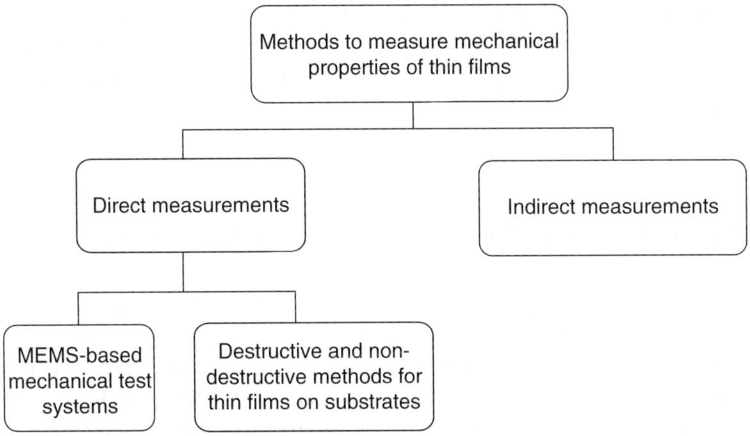


***Figure 10.1*** *Schematic overview of different methods to measure the mechanical properties of thin films*

methods are techniques that depend on changes in a measurable parameter by mechanical response, such as Raman shifts due to strain. Direct methods can be either techniques suitable for mechanical testing of thin films on substrates or MEMS-based mechanical test methods. Techniques suitable for testing thin films on substrates involve minimal sample preparation, but often measure both film and substrate properties. Therefore, significant post-processing effort may be needed to extract intrinsic film properties. In this category, nanoindentation is the most widely used technique in the semiconductor industry to determine Young's modulus, hardness, interfacial adhesion and fracture toughness of low-*k* dielectric films [13–15]. The main advantage of this technique is that it requires no special sample preparation and tests can be performed quickly and inexpensively. Additionally, nanoindentation combined with frequency-specific dynamic analysis also provides access to viscoelastic measurement of near-surface properties of polymeric low-*k* materials. However, several factors may influence the extracted mechanical properties of thin porous low-*k* films. Most notable among these are effects due to the presence of the substrate, which may lead to overestimation of the Young's modulus and hardness values [16–18], issues with tip calibration [19], surface roughness, densification of the film as a result of large hydrostatic stresses [20, 21] and size effects as a result of the inhomogeneous strain field [19]. In order to obtain reliable data, all these factors need to be taken into account. Considerable effort has been invested to understand these issues and relate nanoindentation results to intrinsic material properties [16, 17, 22–26]. Besides nanoindentation, a number of nondestructive dynamic techniques are available to determine the mechanical properties of thin films on substrates, including surface acoustic wave spectroscopy (SAWS) [27], surface Brillouin scattering (SBS) [28], the picosecond ultrasonic method [29], among others. These techniques typically require knowledge of the density of the film and only provide the elastic behavior of the films. MEMS-based techniques require careful specimen preparation and sample handling, but typically yield explicit and accurate film properties. The microtensile test [30, 31], microbeam cantilever deflection technique

[32] and plane-strain bulge test [33, 34] are the most popular techniques in this category and have been used extensively to study the mechanical behavior of thin metal films. However, these techniques become less attractive in the case of low stiffness materials such as low-*k* dielectrics due to difficulties related to fabricating freestanding membranes out of such materials.

### 10.2.2 Effect of Porosity on the Stiffness of Organosilicate Glass Films

The overall mechanical properties of porous materials, including nanoporous low-*k* dielectrics, are governed by the properties of the matrix and the pore structure. Extensive theoretical studies have been directed to modelling porosity effects on the elastic properties of materials [35–43]. Micromechanics models based on Eshelby's inclusion solution can capture the effect of porosity and pore shape explicitly, but are generally limited to structures with simple pore geometries [36–38]. Semi-empirical formulae have also been derived from the minimum solid area model [40]. The microstructure that corresponds to this model is, however, not clearly defined, which limits the predictive and interpretative power of this method. Torquato describes two-phase media using an *n*-point probability function [41]. The effective modulus tensor can be calculated to different degrees of accuracy depending on the order of the series expansion. However, application of this approach requires that the probability functions for a given porous material be determined independently. With the rapid progress of computers, finite element simulations represent a promising approach towards understanding the effects of porosity and pore geometry, as demonstrated by Roberts and Garboczi [42, 43]. Various geometrical factors can be tuned separately, to facilitate the study of their roles in the structure–property relationship for, in principle, arbitrary pore structures.

From the experimental perspective, studying the effect of porosity in nanoporous low-*k* dielectrics is often complicated by the difficulty to control porosity and matrix properties separately. For example, to tune the porosity of organosilicate glass (OSG) thin films using the porogen approach [10], one needs to change the amount of porogen loading during film deposition. Accordingly, a different post-deposition annealing/ultraviolet (UV) curing process is usually required to remove the porogen phase after film deposition. However, the bond network of the matrix is very sensitive to the condition of UV irradiation [44, 45]. Consequently, the porogen-removal step may modify the matrix materials to a different extent for films that contain different amounts of porogen phase. As an example, Figure 10.2 shows some representative Fourier transform infrared (FTIR) absorption spectra from a series of OSG films with increasing porosity [7]. These films are deposited by a PECVD technique using diethoxymethylsilane (DEMS) precursors with different porogen loadings. While qualitatively similar in shape, the relative intensity of the Si–O–Si peak with respect to that of the SiC–$H_3$ peak decreases with increasing film porosity (or porogen loading), indicating a change in the relative amount of the terminal groups compared to the networking bonds in the matrix. To study the intrinsic porosity effect of nanoporous dielectrics, Li and co-authors have demonstrated an experimental procedure to separate out the effects caused by the matrix at different levels of porosity [7]. Such a separation relies on (1) the characterization of the density of various molecular bonds in the dielectric film and (2) the experimentally established correlation between bond

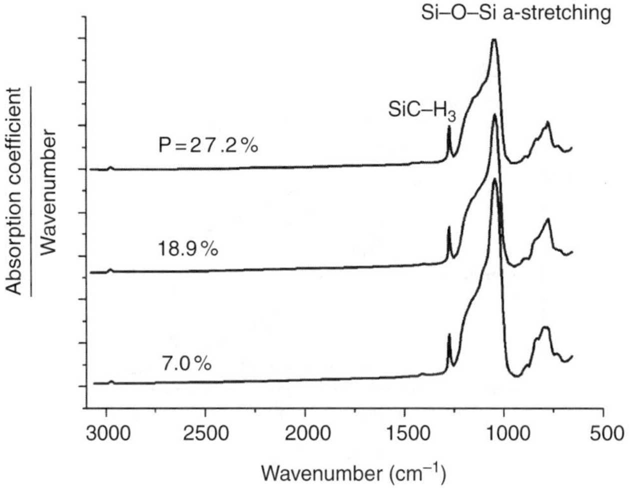


***Figure 10.2*** *FTIR absorption spectra of the OSG films with different levels of porosity (P); spectra are offset for comparison purposes*

densities and material properties for similar but dense dielectrics. Briefly, the various bond densities (Si–O–Si, Si–$CH_2$–Si, Si–$CH_3$, etc.) in the porous films are determined from their FTIR absorption spectra using the infrared absorption cross-sections – a parameter that relates the peak area in the IR absorption spectrum to the number density of the corresponding bond in the material [46]. The bond densities in a porous film are readily converted into the bond densities in the matrix through film porosity. For dense OSG dielectrics, it has been demonstrated experimentally that Young's modulus ($E$) correlates strongly with the densities of networking bonds (Si–O–Si and Si–$CH_2$–Si) of the materials over a broad range of process parameters [47]; this correlation is adopted to model the matrix material of porous OSG due to its structural similarity with the dense OSG films. With that, one can estimate Young's modulus of the matrix, $E_{matrix}$, of a porous film from knowledge of the network bond densities in the matrix. The intrinsic porosity effect on Young's modulus can thus be expressed using a dimensionless function as $f(P) = E_{film}(P)/E_{matrix}(P)$, where $P$ denotes film porosity. For demonstration, the function $f(P)$ determined for the same series of OSG films is shown in Figure 10.3. The experimental results are compared with finite element simulations of some idealized structures and the theoretical Hashin–Shtrikman upper bound [48]. At low porosity, the experimental results agree with the model that consists of randomly distributed spherical pores. As the porosity of the film increases, the experimental results start to track the solid sphere model that mimics a microstructure of sintered ceramic particles. This observation suggests that the pore structure of the OSG films evolves with increasing porosity – going from a structure with isolated pores at low porosity to a structure with greatly interconnected pores at higher porosity. This evolution has indeed been observed in similar dielectrics using positronium annihilation lifetime spectroscopy [49]. Finally, it is noted that this decoupling procedure is generic and can be adapted to treat other material properties as well. For example,

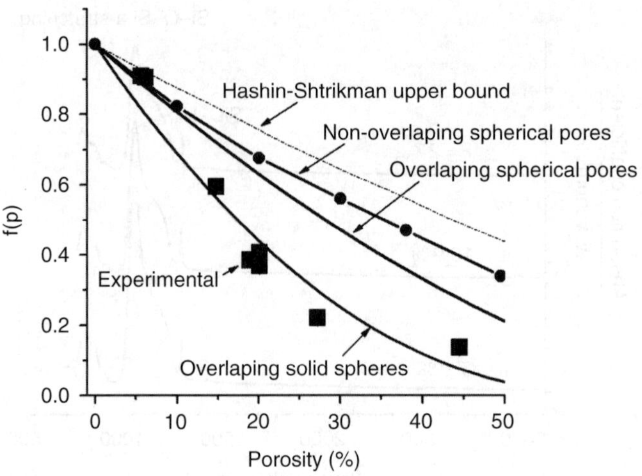


***Figure 10.3*** *Experimental values of the function f(P), compared with finite element simulations of various pore microstructures and theoretical upper bounds. The overlapping spherical pore model and solid spheres model are from Reference* [42]

the cohesive fracture energy of porous dielectrics can be analyzed in a similar way, as discussed in Section 10.3.3.

### 10.2.3 Hybrid Dielectrics Containing Organic/Inorganic Bridging Units

Common to most conventional organosilicate dielectrics is the fact that the low dielectric constants are achieved by introducing organic terminal groups into the silica network [50]. This structure modification decreases the OSG density and lowers the bond polarizability, both beneficial for the dielectric performance. From a mechanical point of view, however, these terminal groups disrupt the covalent molecular network and degrade the mechanical integrity of the materials [2].

The mechanical properties of as-deposited films can be considerably improved by ultraviolet (UV) irradiation and thermal annealing. The property improvements are associated with a network of increased connectivity through the removal of terminal groups and the formation of more networking bonds [44, 45, 51]. Recently, a new strategy has been proposed to strengthen films during film deposition. By engineering the precursor monomer structure, hydrocarbon molecules can be incorporated into the material matrix as network-forming units rather than as terminal groups, resulting in a network that is less disrupted and has improved strength [52–56]. For example, an organic bridging unit such as Si–$CH_2$–Si can be introduced by using a conventional precursor combined with carbosilane precursors in the form of $R_{1,2,3}$Si–$CH_2$–Si$R_{1,2,3}$ [56], where $R_{1,2,3}$ represent identical or different terminal groups.

As the material structure becomes more complex, it is important to understand how the different bonds and terminal groups affect the properties of the hybrid dielectrics. To this end, atomistic simulations have been increasingly applied to complement experimental investigations [57–61]. The molecular network can be systematically varied with precise

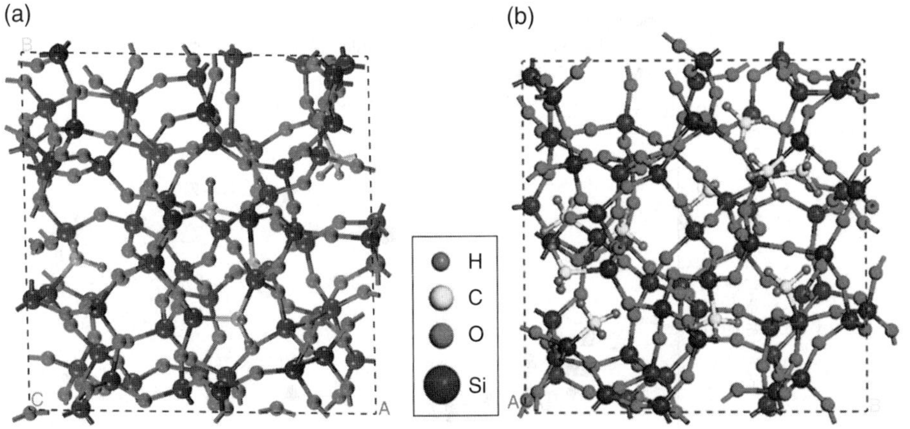


*Figure 10.4* *Unit cell of (a) type-I OSG with five methylene cross-links and (b) type-II OSG with ten methyl terminal groups after relaxation (see color Plate 11)*

control of the composition and bonding structure. To delineate the role of the bridging units and terminal groups, two idealized classes of molecular models have been considered that are built out of the backbone structure of amorphous silica [60, 61]. The first class of models consists of an Si–O–Si and Si–$CH_2$–Si network without terminal groups, and are referred to as type-I OSG models (Figure 10.4 (a)). Type-I models are created by replacing a fraction of the bridging oxygen atoms in the silica backbone structure with methylene groups. The second class of models, or type-II OSG models, have only Si–O–Si cross-linking bonds in the network, but terminal groups (–OH or –$CH_3$) are allowed (Figure 10.4 (b)). Type-II models are created by breaking the Si–O–Si bonds and replacing the oxygen atom with two 'T' groups, where 'T' denotes a terminal group that can be either –OH or –$CH_3$. Molecular dynamics simulations are performed to relax, equilibrate and sample these atomistic models to extract the elastic moduli and mass density of the material as a function of the network structure. The interatomic interactions are described using either a force-field potential (COMPASS) or ab initio calculations based on density functional theory and the tight-binding (DFTB) method [60, 61]. Figure 10.5 shows the bulk modulus and shear modulus of the type-I OSG as a function of the –$CH_2$– concentration in the material. The results show that the incorporation of –$CH_2$– as a bridging unit into the silica network affects the elastic properties of the resultant organosilicates considerably. Namely, the bulk modulus gradually increases from the value of pure silica (~39.5 GPa) to 59.3 GPa as half of the oxygen atoms are replaced. Further increasing the –$CH_2$– content continues the trend, but at a lower rate. The shear modulus, however, depends only slightly on the –$CH_2$– population. The COMPASS results are largely consistent with the calculation from density function theory, although a more significant increase in the elastic modulus (bulk and shear modulus) is obtained for the latter. The mechanism responsible for the enhanced stiffness can be understood by tracking the deformation of bonds of the network under external loading. For instance, the evolution of the silica network structure under hydrostatic loading is summarized in the left panels of Figure 10.6, which indicates that the volumetric

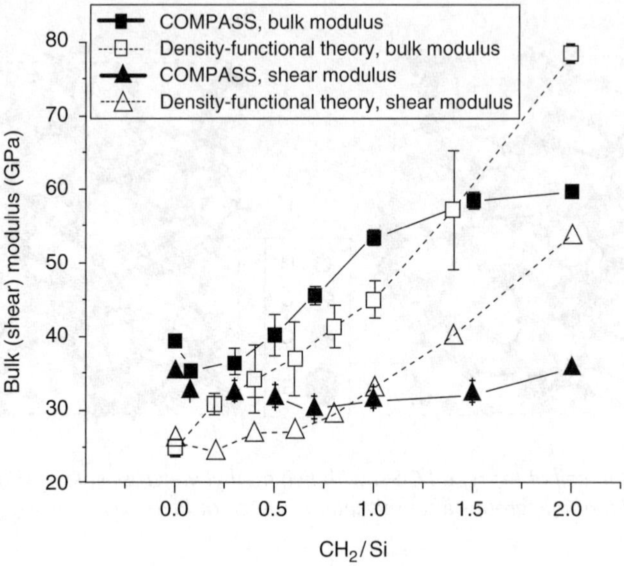


***Figure 10.5*** *Bulk modulus and shear modulus of type-I OSG as a function of $CH_2/Si$ ratio* [60]

deformation of $SiO_2$ is mostly associated with a change in the Si–O–Si bond angle distribution; the relative variation of the mean Si–O–Si angle per unit strain is much larger than the corresponding variation for the O–Si–O angle or the Si–O length (Figure 10.6, left-bottom). Thus deformation of $SiO_2$ under hydrostatic loading is accommodated mainly by the bending of the Si–O–Si bonds connecting nearly rigid $SiO_4$ tetrahedra. In $Si(CH_2)_2$, the elastic deformation of the network is accommodated primarily by bending of the Si–C–Si bonds and by stretching of the Si–C bonds, as illustrated in the right panels of Figure 10.6. The Si–C stretching is invoked in response to external loading because the bending rigidity of Si–C–Si is higher than the Si–O–Si bonds in silica. This greater bending rigidity of Si–C–Si relative to Si–O–Si is responsible for the increased bulk modulus of type-I OSG with increasing $–CH_2–$ population. Indeed, according to density functional theory calculation, the Si–C–Si bonds have an even greater bending rigidity than that described by the COMPASS force field [61], thus causing a steeper rise in the network stiffness. Type-II OSG materials contain a number of terminal groups (–OH, $–CH_3$) interrupting the network structure of amorphous silica. The degree of interruption can be described by the network connectivity number $<r>$, defined as the average coordination number per network forming atom [62]. For type-II OSG, $<r>$ is related to the number of terminal groups through [60]

$$<r> = \frac{16}{6 + N_T / N_{Si}} \tag{10.1}$$

where $N_{Si,T}$ is the number of silicon atoms or terminal groups as denoted by the subscript. The molecular dynamics (MD) simulations show that the bulk modulus and the shear modulus of type-II OSG are reduced significantly with increasing fraction of terminal

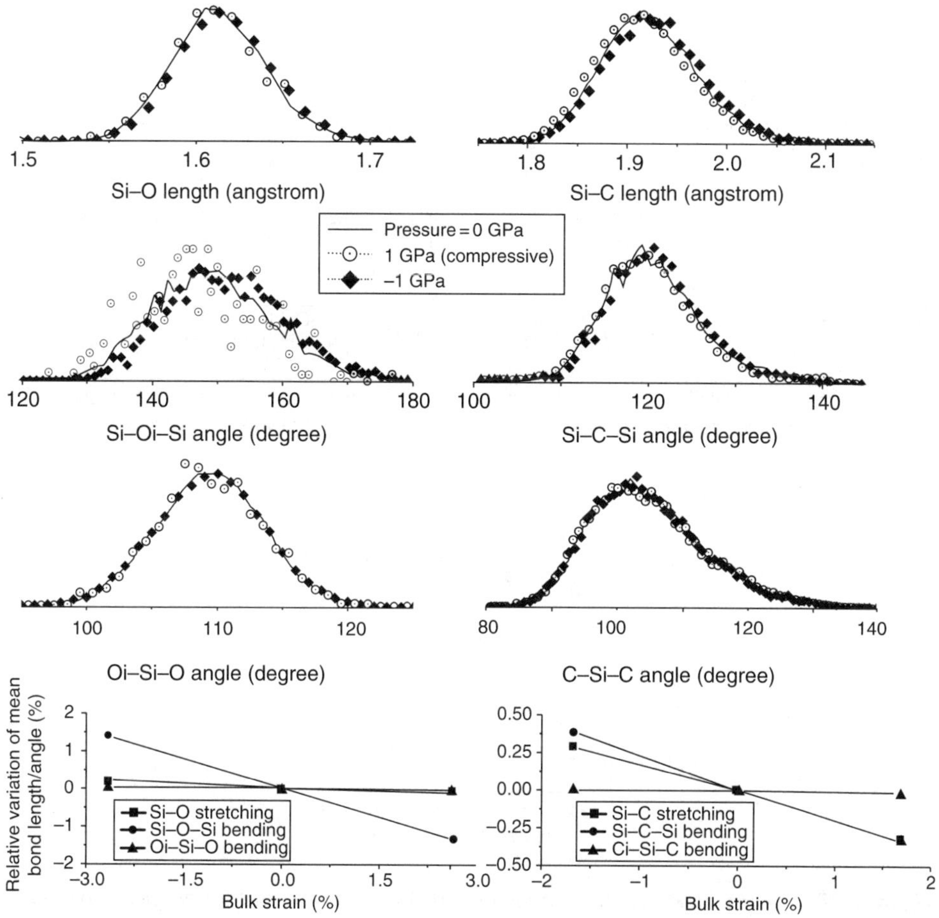


***Figure 10.6*** *The evolution of bond length and angle distribution in $SiO_2$ (left panels) and $Si(CH_2)_2$ (right panels) under hydrostatic loading* [60]

groups or decreasing <r> (Figure 10.7 (a)), while Poisson's ratio increases (Figure 10.7 (b)). Figure 10.7 (c) further shows that the density decreases with decreasing <r>, because more free volume is generated by the terminal groups; that the density of OSG decreases faster with <r> for $-CH_3$ terminal groups than for –OH groups is attributed to the relatively larger free volume associated with the methyl group [60].

These simulation results reveal property changes consistent with those of OSG materials with gradually modified structure due to UV irradiation [44, 49, 51], and a more quantitative comparison has been made in Reference [60]. It is anticipated that atomistic simulations will make meaningful contributions to the rational design of advanced dielectrics for next-generation interconnect technologies. Amorphous solids with covalent bonds have long been modeled as continuous random networks using constraint theory [62, 63]. Numerical simulations based on constraint theory show that the elastic properties of a

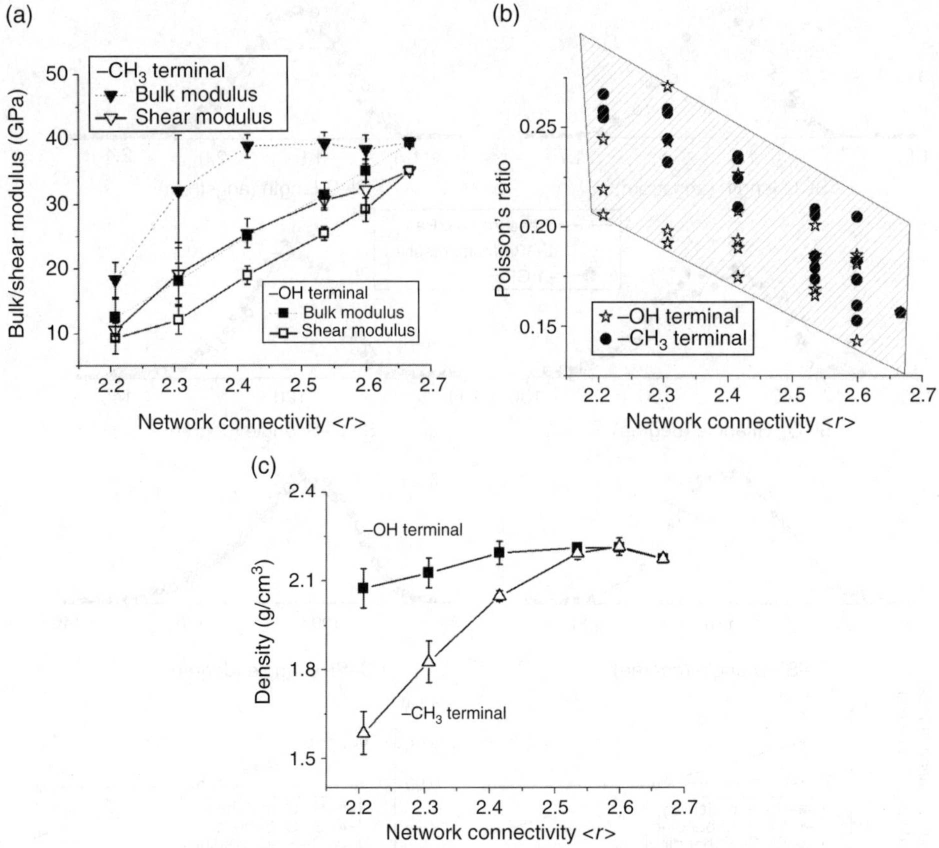


***Figure 10.7*** *(a) The bulk and shear modulus, (b) Poisson's ratio and (c) density of type-II OSG as a function of network connectivity*

continuous random network are determined almost entirely by the network connectivity [63]. The MD simulations demonstrate that the specific terminal groups also play a significant role. Another important difference between the MD simulations and constrained random network theory relates to rigidity percolation. According to classical constraint theory, a network is overconstrained when the number of constraints imposed by the network bonds exceeds the number of degrees of freedom. If the number of constraints is smaller than the degrees of freedom, the network is underconstrained and the stiffness vanishes rapidly. This transition occurs at the rigidity percolation threshold. MD simulations, however, show no sudden change in stiffness near the theoretical percolation threshold. Instead, the stiffness of the OSG networks decreases gradually over a range of *<r>* values. These differences between MD simulation and constraint theory are probably due to the long-range nonbond interactions that are present in the MD simulations but are not considered in the continuous random network model [60].

### 10.2.4 Effect of UV Wavelength and Porogen Content on the Hardening Process of PECVD Low-*k* Dielectrics

As discussed in Section 10.1, the introduction of nanometer-scale porosity into organosilicate glass films degrades their mechanical properties, thereby presenting a real challenge for their integration into ultra-large-scale microelectronic devices. Porosity in these materials is typically introduced by codeposition of the organosilica matrix with organic porogen by oxidation of alkylsilanes in a PECVD process. Subsequently, the films are subjected to a UV-assisted thermal curing process, in which the porogen molecules are photodissociated by the UV light and a reinforced Si–O–Si network is formed through the thermo-induced and photo-induced rearrangement in the bonding structure (cross-linking) of the SiCO:H-based low-*k* material [44, 64, 65]. However, not all porogen is removed during UV curing, as it is partially converted by UV light into nonvolatile graphitized carbon residues (porogen residues). Although porogen is needed to introduce porosity in these films, the presence of porogen residues can have a negative impact on the fundamental film properties and their industrial process compatibility. First of all, the presence of porogen residues with conjugated C=C bonds increases the leakage current and decreases the breakdown voltage of these materials [66, 67]. Secondly, industrial processes such as the zero-damage (no Si–$CH_3$ scission) removal of an organic photoresist mask from OSG low-*k* films also removes porogen residues from the OSG films because porogen contains similar types of bonds (C–C, C=C and C–H) as the organic photo masks [68, 69]. This in turn results in increased film porosity and subsequent degradation of mechanical properties [70, 71–73]. Finally, the amount of porogen residues has a significant impact on optical, chemical and mechanical properties of PECVD low-*k* materials [70, 71, 74, 75]. It is shown that the presence of porogen residues increases the extinction coefficient of organosilica glass films in the UV range [71, 76]. This is illustrated in Figure 10.8, which compares the extinction coefficient of PECVD and spin-on organosilica glass (OSG) with $k = 2.3$, PECVD deposited matrix material without porogen and with $SiO_2$ films. As can be seen from this figure, the extinction coefficients of the spin-on film and PECVD matrix material deposited without porogen are closer to that of $SiO_2$. In contrast, the PECVD glass prepared by standard codeposition of organosilica matrix and porogen followed by UV curing shows significantly higher extinction than other films. Moreover, the low-energy shoulder with a maximum at 4.5 eV in the extinction coefficient reflects the presence of $sp^2$ hybridized carbon [74, 77]. The amorphous $sp^2$ carbon originates mainly from porogen residues, as was proven by Raman spectroscopy [66]. When comparing the mechanical properties of these films, it was found that organic-free OSG deposited by spin-on showed a higher Young's modulus of 6.5 GPa than OSG deposited by PECVD, which has a Young's modulus of 4.6 GPa for the same level of porosity (32 %). This suggests that the porogen residues present in PECVD glass are responsible for its lower Young's modulus compared to spin-on glass films. Moreover, recent work of Urbanowicz and co-authors showed that the initial porogen content in OSG films prior to a UV cure has a severe impact on both the UV-curing mechanism and the obtained mechanical properties after the UV cure [72, 78]. It was demonstrated that the UV-curing mechanism of porogen-free OSG films significantly differs from the conventional UV-curing schemes with porogen. Furthermore, improved mechanical properties and an ultra-low dielectric constant could be obtained by controlled

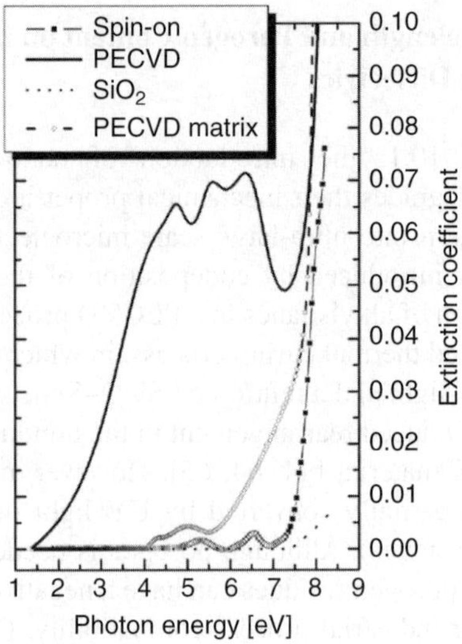


***Figure 10.8*** *Extinction coefficient as measured by UV SE of $SiO_2$ deposited by PECVD using $SiH_4$ as a precursor, PECVD matrix material, standard spin-on nanoclustered silica (k = 2.3) and PECVD low-k films* [78]

decomposition of the porogen molecules prior to the UV hardening step. The controlled removal of porogen was obtained by exposing the PECVD films to an $H_2$-based downstream plasma (DSP) at a temperature of 280 °C. The latter process is similar to the zero-damage photoresist mask removal process as reported by Berry *et al.* [68] and Chen *et al.* [69]. Figure 10.9 illustrates the impact of porogen removal on the optical properties of PECVD low-*k* films. The extinction coefficient of the as-deposited film with porogen was found to be the highest due to the largest porogen content. The $H_2$-based DSP treatment resulted in complete removal of the porogen and porogen residues, and the absorption spectrum becomes similar to the spectrum of the low-*k* matrix material (Figure 10.8). The standard UV-curing process also results in porogen residue creation, as reflected by the increased extinction coefficient compared to $H_2$-based DSP-treated films. According to the Grotthus–Draper law, a reduced film absorption in the UV range after porogen removal leads to less initiation of photochemical reactions during the UV curing [79]. Therefore, lower film absorption of UV light should result in a lower UV-curing effect. However, not only the UV absorption, but also the direct photodissociation of the chemical bonds and bond rearrangement by UV light determines the mechanical robustness of these low-*k* films after UV curing. In the first approximation, one can say that OSG low-*k* materials contain energetically strong Si–O bonds as well as weaker C–H and Si–$CH_3$ bonds. The Si–O bonds are mostly responsible for the mechanical stability of the films, while the less polar Si–$CH_3$ bonds provide the hydrophobic properties and their lower dielectric constant. The direct Si–$CH_3$ bond photodissociation has been reported to be determined by excitation to a singlet state that requires energy in the range of 190–200 nm [80, 81]. By selecting two different UV sources (narrowband ~ 172 nm; broadband > 200 nm), the amount of Si–$CH_3$ bonds can be controlled. The broadband UV source (> 200 nm) emits light that is less

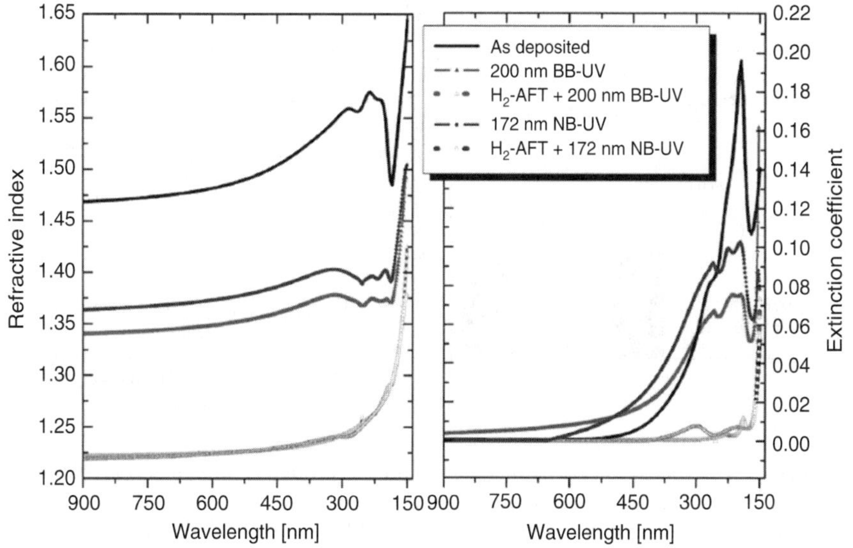


***Figure 10.9*** *Optical properties (150–900 nm) of differently prepared PECVD low-k dielectrics as measured by UV-SE: as-deposited matrix-porogen films cured with and without porogen by narrowband UV light (172 nm) and broadband UV light (>200 nm)* [72]

absorbed in the porogen-free low-*k* film in comparison with porogen-rich low-*k* film and cannot cause direct Si–$CH_3$ photodissociation. In contrast, the light generated by the narrowband UV source (~172 nm) is more efficiently absorbed in comparison with the broadband UV source and direct photodissociation of Si–$CH_3$ can occur due to more energetic light generated by the narrowband UV source (below the Si–$CH_3$ photodissociation threshold). The direct Si–$CH_3$ photodissociation results in more cross-linking opportunities for the film skeleton due to generation of active silyl states, which can react with another silyl state or silanol (Si–OH), which then leads to the creation of Si–O–Si links in the skeleton. This mechanism will have more importance for films cured without porogen, because decomposed porogen fragments such as C and H or $CH_x$ radicals can participate in the photochemical reaction with low-*k* skeleton bonds and lead to the termination of active sites preventing formation of Si–O–Si cross-links. These statements were verified by comparing the mechanical properties of all films with the degree of film cross-linking, which is reflected by its thickness loss due to film shrinkage and corresponds to the improvement of Young's modulus (Figure 10.10). One can see that irradiation of the porogen-containing and porogen-free films by the narrowband (~172 nm) UV source leads to thickness loss that is higher than for the broadband (>200 nm) UV source, which indicates that the degree of cross-linking is higher for the narrowband UV source. Moreover, there is a clear difference in the film thickness loss depending on the UV wavelength used for curing. This difference is smaller for the films cured with porogen and larger for films cured without porogen. In the case of $H_2$-based DSP treatment without a UV cure, almost no thickness loss is observed, proving that $H_2$-based DSP does not cause the film

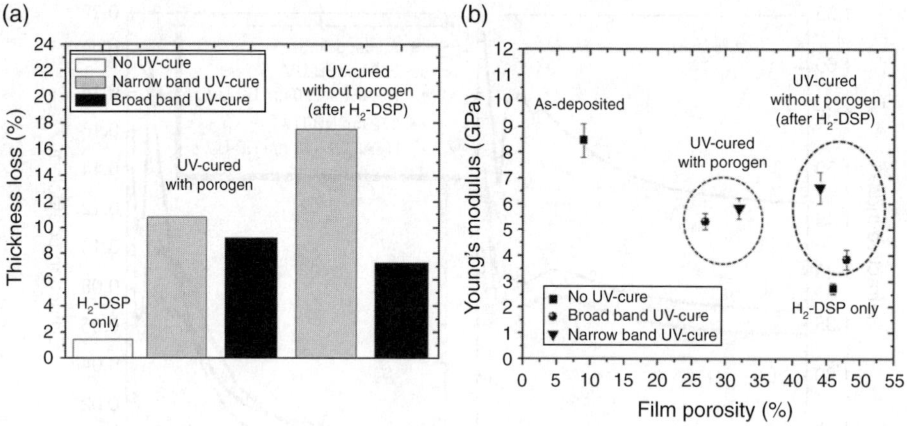


*Figure 10.10* *(a) Thickness loss (shrinkage) as measured by UV-spectroscopic ellipsometry; (b) Young's modulus as-measured nanoindentation versus film porosity as measured by ellipsometric porosimetry of porogen-free film ($H_2$-DSP treated) and films cured with and without porogen* [72]

cross-linking. The achieved values for Young's modulus are slightly higher when porogen-containing films are cured with the narrowband UV source, while in the case where the films were cured without porogen, the improvement in Young's modulus is almost twice as high for the narrowband UV-cured film compared to the broadband UV-cured film. This proves that the presence/absence of porogen during UV curing has a strong impact on the final result due to changes in absorption induced by porogen removal and an additional cross-linking mechanism due to direct Si–$CH_3$ bond photodissociation.

## 10.3 Fracture Properties of Porous Low-*k* Materials

### 10.3.1 Adhesion Measurement Methods

From a thermodynamic point of view, the true work of adhesion of an interface between two materials is the amount of energy needed in order to create free surfaces from the bonded materials. This is an intrinsic property of a specific pair of materials that depends on the type of bonding between both materials and is often measured by contact angle measurements. A wide variety of techniques to study thin-film adhesion are reported in the literature [82]. The majority of all available methods nowadays are techniques by which a controlled external load is applied to the film to induce interfacial failure and depends on the measurement of the critical load value. Some of these tests use continuous films and some require patterning, but all of them use some driving force or stored energy to achieve thin-film delamination. This energy may come from the external mechanical force that is applied or it can be stored inside the film itself through the internal film stress. Typically, these tests measure critical values of applied stress intensity, $K_i$, or strain energy release

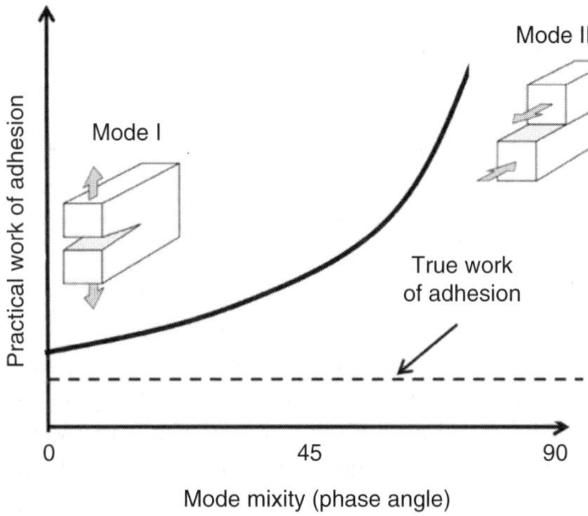


***Figure 10.11*** *Practical work of adhesion as a function of the mode*

rate, $G_i$, where *i* can be mode I (opening mode), mode II (sliding mode), mode III (tearing mode) or of a mixed mode character. However, most of these techniques cannot yield a direct, quantitative and reproducible evaluation of the true work of adhesion. This is because, while debonding, usually both bonded materials experience plastic deformation, which makes it very difficult to extract the true work of adhesion from the total energy measured. Instead, what is measured is the practical work of adhesion, which is a combination of the true work of adhesion, the energy spent in plastic deformation of both bonded materials and the energy loss due to friction. Depending on the nature and thickness of the considered films, plastic dissipation can account for a considerable part of the practical work of adhesion. On the other hand, the amount of plastic energy dissipation also depends on the mode mixity, which is a relative measure of the amount of shear and normal stress components. The concept of mode mixity is presented in Figure 10.11, which shows that the amount of energy dissipation is higher in pure shear (mode II) compared to the pure opening fracture mode (mode I).

To date, the four-point beam bending test is by far the most popular method in the microelectronics industry that can provide quantitative information on the interfacial adhesion energy [82]. This is mainly due to its simple geometry, easy sample preparation and the relatively easy beam mechanics involved in the fracture analysis. However, having a simple geometry also leads to the main disadvantage of these tests, which is the very limited sample geometries that can be investigated. The mode mixity (phase angle) for this test under steady-state crack growth conditions is approximately 43 °. In these tests, the interface of interest is sandwiched between two elastic substrates, after which the upper substrate is notched. During the four-point bending tests, a pre-crack initiates from the notch root and propagates vertically into a specific position in the film stack. At a critical load, the pre-crack initiates a new crack on the interface, The subsequent load that extends the interfacial crack in a stable way then gives a measure of the extension energy of an interfacial crack.

### 10.3.2 Fracture Toughness Measurement Techniques

Toughness is the ability of a material to absorb energy during deformation before rupture occurs, while fracture toughness is the ability of a material to resist the growth of a pre-existing crack [83, 84]. In this way, toughness encompasses the energy required to both create a crack and to enable the crack to propagate until rupture, whereas fracture toughness takes only account of the energy required to facilitate crack propagation to rupture. These are two different concepts that should not be confused and interchangeably used. In the case of bulk materials and thick films, fracture toughness can be easily measured according to ASTM standards [85, 86]. However, for thin film on substrates, there is neither a standard procedure nor commonly accepted methodology to follow. Quantitative measurements for thin films fall into different groups of test methods, including bending tests [87–91], scratch tests [92–98], indentation techniques [99–104], buckling tests [105–107] and tensile tests [108–112]. These test methods can either be stress-based or energy-based. Stress-based test methods examine the stress state near the tip of the crack and include bending tests with pre-crack, scratch in consideration of a critical tensile stress, crack length and spacing, and indentation in consideration of the critical load and corresponding crack length, among others. However, difficulties in these methods lie in the formation of the pre-crack, the determination of the crack length and the critical stress state. These problems are not easily resolved due to the thickness dimension involved. On the other hand, energy-based methods focus on the system's energy state before and after fracture of the film occurs. This energy difference ($\Delta U$) is considered to be consumed in order to create a new crack area ($\Delta A$) and toughness measurements are obtained through the critical energy release rate ($G_c \sim \Delta U/\Delta A$). This category comprises bending tests without a pre-crack in the film, buckling, indentation with chipping, tensile tests, and so on. In order not to confuse this with the classical concept of fracture toughness, it is strongly suggested that the term fracture toughness should not be used in a thin-film toughness description where a pre-crack is not involved.

Among the various available test methods, indentation is by far the most popular method to assess fracture toughness of brittle materials. To avoid the difficulties in making the pre-crack, many researchers directly indent without a pre-crack. When the stress upon indentation exceeds a critical value, radial cracking occurs, emanating from stress concentrations at the facet edges of the indenter tip. For a certain indenter tip geometry, the fracture toughness can be calculated based on the mechanical properties of the investigated material, the maximum applied indentation load and the observed crack length [97, 98, 100]. The mechanical properties can be determined from indentation tests at much smaller loads, while the crack length can be obtained using scanning electron microscopy (SEM). However, difficulty lies in the existence of a crack formation threshold, locating the indent and the determination of the crack length. Moreover, in the case of thin films on substrates, an appropriate theoretical model should be developed to account for both film thickness and stress, and preferably also film porosity. This is because typically the crack generated in these thin films by indentation does not show a half penny shape, as is typically seen for bulk materials. Lawn and Wilshaw have provided a detailed review of the indentation-induced cracking in bulk brittle materials [104].

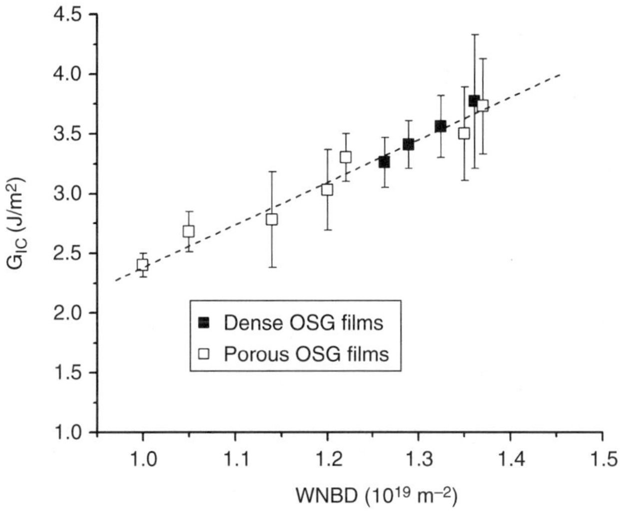


***Figure 10.12*** *Fracture energy of OSG as a function of overall WNBD* [114]. *Data for both dense and porous OSG films are shown*

### 10.3.3 Effect of Porosity and Network Structure on the Fracture Toughness of Organosilicate Glass Films

In Sections 10.2.2 and 10.2.4, we demonstrated that the mechanical properties of OSG materials depend sensitively on their porosity and that the amount of porogen during their fabrication changes both porosity and network structure of the dielectric. This section resolves the contributions of both network structure and porosity on the fracture energy of OSG films. Organosilicate low-*k* dielectrics are intrinsically brittle materials and are prone to cohesive fracture and interfacial delamination during chip fabrication and packaging [113, 114]. The fracture properties of the materials are closely related to the network structure where fracture occurs through atomic bond breaking. For porous dielectrics, the fracture properties are further modulated by porosity [115]. To quantify the amorphous network structure, Lin and Vlassak proposed the use of weighted network bond density, or WNBD, as a structure descriptor [47]. WNBD is the total number density of the two main networking bonds in an OSG material, i.e. the siloxane bond and the methylene bond. The density of the network bonds is normalized for its bond energy, so that an Si–O bond is counted as one network bond and an Si–$CH_2$ bond counts for approximately half a network bond due to its lower bond energy. The WNBD parameter is readily determined from the FTIR spectra of the films using inverse infrared cross-sections and shows a strong correlation with the cohesive fracture energy of OSG films over a wide range of deposition parameters (precursors, UV curing, etc.) [47, 115]. For example, Figure 10.12 plots the fracture energy of a series of dense and porous OSG coatings as a function of the weighted network bond density, WNBD, of the OSG [115]. The dense series of OSG are deposited using octamethylcyclotetrasiloxane (OMCTS) precursor [47] and the porous series are

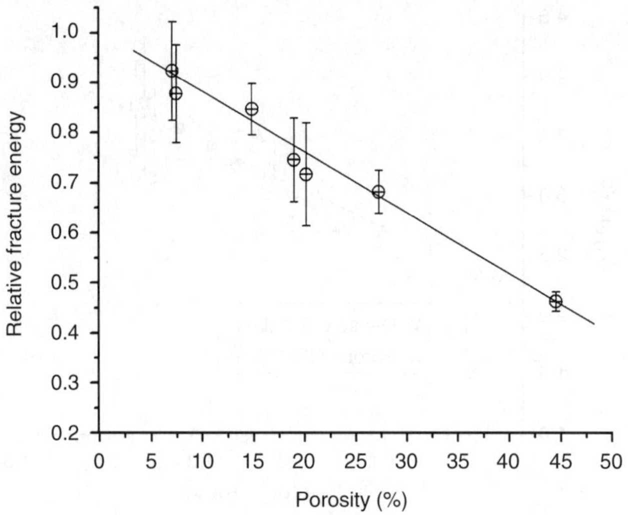


***Figure 10.13*** *Effect of porosity on the fracture energy of OSG as expressed by g(p)*

through the use of DEMS precursor together with porogen precursor [115]. The fracture energy is measured under a mode I condition using the double-cantilever beam technique. Evidently, the fracture energy of the OSG materials scales linearly with WNBD. This linear relationship is not surprising because fracture of brittle materials such as OSG is a bond-breaking process and the energy dissipated during fracture should scale with the number of broken networking bonds and with the energy of the bonds. This structure–property correlation was used by Li *et al.* to compensate any changes in the matrix structure of the porous OSG film as a result of different processing conditions, to facilitate the study of the effect of porosity on fracture energy [115]. The procedure is identical to what has been demonstrated in Section 10.2.2 for the stiffness of porous OSG, and a dimensionless function $g(p)$ can be extracted to represent the intrinsic porosity effect [115]. Experimental values of the function $g(p)$ are plotted in Figure 10.13, which shows that the fracture energy of the OSG coatings decreases linearly with increasing porosity for porosity values up to 45 %. This linear scaling is consistent with a planar through-pore fracture mechanism, where the solid area that must be fractured changes linearly with film porosity; effects associated with pore shape, size or distribution are screened out for a random pore microstructure. A different porosity dependence of fracture energy has been reported in the literature by Guyer *et al.* for methylsilsesquioxane (MSSQ)-based OSG coatings [117]. The cohesive fracture energy of the MSSQ films was measured under a mixed mode using the four-point bend flexure technique, which was shown to scale with average film density (porosity) through an empirical power law relation with an exponent equal to two. This nonlinear behavior was attributed to several factors, including the nonuniformity of the MSSQ density through the coating thickness and contributions from frictional contact between the crack faces [117]. These factors may be important under the mixed-loading conditions of the four-point flexure test, but should have a minimal effect under the mode

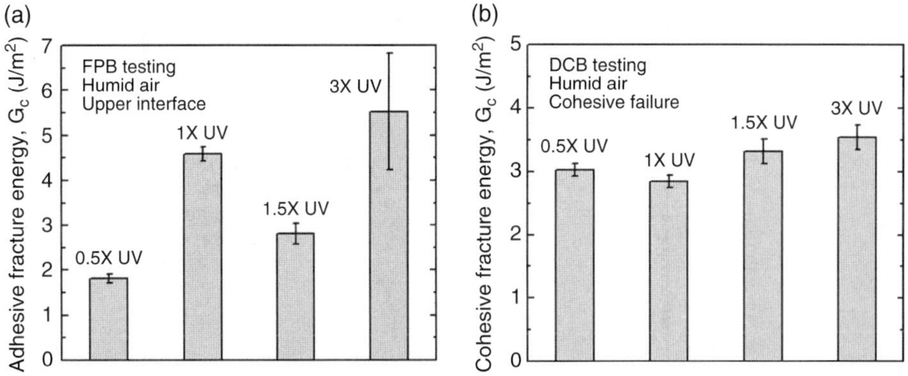


***Figure 10.14*** *(a) Interfacial fracture energies measured by four-point bending (FPB) and (b) cohesive fracture energy measured by the double cantilever beam (DCB) test for different cure durations* [45]

I loading conditions of the double cantilever beam (DCB) technique. In the absence of these complicating factors, the results shown in Figure 10.13 indicate that cohesive energy does scale linearly with the density of the atomic bonds on the fracture plane and decrease linearly with film porosity.

### 10.3.4 Effects of UV Cure on Fracture Properties of Carbon-Doped Oxides

Fracture of organosilicate glass films is a significant challenge for the integration of these mechanically fragile materials in future microelectronics technologies. Their integration has been limited by accelerated cracking rates in reactive or moist environments, leading to premature failure. In Section 10.2.4 it was shown that UV curing significantly influences the glass structure and improves mechanical performance without significantly detracting from the film's low-*k* nature. However, by introducing porosity into these films, the mechanical and fracture properties of organosilicates typically display a strong dependence on film density [117]. Recently, Gage *et al.* have demonstrated the impact of UV cure exposure time on the glass structure and fracture properties of PECVD carbon-doped oxide films through a detailed structural analysis and fracture mechanics based characterization [45]. The structural characterization indicated a significant change in the glass structure with increasing UV curing time, marked by the removal of terminal functional groups and increased network-forming bonds following the initial removal of porogen material. It was further demonstrated that the higher degree of film connectivity for longer UV cure exposure times significantly enhances the interfacial adhesion strength (Figure 10.14 (a)), but only slightly enhances the fracture toughness (Figure 10.14 (b)). Moreover, only minor effects were found on the resistance to subcritical cohesive crack growth rates measured in a moist air environment (Figure 10.15). The slopes of the crack velocity versus strain energy release rates are very similar for different cure conditions. Moreover, within the level of the expected experimental scatter, the different curves did not shift in strain energy

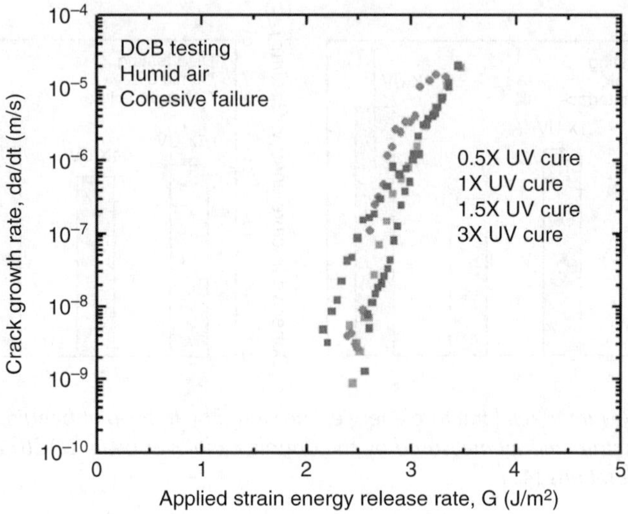


***Figure 10.15*** *Environmentally assisted cohesive cracking measured by the DCB test for the different cure durations* [45]

release rate value, indicating that the reactivity of the bonds being broken at the crack tip are not appreciably changed with UV curing. This was supported by FTIR data that showed no appreciable change in network bond angle. The difference in adhesive or cohesive fracture resistance was explained in terms of a small-scale crack tip deflection phenomenon. During cohesive cracking under DCB conditions, the crack is not constrained and therefore is free to seek out a path of least resistance through the carbon-doped oxide film. This means that the crack tip is able to undergo small-scale deflections toward regions of low connectivity and low density within the glass network, illustrated in Figure 10.16. If such small-scale deflection phenomenon occurs, this means that the crack tip would be more likely to undergo deflection to avoid a significant fraction of the network bonds that are formed during the UV process (encircled). In this way, only a small increase in cohesive strength with UV curing is expected. On the other hand, under four-point bending loading (FPB) conditions, both tensile opening stresses (mode I) and shear stresses (mode II) are present at the crack tip, which tends to constrain the fracture to a more planar path through the OSG film. Hence, it is expected that the interfacial fracture energies would display much more sensitivity to UV curing, since the crack tip does not have the same degree of freedom and cannot easily deflect to avoid new network bonds formed during the UV curing process. This explanation was supported by AFM morphology characterization of the fracture surface. The AFM surface roughness scans on the DCB fracture surfaces revealed rms roughness values around 1 nm, which was somewhat larger than the characteristic size of the glass network. In contrast, the FPB fracture surfaces were in the order of ~0.35 nm, which suggests that the small-scale deflections that occurred under DCB conditions did not occur under FPB conditions.

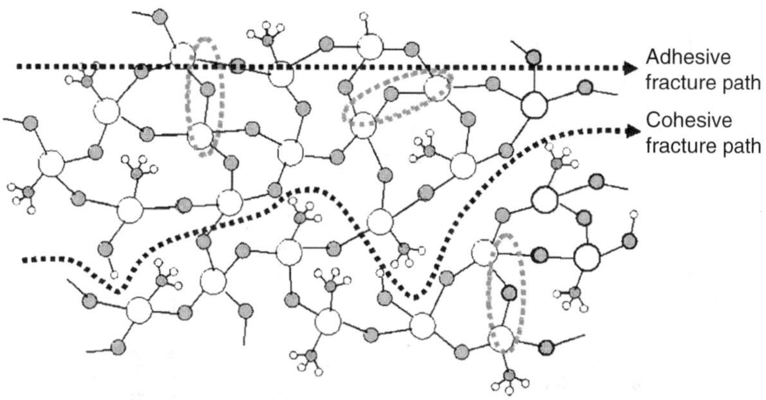


*Figure 10.16* *Schematic illustration of crack path selection under both cohesive failure (double cantilever beam load conditions) and adhesive failure (four-point bending load conditions). The circled network bonds represent bonds created during the UV curing process* [45]

### 10.3.5 Water Diffusion and Fracture Properties of Organosilicate Glass Films

While the intrinsic fracture properties of low-*k* dielectrics are certainly important for the mechanical integrity of the interconnect structures, time-dependent cracking is another concern for chip fabrication and long-term reliability [118, 119]. Organosilicate dielectrics are well known to be susceptible to environmentally assisted cracking because of the Si–O bonds in their backbone structures. The strained Si–O bonds at crack tip can react chemically with water molecules to assist the bond rupture process. As a result, the fracture resistance of the dielectrics can be significantly degraded due to the ingress of water and other reactive species [120, 121]. More recently, water diffusion and the impact on the fracture behavior of nanoporous dielectric film stacks were studied for various organosilicate materials and their interfaces [122, 123]. For example, Li and Vlassak studied various nanoporous OSG materials following the 'soak-then-test' (STT) approach [122]. The OSG layer was sandwiched between a silicon substrate and an SiCN diffusion barrier. The samples were immersed in water to allow water to diffuse in from the sample edge for a controlled amount of time prior to the fracture measurement. The fracture measurements show that the adhesion of the OSG/SiCN interface decreases over time (Figure 10.17), while the cohesive fracture energy of the nanoporous OSG layer remains largely constant. This somewhat unexpected result can be understood by tracking the diffusion path of water in the multilayered structure. To this end, $^{18}O$-labeled water tracer diffusion experiments were carried out combined with dynamic secondary ion mass spectroscopy (SIMS). The depth profiling of $^{18}O$ through the thickness of the film stack revealed sharp peaks at the OSG/SiCN interfaces of the soaked film stacks (Figure 10.18 (a)), while in the bulk of the OSG film the signals were fairly uniform. The magnitude of the $^{18}O$ peaks decreased with increasing distance from the sample edge, following the complementary error function as for one-dimensional diffusion (Figure 10.18 (b)). These observations indicate that water molecules tend to diffuse along the OSG/SiCN interface rather than through the bulk of the

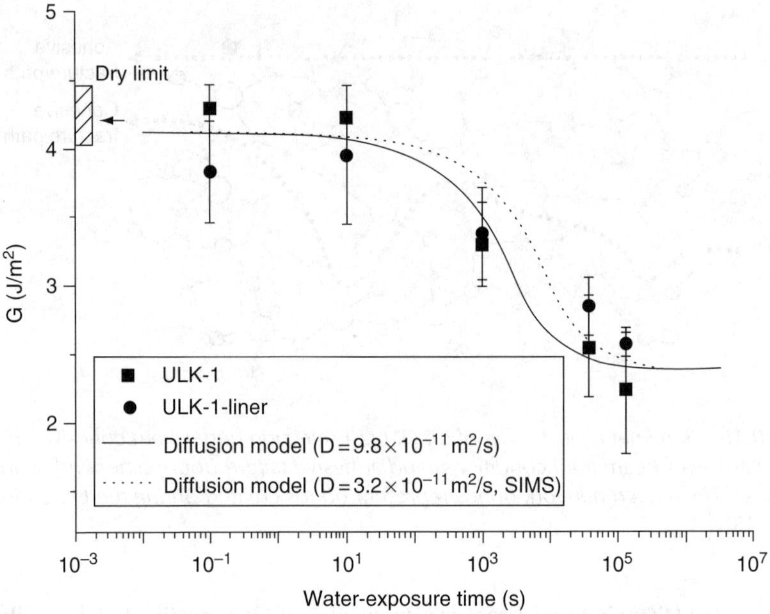


***Figure 10.17*** *Adhesion energy of the OSG/SiCN interface as a function of the water immersion time for samples ULK-1 (square) and ULK-1-liner (circle)*

OSG, consistent with the fracture measurement. This preference of water for the OSG/SiCN interface was attributed to the less hydrophobic character of the interface, as verified by the contact angle measurement. Various processing steps may cause a loss of hydrophobicity to a certain extent, as identified, including demethylation of the OSG surface during the initial phase of the SiCN deposition [124] and/or during the interfacial He plasma treatment [125]. It should be pointed out that common organic additives such as isopropyl alcohol or acetic acid to water can drastically change the diffusion behaviour. Indeed, degradation of the cohesive fracture energy is observed if the OSG film stack is exposed to aqueous solutions with organic additives that enhance wetting of the OSG [122, 126]. The adhesion degradation result can be analyzed quantitatively using the model developed by Vlassak and co-authors [120, 121]. In this model, the water concentration profile along the crack front is calculated as a function of soaking time, and correlated with the overall energy release rate required to drive the crack at a given velocity using subcritical crack growth data. By fitting the experimental data with the analytical function derived from the model, water diffusivity can be extracted. In fact, the diffusivity of water along the OSG/SiCN interface is calculated to be $9.8 \times 10^{-11}$ $m^2/s$, comparable to a value of $3.32\times10^{-11}$ $m^2/s$ obtained from the SIMS measurements. For comparison, the diffusion coefficients of various aqueous solutions through a nanoporous MSSQ measured by Guyer *et al.* [126] using an optical technique were approximately three orders of magnitude smaller than the diffusivity for the OSG/SiCN interface reported [120, 121].

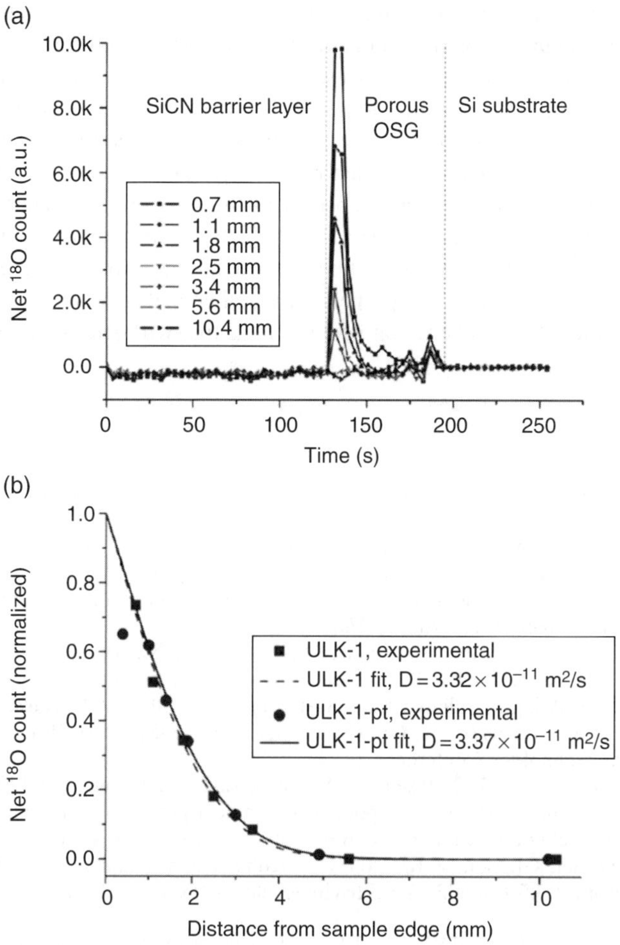


***Figure 10.18*** *(a) Subcritical delamination behavior at the OSG/TaN interface under different environmental humidity conditions and (b) normalized peak intensity of the $^{18}O$ signal along the OSG/SiCN interface after subtracting the reference signal. The diffusion coefficient of water is calculated by fitting the experimental data with the complementary error function*

## 10.4 Conclusion

Integration of porous materials into microelectronic devices poses a number of challenges for the patterning and deposition technologies, as the materials meet strict requirements in terms of properties and reliability. One of the major reliability issues for microelectronic devices is related to mechanical failure and interfacial debonding driven by stresses in multilayer structures. During the various integration processes, porous low-*k* dielectrics are subjected to different thermo-mechanical stresses, which can cause fracture of the low-*k* and/or interfacial delamination due to insufficient mechanical and fracture strength.

This chapter has provided an overview of the most commonly used techniques to characterize mechanical and fracture properties. Furthermore, it was demonstrated that the mechanical and fracture properties of porous ultra-low-*k* dielectrics are intimately linked with porosity, network structure and deposition technologies, while fracture properties of low-*k* are also sensitive to reactive species in the environment. Therefore, the low-*k* survivability upon integration is not just a simple factor of mechanical properties (Young's modulus, hardness) and fracture properties (interfacial adhesion, fracture toughness) but is rather a combination of all these properties, together with a relatively high mechanical or thermomechanical load and environmental effects that will determine its fate.

## References

[1] Moore, G.E. (1965) Cramming more components into integrated circuits, *Electronics*, **38** (8), April.

[2] Maex, K., Baklanov, M.R., Shamiryan, D., Iacopi, F., Brongersma, S.H. and Yanovitskaya, Z.S. (2003) Low dielectric constant materials for microelectronics. *J. Appl. Phys.*, **93**, 8793.

[3] Fan, H.Y. *et al.* (2007) Modulus-density scaling behaviour and framework architecture of nano-porous self-assembled silicas. *Nature Mater.*, **6**, 418.

[4] Ikeda, M., Nakahira, J., Iba, Y., Kitada, H., Nishikawa, N., Miyajima, M., Fukuyama, S., Shimizu, N., Ikeda, K., Ohba, T., Sugiura, I., Suzuki, K., Nakata, Y., Doi, S., Awaji, N. and Yano, E. (2003) A highly reliable nano-clustering silica with low dielectric constant ($k < 2.3$) and high electric modulus ($E = 10\,\text{GPa}$) for copper damascene process. Proceedings of the IEEE International Technology Conference, p. 71.

[5] Dubois, G., Magbitang, T., Volksen, W., Simonyi, E.E. and Miller, R.D. (2005) New spin-on low-*k* dielectric materials with exceptional mechanical properties. Presented at International Interconnects Technology Conference, Burlingame, California, 6–8 June 2005.

[6] Burkey, D.D. and Gleason, K.K. (2003) Structure and mechanical properties of thin films deposited from 1,3,5-trimethyl-1,3,5-trivinylcyclotrisiloxane and water. *J. Appl. Phys.*, **93**, 5143.

[7] Li, H., Lin, Y., Tsui, T.Y. and Vlassak, J.J. (2009) The effect of porogen loading on the stiffness and fracture energy of brittle organosilicates. *J. Mater. Res.*, **24**, 107.

[8] Baklanov, M., Green, M. and Maex, K. (eds) (2007) *Dielectric Films for Advanced Microelectronics*. John Wiley & Sons, Ltd, West Sussex, Ch. 1.

[9] Sreenivasan, R. and Gleason, K.K. (2009) Overview of strategies for the CVD of organic films and functional polymer layers. *Chemical Vapor Deposition*, **15**, 77.

[10] Grill, A. and Patel, V. (2001) Ultralow-*k* dielectrics prepared by plasma-enhanced chemical vapor deposition. *Appl. Phys. Lett.*, **79**, 803.

[11] Nix, W.D. (1989) Mechanical properties of thin films. *Metall. Trans.*, **A 20**, 2217.

[12] Vinci, R.P. and Vlassak, J.J. (1996) Mechanical behavior of thin films. *Annu. Rev. Mater. Sci.*, **26**, 431.

[13] Volinsky, A.A., Vella, J.B. and Gerberich, W.W. (2003) Fracture toughness, adhesion and mechanical properties of low-*k* dielectric thin films measured by nanoindentation. *Thin Solid Films*, **429**, 201.

[14] Vella, J.B., Adhihetty, I.S., Junker, K. and Volinsky, A.A. (2003) Mechanical properties and fracture toughness of organo-silicate glass (OSG) low-*k* dielectric thin films for microelectronic applications. *Int. J. Fract.*, **119**, 487.

[15] Malzbender, J., den Toonder, J.M.J., Balkenende, A.R. and de With, G. (2002) Measuring mechanical properties of coatings: a methodology applied to nano-particle-filled sol-gel coatings on glass. *Mater. Sci. Engng R–Rep.*, **36**, 47.

[16] Chen, X. and Vlassak, J.J. (2001) Numerical study on the measurement of thin film mechanical properties by means of nanoindentation. *J. Mater. Res.*, **16**, 2974.
[17] King, R.B. (1987) Elastic analysis of some punch problems for a layered medium. *Int. J. Solids Struct.*, **23**, 1657.
[18] Bhattacharya, A.K. and Nix, W.D. (1988) Analysis of elastic and plastic deformation associated with indentation testing of thin-films on substrates. *Int. J. Solids Struct.*, **24**, 1287.
[19] McElhaney, K.W., Vlassak, J.J. and Nix, W.D. (1998) Determination of indenter tip geometry and indentation contact area for depth sensing indentation experiments. *J. Mater. Res.*, **13**, 1300.
[20] Xiang, Y., Tsui, T.Y., Vlassak, J.J. and McKerrow, A.J. Measuring the elastic modulus and ultimate strength of low-*k* dielectric materials by means of the bulge test. Proceedings of the IEEE International Interconnect Technology Conferance (edited by IEEE), IEEE, Piscataway, New Jersey, p. 133.
[21] Fleck, N.A., Otoyo, H. and Needleman, A. (1992) Indentation on porous solids. *Int. J. Solids Struct.*, **29**, 1613.
[22] Tsui, T.Y., Vlassak, J.J. and Nix, W.D. (1999) Indentation plastic displacement field: Part I. The case of soft films on hard substrates. *J. Mater. Res.*, **14**, 2196.
[23] Tsui, T.Y., Vlassak, J.J. and Nix, W.D. (1999) Indentation plastic displacement field: Part II. The case of hard films on soft substrates. *J. Mater. Res.*, **14**, 2204.
[24] Saha, R. and Nix, W.D. (2002) Effects of the substrate on the determination of thin film mechanical properties by nanoindentation. *Acta Mater.*, **50**, 23.
[25] Bolshakov, A. and Pharr, G.M. (1998) Influences of pileup on the measurement of mechanical properties by load and depth-sensing indentation techniques. *J. Mater. Res.*, **13**, 1049.
[26] Chen, X., Xiang, Y. and Vlassak, J.J. A novel technique of measuring the mechanical properties of porous materials by nanoindentation: with application to low-*k* dielectric thin films. *J. Mater. Res.*, **21** (3), 715–724.
[27] Farnell, G.W. and Adler, E.L. (1972) *Physical Acoustics*, Academic Press, New York and London.
[28] Carlotti, G., Doucet, L. and Dupeux, M. (1996) Comparative study of the elastic properties of silicate glass films grown by plasma enhanced chemical vapor deposition. *J. Vac. Sci. Technol. B*, **14**, 3460.
[29] Chapelon, L.L., Vitiello, J., Neira, D., Torres, T., Royer, J.C., Barbier, D., Naudin, F., Tas, G., Mukundhan, P. and Clerico, J. (2006) Measuring the Young's modulus of ultralow-*k* materials with the non destructive picosecond ultrasonic method. *Microelectron. Engng*, **83**, 2346–2350.
[30] Huang, H.B. and Spaepen, F. (2000) Tensile testing of free-standing Cu, Ag and Al thin films and Ag/Cu multilayers. *Acta Mater.*, **48**, 3261.
[31] Read, D.T. and Dally, J.W. (1993) A new method for measuring the strength and ductility of thin-films. *J. Mater. Res.*, **8**, 1542.
[32] Schwaiger, R. and Kraft, O. (2004) Analyzing the mechanical behavior of thin films using nanoindentation, cantilever microbeam deflection, and finite element modeling. *J. Mater. Res.*, **19**, 315; original work was by T.P. Wechs, in the late 1980s, early 1990s.
[33] Vlassak, J.J. and Nix, W.D. (1992) A new bulge test technique for the determination of Young's modulus and Poisson's ratio of thin films. *J. Mater. Res.*, **7**, 3242.
[34] Xiang, Y., Chen, X. and Vlassak, J.J. (2005) The plane-strain bulge test for thin films. *J. Mater. Res.*, **20**, 2360.
[35] Nemat-Nasser, S. and Hori, M. (1999) *Micromechanics: Overall Properties of Heterogeneous Materials*, 2nd revised edn, Elsevier, Amsterdam New York.
[36] Roscoe, R. (1952) The viscosity of suspensions of rigid spheres. *Br. J. Appl. Phys.*, **3** (8), 267–269.
[37] Christensen, R.M. and Lo, K.H. (1979) Solutions for effective shear properties in 3 phase sphere and cylinder models. *J. Mechanics and Physics of Solids*, **27** (4), 315–330.
[38] Benveniste, Y. (1987) A new approach to the application of Mori–Tanaka theory in composite-materials. *Mechanics of Mater.*, **6** (2), 147–157.
[39] Eshelby, J.D. (1957) The determination of the elastic field of an ellipsoidal inclusion, and related problems. *Proc. R. Soc. Lond. Ser. A – Math. Phys. Sci.*, **241** (1226), 376–396.
[40] Rice, R.W. (1996) Evaluation and extension of physical property – porosity models based on minimum solid area. *J. Mater. Sci.*, **31** (1), 102–118.

[41] Torquato, S. (1997) Effective stiffness tensor of composite media. 1. Exact series expansions. *J. Mechanics and Physics of Solids*, **45** (9), 1421–1448.

[42] Roberts, A.P. and Garboczi, E.J. (2000) Elastic properties of model porous ceramics. *J. Am. Ceram. Soc.*, **83** (12), 3041–3048.

[43] Roberts, A.P. and Garboczi, E.J. (2002) Computation of the linear elastic properties of random porous materials with a wide variety of microstructure. *Proc. R. Soc. Lond. Ser. A – Math. Phys. Engng Sci.*, **458** (2021), 1033–1054.

[44] Iacopi, F., Travaly, Y., Eyckens, B., Waldfried, C., Abell, T., Guyer, E.P., Gage, D.M., Dauskardt, R.H., Sajavaara, T., Houthoofd, K., Grobet, P., Jacobs, P. and Maex, K. (2006) Short-ranged structural rearrangement and enhancement of mechanical properties of organosilicate glasses induced by ultraviolet radiation. *J. Appl. Phys.*, **99**, 053511.

[45] Gage, D.M., Stebbins, J.F., Peng, L., Cui, Z., Al-Bayati, A., MacWilliams, K.P., M'Saad, H. and Dauskardt, R.H. (2008) Effects of UV cure on glass structure and fracture properties of nanoporous carbon-doped oxide thin films. *J. Appl. Phys.*, **104** (4), 043513.

[46] Lin, Y.B., Tsui, T.Y. and Vlassak, J.J. (2006) Octamethylcyclotetrasiloxane-based, low-permittivity organosilicate coatings – composition, structure, and polarizability. *J. Electrochem. Soc.*, **153** (7), F144–F152.

[47] Lin, Y.B. and Vlassak, J.J. (2008) PECVD low-permittivity organosilicate glass coatings: adhesion, fracture and mechanical properties. *Acta Mater.*, **56** (17), 4932–4943.

[48] Hashin, Z. and Shtrikman, S. (1963) A variational approach to the theory of the elastic behaviour of multiphase materials. J. Mechanics and Physics of Solids, **11** (2), 127–140.

[49] Wang, C.L. *et al.* (2002) Nanometer-scale pores in low-*k* dielectric films probed by positron annihilation lifetime spectroscopy. *Appl. Phys. Lett.*, **81** (23), 4413–4415.

[50] Grill, A. and Neumayer, D.A. (2003) Structure of low dielectric constant to extreme low dielectric constant SiCOH films: Fourier transform infrared spectroscopy characterization. *J. Appl. Phys.*, **94** (10), 6697–6707.

[51] Iacopi, F., Beyer, G., Travaly, Y., Waldfried, C., Gage, D.M., Dauskardt, R.H. Houthoofd, K., Jacobs, P., Adriaensens, P., Schulze, K., Schulz, S.E., List, S. and Carlotti, G. (2007) Thermomechanical properties of thin organosilicate glass films treated with ultraviolet-assisted cure. *Acta Mater.*, **55** (4), 1407–1414.

[52] Kim, S. *et al.* (2004) Organosilicate spin-on glasses – I. Effect of chemical modification on mechanical properties. *J. Electrochem. Soc.*, **151** (3), F37–F44.

[53] Rathore, J.S., Interrante, L.V. and Dubois, G. (2008) Ultra low-*k* films derived from hyperbranched polycarbosilanes (HBPCS). *Advd Functional Mater.*, **18** (24), 4022–4028.

[54] Dubois, G. *et al.* (2008) Superior mechanical properties of dense and porous organic/inorganic hybrid thin films. *J. Sol-Gel Sci. Technol.*, **48** (1–2), 187–193.

[55] Dubois, G. *et al.* (2007) Molecular network reinforcement of sol-gel glasses. *Advd Mater.*, **19** (22), 3989.

[56] Gates, S.M. *et al.* (2009) Adjusting the skeleton and pore structure of porous SiCOH dielectrics. *J. Electrochem. Soc.*, **156** (10), G156–G162.

[57] Bazant, M.Z., Kaxiras, E. and Justo, J.F. (1997) Environment-dependent interatomic potential for bulk silicon. *Phys. Rev. B*, **56** (14), 8542.

[58] Van Ginhoven, R.M., Jonsson, H. and Corrales, L.R. (2005) Silica glass structure generation for ab initio calculations using small samples of amorphous silica. *Phys. Rev. B*, **71** (2), 024208.

[59] Guttman, L. and Rahman, S.M. (1998) Simulation of the structure of amorphous silicon dioxide. *Phys. Rev. B*, **37** (5), 2657.

[60] Li, H. *et al.* (2011) Stiffening of organosilicate glasses by organic cross-linking. *Acta Mater.*, **59** (1), 44–52.

[61] Knaup, J.M. *et al.* (2011) Influence of $CH_2$ content and network defects on the elastic properties of organosilicate glasses. *Phys. Rev. B*, **83** (5), 054204.

[62] Thorpe, M.F. (1983) Continuous deformations in random networks. *J. Non-Crystalline Solids*, **57** (3), 355–370.

[63] He, H. and Thorpe, M.F. (1985) Elastic properties of glasses. *Phys. Rev. Lett.*, **54** (19), 2107–2110.

[64] Chapelon, L.L., Vitiello, J., Gonchond, J.P., Barbier, D. and Torres, J. (2006) UV curing effects on mechanical and electrical performances of a PECVD non-porogen porous SiOC:H films (in *k* [2.2–2.4] range) for 45 nm node and below. *Microelectron. Engng*, **83**, 2136.
[65] Zenasni, A., Jousseaume, V., Holliger, P., Favennec, L., Gourhant, O., Maury, P. and Gerbaud, P. (2007) The role of ultraviolet radiation during ultralow *k* films curing: strengthening mechanisms and sacrificial porogen removal. *J. Appl. Phys.*, **102**, 094107.
[66] Masuura, M., Goto, K., Miura, N., Haag, J.M., Hashii, S. and Asai, K. (2006) Film characterization of ultra low-*k* dielectrics modified by UV curing with different wavelength bands. *Mater. Res. Soc. Symp. Proc.*, **914**, F01–F06.
[67] Aksenov, G., De Roest, D., Verdonck, P., Dultsev, F.N., Marsik, P., Shamiryan, D., Arai, H., Takamure, N. and Baklanov, M.R. (2009) Optimization of low-*k* UV curing: effect of wavelength on critical properties of the dielectric. *Mater. Res. Soc. Symp. Proc.*, **1156**, D02.
[68] Berry, I.L., Han, Q., Waldfried, C., Escorcia, O. and Becknell, A. (2004) Use of plasma assisted sublimation to remove photoresist from porous low-*k* dielectrics. SEMI® Technical Symposium: Innovations in Semiconductor Manufacturing.
[69) Chen, W., Han, Q.Y., Most, R., Waldfried, C., Escorcia O. and Berry, I. (2004) Plasma impacts to an O–SiC low-*k* barrier film. *J. Electrochem. Soc.*, **151**, F182–F188.
[70] Urbanowicz, A.M., Shamiryan, D., Marsik, P., Travaly, Y., Jonas, A., Verdonck, P., Vanstreels, K., Ferchichi, A., De Roest, D., Sprey, H., Matshushita, K., Kaneko, S., Tsuji, N., Luo, S., Escorcia, O., Berry, I.L., Waldfried, C., De Gendt, S. and Baklanov, M.R. (2009) Improved low-*k* dielectric properties using He/$H_2$ plasma for resist removal. MRS Symposia Proceedings XXIV (eds M. Naik, R. Shaviv, T. Yoda and K. Ueno), Materials Research Society, Pittsburgh, Pennsylvania, pp. 594–598.
[71] Urbanowicz, A.M., Vanstreels, K., Shamiryan, D., De Gendt, S. and Baklanov, M.R. (2009) Effect of porogen residue on chemical, optical, and mechanical properties of CVD SiCOH low-*k* materials. *Electrochem. Solid-State Lett.*, **12**, H292.
[72] Urbanowicz, A., Vanstreels, K., Verdonck, P., Van Besien, E., Christos, T., Shamiryan, D., De Gendt, S. and Baklanov, M.R. (2011) Effect of UV wavelength on the hardening process of porogen-containing and porogen-free ultralow-*k* plasma-enhanced chemical vapor deposition dielectrics. *J. Vac. Sci. Technol. B*, **29** (3), 032201.
[73] Vanstreels, K. and Urbanowicz, A.M. (2010) Nanoindentation study of thin plasma enhanced chemical vapor deposition SiCOH low-*k* films modified in He/$H_2$ downstream plasma. *J. Vac. Sci. Technol. B*, **28** (1), 173.
[74] Marsik, P., Urbanowicz, A., Verdonck, P., Ferchichi, K., De Roest, D., Prager, L. and Baklanov, M.R. (2009) Engineering of chemical and physical properties of low-*k* materials by different wavelength of UV light. Proceedings of the 25th Advanced Metallization Conference, 2008, MRS Symposia Proceedings XXIV (eds M. Naik, R. Shaviv, T. Yoda and K. Ueno), Materials Research Society, Pittsburgh, Pennsylvania, pp. 543–549.
[75] Zenasni, A., Ciaramella, F., Jousseaume, V., Le Cornec, C. and Passemard, G. (2007) Investigation of porogen behavior during the curing process of ultralow-*k* spin-on materials. *J. Electrochem. Soc.*, **154**, G6.
[76] Marsik, P., Verdonck, P., De Roest, D. and Baklanov, M.R. (2010) Porogen residues detection in optical properties of low-*k* dielectrics cured by ultraviolet radiation. *Thin Solid Films*, **518**, 4266.
[77] Eslava, S., Eymery, G., Marsik, P., Iacopi, F., Kirschhock, C.E.A., Maex, K., Martens, J.A. and Baklanov, M.R. (2008) Optical property changes in low-*k* films upon ultraviolet-assisted curing. *J. Electrochem. Soc.*, **155**, G115–G120.
[78] Urbanowicz, A.M., Vanstreels, K., Verdonck, P., Shamiryan, D., De Gendt, S. and Baklanov, M.R. (2010) Improving mechanical robustness of ultralow-*k* SiOCH plasma enhanced chemical vapor deposition glasses by controlled porogen decomposition prior to UV-hardening. *J. Appl. Phys.*, **107**, 104122.
[79] Senapati, M.R. (2006) *Advanced Engineering Chemistry*, Laxmi Publications, p. 56.
[80] Eslava, S., Iacopi, F., Urbanowicz, A.M., Kirschhock, C.E.A., Maex, K., Martens, J.A. and Baklanov, M.R. (2008) Ultraviolet-assisted curing of organosilicate glass low-*k* dielectric by excimer lamps. *J. Electrochem. Soc.*, **155**, G231–G236.

[81] Prager, L., Marsik, P., Wennrich, L., Baklanov, M.R., Naumov, S., Pistol, L., Schneider, D., Gerlach, J.W., Verdonck, P. and Buchmeiser, M.R. (2008) Effect of pressure on efficiency of UV curing of CVD-derived low-*k* material at different wavelengths. *Microelectron. Engng.*, **85** (10), 2094.

[82] Lacombe, R. (2006) *Adhesion Measurement Methods – Theory and Practice*, CRC Press, Boca Raton, Florida.

[83] Dieter, G.E. (1976) *Mechanical Metallurgy*, 2nd edn, McGraw-Hill.

[84] Callister Jr, W.D. (2003) *Materials Science and Engineering: An Introduction*, 6th edn, John Wiley & Sons, Inc., New York.

[85] *Standard Test for Plane Strain Fracture Toughness of Metallic Materials*, ASTME-399, American Society for Testing and Materials, Philadelphia, Pennsylvania, 1987.

[86] Cherepanov, G.P. (1979) *Mechanics of Brittle Fracture*, McGraw-Hill.

[87] Jiang, Z., Lu, F.X., Tang, W.Z., Wang, S.G., Tong, Y.M., Huang, T.B. and Liu, J.M. (2000) Accurate measurement of fracture toughness of free standing diamond films by three-point bending tests with sharp pre-cracked specimens. *Diamond Relat. Mater.*, **9**, 1734.

[88] Jaeger, G., Endler, I., Heilmaier, M., Bartsch, K. and Leonhardt, A. (2000) A new method of determining strength and fracture toughness of thin hard coatings. *Thin Solid Films*, **377–378**, 382.

[89] Gille, G. and Wetzig, K. (1983) Investigations on mechanical behaviour of brittle wear-resistant coatings. I. Experimental results. *Thin Solid Films*, **110** (1), 37.

[90] Gille, G. (1984) Investigations on mechanical behaviour of brittle wear-resistant coatings. II: Theory. *Thin Solid Films*, **111** (3), 201.

[91] Cox, L.C. (1988) The four-point bend test as a tool for coating characterization. *Surf. Coat. Technol.*, **36**, 807.

[92] Bellido-Gonzalez, V., Stefanopoulos, N. and Deguilhen, F. (1995) Friction monitored scratch adhesion testing. *Surf. Coat. Technol.*, **74–75**, 884.

[93] Li, X. and Bhushan, B. (1999) Micro/nanomechanical and tribological characterization of ultrathin amorphous carbon coatings. *J. Mater. Res.*, **14** (6), 2328.

[94] Sundararajan, S. and Bhushan, B. (2001) Development of a continuous microscratch technique in an atomic force microscope and its applications to study scratch resistance of ultrathin hard amorphous carbon coatings. *J. Mater. Res.*, **16** (2), 437.

[95] Panjan, P., Cekada, M. and Navinsek, B. (2003) A new experimental method for studying the cracking behaviour of PVD multilayer coatings. *Surf. Coat. Technol.*, **174–175**, 55.

[96] Voevodin, A.A., Rebholz, C., Schneider, J.M., Stevenson, P. and Matthews, A. (1995) Wear resistant composite coatings deposited by electron enhanced closed field unbalanced magnetron sputtering. *Surf. Coat. Technol.*, **73**, 185.

[97] Harrry, E., Rouzaud, A., Juliet, P., Pauleau, Y. and Ignat, M. (1999) Failure and adhesion characterization of tungsten–carbon single layers, multilayered and graded coatings. *Surf. Coat. Technol.*, **116–119**, 172.

[98] Shum, P.W., Li, K.Y., Zhou, Z.F. and Shen, Y.G. (2004) Structural and mechanical properties of titanium–aluminium–nitride films deposited by reactive close-field unbalanced magnetron sputtering. *Surf. Coat. Technol.*, **185**, 245.

[99] Marshall, D.B. and Evans, G. (1984) Measurement of adherence of residually stressed thin films by indentation. I. Mechanics of interface delamination. *J. Appl. Phys.*, **56**, 2632–2638.

[100] Kriese, M.D. and Gerberich, W.W. (1999) Experimental considerations for indentation-induced adhesion measurement of multilayered thin films. *J. Mater. Res.*, **14** (7), 3007.

[101] Volinsky, A.A., Tymiak, N.I., Kriese, M.D., Gerberich, W.W. and Hutchinson, J.W. (1999) Quantitative modeling and measurement of copper thin film adhesion. *Mater. Res. Soc. Symp. Proc.*, **539**, 277.

[102] Volinsky, A.A., Moody, N.R. and Gerberich, W.W. (2002) Interfacial toughness measurements for thin films on substrates. *Acta Mater.*, **50** (3), 441.

[103] Lawn, B.R., Evans, A.G. and Marshall, D.B. (1980) Elastic/plastic indentation damage in ceramics: median/radial crack system. *J. Am. Ceram. Soc.*, **63** (9–10), 574–581.

[104] Lawn, B. and Winslaw, R. (1975) Indentation fracture: principles and applications. *J. Mater. Sci.*, **10** (6), 1049.

[105] Cotterell, B. and Chen, Z. (2000) Buckling and cracking of thin films on compliant substrate under compression. *Int. J. Fract.*, **104**, 169–179.

[106] Chen, Z., Cotterell, B. and Wang, W. (2002) The fracture of brittle thin films on compliant substrate in flexible displays. *Engng Fract. Mechanics*, **69**, 597–603.

[107] Britvec, S.J. (1973) *The Stability of Elastic Systems*, Pergamon Press, New York.

[108] Drory, M.D., Dauskardt, R.H., Kant, A. and Ritchie, R.O. (1995) Fracture of synthetic diamond. *J. Appl. Phys.*, **78** (5), 3083.

[109] Harry, E., Rouzaud, A., Ignat, M. and Juliet, P. (1998) Mechanical properties of W and W(C) thin films: Young's modulus, fracture toughness and adhesion. *Thin Solid Films*, **332**, 195.

[110] Harry, E., Ignat, M., Pauleau, Y., Rouzaud, A. and Juliet, P. (2000) Mechanical behaviour of hard PVD multilayered coatings. *Surf. Coat. Technol.*, **125**, 185.

[111] Harry, E., Ignat, M., Rouzaud, A. and Juliet, P. (1999) Cracking investigation of W and W(C) films deposited by physical vapor deposition on steel substrates. *Surf. Coat. Technol.*, **111**, 177.

[112] Hu, M.S. and Evans, A.G. (1989) The cracking and decohesion of thin films on ductile substrates. *Acta Metall.*, **37** (3), 917.

[113] Mercado, L.L. *et al.* (2003) Analysis of flip-chip packaging challenges on copper low-*k* interconnects. Proceedings of the 53rd Electronic Components and Technology Conference.

[114] Dauskardt, R. *et al.* (1998) Adhesion and debonding of multi-layer thin film structures. *Engng Fract. Mechanics*, **61** (1), 141–162.

[115] Li, H., Lin, Y., Tsui, T.Y. and Vlassak, J.J. (2009) The effect of porogen loading on the stiffness and fracture energy of brittle organosilicates. *J. Mater. Res.*, **24** (1), 107–116.

[116] Vandeperre, L.J., Wang, J. and Clegg, W.J. (2004) Effects of porosity on the measured fracture energy of brittle materials. *Phil. Mag.*, **84** (34), 3689–3704.

[117] Guyer, E.P., Patz, M. and Dauskardt, R.H. (2006) Fracture of nanoporous methyl silsesquioxane thin-film glasses. *J. Mater. Res.*, 21 (4), 882.

[118] Guyer, E.P. and Dauskardt, R.H. (2004) Fracture of nanoporous thin-film glasses, *Nature Mater.*, **3** (1), 53–57.

[119] Vlassak, J.J., Lin, Y. and Tsui, T.Y. (2005) Fracture of organosilicate glass thin films: environmental effects. *Mater. Sci. Engng A – Structural Materials Properties Microstructure and Processing*, **391** (1–2), 159–174.

[120] Lin, Y.B., Tsui, T.Y. and Vlassak, J.J. (2007) Water diffusion and fracture in organosilicate glass film stacks, *Acta Mater.*, **55** (7), 2455–2464.

[121] Tsui, T.Y., McKerrow, A.J. and Vlassak, J.J. (2006) The effect of water diffusion on the adhesion of organosilicate glass film stacks. *J. Mechanics and Physics of Solids*, **54** (5), 887–903.

[122] Li, H., Tsui, T.Y. and Vlassak, J.J. (2009) Water diffusion and fracture behavior in nanoporous low-*k* dielectric film stacks. *J. Appl. Phys.*, **106** (3), 033503.

[123] Lin, Y.B., Tsui, T.Y. and Vlassak, J.J. (2010) Adhesion degradation and water diffusion in nanoporous organosilicate glass thin film stacks. *J. Electrochem. Soc.*, **157** (2), G53–G56.

[124] Shaw, T. (2009) Private communication, 2 March 2009.

[125] Bao, J. *et al.* (2008) Mechanistic study of plasma damage of low *k* dielectric surfaces. *J. Vac. Sci. Technol. B*, **26** (1), 219–226.

[126] Guyer, E.P., Gantz, J. and Dauskardt, R.H. (2007) Aqueous solution diffusion in hydrophobic nanoporous thin-film glasses. *J. Mater. Res.*, **22** (3), 710–718.

# 11

# Electrical Breakdown in Advanced Interconnect Dielectrics

*Ennis T. Ogawa*[1] *and Oliver Aubel*[2]

[1]*Broadcom Corporation, Irvine, California, USA*

[2]*GLOBALFOUNDRIES, Dresden Module One LLC & Co. KG, Dresden, Germany*

## 11.1 Introduction

Prior to the start of this century, generating a fully integrated back-end-of-line (BEOL) interconnect process flow for advanced CMOS technology was viewed as an evolutionary effort, because much of the work in advanced CMOS integration focused on meeting only the technological demands of dimensional scaling. In essence, the critical dimensions continued to scale downward by a factor of about 0.7 per technology generation, but the materials and processes that were used to build up the integrated stack were essentially the same; namely Al (with $Cu_x$ doping) for the metallization and $SiO_2$ as the interlevel dielectric. This situation, however, changed toward the end of the last decade of the 20th century after it became apparent that new BEOL materials would be needed. The introduction of new materials (Cu for electrical resistivity and low-*k* dielectrics for capacitance) was needed to ameliorate the effects of increased RC interconnect delay due to interconnect dimensional shrinkage [1, 2]. Thus, dimensional scaling still progressed at roughly the same rate as before, but new materials were introduced at a sub-250 nm critical (interconnect) dimension to provide an additional performance benefit to dimensional scaling. First, Cu metallization was introduced into manufacturing in 1997 [3] and, recently, low-*k* (LK) dielectrics of

*Advanced Interconnects for ULSI Technology*, First Edition. Edited by Mikhail R. Baklanov, Paul S. Ho and Ehrenfried Zschech.

different types have been incorporated into the BEOL stack [4, 5], starting with the incorporation of F-doped silica dielectric around the year 2000 [6–10] at around the 180 nm node. The introduction of these new materials, however, has greatly complicated the efforts to retain the benefits of interconnect performance scaling from both process/integration and reliability perspectives. Since Cu (with maybe some alloying) is likely to remain the interconnect metallization of choice for the foreseeable future (at least down to the 15 nm node), the main challenge for new materials integration has shifted toward the implementation of progressively lower-*k* dielectrics, where *k* refers to the relative permittivity or 'dielectric constant' in the equation for the dielectric permeability ($\varepsilon = k\varepsilon_0$, where $\varepsilon$ is the dielectric permittivity and $\varepsilon_0$ is the permittivity of free space).

### 11.1.1 Dual-Damascene Integration of Low-*k* Dielectrics

A typical BEOL stack (see Figure 11.1) found in advanced CMOS applications consists of several stacked layers of Cu metal lines or interconnects, running along a given layer, and dielectric materials, insulating these metal lines from one another. These interconnects can connect to other metal lines in other layers directly above or below them through Cu via structures. At the bottom interconnect layer, known as M1, these Cu lines will eventually be connected to even lower levels through W-metal plugs, also commonly called 'contact metal'. These W-contacts either connect (not shown) to silicided well layers, maybe poly-Si interconnect, or silicided poly-silicon gate contacts for conventional $SiO_xN_y$ gate oxide transistors (or, for high-*k*/metal gates, the metal gate electrode) [11]. At M1, the minimum metal-to-metal pitch is at its narrowest compared to other interconnect levels (so that the half-pitch defines the critical interconnect dimension for a given technology node). Above M1 (known as 'intermediate metallization'), the metal pitch and metal thickness are usually similar to those at M1 but might be somewhat relaxed (wider or thicker, respectively, depending on the design practice or product application). At the highest metal levels (still Cu-based) – where global interconnect, signal routing, or power connections are usually found – both the minimum metal pitch and metal thickness will be much larger so that they can be used ultimately to make contact to wirebond pads (usually Al-metal based) or pads with proprietary underbump metallurgy for flip-chip applications. At the level of the wirebond pads (not shown), some routing metallization (Al-based) may be used.

The integration approach used to fabricate Cu-based interconnects is the dual-damascene (DD) process. A dual-damascene integrated via is shown in Figure 11.2. This process is quite complicated and will have different variants – depending on the thin film materials being integrated and the specific critical dimensions (technology node) – and will be explained only at a rudimentary level. In dual-damascene integration, a dielectric film (a low-*k* dielectric) is deposited on the dielectric layer covering the transistor gate/contact area (mostly oxide-based materials), and suitable patterns are etched within this layer. These patterns will be used to build up the Cu interconnects of the first metal layer, which will be described later. After the first metal layer is completed, a second metal layer can be fabricated. From the second metal layer, connections to the first metallization level directly underneath are formed by via holes, which are patterned over specific areas of the

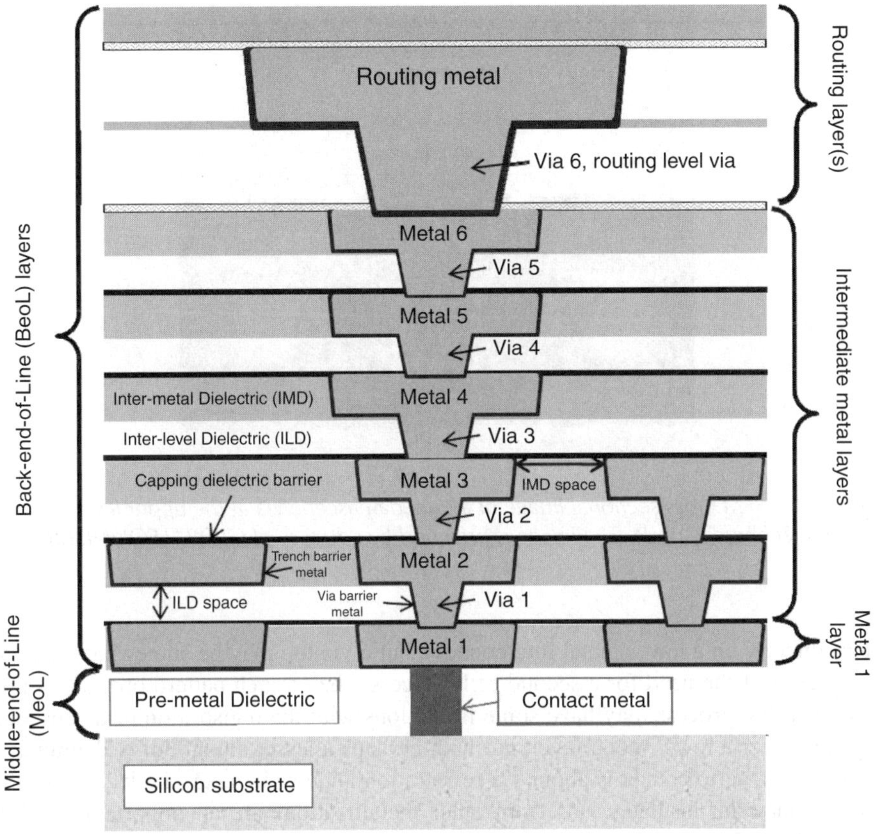


***Figure 11.1*** *Cross-sectional graphic of a typical BeoL stack. Cu metallization must be fully encapsulated by metal barrier on the sides and bottom and on top by capping the dielectric barrier (or maybe by capping the metal liner deposited by the electroless plating method). The intermetal dielectric (IMD) and intermetal dielectric (ILD) are usually the same dielectric material. Above routing metallization, pads to the outside contact are present (not shown) Reprinted with permission from [4].Copyright 2003, American Institute of Physics*

damascene pattern (typically at line ends) so that continuous electrical connection can be made to a lower metal. Two approaches for via formation can be used: either the via is developed before the interconnect line (lithography step) ('via first' process) or vice versa ('trench first' process). It is worth noting that trench and via patterns typically have some taper to them when viewed in cross-section [12, 13] so that the trench-to-trench dielectric spacing at the trench top will be slightly smaller than that found at the trench bottom. Such a taper, if not well controlled, will exacerbate line-edge-roughness (LER) effects at the trench top, which is detrimental to LK/ULK dielectric reliability [14–16]. Furthermore, the opening of a via hole will end up widening the trench feature significantly at the trench top so that via-to-adjacent trench spacing or via-to-adjacent-via spacing will be smaller at the trench top in that vicinity relative to the nominal trench-to-trench spacing at the trench top. The via first process may have some benefit in terms of direct ability to register a via

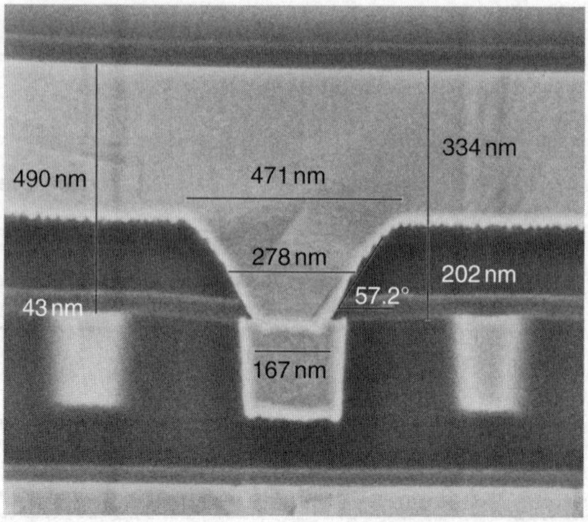


*Figure 11.2* SEM cross-sectional image of a dual-damascene via at the upper level for an advanced technology node. The image is provided by courtesy of GLOBALFOUNDRIES, Inc.

pattern directly on a lower metal interconnect, but a via top may be somewhat wider as a consequence of the need for a second etch process when trench pattern opening is done. The trench first process may have some limitations with the registration of a subsequent via pattern over a lower metal layer, but with perhaps a lesser amount of widening at the top of a via. Due to concerns about via registration and/or via taper, the locations where dielectric material insulates vias from other metallizations are an important reliability concern [17–19].

After trench/via patterning and opening of any capping dielectric to a lower metal, a metal diffusion barrier (typically Ta- or Ti-based) is deposited within the trench and via (trench only if it is M1). The barrier thickness is generally thicker at the bottom of the feature than on the sidewalls, but the sidewall may be thickened at the expense of the bottom by a barrier resputtering method. A Cu seed layer is then deposited using physical vapor deposition (PVD). The trench/via is then filled completely with Cu metal using electrochemical deposition (ECD) until an overburden layer lies on top. This overburden is then removed using chemical mechanical polishing (CMP) to form the trench pattern. Finally, after some cleaning steps, the top surface of the damascene trench is capped by a dielectric diffusion barrier (typically a variant of $SiC_xN_y$), or maybe as we proceed to a more advanced BEOL, a metal liner (such as $CoW_xP_y$) could be used to cap the Cu-filled damascene trench by selective electroless plating (SEP) methods [19, 20]. This metal liner may or may not be capped by an additional dielectric barrier layer. The Cu metallization must be completely surrounded by diffusion barrier material to keep it from out-diffusing into low-*k* dielectric and prematurely shorting interconnects [21–25]. Cu is known to be a ready diffuser into such dielectrics under electrical bias and greatly reduces dielectric reliability [26, 27], although injection of Cu into dielectric should be an activated process [28]. The next dual-damascene metallization is then similarly processed until a full BEOL stack is completed (uppermost metallization process flow is not described for space limitation reasons).

Depending upon the application at a given technology node, the lowest-*k* dielectrics will be found at the intermediate metallization levels above M1. At M1, the lowest-*k* dielectric may or may not be used, depending on the performance margin and/or reliability concerns. At the level where power connections are running, a somewhat higher-*k* dielectric may be used, but a lower-*k* dielectric may be used there too (if reliability concerns for chip-package interaction or CPI, are not warranted). If, at these upper levels, a signal is also transferred over longer distances, then a lower-*k* dielectric will be the most efficient with respect to signal delay reduction. Above the highest metallization using Cu, however, more conventional $SiO_x$-based and nitride liner dielectrics are used to serve as protection layers against low-*k* exposure to any ambient corrosive gases or moisture and to ensure better mechanical strength in the BEOL stack against external stresses.

### 11.1.2 Low-*k* Types and Integrating Low-*k* Dielectrics

Although there are a variety of materials that have a low dielectric constant [4, 5], the semiconductor industry has focused on materials based on $SiO_2$ for advanced interlevel (ILD) or intermetal (IMD) dielectrics, presumably because of their compatibility with existing BEOL processes and also their superior mechanical properties in comparison to organic low-*k* varieties [29]. The typical CVD $SiO_2$ used in earlier BEOL processes has a dielectric constant of around 4.2 as compared to thermal oxide ($k = 3.9$) used in gate oxides because CVD oxide has a higher concentration of –OH bonds. More generally, the nominally higher dielectric constant is related to the relatively high polarizability of Si–O bonds. Consequently, a lower *k* dielectric based on an Si–O backbone must be achieved through individual atomic replacement of Si–O bonds with generally less polarizable species (by lowering the ionic and/or electronic contributions to the polarizability). An early example was to substitute the Si–F bonds for Si–O in F-doped silica glasses (FSG), which has a dielectric constant of about 3.6 [30–33]. In this case, however, the Si–F bond is actually more polar than Si–O, but the now-missing electronic contribution from the replaced $O^{2-}$ ions results in a lower *k* value. A further decrease in the $SiO_2$ dielectric constant is achieved by substitution of Si–O with less-polar Si–$CH_3$ bonds in so-called 'carbon-doped oxides' (CDO or OSG for organosilicate glass); see Figure 11.3 for a stick representation of an $SiO_2$ tetrahedral structure [34], networked CDO, and porous CDO. Table 11.1 shows a list of some of the relevant materials properties. CDOs are typically deposited using a plasma-enhanced chemical vapor deposition (PE-CVD) reactor system. Spin-on low-*k* dielectrics are also available and can generally have similar compositions to PE-CVD CDOs with Si–O–C–H (or SiCOH) making up the basic atomic ingredients. In CDOs, methyl groups (–$CH_3$) and some hydrogen atoms replace some 10–25% of the oxygen atoms, and the resulting $SiO_4$ network packs at a substantially lower density (1.2–1.4 g/cm$^3$). Thus, CDOs intrinsically have a naturally increased interatomic distance or intrinsic 'free volume' (or ultramicroporosity, meaning dielectrics with pore size < 10 Å [5]) within the silica matrix (CDO effective pore size is ~ 5 Å), which further reduces the dielectric constant. The resulting dielectric has a *k* value ranging from about 2.6 to 3.3. Dielectrics with *k* values in this range, where doping of the Si–O matrix with methyl groups is used to lower the overall polarizability, are considered to be conventional 'low-*k*' or, simply, LK dielectrics. It should be noted, however, that potentially useful SiCOH dielectrics exist that possess an

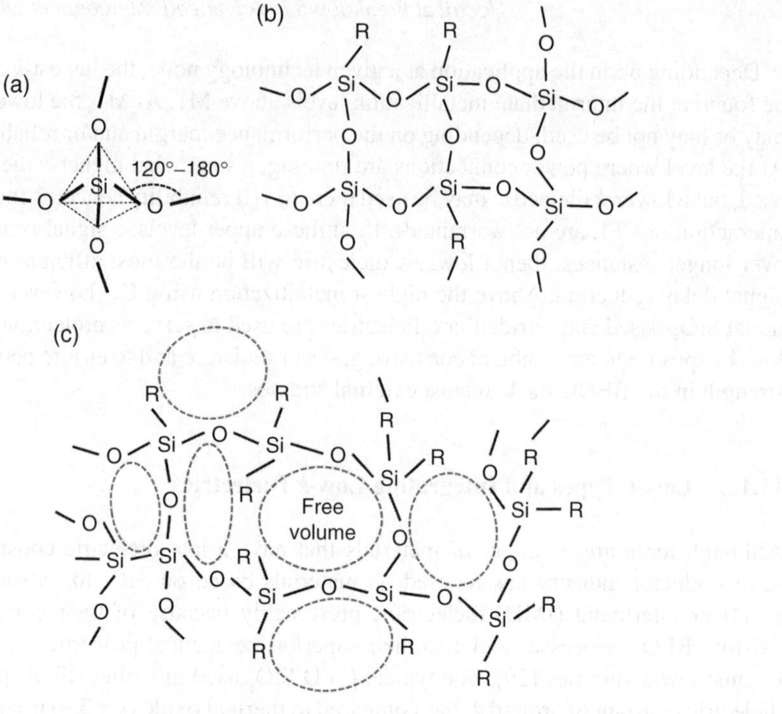


***Figure 11.3*** *(a) Local tetrahedral structure of $SiO_2$. The ideal O–Si–O bond angle is 109°; however, in an amorphous silica network, the bond angle can range between 120° and 180°. (b) Generic structure of CDO. 'R' groups represent replacement groups for O atom and can be $CH_3$, $CH_2$ (bridging), OH, and H atoms, etc. (c) Porosity incorporation occurs by a looser network formation so that a three-dimensional free space is generated within the dielectric*

***Table 11.1*** *Some nominal dielectric properties for the different dielectrics used in CMOS technology are shown for comparison. Porosity range estimates are approximate only but roughly follow K. Maex et al., J. Appl. Phys., 93(11), 8793 (2003). A low-k classification scheme is included for referencing within this work. Classifications include 'LK' ≡ 'low-k'; 'VLK' ≡ 'very low-k'; 'ULK' ≡ 'ultra-low-k'; 'hp-ULK' ≡ 'hyper-porous ULK' [or equivalently, ELK ≡ Extreme LK]. Other authors will have alternative definitions. The terms 'CDO' ≡ carbon-doped oxide and 'OSG' ≡ organo-silicate glass are actually interchangeable; 'SO' ≡ 'spin-on'. Generally, we will refer to k ≥ 2.7 as 'LK' and k < 2.7 as 'ULK' to distinguish those that do, respectively, not have and have porosity incorporation*

| Material | Low-*k* classification | Formation | $k_{eff}$ | Porosity (%) |
|---|---|---|---|---|
| $SiO_2$ (gate oxide) | – | Thermal anneal | 3.9 | 0 |
| TEOS (BeoL silica) | – | PE-CVD | 4.2 | 0 |
| $SiO_xF_y$ (FSG) | LK | PE-CVD | 3.6 | 0 |
| $SiO_xC_yH_z$ (CDO) | LK | PE-CVD | 2.7–2.9 | 0 |
| p-$SiO_xC_yH_z$ (CDO) | VLK | PE-CVD | 2.3–2.7 | 0–10 |
| Very p-$SiO_xC_yH_z$ (CDO) | ULK | PE-CVD | 2.0–2.3 | 10–40 |
| p-SO-OSG (porous OSG) | ULK | Spin-on | 2.0–2.3 | 10–40 |
| Hyper p-$SiO_xC_yH_z$ (CDO) | hp-ULK | PE-CVD | ≤2.0 | > 40 |
| Hyper p-SO-OSG (porous OSG) | hp-ULK | Spin-on | ≤2.0 | > 40 |

additional backbone structure to O–Si–O; namely a methylene bridge (Si–$CH_2$–Si). The different backbone structures coexist and interlock within a SiCOH low-*k* dielectric [35, 36]. CDOs were introduced into manufacturing at around 90 nm [37–44] and have been a mainstay low-*k* dielectric in high-volume manufacturing to this point (90 nm, 65 nm, and 45 nm nodes) [29, 45].

More substantial decreases in the dielectric constant require more aggressive methods; namely through explicit porosity incorporation. Porosity incorporation is a very effective way to decrease the dielectric constant. By simply lowering the atom density, less polarizable bonds per unit volume are obtained. The Si–O network, however, must remain sufficiently rigid so that any additional rotational polarizability component is not enabled. By using the Clausius–Mossotti equation [4, 46], the relationship between the resulting effective dielectric constant for the composite dielectric ($k_{eff}$) and the dielectric constants of two mixed phases ($k_1$ and $k_2$) can be described as follows:

$$\frac{k_{eff}-1}{k_{eff}+2}=\frac{4\pi}{3}\sum_i N_i\alpha_i = f_1\frac{k_1-1}{k_1+2}+f_2\frac{k_2-1}{k_2+2} \tag{11.1}$$

where $N_i$ is the number of molecules per unit volume of type '*i*' in the dielectric, $\alpha_i$ is the *i*th molecular polarizability, and $f_1$ and $f_2 = 1 - f_1$ are the volume fractions of the two components with dielectric constants $k_1$ and $k_2$, respectively. Equation (11.1) shows that, by introducing 50% porosity (with pores having $k_1 = 1$) into silica ($k_2 = 4.2$), $k_{eff} \sim 2$ is found. Low-*k* dielectrics with $k \leq 2.6$ are generically referred to as 'ultra-low-*k*' (or ULK) dielectrics because further effort to incorporate porosity into the Si–O matrix is required. In other words, any given SiCOH-based ULK with $k \leq 2.6$ is more than likely to have pores purposely incorporated into it. In such a dielectric, however, maintaining structural integrity and hardness is a significant materials challenge [47, 48], and this requirement implies that the pore size must be necessarily small even as the density of such pores must increase. Such a requirement, however, may not be physically achievable for silica-based porous dielectrics with $k \leq 2$ (which from this point is referred to as the extreme low-*k*, or ELK, regime).

The explicit incorporation of free space or pores within the dielectric matrix of a low-*k* dielectric can occur through two basic routes [49]: (1) subtractive or templated porosity incorporation [4, 5, 50–54] and (2) structural incorporation of pores [5, 49]. In the subtractive method, a templating sacrificial organic phase (called a 'porogen') is used to form pockets within the $SiO_2$ host matrix. The porogen must be sufficiently decomposed [55, 56] into volatile components at a temperature above the cross-linking or vitrification temperature of the host network but below the glass transition temperature at which the matrix can collapse [57, 58]. With a structural approach, a mixture of precursors [59, 60] is used that can form a network of lower density (free-space generation) because the local network structure enables it but only under optimized processing conditions [61–63]. These types of porous LKs are typically generated using a spin-on-then-cure approach, although not exclusively [64]. A post-deposition treatment may be used to chemically modify and increase the amount of porosity within the network [65], but may be at the risk of generating extended pores, which are undesirable.

Whatever the deposition method used, it is critical that the formed free volume is localized and sufficiently small (ideally, 'microporous' ULK with a pore size <2 nm [4]) to minimize the effect of clustering [66]. In order to strengthen the host matrix, the ULK is treated either with a post-deposition cure, typically using a combination of heat-based curing [67–69] and ultraviolet (UV) beam [70–77] or possibly e-beam curing [78–80]. An optimized UV cure [81–87] is the preferred supplemental approach compared to e-beam curing because there is some perceived risk of charging damage (similar to plasma-induced damage, or PID, observed during certain thin-film deposition processes) to underlying N/PMOS gate oxides under certain e-beam conditions [88–90].

Formation of continuous channels of free space [91, 92] within the ULK (so-called 'open' pores in 'mesoporous' ULK, which have a pore size in the range of 2 nm to 50 nm [4]) is considered a major process headache and potential reliability concern [93–97]. When trench and via structures are being formed in a dual-damascene process [98, 99], difficulties can been found with etch and ash processes [100–102], metal barrier conformality [103–106], moisture absorption [107–109], chemical mechanical polish (CMP) [110, 111], and thermomechanical integrity [112], etc. It should be stated that these concerns also pertain to low-*k* dielectrics in general [113], and that surface chemistry of the pore structure is also critical to how extended pores interact with a given processing environment [114]. An extended pore network must then be addressed by either very careful process optimization or, more likely, 'pore sealing' through chemical/plasma treatment of the pore network prior to processing [115–124]. The open-pore issue will likely be an even greater concern for ELK dielectrics.

New low-*k* materials that will be suitable for BEOL integration must meet a number of materials and performance metrics [125–128]. First, the material must possess thermomechanical properties that will be sufficiently robust against the expected thermomechanical stresses and strains from BEOL integration and packaging; i.e. whether it shows too divergent a coefficient of thermal expansion relative to Si or Cu, or whether it is sufficiently thermally stable under high-temperature processes [129, 130], or whether it will delaminate or fracture during BEOL processing, wafer dicing, or die packaging [131–137]. Second, the material must not suffer any significant degradation in its fundamental properties (chemical/physical/materials, thermal, and electrical properties) during process and integration [127, 138–143]. For example, if its dielectric constant is permanently altered during processing [144–146] or degrades over time in chip ambient conditions or under electrical stress, then it will be useless as a high-performance BEOL dielectric [147–151]. Unfortunately, dielectric degradation during certain process steps is a major concern for both LKs (major concern) and porous ULKs (much greater concern) because they are susceptible to damage during the ashing process (in particular, $O_2$-based plasmas) to remove photoresist residues [152–161]. Such damage causes the SiCOH LK/ULK to lose its natural surface hydrophobicity through carbon (or methyl, $-CH_3$) depletion [162–166] so that a discernable increase in dielectric constant is found [167]. The decomposed Si–$CH_3$ and C–H bonds [168] can form surface silanol (Si–OH) or Si–H bonds [169] and subsequently enable bulk moisture absorption [170, 171]. The damaged area represented by C loss can extend rather deeply (~40 nm) into the bulk low-*k* dielectric [172] and can reduce the process margin [173–176] and reliability [177–179]. Consequently, porous SiCOH ULK integration requires gentler plasma exposure processes [180, 181] and, quite likely, novel methods (such as surface silylation by various means) of localized repair [182–191] to damaged regions (particularly trench-via sidewalls) of the LKs/ULKs to successfully integrate these materials. Alternatively, surface capping methods may be employed [192, 193], although it is natural to ask how thick such a cap needs to be as critical interconnect dimensions trend below 32 nm. Another major concern is damage to the ULK top surface by direct chemical mechanical polishing (CMP) used during dual-damascene processing [194–197] to form the on-chip metallization patterns. Such damage also results in significant loss of reliability margin against early intermetal dielectric (IMD) breakdown [198, 199]. Another form of surface damage is the presence of Cu residues on top of the LK/ULK surface, which can also further degrade the reliability margin against IMD breakdown [200]. Although SiCOH-based ULKs are a predominant focus for advanced BEOL integration, alternatives based on organic ULKs are still being investigated [201–204], precisely because of such concerns. Despite these difficulties, reasonably successful integration of SiCOH-based ULK BEOL has been demonstrated [205–212], even with BEOL stacks that include CoWP metal capping over Cu metallization [20, 213, 214]. Regardless of the types of ULKs used, the robustness of such dielectrics in a full-stack configuration (multiple metal levels) against the thermomechanical stresses generated by wafer dicing and packaging needs to be demonstrated [215, 216]. Such full stacks (and actual products) must be verified electrically after package-level stress assessments [217], which would include high-temperature operational life (HTOL) testing [218]. ULKs are expected to have large-scale implementation in manufacturing somewhere around the 45 nm node or somewhat below [45].

The third consideration for porous ULK integration is that the resulting BEOL stack must be sufficiently reliable against the expected wear and tear that a given semiconductor device will endure during its expected operating life (EOL) [26, 219–221]. Typically, a given population of semiconductor devices in the field should not exhibit an excessive rate of device failures within a given target lifetime or EOL. Within the context of the well-known 'bathtub curve', the EOL[1] should lie within or just at the end of a period of low failure rate that is known as the 'intrinsic' or 'useful' life, where the failure rate is relatively constant [222–225]. The failures found during this period will not be specific to any given failure mechanism but are characterized by random failures due to either a decreasingly low level of defect-related failures or gradually increasing, but very low-probability, intrinsic materials failures (through multiple failure pathways). After the period of useful life, an onset of increased failure rate due to device degradation will commence and is known as the 'wearout' phase, where the rate of intrinsic materials failures rises above a tolerable level. During wearout, physical aging of the materials used within devices will be more evident, such as the increasing presence of voids within the metallization or higher or unstable leakage current through a gate or backend dielectric. Because the EOL is a fixed requirement to suit the needs of humans, the reliability performance of the materials used within semiconductor devices must necessarily exceed this EOL. The reliability performance of a given material, such as a low-*k* dielectric, must meet not just thermomechanical but also electrical requirements. Although low-*k* failure is only one possible mechanism by which a product can fail during or beyond EOL, a low-*k* dielectric is so fundamental to a microelectronics device that its failure should not be the weakest intrinsic link within

---

[1] Generically, 100 000 power-on-hours (100 kPOH) or more is a common EOL standard, depending on the application. A suitable failure rate metric may be 10 FIT, where FIT means 'failures in time' and 1 FIT means 1 failure in $10^9$ device-operating hours. Thus, for $10^4$ devices operating for $10^5$ hours, we would expect no more than 10 failures. Note that 10 FIT may be a rather relaxed metric for certain customer requirements.

a given device; i.e. the LK/ULK dielectric should not suffer premature electrical breakdown during EOL nor possess intrinsic materials reliability (in an electrical sense) that is too close to the expected EOL. Under very long-term electrical bias, LK/ULKs should eventually break down, and understanding mechanistically the inherent failure mechanism(s) and the additional mechanisms by which intrinsic materials reliability is worsened is critical to assess their implications for large-scale manufacturing of CMOS-based microelectronics [226]. This latter concern about the electrical reliability of low-*k* dielectrics is the focus of the bulk of this chapter.

## 11.2 Reliability Testing

### 11.2.1 Measurement of Dielectric Degradation

*Interconnect Layout and Test Structures*

From a BEOL dielectric reliability perspective, electrical stresses to LK/ULK dielectrics are applied in two basic ways: (1) within a given metal level (intermetal dielectric, or IMD; e.g. between adjacent M1 interconnects) and (2) between level-to-level (interlevel dielectric, or ILD; e.g. between M1 and M2 interconnects). In typical CMOS applications, the actual product will include both NMOS and PMOS transistors. Hence, positive, negative, and/or ground-level biases will be applied to these interconnects – some being statically biased, others alternating between $V_{dd}$ and ground (swings of $+V_{dd}$ to $-V_{dd}$ would be used in special applications only). Bias differences between adjacent metallization will be typically at most $V_{dd}$ to ground, where the actual $V_{dd}$ value depends upon the types of transistors used (core, I/O, native, power) and should include allowed-for variations from the external power supplies. It is then reasonable to ask about the magnitude of the electric field strengths that an interconnect dielectric endures in an actual product. If, for example, a pair of interconnects spaced apart by 32 nm has a continuous bias of 5 V applied across it, then the ULK dielectric will need to withstand an electric field of ~1.5 MV/cm, and in a real product, the ULK will need to maintain its properties at such a bias over a fixed EOL (say, 100 kPOH or ~11.4 years). For reference, polyethylene shielded cables may be required to withstand a comparatively lesser electric field (0.085 MV/cm or 1700 V for 5 mm thick insulation) for only one hour [227] (although the maximum total cable length is not noted here, which will have a significant impact on the analysis, as will be demonstrated in a later section). IMD or ILD reliability is typically a concern in those applications where high concentrations of interconnects are located with opposing biases, such as in capacitors or dense circuitry. Additionally, very tight layouts (such as in metallization associated with SRAMs) and locations where unusual interconnect features are present (e.g. where dense and isolated metallization or narrow or wide metal features lie adjacent to one another within a small area [228, 229]) are potential LK/ULK reliability concerns. Finally, usage of LK dielectrics in high-voltage applications needs to be handled with care [230].

Understanding the reliability implications of using LK/ULK dielectrics requires that suitable test structures be utilized to properly assess reliability. As one might expect, the layouts found on real products are rather complicated so that not all possible layouts can be monitored practically; however, it is possible to devise structures that are both representative of the structures found in real designs and also allow for intrinsic studies of the materials reliability issues of LK/ULK dielectrics. Regardless of the reasoning used to devise a given test structure layout, the reliability test structures should represent the worst-case (most critical) design aspects allowed in the prospective technology/product. Concerning Cu/low-*k* interconnect features and layouts, certain general characteristics are understood and are listed below:

(1) Damascene trenches have a taper (yielding a trapezoidal cross-section that is wider at the top) so that the minimum spacing and nominally largest local electric is found at the trench top [12, 13].
(2) In addition to the trench taper, significant field enhancement can be found at the top of the trench due to the barrier deposition and shaping of the trench surface during CMP. Cross-sectional analysis also reveals pronged extensions of the barrier metal, which bend outward to shorten slightly the interconnect spacing at the trench top and can also run along the top of the trench sidewall through the entire length of the interconnect [231–233].
(3) The LK/ULK dielectric at the trench is typically capped by the dielectric barrier so that a leaky interfacial pathway potentially exists [231, 234, 235] right where the highest local field between two adjacent interconnects is found. This interface is also prone to localized damage [236] from CMP, plasma pre-treatment prior to capping dielectric deposition, and possible surface contamination from Cu defects [237] or from surface corrosion of Cu metallization [238]. This situation may change somewhat, however, if metal trench capping approaches are used, but without a secondary dielectric capping layer. A discernable interface layer will likely still exist on top of the ULK portion of the trench top surface because subsequent ULK dielectric deposition must occur on top of the metal cap.
(4) Line terminations [239, 240] and turns within interconnect layouts have particularly high local fields due to electric field enhancement at corners [26, 27, 241].
(5) Vias tend to form distended regions around the top of the via at the trench/capping layer interface that arise from taper effects and via misalignment [242]. Thus, via spacing to adjacent metallization is smaller than that normally at the trench top [17, 18, 243, 244].
(6) A related concern is the dielectric breakdown performance when there exists the potential for misalignment of the via to the underlying trench, yielding an additional reduction of minimum spacing at the lower portion of the trench to adjacent metallization. Such concern might also be registered at M1 if M1 misalignment over contact potentially impacts M1 spacing to unrelated contact.
(7) Metal barrier thickness at the bottom and sides of trenches must be necessarily thinned to reduce the interconnect line resistance increase by downward technology scaling of interconnect dimensions. This decrease in the barrier thickness will increase the risk that Cu out-diffusing will cause barrier failure along the trench and via sidewalls and trench bottom [245]. Barrier weakness and high electric field strength from the bottom corners at trenches may pose a combined reliability risk for trench bottom failure from Cu drift-diffusion under electrical stress [245, 246]. If capping dielectrics are not used in more advanced nodes that use selectively electroless plated (SEP) metal capping over Cu trench, then risk of IMD and ILD failure by Cu drift-diffusion increases.

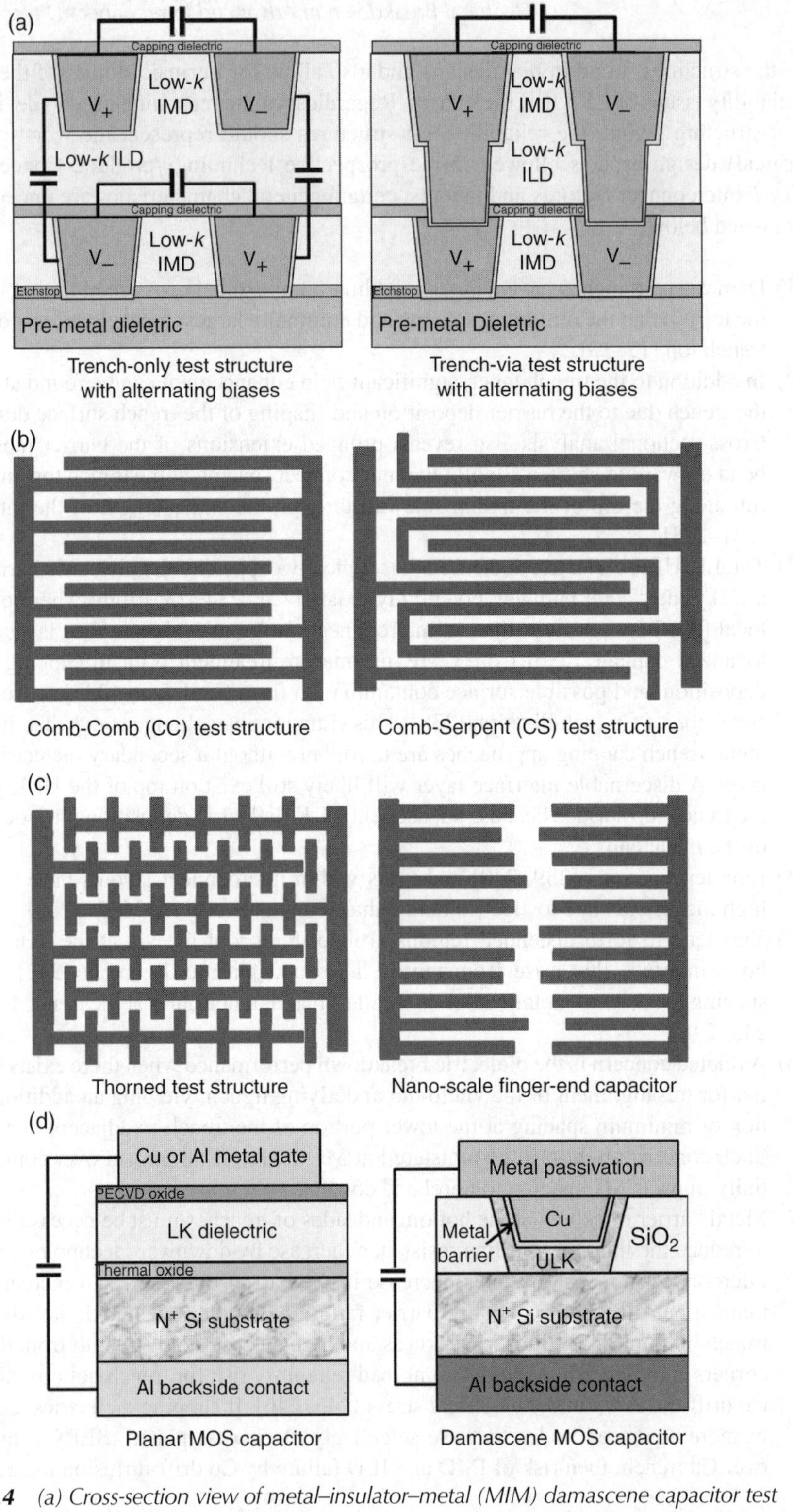


***Figure 11.4*** *(a) Cross-section view of metal–insulator–metal (MIM) damascene capacitor test structures for dielectric breakdown studies. Damascene test structures can consist of several types:*

(8) Similarly, metal barrier thickness reduction will affect the thickness of metal barrier sidewalls (and bottom) within vias, thus potentially worsening the reliability margin for vias against Cu out-diffusion failures (regardless of whether new advanced barrier metallizations are used) [21–25].

Because the trench top taper and interfacial nature of the dielectric at the trench top are unavoidable, they should be considered as regular features of test structures needed to assess IMD reliability. Layout features associated with via-controlled dielectric spacing or pattern density effects can be considered separately as long as a 'standard' type test structure can be used to compare differences. Metal barrier conformality issues should also be addressed with similar structures, but may be under different ambient stressing conditions to activate Cu mobility more effectively. From the above list, it is clear that, instead of an 'intrinsic' materials-limited reliability of bulk LK/ULK, an 'integrated intrinsic' reliability of damascene or dual-damascene LK/ULK dielectrics is being assessed by such test structures.

The types of test structure used to assess LK/ULK IMD reliability and process quality [247, 248] are either comb–comb (CC, or fork–fork, FF) or comb–serpent (CS, or meander–fork, MF) structures (see Figures 11.4 (a) and (b)). For the designs of such structures (sometimes referred to as damascene 'MIM' structures for metal–insulator–metal), the interconnect spacing between the comb finger and serpent or adjacent comb fingers can be varied as well as the lengths and widths of the individual fingers or serpent metal lines. Also, the number of repeats of these paired lines can be varied, and the main limitation to these structures may be the permissible test structure size on a given test chip. Because both CC and CS structures are necessarily terminated with ends or turnabouts (within the serpent meander), there is some concern that they might limit the reliability of a given test structure [27]. Thus, parallel-only line structures have been proposed [249] as a better alternative to more intrinsic breakdown mechanism analysis. The effect of having line terminations and turnabouts may be partially mitigated by ensuring that adequate space exists between the line end and the opposing comb spine or serpent turnabout, although one might expect that making adequate adjustments will be more difficult with newer BEOL integration approaches such as double-patterning [27]. Also, the lengths of the comb teeth (also called 'comb fingers') and straight portions of the serpent can be increased to increase the probability of failure along the length of the test structure. Some analysis should give adequate insights about the proper end spacing [250] and straight line length needed for a test structure that can assess more intrinsic reliability properties of the low-*k* dielectric, and it is a simple matter (but perhaps costly because it is done after the mask layers have already been fabricated) of checking the spatial distribution of IMD breakdown failures [251] (i.e. are the failure sites randomly dispersed?) to confirm test structure effectiveness against layout bias. Specialized structures that can test nanoscale ('end-to-end' or 'tip-to-tip') dielectric breakdown [252–254] or consist entirely of line terminations or meanders (see Figure 11.4 (c)) can be constructed as well [239, 255].

---

***Figure 11.4*** *(continued) a single-level metal layer, multilayer trench structures, or multilevel trench-via structures. Test structures can be designed to test for dielectric breakdown within a metal layer or between metal layers or both simultaneously. Simpler structures, however, are generally better for a more fundamental study. (b) Top view image of comb–comb and comb–serpent MIM capacitor test structures. (c) Specialized test structures for additional dielectric breakdown characterizations. (d) Vertical metal-oxide–semiconductor (MOS) capacitor structures*

Other test structure types exist that are often more suitable for research applications on materials-limited reliability assessment, rather than for assessing process/integration-limited reliability. An example is a planar metal–oxide–semiconductor (MOS) capacitor (see examples in Figure 11.4 (d); some refer to these structures as 'MIS' for metal–insulator–semiconductor), where a dielectric is deposited upon a heavily doped Si wafer and then capped by a suitable metal layer [256–260]. A specialized damascene vertical capacitor has also been effectively demonstrated by Zhao *et al.* [261]. The advantage of such a structure is that the LK/ULK is not damaged by the damascene process flow so that an intrinsic reliability study of bulk LK/ULK is possible, and if Cu is chosen as the metal layer, then a Cu drift-diffusion reliability assessment can be done [256–259].

*Reliability Testing of Low-*k *Dielectrics*

In general terms, reliability testing is used to accelerate the failures in time for a given device by either (a) increasing the stressing forces and/or by (b) increasing the kinetic activity of the nanoscopic elements responsible for the actual device failure [262, 263]. Proper reliability testing requires that the accelerated testing conditions should still represent the actual failure modes found under actual use conditions; otherwise, making reliability projections to at-use conditions for a given failure mechanism will be a meaningless exercise. The stress test must also isolate the particular failure mechanism of interest so that observed failures are not mixed with other potential competing failure mechanisms. This latter condition is actually quite difficult to verify, especially if the effect and scope of competing mechanisms are not well understood. For the case of LK/ULK dielectric reliability assessment, the accelerating conditions are simply the applied forcing stress, given by the applied voltage (or electric field), and the main environmental stress, given by the temperature during stress. Constant current stressing (CCS) or current-density stressing rather than bias stressing is possible too (for doing so-called $Q_{BD}$ or charge-to-breakdown experiments), but it is not as popular for BEOL reliability assessment because actual microelectronics devices usually see constant voltage stress (CVS).

The major complications with regard to testing the integrated intrinsic reliability of LK/ULK dielectrics (and not extrinsic factors) lie with free Cu and trapped moisture due to process-related damage such as Cu corrosion [264] or CMP/etch damage to LK/ULK. Because Cu metallization is involved, any free Cu from either corrosion defects or weak barrier confinement will be activated at higher temperatures. Typically, testing is done at a range of about 100 °C to 150 °C for IMD-TDDB applications, although higher temperatures may be used for kinetic energy determination. Too high a testing temperature cannot be used [265, 266], unless one is testing to directly assess the impact of such free Cu. Dielectric breakdown is generally expected to have low activation energy ($E_a \leq 0.5$ eV) [221, 267–269] as opposed to Cu drift-diffusion controlled breakdown ($E_a \geq 1.0$ eV) [259] so that their individual contributions to dielectric breakdown can be determined (rather painfully) by testing over a wide temperature range [270]. The low activation energy of oxide breakdown also justifies the notion stated above about testing at a moderately high temperature for intrinsic (or integrated intrinsic) reliability assessment. Only modest benefit in testing efficiency is found by testing at much higher temperatures for a failure mechanism with intrinsically low activation energy. On the other hand, the presence of moisture is more complicated because the original locations where moisture may or may not be concentrated is not well defined and is inherently process- and integration- and layout-dependent [271, 272]. Moisture ingress in LK/ULK dielectrics is known to have a deleterious effect on low-*k* leakage [273–275], dielectric breakdown strength, and field acceleration [276–282] and may be made worse when interfaces are present that lie along the electric field lines, as is the case for IMD capacitors [234]. Volatizing adsorbed moisture with heat treatment [283] certainly helps (especially if done during chip fabrication), but chemisorbed and hydrophilic species are not removed by heat treatment and can reabsorb moisture during subsequent processing [284]. Also, in fully integrated capacitors, such structures are sealed to prevent moisture ingress into the die so that whatever moisture is trapped during process and integration remains. Thus, moisture incorporated during processing may diffuse rapidly at higher temperatures to or from a vulnerable region during stress testing. It is reasonable to assert that whatever moisture is trapped within the BEOL stack before stressing must be considered as an inherent property of the given BEOL integration. The testing, however, must not incorporate *additional* moisture during sample preparation for testing or during actual testing so that testing is representative of the actual amount of LK/ULK in the product. Thus, it is advisable to use local seal-ring protection around test structures to minimize ambient exposure effects when package-level testing is used or when extensive wafer-level dielectric breakdown testing is done with several test structures located on the same test die [285, 286].

A major factor associated with the testing of LK/ULK dielectrics is time, because in process development it is usually in short supply. For a given technology that must satisfy the process development timelines established by Moore's law [287], very long-term testing of low-*k* dielectrics is impractical. Therefore, applied voltages are necessarily significantly higher than would be found under use conditions, but how high is too high a stressing bias is an important consideration as well. If a current transport mechanism or pathway is an intrinsic part of the degradation process, then too high a bias, where the current transport mechanism changes from one dominant mode to another, will also alter the reliability assessment picture. For example, it is clear that at very high bias (~5 to 6 MV/cm), leakage current in (relatively undamaged) damascene-integrated LK/ULK dielectrics becomes dominated by Fowler–Nordheim (FN) tunneling, whereas, at a lower bias, the dominant transport mechanism appears to be Schottky emission (SE) [288] or Poole–Frenkel (PF) conduction [221, 289, 290], with some dependence noted on the type of capping dielectric used [291, 292]. Generally, Poole–Frenkel hopping has been identified as the more likely dominant conduction mechanism – at least initially. Because the role of leakage current mechanism in LK/ULK during breakdown is not explicitly clear as yet [293], the role of testing bias range on reliability assessment needs careful analysis [294].

### *Time-Dependent Dielectric Breakdown*

Dielectric reliability in microelectronics is assessed usually by the method of time-dependent dielectric breakdown (TDDB) testing. In TDDB, a fixed bias to a collection of identical test structures is applied at a fixed temperature. The leakage current as a function of time is monitored (the necessary current resolution $\leq 1\,\text{pA}$ for signal-to-noise ratio

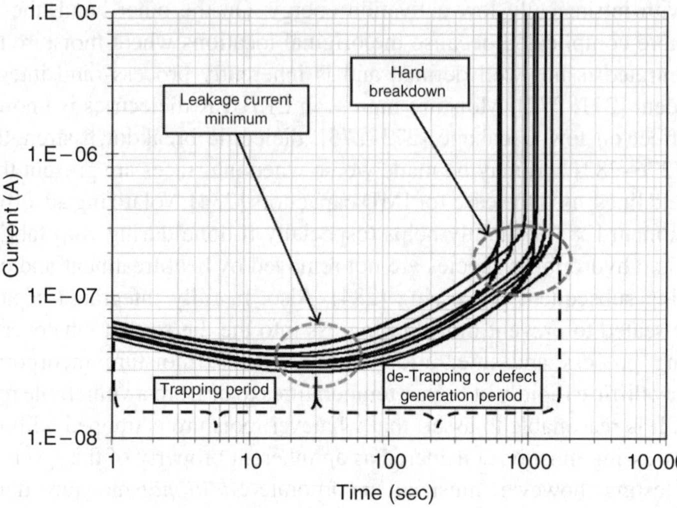


*Figure 11.5 TDDB leakage current trace for several test structures with a low-k dielectric. Leakage trace shows characteristic trapping and de-trapping periods followed by an abrupt leakage increase, indicating hard breakdown failure. The leakage trace data are by courtesy of GLOBALFOUNDRIES, Inc.*

reasons) and the time-to-fail ($t_F$) of each device under stress is tracked. The voltage/electric field dependence is determined by running TDDB tests at different biases on multiple sets of identical test structures but at a fixed temperature. Similarly, the temperature dependence is found by testing multiple sets of identical test structures at a fixed bias but different temperatures. Thus, the testing matrix is similar to how electromigration (EM) is assessed – a driving force testing matrix at a fixed temperature and a kinetic testing matrix at a fixed driving force.

IMD leakage-current traces found during TDDB testing are characteristically similar: a decrease in leakage current over time before the leakage increases and finally shows a breakdown (see Figure 11.5). The initial period – when the leakage current decreases over time – is typically identified as the 'trapping phase' when defects within the LK/ULK trap free charges and inhibit the leakage. Simultaneously during TDDB, new defects are being generated due to the high stress conditions, so that at some point, the leakage begins to increase as the 'de-trapping' or defect generation phase begins to dominate. Thus, there is a clear 'minimum' observed between the two phases during TDDB stressing of LK/ULK dielectrics. As the leakage continues to increase, the dielectric ultimately fails. Typically, this failure is observed to be a 'hard breakdown' (HB), where the dielectric shows a sudden and catastrophic leakage change. If test structures are taken out and then examined using physical failures analysis techniques (hot-spot detection using techniques such as optical beam induced resistance change (OBIRCH), cross-sectioning by focused ion beam (FIB), scanning electron microscropy (SEM), etc.), heavy physical damage to the test structure will be found.

Test structure damage from dielectric breakdown is usually localized compared to the total area occupied by the test structure (e.g. CC type), but within that microscopic region,

***Figure 11.6*** *Typical damage failure after TDDB stress obtained by focused ion beam cross-sectioning and SEM imaging. The image is provided by courtesy of GLOBALFOUNDRIES, Inc.*

the damage will be extensive, making it difficult to definitively identify the actual initial site for the breakdown event. At damage sites (see Figure 11.6 for an example), Cu stringers or filaments can be seen extending from one interconnect line (where a void may be present) to an adjacent line [295–297]. Also, voids within the IMD or voids on top within the capping dielectric layer have been seen, but their appearance after breakdown may depend upon the capping dielectric composition [298]. The damage zone under TDDB stress is usually smaller than that found by RVDB because the dissipated charge during breakdown is somewhat smaller. Because the electric field is typically highest at the trench top, damage is usually found to be concentrated in that area of the test structure cross-section, unless there is a case where another damage path has triggered failure, such as when metal barrier failure has occurred [93, 299]. In that case, shorting damage will be evident where the barrier has failed [300], although, again, unambiguously identifying that site as the trigger site is difficult at best. Specialized failure analysis techniques – such as transmission electron microscopy (TEM) with energy-dispersive X-ray spectroscopy (EDX or EDS), or electron energy loss spectroscopy (EELS), or secondary ion mass spectroscopy (SIMS) line scans – may be used to assess the presence of unusual nanoscopic defects found after TDDB stress [238, 301, 302]. Interestingly, unpublished work has shown that, even in the tiniest of IMD capacitors ('end-to-end' pointed finger capacitors for a 65 nm node interconnect [252] using an LK dielectric, which is somewhat similar to that shown by Tan *et al.* [253]) undergoing IMD-TDDB stress, obvious physical damage to the dielectric will only be evident after an actual HB event occurs but not prior to that, even during the latter stages of the defect-generation period [303]. It is notable that no such evidence of obvious Cu defects permeating through the low-*k* to capping dielectric interface was observed for these structures. This result suggests that, in cases where failure analysis has been able to identify sites where obvious Cu residues are found, such damage sites are either a result of the residual damage from the dielectric breakdown event or of a defect-related failure mechanism (i.e. barrier failure, CMP issues, post-CMP corrosion effects, etc.) that is not necessarily native to the actual failure of a pristine LK dielectric. Of course, if atomic-level concentrations of Cu ions are the root cause for IMD breakdown, imaging such ionic concentrations near the point of dielectric breakdown would be a particularly interesting technical challenge. Recent work by Beuer *et al.* [304] has used similar structures under ramped breakdown conditions to characterize damage just before ULK dielectric failure. Damage that occurs just before failure shows evidence of bubble formations of the Ta barrier extruding into the dielectric space and is explained as a consequence of sudden shrinkage of the ULK due to rapid localized densification (and maybe expulsion of gases formed by carbon depletion) during ramped stress. Such damage could then be considered as precursor damage to the catastrophic damage to follow as Cu metallization is expelled during complete breakdown.

The time-to-fail ($t_F$) criterion is predetermined and is typically related to a distinct change in the leakage current trace over time, usually an HB condition. During stressing, the leakage is constantly measured to detect changes over time. Generally, two different kinds of failure criteria may be applied:

(1) Leakage current ($I_{meas}$) increase over an absolute leakage level, such as $I_{meas} > 1\,\mu A$.
(2) A sudden jump in the leakage current magnitude, e.g. by one decade.

Whatever HB failure criterion is used, it must be used consistently. In some cases, a soft breakdown (SB), where monitored leakage at stressing bias shows obvious instability, may be evident. Such instability during TDDB (or BTS; see below) is typically associated with a substantial concentration of mobile Cu ion defects penetrating into the LK/ULK [26, 235]. Verification of breakdown is done by leakage measurement at a low voltage level, typically near an operational voltage. This type of measurement, known as 'stress-induced leakage current', or SILC (see Figure 11.7), is a commonly used method for gate oxide breakdown detection and permits verification of real impact on real product performance [305], a sort of sanity check to distinguish between potential reliability implications of soft versus hard breakdown. The latter is easily detected, but the former may be represented only by a small spike or 'wiggles' in the leakage trace and may not be detected by the HB criterion above without carefully tightening of the specifications. Tightening the criteria, however, may lead to a significant number of detected breakdowns without a physical breakdown so that SILC investigations become necessary for verification. SILC investigations will thus drive the test measurement equipment requirements. Because switching between the two voltage levels is needed for SILC, any significant overshoot above the stress level would negatively influence the measurement accuracy and must be adequately controlled. With current limits placed upon higher voltage stress, it has been shown that, in nonideal ULK, soft breakdown may be evident during SILC stress but would not necessarily lead to outright device failure during chip operation [305].

TDDB testing can be done at wafer level using probe stations or at the packaging level using package-level testing systems. In wafer-level systems, a wafer prober makes contact with the probe pads of the given test structure and an SMU (source measurement unit) performs both leakage measurement and bias stressing simultaneously. Wafer-level probing systems have the capability to test a single device using a micromanipulator or multiple devices serially using semiautomatic and fully automatic stepping systems. Automatic systems often come with multisite testing setups to enable a limited

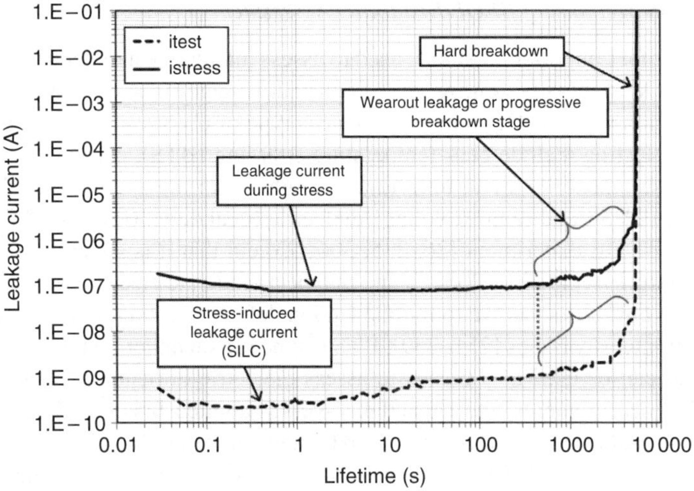


***Figure 11.7*** *Leakage trace of a TDDB measurement supported by SILC investigation. The solid curve, istress, represents the leakage current at the stress voltage; the dashed curve, itest, is the current measured at the sensing voltage. Note that the SILC leakage trace appears to show different interval lengths for the trapping versus de-trapping periods in comparison to at-stress leakage: the trapping period appears shorter while the de-trapping period has onset at an earlier point than that seen at-stress. Toward the test structure end-of-life, leakage degradation may become evident (progressive breakdown) until hard failure is observed. The image is provided by courtesy of GLOBALFOUNDRIES, Inc.*

degree of parallel testing capabilities, but such systems are usually in high demand and cannot be used indefinitely. Thus, wafer-level testing is usually done at a higher stressing bias with a relatively small sample size (10 to 20 units per field condition) to reduce the testing time. In package-level systems, stressing and testing are not done in a fully parallel manner because measurement is done over a sequential switching matrix. This reduces the time resolution of package-level systems compared to wafer-level systems, but it is less of an issue if longer-term stress assessment at lower applied stressing bias is needed. In package-level systems, individual test structures must be diced from the wafer and packaged and may more appropriately simulate some of the harsher conditions found in product fabrication, such as those found during die attach preparation and wirebonding. For appropriately designed test structures, it is then possible to characterize the impact of package processing better on the mechanically weak low-*k* dielectrics to test so-called chip–package interaction (CPI). In such a case, damage to the underlying LK/ULK may show evidence of leakage degradation or premature dielectric breakdown. Also, higher temperatures may be allowable in such testing systems compared to what is practical on a wafer-level probe station. Thus, TDDB assessment in BEOL dielectrics can track a wide variety of potential reliability issues and requires only good test structure designs and adequate testing support. To cover a larger range of testing conditions, both types of systems can be combined to piece together a fuller BEOL lifetime projection picture [285].

*Complementary Methods of BEOL Reliability Assessment*

There are also other important methods used to assess interconnect dielectric reliability and most utilize (at least in part) some aspect of constant voltage stress (CVS) to accelerate the damage formation process. Of course, TDDB is a type of CVS test. These complementary tests to TDDB testing include bias–temperature stressing (BTS), capacitance versus voltage (CV) measurements (with an unfortunate similarity to the acronym CVS), triangular voltage sweep (TVS), and ramped voltage dielectric breakdown (RVDB). BTS is actually pretty much a TDDB experiment [26, 235, 306], except that it is generally run at a much higher temperature (typically ≥ 200 °C) and maybe a moderate electric field strength (~ 0.5 to 2 MV/cm). Thus, it is sensitive to mobile Cu ion defects that might arise from Cu corrosion or barrier integrity issues [256–259]. BTS can be run on either metal-to-metal or MOS structures and can be used as a means to assess dielectric diffusion barrier effectiveness [270]. In CV measurements, the capacitance-versus-applied bias on MOS capacitors is measured at a fixed frequency (typically 1 MHz) and can be used in conjunction with BTS to monitor the rate of mobile Cu incursion into the BEOL dielectric by monitoring the flat-band voltage shift before and after BTS [256–258, 307]. Capacitance measurement can also be run during BTS/TDDB (known as 'time-dependent dielectric-constant increase', or TDDI) to track as-stressed changes to dielectric properties [147, 148, 150].

Like BTS, the triangular voltage sweep (TVS) technique [308] is a technique that detects mobile ionic species (including, but not limited to, Cu and moisture). The procedure is as follows:

(1) Apply a positive bias (the magnitude depending on how much time you have available) to the gate (of an MOS capacitor) at a temperature high enough to enable mobile ions – which are present within the capacitor or are injected from the gate metal – to accumulate at the opposite (substrate) interface.
(2) Slowly run a linear voltage sweep from the positive to the negative gate bias. A net displacement current, which discharges the capacitor, will be directly proportional to the DC-like capacitance. During the sweep to a high negative gate bias, the collection of accumulated mobile ions will register a peak in the low-frequency C–V characteristic at some voltage that is characteristic of the mobility of the ion in question and the thickness of the dielectric layer.
(3) Calculate the net amount of charge that has drifted back through the dielectric by integrating the area under this peak. Note also that TVS can be applied to both planar [309, 310] and damascene capacitors [302, 311, 312]. The technique, depending on the test structure used, can sense charge densities on the order of ~ $10^9/cm^2$, which is said to be about 10 times more sensitive than the BTS method. As a frame of reference, it is estimated that on the order of a Cu particle a density of ~$10^{12}$ at/$cm^2$ or more is needed so that copper-induced degradation to dielectric quality is notable [27, 313]. Thus, BTS, CV, and TVS methods are necessary methods to detect Cu mobile ion defect issues in advanced process development.

RVDB is used primarily as an assessment of quality rather than reliability (see Figure 11.8). In RVDB (other acronyms used are VBD or EBD for voltage or E-field breakdown, RVBD or VRDB for 'ramped voltage breakdown' or 'voltage ramped dielectric breakdown', respectively), the voltage applied to a test structure is ramped at a fixed rate (typically in discrete voltage steps) until the dielectric fails. Because RVDB is a rather fast test in

(a)

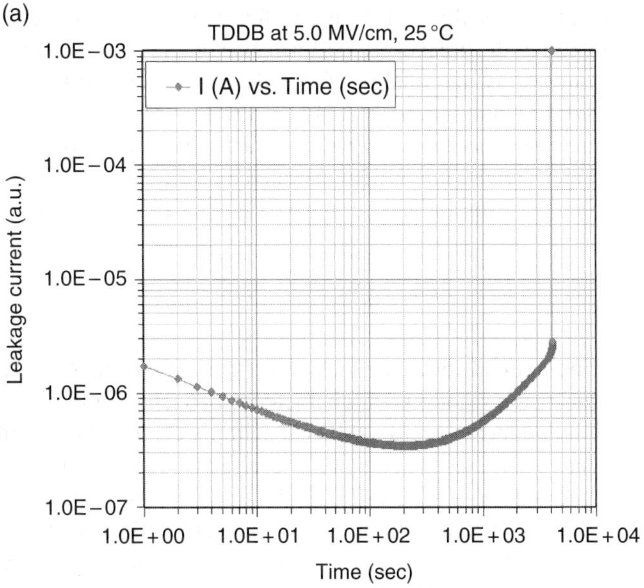


(b)

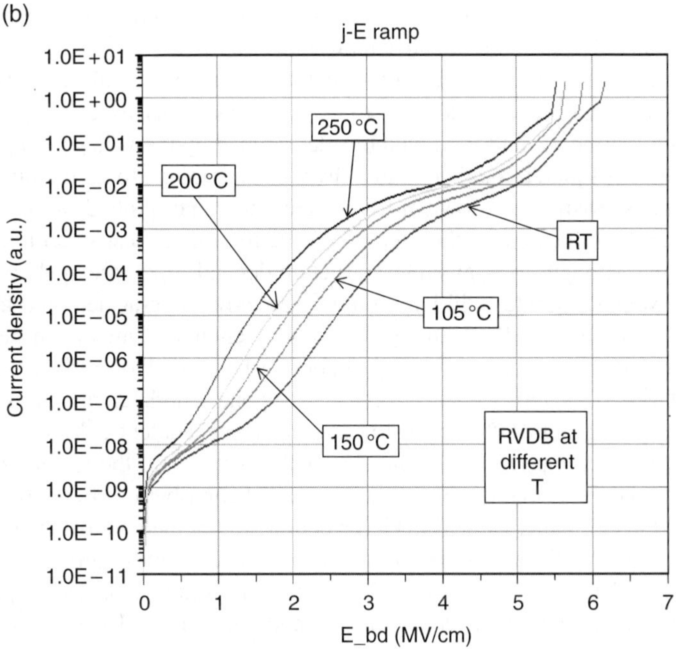


***Figure 11.8*** *(a) Comparison between TDDB and RVDB is shown. TDDB is generally used for reliability prediction and provides detailed information about the degradation process. It suffers from limited sampling statistics and has relatively slow test turnaround time. (b) RVDB is more a measure of device quality for defect detection and is limited in terms of reliability projection. It can provide very good statistical sampling because it generally has rapid test turnaround time. Image was adapted from E. T. Ogawa,* Semiconductor International/ SemiTracks Tutorial, *11July 2005.*

comparison to TDDB and is typically done at the wafer level, larger sample sizes can be monitored. In fact, mapping of the breakdown performance on the entire wafer can be done to monitor the overall uniformity and quality of the LK/ULK stack and has been used to refine quality of the LK/ULK integration process [12, 13, 314–318]. Even though RVDB is not a CVS-type test like the other tests described so far, there is a sort of conjugate connection between RVDB and TDDB, as was demonstrated by Berman and others [319–322]. The effective-time-at-field, $t_{EFF}$, for a given ramp test is given by the following:

$$t_{EFF} = \frac{\Delta\tau}{1 - e^{-\gamma \Delta E}} \tag{11.2}$$

where $\Delta\tau$ is the hold-time interval for a given electric field increment, $\Delta E = \Delta V/s$ ($\Delta V$ is the voltage increment during ramp and $s$ is the characteristic spacing for the capacitor), and $\gamma$ is the field acceleration parameter. For an IMD capacitor, $s$ can be interpreted to be the minimum line-to-line spacing. The above equation is derived by assuming that the $E$-model [319, 323, 324] represents the intrinsic LK/ULK TDDB; however, versions using other lifetime models, such as the $\sqrt{E}$-model, have also been similarly constructed [254, 325, 326] so that the effective-time-at-field expression is modified from Equation (11.2). Because all lifetime models generally agree at higher fields, where ramped breakdown is more relevant, the results among such models (at least empirically) should be similar. Therefore, Berman's original analysis should be fairly representative.

If TDDB is a measurement of time-to-breakdown at fixed bias, the RVDB is a measurement of the voltage or electric field strength at breakdown (respectively, $V_{BD}$ or $E_{BD}$) for a fixed effective-time-at-field, $t_{EFF}$, which is determined by the voltage ramp rate. This breakdown strength is found through a 'search' by voltage ramping to such failure. Thus, in Figure 11.9, a characteristic breakdown bias at a fixed time can be plotted along with TDDB data to determine the full extent of the lifetime profile for a given test structure. Note also that the breakdown strength for a given test structure depends upon the applied bias ramp rate. Thus, very slow ramp rates will have a higher effective-time-at-field and faster ramp rates will have a shorter $t_{EFF}$. If RVDB is done slowly enough, comparison between the characteristic breakdown strength at fixed $t_{EFF}$ under RVDB and characteristic time-to-failure at fixed TDDB bias can be done. RVDB can also be done at high temperatures ($\geq$250 °C) as a sort of sanity check on IMD/ILD test structures against Cu defects, although sensitivity will be limited by the ramp rate used (the slower the better) and the test structure area, which needs to be larger so that extrinsic process-related contributions to poorer reliability margin are detectable. Finally, while RVDB is a conjugate test method to TDDB, it is not a replacement because TDDB is necessary to identify the long-term lifetime behavior under electrical stress of LK/ULK dielectrics [327]. Instead, complementary usage of both methods is needed to ensure both adequate quality and reliability of LK/ULK dielectrics.

### 11.2.2 Reliability Analysis

*Statistical Analysis of Dielectric Breakdown*

To assess both the driving force and kinetic effects, statistical analysis is used. Typically, Weibull statistics [221] or maybe lognormal statistics [328] are used to assess the lifetime distributions found. An example of TDDB data plotted as a Weibull distribution is shown

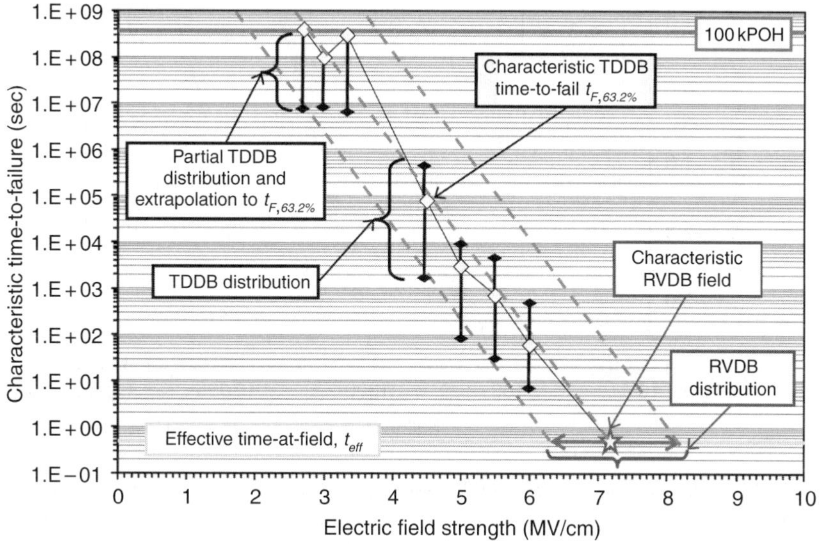


***Figure 11.9*** *This graphic explains the conjugate relationship between RVDB and TDDB using the Berman model. The effective-time-at-field,* $t_{eff}$, *can be viewed as a fixed time in which a population of capacitors shows variable breakdown voltage. On a TDDB lifetime projection plot (here using an E-model), the RVDB distribution can be plotted on the horizontal axis at the value* $t_{eff}$. *Note that a slower ramp rate will correspond to a higher* $t_{eff}$. *This RVDB distribution will then provide expected time-bounds within which capacitors should fail under TDDB stress at a fixed field. Note that in TDDB, the higher lifetimes do not reach the upper bounds expected from an RVDB projection. A rapid RVDB will cause a capacitor to fail in essentially a 'frozen state' so that it will be sensitive to poorly fabricated dielectrics; however, TDDB would be expected to allow defects to form in places within the capacitor that are advantageous for earlier breakdown. Therefore, it might be expected that TDDB will show a tighter breakdown distribution for higher lifetime devices than would be expected from a rapid RVDB projection. Much slower RVDB may provide better agreement for the entire distribution because some element of TDDB-like damage generation will take place. Of course, the benefit of rapid testing turnaround provided by the RVDB will be lost. This discussion demonstrates that the RVDB distribution is a clear measure of the 'initial quality' of the capacitor population that can fail under TDDB stress, so that it is a necessary condition that RVDB distributions be clean and true. The graphic is adapted from J. Kim* et al., *International Relative Physics Symposium (2007) (© 2004 IEEE; permission by IEEE) (see color Plate 12)*

in Figure 11.10. The $x$ axis represents the time-to-failure found for identical test structures tested at a given condition. The $y$ axis represents the statistical plotting function. In this case, a Weibit function ($\mathrm{Wbt}(i) = -\mathrm{Ln}[\mathrm{Ln}(1-F_i)]$) is used, where $F_i$ is the estimated value of the cumulative distribution function (CDF) for the $i$th device. For a given monomodal lifetime distribution, two parameters are important [222]: (a) the characteristic lifetime for each stress condition and (b) the shape factor. In Weibull statistics, the characteristic lifetime is conveniently tracked at $t_{F,63.2\%}$ (where the Weibit function equals zero) and the shape factor, $\beta$, is given by the slope found in a Weibit plot. To identify the lifetime trending with applied bias or electric field, the characteristic lifetime at different electric fields are plotted

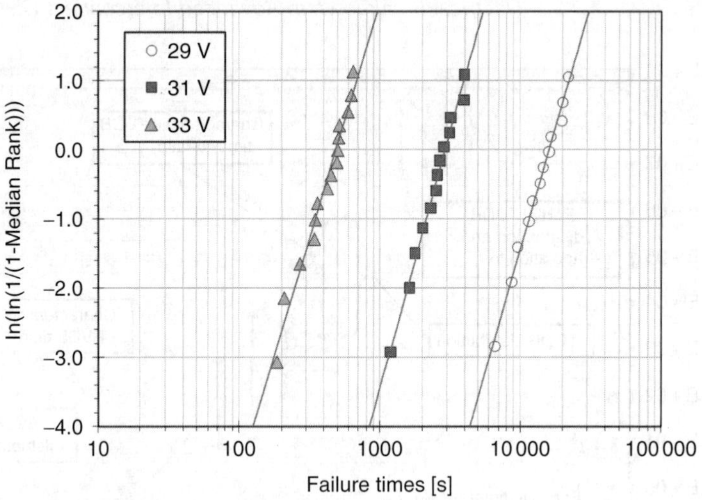


***Figure 11.10*** *Weibull plots of IMD-TDDB results are shown for three different bias conditions. The y-axis function is known as a 'Weibit' and 'Median Rank' is the empirical estimate for the proper CDF value for the test sampling (see the details in Tobias and Trindade, 1986, Reference [223]) . Distributions are offset based on the strength of applied bias stress, but the slopes of the distributions are similar and shows that no abnormal failures are found for the given sampling population*

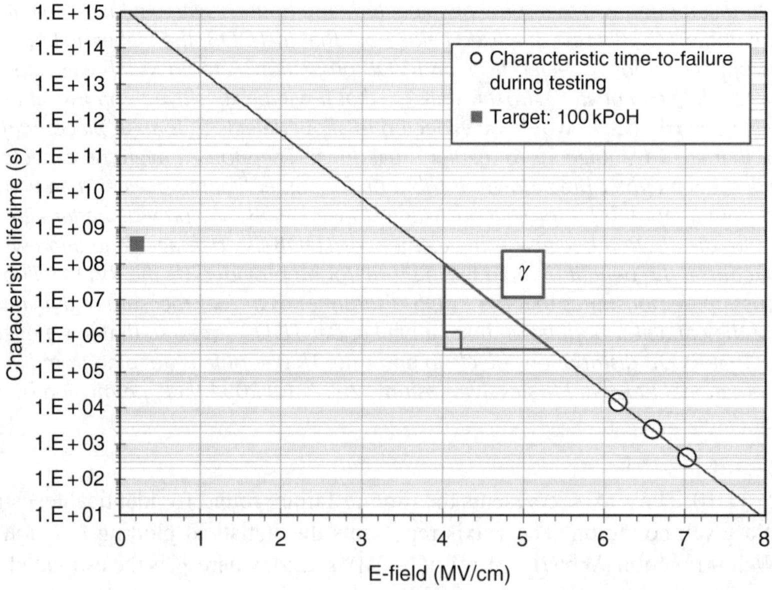


***Figure 11.11*** *TDDB lifetime extrapolation is shown using a linear-E model, where the slope γ is the field acceleration parameter. Extrapolation over a wide range of E-fields results in a larger lifetime at operation voltage (e.g. 1 V) than the required target (e.g. 100 000 hour on power, or PoH)*

(see Figure 11.11). This approach is the standard method by which lifetime extrapolation models are compared.

There has been some discussion as to what sort of distribution (Weibull versus lognormal) is appropriate to assess LK/ULK reliability [329], but a simple argument can be made that suggests that Weibull statistics (or a similar equivalent) is better suited to assess reliability scope (see Figure 11.12). The lifetime distribution of a set of capacitors with total area, $A$, should be identical to a set of capacitors consisting of two parallel capacitors, each with area $A/2$. Likewise, a single capacitor can be viewed as a collection of multiple parallel capacitors. During TDDB, the weakest of these parallel units will fail. With a Weibull distribution, a combination of two Weibull distributions yields a Weibull distribution; however, for a lognormal distribution, a combination of two lognormal distributions does not. Thus, Weibull statistics are ideally suited to assessing the reliability of capacitor-based failures [221]. It should be noted that both Weibull and lognormal statistics can usually fit data obtained using small statistical samplings. The differences would really only be discernable at the percentile extremums (very small or large CDF values), where a Weibull statistics project would yield

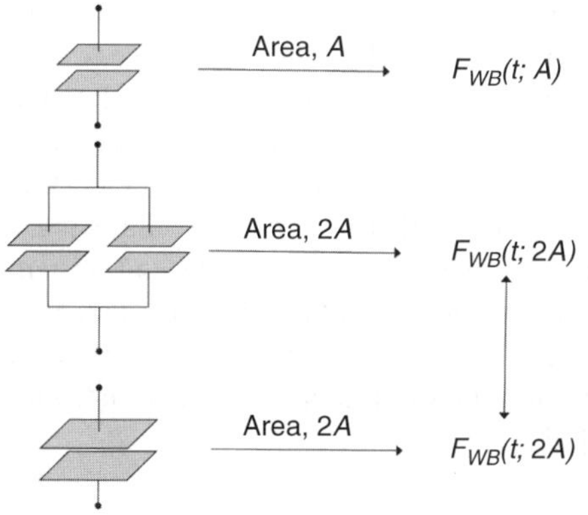


***Figure 11.12*** *If a capacitor of size A obeys a given distribution, then a similar capacitor but of size 2A should also obey the same type of intrinsic distribution ($F_{WB}(t; A_{cap})$, where $F_{WB}$ is the Weibull cumulative distribution function at time t with a capacitor of given area, $A_{cap}$), except that its lifetime may be different but not its shape factor. Yet, a size 2A capacitor is essentially two size A capacitors in parallel, where both size A capacitors should follow the same distribution and failure is determined by the weaker of the two capacitors. Such subdividing of a given capacitor into multiple numbers of smaller capacitors can conceptually go on indefinitely with each smaller capacitor obeying the same essential distribution and failure occurring when the weakest of the group of smaller-sized capacitors fails. At some point, a 'smallest capacitor' can be defined where its distribution serves as a generator for all larger capacitors by an appropriate area-scaling relationship, which defines the shifts observed in the characteristic lifetimes for different-sized capacitors. Since Weibull distributions are of a class of distributions that are self-similar – as would be expected for identical groups of capacitors differing only by their relative area – then a Weibull distribution is an appropriate distribution to use for capacitor reliability statistics. The graphic was adapted from E. T. Ogawa,* Semiconductor International/SemiTracks Tutorial, *11 July 2005*

more conservative lifetime projections (earlier time-to-failure) [330]. There may be instances where lognormal statistics may provide a better fit to the data, and such instances are likely to reflect the impact of 'pre-damaged' regions, which are not represented by Weibull area scaling. The implications for such data, however, are that area scaling to product level would not follow conventional area-scaling relations and would require additional detailed characterization. Such characterization would need to specify whether there are limits to be found as to the extent of such pre-damage so that some form of area-scaling analysis can be done. If not, then a variant of lognormal scaling analysis may be needed, an approach that is analogous to what is done in electromigration multilink reliability failure analysis [331–337]. In such an approach, IMD-TDDB of test capacitors following lognormal statistics would be analyzed as units that can be multilinked for area-scaling projection to product level.

There are instances when the IMD-TDDB (or ILD-TDDB) lifetime or RVDB distributions do not exhibit a straight line when graphed on a suitable Weibit (or lognormal or even normal) plot. The simplest case to understand is one where two separate but parallel distributions appear to coexist and merge at some intermediate CDF value [12]. In that case, spacing variation across the test sampling (likely across-wafer) is evident. When the sampling is tested under different field stress conditions (at a fixed temperature), the observed field acceleration parameter, $\gamma$, should remain the same when extrapolating at a fixed CDF value; otherwise, clear differences in $\gamma$ at different CDF values may mean that there are process-related issues that need to be resolved [338]. Unusually wide but monomodal-looking distributions also indicate problems because the expected area scaling behavior (see the next section) will be poor. If the distribution exhibits a distinct 'tail' – where the lifetime or breakdown strength trend to a lower value is seemingly without end – then a clear defect issue is at hand [314]. Again, process improvements will be needed. Similar problems found at high temperatures are likely indicators of mobile Cu defects. For a more detailed examination of the impact of nonmonomodal statistical distributions, refer to the reference by Degraeve *et al.* [339].

*Area Scaling and Test Structure Size*

When demonstrating the intrinsic robustness of LK/ULK dielectrics, it is often shown that integrated test structures using such dielectrics exhibit a long intrinsic lifetime; however, simply demonstrating a lifetime that exceeds 10 or 11 years of life under stress conditions does not necessarily mean that such performance is adequate for products using integrated LK/ULK dielectrics. The obvious reason is that an actual product based on advanced CMOS technology is not the size of a typical IMD test structure. Typically, the capacitor length of an IMD test structure can range between 0.0001 m and 0.1 m. Assuming a CC capacitor with total length = 0.01 m, an individual segment length = 100 μm, and a pitch of 64 nm, the total capacitor surface area would be only about $6.4\times10^{-3}$ mm$^2$. A typical microchip, however, might be $5\times5$ mm$^2$ or more. If one assumes 100 % areal coverage at a minimum pitch of 64 nm at M1 for a $5\times5$ mm$^2$ device, then the total capacitor length would be 390 m! In fact, if the die size remains fixed, then the total interconnect length will continue to increase geometrically (divide by 0.7 per generation). Thus, if a fixed test structure length is used, then the sensitivity of this structure to any process nonuniformity that impacts reliability will decrease proportionately. If only small test structures are used, then process marginality cannot be detected and the impact on larger area devices is less certain. Such process marginality may lead to early failures at product level under high-temperature operating life (HTOL) stress, suggesting that, although the characterization of large area structures is necessary, it may not be sufficient because even large test structures are generally smaller than the actual product. A counterpoint to the above argument is that the focus on small structures will allow for more accurate evaluation of intrinsic reliability behavior. Reliability evaluation exclusively done on larger structures may superpose extrinsic (defect) contributions, which generally increase the error bar for kinetics evaluation and determination of the field/voltage acceleration parameter. Of course, if extrinsic effects are dominant at product level, then any intrinsic analysis will have little practical value. Given the trade-offs, the moral here is not to focus just on a single type and size of test structure.

Because Weibull area scaling is observed by IMD capacitors [221, 269] (at least, for integrated intrinsic failures), the TDDB lifetime of a test structure can be related to that at product level [340]:

$$\frac{t_{Product}}{t_{TestStructure}} = \left(\frac{A_{TestStructure}}{A_{Product}}\right)^{1/\beta} \left[\frac{\mathrm{Ln}\left(1-F_{Product}\right)}{\mathrm{Ln}\left(1-F_{TestStructure}\right)}\right]^{1/\beta} \cong \left[\left(\frac{A_{TestStructure}}{A_{Product}}\right)\left(\frac{F_{Product}}{F_{TestStructure}}\right)\right]^{1/\beta} \tag{11.3}$$

$$\frac{t_{Product}}{t_{TestStructure}} = \left(\frac{A_{TestStructure}}{A_{Product}}\right)^{1/\beta}, \quad \text{for} \quad F_{Product} = F_{TestStructure} \tag{11.4}$$

where $t_X$ is the time-to-failure ($X$ = *Product* or *TestStructure*), $A_X$ is the capacitor area of $X$, $F_X$ is the CDF value of $X$, and $\beta$ is the Weibull shape factor. Equation (11.3) is the more general form, for different $F_{Product} \neq F_{TestStructure}$, and Equation (11.4) is the simplified form when they are equal. Equation (11.4) tells us immediately that the product lifetime scales roughly as a ratio of the test structure to product areas. Thus, if $\beta = 1$ (which is actually a poor capacitor shape factor) and the area ratio = 0.01/390 = $2.6\times10^{-5}$, then the test structure time-to-fail needs to be at least $4.4\times10^{3}$ years, which means that the product will last 11.4 years. Additionally, the value of the CDF used is important because it determines the expected failure fraction for such devices (1 %, 0.1 %, 10 ppm, 1 ppm, etc.). Of course, this is just an example (e.g. the use-temperature is not considered), but similar exercises are needed to make an appropriate assessment about low-*k* reliability for a given technology.

The importance of area scaling can also be applied to RVDB using the Berman formulation [319]. Using a certain representation of the *E*-model ($\mathrm{Ln}[t(F)] = B(F) - \gamma_{TDDB} E_{stress}$, where $t(F)$ is the time-to-fail for a given CDF $F$, $B(F)$ is an integration constant, $\gamma_{TDDB}$ is the field acceleration parameter, and $E_{stress}$ is the applied electric field), the TDDB lifetime can be represented in terms of a given effective-time-at-field, $t_{EFF}$, as follows:

$$t_F(F) = t_{EFF} \exp\left\{\gamma_{TDDB}\left[E_{BD}(F) - E_{stress}\right]\right\} \tag{11.5}$$

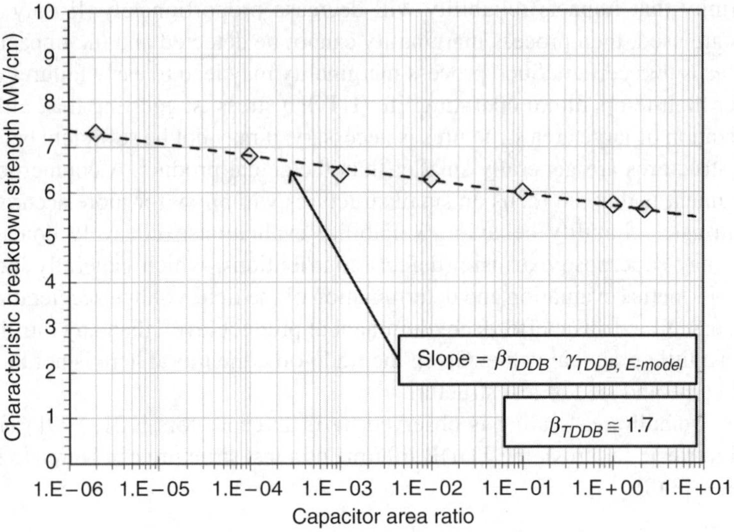


***Figure 11.13*** *Voltage ramp data for capacitors with different areas obey Weibull area scaling over six decades in capacitor area. In fact, the smallest capacitor used was an end-to-end capacitor at a 65 nm node. The graphic was adapted from E. T. Ogawa,* Semiconductor International/SemiTracks Tutorial, *11 July 2005*

where $E_{BD}(F)$ is the breakdown strength for a given CDF, $F$. If Equation (11.5) is substituted into Equation (11.4) for two different test structures with different areas, then we have the following result:

$$E_{BD,2} = E_{BD,1} + \left(\frac{1}{\beta_{TDDB}\,\gamma_{TDDB}}\right)\mathrm{Ln}\left(\frac{A_1}{A_2}\right) \tag{11.6}$$

where $E_{BD,X}$ is the breakdown strength of the capacitor with area $A_X$ and $X \in \{1,2\}$, $\beta_{TDDB}$ is the TDDB Weibull shape factor, and $\gamma_{TDDB}$ is the TDDB field acceleration parameter for an $E$-model. Thus, unlike TDDB, the difference in the breakdown strengths between two capacitors of different area has a logarithmic relationship to the capacitor area ratio. This relationship is shown more explicitly in Figure 11.13, where a shape factor $\beta_{TDDB} \sim 1.7$ is determined for a known value of $\gamma_{TDDB}$ at the same temperature [322]. The relationship shown in Equation (11.6) is also different from that described elsewhere [269], where, using Poisson area scaling arguments, a relation of the form $E_{BD,2} = E_{BD,1}(A_1/A_2)^{1/\beta}$ is generated. Note, however, that the shape factor $\beta$ obtained using this equation will be very different from that which would be obtained from TDDB because no explicit connection can be made between the breakdown strength shape factor and the shape factor obtained by TDDB. In Figure 11.13, a roughly equivalent fit to the observed breakdown strength with an area ratio using the Poisson area-scaling relation is $\beta \sim 53.5$. In the case where the TDDB distribution is wide ($\beta \sim 1$), Chen *et al.* [305] have proposed a 'shift and compare' method to take into account a significant negative impact of a smaller area within a test structure of a suboptimal interconnect dielectric that is dominating the failure behaviors so that its area impact can be properly assessed.

## 11.3 Lifetime Extrapolation and Models

Reliability physics concepts in microelectronics date back almost to the beginnings of the industry. Although 'reliability' was not a specific topic at the inaugural meeting of the International Electron Devices Meeting (IEDM) in 1955, Robillard [341] did mention that 'reasonable aging trends' were found when devices were encapsulated in dry ambient conditions, which is among the earliest records of moisture issues in microelectronics. In 1960, Howard and Dodson [342] noted a generalized stress equation of the form: $f(S) = A \log(t) + B$, where $f(S)$ is a function of an applied stress, $S$ (for '50% failure'), $t$ is the time, and $A$ and $B$ are characteristic device parameters, and of which the temperature Arrhenius equation ($t_F \sim \exp(\Delta H/k_B T)$) is a special case. Descriptions of the methods of biased temperature stressing and voltage breakdown on (3 μm thick!) silicon dioxide films can be found in the paper by Greenough [343] in the inaugural International Reliability Physics Symposium (then known as the 'Symposium on the Physics of Failure'). Also shown was maybe the earliest reliability comparison between 'low-*k* and high-*k*' dielectrics, although, in that case, high-*k* meant one with a *k* value that ranged between 1000 to 5000 [344].

In 1979, when the empirical versions of the *E*-model were first described in papers by Crook [323] and also Anolick and Nelson [324], the gate oxide thickness characterized at that time was about 70 nm, which is about twice as large as the M1 half-pitch needed at the 32 nm node. Thus, even though the dielectric thickness is comparable to that found in gate oxide technologies some 30 years ago, the intrinsic reliability of the interconnect dielectrics is not nearly as robust. This point is made clearer by a comparison shown in Figure 11.14, which compares the relative levels of TDDB performance found among different dielectrics ranging from gate oxides to interconnect silica and to low-*k* dielectrics for roughly the same capacitor area [221]. The TDDB data for the $SiO_2$ gate oxide shown in Figure 11.14 stands out from an earlier era because it represents TDDB done over a three-year period [345]. If one compares the relative applied electric fields needed to cause failure at a fixed time (as if comparing breakdown strength), it is clear that the breakdown strength gets progressively worse as the intrinsic dielectric constant for silica-based dielectrics is decreased. Furthermore, the slopes for lifetime extrapolation are similar, which suggests that the lifetime projection models have similarities (more on that later). Percolation modeling has indicated that treating pores as basically already a nonfunctional dielectric (i.e. with preexisting defects) explains the trends found for breakdown strength, a worsened Weibull shape factor, and consistency with the observed degradation physics (at least at higher bias) [221, 346]. Therefore, the trade-off of improved interconnect *RC* performance has come at the expense of a decreasing reliability margin against dielectric breakdown.

The primary lifetime models for thicker gate oxide (≥4 nm; a hyper-thin gate oxide reliability discussion can be found in the review by Suehle [347]) dielectric breakdown are known as the 1/*E*- [267, 348–350] and *E*-models [319, 323, 324, 351–358]. The two models also define the strategic differences taken to identify a root cause explanation for dielectric breakdown: for the 1/*E*-model, a fluence model leading to impact ionization damage [359];

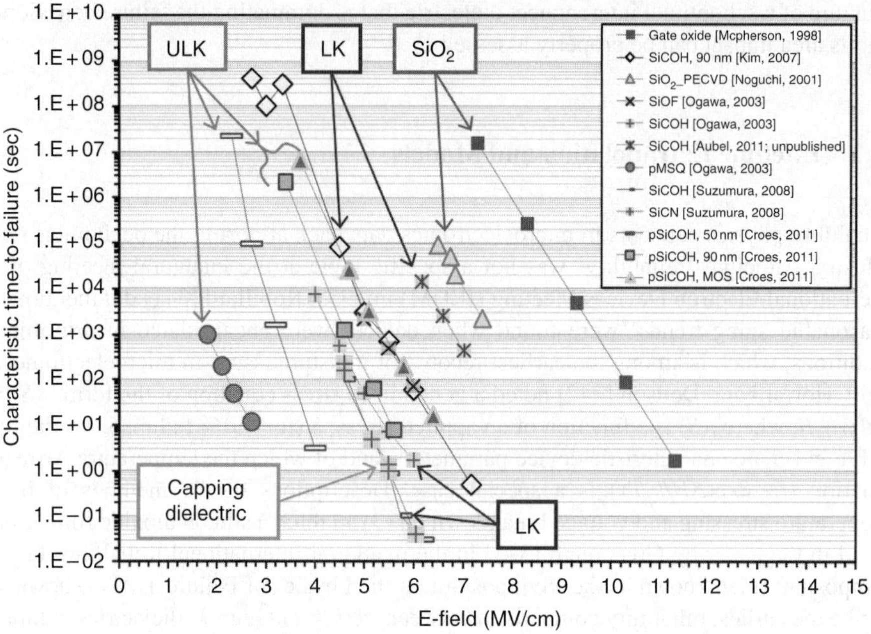


***Figure 11.14*** *A dielectric breakdown comparison is shown for different dielectrics. Note that there is inevitable overlap because of differences in the test structure area and layout, integration approach, process maturity, etc.; however, a clear trend to lower breakdown performance with a decreasing k value is evident. There are also a couple of interesting observations that demonstrate the importance of process/integration quality: (1) for similar LKs, a more recent process shows superior TDDB performance compared to earlier results; (2) in comparatively integrated structures, both LK with* $SiC_xN_y$ *capping dielectric and* $SiC_xN_y$*-only structures show very similar TDDB, indicating that capping dielectric breakdown is also important; (3) TDDB in a MOS structure with ULK only shows better nominal TDDB (it needs an area-scaling correction, however) than integrated ULK and a tighter space integrated-ULK shows a worse performance than a wider space ULK (see color Plate 13)*

for the *E*-model, a field-driven model, where dipole-field coupling reduces the activation energy for thermally induced bond breakage [34]. In the 1/*E*-model, it is hypothesized that the electrons injected (due to Fowler–Nordheim (FN) conduction, even at low-field) into the gate oxide from the substrate cathode end (i.e. assuming NMOS in strong inversion) will generate impact ionization damage as they are accelerated through the dielectric. Subsequently, when electrons reach the cathode, holes can be created that can tunnel back and wreak additional damage (the so-called 'anode-hole injection model') [360, 361]. Because FN tunneling is explicitly dependent upon the exponential of the inverse *E*-field, the time-to-failure, $t_F$, is expected to have the following form:

$$t_F = \tau_0(T)\exp\left[\frac{G(T)}{E_{diel}}\right] \tag{11.7}$$

where $\tau_0(T)$ has the form of an Arrhenius term, $G(T)$ to first order is a constant, and $E_{diel}$ is the applied electric field. In the thermochemical $E$-model [34, 362], the applied field serves to distort and stretch the polar molecular bonds (Si–O) so that these weakened bonds are more susceptible to breakage by thermal effects. Thus, because the degradation rate would be expected to show exponential dependence on the field [342], then the time-to-fail, $t_F$, is expected to exhibit the following form:

$$t_F = A_0 \exp\left[\frac{\Delta H}{k_{Bl}T} - \gamma(T)E_{diel}\right] \tag{11.8}$$

where $A_0$ is a process/materials coefficient, $\Delta H$ is the activation enthalpy (part of the Arrhenius term), $k_B$ is Boltzman's constant, $T$ is the temperature, $\gamma(T)$ is the field acceleration parameter, and $E_{diel}$ is the applied electric field. At or near use-temperatures (say, 105 °C), the $\gamma$ in dielectrics with stretched Si–O bonding [363] is found to be about 4.0 to 4.5 cm/MV [221, 322, 345], which means that for every 0.5 MV/cm change in electric field strength, the characteristic lifetime changes roughly by 10 times. The field acceleration parameter with regard to Si–O bond breakage is shown to have the following temperature dependence [363]:

$$\gamma(T) = \frac{p_{eff}}{k_{Bl}T} \tag{11.9}$$

where $p_{eff}$ is the effective dipole moment. Note that in Equation (11.8), the lifetime for finite temperature at $E_{diel} = 0$ is finite and determines the kinetic term. That is the most significant qualitative difference between the classical $E$-model and fluence-dependent models because it points to a finite lifetime constraint in the low-field behavior of the $E$-model, even as the electrical stress weakens. More recently, attempts have been made to provide a more complementary picture of current injection effects and thermochemical $E$-model impact on TDDB [362, 364–366], but full understanding of dielectric breakdown based on such models remains out of reach.

Early work on lifetime projection of low-$k$ dielectrics focused on the $E$-model because of its ready application to TDDB of simple capacitors [26, 219, 221, 285, 320]; it was clear, however, that the margins afforded by the $E$-model would have eroded significantly as technology went beyond the 45 nm node. Thus, alternative models have been proposed that show an additional lifetime margin in comparison to the $E$-model. One example was the assertion that the Cu drift into low-$k$ constituted the dominant mechanism. The TDDB lifetime would be determined by the increased concentration of Cu defects being incorporated into the low-$k$ dielectric [367, 368] and would be especially problematic in porous ULK dielectrics [369]. Although this situation may be relevant when barrier material is nonexistent or defective, it is not clear how such a model would apply if Cu penetration into LK/ULK were prevented and dielectric breakdown still occurred [370].

Because FN tunneling was not expected to be a dominant transport mechanism during low-field TDDB, a $1/E$-model was not considered suitable for low-$k$ dielectrics. Instead, researchers focused on fluence models that used the leakage–current mechanisms expected to be dominant at low-field. Hence, lifetime models based on a Poole–Frenkel (PF) leakage mechanism have recently emerged, where the time-to-fail has the form:

$$t_F = B_0 \exp\left[\frac{\Delta H}{k_{Bl}T} - \zeta(T)\sqrt{E_{diel}}\right] \tag{11.10}$$

where $B_0$ is a process/materials coefficient, $\Delta H$ is the activation enthalpy (part of the Arrhenius term), $k_B$ is Boltzman's constant, $T$ is the temperature, $\zeta(T)$ is the field acceleration parameter, and $E_{diel}$ is the applied electric field. Not abandoning the notion that Cu ions need to be present for dielectric breakdown to occur, Suzumura *et al.* [371] proposed that TDDB would have the form of Equation (11.10) because injected Cu ions from the anode-side electrode drift along the field direction, especially along the capping dielectric/low-*k* interface, and eventually accumulate at the cathode side at the low-*k*/barrier metal interface. At that point, the conduction band is modified at that interface to allow FN tunneling to dominate and accelerates the rate of Cu accumulation. Once a critical concentration of Cu ($Q_{crit}$) is reached, failure occurs. Because the majority of the mass transport is rate-limited by the PF current [372], the lifetime model would follow the form of Equation (11.10), but possibly with some modifications to the Arrhenius factor. In a follow-up discussion [373], Suzumura *et al.* noted that high-field stress significantly affects the capping $SiC_xN_y$ reliability (which should not be ignored [374–376]), but at lower field, the IMD reliability can be substantially different. Thus, either $E$- or $\sqrt{E}$-models may be both relevant under lower-field stress. In their most recent update, Suzumura *et al.* [377] assert that an $\sqrt{E}$-model fits their data best.

In a separate discussion, Yiang *et al.* [378] showed that the time-dependent generation of traps is consistent with the rate at which the barrier height for PF conduction is lowered with dielectric aging under stress. Because the observed value of the lowered barrier height is consistent with the determined activation energy, such degradation would be consistent with an $\sqrt{E}$-model. Achanta and McLaughlin [379] have demonstrated a charge transport model to account for Cu mobility under bias and field buildup due to such charge accumulation with substantially different (and more pessimistic) trending. They have also matched their model to results for very long-term TDDB data on the ULK dielectric, although a fit of the present lifetimes to an $E$-model is still not out of the question. It is also clear that Cu ingress into the ULK dielectric is quite complicated and depends upon the surface chemistry of the ULK, including the pore surfaces [28, 380]. The implications of surface interaction and pore structure to reliability modeling are unclear. To confuse things even further, Zhao and coworkers have noted that their data on Cu TDDB using MOS capacitors shows a better match to a $1/E$ relation [381, 382].

Chen and coworkers have observed in their TDDB data [383–385] results consistent with an $\sqrt{E}$-model in which the authors also assert that Cu ion incursion into the low-*k* dielectric is at root cause. The TDDB result matching the $\sqrt{E}$-model is distinctly different to the trending observed in a long-term study by Kim *et al.* [285]; as Chen and coworkers noted [383–385], however, the coexistence of more than one failure mechanism is not implausible and that is also the perspective taken by us. Raghavan and Prasad have used probability arguments to construct such a model for low-*k* dielectric breakdown [386]. In fact, coexistence of more than a single failure path has been seen in a very interesting long-term gate oxide TDDB study by Pompl *et al.*, where the BEOL anneal (hydrogen incorporation) markedly changed lifetime acceleration behavior in array MOSFET TDDB [387] so that $1/E$-model, $E$-model, and transitional behavior between both was observed. It is notable that $E$-model behavior was observed at the extremes of the stressing fields where coexistence field acceleration behavior was found, while $1/E$-model behavior was found at intermediate stress. Thus, although this is conjecture at this point, a transition from an earlier failure mode at intermediate stress during IMD-TDDB may trend back toward $E$-model behavior at much lower fields. If that is the case, then accurate understanding of LK/ULK dielectric breakdown strength behavior may be needed to put a lower bound on $E$-model lifetime trending for both test structure and product.

A different kind of fluence-based model (proposed by Lloyd *et al.* and known as the 'impact damage model' or ID-model [388, 389]) has been used to explain IMD-TDDB in LK/ULK dielectrics. This model is basically an impact-ionization type model where electrons within the dielectric undergo acceleration within an electric field and suffer collision with the dielectric matrix after a suitable mean free path, $\lambda_t$. Although the explicit damage mechanism is not put forth, the rate at which damage is generated within the dielectric is proportional to the product of the dominant leakage mechanism within the dielectric (PF conduction) and the probability that an electron will cause damage after some distance $\lambda > \lambda_t$. After integrating over the time up to a critical number of defects, the time-to-fail is shown to have the following form:

$$t_F = \left[\frac{\left(N_f - N_0\right)}{C_0\left(T\right)E_{diel}}\right]\exp\left[-\xi\left(T\right)\sqrt{E_{diel}} + \frac{\alpha}{E_{diel}}\right] \qquad (11.11)$$

where $N_f$ is the critical density of defects at failure, $N_0$ is the density of time-zero defects of similar type, $E_{diel}$ is the electric field strength, $C_0(T)$ is a term incorporating factors from the PF current density, $\xi(T)$ is the field acceleration parameter term that involves the dielectric constant, and $\alpha = U_t/(\mu\ e)$, where $U_t$ is the threshold energy needed to create a defect, $\mu$ is the mean free path between collisions, and $e$ is the electron charge. Because of the extra $1/E_{diel}$ term, the ID-model exhibits a hybrid character between $\sqrt{E}$- and $1/E$-models so that low-field lifetime projections are much more optimistic than both $E$- and $\sqrt{E}$-models. Recently, Lloyd [389] has argued that the ID-model is primarily suited to addressing impact displacement and defect generation of –H ionic bonds. Because LK/ULK materials are expected to contain a significant amount of H atoms, the ID-model is well suited to these kinds of very defective dielectrics in comparison to silica gate oxide, where the H content is (or should be) much more limited.

Haase [286, 390] takes the same problem posed by Lloyd but from a different line of reasoning by posing the dielectric breakdown problem as a trap generation problem instead of a defect creation problem. Through his analysis of the trap generation rate through a series of elaborate trap and detrapping experiments, he notes that the rate of trap generation is highest at the beginning of the stress before even the first measurement is made. Thus, the leakage current decreases as a consequence of such trap generation and subsequent occupancy of such traps; once the trap density increases to sufficient density, however, then overlap between trapped electron wave functions permits tunneling between adjacent traps. After the leakage current passes through a minimum, the leakage current will then increase. This result is generally consistent with commonly observed leakage trace characteristics of

IMD-TDDB. This result is also consistent with observations made by Gischia *et al.* [290], where leakage current during TDDB appears more consistent with FN tunneling than PF leakage at a sufficiently strong field [221]. Based on such an analysis, a trap generation model was created to predict the time-to-minimum-current (TTMC) and is shown to have a dramatically more optimistic lifetime projection than that found by *E*- or $\sqrt{E}$-models. Because a close-form solution is not given due to the fact that the TTMC is obtained computationally, it is difficult to compare the models side-by-side; it is clear, however, that this trap generation model increases at a very fast rate with a lower electric field because a suitable trap generation density cannot be achieved. In fact, the lifetime trending shown by Haase is found to coincide well with the 1/*E*-model (actually, the same can be said about the ID-model). Such a result is actually not so surprising because the original analysis of the 1/*E* used a charge-to-breakdown analysis [348], whereas Haase's treatment tracks the leakage current evolution. Thus, as the electric field decreases, FN tunneling will be minimized to limit the extent of damage to the dielectric under stress. Under very low-field and high-temperature conditions, however, where the thermal energy contributions become critical to controlling the trap creation rate, Haase has argued that deviations from the 1/*E* trending might be expected. Because the thermochemical *E*-model [345, 362] is also a thermally driven process but limited by the effectiveness of localized bond distortions to the Si–O bond, it would be interesting to see how the two models compare under such conditions. What is not explained, however, is the fact that the initial trap generation rate is so high, although low-*k* dielectrics are known to have a significant density of active defects [391–396]. It is not clear whether the initial trap generation is created by a separate pathway from the leakage path and whether this pathway remains a viable pathway to further trap generation as the dielectric ages under stress.

Finally, Croes, Tőkei and others [397, 398] have recently argued through experimental observation that the both *E*- and $\sqrt{E}$-models are far too conservative and that the power-law ($E^n$) model [399] ($t_F \propto E^n$, where *E* is the electric field and *n* is a large exponent ≥ 15) better describes their TDDB results to date. Although there is, at present, no physical justification for such a model for LK/ULK dielectrics, it is interesting that lifetime models for both ultra thin gate oxide and LK/ULK models show evidence of convergence toward some kind of power-law dependence. It is thus clear that the state of knowledge about LK/ULK TDDB reliability is a continually evolving process.

Perceptibly, the standard model used for low-*k* TDDB has shifted away from the *E*-model to the $\sqrt{E}$-model, although its status is under regular review (see Figure 11.15 where the lifetime extrapolations for the different lifetime models are demonstrated). Whatever the model ultimately deemed 'most correct', it is clear that marginal processes that impact general critical dimension uniformity worsen line-edge-roughness (LER) effects, or degrade low-*k* quality through excessive moisture incorporation will worsen reliability margin and cannot be tolerated in high-volume manufacturing. To assess the quality of such processes, it is actually convenient to use the *E*-model as the criterion for process quality because the lifetime extrapolations are so conservative relative to the other models. What this means is that breakdown distributions will need to be clean, and known extrinsic factors such as Cu defects and moisture degradation need to be limited so that *E*-model extrapolation is clean and simple. Unfortunately, as technology moves below the 32 nm node, meeting 100 kPOH requirements at the product level using the *E*-model will be problematic at best. If an area-scaled product lifetime projection falls below 100 kPOH when

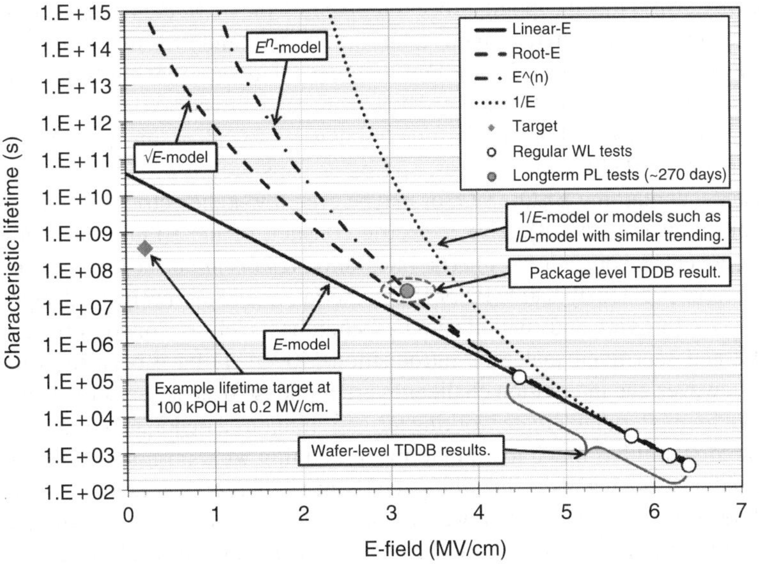


***Figure 11.15*** *Comparison among different lifetime extrapolation models. In practice, wafer-level (WL) testing cannot distinguish among models. Only long-term assessment (mostly a package-level test) can permit additional insight into an appropriate physical model. In this example, a root-E or even a less stringent model is shown to match an additional long-term data point added to the data in Figure 11.11; however, a test time of ~270 days is not suitable for more generic reliability evaluations*

using the $E$-model, the question becomes a matter of what lifetime metric is then considered suitable with an $E$-model, especially because the lifetime difference at use-voltage between the $E$-model and other models will be substantial. For example, the extrapolated lifetime difference between an $E$- and $\sqrt{E}$-model at 1 MV/cm is probably of the order of $10^4$. Thus, comparisons among the different lifetime models will be needed to make judgments for especially at-risk interconnect areas. In that situation, the use of the $E$-model will need to be weighed against competitive concerns. This analysis points to the critical nature of lifetime models because real-world designs can be impacted by them, especially as we go forward with advanced technology nodes with perceptibly smaller lifetime margins.

## 11.4 Future Trends and Concerns

In an earlier section, the problems found with integration of ULK or ELK dielectrics have been shown to be a significant challenge and will remain a major area of research focus for upcoming technology nodes. More detailed questions about the manufacturing feasibility of presently targeted process and integration approaches to ULK materials and their ultimate impact on reliability need to be addressed. Such process challenges include direct

CMP on ULK, SED metal capping process uniformity on ULK, etch damage to ULK, etc., as discussed previously. There remains a fair bit of uncertainty regarding the suitability of ULK/ELK dielectrics with $k < 2.3$ for high-volume manufacturing, but further advances in process and integration optimization of ULK/ELK can be expected.

Looking further ahead, however, efforts to integrate so-called 'air-gap' technologies lie on the horizon [400–405]. In air-gaps, spaces or gaps are purposely integrated into the BEOL stack – containing BEOL $SiO_2$ or maybe even low-$k$ dielectrics – for the purpose of dramatically lowering the effective dielectric constant of the BEOL stack (a gap has a minimum dielectric constant $k \cong 1$). The integration results are basically open areas where one might have previously had IMD or ILD material, and local mechanical support is provided only by the vias and metallization. Techniques such as nonconformal deposition [406–408] and sacrificial materials introduction [409–411] have been used to fabricate air-gaps; however, the integration challenges are immense because selectively and controllably removing material to create working air-gaps remains a major area of focus. Although this technology still has much to do in terms of defining what portions of the interconnect stack should have such gaps and defining the proper integration strategy for such interconnect architecture (especially with regards to mechanically stability), one interesting result found with present approaches is that the observed IMD-TDDB result for an air-gap does not appear to be a 'show-stopper' [412, 413], at least at a larger critical dimension. As long as the Cu barrier structure remains intact, the electrical field lines on the dielectric outside a gap will necessarily consist of those from the fringing field. Thus, the electric field strength on the remaining dielectric will be significantly reduced.

The presence of the gap does bring to mind what remains within the gap after final integration. If the air-gap is filled with actual air, then the breakdown strength of air (according to spark gap tables) is maybe $3.85 \times 10^4$ V/cm for a spark gap with a diameter of 2.5 cm and a gap = 0.13 cm [414]. This rather low field is much smaller than the field that will be seen for microelectronics devices used at even a 65 nm technology node, such that it is natural to ask about the intrinsic breakdown strength for a given capacitor (note that the effective-time-at-field is not specified here). Assuming that macroscopic air breakdown translates well to nanometer spacings (it always does, right?), the capacitor area for this spark gap is a massive (to a nanometer-sized person) $4.9 \times 10^{14}$ nm$^2$. A 1 cm long IMD capacitor with a trench thickness of 500 nm will have a capacitor area of $5.0 \times 10^{-5}$ nm$^2$. If we assert that Poisson area scaling applies [269] so that $E_{BD,2} = E_{BD,1}(A_1/A_2)^{1/\beta}$ (assume $\beta = 53.5$), then the breakdown strength of an IMD air-filled and gapped capacitor increases to only $8.72 \times 10^4$ V/cm. If the gap is closer to vacuum, the breakdown strength should increase roughly as the vacuum pressure decreases so that a value of 10 MV/cm is plausible for large-area capacitors [415, 416]. Then, using area-scaling arguments again, such *vacuum*-gapped IMD capacitors will have a somewhat larger breakdown strength than that of a macroscopic capacitor.

Because there are no bonds to break during TDDB stress in a vacuum, and a vacuum does not age *per se* in any practical sense, the lifetime should be limited only by the extent of vacuum preservation within the gap and whatever magnitude of 'dark current' (spontaneous charge generation within the vacuum) is present. In a sense, maybe the notion of breakdown, as conventionally used for low-$k$ reliability assessment, will need modification because an arcing event across a gap will be transitory and not necessarily detectable by conventional TDDB experiments. If BEOL dielectric breakdown is not limited by the dielectric that is surrounding the gap, then the frequency of gap arcing will presumably depend upon how easily the gas components within the gap are broken down and recovered under a fixed electrical bias. Reliability implications of such transitory breakdown – such as the impact on transistor reliability and of residual damage to interconnects left from repeated arcing strikes within a localized region of the gap – may need further investigation. Still, as long as the macroscopic breakdown physics remains similar down to nanometer dimensions and such vacuum-gaps are not filled with extremely low breakdown-strength gases, the intrinsic reliability against air-gap breakdown looks promising. If such promise holds, then air-gaps may help circumvent to some degree (because air-gaps may not be everywhere on-chip) the clearly decreasing breakdown-strength margin observed with increasing porosity incorporation and limit some of the uncertainty associated with how much integrated intrinsic reliability margin truly exists for interconnects using porous low-*k* dielectrics.

## Acknowledgments

The authors would first like to thank the editors, Prof. Mikhail R. Baklanov, Prof. Paul S. Ho, and Prof. Ehrenfried Zschech, for being given the opportunity to write this manuscript. E. Ogawa acknowledges Dr J. K. Wang, R. Lutze, N. Kim, K. Bult, and Dr H. Samueli for their kind support at Broadcom, and he would also like to extend his gratitude to Prof. P. S. Ho and Ms J. A. Smith (at The University of Texas at Austin) and R. Carter and R. Alden (both at Broadcom) for additional technical support. O. Aubel expresses his thanks to his colleagues K. Y. Yang and A. Beyer (at GLOBALFOUNDRIES) as well as to A. Fischer (at Infineon Technologies AG) for helpful discussion and support. Finally, the authors would like to thank their respective families for their great patience and understanding while they struggled together with the preparation of this manuscript.

## References

[1] Meindl, J.D., Davis, J.A., Zarkesh-Ha, P., Patel, C.S., Martin, K.P. and Kohl, P.A. (2002) Interconnect opportunities for gigascale integration. *IBM J. Res. Develop.*, **46** (2/3), 245–263.

[2] List, S., Bamal, M., Stucchi, M. and Maex, K. (2006) A global view of interconnects. *Microelectron. Engng*, **83** (11/12), 2200–2207.

[3] Edelstein, D., Heidenreich, J., Goldblatt, R., Cote, W., Uzoh, C., Lustig, N., Roper, P., McDevitt, T., Motsiff, W., Simon, A., Dukovic, J., Wachnik, R., Rathore, H., Schulz, R., Su, L., Luce, S. and Slattery, J. (1997) Full copper wiring in a sub-0.25 μm CMOS ULSI technology. Technical Digest, IEEE International Electron Devices Meeting, Piscataway, New Jersey, pp. 773–776.

[4] Maex, K., Baklanov, M.R., Shamiryan, D., Iacopi, F., Brongersma, S.H. and Yanovitskaya, Z.S. (2003) Low dielectric constant materials for microelectronics. *J. Appl. Phys.*, **93** (11), 8793–8841.

[5] Volksen, W., Miller, R.D. and Dubois, G. (2010) Low dielectric constant materials. *Chem. Rev.*, **110** (1), 56–110.

[6] Liu, C.C., Tsai, C.Y., Huang, Y.M., Wu, J.Y. and Lur, W. (1999) Integration of Cu/FSG dual damascene technology for sub-0.18 μm ULSI circuits, in *Advanced Metallization Conference (AMC)* (eds M. E. Gross, T. Gessner, N. Kobayashi and Y. Yasuda), Materials Research Society, Warrendale, Pennsylvania, pp. 403–408.

[7] Barth, E., Ivers, T., McLaughlin, P., McDonald, A., Levine, E., Greco, S., Fitzsimmons, J., Melville, I., Spooner, T., DeWan, C., Chen, X., Manger, D., Nye, H., McGahay, V., Biery, G., Goldblatt, R. and Chen, T. (2000) Integration of copper and fluorosilicate glass for 0.18 μm interconnections. Proceedings of the IEEE 2000 International Interconnect Technology Conference (IITC), IEEE Service Center, Piscataway, New Jersey, pp. 219–221.

[8] Martin, J.S., Taylor, K.J., Luttmer, J.D., Ralston, A.K.R., West, J.A., Bonifield, T.D., Mickler, E.M., Bolnedi, S., Adams, C.T., Chew, K.-H., Bayman, A. and van Schravendijk, B. (2001) FSG process development for copper/damascene technology. Proceedings of the IEEE 2001 International Interconnect Technology Conference (IITC), pp. 39–41.

[9] Wistrom, R., Bomberger, G., Cohen, S., Hazel, S., Lavoie, M., Gambino, J., Poley, D. and Dokumaci, O. (2001) Film properties and integration of a variety of FSG films. Proceedings of the IEEE 2001 International Interconnect Technology Conference (IITC), pp. 168–170.

[10] Stamper, A.K., Adams, C., Chen, X., Christiansen, C., Cooney, F., Cote, W., Gambino, J., Gill, J., Luce, S., McDevitt, T., Porth, B., Spooner, T., Winslow, A. and Wistrom, R. (2002) 0.13 μm generation integration and manufacturing of dual damascene copper in FSG, in *Advanced Metallization Conference (AMC)* (eds B.M. Melnick, T.S. Cale, S. Zaima and T. Ohta), Warrendale, PA, Materials Research Society, Warrendale, Pennsylvania, pp. 485–491.

[11] Yokogawa, S., Uno, S., Kato, I., Tsuchiya, H., Shimizu, T. and Sakamoto, M. (2011) Statistics of breakdown field and time-dependent dielectric breakdown in contact-to-poly modules. Proceedings of the 49th Annual IEEE International Reliability Physics Symposium (IRPS), Piscataway, New Jersey, pp. 149–154.

[12] Haase, G.S. and McPherson, J.W. (2007) Modeling of interconnect dielectric lifetime under stress conditions and new extrapolation methodologies for time-dependent dielectric breakdown. Proceedings of the 45th Annual IEEE International Reliability Physics Symposium (IRPS), Piscataway, New Jersey, pp. 390–398.

[13] Vilmay, M., Roy, D., Monget, C., Volpi, F. and Chaix, J.-M. (2009) Copper-line topology impact on the reliability of SiOCH low-*k* for the 45-nm technology node and beyond. *IEEE Trans. Device and Mater. Reliab.*, **9** (2), 120–127.

[14] Noguchi, J. (2005) Dominant factors in TDDB degradation of Cu interconnects. *IEEE Trans. Electron Devices*, **52** (8), 1743–1750.

[15] Tőkei, Zs., Roussel, Ph., Stucchi, M., Versluijs, J., Ciofi, I., Carbonell, L., Beyer, G.P., Cockburn, A., Agustin, M. and Shah, K. (2009) Impact of LER on BEOL dielectric reliability: a quantitative model and experimental validation. Proceedings of the IEEE 2009 International Interconnect Technology Conference (IITC), IEEE Service Center, Piscataway, New Jersey, pp. 228–230.

[16] Siew, Y.K., Stucchi, M., Versluijs, J., Roussel, P., Kunnen, E., Pantouvaki, M., Beyer, G.P. and Tokei, Zs. (2011) Line edge roughness (LER) correlation and dielectric reliability with spacer-defined double patterning (SDDP) at 20 nm half pitch. Proceedings of the IEEE 2011 International Interconnect Technology Conference (IITC), IEEE Service Center, Piscataway, New Jersey, paper 6.5.

[17] Chen, F., Lloyd, J.R., Chanda, D., Achanta, R., Bravo, O., Strong, A., McLaughlin, P.S., Shinosky, M., Sankaran, S., Gebreselasie, E., Stamper, A.K. and He, Z.X. (2008) Line edge roughness and spacing effect on low-*k* TDDB characteristics. Proceedings of the 46th Annual IEEE International Reliability Physics Symposium (IRPS), pp. 132–137.

[18] Stucchi, M. and Tőkei, Zs. (2008) Impact of LER and via-misalignment on the electric field between nanometer-scale wires. Proceedings of the IEEE 2008 International Interconnect Technology Conference (IITC), IEEE Service Center, Piscataway, New Jersey, pp. 174–176.

[19] Aubel, O., Thierbach, S., Seidel, R., Freudenberg, B., Meyer, M.A., Feustel, F., Poppe, J., Nopper, M., Preusse, A., Zistl, C. and Weide-Zaage, K. (2008) Comprehensive reliability analysis of CoWP metal cap unit processes for high volume production in sub-5m dimensions. Proceedings of the 46th Annual IEEE International Reliability Physics Symposium (IRPS), Piscataway, New Jersey, pp. 675–676.

[20] Nopper, M., Bömmels, J., Aubel, O., Hartz, H. and Preusse, A. (2009) Key aspects of CoWP and ULK implementation into high-volume manufacturing for future technologies, in *Advanced Metallization Conference (AMC)* (eds A.J. McKerrow, Y. Shacham-Diamand, S. Shingubara and Y. Shimogaki), Materials Research Society, Warrendale, Pennsylvania, pp. 157–165.

[21] Michael, N.L., Kim, C.U., Gillespie, P. and Augur, R. (2003) Electromigration failure in ultra-fine copper interconnects. *J. Electron. Mater.*, **32** (10), 988–993.
[22] Michael, N.L., Kim, C.U., Gillespie, P. and Augur, R. (2003) Mechanism of reliability failure in Cu interconnects with ultralow-*k* materials. *Appl. Phys. Lett.*, **83** (10), 1959–1961.
[23] Augur, R.A., Kim, C.U., Blaschke, V., Michael, N.L., Gillespie, P., Rasco, M., Lin, J.C., Kim, S.Y. and Pfeifer, K. (2003) New reliability failure mechanism in porous low-*k* dual damascene interconnects, in *Advanced Metallization Conference (AMC)* (eds G.W. Ray, T. Smy, T. Ohta and M. Tsujimura), Materials Research Society, Warrendale, Pennsylvania, pp. 277–281.
[24] Lu, X., Pyun, J.W., Li, B., Henis, N., Neuman, K., Pfeifer, K. and Ho, P.S. (2005) Barrier layer effects on electromigration reliability of Cu/low *k* interconnects. Proceedings of the IEEE 2005 International Interconnect Technology Conference (IITC), IEEE Service Center, Piscataway, New Jersey, pp. 033–035.
[25] Pyun, J.W., Lu, X., Yoon, S., Henis, N., Neuman, K., Pfeifer, K. and Ho, P.S. (2005) Scaling effect on electromigration reliability for Cu/low-*k* interconnects. Proceedings of the 43rd Annual IEEE International Reliability Physics Symposium (IRPS), Piscataway, New Jersey, pp. 191–194.
[26] Tsu, R., McPherson, J.W. and McKee, W.R. (2000) Leakage and breakdown reliability issues associated with low-*k* dielectrics in a dual-damascene Cu process. Proceedings of the 38th Annual IEEE International Reliability Physics Symposium (IRPS), Piscataway, New Jersey, pp. 348–353.
[27] Tőkei, Zs., Croes, K. and Beyer, G.P. (2010) Reliability of copper low-*k* interconnects. *Microelectron. Engng*, **87** (3), 348–354.
[28] Takeda, K., Ryuzaki, D., Mine, T., Hinode, K. and Yoneyama, R. (2003) Copper-induced dielectric breakdown in silicon oxide deposited by plasma-enhanced chemical vapor deposition using trimethoxysilane. *J. Appl. Phys.*, **94** (4), 2572–2578.
[29] Matsuura, M., Nishioka, Y., Matsumoto, S., Goto, K., Satake, T., Yuasa, H., Okazaki, G., Tomita, K., Hashimoto, K., Tomohisa, S., Yamashita, T., Fukada, T. and Ohsaki, A. (2002) Robust low-*k* SiOC integration in Cu dual damascene interconnect for 90 nm node SoC technology, in *Advanced Metallization Conference (AMC)* (eds B.M. Melnick, T.S. Cale, S. Zaima and T. Ohta), Materials Research Society, Warrendale, Pennsylvania, pp. 493– 499.
[30] Lim, S.W., Shimogaki, Y., Nakano, Y., Tada, K. and Komiyama, H. (1997) Way to obtain low-dielectric-constant F-doped $SiO_2$ films, in *Advanced Metallization and Interconnect Systems for ULSI Applications (AMC)* (eds R. Cheung, J. Klein, K. Tsubouchi, M. Murakami and N. Kobayashi), Materials Research Society, Warrendale, Pennsylvania, pp. 317–323.
[31] Kukuda, T., Sasaki, E., Hosokawa, T. and Kobayashi, N. (1997) Highly reliable SiOF film formation by HDP-CVD using fluorosilanes, in *Advanced Metallization and Interconnect Systems for ULSI Applications (AMC)* (eds R. Cheung, J. Klein, K. Tsubouchi, M. Murakami and N. Kobayashi), Materials Research Society, Warrendale, Pennsylvania, pp. 391–395.
[32] Martin, J.S., Taylor, K.J., Luttmer, J.D., Ralston, A.K.R., West, J.A., Bonifield, T.D., Mickler, E.M., Bolnedi, S., Adams, C.T., Chew, K.-H., Bayman, A. and van Schravendijk, B. (2001) FSG process development for copper/damascene technology. Proceedings of the IEEE 2001 International Interconnect Technology Conference (IITC), IEEE Service Center, Piscataway, New Jersey, pp. 39–41.
[33] Wistrom, R., Bomberger, G., Cohen, S., Hazel, S., Lavoie, M., Gambino, J., Poley, D. and Dokumaci, O. (2001) Film properties and integration of a variety of FSG films. Proceedings of the IEEE 2001 International Interconnect Technology Conference (IITC), IEEE Service Center, Piscataway, New Jersey, pp. 168–170.
[34] McPherson, J.W. and Mogul, H.C. (1998) Underlying physics of the thermochemical *E* model in describing low-field time-dependent dielectric breakdown in $SiO_2$ thin films. *J. Appl. Phys.*, **84** (3), 1513–1523.
[35] Gates, S.M., Grill, A., Dimitrakopoulos, C., Patel, V., Chen, S.T., Spooner, T., Ryan, E.T., Cohen, S.A., Simonyi, E., Liniger, E., Ostrovski, Y. and Bhatia, R. (2008) Integration compatible porous SiCOH dielectrics from 45 to 22 nm, in *Advanced Metallization Conference (AMC)* (eds M. Naik, R. Shaviv, T. Yoda and K. Ueno), Materials Research Society, Warrendale, Pennsylvania, pp. 531–536.

[36] Shimada, M., Watanabe, K., Gawase, A., Miyajim, H. and Yano, H. (2008) Mechanism of plasma damage resistance on porous MSX film with methylene bridge, in *Advanced Metallization Conference (AMC)* (eds M. Naik, R. Shaviv, T. Yoda and K. Ueno), Materials Research Society, Warrendale, Pennsylvania, pp. 559–563.

[37] Yau, W.-F., Lu, Y.-C., Liu, K., Chopra, N., Poon, T., Willecke, R., Lee, J.-H., Matthews, P., Huang, T., Mandal, R., Lee, P., Lang, C.-I., Sugiarto, D., Lou, I.-S., Ma, J., Pang, B., Naik, M., Yost, D. and Cheung, D. (1999) Black Diamond™ – a low-*k* dielectric for Cu damascene applications, in *Advanced Metallization Conference (AMC)* (eds M.E. Gross, T. Gessner, N. Kobayashi and Y. Yasuda), Materials Research Society, Warrendale, Pennsylvania, pp. 379–386.

[38] Donaton, R.A., Struyf, H., Lepage, M., Coenegrachts, B., Stucchi, M., De Roest, D., Baklanov, M.R., Vanhaelemeersch, S., Maex, K., Gaillard, F., Xia, L.-Q., Lim, T.-H., Gotuaco, M., Yieh, E. and Van Autryve, L. (2000) Characterization and integration of a new Si–O–C film deposited by CVD, in *Advanced Metallization Conference (AMC)* (eds D. Edelstein, G. Dixit, Y. Yasuda and T. Ohba), Materials Research Society, Warrendale, Pennsylvania, pp. 595–601.

[39] Kim, T.S., McKerrow, A.J., Hong, Q.-Z., Kirkpatrick, B., Park, H., Hong, H., Xing, G., Newton, K., Permana, D., Eissa, M., Chatterjee, B. and Bonifield, T.D. (2001) Integration of organosilicate glasses (OSGs) in high performance copper interconnects, in *Advanced Metallization Conference (AMC)* (eds A.J. McKerrow, Y. Shacham-Diamand, S. Zaima and T. Ohba), Materials Research Society, Warrendale, Pennsylvania, pp. 25–31.

[40] Mountsier, T., Grief, M., Goonetilleke, C. and Lassig, S. (2001) Integration of Cu with SiOCH (coral) and SiC:H low *k* dielectric films in a dual damascene scheme, in *Advanced Metallization Conference (AMC)* (eds A.J. McKerrow, Y. Shacham-Diamand, S. Zaima and T. Ohba), Materials Research Society, Warrendale, Pennsylvania, pp. 295–299.

[41] Sánchez, H., Melnick, B. and Smith, B. (2002) Productization of Cu/low *k* at the 130 nm and 90 nm technology nodes, in *Advanced Metallization Conference (AMC)* (eds B.M. Melnick, T.S. Cale, S. Zaima and T. Ohta), Materials Research Society, Warrendale, Pennsylvania, pp. 471–483.

[42] Edelstein, D., Rathore, H., Davis, C., Clevenger, L., Cowley, A., Nogami, T., Agarwala, B., Arai, S., Carbone, A., Chanda, K., Chen, F., Cohen, S., Cote, W., Cullinan, M., Dalton, T., Das, S., Davis, P., Demarest, J., Dunn, D., Dziobkowski, C., Filippi, R., Fitzsimmons, J., Flaitz, P., Gates, S., Gill, J., Grill, A., Hawken, D., Ida, K., Klaus, D., Klymko, N., Lane, M., Lane, S., Lee, J., Landers, W., Li, W.-K., Lin, Y.-H., Liniger, E., Liu, X.-H., Madan, A., Malhotra, S., Martin, J., Molis, S., Muzzy, C., Nguyen, D., Nguyen, S., Ono, M., Parks, C., Questad, D., Restaino, D., Sakamoto, A., Shaw, T., Shimooka, Y., Simon, A., Simonyi, E., Swift, A., Van Kleeck, T., Vogt, S., Wang, Y.-Y., Wille, W., Wright, J., Yang, C.-C., Yoon, M. and Ivers, T. (2004) Comprehensive reliability evaluation of a 90 nm CMOS technology with Cu/PECVD low-*k* BEOL. Proceedings of the 42nd Annual IEEE International Reliability Physics Symposium (IRPS), Piscataway, New Jersey, pp. 316–319.

[43] Clevenger, L.A., Yoon, M., Edelstein, D., Cowley, A., Davis, C., Agarwala, B., Biolsi, P., Chanda, K., Cohen, S., Cote, W., Cullinan, M., Dalton, T., Das, S., Davis, P., Demarest, J., Dunn, D., Dziobkowski, C., Economikos, L., Filippi, R., Fitzsimmons, J., Flaitz, P., Fuller, N., Gates, S., Gill, J., Grill, A., Hawken, D., Joy, R., Klymko, N., Lane, M., Law, S., Landers, W., Le, N., Li, W.-K., Lin, Y.-H., Liniger, E., Liu, H., Liu, W., Liu, X., Llera-Hurlburt, D., Low, C.H., Malhotra, S., Matusiewicz, G., Moskowitz, S., Muzzy, C., Nguyen, D., Pal, S., Parks, C., Ramachandran, T., Rathore, H., Restaino, D., Rhoads, B., Shaw, T., Schmitt, J., Simon, A., Tan, J., Tang, T., Tempest, S., Van Kleeck, T., Vogt, S., Wagner, R., Wang, X., Wang, Y.-Y., Wille, W., Wright, J., Yang, C.-C. and Ivers, T. (2004) 90 nm SiCOH technology in 300 mm manufacturing, in *Advanced Metallization Conference (AMC)* (eds D. Erb, P. Ramm, K. Masu and A. Osaki), Materials Research Society, Warrendale, Pennsylvania, pp. 27–36.

[44] Herrick, M., Leung, P., Wilson, B., Nallapati, G., Guvenilir, A., Nagy, A., Carter, R., Smith, D., Legg, J., Rose, D., Sheth, G., Terpolilli, L., Cain, J., Demircan, E., Junker, K., Martinez, H., Cochran, D., Bolton, S., Filipiak, S., Justison, P., Gall, M., Kawasaki, H., Sheth, V., Huang, F. and Melnick, B. (2004) Integration solutions for 90 nm BEOL, in *Advanced Metallization*

*Conference (AMC)* (eds D. Erb, P. Ramm, K. Masu and A. Osaki), Materials Research Society, Warrendale, Pennsylvania, pp. 89–95.
[45] *International Technology Roadmap for Semiconductors (ITRS)*, 2009 edn, Interconnect; Available from: http://www.itrs.net [Accessed 22 October, 2010].
[46] Robertson, W.M., Arjavalingam, G. and Shinde, S.L. (1991) Microwave dielectric measurements of zirconia–alumina ceramic composites: a test of the Clausius–Mossotti mixture equations. *J. Appl. Phys.*, **70** (12), 7648–765.
[47] Morgen, M., Ryan, E.T., Zhao, J.-H., Hu, C., Cho, T. and Ho, P.S. (2000) Low dielectric constant materials for ULSI interconnects. *Annu. Rev. Mater. Sci.*, **30**, 645–680.
[48] Ryan, E.T., Ho, H.-M., Wu, W.-L., Ho, P.S., Gidley, D.W. and Drage, J. (1999) Material property characterization and integration issues for mesoporous silica. Proceedings of the IEEE International Interconnect Technology Conference (IITC), pp. 187–189.
[49] Grill, A. (2009) Porous pSiCOH ultralow-*k* dielectrics for chip interconnects prepared by PECVD. *Annu. Rev. Mater. Sci.*, **39**, 49–69.
[50] Hedrick, J.L., Miller, R.D., Hawker, C.J., Carter, K.R., Volksen, W., Yoon, D.Y. and Trollsas, M. (1998) Templating nanoporosity in thin-film dielectric insulators. *Advd Mater.*, **10** (13), 1049–1053.
[51] Miller, R.D., Volksen, W., Hedrick, J.L., Hawker, C.J., Remenar, J.F., Furuta, P., Nguyen, C.V., Yoon, D., Toney, M., Rice, D.P. and Hay, J. (1999) Sacrificial macromolecular porogens: a route to porous organosilicates for on-chip insulator applications, in *Advanced Metallization Conference (AMC)* (eds M.E. Gross, T. Gessner, N. Kobayashi and Y. Yasuda), Materials Research Society, Warrendale, Pennsylvania, pp. 327–333.
[52] Grill, A. (2001) From tribological coatings to low-*k* dielectrics for ULSI interconnects. *Thin Solid Films*, **398/399**, 527–532.
[53] Grill, A. and Patel, V. (2001) Ultralow-*k* dielectrics prepared by plasma-enhanced chemical vapor deposition. *Appl. Phy. Lett.*, **79** (6), 803–805.
[54] Zheng, Y., Zhu, W., Huang, T.F., Nemani, S., Yim, K.S., Amer, H., Xia, L.-Q. and M'Saad, H. (2003) Plasma enhanced CVD low-*k* Black Diamond™ film formation process for 65 nm technology node, in *Advanced Metallization Conference (AMC)* (eds G.W. Ray, T. Smy, T. Ohta and M. Tsujimura), Materials Research Society, Warrendale, Pennsylvania, pp. 543–547.
[55] Jousseaume, V., Beynet, J., Remiat, B., Gaillard, F., Lumia, G., Didier, G., Fourel, B., Rebiscoul, D., Louis, D. and Passemard, G. (2005) Total porogen extraction by supercritical carbon dioxide to produce porous ULK thin film, in *Advanced Metallization Conference (AMC)* (eds S.H. Brongersma, T.C. Taylor, M. Tsujimura and K. Masu), Materials Research Society, Warrendale, Pennsylvania, pp. 317–323.
[56] Urbanowicz, A.M., Vanstreels, K., Verdonck, P., Shamiryan, D., Cremel, M., De Gendt, S. and Baklanov, M.R. (2009) Fabrication of porogen residue free ultra low-*k* PECVD material by subsequent $H_2$-afterglow plasma treatment and UV curing, in *Advanced Metallization Conference (AMC)* (eds A.J. McKerrow, Y. Shacham-Diamand, S. Shingubara and Y. Shimogaki), Materials Research Society, Warrendale, Pennsylvania, pp. 65–71.
[57] Maitrejean, S., Jousseaume, V., Trouve, H., Gallagher, M., Fayolle, M., Le Cornec, C., Morel, T., Haumesser, P.H., Minoret, S., Zenasni, A., Ciaramella, F., Prokopowicz, G., Zhen, H. and Calvert, J. (2004) Study of ALD and CVD metallization compatibility with ULK using a post integration porogen removal approach, in *Advanced Metallization Conference (AMC)* (eds D. Erb, P. Ramm, K. Masu and A. Osaki), Materials Research Society, Warrendale, Pennsylvania, pp. 83–88.
[58] Tsuji, N., Matsushita, K., Kaneko, S., Takamure, N. and Kobayashi, N. (2008) UV cure control of pore interconnectivity for porogen-based SiOC films, in *Advanced Metallization Conference (AMC)* (eds M. Naik, R. Shaviv, T. Yoda and K. Ueno), Materials Research Society, Warrendale, Pennsylvania, pp. 619–623.
[59] Ashraf, A., Hwang, B.K., Loboda, M.J., Schneider, R. and Perz, S. (2004) Development of PECVD processes for low-*k* ($k \leq 2.5$) and Cu barrier ($k \leq 3.2$) dielectric films using novel organosilanes as CVD precursors for 65 nm technology node and beyond, in *Advanced Metallization Conference (AMC)* (eds D. Erb, P. Ramm, K. Masu and A. Osaki), Materials Research Society, Warrendale, Pennsylvania, pp. 387–391.

[60] Shinriki, M., Tajima, N., Hamada, T., Ohno, T., Kobayashi, N., Hasaka, S. and Inoue, M. (2005) Mechanical-property improvement for PECVD SiCOH film by hydrocarbon substitution effect using molecular modeling, in *Advanced Metallization Conference (AMC)* (eds S.H. Brongersma, T.C. Taylor, M. Tsujimura and K. Masu), Materials Research Society, Warrendale, Pennsylvania, pp. 303–308.

[61] Nakahira, J., Suzuki, K., Iba, Y., Sugiura, I., Suzuki, K., Nakata, Y., Fukuyama, S., Yano, E. and Ohba, T. (2002) A novel porous silica film for Cu/low-*k* multilevel interconnects, in *Advanced Metallization Conference (AMC)* (eds. B.M. Melnick, T.S. Cale, S. Zaima and T. Ohta), Materials Research Society, Warrendale, Pennsylvania, pp. 649–654.

[62] Nakahira, J., Sugiura, I., Nakata, Y., Misawa, N., Iba, Y., Hasegawa, A., Kitada, H., Sugimoto, F., Nishikawa, N., Mizushima, Y., Oshima, M., Ehara, H., Shimizu, N., Miyajima, M., Fukuyama, S., Matsuyama, H., Minamizawa, M., Yano, E. and Ohba, T. (2003) Highly reliable multilevel interconnects with nano-clustering silica, in *Advanced Metallization Conference (AMC)* (eds G.W. Ray, T. Smy, T. Ohta and M. Tsujimura), Materials Research Society, Warrendale, Pennsylvania, pp. 477–482.

[63] Iba, Y., Misawa, N., Sugiura, N., Nishikawa, N., Sugimoto, F., Setta, Y., Kitada, H., Koura, Y., Nakata, Y., Nakahira, J., Nakano, K., Hasegawa, A., Nakaishi, M., Shimizu, N., Fukuyama, S., Nakai, S., Nakamura, T., Yano, E. and Miyajima, M. (2004) Dual damascene formation technique using nano-clustering silica (NCS) for 65 nm node interconnects, in *Advanced Metallization Conference (AMC)* (eds D. Erb, P. Ramm, K. Masu and A. Osaki), Materials Research Society, Warrendale, Pennsylvania, pp. 97–102.

[64] Kinoshita, D., Nakano, A., Kunimi, N., Shimoyama, M., Kawahara, J., Miyoshi, H., Seino, Y., Takasu, Y., Ichikawa, R., Komatsu, M., Hayashi, Y. and Kikkawa, T. (2004) Reduction of dielectric constant in organic-based low-*k* films deposited by plasma copolymerization, in *Advanced Metallization Conference (AMC)* (eds D. Erb, P. Ramm, K. Masu and A. Osaki), Materials Research Society, Warrendale, Pennsylvania, pp. 451–456.

[65] Shamiryan, D.G., Baklanov, M.R., Vanhaelemeersch, S. and Maex, K. (2000) Controllable change of porosity of SiOCH low-*k* dielectric film, in *Advanced Metallization Conference (AMC)* (eds D. Edelstein, G. Dixit, Y. Yasuda and T. Ohba), Materials Research Society, Warrendale, Pennsylvania, pp. 635–639.

[66] Kondoh, E., Aruga, S., Ito, F. and Hayashi, Y. (2008) A fast new approach for evaluating the connectivity of micropores (<2 nm) and supermicropores (<0.5 nm) in low-*k* thin films, in *Advanced Metallization Conference (AMC)* (eds M. Naik, R. Shaviv, T. Yoda and K. Ueno), Materials Research Society, Warrendale, Pennsylvania, pp. 347–352.

[67] Matsuda, K., Miyoshi, R., Seki, H., Takeda, M., Inoue, K., Takahashi, K., Ajioka, T., Sugie, R., Matsunobe, T., Hashimoto, H. and Yoshikawa, M. (2005) Characterization of structural changes of SiOC films by heat treatment, in *Advanced Metallization Conference (AMC)* (eds S.H. Brongersma, T.C. Taylor, M. Tsujimura and K. Masu), Materials Research Society, Warrendale, Pennsylvania, pp. 285–290.

[68] Arao, H., Egami, M. and Nakashima, A. (2007) Mechanical properties of high-temperature vapor-cured ultra low-*k* film (nanoclustering silica $k = 2.0$), in *Advanced Metallization Conference (AMC)* (eds A.J. McKerrow, Y. Shacham-Diamand, S. Shingubara and Y. Shimogaki), Materials Research Society, Warrendale, Pennsylvania, pp. 527–531.

[69] Gourhant, O., Jousseaume, V., Favennec, L., Zanasni, A., Gonon, P., Gerbaud, G. and Vincent, G. (2008) Dielectric constant contributions and structural properties of PECVD porous ULK: impact of the elaboration process, in *Advanced Metallization Conference (AMC)* (eds M. Naik, R. Shaviv, T. Yoda and K. Ueno), Materials Research Society, Warrendale, Pennsylvania, pp. 537–542.

[70] Chapelon, L.L., Vitiello, J., Mastromatteo, E., Barbier, D., Ferrieu, F. and Torres, J. (2004) Characterisation of a CVD porous SiOCH (in *k* [2.2–2.5] range) and evaluation of different post-treatments improving mechanical properties for 45 nm node and below, in *Advanced Metallization Conference (AMC)* (eds D. Erb, P. Ramm, K. Masu and A. Osaki), Materials Research Society, Warrendale, Pennsylvania, pp. 393–403.

[71] Iacopi, F., Waldfried, C., Houthoofd, K., Guyer, E.P., Gage, D.M., Carlotti, G., Travaly, Y., Abell, T., Escorcia, O., Beyer, G., Berry, I., Dauskardt, R.H. and Maex, K. (2005) UV-assisted

curing: an effective technique for toughening low-*k* organosilicate films, in *Advanced Metallization Conference (AMC)* (eds S.H. Brongersma, T.C. Taylor, M. Tsujimura and K. Masu), Materials Research Society, Warrendale, Pennsylvania, pp. 247–254.

[72] Kanto, M., Yoneda, K., Nakao, S., Kondo, S. and Kobayashi, N. (2005) Effects of UV-cure time on electrical properties of Cu/porous SiOC interconnects, in *Advanced Metallization Conference (AMC)* (eds S.H. Brongersma, T.C. Taylor, M. Tsujimura and K. Masu), Materials Research Society, Warrendale, Pennsylvania, pp. 271–276.

[73] Goto, K., Hashii, S., Matsumoto, M., Miura, N., Furusawa, T., Matsuura, M., Ohsaki, A., Ohara, N., Tsuji, N. and Matsushita, K. (2005) Film characterization and integration of UV cured ultra low-*k* for 45 nm node Cu/low-*k* interconnects, in *Advanced Metallization Conference (AMC)* (eds S.H. Brongersma, T.C. Taylor, M. Tsujimura and K. Masu), Materials Research Society, Warrendale, Pennsylvania, pp. 277–283.

[74] Nakao, S.-I., Ushio, J., Ohno, T., Hamada, T., Kamigaki, Y., Kato, M., Yoneda, K., Kondo, S. and Kobayashi, N. (2006) UV/EB cure mechanism for porous PECVD/SOD low-*k* SiCOH materials. Proceedings of the IEEE 2006 International Interconnect Technology Conference (IITC), IEEE Service Center, Piscataway, New Jersey, pp. 66–68.

[75] Iacopi, F., Travaly, Y., Eyckens, B., Waldfried, C., Abell, R., Guyer, E.P., Gage, D.M., Dauskardt, R.H., Sajavaara, T., Houthoofd, K., Grobet, P., Jacobs, P. and Maex, K. (2006) Short-ranged structural rearrangement and enhancement of mechanical properties of organosilicate glasses induced by ultraviolet radiation. *J. Appl. Phys.*, **99** (5), 053511-1-7.

[76] Kim, T.-S., Tsuji, N., van der Hilst, J., Kemeling, N., Matsushita, K., Kobayashi, N., Chumakov, D., Geisler, H., Zschech, E. and Dauskardt, R.H. (2007) Depth dependence of UV curing of organosilicate low-*k* thin films, in *Advanced Metallization Conference (AMC)* (eds A.J. McKerrow, Y. Shacham-Diamand, S. Shingubara and Y. Shimogaki), Materials Research Society, Warrendale, Pennsylvania, pp. 455–458.

[77] Gage, D.M., Stebbins, J.F., Peng, L., Cui, Z., Al-Bayati, A., MacWilliams, K.P., M'Saad, H. and Dauskardt, R.H. (2008) Effects of UV cure on glass structure and fracture properties of nano-porous carbon-doped oxide thin films. *J. Appl. Phys.*, **104** (4), 043513-1-8.

[78] Ito, F., Takeuchi, T. and Hayashi, Y. Improvement of mechanical properties of porous SiOCH films by post-cure treatments, in *Advanced Metallization Conference (AMC)* (eds S.H. Brongersma, T.C. Taylor, M. Tsujimura and K. Masu), Materials Research Society, Warrendale, Pennsylvania, pp. 291–296.

[79] Miyajima, H., Masuda, H., Idaka, T., Shimayama, T., Kagawa, Y., Tabuchi, K., Yano, H., Hasegawa, T., Kadomura, S. and Yoda, T. (2005) Material design of balance between high mechanical strength and high plasma resistance for porous PE-CVD SiOC film ($k = 2.3$), in *Advanced Metallization Conference (AMC)* (eds S.H. Brongersma, T.C. Taylor, M. Tsujimura and K. Masu), Materials Research Society, Warrendale, Pennsylvania, pp. 297–302.

[80] Gage, D.M., Yim, K.S., Al-Bayati, A., Demos, A., M'Saad, H. and Dauskardt, R.H. (2007) UV, E-beam, and thermal curing of low-*k* organosilicates: effects on glass structure and mechanical properties, in *Advanced Metallization Conference (AMC)* (eds A.J. McKerrow, Y. Shacham-Diamand, S. Shingubara and Y. Shimogaki, Materials Research Society, Warrendale, Pennsylvania, pp. 433–438.

[81] Moore, D. and Waldfried, C. (2007) Examining the effects of UV sources on curing of low-*k* dielectrics using Si29 NMR, in *Advanced Metallization Conference (AMC)* (eds A.J. McKerrow, Y. Shacham-Diamand, S. Shingubara and Y. Shimogaki), Materials Research Society, Warrendale, Pennsylvania, pp. 489–494.

[82] Chan, K., Yim, K.S., Nguyen, V., Yi, S.I., Elsheref, K., Nowak, T., Rocha, J., Demos, A., Witty, D. and M'Saad, H. (2007) Structural evolution of nano-porous ultra-low-*k* dielectrics under broadband UV curing, in *Advanced Metallization Conference (AMC)* (eds A.J. McKerrow, Y. Shacham-Diamand, S. Shingubara and Y. Shimogaki), Materials Research Society, Warrendale, Pennsylvania, pp. 489–494.

[83] Matsushita, K., Tsuji, N., Kagami, K., Kato, M., Kaneko, S. and Kobayashi, N. (2007) Effect of UV cure on pore interconnectivity of PECVD SiOC films, in *Advanced Metallization Conference (AMC)* (eds A.J. McKerrow, Y. Shacham-Diamand, S. Shingubara and Y. Shimogaki), Materials Research Society, Warrendale, Pennsylvania, pp. 533–537.

[84] Nakahira, J., Tomioka, K., Soda, K., Kato, M. and Nakao, S. (2007) Robust porous PE-CVD SiOC film ($k$ < 2.4) with ultraviolet curing for Cu/low-$k$ multilevel interconnect fabrication, in *Advanced Metallization Conference (AMC)* (eds A.J. McKerrow, Y. Shacham-Diamand, S. Shingubara and Y. Shimogaki), Materials Research Society, Warrendale, Pennsylvania, pp. 545–549.
[85] Marsik, P., Urbanowicz, A., Verdonck, P., Ferchichi, K., DeRoest, D., Prager, L. and Baklanov, M. (2008) Engineering of chemical and physical properties of low-$k$ materials by different wavelengths of UV light, in *Advanced Metallization Conference (AMC)* (eds M. Naik, R. Shaviv, T. Yoda and K. Ueno), Materials Research Society, Warrendale, Pennsylvania, pp. 543–549.
[86] Ahner, N., Schulz, S., Schaller, M. and Prager, L. (2008) UV assisted curing of plasma damaged porous ultralow-$k$ material after a $k$-restore process: influence of UV-wavelength and curing ambient, in *Advanced Metallization Conference (AMC)* (eds M. Naik, R. Shaviv, T. Yoda and K. Ueno), Materials Research Society, Warrendale, Pennsylvania, pp. 599–604.
[87] Ee, Y.C., Kioussis, D., Angyal, M.S., Kapur, A., Feng, Y., Engel, B.H., Passano, A., Klymko, N., Madan, A., Wang, J.L., Fitzsimmons, J.A., Yusuff, H., Chong, S.Y., Permana, D., Ogunsola, O.O., Child, C., Cheng, T.-J., Sun, Z.G., Choi, Y.J., Sieg, S.A., Molis, S., Parks, C., Linville, J., Grunow, S. and Biery, G. (2009) UV curing-induced photoresist poisoning in advanced ULK BEOL integration schemes, in *Advanced Metallization Conference (AMC)* (eds A.J. McKerrow, Y. Shacham-Diamand, S. Shingubara and Y. Shimogaki), Materials Research Society, Warrendale, Pennsylvania, pp. 185–192.
[88] Rasco, M., Rauf, S., Leung, P., Chatterjee, R., Junker, K., Turner, M., Grudowski, P., Moosa, M. and Ventzek, P. (2005) Process induced damage evaluation of electron-beam cured low-$k$ material on SOI substrates, in *Advanced Metallization Conference (AMC)* (eds S.H. Brongersma, T.C. Taylor, M. Tsujimura and K. Masu), Materials Research Society, Warrendale, Pennsylvania, pp. 333–339.
[89] Mehta, S., Dimitrakopoulos, C., Augur, R., Gambino, J., Chou, A., Hook, T., Linder, B., Tseng, W., Bolam, R., Harmon, D., Massey, D., Gates, S. and Nye, H. (2005) Evaluation of device damage from e-beam curing of ultra low-$k$ BEOL dielectrics, in *Advanced Metallization Conference (AMC)* (eds S.H. Brongersma, T.C. Taylor, M. Tsujimura and K. Masu), Materials Research Society, Warrendale, Pennsylvania, pp. 361–367.
[90] Al-Bayati, A., Shah, A., El Sheref, K., Shimizu, T., Iwasaki, N., Kiyohara, T., Demos, A., Rocha, J.C., M'Saad, H., Tabuchi, K., Ida, K., Kawanami, K., Iemura, S., Nobe, Y., Ishikawa, Y. and Yamane, T. (2007) Eliminating device charging damage in electron beam curing (EBK™) of low $k$ SiCOH films, in *Advanced Metallization Conference (AMC)* (eds A.J. McKerrow, Y. Shacham-Diamand, S. Shingubara and Y. Shimogaki, Materials Research Society, Warrendale, Pennsylvania, pp. 699–704.
[91] Gidley, D.W., Frieze, W.E., Dull, T.L., Sun, J., Yee, A.F., Nguyen, C.V. and Yoon, D.Y. (2000) Determination of pore-size distribution in low-dielectric thin films. *Appl. Phys. Lett.*, **76** (10), 1282–1284.
[92] Peng, H.G., Frieze, W.E., Vallery, R.S., Gidley, D.W., Moore, D.L. and Carter, R.J. (2005) Revealing hidden pore structure in nanoporous thin films using positronium annihilation lifetime spectroscopy. *Appl. Phys. Lett.*, **86** (12), 121904-01-03.
[93] Chen, Z., Prasad, K., Li, C.Y., Lu, P.W., Su, S.S. and Tang, L.J. (2004) Highly reliable dielectric/metal bilayer sidewall diffusion barrier in Cu/porous organic ultra low-$k$ interconnects. Proceedings of the 42nd Annual IEEE International Reliability Physics Symposium (IRPS), Piscataway, New Jersey, pp. 320–325.
[94] Hwang, S.-S., Lee, H.-C., Ro, H.W., Yoon, D.Y. and Joo, Y.-C. (2005) Porosity content dependence of TDDB lifetime and flat band voltage shift by Cu diffusion in porous spin-on low-$k$. Proceedings of the 43rd Annual IEEE International Reliability Physics Symposium (IRPS), Piscataway, New Jersey, pp. 474–477.
[95] Hong, C. and Milor, L. (2006) Porosity-induced electric field enhancement and its impact on charge transport in porous inter-metal dielectrics. Proceedings of the 44th Annual IEEE International Reliability Physics Symposium (IRPS), Piscataway, New Jersey, pp. 679–680.

[96] Li, Y., Bruynseraede, C., Groeseneken, G., Maex, K. and Tőkei, Zs. (2007) On the interaction between inter-metal dielectric reliability and electromigration stress. Proceedings of the 45th Annual IEEE International Reliability Physics Symposium (IRPS), Piscataway, New Jersey, pp. 642–643.
[97] Kikkawa, T., Kayaba, Y., Kohmura, K. and Chikaki, S. (2011) Electrical reliabilities of porous silica low-*k* films. Proceedings of the 49th Annual IEEE International Reliability Physics Symposium (IRPS), Piscataway, New Jersey, pp. 130–133.
[98] Passemard, G. and Demolliens, O. (2000) Integration issues of low *k* materials in damascene structure, in *Advanced Metallization Conference (AMC)* (eds D. Edelstein, G. Dixit, Y. Yasuda and T. Ohba), Materials Research Society, Warrendale, Pennsylvania, pp. 77–83.
[99] Andideh, E., Moon, P., Jain, A., Kloster, G. and Jan, C.-H. (2003) Carbon doped oxide (CDO) film development and integration challenges, in *Advanced Metallization Conference (AMC)* (eds G.W. Ray, T. Smy, T. Ohta and M. Tsujimura), Materials Research Society, Warrendale, Pennsylvania, pp. 439–447.
[100] Lin, S., Wetzel, J.T., Monnig, K.A. and Winebarger, P.A. (2001) Effective porous low-*k* in single damascene integration, in *Advanced Metallization Conference (AMC)* (eds A.J. McKerrow, Y. Shacham-Diamand, S. Zaima and T. Ohba), Materials Research Society, Warrendale, Pennsylvania, pp. 267–272.
[101] Sekiyama, S., Nakamura, K., Ynazawa, H. and Matsunaga, H. (2001) Mechanisms producing bowed profiles in the etching of low-*k* organic films, in *Advanced Metallization Conference (AMC)* (eds A.J. McKerrow, Y. Shacham-Diamand, S. Zaima and T. Ohba), Materials Research Society, Warrendale, Pennsylvania, pp. 341–344.
[102] Sankaran, A., Vasenkov, A.V. and Kushner, M.J. (2002) Fluorocarbon etching of porous silicon dioxide: plasma chemistry and surface kinetics, in *Advanced Metallization Conference (AMC)* (eds B.M. Melnick, T.S. Cale, S. Zaima and T. Ohta), Materials Research Society, Warrendale, Pennsylvania, pp. 551–556.
[103] Ryan, E.T., Lin, S., Nelsen, D., Wetzel, J., Gidley, D. and Drage, J. (2000) The impact of mesoporous silica material properties on Cu/damascene integration, in *Advanced Metallization Conference (AMC)* (eds D. Edelstein, G. Dixit, Y. Yasuda and T. Ohba), Materials Research Society, Warrendale, Pennsylvania, pp. 555–563.
[104] Baklanov, M.R., Jehoul, C., Flannery, C.M., Mogilnikov, K.P., Gore, R., Gronbeck, D., Prokopowicz, G., Sullivan, C., You, Y., Pugliano, N. and Gallagher, M. (2001) Properties of mesoporous low-*k* MSSQ based film prepared using macromolecular porogen, in *Advanced Metallization Conference (AMC)* (eds A.J. McKerrow, Y. Shacham-Diamand, S. Zaima and T. Ohba), Materials Research Society, Warrendale, Pennsylvania, pp. 273–278.
[105] Shamiryan, D., Baklanov, M.R., Tőkei, Z., Iacopi, F. and Maex, K. (2001) Evaluation of Ta(N) diffusion barrier integrity on porous low-*k* films, in *Advanced Metallization Conference (AMC)* (eds A.J. McKerrow, Y. Shacham-Diamand, S. Zaima and T. Ohba), Materials Research Society, Warrendale, Pennsylvania, pp. 279–285.
[106] Gorman, B.P., Mueller, D.W., Chyan, O. and Reidy, R.F. (2005) Analytical TEM characterization of metal penetration and supercritical pore-sealing of ash-damaged porous low-*k* dielectrics, in *Advanced Metallization Conference (AMC)* (eds S.H. Brongersma, T.C. Taylor, M. Tsujimura and K. Masu), Materials Research Society, Warrendale, Pennsylvania, pp. 393–397.
[107] Ito, F., Hijioka, K., Takeuchi, T. and Hayashi, Y. (2004) Nanostructure control of porous low-*k* dielectric films with high water resistance, in *Advanced Metallization Conference (AMC)* (eds D. Erb, P. Ramm, K. Masu and A. Osaki), Materials Research Society, Warrendale, Pennsylvania, pp. 381–386.
[108] Capani, P.M., Gorman, B.P., Reidy, R.F., Mueller, D.W., Walter, E.R., Matz, P.D., Rhoad, J.T. and Busch, E.L. (2004) Drying methods for low-*k* films and their effects on dielectric constants, in *Advanced Metallization Conference (AMC)* (eds D. Erb, P. Ramm, K. Masu and A. Osaki), Materials Research Society, Warrendale, Pennsylvania, pp. 381–386.
[109] Ito, F., Takeuchi, T., Yamamoto, H., Ohdaira, T., Suzuki, R. and Hayashi, Y. (2007) Pore-connectivity dependence of moisture absorption into porous low-*k* films by positron-annihilation lifetime spectroscopy, in *Advanced Metallization Conference (AMC)* (eds

A.J. McKerrow, Y. Shacham-Diamand, S. Shingubara and Y. Shimogaki), Materials Research Society, Warrendale, Pennsylvania, pp. 465–470.

[110] Hanahata, H., Matsuno, S., Miyamoto, M., Ioka, T. and Tanabe, T. (2000) Dependency of basic properties of porous silica ILD thin films in the $k$ range 1.6–2.7 on CMP compatibility, in *Advanced Metallization Conference (AMC)* (eds D. Edelstein, G. Dixit, Y. Yasuda and T. Ohba), Materials Research Society, Warrendale, Pennsylvania, pp. 629–633.

[111] Hsu, C.-L., Lu, K.-T., Lin, W.-C., Lin, J.-C., Chen, C.-H., Tsai, T.-C., Huang, C., Wu, J.Y. and Perng, D.-C. (2010) The TDDB failure mode and its engineering study for 45 nm and beyond in porous low $k$ dielectrics direct polish scheme. Proceedings of the 48th Annual IEEE International Reliability Physics Symposium (IRPS), Piscataway, New Jersey, pp. 918–921.

[112] Lin, J.C., Augur, R.A., Daniels, B.J., Shue, S.L., Yu, C.H. and Liang, M.S. How to eliminate voiding in porous low-$k$ dielectrics and the mechanism of void formation, in *Advanced Metallization Conference (AMC)* (eds B.M. Melnick, T.S. Cale, S. Zaima and T. Ohta), Materials Research Society, Warrendale, Pennsylvania, pp. 637–642.

[113] Maitrejean, S., Mourier, T. and Morel, T. (2001) Cu metallization issues in low $k$ SiOC integration, in *Advanced Metallization Conference (AMC)* (eds A.J. McKerrow, Y. Shacham-Diamand, S. Zaima and T. Ohba), Materials Research Society, Warrendale, Pennsylvania, pp. 301–305.

[114] Shamiryan, D., Yanovitskaya, Z.S., Iacopi, R. and Maex, K. (2002) Barrier deposition on porous low-$k$ films, in *Advanced Metallization Conference (AMC)* (eds B.M. Melnick, T.S. Cale, S. Zaima and T. Ohta), Materials Research Society, Warrendale, Pennsylvania, pp. 829–833.

[115] Fruehauf, S., Streiter, I., Schulz, S.E., Brendler, E., Himcinschi, C., Friedrich, M., Gessner, T. and Zahn, D.R.T. (2001) Hydrophobisation process for porous low $k$ dielectric silica layers, in *Advanced Metallization Conference (AMC)* (eds A.J. McKerrow, Y. Shacham-Diamand, S. Zaima and T. Ohba), Materials Research Society, Warrendale, Pennsylvania, pp. 287–294.

[116] Kastenmeier, B.E.E., Ahlburn, B.T., Gallagher, M., Gidley, D. and Lee, J. (2002) Improvement of film properties by plasma modification of a porous low-$k$ organosilicate glass material, in *Advanced Metallization Conference (AMC)* (eds B.M. Melnick, T.S. Cale, S. Zaima and T. Ohta), Materials Research Society, Warrendale, Pennsylvania, pp. 589–593.

[117] Reidy, R.F., Gorman, B.P., Orozco-Teran, R.A., Zhang, Z., Chang, S. and Mueller, D.W. (2002) Effects of supercritical processing on ultra low-$k$ films, in *Advanced Metallization Conference (AMC)* (eds B.M. Melnick, T.S. Cale, S. Zaima and T. Ohta), Materials Research Society, Warrendale, Pennsylvania, pp. 601–605.

[118] Abell, T., Shamiryan, D., Schuhmacher, J., Besling, W., Sutcliffe, V. and Maex, K. Precursor penetration and sealing of porous CVD SiCOH low $k$ dielectric for atomic layer deposition of $WC_xN_y$ barrier, in *Advanced Metallization Conference (AMC)* (eds B.M. Melnick, T.S. Cale, S. Zaima and T. Ohta), Materials Research Society, Warrendale, Pennsylvania, pp. 717–723.

[119] Abell, T., Shamiryan, D., Patz, M. and Maex, K. (2003) Correlation between solvent diffusion, porosity and pore sealing for low $k$ dielectrics, in *Advanced Metallization Conference (AMC)* (eds G.W. Ray, T. Smy, T. Ohta and M. Tsujimura), Materials Research Society, Warrendale, Pennsylvania, pp. 549–553.

[120] Brongersma, S.H., Sherman, S., Tabat, M., Patz, M., Beyer, G., Travaly, Y., Le, T.Q., Caluwaerts, R. and Hautala, J. (2004) Evaluation of gas cluster ion beam processing for sealing of a porous dielectric, in *Advanced Metallization Conference (AMC)* (eds D. Erb, P. Ramm, K. Masu and A. Osaki), Materials Research Society, Warrendale, Pennsylvania, pp. 463–468.

[121] Yim, K.S., Nguyen, V., Kim, B.H., Dixit, G., Demos, A., Witty, D., and M'Saad, H. (2004) Development of hermetic oxide films for low-$k$ pore sealing, in *Advanced Metallization Conference (AMC)* (eds D. Erb, P. Ramm, K. Masu and A. Osaki), Materials Research Society, Warrendale, Pennsylvania, pp. 469–474.

[122] Xie, B., Muscat, A.J., Busch, E. and Rhoad, T. (2004) The reaction of octyldimethylchlorosilane and supercritical $CO_2$ mixtures with porous methylsilsesquioxane thin films, in *Advanced Metallization Conference (AMC)* (eds D. Erb, P. Ramm, K. Masu and A. Osaki), Materials Research Society, Warrendale, Pennsylvania, pp. 475–479.

[123] Reidy, R.F., Nerusu, P.K., Chaung, E.C., Orozco-Teran, R.A., Kadam, P.P., Matz, P.D., Rhoad, J.T., Busch, E.L. and Mueller, D.W. Supercritical pore sealing of porous MSQ, in *Advanced*

*Metallization Conference (AMC)* (eds D. Erb, P. Ramm, K. Masu and A. Osaki), Materials Research Society, Warrendale, Pennsylvania, pp. 493–497.
[124] Michelon, J., Waeterloos, J., Bancken, P.H.L., Nguyen, V.H., Caluwaerts, R., Beyer, G., Rozeveld, S., Beach, E. and Hoofman, R.J.O.M. (2006) Reliability characterization of different pore sealing techniques on porous SILK™ dielectric films. Proceedings of the 44th Annual IEEE International Reliability Physics Symposium (IRPS), Piscataway, New Jersey, pp. 496–501.
[125] Russell, S.W., McKerrow, A.J., Shih, W.-Y., Singh, A., List, R.S., Ralston, A.R.K., Lee, W.W., Newton, K.J., Chang, M.-C. and Havemann, R.H. (1997) Integration challenges of ultra low-*k* dielectrics, in *Advanced Metallization and Interconnect Systems for ULSI Applications (AMC)* (eds R. Cheung, J. Klein, K. Tsubouchi, M. Murakami and N. Kobayashi), Materials Research Society, Warrendale, Pennsylvania, pp. 289–300.
[126] Homma, T. (1998) Low dielectric constant materials and methods for interlayer dielectric films in ultralarge-scale integrated circuit multilevel interconnections. *Mater. Sci. Engng*, **R23**, 243–285.
[127] Andideh, E. (2001) Evaluation and evolution of ILD materials, in *Advanced Metallization Conference (AMC)* (eds A.J. McKerrow, Y. Shacham-Diamand, S. Zaima and T. Ohba), Materials Research Society, Warrendale, Pennsylvania, pp. 79–84.
[128] Lu, W., Lee, T.J., Widodo, J., Lim, Y.K., Zhang, F. and Hsia, L.C. (2004) Process, integration and reliability of TMCTS based low-*k* dielectric film, in *Advanced Metallization Conference (AMC)* (eds D. Erb, P. Ramm, K. Masu and A. Osaki), Materials Research Society, Warrendale, Pennsylvania, pp. 71–82.
[129] Wu, Z.-C., Shiung, Z.-W., Chiang, C.-C., Wu, W.-H., Chen, M.-C., Jeng, S.M., Chang, W., Chou, P.F., Jung, S.-M., Yu, C.-H. and Liang, M.-S. (2000) Comparative study of physical and electrical characteristics of F- and C-doped low-*k* CVD oxides, in *Advanced Metallization Conference (AMC)* (eds D. Edelstein, G. Dixit, Y. Yasuda and T. Ohba), Materials Research Society, Warrendale, Pennsylvania, pp. 603–609.
[130] Tseng, C.H., Liu, C.C., Hsieh, T.T., Chiang, Y.F., Chen, J.M., Hsieh, W.Y. and Yen, P.W. (2002) Integratibility evaluation of ultra-low-*k* dielectric material for 65 nm technology, in *Advanced Metallization Conference (AMC)* (eds B.M. Melnick, T.S. Cale, S. Zaima and T. Ohta), Materials Research Society, Warrendale, Pennsylvania, pp. 625–628.
[131] Goldberg, C., Freeman, M., Kirksey, S., Sieloff, D., Filipiak, Mercado, L., Braeckelmann, G., Junker, K.H., Grove, N., Pozder, S., Nguyen, T., Prindle, C., Martin, J. and Wang, V. (2002) Interface reliability of higher performance interconnects, in *Advanced Metallization Conference (AMC)* (eds B.M. Melnick, T.S. Cale, S. Zaima and T. Ohta, Materials Research Society, Warrendale, Pennsylvania, pp. 127–137.
[132] Liu, X.H., Lane, M.W., Shaw, T.M., Liniger, E.G. Rosenberg, R.R. and Edelstein, D.C. (2004) Low-*k* BEOL mechanical modeling, in *Advanced Metallization Conference (AMC)* (eds D. Erb, P. Ramm, K. Masu and A. Osaki), Materials Research Society, Warrendale, Pennsylvania, pp. 361–367.
[133] Restaino, D, Molis, S., Grill, A., Patel, V., Lane, M.W., Cheng, T., Demos, A., Gates, S.M., Kioussis, D., Klymko, N., Liu, H., Nguyen, S., Nguyen, V., Patel, J.C., Reiter, S., Shobha, H., Widodo, J., Tim, K. and Edelstein, D. (2007) Optimized interfacial strength for dense and porous SiCOH, in *Advanced Metallization Conference (AMC)* (eds A.J. McKerrow, Y. Shacham-Diamand, S. Shingubara and Y. Shimogaki), Materials Research Society, Warrendale, Pennsylvania, pp. 419–426.
[134] Smith, R.A., Tsui, T. and Ho, P. (2007) Effect of UV cure on the molecular structure and fracture properties of an ultra-low-*k* dielectric, in *Advanced Metallization Conference (AMC)* (eds A.J. McKerrow, Y. Shacham-Diamand, S. Shingubara and Y. Shimogaki), Materials Research Society, Warrendale, Pennsylvania, pp. 483–487.
[135] Maitrejean, S., Braccini, M., Rousseau, M., Verdier, M. and Passemard, G. (2007) Cracking behavior of porous C doped ultra low *k* dielectrics, in *Advanced Metallization Conference (AMC)* (eds A.J. McKerrow, Y. Shacham-Diamand, S. Shingubara and Y. Shimogaki), Materials Research Society, Warrendale, Pennsylvania, pp. 717–721.

[136] Yamamoto, H., Ito, F., Takeuichi, T., Furutake, N. and Hayashi, Y. (2008) Highly-reliable seamless low-*k* SiOCH stacks (SEALS) with high adhesion strength interface by advanced plasma co-polymerization process, in *Advanced Metallization Conference (AMC)* (eds M. Naik, R. Shaviv, T. Yoda and K. Ueno), Materials Research Society, Warrendale, Pennsylvania, pp. 33–39.

[137] Kim, T.-S., Tsuji, N., van der Hilst, J., Matsushita, K., Kobayashi, N., Chumakov, D., Geisler, H., Zschech, E. and Dauskardt, R.H. (2008) Tuning depth profiles of low-*k* dielectrics with UV curing, in *Advanced Metallization Conference (AMC)* (eds M. Naik, R. Shaviv, T. Yoda and K. Ueno), Materials Research Society, Warrendale, Pennsylvania, pp. 337–340.

[138] Mountsier, T., Grief, M., Goonetilleke, C. and Lassig, S. (2001) Integration of Cu with SiOCH (coral) and SiC:H low *k* dielectric films in a dual damascene scheme, in *Advanced Metallization Conference (AMC)* (eds A.J. McKerrow, Y. Shacham-Diamand, S. Zaima and T. Ohba), Materials Research Society, Warrendale, Pennsylvania, pp. 295–299.

[139] Fang, L.-Y., Chen, H.-C., Hu, S.-C., Wang, C.-M., Wu, J.-Y., Hsieh, W.-Y. and Yew, T.-R. (2002) A study on the integration of organosilicate glasses for advanced copper/low-*k* interconnections, in *Advanced Metallization Conference (AMC)* (eds B.M. Melnick, T.S. Cale, S. Zaima and T. Ohta), Materials Research Society, Warrendale, Pennsylvania, pp. 501–506.

[140] Chou, T.J., Chang, S.Y., Chen, Y.H., Lee, S.N., Jang, S.M., Yu, C.H. and Liang, M.S. (2002) Curing process window for MSQ based low-*k* dielectrics, in *Advanced Metallization Conference (AMC)* (eds B.M. Melnick, T.S. Cale, S. Zaima and T. Ohta), Materials Research Society, Warrendale, Pennsylvania, pp. 671–676.

[141] Colburn, M., Nitta, S., Zinter, J., Chakrapani, N., Purushothaman, S., Simonyi, E., Cohen, S., Shi, L., Milkove, K., Volksen, W. and Miller, R. (2003) Analysis of an extreme low-*k* organosilicate dielectric film exposed to representative plasmas and to a restorative process, in *Advanced Metallization Conference (AMC)* (eds G.W. Ray, T. Smy, T. Ohta and M. Tsujimura), Materials Research Society, Warrendale, Pennsylvania, pp. 489–493.

[142] Chikaki, S., Kubota, T., Tanaka, H., Kohmura, K., Hirakawa, M., Nakayama, T., Seino, Y., Oda, N. and Saito, S. (2008) Novel self-hydrophobication and pre-reinforcement of fine pore self-assembled scalable porous silica film for quick turn-around-time low-*k*/Cu interconnect process, in *Advanced Metallization Conference (AMC)* (eds M. Naik, R. Shaviv, T. Yoda and K. Ueno), Materials Research Society, Warrendale, Pennsylvania, pp. 75–81.

[143] Lin, W.C., Tsai, T.C., Hsu, H.K., Lin, J., Tsao, W.C., Chen, W., Cheng, C.M., Hsu, C.L., Liu, C.C., Hsu, C.M., Lin, J.F., Huang, C.C. and Wu, J.Y. (2011) A comprehensive process engineering on TDDB for direct polishing ultra-low *k* dielectric Cu interconnects at 40 nm technology node and beyond. Proceedings of the 49th Annual IEEE International Reliability Physics Symposium (IRPS), Piscataway, New Jersey, pp. 160–164.

[144] Tsai, C.-Y., Yang, N.-H., Hsieh, W.-Y. and Yen, P.W. (2002) Study of process impact on low-*k* dielectric in copper dual damascene interconnect process, in *Advanced Metallization Conference (AMC)* (eds B.M. Melnick, T.S. Cale, S. Zaima and T. Ohta), Materials Research Society, Warrendale, Pennsylvania, pp. 619–623.

[145] Shamiryana, D., Weidner, K., Gray, W.D., Baklanov, M.R., Vanhaelemeersch, S. and Maex, K. (2002) Comparative study of PECVD SiOCH low-*k* films obtained at different deposition conditions. *Microelectron. Engng*, **64** (01–04), 361–366.

[146] Kirimura, T., Iba, Y., Ozaki, S., Kobayashi, Y. and Nakata, Y. (2008) Effect of surface roughness on plasma-induced degradation of porous low-*k* material, in *Advanced Metallization Conference (AMC)* (eds M. Naik, R. Shaviv, T. Yoda and K. Ueno), Materials Research Society, Warrendale, Pennsylvania, pp. 565–571.

[147] Ryuzaki, D., Ishida, T. and Furusawa, T. (2002) Time-dependent dielectric-constant increase – novel criteria for low-*k* materials reliability, in *Advanced Metallization Conference (AMC)* (eds B.M. Melnick, T.S. Cale, S. Zaima and T. Ohta), Materials Research Society, Warrendale, Pennsylvania, pp. 259–264.

[148] Yoshino, T., Hata, N. and Kikkawa, T. (2003) Evaluation of copper ion drift in low-dielectric-constant interlayer films by transient capacitance spectroscopy, in *Materials, Technology and Reliability for Advanced Interconnects and Low-k Dielectrics – 2003, Symposium E Materials*

*Research Society Symposium Proceedings*, vol. 766, Materials Research Society, Warrendale, Pennsylvania, pp. E1.7.1–E1.7.6.

[149] Ryuzaki, D., Ishida, T. and Furusawa, T. (2003) Time-dependent dielectric-constant increase reliability issue for low dielectric-constant materials. *J. Electrochem. Soc.*, **150** (12), F203–F205.

[150] Ryuzaki, D., Sakurai, H., Abe, K., Takeda, K. and Fukuda H. (2004) Enhanced dielectric-constant reliability of low-*k* porous organosilicate glass ($k = 2.3$) for 45-nm-generation Cu interconnects. Technical Digest, IEEE International Electron Devices Meeting, Piscataway, New Jersey, pp. 949–952.

[151] Hung, C.-C., Oates, A.S., Lin, H.C., Chang, P., Wang, J.L., Huang, C.C. and Yau, Y.W. (2007) New understanding of metal-insulator-metal (MIM) capacitor degradation behavior. Proceedings of the 45th Annual IEEE International Reliability Physics Symposium (IRPS), Piscataway, New Jersey, pp. 630–631.

[152] Joseph, E.A., Goeckner, M.J., Overzet, L.J., Gidley, D.W. and Kastenmeier, B.E.E. (2003) Correlation of pore structure to integration issues for low-*k* dielectrics, in *Advanced Metallization Conference (AMC)* (eds G.W. Ray, T. Smy, T. Ohta and M. Tsujimura), Materials Research Society, Warrendale, Pennsylvania, pp. 537–541.

[153] Dalton, T.J., Fuller, N., Tweedie, C., Dunn, D., Labelle, C., Gates, S., Colburn, M., Chen, S.T., Tai, L., Dellaguardia, R., Petrarca, K., Dziobkowski, C., Kumar, K. and Siddiqui, S. (2004) Ash-induced modification of porous and dense SiCOH inter-level-dielectric (ILD) materials during damascene plasma processing. Proceedings of the IEEE 2004 International Interconnect Technology Conference (IITC), IEEE Service Center, Piscataway, New Jersey, pp. 154–156.

[154] Seki, H., Inoue, K., Nagai, N., Shimada, M., Inukai, K., Hashimoto, H. and Ogawa, S. (2004) Characterization of ashing damage in depth in low-*k* dielectric films by micro beam IR method, in *Advanced Metallization Conference (AMC)* (eds D. Erb, P. Ramm, K. Masu and A. Osaki), Materials Research Society, Warrendale, Pennsylvania, pp. 375–380.

[155] Ryu, S., Kim, S., Shin, K., Kim, S., Kim, H.J. and Lee, J.G. (2004) Low damage ashing of organosilicate glass low-*k* materials using oxygen plasma in inductively coupled plasma source, in *Advanced Metallization Conference (AMC)* (eds D. Erb, P. Ramm, K. Masu and A. Osaki), Materials Research Society, Warrendale, Pennsylvania, pp. 639–644.

[156] Tsai, J.S., Su, Y.N., Hsu, J.W., Yang, J.L., Shieh, J.H., Jang, S.M. and Liang, M.S. (2005) Effect of ash process on leakage mechanism of Cu/ELK ($k = 2.5$) interconnect for 65/45 nm generation. Proceedings of the IEEE 2005 International Interconnect Technology Conference (IITC), IEEE Service Center, Piscataway, New Jersey, pp. 94–96.

[157] Soda, E., Watanabe, T., Matsuhara, Y., Matsuura, S., Koba, F., Kondo, S. and Kobayashi, N. (2005) Low-damage etching process for hp-45 Cu/low-*k* interconnects, in *Advanced Metallization Conference (AMC)* (eds S.H. Brongersma, T.C. Taylor, M. Tsujimura and K. Masu), Materials Research Society, Warrendale, Pennsylvania, pp. 263–269.

[158] Tomioka, K., Soda, E. and Kobayashi, N. (2005) Low-damage ash by atomic hydrogen for porous low-*k*/Cu interconnects, in *Advanced Metallization Conference (AMC)* (eds. S.H. Brongersma, T.C. Taylor, M. Tsujimura and K. Masu), Materials Research Society, Warrendale, Pennsylvania, pp. 399–403.

[159] David, T., Possémé, N., Chevolleau, T., Darnon, M., Louveau, O., Louis, D., Passemard, G. and Joubert, O. (2005) A study of plasma treatments limiting metal barrier diffusion into porous low-*k* materials, in *Advanced Metallization Conference (AMC)* (eds S.H. Brongersma, T.C. Taylor, M. Tsujimura and K. Masu), Materials Research Society, Warrendale, Pennsylvania, pp. 405–410.

[160] Takahashi, H., Takimoto, Y., Tada, M. and Ando, Y. (2007) Evaluating actual plasma damage in ultra low-*k* materials for 45-nm nodes, in *Advanced Metallization Conference (AMC)* (eds A.J. McKerrow, Y. Shacham-Diamand, S. Shingubara and Y. Shimogaki), Materials Research Society, Warrendale, Pennsylvania, pp. 551–555.

[161] Braginsky, O.V., Kovalev, A.S., Lopaev, D.V., Malykhin, E.M., Mankelevich, Y.A. Rakhimova, T.V., Rakhimov, A.T., Vasilieva, A.N., Zyryanov, S.M. and Baklanov, M.R. (2008) Recombination probabilities of O and H atoms on the surface of nanoporous low dielectric

constant SiOCH films, in *Advanced Metallization Conference (AMC)* (eds M. Naik, R. Shaviv, T. Yoda and K. Ueno), Materials Research Society, Warrendale, Pennsylvania, pp. 573–579.

[162] Shi, H., Bao, J., Liu, J., Huang, H., Smith, R.S., Zhao, Q., Ho, P.S., Goodner, M.D., Moinpour, M. and Kloster, G.M. (2007) Dielectric recoveries on $O_2$ plasma damaged organosilicate low-*k* dielectrics, in *Advanced Metallization Conference (AMC)* (eds A.J. McKerrow, Y. Shacham-Diamand, S. Shingubara and Y. Shimogaki), Materials Research Society, Warrendale, Pennsylvania, pp. 447–453.

[163] Kurihara, K., Ono, T., Kohmura, K., Tanaka, H., Fujii, N., Hata, N. and Kikkawa, T. (2007) Carbon loss induced by plasma beam irradiation in porous silica films. *J. Appl. Phys.*, **101** (11), 113301-1-6.

[164] Huang, H., Bao, J., Shi, H., Ho, P.S. McSwiney, M.L., Goodner, M.D., Moinpour, M. and Kloster, G.M. (2008) $O_2$ plasma damage and dielectric recoveries to patterned CDO low-*k* dielectrics, in *Advanced Metallization Conference (AMC)* (eds M. Naik, R. Shaviv, T. Yoda and K. Ueno), Materials Research Society, Warrendale, Pennsylvania, pp. 111–116.

[165] Shimada, M., Kojima, A., Otsuka, Y., Shimizu, Y., Seki, H., Nakamura, N. and Yoda, T. (2008) Structural analysis of low-*k* porous SiOC film by electron energy loss spectroscopy (EELS), in *Advanced Metallization Conference (AMC)* (eds M. Naik, R. Shaviv, T. Yoda and K. Ueno), Materials Research Society, Warrendale, Pennsylvania, pp. 353–357.

[166] Shi, H., Huang, H., Bao, J., Smith, R.S., Sun, Y., Ho, P.S., Li, A., Pender, J.T., Armacost, M.D. and Kyser, D. (2008) Carbon dioxide and oxygen plasma induced damage to porous OSG low-*k* dielectrics, in *Advanced Metallization Conference (AMC)* (eds A.J. McKerrow, Y. Shacham-Diamand, S. Shingubara and Y. Shimogaki), Materials Research Society, Warrendale, Pennsylvania, pp. 605–611.

[167] Yonekura, K., Sakamori, S., Goto, K., Matsuura, M., Fujiwara, N. and Yoneda, M. (2004) Investigation of ash damage to ultralow-*k* inorganic materials. *J. Vacuum Sci. Technol. B*, **22** (2), 548–553.

[168] Hoyas, A.M., Schuhmacher, J., Whelan, C.M., Baklanov, M.R., Carbonell, L., Celis, J.P. and Maex, K. (2005) Effect of plasma treatments on a low-*k* dielectric polymer surface. *J. Vacuum Sci. Technol. B*, **23** (4), 1551–1557.

[169] Furusawa, T., Ryuzaki, D., Yoneyama, R., Homma, Y. and Hinode, K. (2001) Oxygen plasma resistance of low-*k* organosilica glass films. *Electrochem. Solid-State Lett.*, **4** (3), G31–G34.

[170] Bao, J.J., Shi, H.L., Liu, J.J., Huang, H., Ho, P.S., Goodner, M.D., Moinpour, M. and Kloster, G.M. (2007) Mechanistic study of plasma damage and $CH_4$ recovery of low *k* dielectric surface. Proceedings of the IEEE 2007 International Interconnect Technology Conference (IITC), IEEE Service Center, Piscataway, New Jersey, pp. 147–149.

[171] Tőkei, Zs., Li, Y.-L., Ciofi, I., Croes, K. and Beyer G.P. (2008) Low-*k* dielectric reliability: impact of test structure choice, copper and integrated dielectric quality. Proceedings of the IEEE 2008 International Interconnect Technology Conference (IITC), IEEE Service Center, Piscataway, New Jersey, pp. 111–113.

[172] Oszinda, T., Schaller, M., Fischer, D., Walsh, C. and Schulz, S.E. (2010) Investigation of physical and chemical property changes of ultra low-*k* SiOCH in aspect of cleaning and chemical repair processes. *Microelectron. Engng*, **87** (3), 457–461.

[173] Nakamura, N., Matsunaga, N., Watanabe, K., Miyajima, H., Enomoto, Y., Okada, N. and Shibata, H. (2007) Influence of moisture uptake in porous PAr film on electrical properties, in *Advanced Metallization Conference (AMC)* (eds A.D. McKerrow, Y. Shacham-Diamand, S, Shingubara and Y. Shimogaki), Materials Research Society, Warrendale, Pennsylvania, pp. 569–573.

[174] Hayashi, Y., Kojima, A., Nakamura, N., Masuda, H., Sato, K., Hayashi, H., Matsunaga, N. and Shibata, H. (2008) Impact of gas silylation process on wire-to-wire leakage current for low-*k*/Cu dual-damascene interconnects, in *Advanced Metallization Conference (AMC)* (eds A.J. McKerrow, Y. Shacham-Diamand, S. Shingubara and Y. Shimogaki), Materials Research Society, Warrendale, Pennsylvania, pp. 25–31.

[175] Zimmermann, S., Blaschta, F., Schaller, M., Rülke, H., Schulz, S.E. and Gessner, T. (2009) Investigation of etch processes of dense and porous low-*k* dielectrics using OES and QMS as in situ diagnostic methods, in *Advanced Metallization Conference (AMC)* (eds A.J. McKerrow,

Y. Shacham-Diamand, S. Shingubara and Y. Shimogaki), Materials Research Society, Warrendale, Pennsylvania, pp. 387–392.

[176] Labelle, C., Arnold, J., Horak, D., Srivastava, R., Lisi, A., Yusuff, H., Lafosse, Linville, J., Choi, Y., Fuller, N., Kumar, K., Patz, R., Darlak, A., Zhou, K., Zhou, Y. and Pender, J. (2008) Porous low-*k* material etch for 32 nm and beyond, in *Advanced Metallization Conference (AMC)* (eds A.J. McKerrow, Y. Shacham-Diamand, S. Shingubara and Y. Shimogaki), Materials Research Society, Warrendale, Pennsylvania, pp. 523–529.

[177] Tőkei, Zs., Van Aelst, J., Waldfried, C., Escorcia, O., Roussel, P., Richard, O., Travaly, Y., Beyer, G.P. and Maex, K. (2005) Role of dielectric and barrier integrity in reliability of sub-100 nm copper low-*k* interconnects. Proceedings of the 43rd Annual IEEE International Reliability Physics Symposium (IRPS), Piscataway, New Jersey, pp. 495–500.

[178] Imada, T., Nakata, Y., Ozaki, S., Kobayashi, Y., Yoshikawa, K. and Nakamura, T. (2007) TDDB characteristics of low-*k* dielectrics after plasma damage recovery process, in *Advanced Metallization Conference (AMC)* (eds A.J. McKerrow, Y. Shacham-Diamand, S Shingubara and Y. Shimogaki), Materials Research Society, Warrendale, Pennsylvania, pp. 557–561.

[179] Pantouvaki, M., Struyf, H., Hendrickx, D., Heylen, N., Richard, O. and Beyer, G. (2008) The impact of ash on TDDB of metal-hard-mask-etched Cu/low-*k* interconnects, in *Advanced Metallization Conference (AMC)* (eds A.J. McKerrow, Y. Shacham-Diamand, S. Shingubara and Y. Shimogaki), Materials Research Society, Warrendale, Pennsylvania, pp. 523–529.

[180] Yanai, K., Hasebe, T., Sumiya, K., Oguni, S. and Koga, K. (2005) Robustness enhancement for subsequent plasma damages utilizing surface modification layer induced by He plasma, in *Advanced Metallization Conference (AMC)* (eds S.H. Brongersma, T.C. Taylor, M. Tsujimura and K. Masu), Materials Research Society, Warrendale, Pennsylvania, pp. 341–346.

[181] Urbanowicz, A.M., Shamiryan, D., Marsik, P., Travaly, Y., Jonas, A., Verdonck, P., Vanstreels, K., Ferchichi, A., De Roest, D., Sprey, H., Matsushita, K., Kaneko, S., Tsuji, N., Luo, S., Escorcia, O., Berry, O., Waldfried, C., De Gendt, S. and Baklanov, M. (2008) Improved low-*k* dielectric properties using He/$H_2$ plasma for resist removal, in *Advanced Metallization Conference (AMC)* (eds A.J. McKerrow, Y. Shacham-Diamand, S. Shingubara and Y. Shimogaki), Materials Research Society, Warrendale, Pennsylvania, pp. 593–598.

[182] Chang, T.C., Mor, Y.S., Liu, P.T., Tsai, T.M., Chen, C.W., Chu, C.J., Pan, F.M., Lur, W. and Sze, S.M. (2002) Trimethylchlorosilane treatment of ultralow dielectric constant material after photoresist removal processing. *J. Electrochem. Soc.*, **149** (10), F145–F148.

[183] Reidy, R.F., Oroaco-Teran, R.A., Zhang, Z., Nerusu, P.K., Matz, P.D. and Mueller, D.W. (2003) Supercritical silylation of ashed Si–O–C low-*k* films to limit changes in critical dimensions, in *Advanced Metallization Conference (AMC)* (eds G.W. Ray, T. Smy, T. Ohta and M. Tsujimura), Materials Research Society, Warrendale, Pennsylvania, pp. 513–517.

[184] Bhanap, A., Ramos, T., Camarena, A. and Naman, A. (2003) Repairing process-induced damage to porous low-*k* ILDs by post-ash treatment, in *Advanced Metallization Conference (AMC)* (eds G.W. Ray, T. Smy, T. Ohta and M. Tsujimura), Materials Research Society, Warrendale, Pennsylvania, pp. 519–523.

[185] Lahlouh, B., Lubguban, J.A., Sivaraman, G., Gale, R. and Gangopadhyay, S. (2004) Silylation using a supercritical carbon dioxide medium to repair plasma-damaged porous organosilicate films. *Electrochem. Solid-State Lett.*, **7** (12), G338–G341.

[186] Xie B. and Muscat, A.J. (2004) Silylation of porous methylsilsesquioxane films in supercritical carbon dioxide. *Microelectron. Engng*, **76** (1–4), 52–59.

[187] Nitta, S.V., Purushothaman, S., Chakrapani, N., Rodriguez, O., Klymko, N., Ryan, E.T., Bonilla, G., Cohen, S., Molis, S. and McCullough, K. (2005) Use of difunctional silylation agents for enhanced repair of post plasma damaged porous low *k* materials, in *Advanced Metallization Conference (AMC)* (eds S.H. Brongersma, T.C. Taylor, M. Tsujimura and K. Masu), Materials Research Society, Warrendale, Pennsylvania, pp. 325–331.

[188] Rajagopalan, T., Lahlouh, B., Lubguban, J.A., Biswas, N., Gangopadhyay, S., Sun, J., Huang, D.H., Simon, S.L., Toma, D. and Butler, R. (2006) Investigation on hexamethyldisilazane vapor treatment of plasma-damaged nanoporous organosilicate films. *Appl. Surf. Sci.*, **252** (18), 6323–6331.

[189] Chaabouni, H., Chapelon, L.-L., Chevolleau, T., Joubert, O. and Torres, J. (2007) Restoration solutions for plasma-damaged ultra low *k* for the 45 nm technology node and beyond, in *Advanced Metallization Conference (AMC)* (eds A.J. McKerrow, Y. Shacham-Diamand, S. Shingubara and Y. Shimogaki), Materials Research Society, Warrendale, Pennsylvania, pp. 293–299.

[190] Kinoshita, K., Chikaki, S., Soda, E., Tomioka, K., Tanaka, H., Kohmura, K., Nakayama, T., Kikkawa, T. and Saito, S. (2007) Process induced damages and recovery by silylation for low-*k*/Cu interconnects with highly-porous self-assembled silica film, in *Advanced Metallization Conference (AMC)* (eds A.J. McKerrow, Y. Shacham-Diamand, S. Shingubara and Y. Shimogaki), Materials Research Society, Warrendale, Pennsylvania, pp. 513–519.

[191] Oszinda, T., Schaller, M., Leppack, S. and Schulz, S.E. (2009) Restoration of plasma damaged porous ultra low-*k* SiOCH films: a coating process with UV activation versus a vapor phase process with thermal activation, in *Advanced Metallization Conference (AMC)* (eds A.J. McKerrow, Y. Shacham-Diamand, S. Shingubara and Y. Shimogaki), Materials Research Society, Warrendale, Pennsylvania, pp. 55–64.

[192] Senkevich, J.J., Carrow, B.P., Woods, B.W., Bae, D.L., Cale, T.S. and Wang, P.-I. (2005) Molecular caulk: enabling aspects for ultra-low *k* dielectric integration, in *Advanced Metallization Conference (AMC)* (eds S.H. Brongersma, T.C. Taylor, M. Tsujimura and K. Masu), Materials Research Society, Warrendale, Pennsylvania, pp. 375–379.

[193] Singh, A.P., Gandhi, D.D., Lipp, E., Eizenberg, M. and Ramanath, G. (2006) Suppression of chemical and electrical instabilities in mesoporous silica films by molecular capping. *J. Appl. Phys.*, **100** (11), 114504-1-5.

[194] Travaly, Y., Sinapi, F., Heylen, N., Humbert, A., Delande, M., Caluwaert, R., deMussy, J.P., Vereecke, G., Baklanov, M.R., Iacopi, F., Hernandez, J.L., Beyer, G. and Fischer, P. (2007) The critical role of the metal/porous low-*k* interface in post direct CMP defectivity generation and resulting ULK surface and bulk hydrophilisation. Proceedings of the IEEE 2007 International Interconnect Technology Conference (IITC), IEEE Service Center, Piscataway, New Jersey, pp. 164–166.

[195] Gallitre, M., Gall, S., Farcy, A., Blampey, B., Icard, B., Jousseaume, V., Guillermet, M., Brun, P., Zenasni, A., Grosgeorges, P., Sbrugnera, V., Bustos, J., Bermond, C., Lacrevaz, T., Fléchet, B., Rivoire, M. and Ancey, P. (2008) Impact of direct CMP on porous ULK properties and interconnect propagation performance on a wide frequency bandwidth, in *Advanced Metallization Conference (AMC)* (eds A.J. McKerrow, Y. Shacham-Diamand, S. Shingubara and Y. Shimogaki), Materials Research Society, Warrendale, Pennsylvania, pp. 423–429.

[196] Choi, J., Tseng, W., Kim, H., Economikos, L., Fang, Q., Zielinski, E., Engbrecht, E., Child, C., Chae, M. and Moon, Y. (2008) Barrier CMP criteria for ULK integration for 32 nm BEOL technology and beyond, in *Advanced Metallization Conference (AMC)* (eds A.J. McKerrow, Y. Shacham-Diamand, S. Shingubara and Y. Shimogaki), Materials Research Society, Warrendale, Pennsylvania, pp. 449–455.

[197] Matz, L.M., Haas, M.K., O'Neill, M.L., Sawayda, R.A. and Meyers, A.M. (2008) Porous ULK dielectric integrity during CMP and post-CMP cleans processing, in *Advanced Metallization Conference (AMC)* (eds A.J. McKerrow, Y. Shacham-Diamand, S. Shingubara and Y. Shimogaki), Materials Research Society, Warrendale, Pennsylvania, pp. 463–467.

[198] Tsutsue, M., Travaly, Y., Ikeda, A., Tokei, Zs., Willegems, M., Struyf, H., Leylen, N., Sinapi, F., Richard, O., Kemeling, N., De Roest, D., Fukazawa, A., Matsuki, N., Sprey, H. and Beyer, G. (2007) Highly reliable Cu/ULK integration scheme using MHM and low-*k* capping film, in *Advanced Metallization Conference (AMC)* (eds A.J. McKerrow, Y. Shacham-Diamand, S. Shingubara and Y. Shimogaki), Materials Research Society, Warrendale, Pennsylvania, pp. 459–463.

[199] Chaabouni, H., Imbert, G., Chapelon, L.L., Mellier, M., Fossati, D., Licitra, C., Chevolleau, T., Joubert, O. and Ancey, P. (2008) Porous SiOCH ultra low-*k* recovery treatments after direct CMP process, in *Advanced Metallization Conference (AMC)* (eds A.J. McKerrow, Y. Shacham-Diamand, S. Shingubara and Y. Shimogaki), Materials Research Society, Warrendale, Pennsylvania, pp. 431–439.

[200] Heylen, N., Li, Y., Kellens, K., Travaly, Y., Vereecke, G., Volders, H., Tokei, Zs., Versluijs, J., Rip, J., Van Besien, E., Carbonell, L., Beyer, G., Fischer, P., Zhao, L., Santoro, G., Cockburn, A. and Nguyen, O. (2008) Post-direct CMP dielectric surface copper contamination: quantitative analysis and impact on dielectric breakdown behavior, in *Advanced Metallization Conference (AMC)* (eds A.J. McKerrow, Y. Shacham-Diamand, S. Shingubara and Y. Shimogaki), Materials Research Society, Warrendale, Pennsylvania, pp. 415–421.

[201] Maeda, N., Arase, S., Homma, Y., Saito, H., Maejima, K. and Kato, I. (2005) Film property of spin-on barrier dielectric, in *Advanced Metallization Conference (AMC)* (eds S.H. Brongersma, T.C. Taylor, M. Tsujimura and K. Masu), Materials Research Society, Warrendale, Pennsylvania, pp. 445–449.

[202] Maeda, N., Takimoto, Y., Sakamoto, Y., Tada, M., Nakajima, M. and Funatsu, K. (2007) PBO/SiC hybrid structure with Ti liner for reliable barrier-free interconnect, in *Advanced Metallization Conference (AMC)* (eds A.J. McKerrow, Y. Shacham-Diamand, S. Shingubara and Y. Shimogaki), Materials Research Society, Warrendale, Pennsylvania, pp. 521–526.

[203] Ono, Y., Miatsutani, M., Harada, T., Tada, M. and Ando, Y. (2008) Spin-on organic low-*k* materials ($k = 2.2$) with thin Ti liner interconnect, in *Advanced Metallization Conference (AMC)* (eds A.J. McKerrow, Y. Shacham-Diamand, S. Shingubara and Y. Shimogaki), Materials Research Society, Warrendale, Pennsylvania, pp. 431–439.

[204] Pantouvaki, M., Zhao, L., Huffman, C., Heylen, N., Ferchichi, A., Ono, Y., Makajima, M., Nakatani, K., Struyf, H., Beyer, B.P. and Baklanov, M.R. (2009) Integration of an organic ultra low-*k* ($k = 2.2$) material, in *Advanced Metallization Conference (AMC)* (eds A.J. McKerrow, Y. Shacham-Diamand, S. Shingubara and Y. Shimogaki), Materials Research Society, Warrendale, Pennsylvania, pp. 173–178.

[205] Van Olmen, J., Al-Bayati, A., Beyer, G., Boelen, P., Carbonell, L., Chan, K., Ciofi, I., Claes, M., Cockburn, A., Druais, G., Hendrickx, D., Heylen, N., Kesters, E., Lytle, S., Noori, A., Shah, K., Op de Beeck, M., Struyf, H., Tőkei, Zs., Versluijs, J. and Zhao, C. (2007) Integration of 50 nm half pitch single damascene copper trenches in Black Diamond® II by means of double patterning 193 nm immersion lithography on metal hardmask, in *Advanced Metallization Conference (AMC)* (eds A.J. McKerrow, Y. Shacham-Diamand, S. Shingubara and Y. Shimogaki), Materials Research Society, Warrendale, Pennsylvania, pp. 355–361.

[206] Naik, M., Dai, H., Ordonio, C., Yoshida, N., Nguyen, P., Fang, H., Li, A., Yang, H.-L., Yu, J., Demos, A., Okazaki, M. and Thothadri, M. (2007) Process flow development and integration of porous low *k* for 45 nm node, in *Advanced Metallization Conference (AMC)* (eds A.J. McKerrow, Y. Shacham-Diamand, S. Shingubara and Y. Shimogaki), Materials Research Society, Warrendale, Pennsylvania, pp. 371–377.

[207] Hamioud, K., Arnal, V., Gall, S., Favennec, L., Assous, M., Farcy, A., Aimadeddine, M., Jousseaume, V., Zenasni, A., Brun, P., Imbert, G., Chapelon, L.-L., Jayet, C., Maitrejean, S., Louis, D., Passemard, G., Barbier, D. and Torres, J. (2007) Development and integration of a porous ULK dielectric ($k$–2.3 and below) for high performaqnce BEOL interconnect, in *Advanced Metallization Conference (AMC)* (eds A.J. McKerrow, Y. Shacham-Diamand, S. Shingubara and Y. Shimogaki), Materials Research Society, Warrendale, Pennsylvania, pp. 439–445.

[208] Kubota, T., Nakahira, J., Tomioka, K., Soda, E., Matsushita, K., Oda, N., Kinoshita, K., Kondo, S., Ogawa, S. and Saito, S. (2007) Integration of 140 mm pitch Cu interconnects featuring template-type PE-CVD SiOC film ($k$ <2.4), in *Advanced Metallization Conference (AMC)* (eds A.J. McKerrow, Y. Shacham-Diamand, S. Shingubara and Y. Shimogaki), Materials Research Society, Warrendale, Pennsylvania, pp. 563–567.

[209] Yoo, H.J., Bielefeld, J.D., Balakrishnan, S., Boyanov, B., Clendenning, S.B., Kloster, G.M., Paluda, P.M., RamachandraRao, V.S., Suri, S. and Younkin, T.R. (2007) Plasma damage and repair of porous low-*k* carbon-doped oxide films, in *Advanced Metallization Conference (AMC)* (eds A.J. McKerrow, Y. Shacham-Diamand, S. Shingubara and Y. Shimogaki), Materials Research Society, Warrendale, Pennsylvania, pp. 705–709.

[210] Travaly, Y., Van Aelst, J., Truffert, V., Verdonck, P., Dupont, T., Camerotto, E., Richard, O., Bender, H., Kroes, C., De Roest, D., Vereecke, G., Claes, M., Le, Q.T., Kesters, E., Van Cauwenberghe, M., Beynet, J., Kaneko, S., Struyf, H., Baklanov, M., Matsushita, K.,

Kobayashi, N., Sprey, H. and Beyer, G. (2008) Key factors to sustain the extension of a MHM-based integration scheme to medium and high porosity PECVD low-*k* materials. Proceedings of the IEEE 2008 International Interconnect Technology Conference (IITC), IEEE Service Center, Piscataway, New Jersey, pp. 52–54.

[211] Galpin, D., Monget, C., Richard, E., Possémé, N., Brun, Ph., Guillermet, M., Verove, C., Ostrovsky, A., Trouiller, Y., Mellier, M., Petitdidier, S., Caubet, V., Rivoire, M., Hinisinger, O., Imbert, G., Le Friec, Y., Favennec, L., Thomas, I., Guillan, J., Kaxaire, K., Caubet, P., Vannier, P., Normandon, Ph., Chhun, S., Arnal, V., Ferreira, P., Berger, T., Besset, C., Arnaud, L. and Roy, D. (2008) Metal hard mask dual damascene architecture: an industrial solution to face scalability challenges, in *Advanced Metallization Conference (AMC)* (eds M. Naik, R. Shaviv, T. Yoda and K. Ueno), Materials Research Society, Warrendale, Pennsylvania, pp. 11–18.

[212] Widodo, J., Ouyang, T., Liu, H., Zhou, M.S., Leong, K.H., Tang, H.S. and Jain, A. (2008) Characterization of pSiCOH for 45 nm BEOL interconnects and beyond, in *Advanced Metallization Conference (AMC)* (eds M. Naik, R. Shaviv, T. Yoda and K. Ueno), Materials Research Society, Warrendale, Pennsylvania, pp. 635–639.

[213] Huang, E., Oh, M., Law, S.B., Petitdidier, S., Ko, T.-M., Sawada, H., Tang, T.J., Hu, C.-K., Cohen, S., Liniger, E., Bonilla, G., Rath, D., Shaw, T., Gignac, L., Edelstein, D., Permana, D., Child, C., Buengener, R., Fitzsimmons, J., Findeis, P., Taft, C., Angyal, M., Ogunsola, O., King, J., Flaitz, P., Baumann, F., Molis, S., Dziobkowski, C., Coffin, J., Davis, R., Zaitz, M., Kapur, A., Kermel, L., Truong, C., Grunow, S., Chen, X., Sankarapandian, M., Li, B., Chen, F., Lee, T., Christiansen, C., Kolics, A., Gilbert, N., Rigoutat, O. and Li, N. (2009) CoWP metal caps for reliable 32 nm 1X Cu interconnects in porous ULK ($k = 2.4$), in *Advanced Metallization Conference (AMC)* (eds A.J. McKerrow, Y. Shacham-Diamand, S. Shingubara and Y. Shimogaki), Materials Research Society, Warrendale, Pennsylvania, pp. 179–184.

[214] Chen, F., Shinosky, M., Li, B., Christiansen, C., Lee, T., Aitken, J., Badami, D., Huang, E., Bonilla, G., Ko, T.-M., Kane, T., Wang, Y., Zaitz, M., Nicholson, L., Angyal, Truong, C., Chen, X., Yang, G., Law, S.B., Tang, T.J., Petitdidier, S., Ribes, G., Oh, M., Child, C., Sawada, H., Kolics, A., Rigoutat, O. and Gilbert, N. (2010) Comprehensive investigations of CoWP metal-cap impacts on low-*k* TDDB for 32 nm technology application. Proceedings of the 48th Annual IEEE International Reliability Physics Symposium (IRPS), Piscataway, New Jersey, pp. 566–573.

[215] Zhang, X., Smith, S., Huang, R. and Ho, P.S. (2008) Study of chip-package interaction and crackstop structures for Cu/ultra low-*k* interconnects, in *Advanced Metallization Conference (AMC)* (eds M. Naik, R. Shaviv, T. Yoda and K. Ueno), Materials Research Society, Warrendale, Pennsylvania, pp. 341–346.

[216] Hatai, M. and Funatsu, K. (2008) Proposal of a simple method for evaluation of wire bonding process with ultra low-*k* materials, in *Advanced Metallization Conference (AMC)* (eds M. Naik, R. Shaviv, T. Yoda and K. Ueno), Materials Research Society, Warrendale, Pennsylvania, pp. 653–657.

[217] Harsányi, G. (1999) Copper may destroy chip-level reliability: handle with care – mechanism and conditions for copper migrated resistive short formation. *IEEE Electron. Device Lett.*, **20** (1), 5–8.

[218] JEDEC/FSA Joint Publication (2004) Foundry process qualification guidelines (wafer fabrication manufacturing sites), Revision JP001.01, JEDEC Solid State Technology Association, Arlington, Virginia; http://www.jedec.org/Catalog/catalog.cfm.

[219] Noguchi, J., Ohashi, N., Yasuda, J., Yamaguchi, H., Owada, N., Takeda, K. and Hinode, K. (2000) TDDB improvement in Cu metallization under bias stress. Proceedings of the 38th Annual IEEE International Reliability Physics Symposium (IRPS), pp. 339–343.

[220] Noguchi, J., Saito, T., Ohashi, N., Ashihara, H., Maruyama, H., Kubo, M., Yamaguchi, H., Ryuzaki, D., Takeda, K. and Hinode, K. (2001) Impact of low-*k* dielectrics and barrier metals on TDDB lifetime of Cu interconnects. Proceedings of the 39th Annual IEEE International Reliability Physics Symposium (IRPS), pp. 355–359.

[221] Ogawa, E.T., Kim, J., Haase, G.S., Mogul, H.C. and McPherson, J.W. (2003) Leakage, breakdown, and TDDB characteristics of porous low-*k* silica-based interconnect dielectrics. Proceedings of the 41st Annual IEEE International Reliability Physics Symposium (IRPS), Piscataway, New Jersey, pp. 166–172.

[222] Ohring, M. (1998) *Reliability and Failure of Electronic Materials and Devices*, Academic Press, San Diego, California.

[223] Tobias, P.A. and Trindade, D. (1986) *Applied Reliability*, Van Nostrand Reinhold Company, Inc., New York.

[224] Reddy, V. (2004) An introduction to CMOS semiconductor reliability. Tutorial Notes, IEEE International Reliability Physics Symposium (IRPS), SAR Associates, Rome, New York.

[225] JEDEC Publication (2010) *Failure Mechanisms and Models for Semiconductor Devices*, Revision JEP122F, JEDEC Solid State Technology Association, Arlington, Virginia; http://www.jedec.org/Catalog/catalog.cfm.

[226] Bashir, M., Kim, D.H., Athikulwongse, K., Lim, S.K. and Milor, L. (2011) Backend low-*k* TDDB chip reliability simulator. Proceedings of the 49th Annual IEEE International Reliability Physics Symposium (IRPS), Piscataway, New Jersey, pp. 65–74.

[227] Dissado, L.A. and Fothergill, J.C. (2008) *Electrical Degradation and Breakdown in Polymers*, The Institution of Engineering and Technology, London, UK.

[228] Milor, L. and Hong, C. (2007) Backend dielectric breakdown dependence on linewidth and pattern density. *Microelectron. Reliab.*, **47** (9–11), 1473–1477.

[229] Bashir, M. and Milor, L. (2010) Analysis of the impact of linewidth variation on low-*k* dielectric breakdown. Proceedings of the 48th Annual IEEE International Reliability Physics Symposium (IRPS), Piscataway, New Jersey, pp. 895–902.

[230] Tao, G. (2004) Reliability issues in advanced monolithic embedded high voltage CMOS technologies. 2004 IEEE International Integrated Reliability Workshop Final Report (IIRW), Piscataway, New Jersey, pp. 174–177.

[231] Yiang, K.Y., Mok, T.S., Yoo, W.J. and Krishnamoorthy, A. (2004) Reliability improvement using buried capping layer in advanced interconnects. Proceedings of the 42nd Annual IEEE International Reliability Physics Symposium (IRPS), Piscataway, New Jersey, pp. 333–337.

[232] Yiang, K.Y., Yoo, W.J., Krishnamoorthy, A. and Tang, L.J. (2005) Impact of buried capping layer on TDDB physics of advanced interconnects. Proceedings of the 43rd Annual IEEE International Reliability Physics Symposium (IRPS), Piscataway, New Jersey, pp. 490–494.

[233] Hsu, C.-L., Lin, W.-C., Tsai, T.-C., Huang, C. and Wu, J.-Y. (2011) A model for post-CMP cleaning effect on TDDB. Proceedings of the 49th Annual IEEE International Reliability Physics Symposium (IRPS), Piscataway, New Jersey, pp. 670–673.

[234] Loke, A.L.S., Wetzel, J.T., Stankus, J.J., Angyal, M.S., Mowry, B.K. and Wong, S.S. (1998) Electrical leakage at low-*k* polyimide/TEOS interface. *IEEE Electron. Device Lett.*, **19** (6), 177–179.

[235] Kim, S.U., Cho, T. and Ho, P.S. (1999) Leakage current degradation and carrier conduction mechanisms for Cu/BCB damascene process under bias temperature stress. Proceedings of the 37th Annual IEEE International Reliability Physics Symposium (IRPS), Piscataway, New Jersey, pp. 277–282.

[236] Yokogawa, S., Oshida, D., Tsuchiya, H., Taiji, T., Morita, T., Tsuchiya, Y. and Takewaki, T. (2008) Prediction of early failure due to non-visual defect on time-dependent dielectric breakdown of low-*k* dielectrics: experimental verification of a yield-reliability model. Proceedings of the 46th Annual IEEE International Reliability Physics Symposium (IRPS), Piscataway, New Jersey, pp. 144–149.

[237] Noguchi, J., Ohashi, N., Yasuda, J., Yamaguchi, H., Owada, N., Takeda, K. and Hinode, K. (2000) TDDB improvement in Cu metallization under bias stress. Proceedings of the 38th Annual IEEE International Reliability Physics Symposium (IRPS), Piscataway, New Jersey, pp. 339–343.

[238] Noguchi, J., Miura, N., Kubo, M., Tamaru, T., Yamaguchi, H., Hamada, N., Makabe, K., Tsuneda, R. and Takeda, K. (2003) Cu-ion-migration phenomena and its influence on TDDB lifetime in Cu metallization. Proceedings of the 41st Annual IEEE International Reliability Physics Symposium (IRPS), Piscataway, New Jersey, pp. 287–292.

[239] Hong, C., Milor, L. and Lin, M.Z. (2004) Analysis of the layout impact on electric fields in interconnect structures using finite element method. *Microelectron. Reliab.*, **44** (9–11), 1867–1871.

[240] Miyazaki H. and Kodama, D. (2009) TDDB lifetime of asymmetric patterns and its comprehension from percolation theory. Proceedings of the 47th Annual IEEE International Reliability Physics Symposium (IRPS), Piscataway, New Jersey, pp. 814–818.

[241] Tőkei, Zs., Li, Y.-L. and Beyer, G.P. (2005) Reliability challenges for copper low-*k* dielectrics and copper diffusion barriers. *Microelectron. Reliab.*, **45** (9–11), 1436–1442.

[242] Yiang, K.-Y., Chin, M., Marathe, A. and Aubel, O. (2010) A simple electrical method for etch bias and process reliability determination. Proceedings of the 48th Annual IEEE International Reliability Physics Symposium (IRPS), Piscataway, New Jersey, pp. 562–565.
[243] Lee, S.-C., Oates, A.S. and Chang, K.M. (2008) Limitation of low-*k* reliability due to dielectric breakdown at vias. Proceeding of the IEEE 2008 International Interconnect Technology Conference (IITC), IEEE Service Center, Piscataway, New Jersey, pp. 177–179.
[244] Fischer, A.H., Lim, Y.K., Riess, Ph., Pompl, Th., Zhang, B.C., Chua, E.C., Keller, W.W., Tan, J.B., Klee, V., Tan, Y.C., Souche, D., Sohn, D.K. and von Glasow, A. (2008) TDDB robustness of highly dense 65 nm BEOL vertical natural capacitor with competitive area capacitance for RF and mixed-signal applications. Proceedings of the 46th Annual IEEE International Reliability Physics Symposium (IRPS), Piscataway, New Jersey, pp. 126–131.
[245] Cheng, C.-M., Hsu, C.-M., Lin, W.C., Huang,H.-F., Liu, Y.-C., Lin, K.-H., Lin, J.-F., Huang, C.C. and Wu, J.Y. (2011) A novel pre-clean process of BEOL barrier-seed process to enhance reliability performance of advanced 40 nm node. Proceedings of the 49th Annual IEEE International Reliability Physics Symposium (IRPS), Piscataway, New Jersey, pp. 661–664.
[246] Lin, T.M.Z., Hsu, W.M., Lee, S.Y., Chiu, C.C. and Wu, K. (2005) Analysis of electric field distribution and its influence on dielectric failures in asymmetric copper interconnect structures. Proceedings of the 43rd Annual IEEE International Reliability Physics Symposium (IRPS), Piscataway, New Jersey, pp. 582–583.
[247] Zhang, W., Zeng, X., Liu, W., Lim, Y.K., Liu, J.F. and Chua, E.C. (2010) Study of electric field-based lifetime projection method in IMD TDDB. Proceedings of the 48th Annual IEEE International Reliability Physics Symposium (IRPS), Piscataway, New Jersey, pp. 938–942.
[248] Xia, F., He, J., Prabhumirashi, P., Schmitz, A., Lowrie, A., Hicks, J., Shusterman, Y. and Brain, R. (2011) Characterization and challenge of TDDB reliability in Cu/low *k* dielectric interconnect. Proceedings of the 49th Annual IEEE International Reliability Physics Symposium (IRPS), Piscataway, New Jersey, pp. 61–64.
[249] Tőkei, Zs., Li, Y.-L., Ciofi, I., Croes, K. and Beyer, G. (2008) Low-*k* dielectric reliability: impact of test structure choice, copper and integrated dielectric quality. Proceedings of the IEEE 2008 International Interconnect Technology Conference (IITC), Piscataway, IEEE Service Center, New Jersey, pp. 111–113.
[250] Lin, T.M.Z., Hsu, W.M., Lin, S.R., Wang, R.C.J., Chiu, C.C. and Wu, K. (2006) Identification and layout modification of copper/low *k* interconnect dielectric reliability assessment by using RVDB test. Proceedings of the 44th Annual IEEE International Reliability Physics Symposium (IRPS), Piscataway, New Jersey, pp. 687–688.
[251] Hwang, N., Micaller-Silvestre, M.C.A., Tsang, C.F., Su, J.Y.-J., Kuo, C.C. and Trigg, A.D. (2004) TDDB reliability assessments of 0.13 μm Cu/low-*k* interconnects fabricated with PECVD low-*k* materials. Proceedings of the 42nd Annual IEEE International Reliability Physics Symposium (IRPS), Piscataway, New Jersey, pp. 338–342.
[252] Ogawa, E.T. (2005) Reliability in Cu/low-*k* advanced interconnects. Semiconductor International/SemiTracks Tutorial, 11 July 2005.
[253] Tan, T.L., Gan, C.L., Du, A.-Y., Cheng, C.K., Ng, C.M. and Chan, L. (2007) Test structure design for precise understanding of Cu/low-*k* dielectric reliability. Proceedings of the 45th Annual IEEE International Reliability Physics Symposium (IRPS), Piscataway, New Jersey, pp. 632–633.
[254] Aubel, O., Feustel, F., Hoffmann, T., Majer, M. and Yiang, K. (2004) Comprehensive characterization of BEOL-TDDB performance using very fast voltage ramp dielectric breakdown tests. Presentation in 2004 IEEE International Integrated Reliability Workshop Final Report (IIRW), Institute of Electrical and Electronic Engineers, Inc., Piscataway, New Jersey, pp. 60–65.
[255] Miyazaki, H. and Kodama, D. (2009) TDDB lifetime of asymmetric patterns and its comprehension from percolation theory. Proceedings of the 47th Annual IEEE International Reliability Physics Symposium (IRPS), Piscataway, New Jersey, pp. 814–818.
[256] Loke, A.L.S., Ryu, C., Yue, C.P., Cho, J.S.H. and Wong, S.S. (1996) Kinetics of copper drift in PECVD dielectrics. *IEEE Electron. Device Lett.*, **17** (12), 549–551.
[257] Loke, A.L.S., Wetzel, J.T., Townsend, P.H., Tanabe, T., Vrtis, R.N., Zussman, M.P., Kumar, D., Ryu, C. and Wong, S.S. (1999) Kinetics of copper drift in low-*k* polymer interlevel dielectrics. *IEEE Trans. Electron. Devices*, **46** (11), 2178–2187.

[258] Mallikarjunan, A., Murarka, S.P. and Lu, T.-M. (2001) Metal drift behavior in low dielectric constant organosiloxane polymer. *Appl. Phys. Lett.*, **79** (12), 1855–1857.
[259] Hwang, S.-S., Jung, S.-Y., Jung J.-K. and Joo, Y.-C. (2006) Study of Cu migration-induced failure of inter-layer dielectric. Proceedings of the 44th Annual IEEE International Reliability Physics Symposium (IRPS), Piscataway, New Jersey, pp. 673–674.
[260] Jung, S.-Y., Kim, B.-J., Lee, N.Y., Kim, B.-M., Yeom, S.J., Kwak, N.J. and Joo, Y.-C. (2009) The characteristics of Cu-drift induced dielectric breakdown under alternating polarity bias temperature stress. Proceedings of the 47th Annual IEEE International Reliability Physics Symposium (IRPS), Piscataway, New Jersey, pp. 825–827.
[261] Zhao, L., Tőkei, Zs., Gischia, G.G., Pantouvaki, M., Croes, K. and Beyer, G. (2009) A novel test structure to study intrinsic reliability of barrier/low-*k*. Proceedings of the 47th Annual IEEE International Reliability Physics Symposium (IRPS), Piscataway, New Jersey, pp. 848–850.
[262] McPherson, J.W. (2010) *Reliability Physics and Engineering – Time-to-Failure Modeling*, Springer Publishing, New York.
[263] McPherson, J.W. (2011) Basics of reliability physics. Tutorial Notes, IEEE International Reliability Physics Symposium (IRPS), SAR Associates, Rome, New York.
[264] Shima, S., Wada, Y., Tokushige, K., Fukunaga, A. and Tsujimura, M. (2010) Analysis on copper photocorrosion induced by illuminance in chemical mechanical planarization equipment using photodiode and quartz crystal microbalance. *Jpn. J. Appl. Phys.*, **49** (05I03), 05FF05-1-6.
[265] Guedj, C., Claret, N., Arnal, V., Aimadeddine, M., Barnes, J.P., Barbe, J.C., Arnaud, L., Reimbold, G., Torres, J., Passemard, G. and Boulanger, F. (2006) Dielectric conduction mechanisms of advanced interconnects: evidence for thermally-induced 3D/2D transition. Proceedings of the 44th Annual IEEE International Reliability Physics Symposium (IRPS), Piscataway, New Jersey, pp. 502–506.
[266] Yiang, K.-Y., Yao, H.W. and Marathe, A. (2008) TDDB kinetics and their relationship with the E- and √E-models. Proceedings of the IEEE 2008 International Interconnect Technology Conference (IITC), IEEE Service Center, Piscataway, New Jersey, pp. 168–170.
[267] Moazzami, R., Lee, J.C. and Hu, C. (1989) Temperature acceleration of time-dependent dielectric breakdown. *IEEE Trans. Electron. Devices*, **36** (11), 2462–2465.
[268] Liu, Z.H., Nee, P., Ko, P.K., Hu, C., Sodini, C.G., Gross, B.J., Ma, T.P. and Cheng, Y.C. (1992) Field and temperature acceleration of time-dependent dielectric breakdown for reoxidized-nitrided and fluorinated oxides. *IEEE Electron. Device Lett.*, **13** (1), 41–43.
[269] Chen, F., McLaughlin, P., Gambino, J., Wu, E., Demarest, J., Meatyard, D. and Shinosky, M. (2007) The effect of metal area and line spacing on TDDB characteristics of 45 nm low-*k* SiCOH dielectrics. Proceedings of the 45th Annual IEEE International Reliability Physics Symposium (IRPS), Piscataway, New Jersey, pp. 382–389.
[270] Takeda, K., Ryuzaki, D., Mine, T., Hinode, K. and Yoneyama, R. (2003) Copper-induced dielectric breakdown in silicon oxide deposited by plasma-enhanced chemical vapor deposition using trimethoxysilane. *J. Appl. Phys.*, **94** (4), 2572–2578.
[271] Okazaki, M., Hatano, M., Yoshida, K., Shibasaki, S., Kaneko, H., Yoda, T. and Hayasaka, N. (2004) 'Sea of Kelvin' multiple-pattern arrangement interconnect characterization for low-*k*/Cu dual damascene and its findings. Proceedings of the IEEE 2004 International Interconnect Technology Conference (IITC), IEEE Service Center, Piscataway, New Jersey, pp. 211–213.
[272] Fujimaki, T., Higashi, K., Nakamura, N., Matsunaga, N., Yoshida, K., Miyawaki, N., Hatano, M., Hasunuma, M., Wada, J., Nishioka, T., Akiyama, K., Kawashima, H., Enomoto, Y., Hasegawa, T., Honda, K., Iwai, M., Yamada, S. and Matsuoka, F. (2005) Mechanism of moisture uptake induced via failure and its impact on 45 nm node interconnect design. Technical Digest, IEEE International Electron Devices Meeting, Piscataway, New Jersey, pp. 183–186.
[273] Devine, R.A.B., Ball, D., Rowe, J.D. and Tringe, J.W. (2003) Irradiation and humidity effects on the leakage current in $SiO_x(CH_3)_y$-based intermetal dielectric films. *J. Electrochem. Soc.*, **150** (8), F151–F155.
[274] Kikkawa, T., Kuroki, S., Sakamoto, S., Kohmura, K., Tanaka, H. and Hata, N. (2005) Influence of humidity on electrical characteristics of self-assembled porous silica low-*k* films. *J. Electrochem. Soc.*, **152** (7), G560–G566.

[275] Kubasch, C., Klaus, C., Ruelke, H., Mayer, U. and Bartha, J.W. (2010) Investigation of moisture uptake in low-*k* dielectric materials. *IEEE Trans. Electron. Devices*, **57** (8), 1865–1872.

[276] Michelon J. and Hoofman, R.J.O.M. (2005) Impact of moisture on porous low-*k* reliability. 2005 IEEE International Integrated Reliability Workshop Final Report (IIRW), Institute of Electrical and Electronic Engineers, Inc., Piscataway, New Jersey, pp. 35–38.

[277] Lloyd, J.R., Shaw, T.M. and Liniger, E.G. (2005) Effect of moisture on the time dependent dielectric breakdown (TDDB) behavior in an ultra-low-*k* (ULK) dielectric. 2005 IEEE International Integrated Reliability Workshop Final Report (IIRW), Institute of Electrical and Electronic Engineers, Inc., Piscataway, New Jersey, pp. 39–43.

[278] Chen, F., Ungar, F., Fischer, A.H., Gill, J., Chinthakindi, A., Goebel, T., Shinosky, M., Coolbaugh, D., Ramachandran, V., Siew, Y.K., Kaltalioglu, E., Kim, S.O. and Park, K. (2006) Reliability characterization of BEOL vertical natural capacitor using copper and low-*k* SiCOH dielectric for 65 nm RF and mixed-signal applications. Proceedings of the 44th Annual IEEE International Reliability Physics Symposium (IRPS), Piscataway, New Jersey, pp. 490–495.

[279] Michelon J. and Hoofman, R.J.O.M. (2006) Moisture influence on porous low-*k* reliability. *IEEE Trans. Device and Mater. Reliab.*, **6** (2), 169–174.

[280] Li, Y., Ciofi, I., Carbonell, L., Groeseneken, G., Maex, K. and Tőkei, Zs. (2007) Moisture related low-*k* dielectric reliability before and after thermal annealing. Proceedings of the 45th Annual IEEE International Reliability Physics Symposium (IRPS), Piscataway, New Jersey, pp. 405–409.

[281] Li, Y., Ciofi, I., Carbonell, L., Heylen, N., Van Aelst, J., Baklanov, M.R., Groeseneken, G., Maex, K. and Tőkei, Zs. (2008) Influence of absorbed water components on SiOCH low-*k* reliability. *J. Appl. Phys.*, **104** (3), 034113-1-6.

[282] Choi, S.Y. (2009) Effects of moisture on the breakdown strength and lifetime of low permittivity dielectric for nanometer scale interconnects. PhD dissertation, University of Texas at Austin, Texas.

[283] Biswas, N., Lubguban, J.A. and Gangopadhyay, S. (2004) Electric field and temperature-induced removal of moisture in nanoporous organosilicate films. *Appl. Phys. Lett.*, **84** (21), 4254–4256.

[284] Yu, S., Wong, T.K.S., Pita, K., Hu, X. and Ligatchev, V. (2002) Surface modified silica mesoporous films as a low dielectric constant intermetal dielectric. *J. Appl. Phys.*, **92** (6), 3338–3344.

[285] Kim, J., Ogawa, E.T. and McPherson, J.W. (2007) Time dependent dielectric breakdown characteristics of low-*k* dielectric (SiOC) over a wide range of test areas and electric fields. Proceedings of the 45th Annual IEEE International Reliability Physics Symposium (IRPS), Piscataway, New Jersey, pp. 399–404.

[286] Haase, G.S. (2008) An alternative model for interconnect low-*k* dielectric lifetime dependence on voltage. Proceedings of the 46th Annual IEEE International Reliability Physics Symposium (IRPS), Piscataway, New Jersey, pp. 556–565.

[287] Moore, G. (1965) Cramming more components onto integrated circuits. *Electronics*, **38** (8), 114–117.

[288] Zhou, H., Shi, F.G., Zhao, B. and Yota, J. (2005) Temperature accelerated dielectric breakdown of PECVD low-*k* carbon doped silicon dioxide dielectric thin films. *Appl. Phys. A: Mater. Sci. Process.*, **81** (4), 767–771.

[289] Chen, F., Huang, E., Shinosky, M., Angyal, M., Kane, T., Wang, Y. and Kolics A. (2010) A comparative study of ULK conduction mechanisms and TDDB characteristics for Cu interconnects with and without CoWP metal cap at 32 nm technology. Proceedings of the IEEE 2010 International Interconnect Technology Conference (IITC), Piscataway, IEEE Service Center, New Jersey, pp. 1–3.

[290] Gischia, G.G., Croes, K., Groeseneken, G. and Tőkei, Zs. (2010) Study of leakage mechanism and trap density in porous low-*k* materials. Proceedings of the 48th Annual IEEE International Reliability Physics Symposium (IRPS), Piscataway, New Jersey, pp. 549–555.

[291] Ngwana, V.C., Zhu, C. and Krishnamoorthy, A. (2004) Dependence of leakage mechanisms on dielectric barrier in Cu–SiOC damascene interconnects. *Appl. Phys. Lett.*, **84** (13), 2316–2318.

[292] Ueki, M., Yamamoto, H., Ito, F., Kawahara, J., Tada, M., Takeuchi, T., Saito, S., Furutake, N., Onodera, T. and Hayashi Y. (2007) Cost-effective and high performance Cu interconnects ($k_{eff}$ = 2.75) with continuous SiOCH stack incorporating a low-*k* barrier cap (*k* = 3.1). Technical Digest, IEEE International Electron Devices Meeting, The Institute of Electrical and Electronics Engineers, Inc., Piscataway, New Jersey, pp. 973–976.

[293] Ngwan, V.C., Zhu, C. and Krishnamoorthy, A. (2004) Analysis of leakage mechanisms and leakage pathways in intra-level Cu interconnects. Proceedings of the 42nd Annual IEEE International Reliability Physics Symposium (IRPS), Piscataway, New Jersey, pp. 671–672.

[294] Aubel, O., Kiene, M. and Yao, W. (2005) New approach of 90 nm low-*k* interconnect evaluation using a voltage ramp dielectric breakdown (VRDB) test. Proceedings of the 43rd Annual IEEE International Reliability Physics Symposium (IRPS), Piscataway, New Jersey, pp. 283–589.

[295] Jow, K., Alers, G.B., Sanganeria, M., Harm, G., Fu, H., Tang, X., Kooi, G., Ray, G.W. and Danek, M. (2003) TDDB and voltage-ramp reliability of SiC-base dielectric diffusion barriers in Cu/low-*k* interconnects. Proceedings of the 41st Annual IEEE International Reliability Physics Symposium (IRPS), Piscataway, New Jersey, pp. 598–599.

[296] Alers, G.B., Jow, K., Shaviv, R., Kooi, G. and Ray, G.W. (2004) Interlevel dielectric failures in copper/low-*k* structures. *IEEE Trans. Device and Mater. Reliab.*, **4** (2), 148–152.

[297] Chatterjee, A., Brewer, F., Lee, S.C. and Oates, A.S. (2010) Investigating the electro-thermal origin of breakdown in low-*k*/Cu dielectrics under short duration over stressed pulsed regime. Proceedings of the 48th Annual IEEE International Reliability Physics Symposium (IRPS), Piscataway, New Jersey, pp. 932–937.

[298] Krishnamoorthy, A., Huang, N.Y. and Chong, S.-Y. (2003) Assessment of reliability of cap layers used in Cu–Black Diamond™ interconnects, in *Materials Research Society (MRS) Symposium Proceedings*, vol. 766, pp. E2.9.1–E2.9.6.

[299] Song, W.S., Kim, T.J., Lee, D.H., Kim, T.K., Lee, C.S., Kim, J.W., Kim, S.Y., Jeong, D.K., Park, K.C., Wee, Y.J., Suh, B.S., Choi, S.M., Kang, H.-K., Suh, K.P. and Kim, S.U. (2002) Pseudo-breakdown events induced by biased-thermal-stressing of intra-level Cu interconnects – reliability and performance impact. Proceedings of the 40th Annual IEEE International Reliability Physics Symposium (IRPS), Piscataway, New Jersey, pp. 305–311.

[300] Chen, Z., Prasad, K., Li, C.Y., Lu, P.W., Su, S.S., Tang, L.J., Gui, D., Balakumar, S., Shu, R. and Kumar, R. (2004) Dielectric/metal sidewall diffusion barrier for Cu/porous ultralow-*k* interconnect technology. *Appl. Phys. Lett.*, **84** (13), 2442–2444.

[301] Lia, Y., Hunter, J. and Tate, T.J. (2004) SIMS study of Cu trapping and migration in low-*k* dielectric films. *Appl. Surf. Sci.*, **231/232**, 791–795.

[302] Chen, F., Shinosky, M., Li, B., Aitken, J., Cohen, S., Bonilla, G., Simon, A., McLaughlin, P., Achanta, R., Baumann, F., Parks, C. and Angyal, M. (2011) Invasion percolation model for abnormal TDDB characteristic of ULK dielectrics with Cu interconnect at advanced technology nodes. Proceedings of the 49th Annual IEEE International Reliability Physics Symposium (IRPS), Piscataway, New Jersey, pp. 134–141.

[303] Haase, G.S. and Ogawa, E.T. (2005) Unpublished work.

[304] Breuer, T., Kerst, U., Boit, C., Langer, E. and Ruelke, H. (2010) Ultra-low-*k* dielectric degradation before breakdown. Proceedings of the 48th Annual IEEE International Reliability Physics Symposium (IRPS), Piscataway, New Jersey, pp. 890–894.

[305] Chen, F., Shinosky, M., Li, B., Gambino, J., Mongeon, S., Pokrinchak, P., Aitken, J., Badami, D., Angyal, M., Achanta, R., Bonilla, G., Yang, G., Liu, P., Li, K., Sudijono, J., Tan, Y., Tang, T.J. and Child, C. (2009) Critical ultra low-*k* TDDB reliability issues for advanced CMOS technologies. Proceedings of the 47th Annual IEEE International Reliability Physics Symposium (IRPS), Piscataway, New Jersey, pp. 464–475.

[306] Bersuker, G., Blaschke, V., Choi, S. and Wick, D. (2000) Conduction processes in Cu/low-*k* interconnection. Proceedings of the 38th Annual IEEE International Reliability Physics Symposium (IRPS), Piscataway, New Jersey, pp. 344–347.

[307] Cluzel, J., Mondon, F., Blachier, D., Morand, Y., Martel, L. and Reimbold, G. (2002) Electrical characterization of copper penetration effects in silicon dioxide. Proceedings of the 40th

Annual IEEE International Reliability Physics Symposium (IRPS), Piscataway, New Jersey, pp. 431–432.

[308] Cohen, S.A., Liu, J., Gignac, L., Ivers, T., Armbrust, D., Rodbell, K.P. and Gates, S.M. (1991) Characterization of thin dielectric films as copper diffusion barriers using triangular voltage sweep, in *Materials Research Society Symposium Proceedings –Symposium N – Advanced Interconnects and Contacts*, vol. 564, Materials Research Society, pp. 551–558.

[309] Book, G.W., Pfeifer, K. and Smith, S. (2002) Barrier integrity testing of Ta using triangular voltage sweep and a novel CV-BTS test structure. *Microelectron. Engng*, **64** (1–4), 255–260.

[310] Mallikarjunan, A., Murarka, S.P. and Lu, T.-M. (2002) Mobile ion detection in organosiloxane polymer using triangular voltage sweep. *J. Electrochem. Soc.*, **149** (10), F155–F159.

[311] Ciofi, I., Tökei, Zs., Visalli, D. and Van Hove, M. (2006) Water and copper contamination in SiOC:H damascene: novel characterization methodology based on triangular voltage sweep measurements. Proceedings of the IEEE 2006 International Interconnect Technology Conference (IITC), IEEE Service Center, Piscataway, New Jersey, pp. 181–183.

[312] Miyazaki, H., Kodama, D. and Suzumura, N. (2008) The observation of stress-induced leakage current of damascene interconnects after bias temperature aging. Proceedings of the 46th Annual IEEE International Reliability Physics Symposium (IRPS), Piscataway, New Jersey, pp. 150–157.

[313] Gonella, R., Motte, P. and Torres, J. (2000) Assessment of copper contamination impact on inter-level dielectric reliability performed with time-dependent-dielectric-breakdown tests. *Microelectron. Reliab.*, **40** (8–10), 1305–1309.

[314] Hartfield, C.D., Ogawa, E.T., Park, Y.-J., Chiu, T.-C. and Guo, H. (2004) Interface reliability assessments for copper/low-*k* products. *IEEE Trans. Device and Mater. Reliab.*, **04** (02), 129–141.

[315] Hong, C., Milor, L., Lin, M.Z., Hsu, W.M. and Peng, Y. (2005) Fast reliability evaluation of backend dielectrics using lifetime prediction techniques at wafer level. Proceedings of the 43rd Annual IEEE International Reliability Physics Symposium (IRPS), Piscataway, New Jersey, pp. 588–589.

[316] Chattopadhyay, K., van Schravendijk, B., Mountsier, T.W., Alers, G.B., Hornbeck, M., Wu, H.J., Shaviv, R., Harm, G., Vitkavage, D., Apen, E., Yu, Y. and Havemann, R. (2006) In-situ formation of a copper silicide cap for TDDB and electromigration improvement. Proceedings of the 44th Annual IEEE International Reliability Physics Symposium (IRPS), Piscataway, New Jersey, pp. 128–130.

[317] Chen, Y.W., Zou, J., Zhang, G., Chen, S., Wang, J., Hu, T., Bei, E., Wang, J., Fan, A. and Chen I.-C. (2006) Reliability improvement in multi-level Cu/SiOC low *k* integration. 13th International Symposium on the Physical and Failure Analysis of Integrated Circuits (IPFA), Institute of Electrical and Electronic Engineers, Inc., Piscataway, New Jersey, pp. 110–113.

[318] Wang, Q., Gan, H., Zhao, L., Zheng, K., Bei, E. and Ning, J. (2008) Low-k breakdown improvement in 65 nm dual-damascene Cu process. 9th International Conference on Solid-State and Integrated-Circuit Technology (ICSICT), pp. 1324–1327.

[319] Berman, A. (1981) Time zero dielectric reliability test by a ramp method. Proceedings of the 19th Annual IEEE International Reliability Physics Symposium (IRPS), Piscataway, New Jersey, pp. 204–209.

[320] Haase, G.S., Ogawa, E.T. and McPherson, J.W. (2005) Breakdown characteristics of interconnect dielectrics. Proceedings of the 43rd Annual IEEE International Reliability Physics Symposium (IRPS), Piscataway, New Jersey, pp. 466–473.

[321] Haase, G.S., Ogawa, E.T. and McPherson, J.W. (2005) Reliability analysis method for low-*k* interconnect dielectrics breakdown in integrated circuits. *J. Appl. Phys.*, **98** (03), 034503-01-10.

[322] Kim, J., Ogawa, E.T. and McPherson, J.W. (2006) A statistical evaluation of the field acceleration parameter observed during time dependent dielectric breakdown testing of silica-based low-*k* interconnect dielectrics. Proceedings of the 44th Annual IEEE International Reliability Physics Symposium (IRPS), Piscataway, New Jersey, pp. 478–483.

[323] Crook, D.L. (1979) Method of determining reliability screens for time dependent dielectric breakdown. Proceedings of the 17th Annual IEEE International Reliability Physics Symposium (IRPS), Piscataway, New Jersey, pp. 1–7.

[324] Anolick, E.S. and Nelson, G.R. (1979) Low field time dependent dielectric integrity. Proceedings of the 17th Annual IEEE International Reliability Physics Symposium (IRPS), Piscataway, New Jersey, pp. 8–12.
[325] Lin, M., Liang, J.W. and Su, K.C. (2010) New voltage ramp dielectric breakdown methodology based on square root E model for Cu/low-*k* interconnect reliability. Proceedings of the 48th Annual IEEE International Reliability Physics Symposium (IRPS), Piscataway, New Jersey, pp. 556–561.
[326] Lin, M. and Su, K.C. (2010) Correlation between TDDB and VRDB for low-*k* dielectrics with square root E model. *IEEE Electron. Device Lett.*, **31** (5), 494–496.
[327] Chen, F., McLaughlin, P., Gambino, J. and Gill, J. (2007) A comparison of voltage ramp and time dependent dielectric breakdown tests for evaluation of 45 nm low-*k* SiCOH reliability. Proceeding of the IEEE 2007 International Interconnect Technology Conference (IITC), IEEE Service Center, Piscataway, New Jersey, pp. 120–127.
[328] Lloyd, J.R., Liniger, E. and Chen, S.T. (2004) Time dependent dielectric breakdown in a low-*k* interlevel dielectric. *Microelectron. Reliab.*, **44** (9–11), 1861–1865.
[329] Tőkei, Zs., Sutcliffe, V., Demuynck, S., Iacopi, F., Roussel, P., Beyer, G.P., Hoofman, R.J.O.M. and Maex, K. (2004) Impact of the barrier/dielectric interface quality on reliability of Cu porous-low-*k* interconnects. Proceedings of the 42nd Annual IEEE International Reliability Physics Symposium (IRPS), Piscataway, New Jersey, pp. 326–332.
[330] McPherson, J.W. and Ogawa, E.T. (2008) *Handbook of Semiconductor Manufacturing Technology*, 2nd edn, CRC Press, Boca Raton, Florida, Chapter 30.
[331] Gall, M., Capasso, C., Jawarani, D., Hernandez, R., Kawasaki, H. and Ho, P.S. (1997) Statistical evaluation of device-level electromigration reliability, in *Stress-Induced Phenomena in Metallization* (eds H. Okabayashi, S. Shingubara and P.S. Ho), vol. 418, American Institute of Physics, Woodbury, New York, pp. 483–494.
[332] Ogawa, E.T., Lee, K.D., Matsuhashi, H., Ko, K.S., Justison, P.R., Ramamurthi, A.N., Bierwag, A.J. and Ho, P.S. (2001) Statistics of electromigration early failures in Cu/oxide dual-damascene interconnects. Proceedings of the 39th Annual IEEE International Reliability Physics Symposium (IRPS), Piscataway, New Jersey, pp. 341–349.
[333] Lee, K.-D. and Ho, P.S. (2004) Statistical study for electromigration reliability in dual-damascene Cu interconnects. *IEEE Trans. Device and Mater. Reliab.*, **04** (2), 237–245.
[334] Oates, A.S. and Lee, S.C. (2006) Electromigration failure distributions of dual damascene Cu/low-*k* interconnects. *Microelectron. Reliab.*, **46** (9–11), 1581–1586.
[335] Tsuchiya, H. and Yokogawa, S. (2006) Electromigration lifetimes and void growth at low cumulative failure probability. *Microelectron. Reliab.*, **46** (9–11), 1415–1420.
[336] Hauschildt, M., Gall, M., Justison, P., Hernandez, R. and Herrick, M. (2007) Large-scale statistical study of electromigration early failure for Cu/low-*k* interconnects, in *Stress-Induced Phenomena in Metallization* (eds S. Ogawa, P.S. Ho and E. Zschech), vol. 945, American Institute of Physics, Melville, New York, pp. 66–81.
[337] Hauschildt, M., Gall, M. and Hernandez, R. (2008) Large-scale statistics for Cu electromigration, in *Stress-Induced Phenomena in Metallization* (eds P.S. Ho, S. Ogawa and E. Zschech), vol. 1143, American Institute of Physics, Melville, New York, pp. 31–46.
[338] Chen, F., Ungar, F., Fischer, A.H., Gill, J., Chinthakindi, A., Goebel, T., Shinosky, M., Coolbaugh, D., Ramachandran, V., Siew, Y.K., Kaltalioglu, E., Kim, S.O. and Park, K. (2006) Reliability characterization of BEOL vertical natural capacitor using copper and low-*k* SiCOH dielectric for 65 nm RF and mixed-signal applications. Proceedings of the 44th Annual IEEE International Reliability Physics Symposium (IRPS), Piscataway, New Jersey, pp. 490–495.
[339] Degraeve, R., Ogier, J.L., Bellens, R., Roussel, P.J., Groeseneken, G. and Maes, H.E. (1998) A new model for the field dependence of intrinsic and extrinsic time-dependent dielectric breakdown. *IEEE Trans. Electron. Devices*, **45** (2), 472–481.
[340] Stathis, J.H. (2001) Physical and predictive models of ultrathin oxide reliability in CMOS devices and circuits. *IEEE Trans. Device and Mater. Reliab.*, **1** (1), 43–59.
[341] Robillard, T.R. (1955) Techniques employed in the fabrication of p-n-p power alloy transistors. Technical Digest, IEEE International Electron Devices Meeting, The Institute of Electrical and Electronics Engineers, Inc., Piscataway, New Jersey, p. 24.

[342] Howard, B.T. and Dodson, G.A. (1960) A method for the rapid evaluation of semiconductor device reliability. Technical Digest, IEEE International Electron Devices Meeting, The Institute of Electrical and Electronics Engineers, Inc., Piscataway, New Jersey, p. 46.
[343] Greenough, K.F. (1962) Progress report on a study of the modes of failure primary found occurring at material interfaces in thin-film solid state devices. First Annual Symposium on the Physics of Failure in Electronics, 1962, pp. 20–33.
[344] Gerson, R. and Berlincourt, D. (1962) Some aspects of dielectric degradation in perovskites. First Annual Symposium on the Physics of Failure in Electronics, 1962, pp. 223–230.
[345] McPherson, J., Reddy, V., Banerjee, K. and Le, H. (1998) Comparison of E and 1/E TDDB models for $SiO_2$ under long-term/low-field test conditions. Technical Digest, IEEE International Electron Devices Meeting, The Institute of Electrical and Electronics Engineers, Inc., Piscataway, New Jersey, pp. 171–174.
[346] Lee, S.-C. and Oates, A.S. (2011) Reliability limitations to the scaling of porous low-k dielectrics. Proceedings of the 49th Annual IEEE International Reliability Physics Symposium (IRPS), Piscataway, New Jersey, pp. 155–159.
[347] Suehle, J.S. (2002) Ultrathin gate oxide reliability: physical models, statistics, and characterization. *IEEE Trans. Electron. Devices*, **49** (6), 958–971.
[348] Chen, I.-C., Holland, S.E. and Hu, C. (1985) Electrical breakdown in thin gate and tunneling oxides. *IEEE Trans. Electron. Devices*, **ED-32** (2), 413–422.
[349] Chen, I.-C., Holland, S.E. and Hu, C. (1985) A quantitative physical model for time-dependent breakdown in $SiO_2$. Proceedings of the 23rd Annual IEEE International Reliability Physics Symposium (IRPS), Piscataway, New Jersey, pp. 24–31.
[350] Lee, J., Chen, I.-C. and Hu, C. (1988) Statistical modeling of silicon dioxide reliability. Proceedings of the 26th Annual IEEE International Reliability Physics Symposium (IRPS), Piscataway, New Jersey, pp. 131–138.
[351] McPherson, J.W. and Baglee, D.A. (1985) Acceleration factors for thin gate oxide stressing. Proceedings of the 23rd Annual IEEE International Reliability Physics Symposium (IRPS), Piscataway, New Jersey, pp. 1–5.
[352] McPherson, J.W. and Baglee, D.A. (1985) Acceleration factors for thin oxide breakdown. *J. Electrochem. Soc.: Solid-State Sci. Technol.*, **132** (8), 1903–1908.
[353] Swartz, G.A. (1986) Gate oxide integrity of NMOS transistor arrays. *IEEE Trans. Electron. Devices*, **ED-33** (11), 1826–1829.
[354] Boyko, K.C. and Gerlach, D.L. (1989) Time dependent dielectric breakdown of 210Å oxides. Proceedings of the 27th Annual IEEE International Reliability Physics Symposium (IRPS), Piscataway, New Jersey, pp. 1–8.
[355] Suehle, J.S., Chaparala, P., Messick, C., Miller, W.M. and Boyko, K.C. (1994) Field and temperature acceleration of time-dependent dielectric breakdown in intrinsic thin $SiO_2$. Proceedings of the 32nd Annual IEEE International Reliability Physics Symposium (IRPS), Piscataway, New Jersey, 120–125.
[356] Chaparala, P., Suehle. J.S., Messick, C. and Roush, M. (1996) Electric field dependent dielectric breakdown of intrinsic $SiO_2$ films under dynamic stress. Proceedings of the 34th Annual IEEE International Reliability Physics Symposium (IRPS), Piscataway, New Jersey, pp. 61–66.
[357] Suehle, J.S. and Chaparala, P. (1997) Low electric field breakdown of thin $SiO_2$ films under static and dynamic stress. *IEEE Trans. Electron. Devices*, **44** (5), 801–808.
[358] Kimura, M. (1997) Oxide breakdown mechanism and quantum physical chemistry for time-dependent dielectric breakdown. Proceedings of the 35th Annual IEEE International Reliability Physics Symposium (IRPS), Piscataway, New Jersey, pp. 190–200.
[359] Shatzkes M. and Av-Ron, M. (1976) Impact ionization and positive charge in thin $SiO_2$ films. *J. Appl. Phys.*, **47** (07), 3194–3202.
[360] Schuegraf, K.F. and Hu, C. (1993) Hole injection oxide breakdown model for very low voltage lifetime extrapolation. Proceedings of the 31st Annual IEEE International Reliability Physics Symposium (IRPS), Piscataway, New Jersey, pp. 7–12.
[361] Schuegraf, K.F. and Hu, C. (1994) Hole injection $SiO_2$ breakdown model for very low voltage lifetime extrapolation. *IEEE Trans. Electron. Devices*, **41** (5), 761–767.

[362] McPherson, J.W., Khamankar, R.B. and Shanware, A. (2000) Underlying physics of the thermochemical E model in describing low-field time-dependent dielectric breakdown in $SiO_2$ thin films. *J. Appl. Phys.*, **88** (9), 5351–5359.

[363] McPherson, J.W. (2004) Determination of the nature of molecular bonding in silica from time-dependent dielectric breakdown data. *J. Appl. Phys.*, **95** (12), 8101–8109.

[364] Hu, C. and Lu, Q. (1999) A unified gate oxide reliability model. Proceedings of the 37th Annual IEEE International Reliability Physics Symposium (IRPS), Piscataway, New Jersey, pp. 47–51.

[365] Cheung, K.P. (1999) A physics-based, unified gate-oxide breakdown model. Technical Digest, IEEE International Electron Devices Meeting, The Institute of Electrical and Electronics Engineers, Inc., Piscataway, New Jersey, pp. 719–722.

[366] McPherson, J.W. (2007) Quantum mechanical treatment of Si–O bond breakage in silica under time dependent dielectric breakdown testing. Proceedings of the 45th Annual IEEE International Reliability Physics Symposium (IRPS), Piscataway, New Jersey, pp. 209–216.

[367] Wu, W., Duan, X. and Yuan, J.S. (2003) A physical model of time-dependent dielectric breakdown in copper metallization. Proceedings of the 41st Annual IEEE International Reliability Physics Symposium (IRPS), Piscataway, New Jersey, pp. 282–286.

[368] Lloyd, J.R., Ponoth, S., Liniger, E. and Cohen, S. (2007) Role of Cu in TDDB of low-*k* dielectrics. Proceedings of the 45th Annual IEEE International Reliability Physics Symposium (IRPS), Piscataway, New Jersey, pp. 410–411.

[369] Lee, S.-C., Oates, A.S. and Chang, K.-M. (2009) Fundamental understanding of porous low-*k* dielectric breakdown. Proceedings of the 47th Annual IEEE International Reliability Physics Symposium (IRPS), Piscataway, New Jersey, pp. 481–485.

[370] Lloyd, J.R., Murray, C.E., Ponoth, S., Cohen, S. and Liniger, E. (2006) The effect of Cu diffusion on the TDDB behavior in a low-*k* interlevel dielectrics. *Microelectron. Reliab.*, **46** (9–11), 1643–1647.

[371] Suzumura, N., Yamamoto, S., Kodama, D., Makabe, K., Komori, J., Murakami, E., Maegawa, S. and Kubota, K. (2006) A new TDDB degradation model based on Cu ion drift in Cu interconnect dielectrics. Proceedings of the 44th Annual IEEE International Reliability Physics Symposium (IRPS), Piscataway, New Jersey, pp. 484–489.

[372] Hwang, S.-S., Jung, S.-Y. and Joo, Y.-C. (2007) Leakage current characteristic of pre-damaged interlayer dielectric during voltage ramp method. Proceedings of the 45th Annual IEEE International Reliability Physics Symposium (IRPS), Piscataway, New Jersey, pp. 628–629.

[373] Suzumura, N., Yamamoto, S., Kodama, D., Miyazaki, H., Ogasawara, M., Komori, J. and Murakami, E. (2008) Electric-field and temperature dependencies of TDDB degradation in Cu/low-*k* damascene structures. Proceedings of the 46th Annual IEEE International Reliability Physics Symposium (IRPS), Piscataway, New Jersey, pp. 138–143.

[374] van Delden, M.H.W.M. and van der Wel, P.J. (2003) Reliability and electric properties for PECVD a-$SiN_x$:H films with an optical bandgap ranging from 2.5 to 5.38 eV. Proceedings of the 41st Annual IEEE International Reliability Physics Symposium (IRPS), Piscataway, New Jersey, pp. 293–297.

[375] Cui H. and Burke, P.A. (2004) Time-dependent dielectric breakdown studies of PECVD H:SiCN and H:SiC thin films for copper metallization. *J. Electrochem. Soc.*, **151** (11), G795–G800.

[376] Remmel, T., Ramprasad, R., Roberts, D., Raymond, M., Martin, M., Qualls, D., Luckowski, E., Braithwaite, S., Miller, M. and Walls, J. (2004) Reliability of silicon nitride dielectric-based metal–insulator–metal capacitors. Proceedings of the 42nd Annual IEEE International Reliability Physics Symposium (IRPS), Piscataway, New Jersey, pp. 673–674.

[377] Suzumura, N., Ogasawara, M., Makabe, K., Kamoshima, T., Ouchi, T., Yamamoto, S., Furusawa, T. and Murakami, E. (2011) Comprehensive lifetime prediction for intrinsic and extrinsic TDDB failures in Cu/low-*k* interconnects. Proceedings of the IEEE 2011 International Interconnect Technology Conference (IITC), IEEE Service Center, Piscataway, New Jersey, paper 6.1.

[378] Yiang, K.-Y., Yao, H.W., Marathe, A. and Aubel, O. (2009) New perspectives of dielectric breakdown in low-*k* interconnects. Proceedings of the 47th Annual IEEE International Reliability Physics Symposium (IRPS), Piscataway, New Jersey, pp. 476–480.

[379] Achanta, R. and McLaughlin, P. (2011) A charge transport based acceleration model for interlevel dielectric breakdown. Proceedings of the 49th Annual IEEE International Reliability Physics Symposium (IRPS), Piscataway, New Jersey, pp. 665–669.
[380] Rodriguez, O.R., Cho, W., Saxen, R., Plawsky, J.L. and Gill, W.N. (2005) Mechanism of Cu diffusion in porous low-*k* dielectrics. *J. Appl. Phys.*, **98** (2), 024108-1-9.
[381] Croes, K., Cannatá, G., Zhao, L. and Tőkei, Zs. (2008) Study of copper drift during TDDB of intermetal dielectrics by using fully passivated MOS capacitors as test vehicle. *Microelectron. Reliab.*, **48** (8–9), 1384–1387.
[382] Zhao, L., Tőkei, Zs., Croes, K., Wilson, C.J., Baklanov, M., Beyer, G.P. and Claeys, C. (2011) Direct observation of the 1/*E* dependence of time dependent dielectric breakdown in the presence of copper. *Appl. Phys. Lett.*, **98** (2), 032107-1-3.
[383] Chen, F., Chanda, K., Gill, J., Angyal, M., Demarest, J., Sullivan, T., Kontra, R., Shinosky, M., Li, J., Economikos, L., Hoinkis, M., Lane, S., McHerron, D., Inohara, M., Boettcher, S., Dunn, D., Fukasawa, M., Zhang, B.C., Ida, K., Ema, T., Lembach, G., Kumar, K., Lin, Y., Maynard, H., Urata, K., Bolom, T., Inoue, K., Smith, J., Ishikawa, Y., Naujok, M., Ong, P., Sakamoto, A., Hunt, D. and Aitken, J. (2005) Investigation of CVD SiCOH low-*k* time-dependent dielectric breakdown at 65 nm node technology. Proceedings of the 43rd Annual IEEE International Reliability Physics Symposium (IRPS), Piscataway, New Jersey, pp. 501–507.
[384] Chen, F., Bravo, O., Chanda, K., McLaughlin, P., Sullivan, T., Gill, J., Lloyd, J., Kontra1, R. and Aitken, J. (2006) A comprehensive study of low-*k* SiCOH TDDB phenomena and its reliability lifetime model development. Proceedings of the 44th Annual IEEE International Reliability Physics Symposium (IRPS), Piscataway, New Jersey, pp. 46–53.
[385] Chen, F. and Shinosky, M. (2009) Addressing Cu/low-*k* dielectric TDDB-reliability challenges for advanced CMOS technologies. *IEEE Trans. Electron. Devices*, **56** (1), 2–12.
[386] Raghavan, N. and Prasad, K. (2009) Statistical outlook into the physics of failure for copper low-*k* intra-metal dielectric breakdown. Proceedings of the 47th Annual IEEE International Reliability Physics Symposium (IRPS), Piscataway, New Jersey, pp. 819–824.
[387] Pompl, T., Allers, K.-H., Schwab, R., Hofmann, K. and Roehner, M. (2005) Change of acceleration behavior of time-dependent dielectric breakdown by the BEOL process: indications for hydrogen induced transition in dominant degradation mechanism. Proceedings of the 43rd Annual IEEE International Reliability Physics Symposium (IRPS), Piscataway, New Jersey, pp. 338–397.
[388] Lloyd, J.R., Liniger, E. and Shaw, T.M. (2005) Simple model for time-dependent dielectric breakdown in inter- and intralevel low-*k* dielectrics. *J. Appl. Phys.*, **98** (8), 084109-01-06.
[389] Lloyd, J.R. (2010) On the physical interpretation of the impact damage model in TDDB of low-*k* dielectrics. Proceedings of the 48th Annual IEEE International Reliability Physics Symposium (IRPS), Piscataway, New Jersey, pp. 943–946.
[390] Haase, G.S. (2009) A model for electric degradation of interconnect low-*k* dielectrics in microelectronic integrated circuits. *J. Appl. Phys.*, **105** (04), 044908-01-10.
[391] Ligatchev, V. and Wong, T.K.S. (2004) On distributions of defect states in low-*k* carbon doped silicon dioxide films in vicinity of Fermi level. *Electrochem. Solid-State Lett.*, **7** (12), F89–F92.
[392] Shamuilia, S., Afanas'ev,_V.V., Somers, P., Stesmans, A., Li, Y.L., Tőkei, Zs., Groeseneken, G. and Maex, K. (2006) Internal photoemission of electrons at interfaces of metals with low-*k* insulators. *Appl. Phys. Lett.*, **89** (20), 202909-1-3.
[393] Atkin, J.M., Song, D., Shaw, T.M., Cartier, E., Laibowitz, R.B. and Heinz, T.F. (2008) Photocurrent spectroscopy of low-*k* dielectric materials: barrier heights and trap densities. *J. Appl. Phys.*, **103** (06), 094104-1-6.
[394] Bittel, B.C., Lenahan, P.M. and King, S. (2009) Reliability and performance limiting defects in low-*k* dielectrics for use as interlayer dielectrics. Proceedings of the 47th Annual IEEE International Reliability Physics Symposium (IRPS), Piscataway, New Jersey, pp. 947–950.
[395] Bittel, B.C., Pomorski, T.A., Lenahan, P.M. and King, S. (2010) Defects in low-*k* dielectrics and etch stop layers for use as interlayer dielectrics in ULSI. 2010 IEEE International Integrated Reliability Workshop Final Report (IIRW), Institute of Electrical and Electronic Engineers, Inc., Piscataway, New Jersey, pp. 37–41.

[396] Atkin, J.M., Laibowitz, R.B., Heinz, T.F., Lloyd, J.R., Shaw, T.M. and Cartier, E. (2010) Effects of photoinduced carrier injection on time-dependent dielectric breakdown. Proceedings of the 48th Annual IEEE International Reliability Physics Symposium (IRPS), Piscataway, New Jersey, pp. 851–854.

[397] Croes, K. and Tőkei, ZS. E- and √E-model too conservative to describe low field time dependent dielectric breakdown. Proceedings of the 48th Annual IEEE International Reliability Physics Symposium (IRPS), Piscataway, New Jersey, pp. 543–548.

[398] Croes, K., Pantouvaki, M., Carbonell, L., Zhao, L., Beyer, G.P. and Tőkei, Zs. (2011) Comparison between intrinsic and integrated reliability properties of low-*k* materials. Proceedings of the 49th Annual IEEE International Reliability Physics Symposium (IRPS), Piscataway, New Jersey, pp. 142–148.

[399] Wu, E.Y., Vayshenker, A., Nowak, E., Suñé, J., Vollertsen, R.P., Lai, W. and Harmon, D. (2002) Experimental evidence of $T_{BD}$ power-law for voltage dependence of oxide breakdown in ultrathin gate oxides. *IEEE Trans. Electron. Devices*, **49** (12), 2244–2252.

[400] Schindler, G., Stich, A., Gabric, Z. and Pamler, W. (2005) Air gaps: can they be a solution to the low-*k* challenge?, in *Advanced Metallization Conference (AMC)* (eds S.H. Brongersma, T.C. Taylor, M. Tsujimura and K. Masu), Materials Research Society, Warrendale, Pennsylvania, pp. 255–261.

[401] Schulze, K., Schulz, S.E., Rennau, M. and Gessner, T. (2005) Formation of air gap structures via wet etch removal of sacrificial dielectrics, in *Advanced Metallization Conference (AMC)* (eds S.H. Brongersma, T.C. Taylor, M. Tsujimura and K. Masu), Materials Research Society, Warrendale, Pennsylvania, pp. 309–316.

[402] Hoofman, R.J.O.M., Bao, T.I., Daamen, R., Chen, H.C., Yang, A., Orain, S., Ernur-Badaroglu, D., Humbert, A., Shue, S.L. and Yu, C.H. (2007) Air gaps as a viable alternative for low-*k* dielectrics in future technology nodes, in *Advanced Metallization Conference (AMC)* (eds A.J. McKerrow, Y. Shacham-Diamand, S. Shingubara and Y. Shimogaki), Materials Research Society, Warrendale, Pennsylvania, pp. 411–417.

[403] Schulze, K., Schulz, S.E., Körner, H. and Gessner, T. (2008) Airgap structures by using sacrifical wet etch: fabrication, thermal and mechanical behavior, reliability, in *Advanced Metallization Conference (AMC)* (eds M. Naik, R. Shaviv, T. Yoda and K. Ueno), Materials Research Society, Warrendale, Pennsylvania, pp. 41–52.

[404] Harada, T., Ueki, A., Kido, S., Tomita, K., Kanda, Y., Sasaki, T., Tsuji, H., Furuhashi, T., Kabe, T., Shibata, J., Iwasaki, A., Izumitani, J., Matsumoto, S., Kawano, Y. and Ueda, T. (2008) Air gap integration of 45 nm node Cu interconnects and evaluations of electrical and mechanical characteristics, in *Advanced Metallization Conference (AMC)* (eds M. Naik, R. Shaviv, T. Yoda and K. Ueno), Materials Research Society, Warrendale, Pennsylvania, pp. 53–59.

[405] Gras, R., Gaillard, F., Bouchu, D.C., Haumesser, P.H., Imbert, G., Vandroux, L., Farcy, A., Chevolleau, T., Passemard, G., Torres, J. and Ancey, P. (2008) Multi levels air gap integration using sacrificial material approach for advanced Cu interconnects technologies, in *Advanced Metallization Conference (AMC)* (eds M. Naik, R. Shaviv, T. Yoda and K. Ueno), Materials Research Society, Warrendale, Pennsylvania, pp. 69–74.

[406] Gabric, Z., Pamler, W., Schindler, G., Steinhögl, W. and Traving, M. (2004) Air gap technology by selective ozone/TEOS deposition. Proceedings of the IEEE 2004 International Interconnect Technology Conference (IITC), IEEE Service Center, Piscataway, New Jersey, pp. 151–153.

[407] Harada, T., Takahashi, M., Murakami, K., Korogi, H., Sasaki, T., Hattori, T., Ogawa, S. and Ueda, T. (2006) Manufacturable low $k_{eff}$ ($k_{eff}$ < 2.5) Cu interconnects by selective/low damage air gap formation. Proceedings of the IEEE 2006 International Interconnect Technology Conference (IITC), IEEE Service Center, Piscataway, New Jersey, pp. 15–18.

[408] Nitta, S., Edelstein, D., Ponoth, S., Clevenger, L., Liu, X. and Standaert, T. (2008) Performance and reliability of airgaps for advanced BEOL interconnects. Proceedings of the IEEE 2008 International Interconnect Technology Conference (IITC), IEEE Service Center, Piscataway, New Jersey, pp. 191–192.

[409] Daamen, R., Verheijden, G.J.A.M., Bancken, P.H.L., Vandeweyer, T., Michelon, J., Nguyen Hoang, V., Hoofman, R.J.O.M. and Gallagher, M.K. (2005) Air gap integration for the 45 nm

node and beyond. Proceedings of the IEEE 2005 International Interconnect Technology Conference (IITC), IEEE Service Center, Piscataway, New Jersey, pp. 240–242.

[410] Gosset, L.G., Gaillard, F., Bouchu, D., Gras, R., de Pontcharra, J., Orain, S., Cueto, O., Lyan, Ph., Louveau, O., Passemard, G. and Torres, J. (2007) Multi-level Cu interconnects integration and characterization with air gap as ultra-low *k* material formed using a hybrid sacrificial oxide/polymer stack. Proceedings of the IEEE 2007 International Interconnect Technology Conference (IITC), IEEE Service Center, Piscataway, New Jersey, pp. 58–60.

[411] Daamen, R., Bancken, P.H.L., Ernur Badaroglu, D., Michelon, J., Nguyen, V.H., Verheijden, G.J.A.M., Humbert, A., Waeterloos, J., Yang, A., Cheng, J.K., Chen, L., Martens, T. and Hoofman, R.J.O.M. (2007) Multi-level air gap integration for 32/22 nm nodes using a spin-on thermal degradable polymer and a SiOC CVD hard mask. Proceedings of the IEEE 2007 International Interconnect Technology Conference (IITC), IEEE Service Center, Piscataway, New Jersey, pp. 61–63.

[412] Noguchi, J., Sato, K., Konishi, N., Uno, S., Oshima, T., Tanaka, U., Ishikawa, K., Ashihara, H., Saito, T., Kubo, M., Aoki, H. and Fujiwara, T. (2004) Reliability of air-gap Cu interconnect and approach to selective W sealing using 90 nm node technology. Proceedings of the IEEE 2004 International Interconnect Technology Conference (IITC), IEEE Service Center, Piscataway, New Jersey, pp. 81–83.

[413] Noguchi, J., Sato, K., Konishi, N., Uno, S., Oshima, T., Ishikawa, K., Ashihara, H., Saito, T., Kubo, M., Tamaru, T., Yamada, Y., Aoki, H. and Fujiwara, T. (2005) Process and reliability of air-gap Cu interconnect using 90-nm node technology. *IEEE Trans. Electron. Devices*, **52** (3), 352–359.

[414] Weast, R.B. (ed.) (1983–1984) Spark gap voltage, in *CRC Handbook of Chemistry and Physics*, 64th edn, CRC Press, Inc., Boca Raton, Florida, p. E-49.

[415] Kobayashi, S., Saito, Y., Nagai, Y. and Yamamoto, Y. (1993) Vacuum breakdown strength of vacuum-degassed oxygen-free copper electrodes. *IEEE Trans. Elect, Insulation*, **28** (4), 500–506.

[416] Cuttone, G., Marchetta, C., Torrisi, L., Della Mea, G., Quaranta, A., Rigato, V. and Zandolin, S. (1997) Surface treatment of HV electrodes for superconducting cyclotron beam extraction. *IEEE Trans. Dielectrics and Elect. Insulation*, **4** (2), 218–223.

# Section IV

## New Approaches

# 12

# 3D Interconnect Technology

*John U. Knickerbocker*[1], *Lay Wai Kong*[2], *Sven Niese*[3], *Alain Diebold*[2] *and Ehrenfried Zschech*[3]

[1]*IBM Research Division, Thomas J. Watson Research Center, New York, USA*
[2]*College of Nanoscale Science and Engineering at the University at Albany, NewYork, USA*
[3]*Fraunhofer Institute for Nondestructive Testing, Dresden, Germany*

## 12.1 Introduction

Three-dimensional interconnected circuits (3DICs) offer great advantages for system applications. Examples of system advantages from silicon integration technologies can include: power reduction, performance improvements, product miniaturization, cost reduction, modular design for heterogeneous die and improved time to market. In order to take advantage of these 3D system benefits, the design and architecture circuits needs to be optimized to take advantage of emerging 3D technology. In addition, the materials, processes and assembly of stacked dies and integration of 3D modules including power delivery, test and cooling need to be able to be manufactured at cost-effective yields and meet system reliability requirements. 3D stacked dies can be integrated using traditional packaging laminate or ceramic packaging or can use emerging silicon and glass packages for higher bandwidth leveraging, higher wiring density between dies and/or stacked dies in multichip modules. Therefore, to optimize the system applications leveraging 3D interconnections, ongoing technical challenges include: (i) 3D system design and architecture, (ii) wafer fabrication with through-silicon vias (TSV), (iii) wafer finishing including wafer thinning, thin wafer handling and wafer backside processing, (iv) assembly using die-to-die, die-to-wafer or wafer-to-wafer stacking, (v) wafer test and burn-in, (vi) power delivery,

*Advanced Interconnects for ULSI Technology*, First Edition. Edited by Mikhail R. Baklanov, Paul S. Ho and Ehrenfried Zschech.

(vii) cooling and (viii) module integration. This chapter will explore the system opportunities and technical challenges related to 3D silicon interconnections in systems. Examples of 3D test vehicles, data and results are reported for the technical challenges used in 3D design, wafer fabrication and module integration. The second part of the chapter analyses application of advanced microscopy techniques for evaluation of 3D structures.

## 12.2 Dimensional Interconnected Circuits (3DICs) for System Applications

*John U. Knickerbocker*

### 12.2.1 Introduction

On-chip interconnection has scaled over time from tens of circuits to billions of circuits over the last six decades [1, 2]. An example of memory density scaling over time is shown in Figure 12.1 [3]. Unlike on-chip integration, off-chip interconnect has not scaled at the same rate or level over the same period of time. Over the last two decades, die stacking using a wirebond interconnect and package-on-package (POP) technology using peripheral connections have been reported. However, wirebond die stacking, SIP, SOP and POP technology is not the focus of this chapter since aspects of these technologies have previously been reported [1, 4, 5] and prior examples typically have limited system interconnect density, bandwidth, frequency, power efficiency and performance. This chapter will focus on emerging three-dimensional (3D) integrated circuits and interconnect using 3D die stacking of thinned silicon with through-silicon vias (TSVs) and 3D advanced packaging system solutions such as silicon packaging and glass packaging, which offer 3D, short vertical wire lengths and high interconnect density for product applications. Figure 12.2 shows a schematic cross-section of examples of 3D integration with TSVs [6].

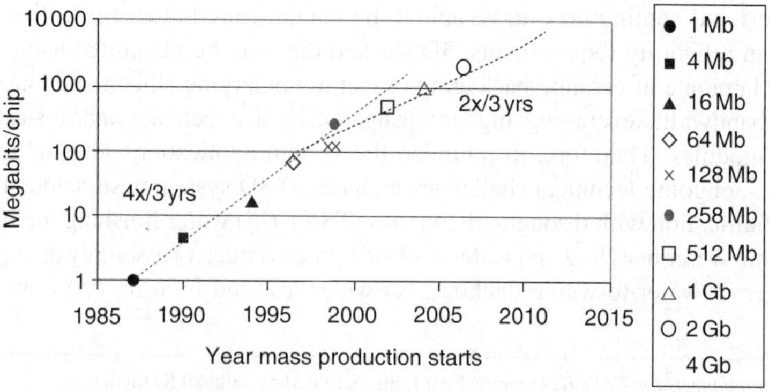


*Figure 12.1* *Evolution of memory density over time* [3]

Figure 12.3 shows a relative comparison of interconnection density comparing traditional packaging such as ceramic and laminate packaging using area array interconnects, higher density silicon and glass package interconnect and highest interconnection density within a stacked die for silicon circuit integration [7].

Universities, consortia and industry have led research and early demonstrations for this emerging technology with TSV, thinned silicon and high-density interconnects. Unlike prior off-die integration technologies, these new 3D structures offer an opportunity for superior electrical characteristics and high-density vertical interconnect between circuits on silicon die or strata levels by reducing interconnection distance and electrical parasitics. This new technology offers potential advantages compared to a traditional system on a chip (SOC), a system in a package (SIP) and peripheral wirebond die stacks and package-on-package (POP) technologies. Moreover, the short distance between circuits can permit silicon die or strata to be customized using homogeneous technology and thereby to simplify wafer processing and reduce wafer costs. For example, manufacture of custom 3D microprocessor wafers, memory wafers, I/O communication wafers, digital wafers, analog

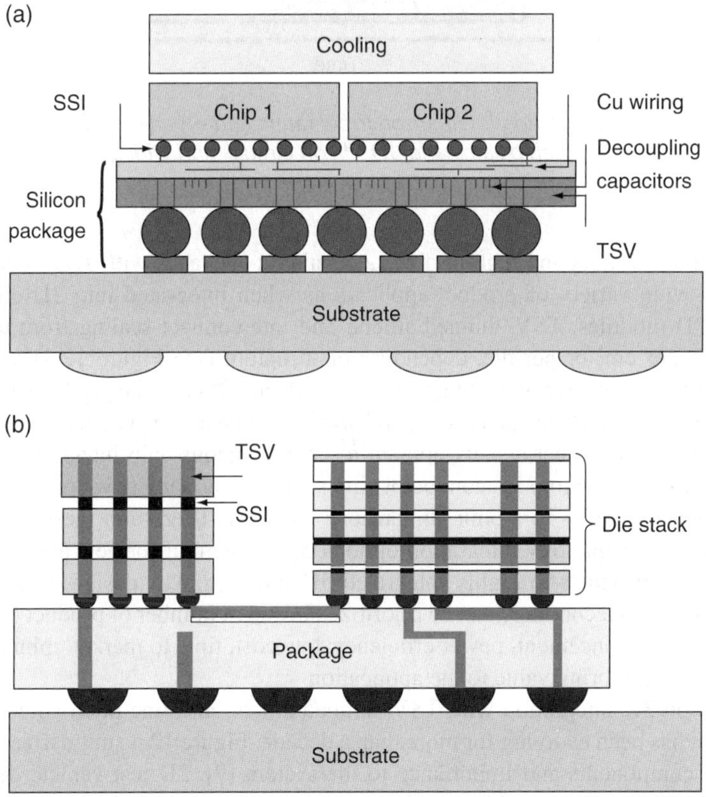


***Figure 12.2*** *Schematic cross-section for (a) 3D silicon package with through-silicon vias (TSVs) interconnecting two die and (b) multiple stacked die using TSVs and silicon–silicon interconnect (SSI) on a base package* [6]

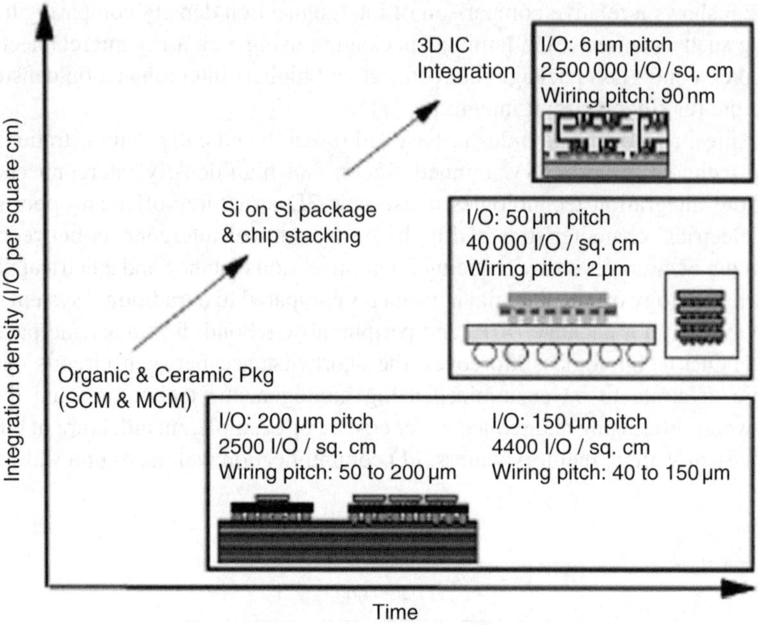


*Figure 12.3* *Silicon integration comparison for ceramic and organic packaging, silicon carrier or glass carrier and chip stack and 3D silicon circuits and wiring*

wafers, photonic wafers and high-performance silicon packages with TSVs could be used to permit a wide variety of product applications when integrated into 3D die stacks or integrated 3D modules. TSV, thinned silicon and interconnect scaling from under 10 to over $10^8$ I/O per $cm^2$ or per die, depending on structure (see Figure 12.3), compared to traditional off-chip integration scaling up to about $10^3$ $cm^2$ I/O fills the gap between 'on-chip' integration density and traditional 'off-chip' I/O interconnection. The wide range of TSVs and die-to-die interconnect density applied to heterogeneous chip integration comes at a time when Moore's law for semiconductor chip scaling is slowing down or reaching an end as the technology scales to atomic dimensions [1, 2, 8]. Therefore, these newly created high-density 3D technology integration options offer potential for new applications from miniature-sized products to highly integrated 3D products. The design, architecture and form factors for 3D technology can be prioritized toward a number of product benefits such as performance enhancement, power efficiency, low cost, time to market, miniature size or other attributes that bring value to the application.

Research on 3D integration with TSV, thinned silicon and fine-pitch silicon-to-silicon interconnect has been evolving for more than a decade. Figure 12.4 summarizes 3D system technology components and importance to the system [9]. 3D test vehicle designs have been followed by build, assembly and characterization studies to provide an understanding of design, architecture, structure and process integration capabilities and limitations. Results from these design, modeling and technology studies provide guidance on 3D performance, power efficiency, design rules, structures, processes, test and reliability, which

| System Technology Component | 3D System Impact / Comments |
|---|---|
| 1.3D Design & Design Library | Design Rules, Performance, Power Density, Thermal Issues, Cost |
| 2. Architecture | System Performance, Heterogeneous Integration |
| 3. 3D Design Tools & Modeling | Floorplanning, Layout, Routing, Interconnect, Power, Thermal & Electrical Models, ... |
| 4. Chip Technology | CMOS, Bulk, SOI, Technology Node, |
| 5. Package & Interconnect | Silicon, Glass, Laminate, Ceramic, Hybrid, Area Array, Peripheral |
| 6. Assembly & Bonding | Chip to Chip, Chip to Wafer, Wafer to Wafer, Solder, Metal, Oxide, Adhesive. |
| 7. Test | Wafer Test, BIST, Module Test, Burn-in, Known Good Die (KGD), KGD stacks, Cost |
| 8. Module Integration & Cooling | Module Integration, Board Interconnect, Power Delivery & Distribution, Thermal & Cooling |
| 9. System Reliability | Robust Materials, Structures, Processes and Modules, Self-repair, Sparing |

***Figure 12.4*** *3D technology integration components and system impact* [9]

can support a growing variety of product requirements. Data from these investigations provide guidance toward robust groundrules and processes, which can support manufacturing and product reliability requirements. Practical design, technology fabrication and interconnect approaches need to be considered for targeted TSV and assembly processes to achieve interconnect density, silicon thickness and power densities of the system application. Options such as TSV conductor material and interconnect material and an assembly approach between die-on-die, die-on-wafer and wafer-to-wafer processes should be taken into consideration. At the same time, one must also consider not only the specific new 3D technology features of TSV, thinned silicon and silicon-to-silicon interconnect but also a range of technology elements needs to be considered around the product requirements. A balance between application design objectives and a high yielding process, such as including feature redundancy for interconnections if needed, die size, manufacturing throughput, cost and test methodology, is also an important consideration for specific applications. In this chapter, we report on examples of test vehicles designs, fabrication and characterization.

### 12.2.2 System Needs

System design, silicon technology and packaging integration for system applications are reaching a time in the semiconductor industry where severe limitations from evolutionary trends are evident [8, 10, 11]. Frequency scaling for microprocessors has been limited by escalating power consumption and leakage current technology nodes below 65 nm. Semiconductor scaling trends are fast approaching atomic limits. As semiconductor manufacturing moves from 32 nm and 28 nm node to 22 nm, 15 nm, 11 nm and perhaps 8 nm dimensions, we might expect that Moore's law [2] and Dennard's semiconductor scaling rules [12] will be slowing down and reaching an end. As we approach semiconductor scaling limits, quantum mechanical properties take over and thus traditional semiconductor CMOS scaling based on silicon node advancements will reach limits and end. As we approach these semiconductor limits, use of new technologies or hybrid integration of technologies could be needed. Custom application accelerators, carbon nanotubes, graphene, use of quantum computing or alternate new technologies may be necessary for future systems [8].

System performance scaling challenges are also growing. Historical system trends have benefited greater than a 60 % system performance advantage due in part to processor chip performance improvement per generation to 65 nm technology. System scaling has also been limited in part by less than 15 % improvements in memory access time, less than 10 % growth in off-chip interconnection density and limitations in chip cooling [10, 13]. It is expected that direct system performance scaling from semiconductor processor node advancements can be < 20 % for each new semiconductor generation. Even with lithography scaling giving circuit density growth for each new technology node, challenges such as leakage power, latency in memory and interconnection bandwidth to memory could impact future systems performance [10, 13]. In order to achieve continued system performance scaling at better than 60 % per generation, the attributes of a system stack including hardware and software are needed. For example, increasing the numbers of software threads combined with an increasing number of cores per processor die is expected to help system performance improvements. In many systems, gate leakage limits power efficiency at higher frequencies but introduction of high-*k* metal gates has helped reduce leakage current. For advanced lithographic nodes, increased numbers of chip repeaters are needed due to smaller wire sizes and long-distance interconnections at these smaller feature sizes, but they can impact total power and power efficiency. As these multicore processors scale in systems usage, so too will they need to scale access to cache and memory with increased bandwidth.

Increased subsystem communications can help system scaling both through the use of 3D electrical interconnection with a shorter interconnect length between circuits and through the use of optical communications. In a 2D logic and memory die, the interconnection length for domains can range in length up to a few or even tens of millimeters. Use of 3D integration can offer domain interconnections at lengths of a few tens of micrometers, thereby providing 2 to 3 orders of magnitude reduction in wire length for reduced latency. 3D also offers the opportunity for significant interconnection density increases between silicon layers in a stack or with the use of silicon packaging, which can directly improve the interconnection bandwidth by 100 times to 10 000 times compared with traditional off-chip interconnection density or bandwidth [7, 10, 14]. Thus system challenges and semiconductor scaling limits need not limit system scaling, power reduction and miniaturization. For example, to compliment semiconductor advancements, subsystem scaling using three-dimensional, high-performance die stacks and 3D integration packages can benefit system advancements [8]. The subsystem scaling can include 3D integration with high-bandwidth electrical and optical interconnection between large numbers of multicore processors and memory [6, 8, 10, 13]. Appropriate architecture and software that can leverage 3D multicore, multithreaded structures will also be critical to optimize the benefit for system scaling for future systems. It is expected that a combination of semiconductor scaling, 3D subsystem scaling (use of 3D electrical and optical interconnection), use of

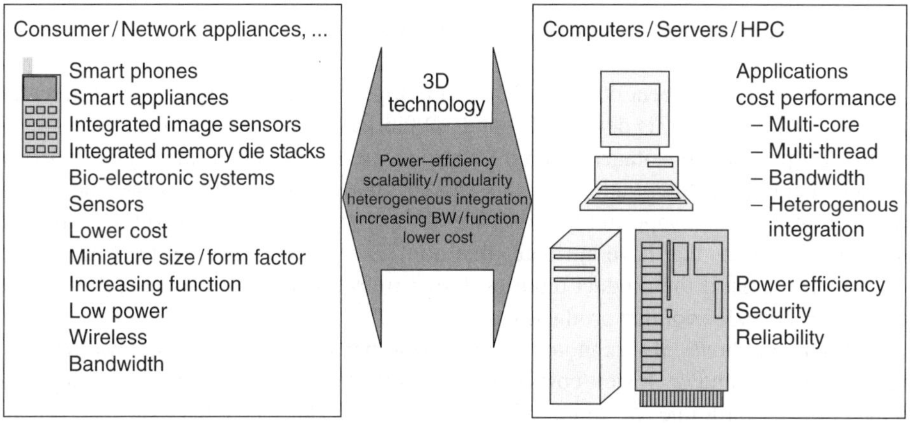


***Figure 12.5*** *3D technology offers system advantages for mobile applications needs and high-performance system needs* [12]

multicore, multithread processors, system accelerators and system software enhancements will be necessary to support system performance scaling at greater than 60 % per generation. In addition, traditional off-chip packaging interconnect input–output (I/O) connections have taken a circuit area that limited scaling to a finer pitch and have required >10 mW/Gbps power levels to connect to on-board circuits. System requirements are increasing demands for more I/O and at much better interconnect power efficiency. For mobile systems, increased I/O are required to support higher data rates such as streaming video for hand-held devices. For high-performance computing, increasing I/O, increasing bandwidth and more power efficient interconnect are needed to support power efficient computing. These requirements are becoming important for both mobile platforms and high-performance systems, respectively driven by 'on demand' access to the internet, limitations in mobile platform battery life versus weight and the rising cost of energy for high-performance computing systems operations.

Emerging three-dimensional (3D) technology can offer close proximity interconnect that can translate into system benefits including higher performance, miniaturization of system hardware, short interconnection lengths offering lower latency and high bandwidth, lower power interconnect, lower cost and other benefits. Figure 12.5 shows how 3D technology can support system needs for mobile and network applications systems as well as satisfy high-performance computing applications [15]. Portable electronics systems such as smart telephones, internet platforms, smart sensors and smart appliances need increasing function in small form factors, wireless connections with high bandwidth, low cost, manufacturing for large volumes and low power consumption for longer battery life to meet the demands of an ever-increasing function in hand-held systems. High-performance computing systems such as servers, cloud computing and supercomputing systems need high performance using multicore, multithread, high-bandwidth interconnects, high power efficiency and efficient cooling solutions, high security and high reliability to meet the demands of 100 % availability, trusted and cost-effective computing systems.

### 12.2.3 3D Interconnect Design and Architecture

The greatest system benefits for 3D technology require design and architecture optimization leveraging 3D design rules and 3D fabrication advancements in order to gain the greatest product advantages when it is introduced. Development of a 3D competitive product requires differentiation in the market place, competitive cost and timely introduction. A great opportunity exists as 3D technology advancements are made to create new products that have attributes that non-3D technology cannot provide. A great risk may also exist if the product requires design rules that do not achieve high yield and competitive costs for volume production.

As discussed above, 3D technology can provide performance benefits, power savings, high bandwidth and/or lower costs where system optimization requires design and architecture toward one or more of the targeted product benefits. Against product requirements or objectives, a comparison of 2D and 3D potential design and architecture options should compare relative advantages and disadvantages of structures that might be considered. Advantages of 3D technology and structures may include shorter wire lengths, higher bandwidth interconnections, low latency, lower power, smaller size and potentially higher performance and lower costs. Depending upon application and the integration structure, disadvantages of 3D structures may include increased power density, which could lead to challenges in power delivery including voltage and current distribution as well as challenges in cooling the 3D structure. Depending on the complexity of a 3D product, challenges may also exist in design tools and manufacturing, including wafer fabrication, wafer thinning, wafer handling and finishing, assembly, test and module integration. Some products may not be best suited to utilize 3D technology and others may have advantages using 3D structures that cannot be achieved with 2D technology.

A 3D integrated circuit (3DIC) design using TSVs can consist of 3DIC fabrication using thinned die or thinned bonded wafers where TSVs may be in peripheral interconnect IC, area-array interconnect IC or alternative IC patterned layouts [1]. A 3D packaging design may consist of system-on-package (SOP), system-in-package (SIP), such as wire-bonded, peripheral interconnects packaging, area-array interconnect packaging, package-on-package (POP) or alternative package layouts [1, 4, 5, 7]. The focus of this chapter is toward 3D integration using vertical interconnect TSVs for integrated circuits such as stacked die in systems and short-length (non-wirebond) vertical interconnections in 3D packaging systems.

*Physical Design*

Efficient floor planning of 3DICs and/or 3D packaged systems may include single- or multistep approaches and single- or multiobjective approaches [15–18]. The objective in 2D floor planning is typically to minimize the total area of the 2D integrated circuit blocks and packaged systems, which can lead to smaller size, shorter wire lengths and lower costs. For 3DIC and 3D packaged systems, circuit blocks may be in the same plane or on another plane. A design with many circuit blocks must then consider a floor plan to interconnect these circuit blocks to achieve small size, short wire length and timing while optimizing placement of these blocks for performance, power and minimum complexity, which impacts cost. The 3D floor plan and trade-offs become more complex for interconnects, timing, wiring and, furthermore, must consider power density for power delivery/distribution and thermal considerations/cooling. 3D layout tools and routing must take into consideration the 3D hierarchy to account for circuits at multiple levels such as 1, 2 and 3 and vertical interconnections between these circuits such as interconnect levels 1 and 2, which may consist of TSVs and an interconnect structure. Thus design tools and design need to accommodate traditional routing techniques along with hierarchy for the third dimension. Furthermore, these structures should take into consideration design for manufacturing and an appropriate test methodology. Design and test must consider either appropriate redundancy, test level repair and/or a built-in self-test to permit known good die (KGD) and/or known good wafers to lead to known good 3D ICs and 3D packaged systems.

*Timing*

Two-terminal and multiterminal interconnect timing across multiple planes or levels of circuits may result in different impedances [1]. The timing is dependent on the TSV and interconnect manufacturing processes (such as using one or more manufacturers) and is dependent upon the 3D structure, such as face-to-face circuits, face-to-back circuits, back-to-back circuits or some interlayer combination. Optimization of timing usually includes optimization of the longest wire interconnections and from modeling the electrical characteristics of all the interconnect segments. Minimizing the interconnect delay requires optimization of wire characteristics as well as placement of vias, and interlayer TSVs and interconnects.

*Clocking*

Symmetrical H-tree and X-tree structures for 2D circuit clocks can permit a clock signal to simultaneously arrive at the multiple branched circuits to provide synchronous data processing [19, 20]. The intralayer and interlayer interconnects associated with 3D clock structures make the challenge of clock distribution to multiple levels within the 3DICs inherently more difficult. Clock distribution can be distributed to a global network and to multiple smaller local networks. Especially for longer distances and higher frequencies, load imbalances, process variations and cross-talk can impact the arrival time of the clock at various locations; this variation is called clock skew. Modeling and measurements of the clock distribution as segments of a transmission line can aid in understanding reflections at branch points and the speed and power consumed by the clock network. 3D clock structures and test circuits compared H-tree structures, local mesh topologies and global rings for 3D structures and showed that the local mesh structures consumed the least power for the structures studied [1].

*3D Circuit Architectures*

3D heterogeneous integration offers great opportunity for enhanced performance and novel systems. Examples such as integration of FPGA with memory, processors with memory, GaAs or SiGe analog circuits with digital circuits, and sensors with microprocessors are but a small list in a growing variety of heterogeneous systems opportunities from 3D technology. Design and optimization of the architecture leveraging the third dimension of integration that vertical TSV's and interconnect between layers of circuits offers, can lead to orders of magnitude of increased interconnects and much shorter connection lengths. These attributes of 3D design and architecture then lend themselves to minimize latency and power consumption for communication channels. These advantages can be leveraged to improve system performance and power efficiency. Proper partitioning of cache memory in the 3D structure and ability to provide multiple layers of memory in close proximity to others layers of circuits such as processors can provide reduced time to access memory. In addition, use of low power memory layers adjacent to processors can increase total memory size and bandwidth while staying within an acceptable thermal budget for a 3D structure. Similarly, FPGA to memory, sensor to processor and other 3D heterogeneous architectures can profoundly lead to revolutionary new systems [10].

*3D Industry Standards*

Over the last decade, design and technology advances in 3D chip integration have been making progress at universities, consortia and in industries from around the world [1, 6, 7, 9, 10, 21–36]. Initial product applications using TSVs and 3D integration have begun to slip since about 2008 including image sensors and power amplifiers for wireless applications [37, 38]. These first applications have had limited numbers of vertical interconnections and have not required design standards and industry compatibility. These initial products have also not required 3D engineering design automation (EDA) tooling, which will emerge as more complex 3D designs are needed for product applications. As complexity increases and integration of heterogeneous technologies begins to emerge, the need for 3D compatibility between multiple wafer and assembly manufacturers will also become more important. It is expected that over the next decade the next generations of complex 3D product applications, industry design and compatibility standards will begin to emerge. Recently, SEMATECH, the Semiconductor Industry Association (SIA) and SRC have announced formation of a 3D Technology Enablement Center to establish 3D industry standards and enable growth using 3D technology by working with SEMI, Semiconductor Research Corporation (SRC) and industry participants [39]. 3D designs and products are beginning to be announced, such as use of a silicon package with TSVs to integrate multiple die such as that from Xilinx [40] using TSMC 3D technology and from Semtech [41] using IBM 3D technology. In addition, wide I/O memory using stacked memory die and a logic controller die are nearing industry readiness, such as that from Elpida [27], Samsung [42] and Micron [43].

### 12.2.4 3D Fabrication and Interconnect Technology

*3D Technology Advancements*

Over the last decade, publications have described research including approaches for wafer fabrication with TSVs and interconnect assembly to form stacked die, bonded wafers and/or integrated 3D modules with silicon packages or high-bandwidth vertical interconnected packages. From the late 1990s to the present, 3D technical publications and research progress from organizations and companies including ASET Consortia of Japan, Fraunhofer Institute of Germany, Massachusetts Institute of Technology (MIT), CEA-Leti, imec, Sematech, Elpida, Samsung, IBM, Amkor and many others [22–36]. Research investigations

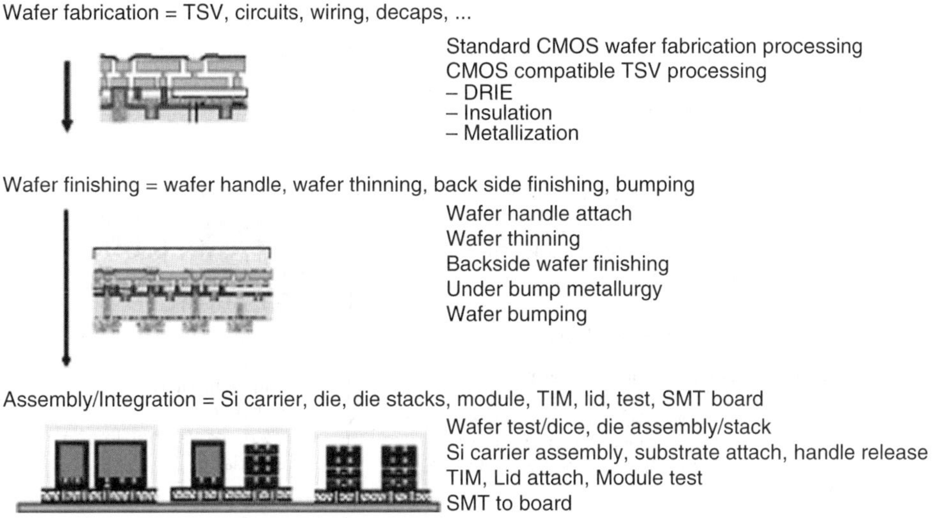


***Figure 12.6*** *An example of a 3D process flow including wafer fabrication with TSVs, wafer finishing and assembly* [14]

have examined a variety of materials, structures and wafer fabrication processes with TSVs, 3D die stacking, wafer bonding and 3D packaging/module integration approaches. Researchers have recognized the importance of developing fine-pitch vertical interconnections using through-silicon vias, developing wafer handling and thinning technology for silicon wafers and interconnection processes for joining thinned silicon die into die stacks, die to silicon or glass packages and developing wafer-to-wafer bonding technologies. Figure 12.6 shows an example of a high-level 3D process flow including wafer fabrication with TSVs, wafer finishing and a 3D die stack assembly [14]. In addition, 3D process and assembly advancements and the increased power density for some 3D applications are driving advancements in materials and structures capable of supporting higher current flow and power delivery per unit area. Advancements in vertical interconnection density demonstrations also support increased signal interconnections for increased bandwidth. Researchers continue to study the best approaches for thermal cooling and modeling of heat removal from thinned silicon and 3D stacked structures with TSVs. Each 3D application will have its own integration challenges. Common requirements for successful 3D product applications include design, fabrication, assembly and timely introduction of competitive products. Figure 12.7 shows the 3D challenges and readiness of the technology that can apply to 3D applications [44, 45]. Advances in 3D technology have begun with 3D products using low TSV interconnection density and pitch. A continuing extension of understanding for design, architecture, wafer processing, wafer finishing, assembly, test and module integration as well as standards to ease the use of die and components from multiple suppliers will also need to compliment 3D technology advancements for more complex products.

Many technical publications have reported on a wide variety of TSV processes and sizes [7, 9, 21–23, 27, 30–32, 46, 47]. TSV diameters, TSV pitch and silicon thickness for

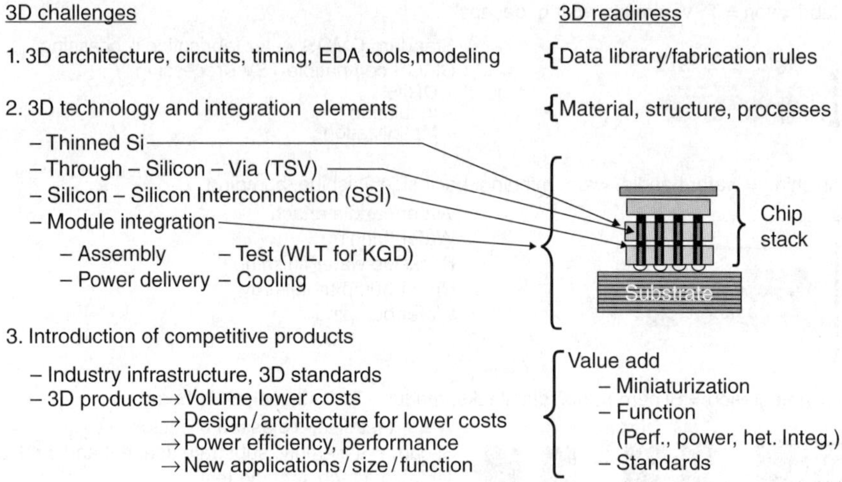


***Figure 12.7*** *3D-technology challenges and readiness including 3D design, technology and introduction of competitive products* [44, 45]

vertical connections have ranged from large sizes to small sizes. Large TSVs have included about 80–100 μm via diameter, over 200 μm pitch and silicon thickness of over 100 μm to 300 μm, and have been targeted for image sensors, power amplifiers, analog devices and advanced packaging technologies. Small TSV sizes include via diameters from under 20 μm to less than 0.5 μm, via pitch of 50 μm to under 3 μm and silicon thickness of 60 μm to under 1 μm and have been targeted for advanced CMOS semiconductor applications, memory die and die stacks, heterogeneous product applications and small form factor mobile electronics. TSV conductors reported include tungsten, copper, nickel, composite, paste, doped polysilicon as well as other electrical conductors. Takahaski *et al.* investigated 10 μm copper conductors utilizing through-silicon vias for electrical interconnection at 20 μm (Figure 12.8) [22]. Another report by Topol *et al.* [48] and one by Guarini *et al.* [49] were on TSVs or vias under 0.2 μm diameter, pitch under 6.7 μm and TSV or via height of under 1 μm to 2 μm. In these silicon-on-insulator (SOI) vertical interconnect examples, the vertical copper via does not need to go through silicon because the bulk silicon is removed. Joseph *et al.* [50] and Stamper *et al.* [51] reported on the use of about 100 μm height TSVs with a high aspect ratio (see Figure 12.9). The high aspect ratio tungsten TSVs supported next-generation SiGe power amplifiers for wireless communications where noninsulated tungsten-filled vertical connections provided a low inductance path to ground to enhance product performance and insulated tungsten TSVs for some stacked die applications.

Numerous technical publications have reported high-density silicon interconnect technologies for die stacking, silicon assembly to silicon packages or thinned silicon assembly including use of anisotropic conductive adhesives, lead containing solder, lead-free solder interconnect, Au–Sn interconnect, thin intermetallic composition interconnect, Cu–Sn to Cu

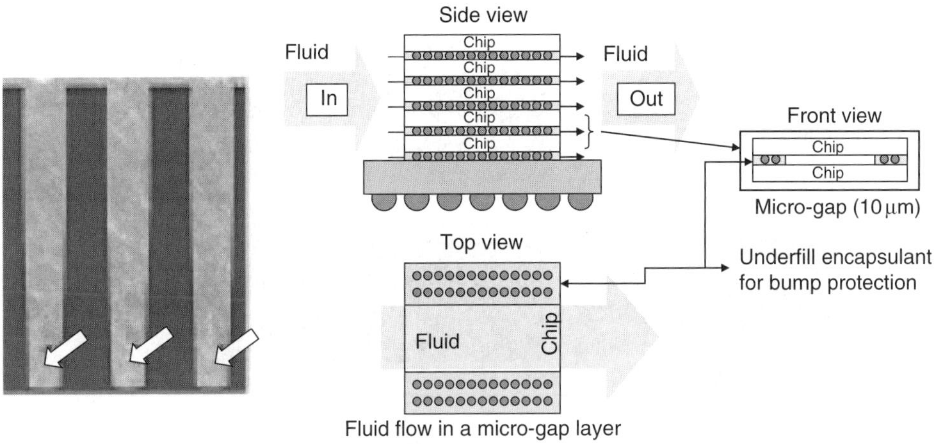


***Figure 12.8*** *Takahaski et al. investigated 10 μm copper conductors utilizing through-silicon vias for electrical interconnection at 20 μm and die stack integration with fluid flow for cooling* [22]

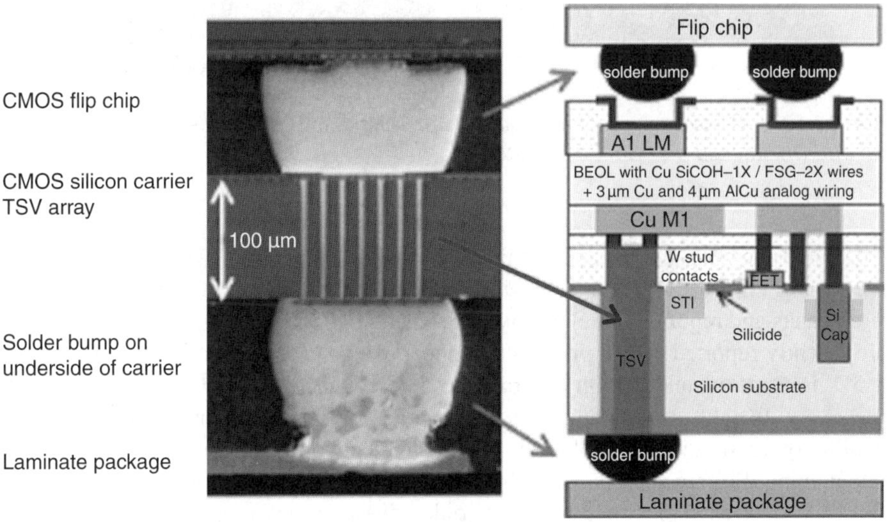


***Figure 12.9*** *3D tungsten TSV cross-section and schematic* [51]

joins and Cu–Cu bonding [7, 24, 25, 29, 52–56]. Each of this interconnects drives a variety of structures and processes to achieve vertical interconnects in combination with TSVs that permit a short-length electrical connection between circuits of adjacent silicon layers. Product applications may drive specific electrical requirements such as a current-carrying capability, capacitance, resistance, inductance, signal data rate, cross-talk, reliability and cost, depending on the targeted application requirements. Feil *et al.* have shown bonding

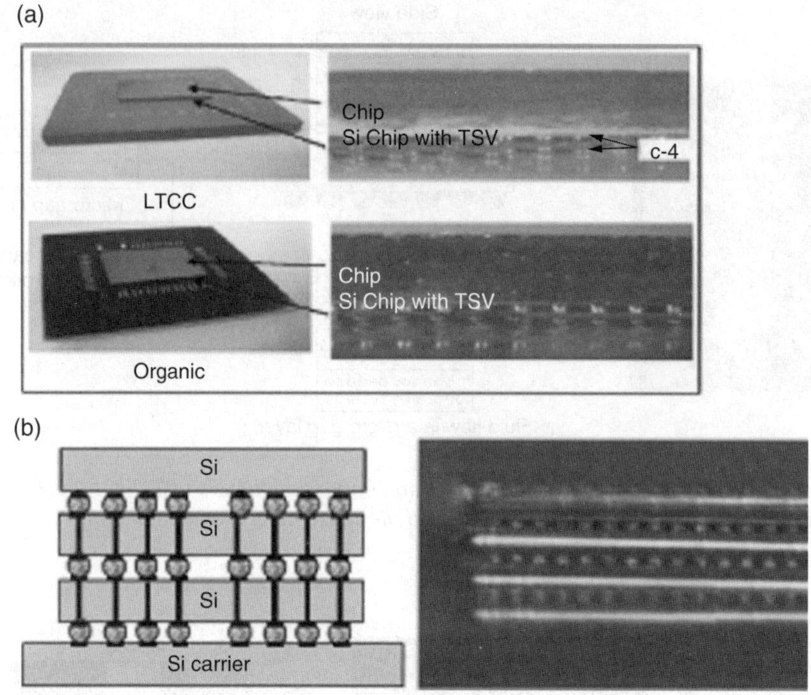


*Figure 12.10* *3D solder interconnection examples including: (a) stacked die with 200 μm pitch and (b) stacked die with 50 μm pitch*

with anisotropic adhesive for 25 μm thin silicon where the bonding requires a continuous force during curing [24]. Hunter *et al.* have assembled silicon 50 μm diameter Au/Sn interconnects and reported on reliability [25]. Reed *et al.* [52] and Agarwal *et al.* [53] have more recently reported 10 μm pitch interconnect using Cu/Sn to Cu or Cu to Cu bonding for 3D integration and Cu/Sn microbump interconnect for 3D TSV chip stacking, respectively. Wright *et al.*, Dang *et al.* and Maria *et al.* have reported on assembly and reliability of 25 μm diameter solder microinterconnect compositions with PbSn, and high Sn-containing, Pb-free solders for high-density silicon integration [54–56]. Figures 12.10 and 12.11 show examples of solder and solder fine-pitch interconnection used for a die stack and 3D multichip package and 3D fine-pitch multichip integration to support high /I/O, wide-bandwidth silicon circuit integration, respectively. These microbump controlled collapse chip connection (uC-4) papers reported on the use of high-density interconnection for silicon package interconnection, reworkability of uC-4, die stack assembly of multiple, thinned silicon layers and reliability of these solder interconnects. Technical challenges for successful fine-pitch interconnect include interconnect alignment, coplanarity of the interconnection, contamination control, assembly yield, adhesive underfill for fine-pitch interconnection or no-flow adhesives, current capacity of the interconnection to avoid electromigration and thermal migration, consistent electrical characteristics of resistance, inductance, capacitance and high thermal conductivity. Ultimately as TSV and

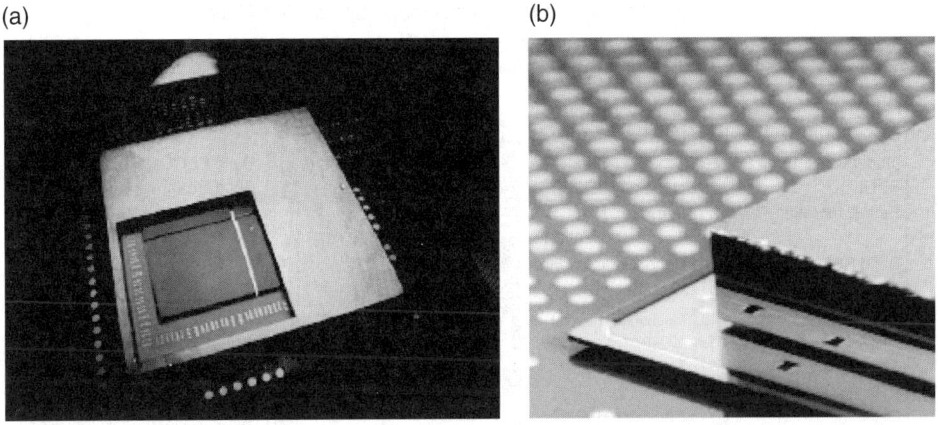

*Figure 12.11* 3D solder interconnection examples including: (a) an integrated silicon 3D package with multiple die with 200 µm pitch and (b) 3D silicon interconnect package with fine-pitch interconnection

interconnections are adopted for products, optimization of the 3D structure, materials and processes will lead to robust solutions that meet product and reliability specifications. As 3D technology adoption leads to higher volume product shipments, more significant statistical data on manufacturing, yield and product reliability will be available. Relative costs for TSV, wafer thinning and 3D silicon interconnection will be reduced through manufacturing continuous yield improvements and productivity improvements in future generations of tooling and associated processes for growing volume production.

This chapter section will provide more information and discussion associated with 3D technologies.

### *Wafer Fabrication with Through-Silicon Vias (TSVs)*

Applications may need only a few, tens, hundreds, thousands or millions of vertical interconnections that are product dependent. Product design, architecture and specifications will determine the appropriate 3D technology, materials, structures and processes needed to manufacture the hardware. As discussed above in Sections 12.2.3 and 12.2.4, use of 3D technology should provide a competitive benefit for the product. Examples of TSV quantities, sizes, structures, fabrication processes, wafer finishing processes and subsequent integration will impact wafer fabrication process choice with TSVs. For example, the range in TSV size includes diameter or *X*–*Y* sizes from <0.5 µm to about 100 µm. The silicon thickness includes SOI or silicon on the thin side of < 1 µm to 10 µm and up to a full wafer thickness of 730 µm to 785 µm for the thickest silicon TSV (although most reports have been at 300 µm or less and more recently at 100 µm or less). TSV height to diameter aspect ratios reported included evaluations from a low aspect ratio of about 2 to 1 to values of over 20 to 1 for copper fill and up to about 40 to 1 for tungsten TSV conductors, respectively. These structures, sizes and aspect ratios impact the physical process as well as resulting electrical, mechanical, thermal and reliability characteristics of the TSV.

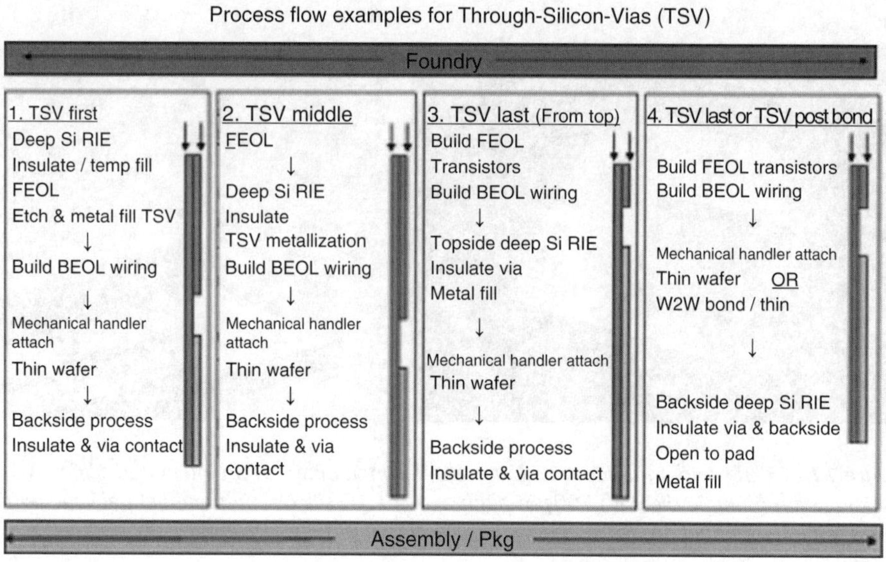


***Figure 12.12*** *3D TSV comparison processes including TSV first, TSV middle and TSV last processes* [6]

### *TSV Process Sequence*

A comparison of process sequences for TSV formation resulting in vertical electrical connections through thinned silicon structures is shown in Figure 12.12 [6]. In this example, the TSV process sequence is compared for TSV first, TSV middle and TSV last processes. In a TSV process, typically an isotropically etched hole is formed in silicon, followed by deposition of a dielectric layer for electrical insulation of the via from the silicon, followed by metallization, which includes a liner deposition, seed and metal fill operations. Figure 12.13 shows an example of via schematics and photographs for TSV formation [31]. When initiated prior to circuit fabrication, the TSV process is often referred to as the TSV first process and when TSV formation follows circuit fabrication but prior to wafer finishing steps, the TSV process is often called a TSV middle process. TSV last processes can be employed from the top side of a wafer, from the wafer backside to a landing pad on the circuit side following wafer thinning or may also refer to creation of a vertical interconnect after bonding two wafers and thinning one wafer, as previously discussed [31, 32, 35]. Next steps for vias first or middle processing would include circuits processing and back-end-of-line (BEOL) or just BEOL processes to complete the top side of a wafer, attaching a mechanical wafer to the top of the wafer using a glass or silicon wafer to provide mechanical support while the active or passive circuit wafer is thinned from a standard wafer thickness of about 730 μm (200 mm) or 785 μm (300 mm) to the desired product thickness. The mechanical wafer can be attached by means of a polymer adhesive, which can later be removed by means of chemical etch, heat, laser or an alternative method. Adhesive properties can limit subsequent process steps depending on the adhesive material and mechanical handle wafer used The wafer can then be thinned by mechanical grinding in one or more

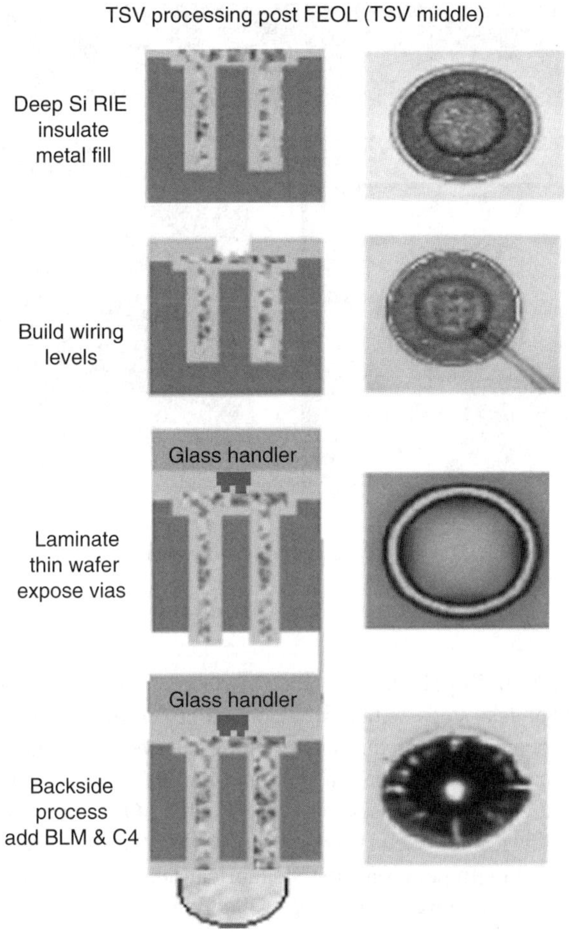


***Figure 12.13*** *An example of via schematics by process step and photographs for TSV formation* [31]

process steps, polishing or chemical–mechanical polishing, chemical etching (wet or dry) or a combination of mechanical and etch methods. Next the backside metallization and redistribution wiring processes for backside electrical insulation around the TSV and electrical interconnection to the TSV would be completed followed by the creation of an interconnect metallurgy for subsequent assembly. Several wafer finishing and thinning processes have been reported using mechanical attach wafers with bonding and debonding processes [57–60].

There are many combinations of TSV formation using wafer processes that may be possible. Fabrication of active circuit wafers with vertical interconnections from TSV first, TSV middle and TSV last sequence steps that can also lead to robust structures are all possible, but in all cases care must be given to avoid damage to active circuits based on design rules, TSV materials and structures and fabrication process temperatures employed

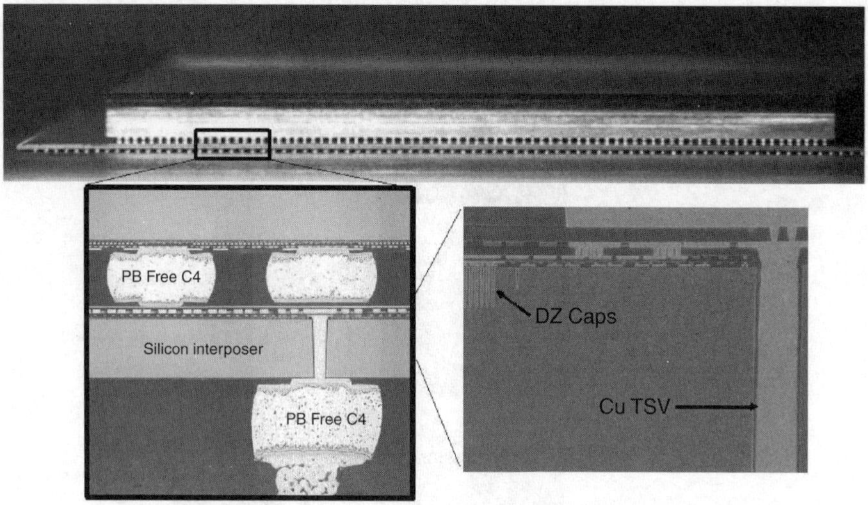


***Figure 12.14*** *3D Sematech press release photograph with cross-section of copper TSV and silicon interposer from IBM* [41]

during wafer fabrication. Simple passive silicon packages can utilize TSV processes such as TSV first or TSV last steps with fewer constraints compared to wafers with active circuits. Additional perspectives on through-silicon vias have been reported [31, 32, 35, 51]. Examples of TSV cross-sections for tungsten and copper vertical conductors are shown in Figure 12.9, Figure 12.14 and elsewhere [32, 35] for SOI bonded wafers. A dielectric insulator, adhesion layer, liner, seed layer and conductor fill and excess conductor removal steps may also be used in preparation for TSV fabrication. A variety of TSV shapes can be considered in order to be compatible with polysilicon, tungsten, copper or composite TSV conductors. Examples include annular TSV, bar-shaped TSV or high aspect ratio TSV, as previously reported [22, 23, 31, 32, 35, 50, 51, 61]. Shape and size of the TSV can be determined by a combination of considerations, such as the cross-sectional area required to meet electrical resistance specifications, current carrying specifications, and taking into consideration the manufacturing yield and thermomechanical stress of TSVs based on shape, conductor and dielectric. For passive silicon packages and thick silicon, low vertical interconnection needs, larger area TSVs may be acceptable; however, for an active circuit die especially requiring high numbers of vertical interconnect, a small cross-sectional area for TSVs is desired. For wafer processing, a thick silicon such as one from 150 μm to 300 μm, many BEOL and wafer handling or dicing process steps, a free-standing silicon wafer can be processed. In some instances, a thick silicon along the wafer perimeter with a thinner silicon thickness across the rest of the wafer may be acceptable. In contrast, a bonded wafer pair consisting of a glass or silicon handle wafer adhesively attached to the wafer to be thinned is typically required to support the active wafer for backside processing. The thin wafer can be released from the mechanical wafer by several different options when thinner silicon structures (such as less than 100 μm thickness) are desired [57–60]. The thinned silicon die or die stack can later be bonded to other die in a die stack, to a package or to other module components for mechanical integrity after releasing from the mechanical support wafer. In the case of wafer-to-wafer permanent bonding, two wafers can be bonded such as with adhesive, oxide or metal bonding and then silicon from one can be removed or thinned while the other wafer provides mechanical support for processing and in the product application. This process can be repeated for multiple high circuit layers [32, 35].

*TSV Etch Process*

TSV structures can be fabricated by a number of methods of silicon removal such as wet processing using a chemical etch or dry etching such as laser drilling, reactive ion etch or deep reactive ion etch (DRIE). Combinations of these etch processes are also possible. Etching often begins with photolithography and use of a photoresist or a 'hardmask', such as patterned oxide or a nitride layer, to define the features to be etched.

Wet etching of silicon can use a mixture of nitric acid ($HNO_3$) and hydrofluoric acid (HF) diluted with water or acetic acid performed at room temperature for isotropic etching. Anisotropic wet etching can be achieved using diluted potassium hydroxide (KOH) or other etch chemistries such as tetramethylammonium hydroxide (TMAH), where in silicon preferential etching in the (110) and (100) crystal planes is compared to the (111) orientation. These etches can lead to many different etch morphologies depending on wafer bulk crystalline orientation, the mask and wafer orientation. However, for many 3D applications isotropic vertical etched shapes are desired and thus wet etching may provide shape adjustment in combination with a dry etch.

Dry etching using a laser can be used to create high aspect ratio holes. Alternatively, silicon can be etched using reactive ion etch (RIE), which uses a plasma etch consisting of source gases broken down in a plasma to create highly reactive electrons, photons, neutral species and positive ions that impinge with surface atoms, both removing them and forming stable compounds. The chemical and physical reactions etch the surface and remaining by-products can be removed, causing the plasma to continue to react. RIE has been used for semiconductor processing and can create high aspect ratio structures or TSV but can be slow. An enhanced process called time multiplexed deep etching or just deep reactive ion etching (DRIE) was patented by Bosch (US Patent 5501892), which combines cycles of deep silicon plasma etching with passivation sequentially to etch very high aspect ratios at high etch rates on the order of up to 10 μm per minute. The sequential operation with plasma radicals such as $CF_x$ from octobluorocyclobutane ($C_4F_8$) followed by passivation can also lead to a scallop surface in the TSV.

*TSV Dielectric*

TSVs may be fabricated with or without dielectric insulation from the silicon. For simple structures with a desired short to the base silicon no dielectric is required. For application requiring electrical isolation from the silicon, a polymer dielectric, oxide, nitride or composite dielectric may be used.

Polymer dielectrics can be processed by means of a spray, spin coat or alternate deposition method where temperature limitations may be required, such as less than 200 °C or less than 300 °C. Robust fill and uniform isolation can be dependent on the TSV diameter and aspect ratio and processing conditions. Following deposition, polymers are cured. For via last, post-FEOL and BEOL wafer finishing for TSV, polymer dielectrics may be appropriate. Similarly, for a larger size TSV, a polymer dielectric may be a low-cost method to isolate the conductor from the silicon.

Oxide and/or nitride depositions can be completed such as thermal oxide processing at temperatures of about 900–1100 °C, whereas plasma-assisted deposition of oxide or nitrides may be completed at temperatures down to about 200–250 °C. Other deposition techniques may also be employed and again TSV diameter, size and aspect ratio can impact the uniformity of dielectric isolation. A thick dielectric with a hard oxide or nitride, thick soft polymer dielectric or combined composite dielectric may be employed to isolate the conductor from silicon and thereby minimize TSV capacitance.

*TSV Conductor*

Deposition of a conductor begins with deposition of a thin conductive liner and seed layer. Similar to BEOL processing of copper lines, TSV can use a physical vapor deposition (PVD) of tantalum nitride tantalum (TaNTa) and copper (Cu) followed by a full metal deposition. The liner–seed deposition should be uniform and thus deposition quality can be dependent on TSV shape and aspect ratio. A top opening that is smaller than the corresponding vertically etched hole should be avoided as this can lead to top of the via fill completion prior to complete fill of the entire TSV, thus leaving a central void or defect in the TSV. Thus a vertically etched wall or one with a minimal taper is desired. Defects in the liner–seed layer can also lead to defects such as voids along the sidewall when filling a TSV and thus surface defects or disruptive morphologies should be avoided. Other liner–seed deposition materials and/or methods may be used to provide the same function of a continuous electrically conducting layer inside the TSV surfaces with uniformity and proper morphology.

Examples of TSV conductor fill materials include polysilicon, copper, tungsten and composite materials. Doped polysilicon used in TSVs can have the advantage of a coefficient of thermal expansion (CTE) match to silicon, compatibility with CMOS processing and simplified process sequences. However, doped polysilicon has the disadvantage of orders of magnitude higher electrical resistance and current-carrying limitations. Multihigh chip stacks, use of thick silicon layers or applications with high power density may have product specifications where polysilicon may not be an acceptable conductor of choice. For lower TSV electrical resistance and higher TSV current-carrying applications, a metal conductor may be a better choice. Conductors such as copper, tungsten or a metal-composite filled conductor may be required. Copper has a desirable low electrical resistance and a high current per unit cross-section. The disadvantages of copper include a high CTE of 18 ppm versus silicon of 3 ppm, which can give rise to high thermomechanical stresses during processing or in product use depending on size and shape of the TSV. Thus use of a copper TSV may be best for smaller diameters and limited length or may require a structure compatible with robust processing and application use. In addition, use of copper conductors must be compatible with active circuit wafer processes. Tungsten can also be considered as a TSV conductor and offers the advantages of a CTE of about 4.5 ppm, which is closer to silicon and can provide lower thermal–mechanical stress compared to copper, but has a disadvantage in electrical resistance. Tungsten electrical resistance is three times higher than copper but far better than doped polysilicon and therefore may satisfy product specifications for many applications. Another benefit for tungsten is that it can be used to fill high aspect ratio vias, even in excess of 40 to 1. Metal composites may also be used for TSV conductors but their properties are dependent on the choice of conductor, matrix fill material and the composite microstructure. Silver-filled pastes, copper annular structures with central filled composites and other TSV structures can also serve some application requirements. A wide variety of TSV electrical parametrics is possible depending on the dielectric, thickness, conductor, size, shape and conductor material of the vertical interconnect. For the TSV conductors, the product TSV specifications are critical in determination of the best conductor, structure and process.

Other TSV conductor considerations include cost, compatibility with the desired semiconductor wafer and package. For each TSV conductor, wafer and 3D process compatibility should be established. Examples include processing the TSV prior to FEOL, post-FEOL or post-FEOL and BEOL from the top or bottom wafer surface. Another question may include other limitations, such as temperature limitations based on the application requirements and reliability, mechanical handler process and compatibility with semiconductor wafer fabrication processes. In addition, subsequent die stacking or wafer bonding process operations, test and module integration compatibility must be acceptable. For example, thermal limitations may preclude the use of some conductors that require high-temperature deposition or annealing steps. Ultimately the TSV material choice is dependent on the resulting microstructure, cost and properties against the application TSV specifications or requirements. In each application, consideration for cost and specifications such as electrical, mechanical, thermal and reliability are necessary.

Silicon test structures have been used to establish robust TSV materials, process, sizes and CMOS- and SiGe-compatible design groundrules. The TSV processes must achieve high yield with desired electrical parametrics and high reliability. Many demonstration test structures were characterized and the best structures established for various TSV objectives. Mechanical modeling and simulations compliment experimental measurements. 3D stress models were able to identify high stress regions in a TSV structure and permit modifications when necessary to improve TSV structures for process compatibility and reliability. Results have shown up to 100 % yields in wafer tests, elimination of defects and establishment of desired groundrules that satisfy compatibility with CMOS or SiGe wafer processing and that meet reliability objectives [31, 41, 50, 51].

### *Back-End-of-Line (BEOL) and Wiring*

For 3D structures and wiring, standard back-end-of-line (BEOL) processes include dielectric insulation, photolithography, metal deposition, chemical–mechanical polish and dielectric deposition. Dielectric insulation layers such as plasma-enhanced chemical vapor deposition (PECVD) or a polymer-based dielectric can be deposited on the wafer surface by a spin-on process and bake. Next photolithographic processes include spin-on photoresists deposited at a given thickness on a wafer. The film is baked to remove solvents and then exposed to ultraviolet light. Post-exposure the film is developed and baked to create a stable film ready for subsequent processing, such as etch of a dielectric film.

Similar to the TSV process described above, BEOL metal wiring and vias are formed using PVD liner–seed deposition and copper electroplating. Electroplating of copper has been used for more than a decade in semiconductor processing, which involves use of an electrolyte against the wafer and use of a voltage potential applied between the seed layer (cathode) and an anode. The voltage potential drives a current in the anode or copper ions contained in a chemical bath to be reduced to copper metal on the surface of the cathode or seed layer. A copper plating bath can have many components including a copper electrolyte such as copper sulfate and sulfuric acid as well as chloride ions. Other additives in the plating bath can be accelerators, suppressors and levelers, which can aid in the deposition rate and quality of the copper film being deposited. Copper from copper anodes can be dissolved into the chemical bath replenishing the ions driven to the cathode seed layer. The copper surface deposition fills the etched photoresist, which later becomes the vias and wires on a wafer following copper removal from the surface of the photoresist, such as by chemical–mechanical planarization (CMP). The quality of the copper deposition is also dependent on the current flow, uniformity, tool and the waveform, such as a pulsed waveform during deposition, which can accelerate deposition rates.

Chemical–mechanical planarization (CMP) can be used to remove the over-plated copper. The copper damascene wiring levels use this CMP process, which has a combination of mechanical abrasion of the surface copper with a mechanical polishing pad and abrasive colloidal slurry, thus causing a combined mechanical material removal and a chemical reaction to remove excess copper. The polishing action of the slurry and polishing pad aids in material removal and creates a more uniform flat surface. During this operation the wafer is held with a backing film and pressure is applied to remove the excess copper uniformly to create a more planar surface. Single damascene processing can create a via or wire layer whereas a via and wire structure can be created using a dual-damascene process.

*Signal Integrity and Electrical Characterization*

For 3D structures, electrical characteristics of TSV, wiring, via and interconnect structures need to be measured and correlated with electrical models and simulation tools. For example, measurements from test vehicles to establish the *X–Y* wire width, thicknesses and spaces, vertical TSVs dimensions, via dimensions and silicon interconnect dimensions help to establish understanding for vertical buses through one or multiple layers of thin stacked silicon. Examples of test vehicle results have been reported [7, 28, 50, 51, 61]. For example, signal integrity has been characterized using frequency and time domain measurements [28, 45, 50, 51, 62–64]. Results were measured for signal transmissions from 2.5 mm to 75 mm line length and for frequencies between 1 and 10 GHz. Further results explored low-power I/O drivers and signal integrity using line lengths from 2 mm to 60 mm. In some test structures, TSVs had 0.2 mΩ of resistance per μm of silicon thickness and 0.15 pH inductance per μm of silicon thickness. Signal integrity measurements for horizontal silicon packages included wire sizes of 0.8 μm to 3 μm width with line lengths of 2.5 mm to 7.5 mm and line widths of 2 μm to 6 μm for transmission lines from 2 mm to 60 mm in length. A variety of line widths and spaces from 0.8 μm to 3 μm and different signal to ground structures showed that open-eye diagrams could be obtained for frequencies from 3 GHz to 8.5 GHz where the signal delay was measured from under 25 ps to 95 ps and far-end cross-talk noise was measured to be from 1.2 % to 35 %. Microbump interconnection characterization showed direct current electrical resistance under 18 mΩ for 25 μm diameter solder interconnects between surface pads on silicon layers. Signal integrity and chip-to-chip I/O link characterization measurements are also important in order to increase bandwidth between the die in a die stack or between the die on a silicon package. I/O driver circuits can be reduced in size and reduced in power for short-reach communications.

*Interconnection Material, Structure and Processes*

Traditional off-chip interconnection has used wirebonding or area array flip chip solder bonding to provide interconnection to a package. For limited I/O wirebond applications, aluminum or gold wirebonds have often used up to four perimeter rows of interconnection with a bond pad pitch down to about 30 μm and total number of connections per die up to about 800 input/output (I/O). Area array off-chip interconnections such as with lead tin solder or more recently Pb-free solder have provided I/O pads at a pitch down to 150 μm where total I/O per die have been to about 10 000 I/O. The wirebond or solder interconnections have provided not only power and signal interconnections but have accommodated many product form factors such as wirebonded die stacking, organic and ceramic packages assemblies at high yield, wafer test compatibility and, where needed, use of overmolds and underfills to improve interconnection reliability against product specifications.

For emerging 3D die stacking and 3D silicon packaging, higher density interconnection densities are desired and possible for this new class of applications. In these applications, fine-pitch interconnection may be necessary for high bandwidth and low latency in either a 3D die stack or 3D package. These TSV and interconnects need to accommodate power delivery and distribution, electrical signal transmission, thermal transmission, mechanical integrity and reliability while also achieving 3D product form factors with a high assembly yield to meet product cost objectives. In many applications, use of the wafer test for known good die (KGD) and creation of known good die stacks will be needed to serve the marketplace. In applications using wafer-to-wafer bonding processes, a small-size die with a high wafer bonding yield is needed. Some products may leverage design redundancy or appropriate repair techniques to ensure good manufacturing yields.

Research and development silicon layer interconnect investigations for 3D circuit integration have included SOI structures, thin die stacks using face-to-face or face-to-back structures and silicon-on-silicon package interconnection investigations [21, 29–34, 45, 51, 54–56]. These investigations have included different bonding approaches including oxide-to-oxide bonding, copper-to-copper interconnection and solder interconnects for vertical silicon–silicon interconnection between silicon strata levels or silicon layers. Experiments included a thinned silicon interposer with TSVs placed between one or more die and either a ceramic or organic package or thinned die stacked and assembled on a base package. For these structures, many options can be considered for module assembly. Figure 12.15 shows a comparison of chip-to-chip, chip-to-wafer and wafer-to-wafer assembly approaches [6].

A 3D interconnect using solder can be used for die stacking, fine-pitch interconnects from die or die stacks to silicon packages and within 3D modules to ceramic, glass or laminate packages. Fine-pitch solder interconnects have used sequential die stack and parallel die stack assembly processes. Similarly, die and die stacks can be assembled to

Interconnection metallurgies, process & chip stacking

| | Chip to chip 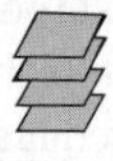 | Chip to wafer  | Wafer to wafer 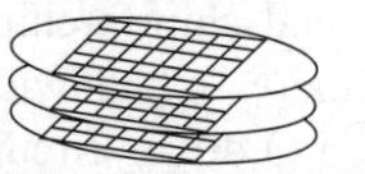 |
|---|---|---|---|
| Pro | Flexible, use of KGD | Flexible, use of KGD | Volume Mfg |
| Con | Handling and bonding | Handling and bonding | Overall yield, same chip size |
| Die or wafer thickness | < 4 µm to > 150 µm | < 4 µm to > 150 µm | < 4 µm to > 150 µm |
| Bonding technology | Solder<br>Metal to Metal<br>Adhesive | Solder<br>Metal to Metal<br>Adhesive | Solder or Metal<br>Oxide bonding<br>Adhesive |
| Demonstrations | 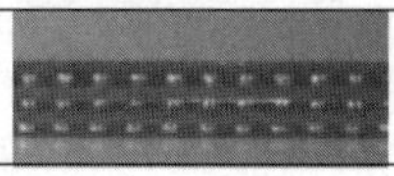 | 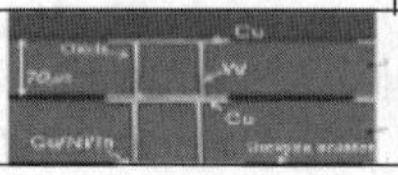 | 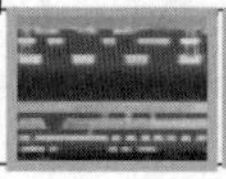  |

***Figure 12.15*** *3D assembly approach comparison including chip-to-chip, chip-to-wafer and wafer-to-wafer integration* [6]

silicon, laminate, ceramic or glass substrates or other base substrates. For both stacked die with TSVs and fine pitch interconnections and 3D integrated packages TSVs and fine-pitch interconnects, coplanarity of the thinned silicon and contamination control are important to achieve high yield. The solder, interconnect metallurgy, pad structure and associated process methodology for assembly and adhesive joining need to be optimized for a wide variety of 3D form factors. As the size and pitch of the electrical interconnect are reduced, alignment, bonding coplanarity, contamination control, mechanical integrity, process handling and adhesive enhancement become more challenging. Adhesive approaches such as capillary underfill, vacuum underfill and no-flow underfill have been investigated. An additional challenge for fine-pitch 3D assembly adhesive interconnect materials can be enhanced thermal conductivity adhesives to support higher thermal conductivity to aid in head removal from higher power density integrated 3D structures.

Test vehicle demonstrations have included 200 µm pitch with die-to-die stack assemblies with approximately 9000 I/O per layer and multiple die on a silicon package again with approximately 200 µm pitch and about 8000 I/O per die for a total of 32 000 I/O per layer. Figure 12.10 (a) shows the stacked die and Figure 12.11 (a) shows an example of the assembled module. In each of these demonstration test vehicles, 100 % assembly yields were demonstrated. Similarly for fine-pitch die stack test vehicles with a 50 µm pitch solder interconnection, 100 % assembly yields have been demonstrated for a range in die sizes and total I/O interconnection including approximately 2000, 5000, 10 000, 42 000 and 80 000 connections per layer. In these studies, either 97-3 Pb–Sn solder, 37-63 Pb–Sn eutectic solder or high Sn, lead-free solders were used for evaluations [51, 54–56, 64]. The 25 µm diameter solder uC-4 was formed using either plating processes or injection molded solder. Die assembly test vehicle demonstrations such as these permit assembly process learning to compare process options, study interconnection metallurgies, define process specifications, support assembly characterization including yield statistics, alignment and physical characterization, electrical parametric characterization, thermal characterization and reliability characterization.

A 3D interconnect for a die stack assembly can also use alternate interconnection approaches such as use of intermetallic compounds or transient liquid phase sintering, copper-to-copper bonding, anisotropic conductive adhesive bonding or alternate approaches, as have been reported previously [24, 25, 29, 52, 53].

A 3D die-to-wafer assembly has been reported as a means to increase productivity [59]. In this assembly approach, known good die can be placed on to known good die sites using rapid pick-and-place tools. Following die pick and place for all good die sites on the wafer, a single-step attachment can join all die to their corresponding die on the wafer with one process. This may be using solder or copper-to-copper bonding or an alternate bonding approach. This approach can also be continued to create multihigh die stacks. An alternate approach to the die-to-wafer assembly is either a die-to-die assembly or die-to-wafer bonding; die can be pre-tested to help assure a high yield for two or more die in the die stack. In addition, the die may be the same size or may vary in size. For thinned die, care must be taken in handling the die to avoid edge damage and breakage. Also for fine-pitch silicon interconnections, a seal band or underfill adhesive may be used to reduce the potential for joint degradation during use due to corrosion or fatigue.

3D wafer-to-wafer bonding can utilize oxide-to-oxide bonding, adhesive bonding, metal-to-metal bonding or hybrid bonding such as combined metal bonding and oxide bonding in one joining process. Wafer-to-wafer permanent bonding can create high-density interconnection using small-diameter TSVs [7, 21, 30, 32, 35]. Wafer-to-wafer bonding and vertical interconnection of circuits includes wafer-to-wafer alignment, wafer bonding, wafer thinning and vertical interconnect formation. Alignment and bonding technical results for the interconnections between silicon levels have been reported with dimensions as small as approximately 0.14 μm diameter, 1.6 μm height, 0.4 μm pitch and density of interconnections of $10^8/cm^2$ [21, 30, 32]. Application requirements along with process integration maturity will expand over time and be expected to support interconnection densities such as $10^8/cm^2$ using wafer-to-wafer bonding for products. For 200 mm and 300 mm wafers, similar to wafer processing, wafer stacking processes will need to be robust to support millions of interconnections between each silicon layer. Wafer-to-wafer bonding and vertical interconnection still have challenges to achieve high yields. Wafer-to-wafer bonding challenges include wafer alignment, coplanarity, permanent bonding and high-yielding TSV or vertical via formation. Design redundancy, integrated repair and/or sparing may aid in the adoption and yield of 3D stacked structures using wafer-to-wafer bonding.

### *Wafer Level Test, Known Good Die and Reliability Testing*

Known good die (KGD) can be identified from pre-testing die with a wafer level test. Statistical sampling may help reduce the need for a 100 % test. The creation of known good die stacks beginning with KGD can substantially improve the total yield. Understanding defect causes and putting in place corrections can improve yields. Redundant interconnections in some applications may aid die stack yields. To demonstrate a path forward for 'known good die' and 'known good die stacks' with fine pitch interconnections, a wafer level test using a built-in self test (BIST) and existing probe technology has been demonstrated. Extensions to fine-pitch probes has begun but will drive learning in probe tips, wafer test and manufacturability in the future. Module level testing can complement the wafer level test to support known good die stacks and for burn-in [7].

Reliability testing for TSVs, fine-pitch interconnections, stacked die and 3D integrated modules has continued to be studied through the use of demonstration test vehicles. For example, stacked die advancing technology nodes, integrated TSVs and interconnects have been designed, fabricated, assembled and put into reliability tests. Test vehicles can range from simple electrical continuity tests with TSVs and uC-4 to active circuit test vehicles. Prior characterization using TSVs and solder microbumps at 50 μm pitch have shown that these robust structures can support reliability stress including electromigration results of over 2000 hours for 50 mA to 100 mA current at 150 °C, deep thermal cycle results of over 1500 to 25 000 cycles of −55 °C to +125 °C, temperature–humidity bias of over 1000 hours for 85°C, 85% relative humidity and 1.5 V, and over 2000 hours of high temperature storage at 150 °C [51, 54, 56, 61]. Results indicate that TSVs and fine-pitch interconnections can be fabricated and meet typical product reliability stress requirements. New test structures and form factors for TSV, interconnect structure, die stacks and 3D integrated packaging and modules continue to be evaluated for semiconductor active and passive applications using copper and tungsten TSVs and with numerous interconnect metallurgies and structures. These further studies of multichip and die stack test structures with increased interconnection densities between $10^3/cm^2$ and $10^8/cm^2$ are at various stages of design, build and characterization. They will permit ongoing experiments and data to be investigated including design rules, process, bonding, test structures/methodology and characterization. These ongoing investigations include die-to-die, die-to-wafer and wafer-to-wafer processing approaches to 3D applications. Assessments provide data on interconnection density, materials, structures and processes to help optimize applications and compatibility with manufacturing. Data collected can provide guidance to meet application reliability objectives for TSV, vertical interconnect and in a variety of 3D integrated package and module form factors. These tests, reliability studies and characterization support system level decisions for product applications.

### *3D Modeling*

Many models and simulation tools exist for 2D applications and many now also exist using extensions of 2D tools for 3D applications. For example, within a chip design, 3D modeling tools exist that permit electrical design, electrical transmission modeling and simulations. However, for 3D structures with multiple levels of silicon strata or layers and module form factors, full 3D designs, models and simulations can require vertical layer identification, and with the expansion of 3D interconnections, the existing tool capacities have needed to be expanded to support the magnitude of modeling and simulations needed for 3D applications. Tools exist for mechanical and thermal modeling in 3D structures but tools for performance simulations, full design and comparison are not broadly available. Combination tools supporting mechanical, thermal and electrical assessments generally are not available at this time. However, extensions from existing tools have permitted great learning for design, modeling and simulations. In time, further release of next generations of 3D design and modeling tools will be able to improve 3D comparisons, 3D electrical transmission models and simulations tools for die stacks, 3D packaging and 3D module structures. System level evaluations to understand performance modeling, power efficiency, cost and form factor trade-offs is more challenging but can be achieved with data analysis. As 3D manufacturing and industry adoption take place, greater experience and data on 3D processes, yield understanding and cost models will help to optimize 3D designs and structures for future applications. Examples of mechanical and thermal modeling comparing 3D die stacks and 3D packaging was summarized by Sri-Jayantha *et al.* [65].

Design, architecture and performance modeling provide a great opportunity to improve system solutions using 3D structures. Examples of architecture considerations and performance benefits for 3D have been reported [1, 10, 50]. Examples of electrical transmission measurements, modeling and simulation has been reported by Patel and Dickson *et al.*, respectively [62, 63]. Mechanical models have been reported for TSV, 3D interconnect using small solder bumps and thermomechanical evaluations of 3D structures [7, 65]. Stress and deformation have been evaluated at each stage of the TSV manufacturing process at the temperature for each operation. Elastic properties were characterized by the elastic modulus and Poisson ratio [31]. For Cu TSVs, the yield strength of the material is likely to be exceeded and the nonlinear properties need to be included. A stress–strain curve can be incorporated but simple yield stress is usually sufficient. The range of process temperatures drove inclusion of the coefficient of thermal expansion, as discussed above. In addition, shear stresses needed to be evaluated at material interfaces and compared with the adhesion strengths between materials. The highest stress conditions are generally seen at the TSV to adjacent wiring and dielectric layers. Understanding mechanical aspects of via structure and process flow can be leveraged to minimize the maximum vertical stress for use in silicon-based technology. From an understanding of structure and stresses, such as for TSV in thinned stacked die or die on 3D silicon packages, the electrical and mechanical design specifications for the product application can be optimized and successfully applied.

For silicon interconnect modeling, initial development of a model to understand stress and strain levels in a solder μ-C4 began with the use of a macro–micro model [7, 65]. In the finite element model, the macro characteristics of the structure could be considered while still providing micro-level detailed understanding for the high volume of small features that were needed to understand mechanical characteristics. For example, the model would address the large quantity of micro joints used in the structure while being able to begin to understand actual stress and strain on an individual solder μ-C4 level. *X* and *Y* displacements could show the relative pressure loads in the macro model and distributed the stress to the solder interconnections for relative comparison. The macro and micro mechanical modeling of stress in the solder μ-C4 can then be evaluated across the various BLM and solder interconnections or compared to alternative fine-pitch interconnections such as with copper-to-copper bonding or when using oxide-to-oxide bonding for fine-pitch interconnections.

3D power delivery, distribution and cooling models and demonstration vehicles are under further investigation and should lead to improved understanding and application of products in time. Similarly, 3D knowledge for wafer build, assembly, yield and cost models is also leading to improved understanding over time.

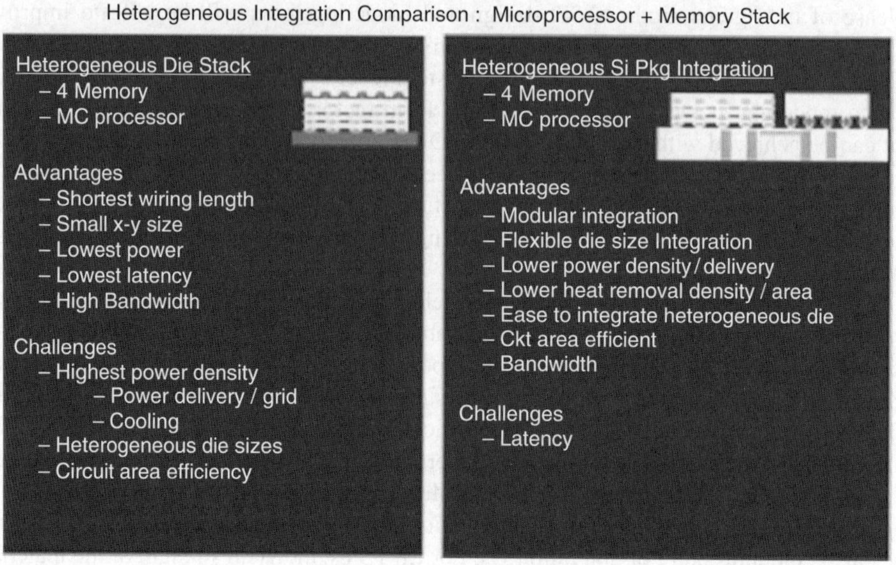


***Figure 12.16*** *Design form factor comparison of a 3D die stack and 3D silicon package interconnection*

### 12.2.5 Trade-offs in Application Design and Product Applications

Leverage of TSV, thinned silicon and vertical silicon interconnects for system integration permits a wide range of products covering varied interconnection densities. For example, simple wireless die with $< 10$ TSVs to applications for high-performance computing that may require $10^8$ TSVs and vertical interconnects between silicon circuit layers may each benefit from emerging 3D technology.

Another consideration toward 3D integration is the form factor in which the product is designed. Figure 12.16 shows a schematic with form factor options of integration that could be considered as part of a 3D system application [44]. One option gains the highest bandwidth for silicon interconnection by means of a vertical stack of silicon die and the other high bandwidth silicon interconnection by means of a silicon package circuit that combines all heterogeneous silicon circuit layers into one die stack. In the die stack option, advantages can include the shortest wire length between heterogeneous die and an opportunity to reduce power for signal communications due to reduced capacitance and resistance in the wire lengths and sizes. Wire lengths for die to die in a die stack may be tens of micrometers. However, the design also has challenges including delivery of power to each level within the stack, the circuit density reduction due to a TSV area takeaway from each silicon layer to support vertical power delivery and signal communications. In addition, for a vertical stack, removing heat from the stack can also lead to power density or operational performance limits depending on the type of die being stacked. Thus this may be the best option provided the design can support power delivery, heat removal and have a minimal loss of circuits to support vertical interconnects. Another option for comparison could be a combination of die stacks with interconnection on a high bandwidth,

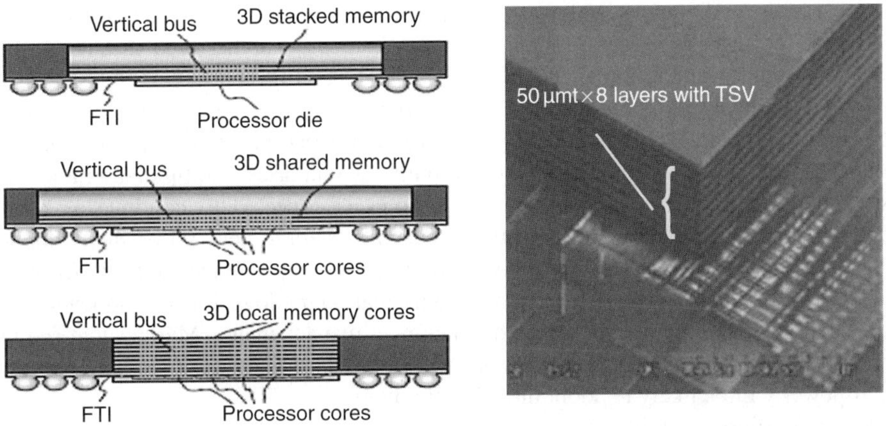


**Figure 12.17** *Multihigh memory die stack from Elpida using a TSV and silicon interconnect* [27, 67]

3D silicon package. In this example, total power delivery to multiple die stacks can be spread over a larger area while maintaining high bandwidth between die stacks. This can help to reduce current flow per area, reduce power density and aid in heat removal as cooling can be across multiple die stacks. However, in this option, increased wire length leads to an increase in latency. A short wire length can be achieved at very high interconnect density for adjacent die stacks on a silicon package. Wire lengths for die-to-die or die-to-die stack interconnection across a silicon package or a glass package can be supported from a few millimeters length to tens of millimeters. Limitations in total length at high bandwidths are limited to power for interconnect at distances in excess of perhaps about 100 mm unless large cross-sectional wires are used, in which case the wire length could be increased.

Press announcements and technical publications have begun to show both first applications of products using TSV and 3D silicon vertical interconnects including image sensors and power amplifiers where these product have utilized low-density vertical interconnects [37, 38, 50, 66]. More recently, two new product application form factors have been announced. The use of a 3D silicon package with TSVs to support multiple die with wiring interconnection has also been announced by Xilinx and Semtech, respectively [37, 38, 40, 41]. In addition, Elpida, Samsung and Micron have had technical publications and/or press announcements about wide I/O bandwidth, stacked die memory applications that utilize TSVs and vertical silicon stacked die [27, 37, 38, 42, 43]. Figure 12.17 shows an example of a memory chip stack from Elpida [27, 67].

Wider industry adoption and acceleration of product application introduction might be expected as wide I/O memory die stacks begin to enter the marketplace during the next two years. Design and enablement tools are available for some applications but more complex products would benefit from expanded capability design tools that can better support 3D applications. Equipment for 3D technology is also available but next generations of 3D tools will need to continue to emerge to support high-productivity manufacturing. Examples of tools that need improvement include bonding and debonding tools, 3D assembly tools for die stacks, wafer-to-wafer permanent bonding tools with high throughput and fine-pitch test probes and associated wafer testbeds. Another area that would help accelerate industry adoption of more complex 3D heterogeneous die, die stacks and integrated 3D packaged solutions is standards for bonded wafer pairs, standards for I/O pitch and floor plans, standards for chip-to-chip communication at low power and technical products that utilize multiple design shops, multiple foundries, multiple assemblers and test companies. Industry growth for 3D products is expected to continue as design, manufacturing and product demonstrations reduce barriers to new 3D product adoption. Like many new technologies, 3D is expected to provide advantages and disadvantages and therefore is not expected to displace traditional technology for all applications but rather will grow into markets where it has a competitive advantage. Markets that can take advantage of high bandwidth for performance advantages, wide I/O memory access and lower power will be likely to adopt the 3D technology.

Beyond traditional applications of 3D microelectronics, application expectations for 3D can be expected to be far reaching with time. Examples might include portable electronics such as cell phones, portable medical products and portable sensors. With the reduced power consumption, portable products may benefit from enhanced battery life, not to mention significantly more compact products with scaling functional capabilities. Additional applications could include military, information technology, communications, automotive and space applications. For computing applications, high bandwidth memory chip stacks for heterogeneous integration to microprocessors and field programmable gate arrays (FPGAs) could provide reduced power, system performance scaling and smaller products. In addition, it is likely that new applications and products will emerge between advancements in these microelectronics and nanoelectronics technologies and emerging biotechnology as well as other nano emerging technologies. It seems clear that the industry is just beginning to consider new applications and products that may take advantage of 3D silicon integration.

### 12.2.6 Summary

Emerging three-dimensional (3D) silicon integration using through-silicon vias (TSVs) and silicon interconnections has the potential to become used in a broad range of applications. Technology advancements and implementation using 200 mm and 300 mm tools is growing in the industry, including first-generation image sensors and power amplifiers to second-generation 3D silicon package integration modules and 3D wide I/O stacked memory die using TSV and fine-pitch interconnect. Further technology advancements toward heterogeneous die stacks and 3D integrated packaging and modules are expected to continue to advance for future applications. 3D technical and business challenges still remain in the industry but continued advancements in design, design tools, 3D manufacturing and manufacturing tools, 3D standards and industry compatibility will lead to more product applications. Future products are expected to take advantage of new 3D design, fabrication, assembly and characterization demonstrations. Heterogeneous integration of 3D products with TSVs and fine-pitch interconnect will benefit from collaboration between research, development, manufacturing, equipment suppliers, materials suppliers and standards defining organizations.

## Acknowledgments

This work has also been partially supported by Maryland Procurement Office (MPO), Contract H98230-04-C-0920 and Contract H98230-07-C-0409.

The author wishes to acknowledge support from the 3D research team including Paul Andry, Bing Dang, Chris Jahnes, Yang Liu, Joana Maria, Robert Polastre, Cornelia Tsang, Bucknell Webb and Stephen Wright. In addition, the author wishes to thank the IBM Research – Materials Research Laboratory, Central Services Organization and collaboration with System and Technology Group.

## 12.3 Advanced Microscopy Techniques for 3D Interconnect Characterization

*Lay Wai Kong, Sven Niese, Alain Diebold and Ehrenfried Zschech*

The rapid development of the 3D IC stacking technology is driving the need for advances in microscopy methods. Overlay measurements after wafer bonding, defect detection and void inspection after copper plating are challenging due to the need to image through silicon and inside copper. Microscopy techniques for which silicon is not opaque, such as scanning acoustic microscopy (SAM) and confocal infrared (IR) microscopy are capable of inspecting the interface between bonded wafer pairs, while high-resolution X-ray computed tomography (XCT) is used to detect voids in metal TSVs. This chapter covers several analytical techniques for process and quality control. The current status of SAM, IR microscopy and nano-XCT, complemented by techniques for microstructure characterization, is discussed in terms of their application to process metrology and failure analysis for 3D IC integration.

### 12.3.1 Scanning Acoustic Microscopy

Scanning acoustic microscopy (SAM) provides the ability to image sub-surface features for bonded wafers and packaged integrated circuits. In SAM, sound waves propagate through elastic media reflecting from regions where the materials properties change. The principles of acoustic microscopy were thoroughly reviewed in Briggs and Kolosov's book [68]. Below, we describe only the basics of acoustic microscopy to aid the discussion of the application of SAM to interconnect materials based on Briggs and Kolosov [68].

*Principles of Scanning Acoustic Microscopy*

A scanning acoustic microscope visualizes changes in the reflected intensity of ultrasonic waves. The intensity of the reflected waves changes across a sample as local changes in the acoustic impedance, $Z$, occur at interfaces between materials. The acoustic impedance is

**Table 12.1** Density, acoustic wave velocity and acoustic impedance of several materials [69]

| Material | Density ($g/cm^3$) | Longitudinal Wave Velocity (m/s) | Acoustic Impedance ($\times 10^6$ $kg/m^2s$) |
|---|---|---|---|
| Water (20 °C) | 1.00 | 1483 | 1.48 |
| Alcohol (20 °C) | 0.79 | 1163 | 0.92 |
| Air (20 °C) | 0.00 | 344 | 0.00 |
| Silicon | 2.33 | 8600 | 20.04 |
| Gold | 19.3 | 3240 | 62.53 |
| Copper | 8.90 | 4700 | 41.83 |
| Aluminium | 2.70 | 6260 | 16.90 |
| Epoxy Resin | 1.20 | 2600 | 3.12 |
| Resin (IC Pkg) | 1.72 | 3930 | 6.76 |
| Glass | 2.7 | 5570 | 15.04 |

the product of the density and the speed of sound in the medium, as given in the following equation:

$$Z = \rho V \tag{12.1}$$

where $Z$ is the acoustic impedance, $\rho$ is the density and $V$ is the acoustic wave velocity.

Since the elastic properties of materials of different composition are different, SAM is an important method of imaging many structures of interest to the semiconductor industry. Another useful relationship is that between the acoustic wave velocity as well as the elastic constant and the density of a material, as indicated by

$$V = (C/\rho)^{1/2} \tag{12.2}$$

The elastic constant $C$ can be calculated from the elastic properties of the material for a particular mode, which yields different speeds of sound, e.g. for transversal and longitudinal waves.

In Table 12.1, density, acoustic wave velocity and acoustic impedance of a variety of materials are listed. The mismatches between the acoustic impedance of air and most other materials results in reflection of the wave with little or no transmission. Therefore, samples are immersed in water, which is further described below. Water is the coupling medium that facilitates the transmission of ultrasonic energy from the transducer into the specimen.

In SAM, the transducer both transmits and detects the reflected ultrasonic waves. The SAM used in many of the studies illustrated in this chapter is equipped with a piezoelectric transducer that transmits and receives longitudinal waves. A high-voltage pulse is used to vibrate the transducer and to generate a plane wave. The plane wave is focused by a lens and then propagates through the coupling medium to the sample. Part of the wave is reflected back to the transducer and part of it transmits through the sample. Inside the sample, the wave is also partially reflected from the structure inside the sample. When the wave reaches the surface of any solid specimen, part of the wave remains as a longitudinal wave and part of the wave propagates as a transverse wave at a typical angle through the specimen.

Figure 12.18 shows a sample immerged in a coupling medium like water with a transducer generating an ultrasound toward the specimen, the initial ultrasound wave. The longitudinal wave from the transducer propagates through the water to the sample. At the boundary between the water and the sample, part of the initial ultrasound wave reflects

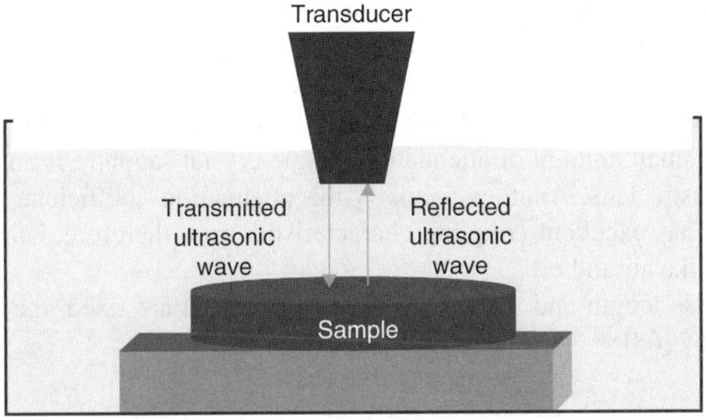


***Figure 12.18*** *Basic setup for scanning acoustic microscopy*

and part transmits. The amount of transmission and reflectance is given by the acoustic impedance of water, $Z_1$ and sample, $Z_2$. The fraction of this initial wave intensity that is reflected can be calculated with the following equation assuming normal incidence [68]

$$R = \frac{(Z_2 - Z_1)}{(Z_2 + Z_1)} \tag{12.3}$$

The term of Equation (12.3) is known as a reflection coefficient. The transmission coefficient $T$ is $1 - R$ or as an expression of the impedances

$$T = \frac{2(Z_1)}{(Z_2 + Z_1)} \tag{12.4}$$

*Resolution*

The spatial resolution, $W$, of an acoustic microscope is based on the Rayleigh criterion and is given by [68]

$$W = \frac{0.51\lambda_0}{\text{NA}} \tag{12.5}$$

where NA is the numerical aperture and the wavelength, $\lambda_0$, is given by $V_0/f$ ($V_0$ is the velocity of sound in that medium and $f$ is the freqency). The resolution is proportional to the wavelength of the ultrasound, which depends on the frequency and the velocity of sound in the material. Equation (12.5) implies that the only limit on resolution is the frequency of the wave. Unfortunately, the attenuation of ultrasonic waves is proportional to the square of the frequency providing practical limits to resolution for a given minimum penetration depth. This is one motivation for thinning samples before measurement, and vice versa; the attenuation is the key to determine the most useful high frequency for imaging the object. The attenuation per unit distance is stated by

$$A = \alpha_0 f^2 \tag{12.6}$$

Here, $A$ is the attenuation, $\alpha_0$ is the attenuation coefficient of the material and $f$ is the frequency of sound in the material.

The attenuation is caused by the combination of scattering and absorption. Scattering causes a change in the direction of propagation of the ultrasonic wave and absorption occurs if the ultrasonic wave is converted into other forms of energy.

The acoustic wave propagates through a lens before entering the coupling medium. Due to the relatively small amount of attenuation, single-crystal sapphire is an excellent material for an acoustic lens. Another factor is the attenuation coefficient of the coupling medium. Water has excellent coupling characteristics and, therefore, it is preferred over other mediums like air and oil.

When the pulse length and the time between pulses ($t_0$) are fixed, the minimum focal length is given by [68]

$$\text{FL} = \frac{V_0 t_0}{2} \tag{12.7}$$

where FL is the focal length, $V_0$ is the velocity of the ultrasonic wave and $t_0$ is the time interval between echoes. From Equation (12.5), it is known that the attenuation is proportional to the square of the frequency [68]. An acceptable attenuation, $\alpha_{acc}$, is needed to determine the maximum frequency that can be used. The frequency limit due to $\alpha_{acc}$ is given by

$$f \leq \sqrt{\frac{\alpha_{acc}}{2\alpha_0 \text{FL}}} \tag{12.8}$$

where $f$ is the frequency of the ultrasonic wave, $\alpha_{acc}$ is the acceptable attenuation and $\alpha_0$ is the attenuation coefficient. This maximum frequency is then defined by the shortest wavelength, $\lambda_{min}$, that can be used. This is stated by

$$\lambda_{\min} \equiv \frac{V_0}{f} = \sqrt{\frac{V_0^3 \alpha_0 t_0}{\alpha_{acc}}} \tag{12.9}$$

This allows us to define a resolution coefficient $R_c$:

$$R_c \equiv \sqrt{V_0^3 \alpha_0} \tag{12.10}$$

The resolution coefficient is proportional to the shortest wavelength that can be used with the given $\alpha_{acc}$ and $t_0$. The resolution coefficient for water is ~ 0.8 and for air is ~ 80 [68].

*Applications of Scanning Acoustic Microscopy in Microelectronics*

SAM characterization of bonded wafer pairs has been proven to be very useful during process development. SAM images of patterned sections of the wafer are not readily comparable to scanning electron microscopy (SEM) images of patterned circuit features, as shown in Figure 12.19 [70]. Several bonding materials have been used to bond wafer pairs. The so-called dendritic structure was observed in wafer pairs during development of the curing step for the bonding material benzocyclobutene (BCB). Figure 12.20 shows defect structures observed for unpatterned wafers bonded using BCB. The combination of SAM and IR microscopy has been proven to be a great aid during process development.

***Figure 12.19*** *SAM image of metal–metal alignment marks and features inside a bonded wafer pair composed of two 775 μm wafers. Figure is courtesy of SEMATECH*

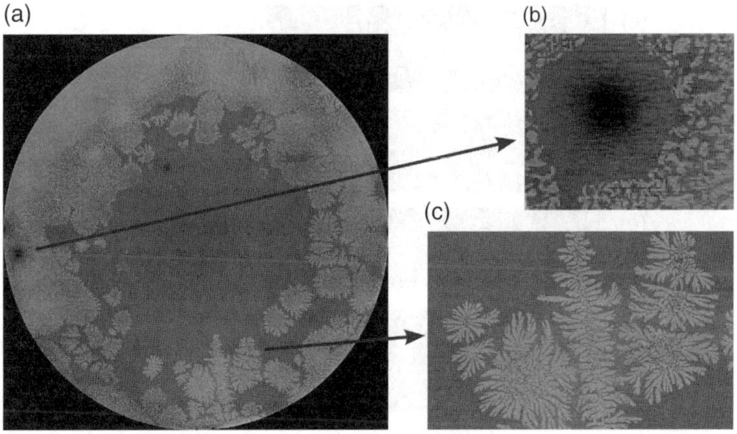


***Figure 12.20*** *(a) SAM image of a wafer pair bonded using BCB. (b) An area where the wafers touch due to missing BCB. (c) Dendritic structures that grow during solidification of the BCB if curing conditions are not optimized. Figure is courtesy of SEMATECH [70]*

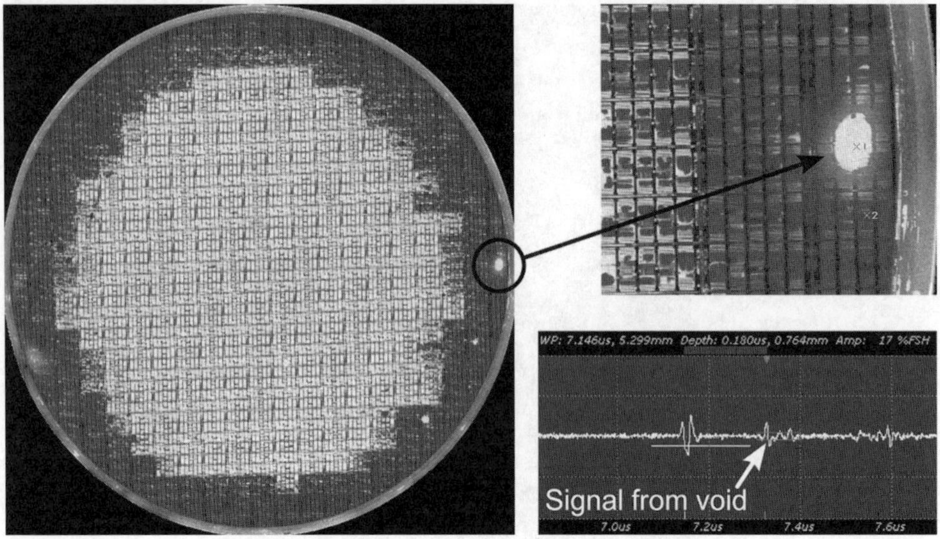


***Figure 12.21*** *SAM image of a bonded pair of patterned wafers (110 MHz, 1.45 h scan time, unbonded area shown in bright). Figure is courtesy of SEMATECH* [70]

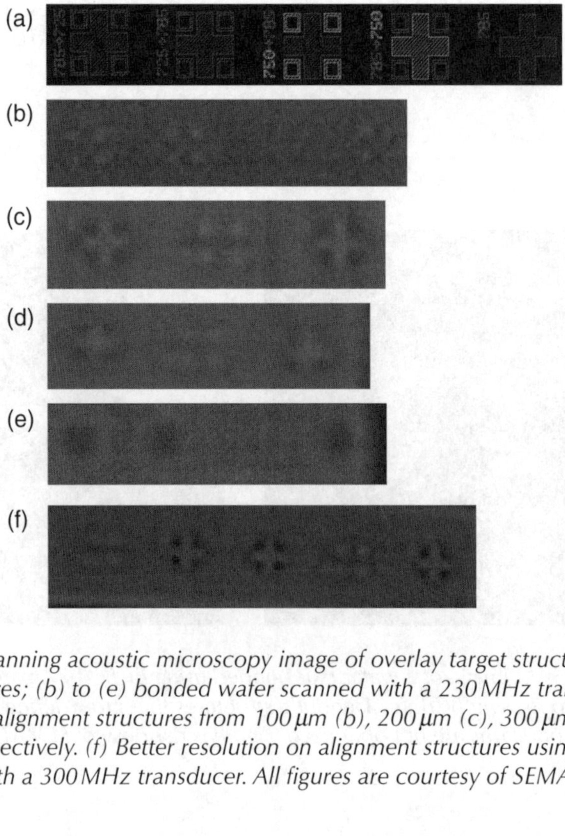


***Figure 12.22*** *Scanning acoustic microscopy image of overlay target structures: (a) gds. of alignment structures; (b) to (e) bonded wafer scanned with a 230 MHz transducer. SAM images show the alignment structures from 100 μm (b), 200 μm (c), 300 μm (d) and 400 μm (e) top wafer respectively. (f) Better resolution on alignment structures using 300 μm top wafer scanned with a 300 MHz transducer. All figures are courtesy of SEMATECH*

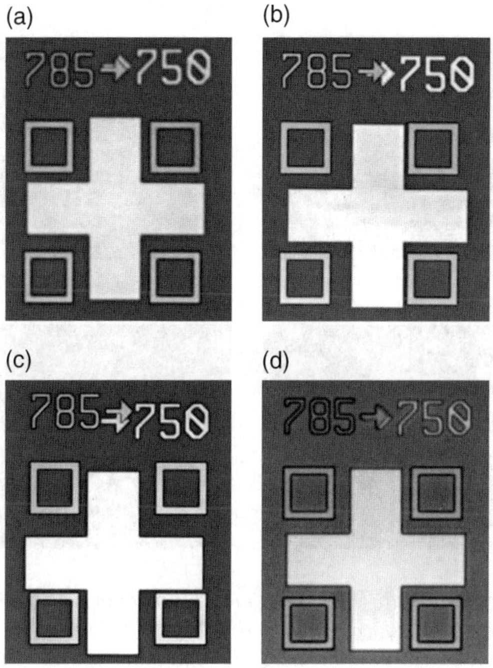


*Figure 12.23* IR microscopy images of overlay target structure: (a) alignment structure on a bonded top wafer with 100 μm thickness; (b) alignment structure on a bonded top wafer with 300 μm thickness; (c) alignment structure on a bonded top wafer with 400 μm thickness; (d) alignment structure on a bonded wafer with 775 μm thickness. All figures are courtesy of SEMATECH

Figure 12.21 shows the presence of an area where two patterned wafers touch. Figure 12.22 clearly shows that IR microscopy is the preferred method for overlay metrology than acoustic microscopy.

### 12.3.2 IR Microscopy

Microscopy of 3D structures is challenged by the opaque nature of silicon for most energies. However, since silicon is transparent at infrared wavelengths from about 1.3 μm to 8 μm, it is possible to examine silicon dies or wafers, and particularly bonded wafers, with IR techniques. Infrared (IR) microscopy is a nondestructive microscopy technique that is used in 3D IC manufacturing to image sub-surface features at the interface of bonded wafers [71]. This practical solution enables a variety of through-silicon metrology, including overlay alignment, review of pre-existing defects from each wafer at the bond interface and detection of new defects created during the bonding process. The use of IR microscopy to measure wafer pair overlay and to study the interface defectivity of bonded wafer pairs is demonstrated in Reference [71].

IR microscopy measurements of overlay target structures at the interface of bonded wafers are demonstrated for bonded top wafers with wafer thicknesses varying from 100 μm to 775 μm in Figures 12.23 (a) to (d).

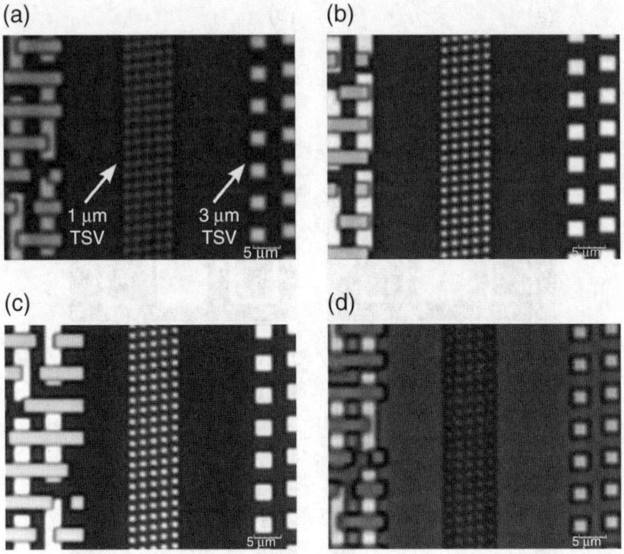


***Figure 12.24*** *(a) 1 µm and 3 µm TSVs in a bonded top wafer with 100 µm thickness; (b) 1 µm and 3 µm TSVs in a bonded top wafer with 300 µm thickness; (c) 1 µm and 3 µm TSVs in a bonded top wafer with 400 µm thickness; d) 1 µm and 3 µm TSV in a bonded top wafer with 775 µm thickness. All figures are courtesy of SEMATECH*

The effect of top wafer thickness on image resolution is demonstrated using TSV SEM bar test structures. We show images of a structure consisting of 1 µm and 3 µm TSVs in Figure 12.24 under an objective lens with specified working distance, NA, and magnification. Consistent with the alignment structure IR images, the bonded wafers with 400 µm top wafer show the sharpest image with the best contrast. These micrographs explain that the 400 µm top wafer has more in-focus reflected signal passing through the pinhole impinging on the detector. Therefore, the intensity using a 400 µm top wafer is higher for bonded wafers with other thickness values. This implies that matching the right IR objective lens to the thickness of the top wafer is important for obtaining the best resolution during inspection.

### 12.3.3 Transmission X-ray Microscopy and Tomography

High-resolution X-ray imaging with a spatial resolution as good as 10 nm on synchrotron-based and 30 nm for lab-based systems offers unique capabilities for process development and failure analysis in the semiconductor industry. Buried metal interconnect structures like copper on-chip interconnects and through-silicon vias (TSVs) for 3D IC integration can be studied with excellent element-specific contrast. Transmission X-ray microscopy (TXM) is done over a wide range of X-ray energies at synchrotron radiation sources. The selection of the X-ray energy is often a matter of convenience. However, selecting a useful energy range can improve resolution, contrast and penetration depth. X-ray tubes provide fixed energies like Cu-$K_\alpha$ radiation at 8.0 keV (characteristic X-rays) with an adequate intensity to limit the flexibility but enable X-ray microscopy to be used in the lab.

Furthermore, it is possible to reconstruct a three-dimensional dataset of the sample, the tomogram, based on the acquisition of a number of radiographs (projections), which is well known from nondestructive testing of macroscopic objects using microfocus X-ray tubes or from medical applications. The resulting radiographs usually show the transmissivity of the X-rays; i.e. regions with a high transmission appear bright and regions with a high absorption appear dark. The tomogram and hence the determined virtual cross-sections represent the distribution of the linear absorption coefficient in the sample. Features with a high absorption, e.g. copper in silicon, appear bright.

*Principles and Implementation of TXM*

This section provides a short review of the physical principles of TXM as well as basic information about synchrotron-based and lab-based microscopes.

***Physical Principles.*** The refractive index of most materials is close to 1 for X-rays; i.e. it is hardly possible to use refractive lenses for imaging as in visible light microscopy. Focusing X-rays for high-resolution imaging requires different X-ray optics as on a Fresnel zone plate. The zone plate consists of a set of concentric rings that decrease in width and increase in radius toward the outermost zone ring, which has the smallest width. The resolution of the image is approximately determined by the outermost zone width and is expressed by

$$\sigma = 1.22\,R_n \tag{12.11}$$

where $\sigma$ is the resolution and $R_n$ is the outermost zone width. The resolution of the image is limited by how small the zone widths can be made as this prevails the Abbe criteria in the case of X-ray imaging [72].

Transmission X-ray microscopy is based on detecting the intensity of transmitted X-rays. The capability of the X-rays to transmit through the material is characterized by the linear attenuation coefficient. A material with a small attenuation coefficient is relatively transparent to X-rays, while a large attenuation coefficient means less transmission for an equal penetrated thickness. This relation can be described by the Beer–Lambert law, given as

$$I = I_0 \exp(-\alpha x) \tag{12.12}$$

The Beer–Lambert law describes the intensity of transmitted X-rays, $I$, after a beam with an initial intensity, $I_0$, passes through a material with a thickness, $x$, and linear attenuation coefficient, $\alpha$ [73]. For X-rays, $\alpha$ only depends on the mass density and the chemical composition, i.e. the present elements, of the sampled region. As the bond energies are usually small compared to the applied photon energies, TXM is not sensitive to chemical bonds. Therefore, the radiographs represent the integrated spatial variation of different elements weighted with the density. The absorption cross-section for a certain element changes drastically at an X-ray absorption edge. As a consequence, the ability to tune the X-ray energy at synchrotron radiation sources can be used to identify certain elements by taking radiographs at energies below and above an absorption edge.

As an X-ray microscope uses Fresnel zone plates that act as lenses to image the object, it is possible to get a Zernike phase contrast as a complementary imaging mode to the absorption contrast described above. For this purpose, a phase ring has to be inserted in the rear focal plane and hollow cone illumination has to be used, e.g. by using a capillary condenser. The non-scattered light gets an additional phase shift of, for example, half a wavelength and interferes with the scattered light of the object that was also shifted in

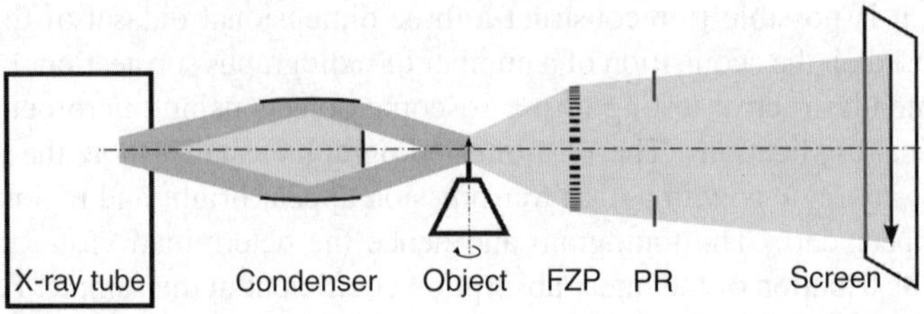


*Figure 12.25* Schematic drawing of the optical path of a lab-based TXM. The focus of the X-ray tube is focused on to the object plane by a capillary condenser. The object is then imaged to the scintillating screen by a Fresnel zone plate (FZP). A phase ring (PR) in the rear focal plane can be inserted to obtain Zernike phase contrast

phase with respect to the refractive index of the transmitted material. Zernike phase contrast especially exaggerates the contrast in areas with gradients of the refractive index, e.g. at phase boundaries, and is overlaying the absorption contrast.

Besides the possibility to obtain single radiographs of the object, the acquisition of 3D data is desirable. There exist well-known algorithms such as filtered back projection (FBP) and various algebraic reconstruction techniques, which are available to reconstruct the 3D information based on a series of projections of the object that are taken at several angles [74]. Typically, the object has to be rotated by 180 or 360 degrees in certain angle steps. This method is called computed tomography. There is a variety of software available that implements these algorithms and performs the reconstruction of the tomogram, which represents the three-dimensional distribution of the linear attenuation coefficient. The availability of the tomogram enables further analysis like virtual cross-sectioning, pore and phase analysis and 3D visualizations.

***Synchrotron-Based X-ray Microscopy.*** The experimental implementation of X-ray microscopy was driven by synchrotron facilities as they provide considerably higher brightness compared to lab-based sources. Thus, the main focus of TXM is still there. The available photon energy range depends on the specific storage ring, beamline optics and the attached microscope. In general, TXM at synchrotron radiation sources is performed with both soft X-rays at minimum photon energies down to 100 eV and hard X-rays up to 20 keV. TXM with soft X-rays is one step ahead in terms of achievable resolutions [75–78].

Synchrotron-based X-ray microscopy provides the full potential of TXM, i.e. using matching photon energies corresponding to the investigated materials and short acquisition times for time-resolved studies [79].

***X-Ray Microscopy in the Laboratory.*** The development of high power rotating anode X-ray sources as well as efficient condenser optics and zone plates have enabled the use of X-ray microscopy in the lab. The beam path of a typical lab-based X-ray microscope is shown on the right-hand side of Figure 12.25. Anode materials used are, for example, copper or chromium. The X-ray beam is focused on to the object by a capillary condenser. The sample stage provides sample rotation during tomography as well as arbitrary translation. Transmitted light from the sample is then imaged on to a scintillating screen by a Fresnel zone plate and subsequently magnified by a folded visible light microscope to a CCD camera [80].

Such lab-based systems are able to provide both individual radiographs and tomography images, e.g. of multiple TSVs. Therefore, lab-based transmission X-ray microscopy

***Table 12.2*** *X-ray transmission of several thicknesses of copper and silicon for Cu-$K_\alpha$ radiation*

| Material | Density (g/cm$^3$) | Absorption length (μm) | Transmission (%) | | |
|---|---|---|---|---|---|
| | | | 50 μm | 30 μm | 5 μm |
| Silicon | 2.33 | 70.8 | 49.4 | 65.5 | 93.2 |
| Copper | 8.96 | 23.0 | 10.6 | 26.0 | 79.9 |

represents the technique of choice to localize defects in copper TSV interconnects for 3D IC integration. The penetration depth for Cu-$K_\alpha$ radiation (8.0 keV) is summarized in Table 12.2. However, the acquisition times are significantly longer compared to synchrotron-based TXM. According to the required magnification, image size and noise level, the acquisition of a single radiograph or tomography might take several minutes or several hours to tens of hours, respectively.

***Sample Preparation.*** As discussed in previous sections, the intensity of the transmitted X-rays depends on the penetrated thickness of the material and on the chosen photon energy. The penetration depth of hard X-rays is in the order of several tens of micrometers in the case of most of the typical materials used in microelectronic products. If soft X-rays are used, the penetration depth is reduced to several micrometers. In most cases sample preparation is needed to provide target prepared samples with a feasible remaining thickness of the region of interest.

Samples for TXM with hard X-rays can usually be prepared by common metallographic methods. With respect to the feature of interest either chip thinning by backside grinding, dimpling and polishing or dicing of a lamellae out of the chip are convenient approaches to prepare a sample with a thickness of, for example, 60 μm. Radiographs of such samples can be directly acquired. If three-dimensional information is desired, any approach to obtain tomography data from *flat samples* like limited angle tomography or tilted rotational axis tomography [81] can be applied. If the available penetration depth demands thinner specimens, more advanced sample preparation has to be done. Analogously to the preparation for transmission electron microscopy (TEM), the region of interest can be milled with focused ion beam (FIB) tools to the desired thickness out of a diced lamella [82]. Furthermore, it should be noted that upcoming methods like laser ablation and plasma FIB [83, 84] are promising techniques for sample preparation as they allow removal of large amounts of material in micrometer scale with the required precision.

*Applications of X-Ray Microscopy in Microelectronics*

The TSV characterization described here has been performed using a lab-based TXM *NanoXCT-100*® from Xradia Inc. as well as the soft X-ray microscope U41-TXM at the BESSY II synchrotron facility. Although the relatively high penetration depth of X-rays allows imaging an entire array of TSVs, some sample preparation is required, as discussed above. Since the region of interest remains intact, X-ray microscopy can be used to study the same sample during a series of process steps (e.g. annealing). If necessary, a subsequent analysis with other methods is possible as well [85].

*Table 12.3* Method comparison for void inspection in TSVs [86]

| | Spatial resolution | Probing depth | Sample preparation | 3D imaging |
|---|---|---|---|---|
| TXM | ~50 nm | ~60 μm | Minimal, no physical cross-sectioning into TSV | Yes |
| SEM | ~1–10 nm | <10 nm | Medium, cross-sectioning into TSV | Yes, e.g. 'Slice and FIB' |
| TEM | ~0.1 nm | <200 nm | Extensive, thinning to <200 nm by FIB | Yes, limited to lamella thickness |

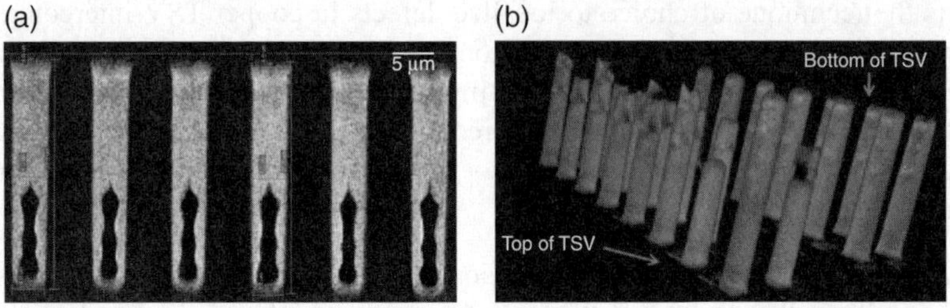


*Figure 12.26* (a) Cross-section of a tomogram of a row of 5 μm TSVs with partially filled vias. (b) 3D visualization of 4 μm (front rows) and 5 μm (rear rows) TSVs. All figures are courtesy of SEMATECH

***Void imaging Using X-ray Microscopy.*** The methods shown in Table 12.3, TXM, SEM and TEM, have the capability of inspecting copper-filled TSVs for voids. XCT has the advantage that it can characterize voids in an array of TSV structures nondestructively [86]. Although SEM and TEM studies provide a better resolution than TXM, the probing depth is very low and physical cross-sectioning of the TSV is required. SEM-based 3D information can be obtained using the FIB technique. This technique allows imaging of the TSV applying a serial-cut method, and the series of images can be assembled into a 3D tomography image. This process is not only time consuming but it also mills the TSV; i.e. a further analysis of detected features is not possible. For electron tomography in a TEM, the sample has to be thinned to ~100 nm.

Figure 12.26 shows a tomography of two arrays of TSVs with diameters of 4 μm and 5 μm, respectively, and 27 μm of depth, filled using the identical process. The photon energy was 8.0 keV and the field of view 67 μm. Defects observed in these TSVs were incomplete filling at the bottom (4 μm TSVs) and large voids at the bottom (5 μm TSVs).

Figure 12.27 shows cross-sections at identical sample positions of TSVs before and after annealing (225 °C/24 h). It is observed that the shape of the voids is changed and copper is extruded at the top of the TSVs.

***Low-Energy Synchrotron-Based TXM Structure Imaging.*** Low-energy synchrotron radiation (<2 keV) has a reduced penetration depth of less than 5 μm for silicon wafers with copper TSVs. Therefore, a lamella with an appropriate thickness was prepared. Figure 12.28 shows radiographs of several TSVs with an $SiO_2$ liner and TaN/Ta/Cu stack before copper

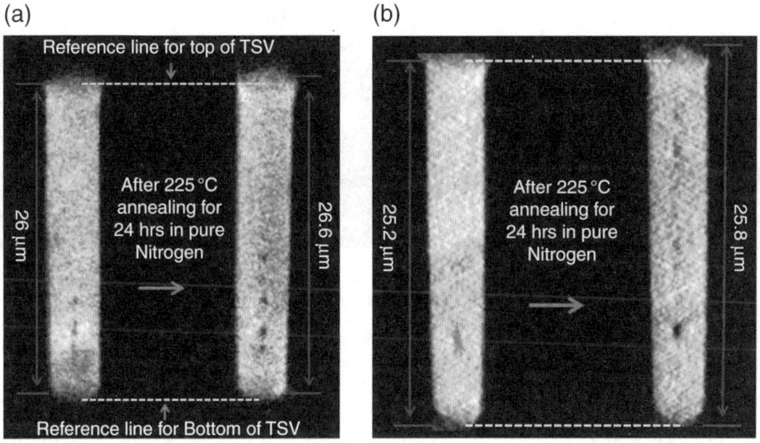


*Figure 12.27* *Changes of the original defect after 225 °C/24 h for (a) 5 µm and (b) 4 µm TSVs* [87]

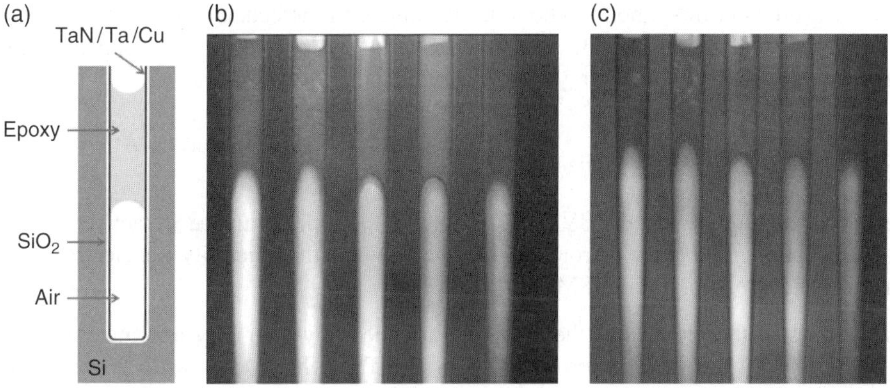


*Figure 12.28* *(a) Scheme of the samples; (b) and (c) radiographs of dummy TSVs at 510 eV and 555 eV photon energy, respectively*

plating at two photon energies, 510 eV and 555 eV, respectively. The experiments had been performed at the U41-TXM at the storage ring BESSY II [88]. The vias are completely embedded in the lamella.

The radiographs offer the possibility to study, for example, the liner thickness and conformity. The influence of absorption edges can be seen by comparing the images at 510 eV and 555 eV. The $SiO_2$ liner appears dark at the higher energy since the X-ray absorption of the $SiO_2$ material increases significantly above the O-*K* absorption edge (543 eV) [86].

***X-Ray Microfocus Tomography.*** As a complementary method, X-ray microfocus tomography provides the opportunity to investigate larger regions with lower resolution. The

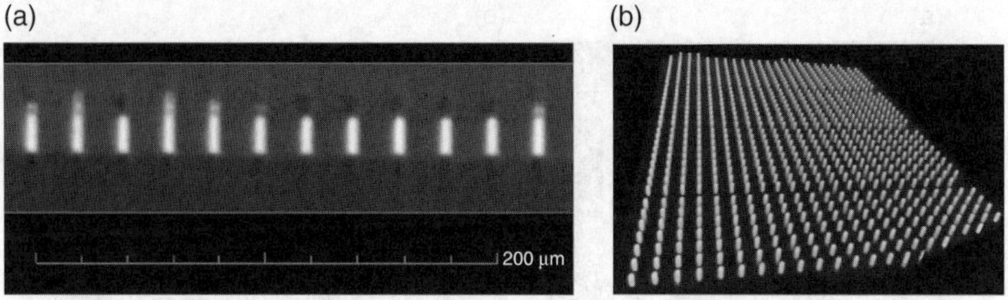


*Figure 12.29* (a) Upside-down cross-section and (b) 3D visualization of an X-ray microfocus tomography of a TSV array

method is well known from nondestructive testing of chip packages and printed circuit boards. Commercial systems offer submicrometer resolution. Hence, they are applicable to the inspection of 3D chip interconnects. The photon energy of several 10 keV can be matched to the specific object to provide sufficient transmission and contrast. In general, no sample preparation is needed.

In Figure 12.29, 5 μm TSVs from via chain structures are shown, based on a tomography study (measurement time about 10 hours). Incomplete fill defects are detected at most of the bottoms of this sample. The 3D visualization offers an overview of the sampled region that includes a relatively large array of TSV structures.

### 12.3.4 Microstructure Analysis

The microstructure of metal TSVs can affect both the performance of active devices through the stress field in the surroundings of the TSV and the reliability, i.e. electromigration (EM) and stress migration (SM) [89]:

- The stress in the vicinity of the TSVs is influenced by geometry and materials, but particularly by the metal microstructure of the TSV. Since the elastic properties of copper are highly anisotropic (the anisotropy factor is 3.2), large copper grains and texture have a significant effect on the stress in the silicon next to the copper TSV.
- Both electromigration and stress migration are phenomena that are based on directed atomic transport, caused by gradients of the electric potential and of the stress, respectively. Interfaces and grain boundaries are fast pathways for atomic transport and affect the lifetime of interconnects.

This means that the characterization of the microstructure of the metal interconnects (e.g. TSVs) is of high importance.

The microstructure of on-chip interconnects has been studied extensively using X-ray diffraction (XRD), electron backscatter diffraction (EBSD) in a scanning electron microscope (SEM) and conical dark-field diffraction (CDF) analysis in a transmission electron microscope (TEM). Since the methodology can be transferred from on-chip interconnects to 3D TSVs, some of the examples shown here are from well-studied copper dual-damascene interconnect structures.

*General Concept of Microstructure Characterization of 3D Interconnects*

The following microstructure parameters of interconnect materials are of importance [90]: grain size, texture, stress and all kinds of defects of filled TSVs. In addition, liner, diffusion barrier and seed layer have to be characterized. The microstructure is completely characterized by the general microstructure function as defined by Bunge [91]:

$$G(x) = \begin{cases} i(x) & \text{phase} \\ g(x) & \text{orientation} \\ D(x) & \text{defects, lattice strain} \end{cases} \tag{12.13}$$

where $G(x)$ specifies phase $i$, crystal orientation $g$, as well as lattice defects $D$ (including local residual stress) in any volume element of the material at the position $x$ of the sample. The first part of this function, $i(x)$, is of concern if the microstructure of the barrier metal has to be included in the characterization of interconnect structures (e.g. TSVs). The whole texture information is contained in the term $g(x)$. Additionally, the mechanical stress is of particular interest, as mentioned above. The first-order strain or macrostrain $\langle\varepsilon\rangle$ is the quantity that is experimentally accessible by common X-ray diffraction methods. The first-order residual stress $\langle\sigma\rangle$ can be considered as one contribution to the term $\langle D\rangle$ rather than $D(x)$ in Equation (12.13), i.e. an average value over a specific volume [92]. Due to this fact, the measured stress is influenced by the texture of the copper inside the probed volume. Hence, a complete determination of the three-dimensional stress state of copper applying X-ray diffraction requires knowledge of the texture as well [93], especially due to the elastic anisotropy of copper (Young's modulus $E_{Cu}\langle 111\rangle =$ 191.1 GPa, $E_{Cu}<100> = 66.7$ GPa). The most complete analysis would imply the simultaneous measurement of both the orientation of single grains (i.e. $g(x)$) and their local stress state $\sigma(x)$.

*X-Ray Texture Analysis – Classical Orientation Distribution*

X-ray diffraction is one of the most suitable nondestructive methods for texture analysis of copper interconnects. Several (*hkl*) pole figures of the material (e.g. (111), (200) and (220) for copper) have to be recorded in order to obtain the required information about the orientation distribution function $f(g)$ (ODF) of the crystallites [94]:

$$f(g) = \frac{dV_g/V}{dg} = f(\varphi_1, \Phi, \varphi_2) \tag{12.14}$$

The crystal orientation $g$ (i.e. the orientation of the crystallographic axes with respect to the chosen sample reference axes) is expressed by the Euler angles $\varphi_1$, $\Phi$, $\varphi_2$ [95]. The ODF considers the volume fractions $dV_g$ of crystallites with orientation $g$ in the irradiated sample volume $V$ with an orientation spread $dg$. Software for the calculation of the ODF is commonly based on one of two general approaches, either the harmonic method or the direct methods, e.g. the ADC algorithm (arbitrarily defined cells method) [96] and the WIMV algorithm (Williams–Imhof–Matthies–Vinel) [97, 98]. The direct methods provide an improved ODF approximation in the case of sharp textures. Quantitative information about the volume fractions of certain texture components can then be extracted from the

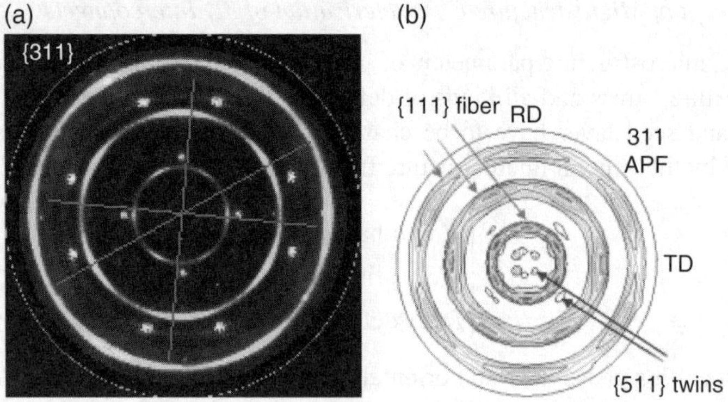


*Figure 12.30* *(a) Cu {311} pole figure of copper lines with w = 180 nm, where the positions for stress measurements are indicated by lines. (b) A reconstructed {311} pole figure, calculated from the ODF, rotated by 90°. Contributions to the {311} pole figure from the {111} oriented grains (fiber) and {511} oriented grains (twins) are indicated* [90]

calculated ODF. Some complementary information is already contained in the pole figures themselves, e.g. the full-width-at-half-maximum (FWHM) of a pole, which gives the orientation spread of a texture component.

From the experimental point of view, the measurement of pole figures has become more convenient and less time consuming since sensitive X-ray area detectors became available [99, 100]. Area detectors with a high quantum efficiency allow very small diffracted signals to be detected, which is beneficial for test structures with a low density of Cu structures like Cu TSV arrays in Si, and reduce the required set of angles $\{\chi; \phi\}$ for the acquisition of a pole figure considerably due to the fact that a whole segment of the reciprocal space is recorded in just one frame. Additionally, the segments of several (*hkl*) diffraction cones (e.g. Cu (111) and (200)) can be captured simultaneously with an area detector [90]. The texture components can basically be described by the FWHM, but a quantitative description needs the ODF to be calculated.

*Texture Characterization for X-Ray Stress Measurements*

The use of an X-ray diffractometer with an area detector and a precise Eulerian cradle makes it possible to perform both stress and pole figure measurements. A triaxial stress data analysis [101] is needed for Cu TSVs for which a biaxial stress model cannot be assumed to be valid. In principle, the pole figure of a higher order (*hkl*) reflection of Cu like {311} has to be measured at first to find the orientation distribution of the {311} crystal planes as a function of the angles ($\chi$; $\phi$). The stress measurements can then be optimized by choosing the optimum angles where the highest possible intensities will be obtained. Results for on-chip copper interconnects that prove this concept are given in Figure 12.30 [90].

Additionally, the (111), (200) and (220) pole figures can be recorded during the same measurement series. This set of pole figures is sufficient to calculate the ODF and to reconstruct the {311} pole figure from this ODF (Figure 12.30 (b)). As an advantage, the texture of the copper structure is characterized at exactly the same position where the stress measurements are performed. During the 2D or 3D stress analysis, the distortions of the segments of the {311} diffraction cones are evaluated for multiple pairs of angles ($\chi$; $\phi$) and the complete stress tensor is calculated, including the shear-stress components. The anisotropy factor of copper is also taken into account. The advantages of this method are obvious. The opportunity exists to correlate the obtained stress values with possibly observed changes of the texture in the investigated copper test structures. Since the {311} pole figure has been measured, it is known which groups of crystal orientations contribute to the stress measurement at a certain set of angles ($\chi$; $\phi$), e.g. if the diffracted intensity includes the fiber as well as the engaged texture component or if it includes the fiber texture only.

*EBSD and CDF – Complementary Orientation Imaging Microscopy*

The main advantage of OIM (orientation imaging microscopy), including EBSD and CDF analysis, is that these methods directly provide the orientation location distribution $g(x)$. This is complementary to X-ray diffraction, which provides integrated information over a vast sample volume (resulting in $\langle g \rangle$ instead of $g(x)$) with penetration depths of at least several micrometers, containing a larger number of crystallites. Particularly for electromigration and stress migration studies, the orientation of single grains in specific TSVs is of interest. To get access to this location and to measure the local texture exactly there, EBSD or CDF analysis can be performed at cross-sections of the Cu TSV structures. Additionally, OIM provides information about the grain sizes. Due to the fact that the relative orientations of neighboring grains are resolved, twin boundaries (i.e. $\Sigma3$) as well as small-angle and high-angle grain boundaries can be identified with EBSD and CDF analysis. Compared to the X-ray texture analysis, OIM provides the distribution of the different types of grain boundaries in copper TSVs. In the case of studies to understand reliability-limiting mechanisms in interconnects, it is most desirable to record OIM images directly on EM or SM test structures before and after the test, and to correlate the EM or SM behavior directly with the grain orientation, grain size distribution and the types of prevailing grain boundaries.

*Microstructure Study Using EBSD Analysis*

EBSD is a proven analytical technique to characterize the microstructure of Cu interconnects, even at FIB-prepared cross-sections. This technique requires minimum grain sizes of about 30 nm. In particular, inverse pole figure (IPF) maps are used to describe the microstructure. Figure 12.31 (a) shows an example for a plan-view EBSD IPF map of an array of 200 nm wide inlaid copper lines after plating, polishing and annealing. Typical experimental parameters are an SEM operating voltage of 20 kV, 70° sample tilt, 3 μm × 10 μm scan size and 40 nm step width. Most of the copper grains have a {111} orientation with respect to the wafer normal, as clearly seen in Figure 12.31 (a). The twin boundaries are marked by solid black lines on the EBSD image. Besides {111} oriented grains, a certain amount of {511} and {611} twins was identified. There is only a very small number of {100} or {101} oriented crystallites. Pole figures can be calculated from the EBSD patterns on inlaid copper lines as well (see, for example, Reference [102]), but their angular resolution is worse than that from X-ray texture analysis (Figure 12.32).

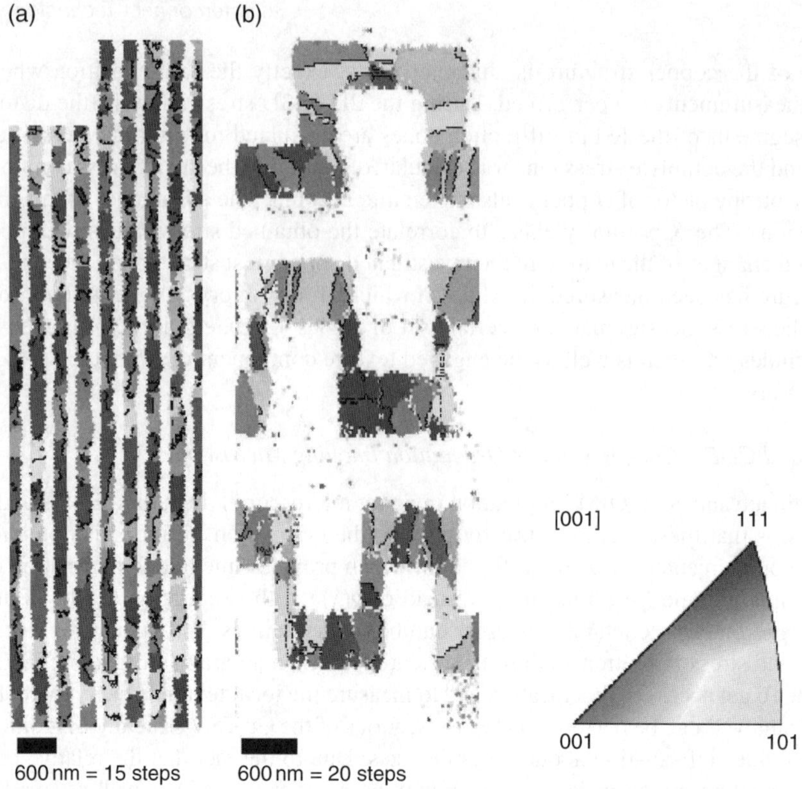


***Figure 12.31*** *EBSD inverse pole figure (IPF) map on 200 nm wide inlaid copper lines. (a) Plan view. (b) EBSD IPF map on cross-sections through via1, via3, via5 chains, showing the orientation of the Cu crystallites inside the vias. Orientations are defined by the coding of the orientation triangle at the bottom margin of the figure and are referenced to the wafer normal for both (a) and (b). Σ3 twin boundaries are marked by black lines. The orientation stereology g(x) is obtained with this technique* [90]

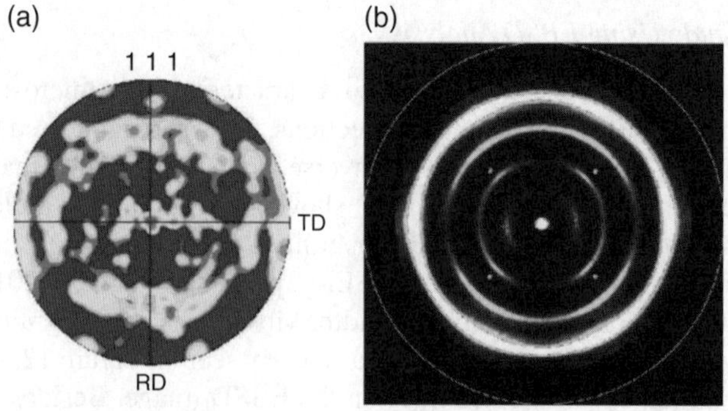


***Figure 12.32*** *(a) {111} pole figure, calculated from a 3 μm × 10 μm EBSD map of an array of 200 nm wide copper lines, like those shown, for example, in Figure 12.31 (a), and (b) {111} pole figure, obtained from X-ray diffraction with an 80 μm beam diameter on an array of 180 nm wide copper metal lines* [90]

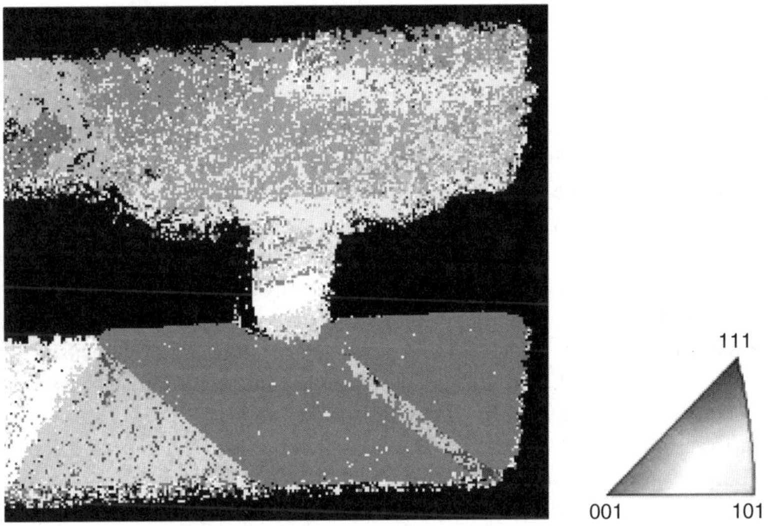


***Figure 12.33*** *[001] inverse pole figure map of a Cu structure* [103]

A typical EBSD scan of an array of copper structures, like the one shown in Figure 12.31 (a), covers an area of several tens of $\mu m^2$, containing a much smaller number of crystallites compared to X-ray diffraction. Therefore, the volume fractions of orientated grains that were obtained with X-ray diffraction are statistically more significant than those from EBSD, leading to more detailed pole figures. On the other hand, EBSD provides the orientation of single crystallites together with their coordinates, and thus directly determines the orientation function $g(x)$ as defined in Equation (12.30).

*Microstructure Study Using CDF Analysis*

In some cases, even for TSVs, the EBSD analysis reaches its limits with respect to texture and grain-size analysis because of grain sizes smaller than 30–50 nm. Therefore, a complete microstructure analysis of TSVs, including the barrier and seed layers, will only be possible using the CDF analysis as a complementary technique, even though it is more time-consuming. Due to a resolution better than 5 nm, the CDF circular scan technique in the TEM provides grain size distributions and complete grain orientation maps for grain sizes <30 nm. Consequently, it allows characterizing the Cu microstructure of regions with small grains [103].

In a first step, centered dark-field images of a particular view of the specimen are recorded in a circular way for several tilt and rotation angles of the parallel electron beam. In a second step, spot diffraction patterns for every point of the field of interest are reconstructed using all recorded dark-field images. Based on the crystallographic structure of the polycrystalline material, the diffraction patterns are indexed. Finally, IPF maps and, hence, the crystallographic orientations of individual Cu grains can be determined [104]. As an example, Figure 12.33 shows a [001] inverse pole figure map of a Cu interconnect structure.

## 12.4 Summary

At the same time as research in 3D TSV technology is advancing quickly, the microscopy techniques used in the evaluation of 3D stacks must also advance in capability. Overlay measurements after wafer bonding, defect detection and void inspection after copper plating are challenging due to the need to image through silicon and inside copper. Microscopy techniques for which silicon is opaque, such as scanning acoustic microscopy (SAM) and confocal infrared (IR) microscopy, are capable of inspecting the interface between bonded wafer pairs, while high-resolution X-ray computed tomography (XCT) is used to detect voids in TSVs. With nano-XCT, voids in copper TSVs with sub-100 nm size can be visualized. For more detailed failure characterization, a target focused ion beam (FIB) cross-section through the localized region of interest (defect) and subsequent scanning electron microscopy (SEM) imaging is proposed [105].

X-ray diffraction and OIM techniques are applicable for TSV metal microstructure analysis. Texture, grain size and the first-order stress of Cu-filled TSVs can be thoroughly characterized using X-ray diffraction, EBSD and CDF analysis. The influence of the TSV geometry, the deposition conditions and the type of liner and barrier layers on the copper texture in filled TSVs can be studied. X-ray diffraction, EBSD and CDF analysis are complementary techniques to X-ray diffraction since the X-ray technique integrates over a larger sample volume, resulting in statistically more significant texture data, whereas EBSD and CDF additionally provide grain boundary distributions, grain sizes and the grain orientation stereology $g(x)$. EBSD and CDF analysis are also suitable methods for studying the local grain orientation, grain boundary distribution and grain size distribution in TSVs.

## References

[1] Pavlidis, V.F. and Friedman, E.G. (2009) *Three Dimensional Integrated Circuit Design*, Elsevier, Morgan Kaufmann, New York.
[2] Moore, G.E. (1965) Cramming more components onto integrated circuits. *Electronics*, **38** (8), 19 April 1965.
[3] Yoon, J.H. (2005) Advanced DRAM technology status and future directions. IBM Global Procurement, JEDEX San Jose, California, 31 March 2005.
[4] Tummala, R.R. (2001) *Fundamentals of Microsystems Packaging*, McGraw-Hill, New York.
[5] Tummala, R.R., Rymaszewski, E.J. and Klopfenstein, A.J. (eds) (1989) *Microelectronics Packaging Handbook*, Van Nostrand Reinhold, New York.
[6] Knickerbocker, J.U. *et al.* (2008) Three-dimensional silicon integration. *IBM J. Res. Develop.*, **52** (6).
[7] Knickerbocker, J.U. *et al.* (2005) Development of next-generation system-on-package (SOP) technology based on silicon carriers with fine pitch chip interconnection. *IBM J. Res. Develop.*, **49** (4/5).
[8] Chen, T.C. (2006) Where Si-CMOS is going: trendy hype vs real technology. Keynote ISSCC.
[9] Bakir, M.S. and Meindl, J.D. (eds) (2009) *Integrated Interconnect Technologies for 3D Nanoelectronic Systems*, Artech House, Boston, Massachusetts.
[10] Emma, P.G. and Kursun, E. (2008) Is 3D chip technology the next growth engine for performance improvement? *IBM J. Res. Develop.*, **52** (6).
[11] Davis, J.A. *et al.* (2001) Interconnect limits on gigascale integration (GSI) in the 21st century. *Proc. IEEE*, 89 (3), 305–324.

[12] Dennard, R.H. *et al.* (1974) Design of ion-implanted MOSFETs with very small physical dimensions. *IEEE J. Solid State Circuits*, **10 SC-9**, October 1974, 256–268.
[13] Agerwala, T. and Gupta, M. (2006) Systems research challenges. A scale-out perspective. *IBM J. Res. Develop.*, **50** (2/3).
[14] Knickerbocker, J.U. *et al.* (2008) Challenges and opportunities for exploiting 3D technology in system designs. AMC & SEMATECH Workshop on 3D Manufacturing and Reliability Challenges for 3D ICs Using TSVs, Del-Mar, California.
[15] Otten, R.H.J.M. (1982) Automatic floorplan design. Proceedings of the IEEE/ACM Design Automation Conference, June 1982, pp. 261–267.
[16] Healy, M. *et al.* (2007) Multiobjective microarchitectural, floorplanning for 2-D and 3-D ICs. *IEEE Trans. on Computer-Aided Design of Integrated Circuits and Systems*, **26** (1), 38–52, January 2007.
[17] Bernstein, K. *et al.* (2007) Interconnects in the third dimension: design challenges for 3-D ICs. Proceedings of the IEEE/ACM Design Automation Conference, June 2007, pp. 562–567.
[18] MacIlrath, L. (2009) CAD tools and design flow for 3D integration. 3D Architectures for Semiconductor Integration and Packaging, December 2009.
[19] Friedman, E.G. (ed.) (1975) *Clock Distribution Networks in VLSI Circuits and Systems*, IEEE Press, New Jersey.
[20] Deutsch, A. and Restle, P.J. (1998) Designing the best clock distribution network. Proceedings of the IEEE Symposium on VLSI Circuits, June 1998, pp. 2–5.
[21] Topol, A.W. *et al.* (2006) Three-dimensional integrated circuits. *IBM J. Res. Develop.*, 50 (4/5).
[22] Takahaski, K. *et al.* (2004) Process integration of 3D chip stack with vertical interconnection. Electronic Components and Technology Conference, pp. 601–609.
[23] Umemoto, M. *et al.* (2004) High performance vertical interconnection for high-density 3D chip stacking package. Electronic Components and Technology Conference, pp. 616–623.
[24] Feil, M. *et al.* (2004) The challenge of ultra thin chip assembly. Electronic Components and Technology Conference.
[25] Hunter, M. *et al.* (2004) Assembly and reliability of flip chip solder joints using miniaturized Au/Sn bumps. Electronic Components and Technology Conference.
[26] Kripesh, V. *et al.* (2005) Three dimensional system-in-package using stacked Si platform technology. *IEEE Trans. on Advanced Packaging*, **28** (3), August 2005.
[27] Ikeda, H., Kawano, M. and Mitsuhashi, T. (2005) Stacked memory chip technology development. SEMI Technology Symposium (STS) 2005 Proceedings, Session 9, pp. 37–42.
[28] Patel, C.S. *et al.* (2005) Silicon carrier with deep through-vias, fine pitch wiring, and through cavity for parallel optical transceiver. 55th Electronic Components and Technology Conference.
[29] Sakuma, K. *et al.* (2007) 3D chip stacking technology with low-volume lead-free interconnections. Electronic Components and Technology Conference, pp. 627–632.
[30] Guarini, K.W. *et al.* (2002) Electrical integrity of state-of-the-art 0.13 μm SOI CMOS devices and circuits transferred for three-dimensional (3D) integrated circuit (IC) fabrication. *IEDM Tech. Dig.*, 943.
[31] Andry, P.S. *et al.* (2008) Design and fabrication of robust through-silicon vias. *IBM J. Res. Develop.*, **52** (6).
[32] Koester, S. *et al.* (2008) Wafer level – three dimension integration technology. *IBM J. Res. Develop.*, **52** (6).
[33] Dang, B. *et al.* (2008) 3D chip stacking with C4 technology. *IBM J. Res. Develop.*, 52 (6).
[34] Sakuma, K. *et al.* (2008) 3D chip-stacking technology with through silicon vias and low-volume lead-free interconnections. *IBM J. Res. Develop.*, **52** (6).
[35] Keast, C. (2009) A SOI-based wafer-scale 3D circuit integration technology. 3D Architectures for Semiconductor Integration and Packaging, December 2009.
[36] Scannell, M. (2009) 3D activities at CEA-Leti. 3D Architectures for Semiconductor Integration and Packaging, December 2009.
[37] Perkins, J. (2009) 3-D TSV markets and technology update. 3D Architectures for Semiconductor Integration and Packaging, December 2009.
[38] Vardamin, J. (2009) 3-D TSV markets: infrastructure requirements for growth. 3D Architectures for Semiconductor Integration and Packaging, December 2009.

[39] http://www.azonano.com/news.asp?newsID=20908, December 2010.
[40] http://eda360insider.wordpress.com/2010/11/16/need-really-big-fpgas-xilinx-will-be-taking-the-%E2%80%9C3d%E2%80%9D-route-for-initial-virtex-7-parts/. Need really big FPGAs? Xilinx will be taking the '3D' route for initial Virtex 7 parts.
[41] http://finance.yahoo.com/news/Semtech-and-IBM-Join-Forces-bw-2431012774.html?x=0&.v=1, December 2010.
[42] www.channelregister.co.uk/2010/12/07/samsng_3d_tsv/. Cached – Samsung readies less sock-ety servers for 3D stacked memory.
[43] http://www.eetimes.com/electronics-news/4212072/Micron-COO-talks-450-mm–3-D–EUV. Micron COO talks 450 mm, 3D, EUV, 1 November 2011.
[44] Knickerbocker, J.U. *et al.* (2007) 3D chip stacking and silicon packaging using through-silicon-vias (TSV) for systems integration. 3DIC, Tokyo, Japan.
[45] Knickerbocker, J.U. *et al.* (2009) Advancements in 3-D technology for system applications. 3D Architectures for Semiconductor Integration and Packaging, December 2009.
[46] Vardaman, J. (2007) 3-D through-siliocn vias become a reality. Semiconductor International, 1 June 2007; see http: //www..semiconductor.net.article/CA6445435.html.
[47] Johnson, S,C. (2007) 3-D TSV chips take off. Semiconductor International, 1 July 2007; see http: //www.semiconductor.net/article//CA6455523.html.
[48] Topol, A.W. *et al.* (2005) Enabling SOI-based assembly technology for three-dimensional (3D) integrated circuits (ICs). Proceedings of the IEEE International Electron Devices Meeting, Washington, DC, pp. 352–355.
[49] Guarini, K.W. *et al.* (2002) Electrical integrity of state-of-the-art 0.13 μm SOI CMOS devices and circuits transferred for three-dimensional (3D) integrated circuit (IC) fabrication. *IEDM Tech. Dig.*, 943–945 (2002).
[50] Joseph, A.J. *et al.* (2008) Through-silicon vias enable next-generation SiGe power amplifiers for wireless communications. *IBM J. Res. Develop.*, **52** (6).
[51] Stamper, A.K. *et al.* (2009) CMOS compatible insulated through silicon vias for 3D silicon. Advanced Metallization Conference, Baltimore, Maryland.
[52] Reed, J.D. *et al.* (2010) High density interconnection at 10 μm pitch with mechanically keyed Cu/Sn–Cu and Cu–Cu bonding for 3-D integration. Electronic Components and Technology Conference, Las Vegas, Nevada, pp. 846–852.
[53] Agarwal, R. *et al.* (2010) Cu/Sn microbumps interconnect for 3D TSV chip stacking. Electronic Components and Technology Conference, Las Vegas, Nevada, pp. 858–863.
[54] Wright, S.L. *et al.* (2006) Characterization of micro-bump C4 interconnections for Si-carrier SOP applications. Electronic Components and Technology Conference, pp. 633–640.
[55] Dang, B. *et al.* (2007) Assembly, characterization, and reworkability of Pb-free ultra-fine pitch C4s for system-on-package. Electronic Components and Technology Conference, pp. 42–48.
[56] Maria, J. *et al.* (2011) 3D chip stacking with 50 μm pitch lead-free micro-C4 interconnections. Electronic Components and Technology Conference, Orlando, Florida.
[57] Tamura, T. *et al.* (2010) Novel adhesive development for CMOS-compatible thin wafer handling. Electronic Components and Technology Conference, Las Vegas, Nevada.
[58] Dang, B. *et al.* (2010) CMOS compatible thin wafer processing using temporary mechanical wafer, adhesive and laser release of thin chips/wafers for 3D integration. Electronic Components and Technology Conference, Las Vegas, Nevada.
[59] Kim, B. and Matthias, T. (2009) Thin wafer handling and chip stacking for 3D integration. 3D Architectures for Semiconductor Integration and Packaging, December 2009.
[60] Bair, B. (2009) Efficient CAPAX utilization for thin wafer handling. 3D Architectures for Semiconductor Integration and Packaging, December 2009.
[61] Andry, P. *et al.* (2006) A CMOS-compatible process for fabricating electrical through-vias in silicon. Electronic Components and Technology Conference.
[62] Patel, C.S. (2006) Silicon carrier for computer systems. Design Automation Conference, 24–28 July 2006.
[63] Dickson,T. *et al.* (2011) IEDM paper submission.
[64] Knickerbocker, J.U. *et al.* (2006) 3D silicon integration and silicon packaging technology using silicon through-vias. JSSC.

[65] Sri-Jayantha, S.M. *et al.* (2008) Thermomechanical modeling of 3D elctronic packages. *IBM J. Res. Develop.*, **52** (6).

[66] Takahashi, K. and M. Sekiguchi, M. (2006) Through silicon via and 3D wafer/chip stacking technology. Symposium on VLSI Circuits, Digest of Technical Papers, Symposium Issue, 2006, pp.89–92.

[67] Kurita, Y. *et al.* (2007) A 3D stacked memory integrated on a logic device using SMAFTI technology. Electronic Components and Technology Conference, Reno, Nevada.

[68] Briggs, A. and Kolosov, O. (2010) *Acoustic Microscopy*, 2nd edn, Oxford University Press, Oxford.

[69] http://www.sonix.com/learning (01 June 2011).

[70] Kong, L.W., Krueger, P., Zschech, E., Rudack, A.C., Arkalgud, S. and Diebold, A.C. (2010) Sub-imaging techniques for 3D-interconnect on bonded wafer pairs. Proceedings of the 11th International Workshop on Stress-Induced Phenomena in Metallization, Dresden, Germany, 12–14 April 2010; *AIP Conf. Proc.*, **1300**, 221–228.

[71] Rudack, A.C., Kong, L.W. and Baker, G.G. (2010) Proceedings of the IEEE Advanced Semiconductor Manufacturing Conference (ASMC), 11–13 July 2010, pp. 347–352.

[72] Michette, A.G. (1993) *Diffractive X-ray Optics in X-Ray Science and Technology*, Institute of Physics Publications, London.

[73] Thompson, A. *et al.* (2001) *X-Ray Data Booklet*, 2nd edn, Lawrence Berkley National Laboratory, Berkeley, California.

[74] Kak, A.C. and Slaney, M. (1988) *Principles of Computerized Tomographic Imaging*, IEEE Press, Piscataway, New Jersey.

[75] Koch, E.-E. (ed.) (1983) *Handbook on Synchrotron Radiation*, vol.1a, North Holland.

[76] Kirz, J. and Jacobsen, C. (2009) The history and future of X-ray microscopy. *J. Phys.: Conf. Ser.*, **186**, 012001.

[77] Chen, J., Li, W., Liu, Y., Yue, Z., Tian, J., Liu, L., Xiong, Y., Liu, G., Wang, C., Wu, Z., Yu, H. and Tian, Y. (2009) Fresnel zone-plate based X-ray microscopy in Zernike phase contrast with sub-50 nm resolution at NSRL. *J. Phys.: Conf. Ser.*, **186**, 012005.

[78] Chu, Y.S., Yi, J.M., De Carlo, F., Shen, Q., and Lee, W.-K. (2008) Hard-X-ray microscopy with Fresnel zone plates reaches 40 nm Rayleigh resolution. *Appl. Phys. Lett.*, **92**, 103119.

[79] Schneider, G., Hambach, D., Niemann, B., Kaulich, B., Susini, J., Hoffmann, N. and Hasse, W. (2001) In situ X-ray microscopic observation of the electromigration in passivated Cu interconnects. *Appl. Phys. Lett.*, **78**, 1936.

[80] http://xradia.com/technology/index.php (1 June 2011).

[81] Krüger, P., Niese, S., Zschech, E., Gelb, J. and Feser, M. (2010) Improved scanning geometry to collect 3D-geometry data in flat samples. Proceedings of the 10th International Conference on X-ray Microscopy, Chicago, in press.

[82] Ritz, Y., Stegmann, H., Engelmann, H.J. and Zschech, E. (2004) Target preparation of samples for 3D-TEM using micromanipulators. *Pract. Metrol.*, **41**, 180–189.

[83] Stegmann, H., Doemer, H., Cai, H., Rosenkranz, R. and Zschech, E. (2011) Efficient target preparation by combining laser ablation and FIB milling in a single tool. Proceedings of the Semiconductor Conference, Dresden SCD, Dresden, to be published.

[84] Kwakman, L., Franz, G., Klumpp, A. and Ramm, P. (2011) Characterization and failure analysis of 3D integrated systems using a novel plasma-FIB system. Proceedings of the International Conference on Frontiers of Characterization and Metrology for Nanoelectronics, Grenoble, to be published.

[85] Niese, S., Krueger, P. and Zschech, E. (2011) NanoXCT – a high-resolution technique for TSV characterization. Proceedings of the International Workshop on Stress Management for 3D ICs Using Through Silicon Vias (eds E. Zschech, R. Radojcic, V. Sukharev and L. Smith); *AIP Conf. Proc.*, **1378**, in press.

[86] Kong, L.W., Krueger, P., Zschech, E., Rudack, A.C., Arkalgud, S. and Diebold, A. (2010) Sub-imaging techniques for 3D-interconnects on bonded wafer pairs. Proceedings of the 11th International Workshop on Stress-Induced Phenomena in Metallization (eds P.S. Ho, S. Ogawa and E. Zschech); *AIP Conf. Proc.*, **1300**, 221.

[87] Kong, L.W., Rudack, A.C., Krüger, P., Zschech, E., Arkalgud, S. and Diebold, A.C. (2011) 3D-interconnect: visualization of extrusion and voids induced in copper-filled through-silicon

vias (TSVs) at various temperatures using X-ray. *Microscopy, Microelectron. Engng*, in press; DOI: 10.1016/j.mee.2011.04.012.

[88] Guttmann, P., Niemann, B., Thieme, J., Hambach, D., Schneider, G., Wiesemann, U., Rudolph, D. and Schmahl, G. (2001) Instrumentation advances with the new X-ray microscopes at BESSY II. *Nucl. Instrums and Meth. in Physics Res. A*, **467–468**, 849–852.

[89] Zschech, E., Ho, P.S., Schmeisser, D., Meyer, M.A., Vairagar, A.V., Schneider, G., Hauschildt, M., Kraatz, M. and Sukharev, V. (2009) Geometry and microstructure effects on electromigration-induced copper interconnect degradation. *IEEE Trans. on Devices and Mater. Reliab.*, **9**, 20–30.

[90] Geisler, H., Zienert, I., Prinz, H., Meyer, M.A. and Zschech, E. (2003) Potential and limits of texture measurement techniques for inlaid copper process optimization. Proceedings of the International Conference on Characterization and Metrology for ULSI Technology, vol. 683, pp. 469–479.

[91] Bunge, H.J. (1999) Texture, microstructure and properties of polycrystalline materials. Proceedings of NASAT-97: First National Seminar on the Application of Textures in Materials Research, Hyderabad, India, 4–5 December 1997 (eds R.K. Ray and A.K. Singh), Oxford & IBH Publishing, New Delhi, pp. 3–44.

[92] Noyan, I.C. and Cohen, J.B. (1987) *Residual Stress*, Springer, New York, pp. 136–140.

[93] Van Houtte, P. and De Buyser, L. (1993) The influence of crystallographic texture on diffraction measurements of residual stress. *Acta Metall. Mater.*, **41**, 323–336.

[94] Bunge, H.J. and Schwarzer, R.A. (2001) Orientation stereology – a new branch in texture research. *Adv. Engng Mater.*, **3**, 25–39.

[95] Bunge, H.J. (1982) *Texture Analysis in Materials Science: Mathematical Methods*, Butterworths, London.

[96] Pawlik, K., Pospiech, J. and Lücke, K. (1991) The ODF approximation from pole figures with the aid of the ADC method. *Textures and Microstructs*, **14–18**, 25–30.

[97] Matthies, S. and Vinel, G.W. (1982) On the reproduction of the orientation distribution function of texturized samples from reduced pole figures using the conception of a conditional ghost correction. *Phys. Stat. Sol.(b)*, **112**, K111–K114.

[98] Kocks, U.F., Tomé, C.N. and Wenk, H.-R. (1998) *Texture and Anisotropy*, Cambridge University Press, Cambridge, pp. 102–124.

[99] Bunge, H. J. and Klein, H. (1996) Determination of quantitative, high-resolution pole figures with the area detector. *Z. Metallkde.*, **87**, 465–475.

[100] Smith, K.L. and Ortega, R.B. (1993) Use of a two-dimensional, position sensitive detector for collecting pole figures, in *Advances in X-Ray Analysis*, vol. 36 (eds J.V. Gilfrich *et al.*), Plenum, New York, pp. 641–647.

[101] He, B.B., Preckwinkel, U. and Smith, K.L. (2000) Advantages of using 2D detectors for residual stress measurements, in *Advances in X-Ray Analysis*, vol. 42, Proceedings of the 47th Annual Denver X-Ray Conference 1998, International Centre for Diffraction Data, Newton Square.

[102] Zienert, I., Besser, P., Blum, W. and Zschech, E. (2001) Crystallographic texture characterization of inlaid copper interconnects, in *Mechanisms of Surface and Microstructure Evolution in Deposited Films and Film Structures* (eds J. Sanchez Jr *et al.*), Material Research Society Symposium Proceedings, vol. 672, MRS, Warrendale, Pennsylvania, pp. O7.9.1–O7.9.6.

[103] Zschech, E., Huebner, R., Potapov, P., Zienert, I., Meyer, M.A., Chumakov, D., Geisler, H., Hecker, M., Engelmann, H.J. and Langer, E. (2008) Process control and physical failure analysis for sub-100 nm Cu/low-*k* structures. Proceedings of the IITC 2008, p. 67.

[104] Huebner, R., Engelmann, H.J. and Zschech, E. (2010) Conical dark field analysis for small grain characterization in narrow Cu interconnect structures: potential and challenges. Proceedings of the 11th Workshop on Stress-Induced Phenomena in Metallization, Dresden, Germany, 12–14 April 2010; *AIP Conf. Proc.*, **1300**, 39–46.

[105] Zschech, E. and Diebold, A. (2011) Metrology and failure analysis for 3D IC integration. Proceedings of the International Conference on Frontiers of Characterization and Metrology for Nanoelectronics, Grenoble, to be published.

# 13

# Carbon Nanotubes for Interconnects

*Mizuhisa Nihei, Motonobu Sato, Akio Kawabata, Shintaro Sato and Yuji Awano*

*MIRAI-Selete, Atsugi, Japan*

## 13.1 Introduction

With shrinking dimensions in LSI circuits, the resistivity and electromigration (EM) reliability of Cu interconnects deteriorates as the line width decreases. Carbon nanotubes (CNTs) have been studied as candidates of emerging research materials. Carbon nanotubes exhibit excellent electrical properties that include a long mean free path up to several micrometers [1], high current carrying capability exceeding $10^9$ A/cm$^2$ [2] and high thermal conductivity [3]. Because of these factors, with their low electrical resistance and large electromigration tolerance, it is expected that CNTs will be used as aggressively scaled interconnects for next-generation LSIs.

We report our trials of using CNT bundles as vertical interconnects such as vias on Cu wire and contact plugs on CMOS, as shown in Figure 13.1. We proposed CNT damascene processes to integrate CNT contact plugs with CMOS transistors. We succeeded in growing vertically aligned multiwalled CNTs with a diameter of 10 nm and a density of $3 \times 10^{11}$ cm$^{-2}$ at temperatures as low as 400 °C in the via and contact hole. The resistance of CNT bundles was found to be of the same order as that of conventional tungsten contact plugs [4]. In order to decrease the resistance as low as that of Cu wire, we need to increase the site density of CNTs. We have already reported a high-density CNT growth of more than $10^{12}$ tubes/cm$^2$ by optimizing catalyst metals [5]. We also demonstrated low-temperature CNT growth at 365 °C, which would be more helpful in realizing reliable CNT vias for

*Advanced Interconnects for ULSI Technology*, First Edition. Edited by Mikhail R. Baklanov, Paul S. Ho and Ehrenfried Zschech.

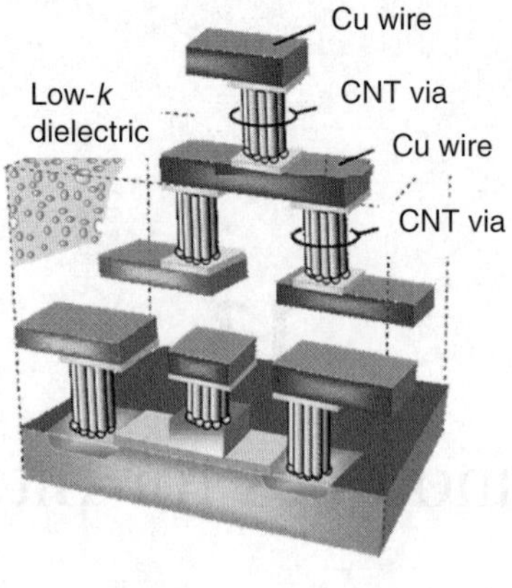


*Figure 13.1* Schematic of future LSI interconnects consisting of CNT vias and low-k materials

future LSIs [6]. At such low temperatures, the ultra-low-*k* dielectric layer and the silicide layer were not damaged.

Moreover, we demonstrated that the CNT was able to sustain a current density as high as $1.7\times10^{8}$ A/cm$^2$ per tube in a CNT via without any deterioration in its properties [7]. Our CMP process, which could improve the contact to the upper metal line, had not only lowered via resistance but also effectively improved the current tolerance. In the same way, several improvements of the contact to both the upper and the lower metal lines could withstand a high current density of a CNT via.

## 13.2 Advantage of CNT Vias

The advantage of CNT bundles is their low resistance, which may be the solution to the problem of high resistance in scaled-down vias. As shown in Figure 13.2, we estimated the resistance of a 50-nm-diameter via depending on the filling rate of CNTs in the via area [4]. In this estimation we assumed that CNTs have the quantum resistance $R_Q = h/4e^2 = 6.45\,\text{k}\Omega$ (conductance $G_Q = 2G_{Q0} = 4e^2/h$, which reaches the maximum conductance limit for ballistic transport in two channels of a CNT), that current flows through each shell of MWNTs and that there is no dependence of ballistic transport on CNT length. In order to lower the resistance of CNT vias, it is necessary to increase the nanotube's density by decreasing its diameter. Regarding the electrical properties, CNTs consist of semiconductive CNTs as well as metallic types. Since the energy gap of a semiconductive CNT is inversely proportional to its diameter, smaller-diameter SWNTs may adversely influence the current conduction property. On the other hand, larger-diameter MWNTs seem to have a vanishing energy gap at room temperature. We are therefore aiming at using metallic MWNTs with their ballistic transport properties as vias.

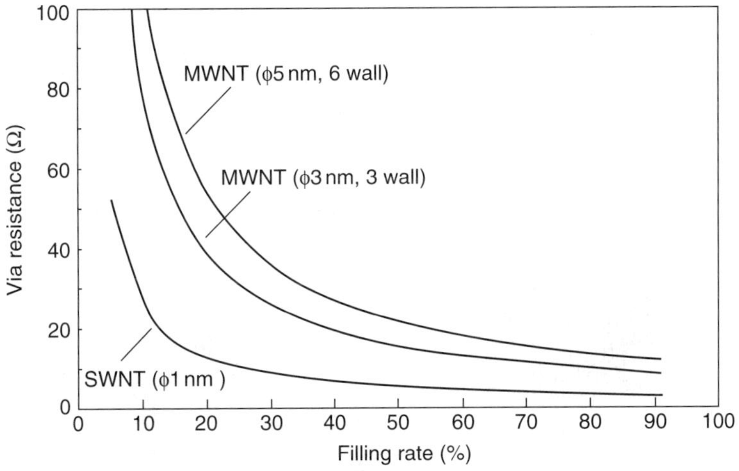


***Figure 13.2*** *Estimated resistance of 50-nm-diameter vias dependent on the filling rate of CNTs in a via hole for a 1-nm-diameter SWNT, 3-nm-diameter three-walled MWNT and 5-nm-diameter six-walled MWNT*

## 13.3 Fabrication Processes of CNT Vias

The fabrication processes are compatible with conventional LSI processes as shown in Figure 13.3 [4]. A substrate with a Cu interconnect covered by a dielectric layer was first prepared. The dielectric layer was SiOC with $k = 3.0$ or $k = 2.6$ (ULK). Via holes with a diameter of 160 nm were made using conventional photolithography followed by dry etching. A TaN/Ta barrier layer and a TiN contact layer were deposited by physical vapor deposition (PVD). Size-controlled Co particles with a mean diameter of about 4 nm were then deposited using a nanoparticles deposition system [8]. MWNTs were grown by thermal CVD with $C_2H_2$ diluted by argon as the source gas. The substrate temperature ranged from 365 °C to 450 °C. The substrate with MWNTs were then coated with spin-on glass (SOG) and planarized by CMP. The CMP condition was similar to the one used for polishing a silicon dioxide layer. Finally, a Ti top contact layer, a Ta barrier layer and a Cu wire were connected to the CNT vias by PVD.

Figures 13.4 (a) and (b) are the cross-sectional scanning electron microscopy (SEM) images of CNT vias fabricated with growth temperatures of 450 °C and 400 °C. We can see in the images that CNTs grown at 400 °C are a little less straight than those at 450 °C, suggesting that CNTs at 400 °C are a little more defective. To investigate the quality of CNTs further, we performed transmission electron microscopy (TEM) analyses, whose results are shown in Figure 13.5. The TEM images indicate that CNTs grown at either temperature are of high quality. However, CNTs at 400 °C appear to be a little more defective.

Moreover, we demonstrate CNT growth at temperatures below 400 °C. In fact, the lower the growth temperature, the better, since some parts of LSIs can be damaged by heat. Figure 13.6 shows the cross-sectional SEM image of CNTs grown by thermal CVD at 365 °C

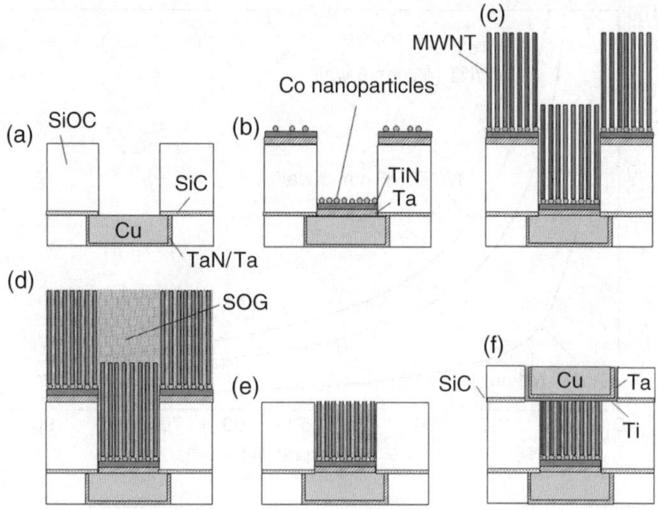


***Figure 13.3*** *CNT damascene via process: (a) via hole formation on bottom Cu interconnect, (b) TaN/Ta barrier layer, TiN contact layer and Co catalyst nanoparticle formation, (c) MWNT growth, (d) SOG coating, (e) CMP planarization and (f) top Cu interconnect formation*

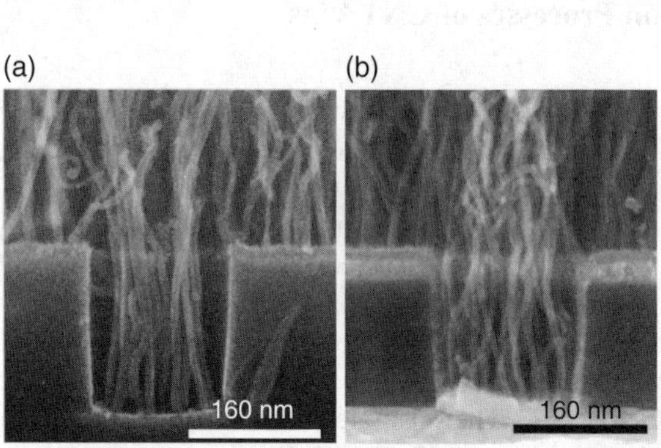


***Figure 13.4*** *Cross-sectional SEM image of the 160-nm-diameter CNT vias at the growth temperatures of (a) 450 °C and (b) 400 °C*

[6]. It is amazing that we can synthesize CNTs even at this low temperature. Also, we have recently reported on CNT growth below 400 °C by plasma-enhanced CVD (PECVD) [9, 10].

Figure 13.7 (a) shows a cross-sectional SEM image of CNTs formed all over the substrate, having 160-nm-diameter via holes, at the growth temperature of 450 °C. We succeeded in growing vertically aligned MWNTs with a diameter of 10 nm, a shell number of 7 and a density of $3\times10^{11}$ cm$^{-2}$. Figure 13.7 (b) shows a cross-sectional SEM image of CNT vias after CMP planarization. MWNT bundles were successfully polished under pressures as low as those in the conventional Cu/low-*k* CMP process. Although SOG is well filled with MWNTs inside the 160-nm-diameter via hole, the filling factor of CNT in via is still low.

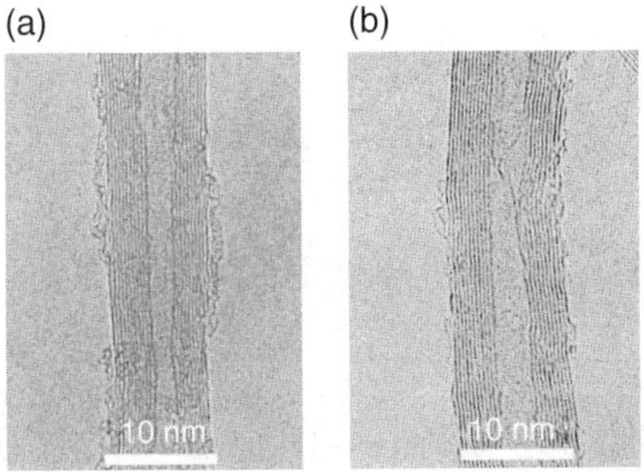


***Figure 13.5*** *TEM image of the CNT at the growth temperatures of (a) 450 °C and (b) 400 °C*

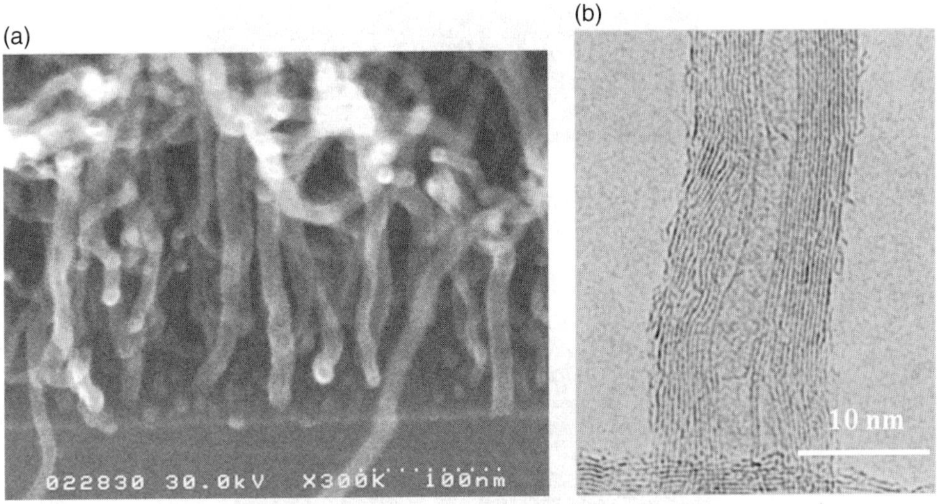


***Figure 13.6*** *(a) Cross-sectional SEM image of the CNT at the growth temperature of 365 °C and (b) TEM image*

We reported the high-density CNT growth of more than $10^{12}$ tubes/cm$^2$ by optimizing catalyst metals [11, 12]. Figure 13.8 shows a cross-sectional SEM image of high-density CNT vias. CNTs were synthesized by the thermal CVD method. A Co catalyst (1.0 nm) was deposited on a TiN contact layer (0.5 nm)/TaN barrier layer/Cu film. The source gas for CVD was $C_2H_2$ diluted by Ar and the substrate temperature was 450 °C. We analyzed the metal structure of the activated catalyst after the CNT growth by using cross-sectional TEM-EELS (electron energy loss spectroscopy) measurement. From the measurement, we found that the peak position of Co and Ti overlaid each other for the activated catalyst. On the other hand, the position of Co and Ti were almost separated for the catalyst that CNT did not grow. We speculated that the Co catalyst that was wrapped with Ti was easy to activate to grow CNTs.

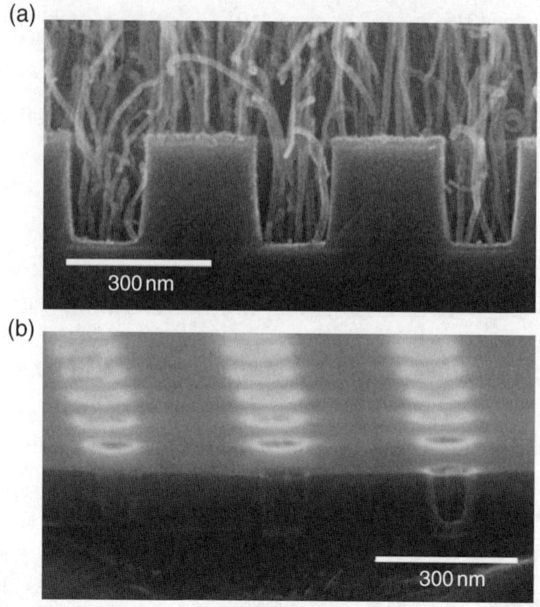


*Figure 13.7* Cross-sectional SEM image of (a) vertically aligned MWNTs formed all over the substrate, having 160-nm-diameter via holes, and (b) 160-nm-diameter CNT vias after CMP planarization

*Figure 13.8* Cross-sectional SEM image of the high-density 160-nm-diameter CNT vias at the growth temperature of 450 °C

## 13.4 Electrical Properties of CNT Vias

We measured the resistance of 160-nm CNT vias with a four-point probe using Kelvin patterns. Figure 13.9 (a) shows the current–voltage characteristics on the low-bias region. It was found that the resistance depended on the growth temperature. The via resistance

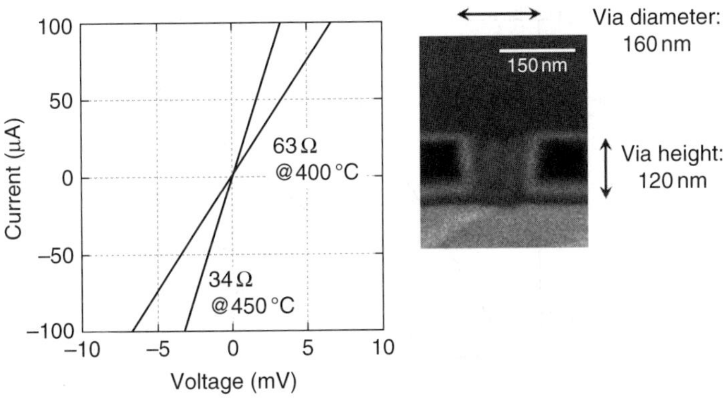


***Figure 13.9*** *(a) Current–voltage characteristic of the 160-nm-diameter CNT vias with a via height of 120 nm and (b) SEM image*

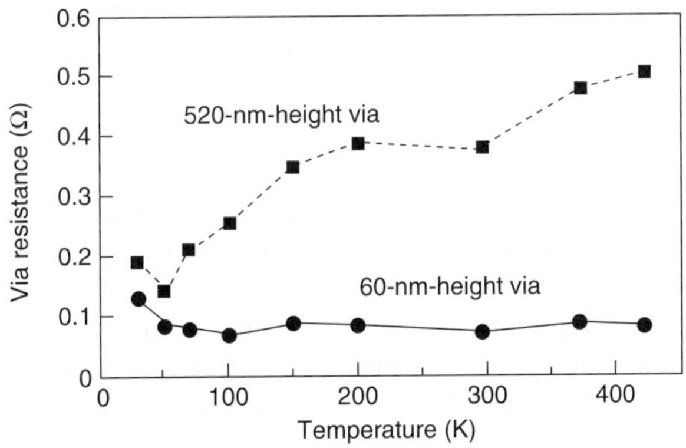


***Figure 13.10*** *Temperature dependence of the via resistance for the 60-nm and 520-nm-height CNT via*

was 34 Ω for a growth temperature of 450 °C and 64 Ω for 400 °C. Since the site density of the CNTs was similar for both temperatures, we speculate that the difference in resistance may have been caused by the difference in the CNT quality.

To investigate the transport mechanism, we measured the temperature dependence of the via resistance. Figure 13.10 shows the temperature dependence of the via resistance for the 60-nm and 520-nm-height CNT via. The 520-nm-height vias show the linear decrease of the resistance by decreasing the temperature. This characteristic is ohmic, which has been attributed to electron–phonon scattering [13]. On the other hand, the resistance of 60-nm-height vias was independent of temperatures as high as 423 K, which suggests that the carrier transport is ballistic. Figure 13.11 shows the via resistance as a function of the via height. The filled circles show the previous results for 2.8-μm vias with a growth temperature of 450 °C [4]. The solid lines indicate the via resistance calculated assuming various electron mean free paths. A solid rectangle or triangle indicates the current results

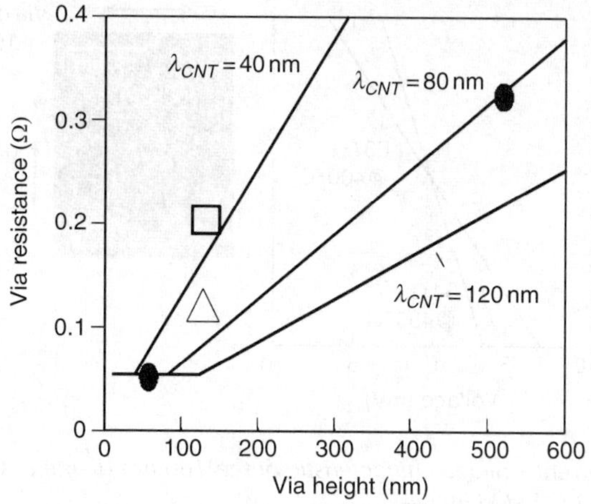


*Figure 13.11* Via resistance dependence as a function of the via height. Solid line: the via resistance calculated assuming various electron mean free paths; ●: 2800-nm-diameter via 450 °C growth; □: 160-nm-diameter via 450 °C growth; △: 160-nm-diameter via 400 °C growth

(shown in Figure 13.9) normalized to a diameter of 2.8 μm. As can be seen in the figure, the current result for 450 °C falls on the line for an electron mean free path of 80 nm. This length is two times larger than that of Cu and about the same as the via height in the half-pitch 32-nm technology node. On the other hand, the resistance for 400 °C falls on the line for an electron mean free path of 40 nm, which suggests the quality of CNTs grown at 400 °C is not as high as that at 450 °C, as also speculated from the SEM and TEM results.

## 13.5 Current Reliability of CNT Vias

The excellent tolerance of CNT vias to a high current density was demonstrated [7]. Figure 13.12 is a schematic cross-sectional structure of a CNT via. We integrated a CNT via with the upper (M2) and the lower (M1) Cu lines. Figure 13.13 is a schematic diagram of the Kelvin pattern for the reliability test of a CNT via. The tested CNT via had a diameter of 160 nm, a length of 275 nm and there were about 60 tubes with a diameter of 10 nm, as shown in Figure 13.14. The test conditions were as follows. The current densities per via ranged from $5 \times 10^6$ to $4 \times 10^7$ A/cm$^2$, i.e. the average current densities per CNT ranged from $2.1 \times 10^7$ to $1.7 \times 10^8$ A/cm$^2$ in the electron downstream direction at room temperature. Resistance increasing to infinity was taken as the fail criterion.

Figure 13.15 shows the cumulative lifetime plots. Plot (A) describes the pre-improvement samples without CMP. Plot (B) shows the TaN improvement samples, which had a TaN barrier layer deposited on the bottom of the via hole using the other PVD equipment. Plot (C) shows the samples with CMP without TaN improvement. Plots (B) and (C) had a longer lifetime than (A). In the case of (B), it was considered that the electrical contact of the

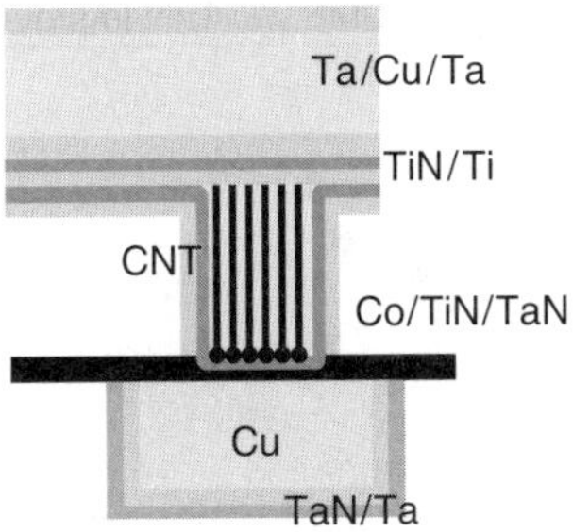


*Figure 13.12* Schematic cross-section of a CNT via

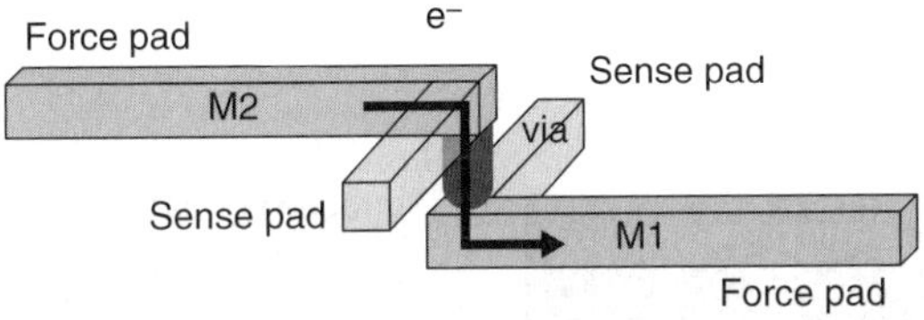


*Figure 13.13* Schematic diagram of a Kelvin pattern for the reliability test of a CNT via. The current was applied in the electron downstream direction

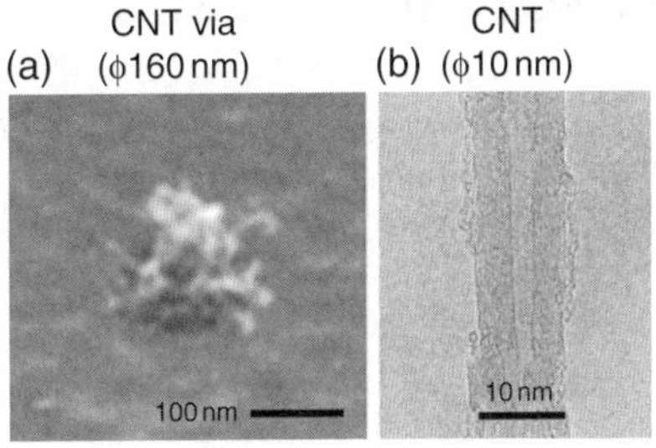


*Figure 13.14* (a) SEM image of the top surface of a CNT via after a CNT growth process. The CNT via had a diameter of 160 nm. The CNT density was $3 \times 10^{11}$ tubes/cm$^2$. There were 60 tubes of CNTs with a diameter of 10 nm in a via. (b) TEM image

via-TaN/Cu-line interface was improved. In the case of (C), the electrical contact of the M2-metal/CNTs interface was improved. As a result, the current density through a CNT decreased because the number of CNTs contributing to the current flow increased. It was considered that the current density per CNT of (A) was up to several times that of (B) and (C).

Figure 13.16 shows a cross-sectional TEM image of sample (A) after the reliability test. It seems that CNTs were cut from the surroundings of the via in the first stage because the majority of current passed the surrounding CNT of a via and the outer layer of the CNT, which was easy to connect to the metal line.

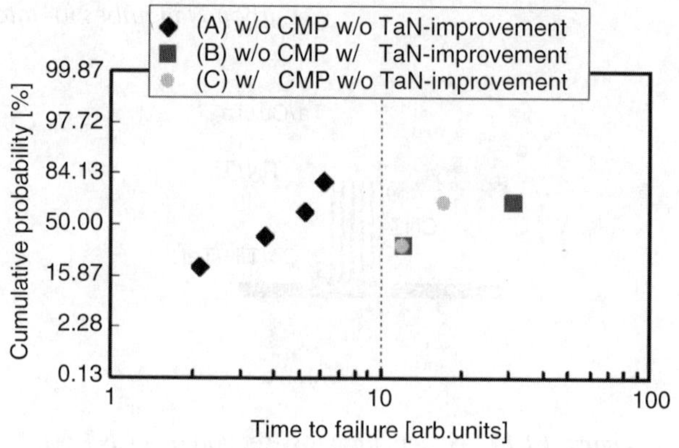


*Figure 13.15* Cumulative lifetime plots. The current density was $5\times10^6$ A/cm$^2$ per via, i.e. an average current density of $2.1\times10^7$ A/cm$^2$ per CNT

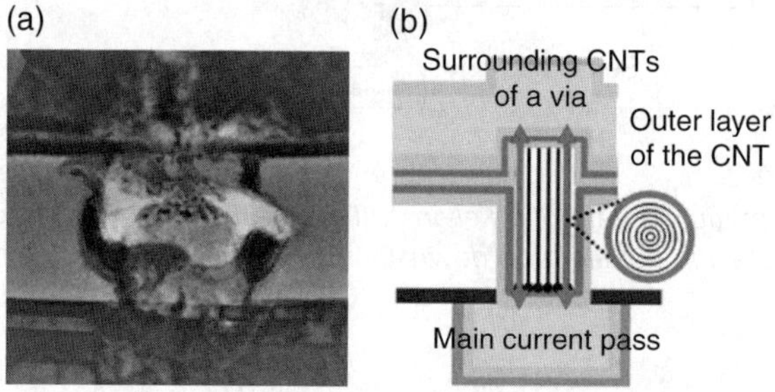


*Figure 13.16* (a) Cross-sectional TEM image of sample (A) after the reliability test. It seems that CNTs were cut from the surroundings of the via. (b) Schematic of the main current pass. The surrounding CNTs of a via and the outer layer of the CNT were easy to connect to the metal line

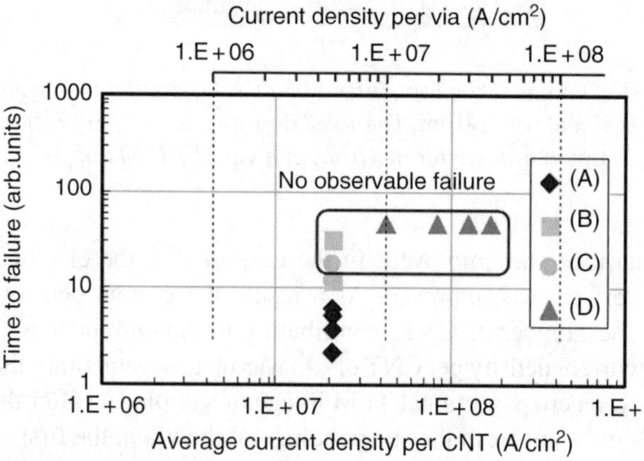


*Figure 13.17* Lifetime versus current density plots. Plots of (A), (B) and (C) are equal to those of Figure 13.15. Plot (D) shows several improvement samples

Figure 13.17 shows the lifetime versus current density plots. Plot (D) had undergone several improvements of electrical contact to both the upper and the lower metal lines. Plot (D) was tested under a higher current density from $1\times10^7$ to $4\times10^7$ A/cm$^2$ per via, i.e. from $4.2\times10^7$ to $1.7\times10^8$ A/cm$^2$ per CNT. Plot (D) had a much better lifetime than (A), (B) or (C). Furthermore, (D) did not have any failures during the plotted test time. In the case of (D), it was considered that the current density through a CNT dropped substantially to the average current density of each test condition. Also, in the test of (A), (B) and (C), it was considered that the current density per CNT was up to about 10 times the average current density of $2.1\times10^7$ A/cm$^2$.

## 13.6 Conclusion

In this chapter, we report our trials of using bundles of CNTs with their high current-carrying capability as via interconnects of LSIs. We proposed CNT damascene processes to integrate CNT vias with CMOS in LSIs, and evaluated its electrical properties and robustness over a high-density current. We demonstrated low-temperature CNT growth at the temperature as low as 365 °C, which would be more helpful in realizing reliable CNT vias for future LSIs. Moreover, the improvement of the contact to both the upper and the lower metal lines, such as the improved performance of the CMP process and the improvement of the TaN barrier layer, results in a high current reliability because the current density per tube decreases.

## Acknowledgments

We would like to thank Dr H. Watanabe of MIRAI-Selete and Dr N. Yokoyama at Fujitsu Laboratories Ltd for their support and useful suggestions. This work was completed as part of the MIRAI Project supported by NEDO.

## References

[1] Kong, J., Yenilmez, E., Tombler, T.W., Kim, W. and Dai, H. (2001) *Phys. Rev. Lett.*, **87**, 106801.
[2] Yao, Z., Kane, C.L. and Dekker, C. (2000) *Phys. Rev. Lett.*, **84**, 2941.
[3] Hone, J., Whitney, M. and Zettl, A. (1999) *Synthetic Metals*, **103**, 2498.
[4] Nihei, M., Hyakushima, T., Sato, S., Nozue, T., Norimatsu, M., Mishima, M., Murakami, T., Kondo, D., Kawabata, A., Ohfuti, M. and Awano, Y. (2007) Proceedings of the IEEE International Interconnect Technology Conference 2007, 204 pp.
[5] Kawabata, A., Nozue, T., Murakami, T., Daidou, T., Sato, M., Hyakushima, T., Kondo, D., Sato, S., Nihei, M. and Awano, Y. (2009) International Proceedings of the Conference on Scientific Applications of Nanotubes (NT09).
[6] Kawabata, A., Sato, S., Nozue, T., Hyakushima, T., Norimatsu, M., Mishima, M., Murakami, T., Kondo, D., Asano, K., Kawarada, H., Sakai, T., Nihei, M. and Awano, Y. (2008) Proceedings of the IEEE International Interconnect Technology Conference 2008, 237 pp.
[7] Sato, M., Hyakushima, T., Kawabata, A., Nozue, T., Sato, S., Nihei, M. and Awano, Y. (2009) Proceedings of the 2009 International Conference on Solid State Devices and Materials, Sendai, pp. 1078–1079.

[8] Sato, S., Nihei, M., Mimura, A., Kawabata, A., Kondo, D., Shioya, H., Iwai, T., Mishima, M., Ohfuti, M. and Awano, Y. (2006) Proceedings of the IEEE International Interconnect Technology Conference 2006, 230 pp.
[9] Yamazaki, Y., Sakuma, N., Katagiri, M., Suzuki, M., Sakai, T., Sato, S., Nihei, M. and Awano, Y. (2008) *Appl. Physics Express*, **1**, 034004.
[10] Yamazaki, Y., Katagiri, M., Sakuma, N., Suzuki, M., Sato, S., Nihei, M., Wada, M., Matsunaga, N., Sakai, T. and Awano, Y. (2010) *Appl. Physics Express*, **3**, 055002.
[11] Kawabata, A., Nakano, H., Murakami, T., Daidou, T., Sato, M., Hyakushima, T., Kondo, D., Sato, S., Nihei, M., Awano, Y. and Yokoyama, N. (2010) Proceedings of the International Conference on Scientific Applications of Nanotubes (NT10).
[12] Awano, Y., Sato, S., Nihei, M., Sakai, T., Ohno, Y. and Mizutani, T. (2010) *Proc. IEEE*, **98**, 2015.
[13] Liu, K., Avouris, Ph., Martel, R. and Hsu, W.K. (2001) *Phys. Rev. B*, **63**, 161404.

# 14

# Optical Interconnects

*Wim Bogaerts*

*Ghent University – IMEC, Department of Information Technology, Ghent, Belgium*

## 14.1 Introduction

Optical communication is no new concept. The introduction of the laser diode and the optical fiber in the 1960s and 1970s has sparked a dramatic breakthrough for long-distance telecommunication. Optical fibers could transport much larger bandwidths over a much larger range, and optical links quickly became the backbone of what later would become the Internet. Since then, fiber communication has steadily increased its share in the communication infrastructure.

From a topological point of view, we can make a relatively clear distinction between optical links and electrical links. Given the fact that in most communication systems people start from an installed base and gradually expand it, it is clear that electrical communication is prevalent inside the closed 'box', while optical links are easier to implement on the edge of the 'box', to connect boxes together. The notion of what exactly is such a box depends on the application field, but in general we can say that the size of the box is shrinking. Forty years ago, optical links were used to connect the telecommunication infrastructure of different continents, but optical communication has since then penetrated the metropolitan networks (box = town), access networks (box = building) and supercomputing infrastructure (box = rack) (see Figure 14.1). The further penetration of optical communication has been

*Advanced Interconnects for ULSI Technology*, First Edition. Edited by Mikhail R. Baklanov, Paul S. Ho and Ehrenfried Zschech.

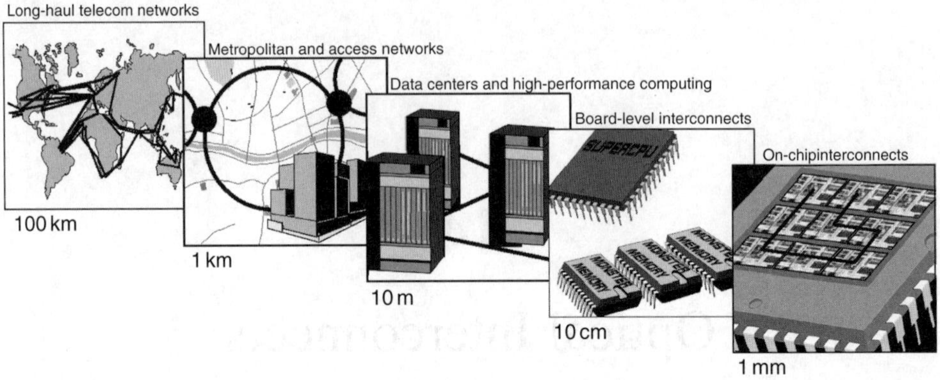


*Figure 14.1* Boxes for optical interconnects: optical links are used between boxes. As the required bandwidth scales up, the boxes shrink

accompanied with an increasing drive to reduce the cost, size and power consumption of the *photonic* components that form the optical link.

The more recent evolution in photonics now brings optical links to even smaller boxes: board-to-board interconnects are now being introduced in the high-performance computing (HPC) scene. There is a substantial amount of research going on in board-level interconnects and with the advent of multicore processors, optical links start to make sense for on-chip interconnects, where cores and memory could be considered as yet another set of boxes [1, 2].

With the advent of the fiber, people quickly realized the merit of optical communication. While light is just another form of electromagnetic radiation, its base frequency is in the order of several hundreds of THz. This means that it can behave much like a radio frequency carrier for a signal with much lower bandwidth, which in the optical case could still cover tens of THz. Also, optical fibers could be engineered to exhibit extremely low propagation losses, allowing links of over 100 km without repeaters. This gives optical communication a potentially huge bandwidth × distance product. Also, being confined in a fiber (for a description of the confinement mechanism see Section 14.2), the optical signals will not radiate to the outside world and cause electromagnetic interference and cross-talk.

Even through the available bandwidth for optical signals is tens of THz, its efficient use is limited by the speed at which signals can be modulated, and this is typically done with electronics. Therefore, most telecommunication links now adopt wavelength division multiplexing (WDM): signals are encoded on to different carrier wavelengths, which are transmitted independently through the same physical channel, not unlike different FM and TV channels sharing the same medium. In WDM, the carriers are typically generated using lasers operating at different wavelengths or colors. For telecommunication, these wavelengths are situated in the near-infrared, around 1300 nm or 1550 nm, as optical fibers exhibit low loss and little dispersion, which is the difference in propagation velocity at different wavelengths. If dispersion is nonzero, signals will be distorted as they travel along the fiber or waveguide. Current WDM systems for telecom can be separated into dense WDM, which supports tens to hundreds of channels with 50 GHz or 100 GHz separation, and coarse WDM, which has a channel spacing of 10 nm (1.2 THz) or 20 nm (2.5 THz).

The bandwidth advantage of optical links has pushed the technology to interconnects between even smaller boxes. Still, the main obstacle for the adoption of optical links on a new level is the physical integration with the existing infrastructure. For telecommunication networks this translates in the installation of fiber lines to replace the copper infrastructure. For board-level and chip-level interconnects, this translates in the integration of the different optical components in the electronic process flow. As we will discuss further, to build optical links, many different components are required, which will add significantly to the cost of the chips, modules or boards. Also, optical components come in many flavors and material systems, which are typically mutually incompatible, as well as incompatible with CMOS electronics manufacturing. The emerging exception is silicon photonics [3], where one of the drivers is exactly its compatibility with existing electronics manufacturing solutions. However, as we will discuss in Section 14.3, not everything is possible in silicon: especially efficient light sources turn out to be a major challenge.

Depending on the required distance and bandwidth, switching to optical interconnects will not necessarily result in reduced power consumption. While the actual transport of the light will most likely induce less power dissipation, there is additional power consumption for the laser(s) and the specific driver electronics. Also, these latter might result in floor space consumption just to support the link itself.

The aim of this chapter is to give people unfamiliar with integrated optics an overview of the relevant technologies that are most likely to be useful for on-chip optical interconnects, as well as references to the current state-of-the-art as it is published in literature. Therefore, many topics will be discussed only qualitatively. We will first describe the anatomy of a link to come to a list of all the required building blocks and then apply that to the most promising technology to implement them, namely silicon photonics. With this technology base, we will cover the possible implementations for optical interconnect networks. Finally, in Section 14.5 we will go deeper into integration strategies.

## 14.2 Optical Links

All communication links consist of a transmitter, a medium through which the signal propagates and a receiver. This also holds for optical links. The anatomy of a basic optical link is sketched in Figure 14.2. In the transmitter, a modulated beam of light is generated from an electrical signal. This beam is sent to the receiver through a medium, which is typically an optical waveguide. At the receiver end, the optical signal is converted back to an electrical signal.

In its simplest form, the transmitter consists of a directly modulated laser, which switches on or off based on the electrical input. However, as we will see further, this solution has disadvantages from an integration point of view. Instead, one can consider the laser light as an external source and use a *modulator* to imprint the signal on to the carrier. The simplest modulation format is on–off keying (OOK), which is also the simplest to decode at the receiver side, where a single photodetector can directly convert the optical signal into an electrical signal. Of course, both the transmitter and receiver will include some electronics for driving and control.

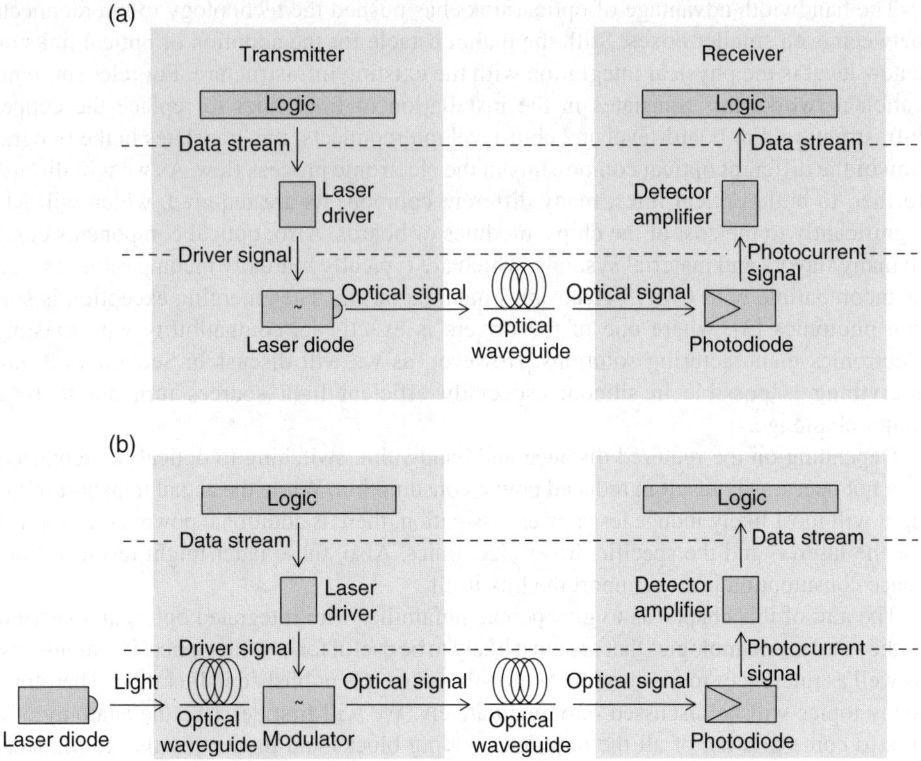


***Figure 14.2*** *Optical links: (a) using direct modulation on the laser and (b) using a continuous wave light source and a separate modulator*

Obviously, the light should be sent from the transmitter to the receiver. While this can be done through free space, we will mainly focus on on-chip waveguides, because waveguide circuits also make it relatively easy to implement WDM (de)multiplexers. For on-chip interconnects, the free space above the chip could be used for controlling the flow of light, using fixed or movable mirrors (MEMS). Such free-space switches already exist for telecom applications [4, 5]. In this chapter we will not concern ourselves with this approach, but rather focus on waveguide-based solutions.

In a WDM link, individual signals are encoded onto different wavelengths, which propagate through a shared physical medium (the waveguide). There are different approaches to accomplish this: the most straightforward is that in the transmitter the wavelengths each occupy their own waveguide, and after the signals are modulated they are joined together using a multiplexer. This is shown in Figure 14.3. At the receiver side the wavelengths are unraveled again using a demultiplexer and converted each with their own photodetector.

Instead of a (de)multiplexer, one can use add/drop filters at the transmitter or receiver side to put a single wavelength signal on to a bus waveguide, or vice versa extract it from the bus. Such filters will be discussed further and give the opportunity to use alternative

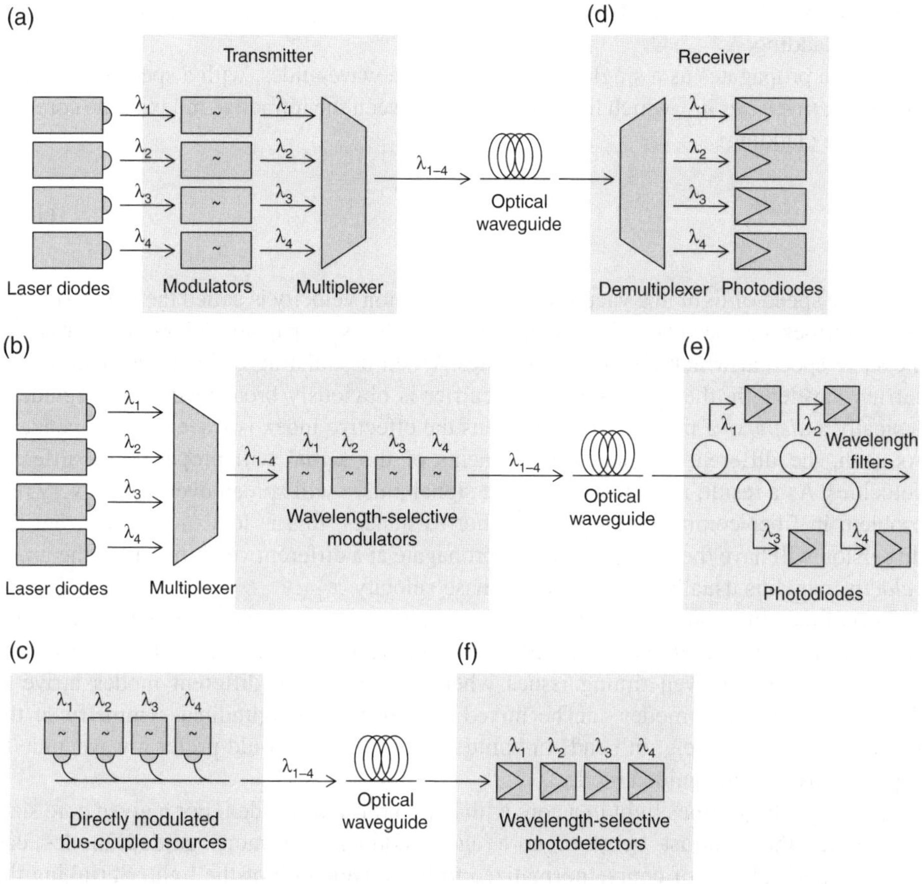


***Figure 14.3*** *Point-to-point links using wavelength division multiplexing. The three transmitter and receiver configurations can be interchanged: (a) using broadband modulators and a wavelength multiplexer, (b) using wavelength-dependent modulators on a single bus, (c) using directly modulated sources on a bus waveguide, (d) receiver using a demultiplexer and broadband photodiodes, (e) using a bus with single-wavelength filters and broadband photodiodes and (f) using wavelength-selective photodetectors on a bus waveguide*

network topologies (see Section 14.4). One can even integrate the wavelength selective filter in the light source, modulator and/or detector.

### 14.2.1 Waveguides

In optical links the waveguide fullfills the core functionality of transporting the signals from the transmitter to the receiver. Optical waveguides confine light in a *core* with a high refractive index surrounded by a cladding with a lower refractive index. Such a waveguide supports optical *modes:* light distributions centered in the core but extending somewhat into the cladding.

A mode propagates as a single entity through the waveguides, with a speed determined by an *effective index* $n_{eff}$, which lies somewhere between the refractive index of the core and that of the cladding:

$$v = \frac{c}{n_{eff}} \tag{14.1}$$

with $c$ the speed of light in a vacuum. This propagation velocity is called the *phase velocity* as it describes the speed at which the optical phase fronts propagate. This is not necessarily the same speed at which a signal propagates. When modulating a signal on to an optical carrier wavelength, the spectrum of that carrier is obviously broadened. A waveguide is typically a *dispersive* medium, which means the effective index is wavelength dependent. As such, the different frequency components of the signal will propagate at different velocities. As a result, in a long waveguide signal pulses will spread over time, a very real problem in fiber communication. Over short distances, this is less of an issue, but the dispersion will have the effect that pulses propagate at a different velocity, called the *group velocity*, which is usually lower than the phase velocity.

Every mode in the waveguide has a different phase and group velocity. Therefore, if we excite different modes in the waveguide, this could result in phase-dependent interference between modes, or even timing issues where the signals in different modes arrive at different times. Also, modes can be mixed wherever the longitudinal symmetry of the waveguide is broken, e.g. in bends or splitters. For this, we would prefer a waveguide to support only a single guided mode.

A waveguide confines light in a core with a high refractive index. For a given core size, the higher the contrast between core and cladding, the more guided modes can be supported. This is of course normalized to the wavelength of the light. Shrinking the core reduces the number of guided modes. If we want a single-mode waveguide, we can reduce the core size (and therefore the footprint) of the waveguide by increasing the refractive index of the core. This is illustrated in Figure 14.4. Single-mode optical fibers typically have a very low index contrast, of less than 1 %, and the core diameter is of the order of 10 μm (for telecom wavelengths between 1300 nm and 1550 nm). Planar waveguides can be defined using implantation or etching. Glass waveguides have a similar contrast as optical fibers. When going to a silicon (oxy)nitride system, core sizes of 1–3 μm are possible. The smallest waveguides can be accomplished using silicon waveguides surrounded by an oxide or air cladding, where one can make single-mode waveguides with a core size of 300–500 nm. In principle it is possible to get even larger index contrast when including metals. However, these so-called plasmon waveguides suffer from very high absorption losses and will not be considered here. The small core of these *photonic wires* significantly reduces the footprint of photonic circuitry. This is one of the key advantages of using silicon for optical waveguides, as will be discussed in more detail in Section 14.3.

A higher refractive index brings another advantage. It allows the bend radius of the waveguides to be reduced. If a waveguide bend is too tight, light will radiate out from the core into the cladding. That is why optical fibers typically have a bend radius of 1 cm, and

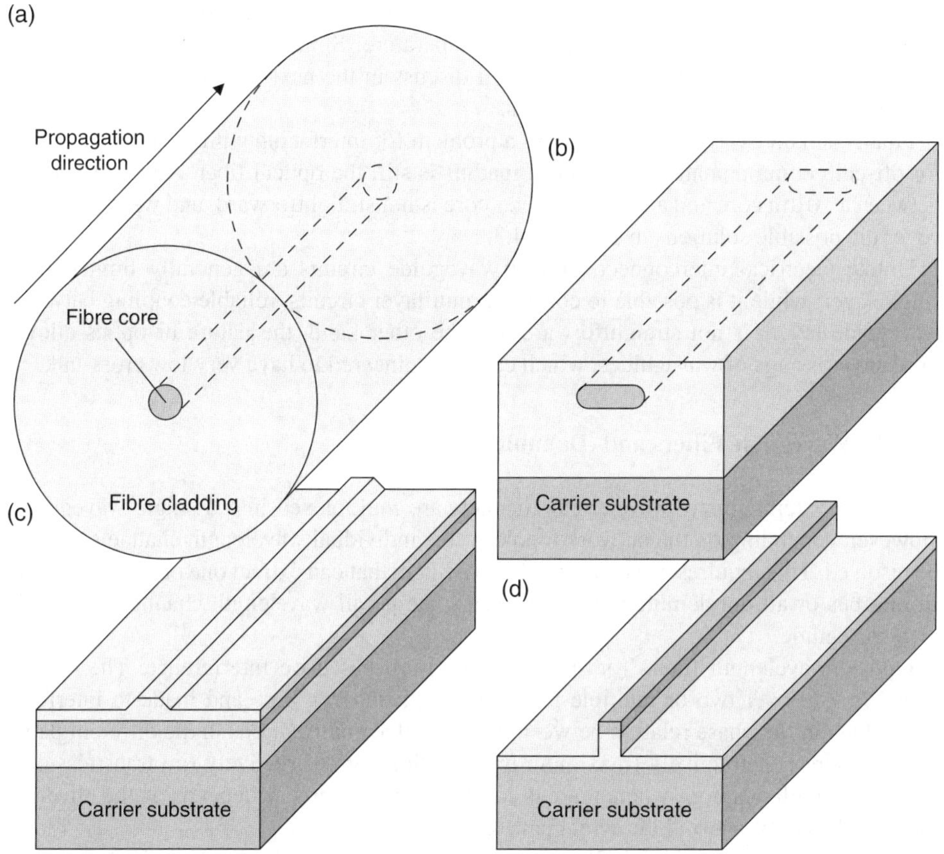


***Figure 14.4*** *Examples of optical waveguides: (a) optical fiber, (b) low-contrast buried waveguides, (c) loaded rib waveguide and (d) high-contrast strip waveguide*

the same is true for glass waveguides on a chip. Silicon photonic wires can have bends with acceptable radii down to a few micrometers.

As already mentioned, waveguide modes are not fully confined in the waveguide core, but a significant fraction can extend into the cladding. This does not cause leakage in its own right, as the entire mode is bound to the waveguide. However, leakage can occur if another high-index material is brought in close proximity, such as another waveguide. Therefore, irrelevant of the index contrast, care has to be taken to keep sufficient spacing between waveguide cores; otherwise leakage can result in cross-talk between waveguide channels.

However, higher index contrast waveguides come at a price. As the index of the core is quite high, both the effective index and the group index of the waveguide are also high. A silicon waveguide can have a group index between 4 and 5, which means signals will encounter a 3 times larger delay than in glass waveguides, where the group index is of the order of 1.5.

Also, high-contrast silicon waveguides are much more sensitive to all kinds of perturbations: linewidth variations, roughness and temperature. Small variations will have their effect on the effective index and, as we will discuss in the next section, this will mainly impact the functionality of WDM systems.

Finally, smaller waveguide cores pose a problem for interfacing with the outside world; for off-chip communication, the typical medius is still the optical fiber. A good interface between a 10 μm core and a submicrometer core is not straightforward, and we will briefly cover the possible solutions in Section 14.3.

Unlike electrical interconnect, optical waveguide circuits are generally limited to a single layer; while it is possible to construct multilayer circuits, reliable coupling between waveguide layers is not straightforward. On the other hand, the nature of optics allows in-plane crossings of waveguides, which can be engineered to have very low cross-talk.

### 14.2.2 Waveguide Filters and (De)multiplexers

In WDM systems, many wavelength channels are multiplexed into a single waveguide. However, depending on the network topology, the individual wavelength channels need to be resolved. This requires wavelength selective filters that can extract one or more channels from a bus or all-out demultiplexers that can separate all wavelength channels into their own waveguide.

Optical wavelength filters (or spectral filters) are based on interference. The optical signal is split over two or multiple paths with a given delay line and made to interfere; depending on the phase relation between the paths, the contributions in the different paths will interfere constructively (maximum transmission) or destructively (no transmission). The wavelength positions where there is constructive interference depends on the effective index in the delay line and the actual path length.

The most simple wavelength filter is the Mach–Zehnder interferometer (MZI), depicted in Figure 14.5. The device has two input ports and two output ports. Light from the bottom input port will end up in the bottom output port if the phases in the arms interfere constructively:

$$\Delta\phi = (\Delta L / \lambda).n_{eff} = m.2\pi \qquad (14.2)$$

This will result in a wavelength-periodic transmission spectrum. By cascading such filters, any filter characteristic can be obtained by controlling the delay line length and the coupling ratios in the splitters and combiners. Both delay and coupling ratio are susceptible to perturbations, which become worse for higher index contrast.

Instead of cascading MZIs, one can also construct multipath interference filters. The most commonly used type is the *arrayed waveguide grating* (AWG), shown in Figure 14.6 [6, 7]. At the entrance, the light is coupled into a *star coupler*; in the *slab area* there is no lateral confinement on the chip and the light will diffract out in a circular phase front. In this way, it is distributed over an array of delay lines with a fixed length difference. At the output star coupler the light will arrive at a circular array of waveguide apertures with a fixed phase difference. When all contributions are in phase $\Delta\phi = 0$, a circular phase front is formed that focuses in the central output waveguide. When $\Delta\phi \neq 0$, this circular phase front is tilted, focusing in adjacent waveguide apertures. As the

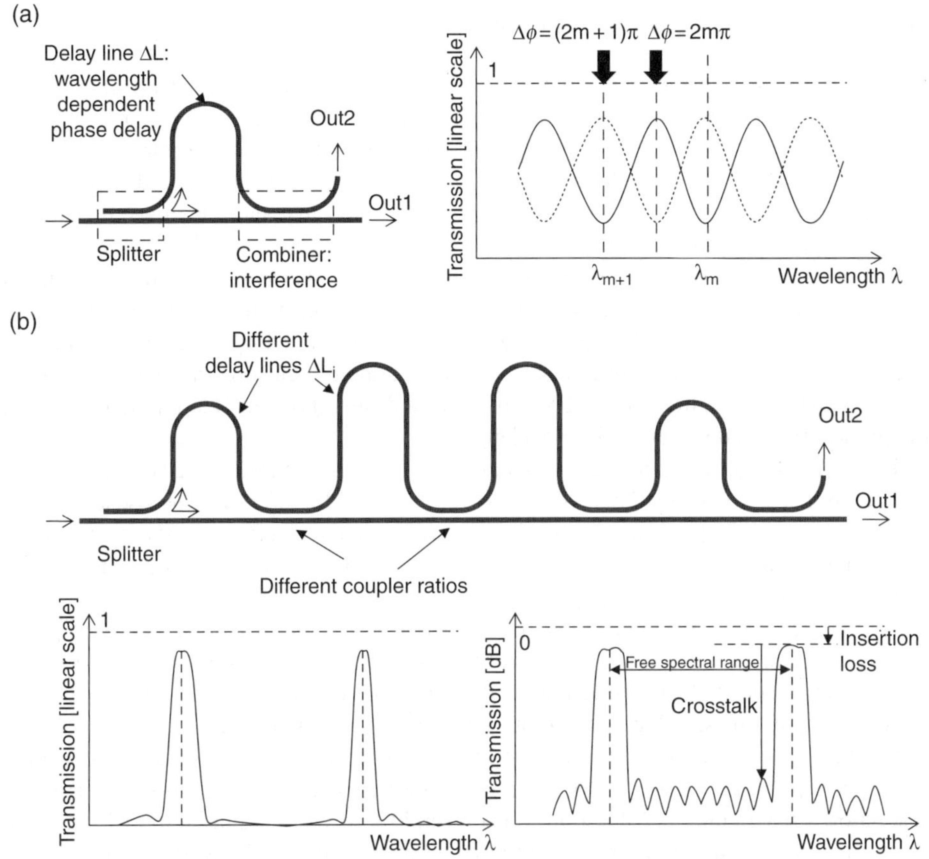


***Figure 14.5*** *Wavelength filters based on Mach–Zehnder interferometers (MZIs). (a) A single MZI, where light from both waveguides interferes constructively and destructively, resulting in a sinusoidal output variation. (b) Higher-order filter based on MZIs, where by choosing the correct coupling factors and delay lengths, sharper filters can be obtained. Transmission of such filters is usually plotted in a logarithmic dB scale*

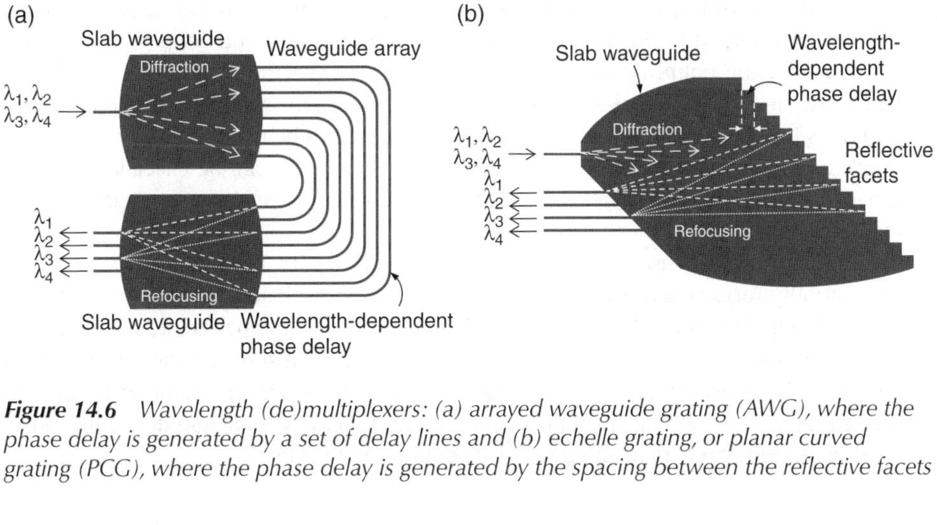


***Figure 14.6*** *Wavelength (de)multiplexers: (a) arrayed waveguide grating (AWG), where the phase delay is generated by a set of delay lines and (b) echelle grating, or planar curved grating (PCG), where the phase delay is generated by the spacing between the reflective facets*

phase difference is wavelength dependent, an AWG will demultiplex different wavelengths into different output waveguides. As in the MZI, the phase delay is dependent on the effective index, which in turn depends on the geometry, temperature, etc., of the waveguide. Perturbations will lead to slight phase errors in the arrayed waveguides, which in turn lead to distortions of the phase front and leakage, or cross-talk into the wrong output waveguides.

Instead of using arrayed waveguides as delay, one can also implement the delay in the slab area (see Figure 14.6) [8, 9]. By implementing reflectors at the correct distances, a set of phase delays is created. This phase delay is now determined by the path length difference and the effective index of the slab (which is a one-dimensional layered waveguide). Just as with the AWG, the light will be refocused at different output positions depending on the phase delay (and therefore the wavelength).

The advantage of such a *planar curved grating* (PCG), sometimes called *echelle grating*, is that the delay is not implemented as a waveguide, and therefore it is less susceptible to fabrication errors. It still remains sensitive to variations in the slab geometry, more specifically the thickness of the guiding layer. Also, implementing good reflectors is not always trivial, depending on the technology [8].

The examples of wavelength filters discussed here are *feedforward* filters: light passes only in one direction and there is no feedback. The alternative is to implement feedback, where light can be made to interfere with itself. This has the advantage that a delay line can be reused many times, reducing the footprint of the filter component.

By introducing feedback into a waveguide, one creates a resonator. The two most common topologies of resonators are illustrated in Figure 14.7 [10–12]. In a ring resonator, a waveguide is folded on to itself, and there will be a buildup of light in the resonator if the wavelength of the light fits a whole number of times in the length of the resonator (corrected for effective index, of course). This is a traveling wave resonator, as the light circulates either clockwise or counterclockwise. One can couple light into a ring resonator using a directional coupler: a bus waveguide that is brought close to the ring waveguide. As the mode in the waveguide extends somewhat into the cladding, the proximity of the waveguides will allow the light to couple from the bus in the ring. On resonance, power will be extracted from the bus waveguide, which will manifest itself as a dip in the transmission spectrum. When adding a second waveguide, the extracted power will be coupled to this new waveguide. In order to get good coupling to the output waveguide, the coupling ratios in the two coupling sections should be well controlled [13]. In this way, a ring resonator can be used as a wavelength channel drop filter, but as the device is reciprocal, the ring filter can also be used to add a wavelength channel to the bus waveguide. As with MZIs, ring resonators can be cascaded to tailor higher-order filter characteristics, by tweaking the ring circumference and the coupling ratios between rings [14].

An alternative to ring resonators is a *Fabry–Perot* (FP) cavity. Here, the cavity is formed by two partially reflecting mirrors: Light will be reflected back and forth in the cavity and power will build up when the round-trip length is a whole number of wavelengths. On resonance, the cavity will have a high transmission, while off-resonance, the cavity will act as a reflector. In a waveguide geometry, the reflected light shares the same waveguide as the incoming light, which makes it difficult to use an FP cavity as an add/drop filter. Also, fabrication of good mirrors in an optical waveguide is not trivial.

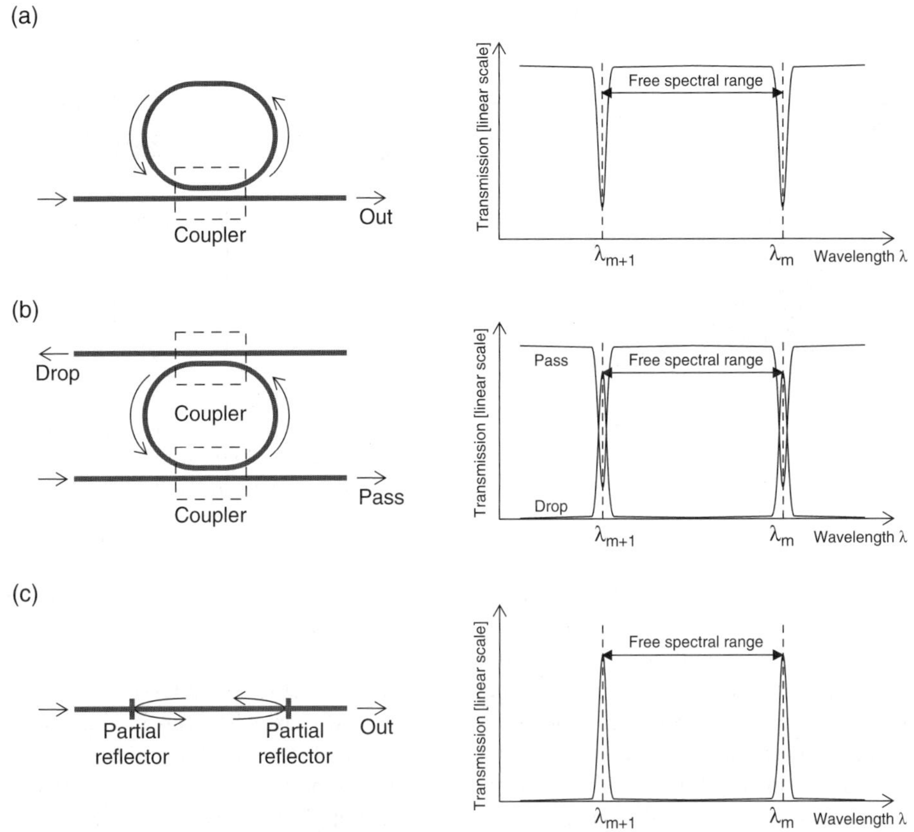


***Figure 14.7*** *Photonic resonators: light resonates in a cavity when the wavelength fits a whole number of times inside the optical length of the cavity. (a) Ring resonator: light is coupled laterally to a ring. (b) Ring drop filter: on resonance, light in the ring is coupled to the second waveguide. (c) Fabry–Perot cavity: light is transmitted when the cavity is on-resonance*

### 14.2.3 Transmitter: Light Source

One of the key components of the transmitter is the light source. This is typically a III–V laser that emits at a single wavelength. Semiconductor lasers for telecommunication can be purchased off the shelf for fixed wavelengths or (at a cost) with a certain tuning range. In a semiconductor laser a gain medium (a double III–V heterojunction or quantum wells/dots) is embedded in a waveguide geometry for optical confinement. The waveguide is embedded in a cavity for feedback. With current injection, the cavity will resonate and lasing light will be emitted from the cavity outputs. The simplest form is the Fabry–Perot laser (see Figure 14.8), which has a cavity of two simple mirrors (e.g. cleaved facets). An FP laser has multiple laser modes, because in the long cavity multiple wavelengths resonate in the bandwidth of the material gain. However, different feedback mechanisms can be used to select a single mode: the most commonly used type is the distributed feedback (DFB) laser, which has a wavelength-selective grating embedded in the waveguide. Alternatively, one can use

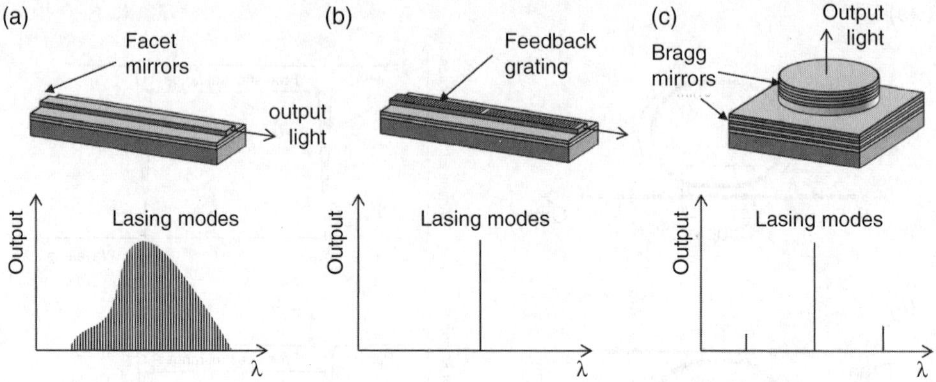


*Figure 14.8* Semiconductor laser sources. (a) Fabry–Perot laser with many longitudinal laser modes. (b) DFB laser, which selects a single longitudinal mode. (c) Vertical cavity surface emitting laser (VCSEL), which has very few (or only one) longitudinal mode because the vertical cavity is so short

a VCSEL (vertical cavity surface emitting laser), which emits light perpendicular to the substrate. The vertical cavity is formed by two Bragg mirrors and the cavity is so short that only one or a few longitudinal modes can exist in the cavity.

In both cases, the laser light needs to be coupled into the chip where it will be used for the interconnect. This coupling problem is not straightforward when considering submicrometer waveguides. The simplest solution is to keep the lasers off-chip and use fibers or large-core board-level waveguides to bring light to the chips as an 'optical power supply'. This automatically solves thermal management issues with lasers (lasers are very sensitive to heat and are also a significant source of heat).

It is possible to co-package the laser with the waveguides in a single module using flip-chip-like technologies, but these typically require a very high alignment accuracy (1 μm) to keep coupling losses to the chip acceptable. Depending on the use of an edge-emitting laser or a surface-emitting laser, different coupling solutions are required. Also, thermal management inside the module now becomes more complicated.

Longer-term solutions are exploring the integration of the laser directly on the chip, where the light is directly coupled to the photonic wires. In this way, alignment and fabrication is handled by the same wafer-scale tools used for the rest of the fabrication. However, the thermal management may become extremely challenging, especially when close integration with electronics is required for on-chip interconnects.

In this close integration scheme, it becomes possible to modulate the signal directly on the laser. This obviates the requirement for a modulator and, as we will discuss in Section 14.4, it can also simplify the network topology. However, the dynamic behavior of the laser is limited by the carrier dynamics, which limits the modulation speed to 5–10 GHz.

### 14.2.4 Transmitter: Modulators

Electrically modulating a signal on a carrier wavelength typically involves a modification of the effective index or the absorption of a waveguide. The mechanism depends on the modulation scheme. As with radio signals, either phase modulation or amplitude

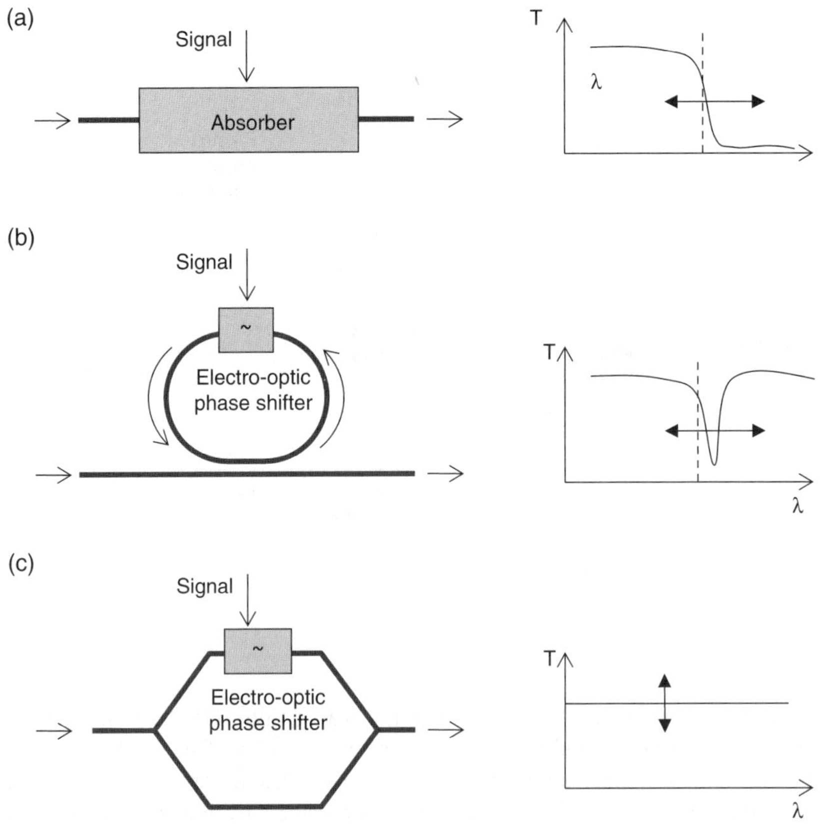


***Figure 14.9*** *Electro-optic amplitude modulation. (a) An electrical signal drives an electro-absorber, where the absorption edge shifts as a function of the electric field. (b) An electro-optic phase shifter changes the optical length of the cavity and thus shifts the absorption wavelength. (c) A phase shifter in the MZI changes the phase difference in the two arms from constructive to destructive interference*

modulation can be used. Amplitude modulation encodes the signal in the amplitude of the light. For a digital signal, this scheme is typically referred to as *on–off keying* (OOK). The advantage of this scheme is that it is very straightforward to decode on the receiver side. Direct amplitude modulators are typically based on direct electroabsorption effects in semiconductors, which rely on a band-edge shift induced by a modulated electric field. However, it will require some normalization of the power if a signal becomes attenuated. Also, OOK does not make efficient use of the spectrum. Even with the huge bandwidth of optical fibers, this is already becoming an issue in long-haul links.

Alternatively, phase modulation will encode a signal in the phase of the light. Different modulation schemes exist that make more efficient use of the spectrum, and also the implementation of the modulator can be simplified. On the other hand, the receiver side becomes more complicated and will require an interferometer and multiple detectors, or even a local oscillator.

Figure 14.9 shows the distinction between modulators. OOK can be achieved using direct absorption modulation. Different implementations based on electroabsorption

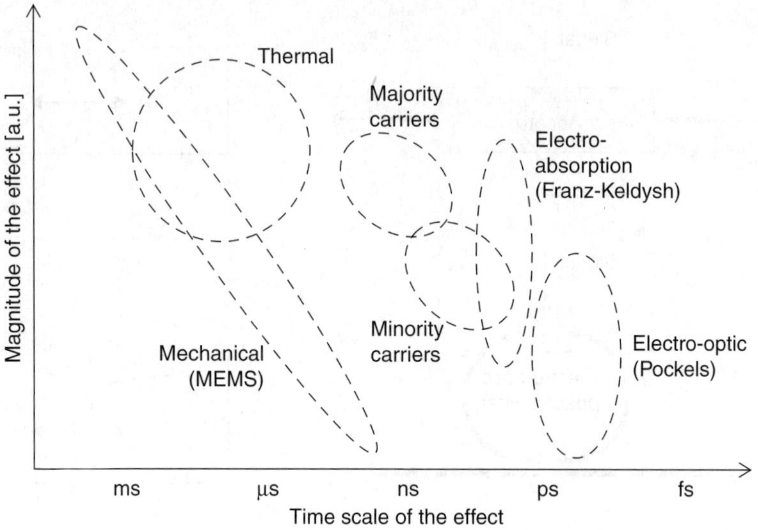


***Figure 14.10*** *Modulator mechanisms*

exist, but these typically work only for a single wavelength. On the other hand, phase modulation just requires a modification of the effective index. This can be done through different mechanisms: mechanical, thermal, using electrical carriers or direct electro-optic (EO) effects (such as the Pockels effect described by the $\chi^{(2)}$ material tensor). Depending on the materials used, these effects have decreasing magnitude, but increasing operation speed, as shown in Figure 14.10. For multi-GHz signals, only the latter two mechanisms are useful. Still, thermal and mechanical effects can be useful for tuning or switching, where ms or μs speeds are required. Such tuning can be essential in WDM links, where the tuning can be used to compensate for fabrication variations and operating conditions.

A phase modulator can also be converted to an amplitude modulator, by combining it with an interferometer or a resonator. If a phase modulator is embedded in an arm of an MZI with equal arm lengths, it can be used to flip the phase difference at the combiner from 0 to $\pi$, effectively modulating the transmission of the MZI from 1 (constructive interference) to 0 (destructive interference). Alternatively, if a modulator is embedded in a resonator, it can modify the optical path length of the resonator, thereby moving the wavelength of the resonance. If the operating wavelength is on the edge of the resonance, the phase modulator will move the cavity in and out of resonance.

Depending on the device type, modulators need special driver electronics. Generally E/O modulators are voltage driven. If the device is small, a simple voltage driver can be used. However, when using a quite weak effect, a modulator can become relatively large compared to the bit length, and this will require distributed drivers or microwave engineering of the electrical contacts. Also, depending on the intrinsic and parasitic load, the drivers might require complex signal generation such as pre-emphasis.

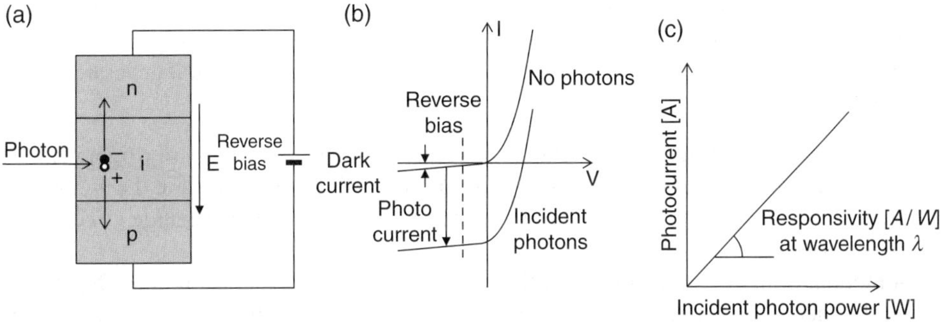


***Figure 14.11*** *Operation of a photodiode. (a) A photon impacts on a reverse-biased p-i-n diode. The resulting carriers generate a photocurrent. (b) The photocurrent depends on the reverse bias and the amount of incident light. (c) In a normal regime, the photocurrent scales linearly with the incident power (for a given wavelength)*

### 14.2.5 Receiver: Photodetector

At the receiver side, the key component is the photodetector, which converts the optical signal into an electrical signal. Photodetectors typically consist of a reverse-biased diode where carriers generated by an absorbed photon generate a photocurrent (see Figure 14.11). In general, the photocurrent scales linear with the power of the incident light (for a given wavelength) and the responsivity is expressed in *A/W*. The other key performance metric is the dark current (i.e. the current generated by the photodiode even when there is no incident light), which should be as low as possible.

For a higher responsivity, *avalanche photodetectors* (APD) can be used. In APDs, the generated carriers are amplified several times using an avalanche effect generated by a strong electric field. However, the avalanche effect typically induces additional noise, which should be traded off with the increased responsivity.

Photodetectors typically need some additional driver electronics. The most commonly used read-out is a transimpedance amplifier [15], which translates the photocurrent to a voltage signal, but read-out by directly interfacing a photodiode to an inverter is becoming more common [16].

### 14.2.6 Power Consumption and Heat Dissipation

When discussing an interconnect link, the usual metric is expressed in power consumption per bit. This is not a straightforward calculation in an optical link, as there are many contributions that need to be taken into account. Published experiments invariably incorporate only a fraction of the total power consumption of the link. The relevant contributions are:

- The electrical power consumed by the laser.
- The power dissipated in the modulator.
- The power dissipated just for tuning the WDM components.
- The power dissipated by the photodiode.
- The power dissipated by the control electronics.

In many cases, the power consumed by the laser is not counted in the case where the laser is used as an external source of light. However, it contributes a very large fraction of the total power consumption. Moreover, a significant portion of the light it generates is not used effectively:

- Coupling the laser light to the chip will result in some loss. The stray light will be absorbed elsewhere. In the best case this absorption will generate heat; in the worst case the stray light will find its way through parasitic paths into the waveguide circuits and cause cross-talk.
- A modulator in on–off keying will effectively drop half of the optical power. Depending on the modulation mechanism, this power is absorbed in the modulator (heating it) or scattered, which will also cause heating and possibly cross-talk.
- Waveguides have propagation loss, either through absorption or scattering.
- Photodetectors intentionally absorb light, but this will also generate heat.

When bringing all this together, optical links can have a significant power consumption as well as pose a considerable thermal challenge. However, it is very difficult to estimate the actual power consumption of an entire link without having detailed knowledge of the exact fabrication technology as well as the operating conditions.

### 14.2.7 Different Materials

Unlike in electronics, the photonics community has developed a variety of material systems over the past decades, each with its respective merits and drawbacks. We compare the most prominent of them them in Table 14.1. For the use in optical interconnects, several metrics stand out. Integration density is a primary requirement, especially for future bandwidth scaling where complex WDM networks on a chip can be used. In that respect, it also becomes important to integrate light sources on the chip, instead of treating them as an external power supply. Also, compatibility with existing CMOS processing technology is important, with respect to contamination and thermal budget.

***Table 14.1*** *Comparison of photonic materials*

| | Glass | Nitride | Polymers | III–V | Silicon |
|---|---|---|---|---|---|
| Index contrast | – | o | – | + | ++ |
| Integration density | – | o | – | + | ++ |
| cost | + | + | ++ | – | + |
| CMOS compatibility | + | ++ | o | – | ++ |
| Stability | ++ | ++ | – | ++ | ++ |
| Thermal operating budget | + | ++ | – | + | ++ |
| Volume scaling | o | ++ | ++ | + | ++ |
| Passives/WDM | ++ | ++ | + | ++ | + |
| Coupling to fiber | ++ | + | ++ | o | + |
| Modulators | – | – | + | ++ | + |
| Detectors | – | – | – | ++ | ++ |
| Lasers | – | – | o | ++ | +[1] |
| Electronics | – | – | – | + | ++ |

[1] Implementing lasers in silicon involves heterogeneous integration of III-V materials. See 14.3.4

In terms of integration density, semiconductors offer the best perspectives, because of their high index contrast. Silicon and III–V compound semiconductors both have their advantages: Silicon can make use of existing CMOS infrastructure, while in III–V materials active–passive integration technologies make it possible to integrate high-quality lasers directly with the waveguides. In view of on-chip optical interconnects, we will go deeper into silicon photonics in the next section.

### 14.2.8 Conclusion

To implement an optical link, many subcomponents are required. Many of these require incompatible material systems or operating conditions, and all are susceptible to imperfections and changing operation conditions. For on-chip interconnects, where a footprint is a prime resource, high-contrast waveguides offer the most benefits, but also the steepest challenges in terms of process control. Silicon photonics might be the technology that addresses these issues.

## 14.3 The Case for Silicon Photonics

Silicon-based waveguide circuits have several advantages for applications in optical interconnects [17, 18]. As already discussed, silicon photonic wire waveguides can have an exceptionally low footprint thanks to their high index contrast. In addition, the material can be processed in using standard CMOS manufacturing tools. As we will discuss further, silicon photonics can also be used to monolithically integrate compact WDM filters, modulators and photodetectors. Also, these subcomponents can be made very compact, which allows integration into larger circuits. Only the light sources are currently not easy to integrate, but we will briefly discuss possibilities to integrate III–V lasers on to the same chip.

We will now discuss the specific silicon subcomponents from Section 14.2 in somewhat more detail.

### 14.3.1 Waveguides and WDM Components

Silicon waveguides come in many forms. The most straightforward waveguide geometry is the photonic wire, which has a high lateral and vertical contrast with the silicon dioxide cladding. Photonic wire dimensions range from 300 nm × 300 nm to 500 nm × 200 nm. Especially the latter dimensions are used by many groups doing research on such waveguide components.

The silicon waveguides need a lower-index cladding on all sides to enable optical confinement.. Silicon dioxide provides a good cladding material and *silicon-on-insulator* (SOI) wafers are especially suited for optical waveguides: they have a high-quality single-crystal top layer. However, there are some restrictions on the SOI specifications. The waveguide layer should be of the right thickness, but also the buried oxide should be sufficiently thick. If the layer is too thin, light will leak out of the waveguide to the silicon substrate below [19]. An alternative to SOI is deposited silicon. Polysilicon will have a

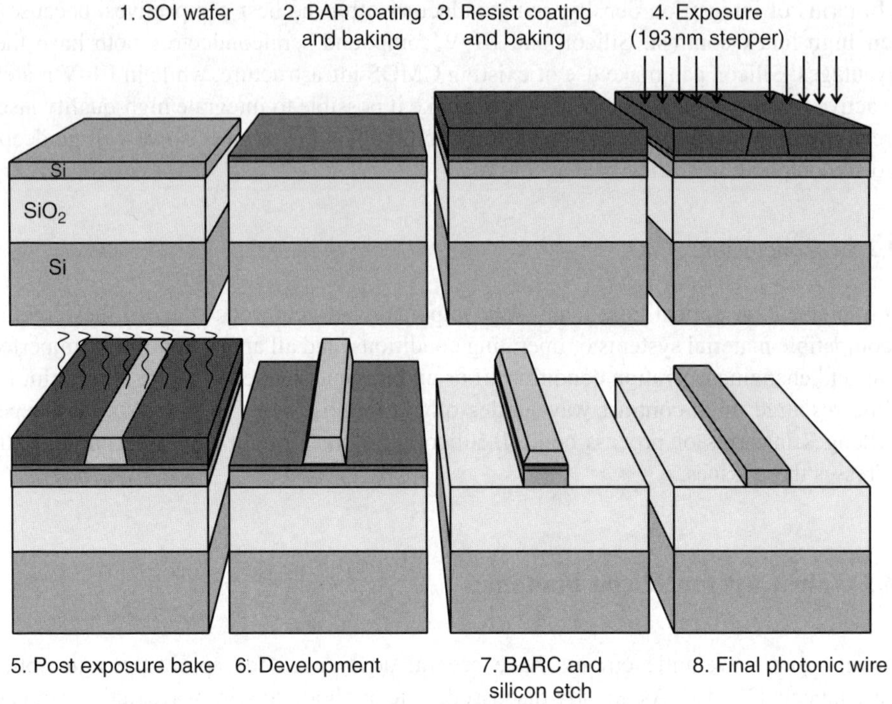


***Figure 14.12*** *Waveguide fabrication*

relatively high loss due to scattering and absorption at grain boundaries [20–22], but amorphous silicon is a good candidate for waveguide circuits [23–25].

A typical fabrication process for silicon waveguides is shown in Figure 14.12. The substrate is usually a custom silicon-on-insulator wafer, with a top silicon layer of the core thickness (around 200 nm) and a buried oxide of at least 1 μm, but preferably 2 μm thickness. First the waveguide patterns are defined in a spun-on resist layer using a high-resolution lithography tool (e-beam or a more industrial deep UV scanner), and subsequently the pattern is transferred to the underlying silicon. Finally, the remaining resist is removed.

One of the key performance metrics for on-chip waveguides is their propagation loss. In high-contrast waveguides the losses are typically induced by scattering of light at sidewall roughness created by the dry etching process. A current state-of-the-art wire waveguide now exhibits losses of about 1 dB/cm. Over a length of 1 cm they lose about 20 % of the power. This is already suitable for many applications. Photonic wire bends can also be very compact with a bend radius of only 3–5 μm.

While we stated that we prefer a waveguide with only a single guided mode, this is not entirely true for photonic wires. These waveguides support a single guided mode for two orthogonal polarizations: the fundamental TE polarization (electric field in the plane of the circuit) and TM (magnetic field in the plane of the circuit). The two polarizations have very different properties, but in general they do not mix. Therefore, silicon photonic circuitry is

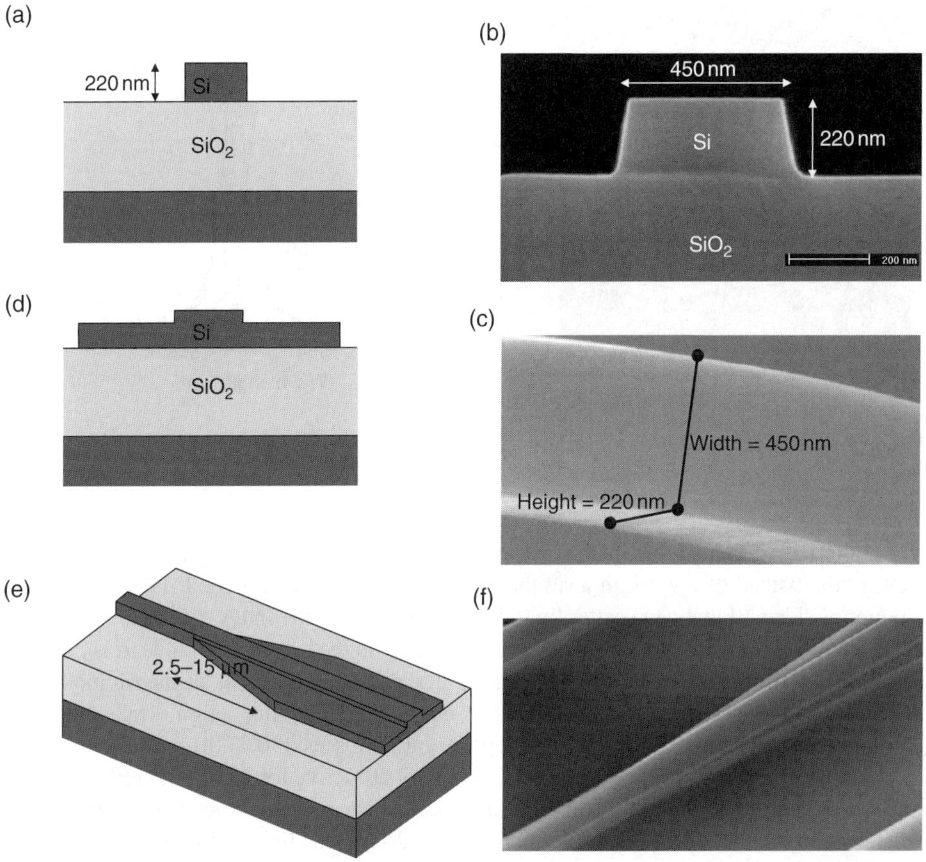


***Figure 14.13*** *Silicon waveguides. (a, b) Cross-section of a deep-etched photonic wire, (c) photonic wire in a birds-eye view on the etched sidewalls, (d) shallow-etched rib waveguide and (e, f) transition between the wire strip waveguide and the rib waveguide*

generally designed to work only for a single polarization; for the remainder of this chapter we will use the TE polarization, unless otherwise specified.

When longer propagation lengths are required, photonic wires can be replaced by alternative waveguide geometries, which can reduce the propagation loss. Figure 14.13 shows a photonic wire waveguide and two alternatives. By widening the waveguide core, the light becomes better confined in the waveguide, which results in a lower intensity on the sidewalls. This in turn reduces the losses by scattering. Waveguide losses as low as 0.3 dB/cm have been demonstrated [26]. However, as discussed in Section 14.2, widening the waveguide while keeping the contrast constant will increase the number of guided modes, which in turn can lead to multimode interference in bends. Therefore, while the broader waveguide can be used in straight sections, it has to be tapered down to a single-mode wire in the bends. This can be done using a gradual, adiabatic transition with a length of a few tens of μm, depending on the width of the straight section.

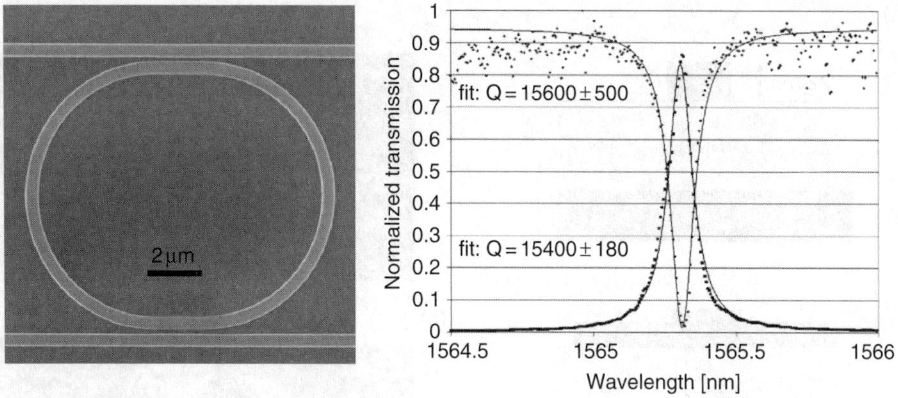


***Figure 14.14*** *Silicon ring resonator: the coupling section is elongated into a racetrack shape to enhance the coupling between the bus and the ring*

An alternative that preserves the single-mode behavior of the waveguide is to use a rib waveguide instead of a wire. In a rib the silicon is only partially etched. This reduces the amount of sidewall that can scatter light, but it also reduces the index contrast. As a result, bends cannot be made as tight as with wires, but, as with the broad waveguide, it is possible to engineer a transition between the rib waveguide and the wire. Also, as the rib mode size is relatively small, such a transition can be made over a shorter length scale. Such rib waveguides have been demonstrated with propagation losses of 0.27 dB/cm [27].

The tight bends make silicon photonic wires very attractive for wavelength filters: delay lines can be curled up in tight spirals and ring resonators can be made extremely small [28], as shown in Figure 14.14. However, as already discussed, the photonic wires are very susceptible to small variations in geometry. In a ring resonator or a delay-line-based filter, this will translate in unpredictable spectral positions of the resonances. Photonic wire-based filters will therefore always need an additional tuning mechanism, but the tuning power consumption can be kept low with good process control.

All the same advantages and disadvantages hold for WDM (de)multiplexers. An arrayed waveguide grating, with its many waveguide delay lines, can be made very compact because of the tight bends, but can suffer from high cross-talk from roughness-induced phase errors. Figure 14.15 shows an example with the transmission spectrum in the different output waveguides [29]. This shows that cross-talk levels are quite low. They could be reduced even further by local tuning of the individual delay lines, but the control algorithms will be much more complicated than those for the tuning of a single ring.

While it is possible thermally to adjust the spectral characteristics of a wavelength filter or multiplexer, this also generates problems, especially in the context of on-chip optical interconnects. Because of the on-chip nature, the interconnect layer will have to be close to the electronics. Because electronics can generate heat in an unpredictable way, with temperature swings of tens of °C, this can seriously disrupt the operation of the WDM components.

There is no clear-cut solution to this problem. Some propose to use waveguides that are athermal by combining silicon cores with polymer cladding, which compensate each

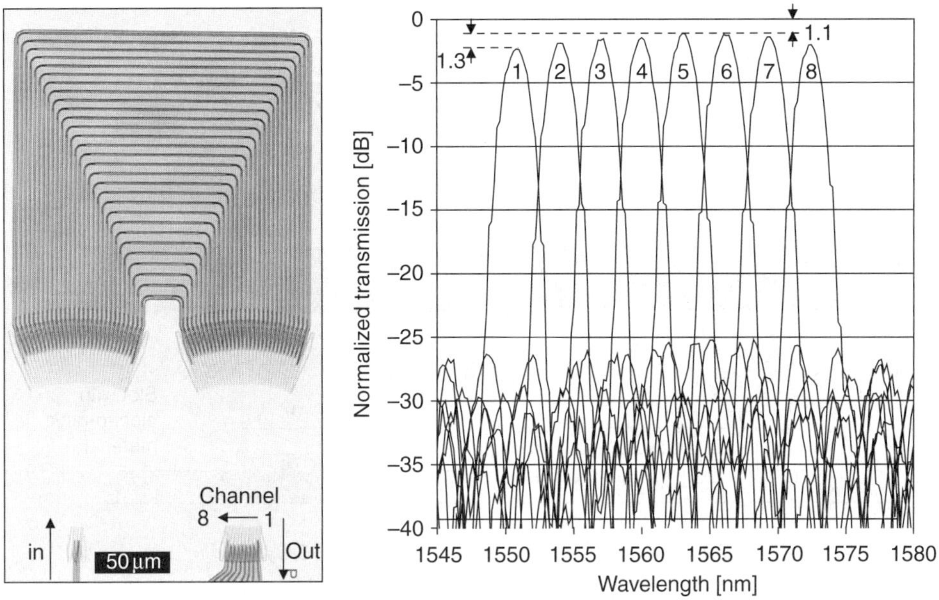


***Figure 14.15*** *Silicon arrayed waveguide grating. This eight-channel device with 400 GHz channel spacing has a footprint of only 300 µm × 450 µm*

other's thermal drift [30]. However, these waveguides have to make compromises in the geometry, which results in much higher propagation losses for a given fabrication technology. Also, polymers are not attractive materials for integration in a CMOS environment, and they typically exhibit thermal hysteresis and reliability issues.

Another alternative is to operate all photonic circuitry at an elevated temperature near the maximum operating point of the CMOS, and include heaters and monitoring circuits to keep the photonics at a stable temperature [31, 17]. This is a working solution, but has a significant detrimental effect on power consumption and might also pose long-term reliability issues.

### 14.3.2 Modulators, Tuners and Switches

As discussed, the fastest and most efficient phase modulators are based on direct electro-optic effects. However, silicon does not have such effects. While it is possible to induce such effects using strain [32], this requires substantial substrate engineering.

The most practical solution to make modulators in silicon is through the use of carriers [33]. The refractive index and the absorption of silicon is influenced by the actual concentration of electrons and holes [34]. Therefore, by injecting or extracting carriers in a waveguide it is possible to modulate a signal. The most commonly used mechanisms for the manipulation of the carrier density in a waveguide are shown in Figure 14.16. The most efficient mechanism in terms of magnitude is carrier injection in a p-i-n diode. The intrinsic zone is located in the center of the waveguide core. By forward biasing the diode, majority carriers are forced into the core, strongly affecting the refractive index [35, 36]. However, this

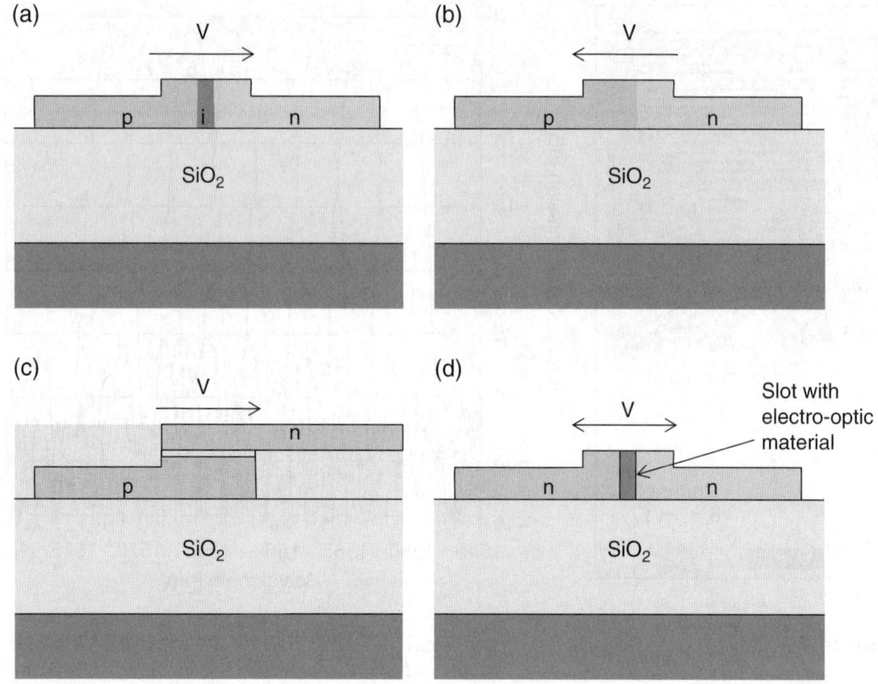


***Figure 14.16*** *Silicon modulator geometries: (a) forward-biased p-i-n diode, (b) reverse-biased p-n diode, (c) vertical silicon–oxide–silicon capacitor, (d) slot-based silicon hybrid modulator*

mechanism is limited in operation speed by the recombination time of the carriers in the core. Alternatively, one can start with a p-n diode in the core, which can be reverse-biased to increase or decrease the depletion zone in the junction. Because the effect moves around less carriers, the effect is much weaker, but it is also potentially much faster, and limited mainly by the capacitance of the junction [37–39]. By using complex junction geometries, or multiple junctions inside the waveguide core, the efficiency can be increased, as a stronger overlap between the carriers and the optical waveguide is achieved [40, 41]. However, as the modulation efficiency is directly linked to the amount of carriers that are moved around, a high modulation efficiency is typically combined with rather high absorption losses.

Instead of a junction, it is also possible to use carrier accumulation in a capacitor [42]. However, it is not straightforward to make a good capacitor with a vertical insulator, so the most promising geometry is the use of a layered capacitor. This involves somewhat more elaborate processing, but it also allows very efficient capacitors with a thin oxide to be made, which can accumulate a lot of carriers for a given operating voltage. The effect of the capacitor can be enhanced by using electro-optic materials. When using a so-called *slot waveguide*, the silicon waveguide is cut in the middle with a narrow slot, which can be filled with another material that has a strong electro-optic effect; an external electric field will change the refractive index of the material. If the material also has a low refractive index compared to the silicon core, the slot waveguide can be engineered such that there is a very high confinement of the light in the slot. Modulators based on this effect have been demonstrated, using an electro-optic polymer filling in the slot [43–46]. Such modulators are also less limited by carrier dynamics, as the electro-optic effect of the filling is much faster than the carrier dynamics.

Even though the carriers induce an absorption, the main effect is a change in the effective index. To make an amplitude modulator, therefore, the junction or capacitor should be integrated in an MZI or a (ring) cavity. Both approaches have been demonstrated by different groups. In an MZI, the modulator could be included in both arms, and the device could be operated in push–pull. For carrier injection modulators, where the effect is quite strong, devices have been demonstrated with a length of only 150 μm [36]. This is sufficiently small to be controlled as a lumped element. However, carrier depletion modulators, which are potentially much faster, are also much weaker. For these modulators the length quickly scales up to 1 mm (for a CMOS compatible operating voltage). At these length scales, for high-speed modulation (up to 40 Gbps), special care has to be taken with the electrical drivers. The usual approach is to use a microwave traveling wave electrode: When properly designed, the electrical wave will co-propagate with the optical mode along the modulator. However, a microwave electrode needs proper termination, which will dissipate a lot of power, enough to dominate the power consumption of the entire link.

In a ring configuration, the modulator is looped in a compact ring [35, 47] or disk [48], which means that is can be driven as a lumped element, which dramatically reduces power consumption, as there is no need for microwave circuitry. However, it is essential that the edge of the transmission dip is spectrally aligned with the operating wavelength of the source. This will require good process control, but ultimately a tuning mechanism to compensate for external influences, such as thermal variations, is needed. The modulator could be tuned by applying a bias to the modulation voltage, but as the modulation effects are typically quite small, in most cases the tuning range will be too small. Thermal tuning is the strongest tuning mechanism available, but it can only be used in one direction (heating not cooling). This means that, to compensate for thermal fluctuations in the chip, the operating temperature must be set near the upper boundary of the operation specs. This means that continuous heating is required, which translates in additional power consumption.

Thermal effects can also be used in their own right. For instance, they can be used spectrally to tune WDM filter components. Such tuning should of course be controlled by a feedback loop, which requires additional electronic circuitry. As already mentioned, thermal tuning operates in only one direction, so the operating point should be carefully chosen, and continuous heating can have a serious effect on the overall power consumption.

Instead of using heaters or carriers for modulation, they can also be used for switching. While the mechanism is the same, the operational requirements for switching are different. Response times are of the order of μs or ms, and power efficiency is important, as switches operate as passive devices. As with WDM components, all dissipation will add to the link power budget. In this respect, thermal switches seem the simplest solution, and thermal switches have been demonstrated [49–51]. Alternatively, one can use carrier injection. This effect is still quite strong, and one of the main drawbacks can now be turned into an advantage. As a modulator, the carrier injection device is limited by the carrier lifetime in the intrinsic region of the junction. However, if the structure can be engineered to increase this lifetime, a switch can maintain its state for longer without additional power consumption [52]. This is an effect that also applies to charge accumulation devices, where the switching action is controlled by charging or draining a capacitor.

### 14.3.3 Photodetectors

Silicon is a very good optical material for waveguide, as it has very little absorption losses at common telecom wavelengths. This transparency makes it unsuitable for photodetectors as a material with a smaller electronic bandgap is required. Germanium and several III–V compound semiconductors fit that bill. Germanium has the key advantage that it is already a material that can be integrated with silicon, and several foundries incorporate Ge or SiGe epitaxy in their fabrication cycle.

In photodetectors, an incident photon will create carriers. For efficient detection, these carriers must be collected at the electrodes, which means that they should not be allowed to recombine at defects. This poses some significant challenges, technology-wise. Germanium epitaxy on silicon typically involves an interface layer with dislocations. Thus a careful process and device design is needed to integrate germanium with silicon into a photodetector where neither the light or the generated carriers come into contact with such layers. Several integrated germanium photodetectors have been demonstrated [31, 53–55] with good responsivity (of the order of 1 A/W at 1550 nm) and reasonably low dark current (a few nA).

Also, avalanche photodetectors have been demonstrated [56, 57], but while these have a gain–bandwidth product of several hundred GHz, they typically suffer from the noise inherent to APDs.

### 14.3.4 Light Sources

One of the key issues with silicon photonics is the integration of the light source. Efficient semiconductor lasers require a direct-bandgap material. Both silicon and germanium have an indirect bandgap. Some recent experiments have shown that germanium can be bandgap engineered to have a direct bandgap, but even then it is far from making an efficient source As already discussed, an external laser can be used as an optical power supply. However, this limits the scalability and the topology of the link network. Also, in WDM systems an external laser for each wavelength is required, which can quickly ramp up the cost.

Therefore, a better long-term solution is to integrate the laser on the chip. Flip-chip or other hybrid integration schemes could already provide this, but this provides little added value compared to separate packaging. Even worse, inclusion of the laser can produce additional thermal issues.

An alternative is to integrate the laser even closer to the optical circuitry. This would require integration of a good laser material: epitaxially grown III–V semiconductors fit the bill. However, while direct epitaxy of III–V materials on silicon has already been demonstrated, this technology is far from the maturity required for laser-quality layer stacks. Therefore, the most commonly explored route is using a bonding approach, where pieces of high-quality III–V material are bonded to a wafer with silicon photonic circuitry and subsequently thinned down to a film (see Figure 14.17). The bonding can be either direct (molecular bonding [58, 59]), using an adhesive layer [60–62] or metallic bonding [63]. This film is then processed into lasers using lithography and etching, the same kind of tools used for the silicon processing. This guarantees very good alignment and small feature sizes. However, the main advantage is that it is now possible to integrate many small lasers

***Figure 14.17*** *III–V integration using die-to-wafer bonding*

directly on the waveguide circuitry and that these lasers can be designed to emit the wavelength of choice (as long as the wavelength within the emission band of the III–V material), thus reducing the WDM source problem significantly.

An additional advantage of having lasers directly integrated on the waveguide circuity is that they could be directly modulated. This obviates the need for an additional modulator, but the modulation speed is limited by the carrier dynamics in the laser, so it is generally limited in speed. The integrated laser also has a topological advantage, as optical feed lines are not required: the laser can be located where the light is needed.

Apart from the additional process requirements (i.e. processing III–V materials in a silicon fab), the integrated lasers also introduce some operational issues. Semiconductor lasers tend to be very temperature sensitive: emission efficiency drops for higher temperatures and, as lasers generate quite a lot of heat themselves, this leads to a severe thermal management problem.

Depending on the application, different types of lasers can be used: Microdisk lasers are extremely small, with a footprint of only 15 μm × 15 μm, and can be directly coupled to a photonic wire waveguide [64, 65]. Also, a multiple of these waveguides can be integrated on the same bus waveguide [66]. However, given their small size, the output power is limited to 100–300 μW, which could well be sufficient for on-chip interconnects.

For higher power levels, alternative laser geometries are needed. These have a larger footprint and are often more complicated to couple to a silicon wire waveguide. This can be solved by constructing geometries where a significant portion of the light remains in the silicon, with a small extension of the optical mode experiencing gain in the III–V material [58] (see Figure 14.18). This implies a relatively large footprint to obtain sufficient power output. Alternative geometries confine as much light as possible in the III–V material, but this complicates the transition to the silicon wire [67].

### 14.3.5 Conclusion

Silicon photonics, with its high index contrast coupled to state-of-the-art silicon process technology, can be used to integrate all building blocks necessary to construct WDM links on a chip. Moreover, the compact photonic wires allow very dense circuitry and more complex functionality. However, there are significant challenges to implement silicon photonic optical interconnects. One is the large sensitivity to fabrication variation and operational

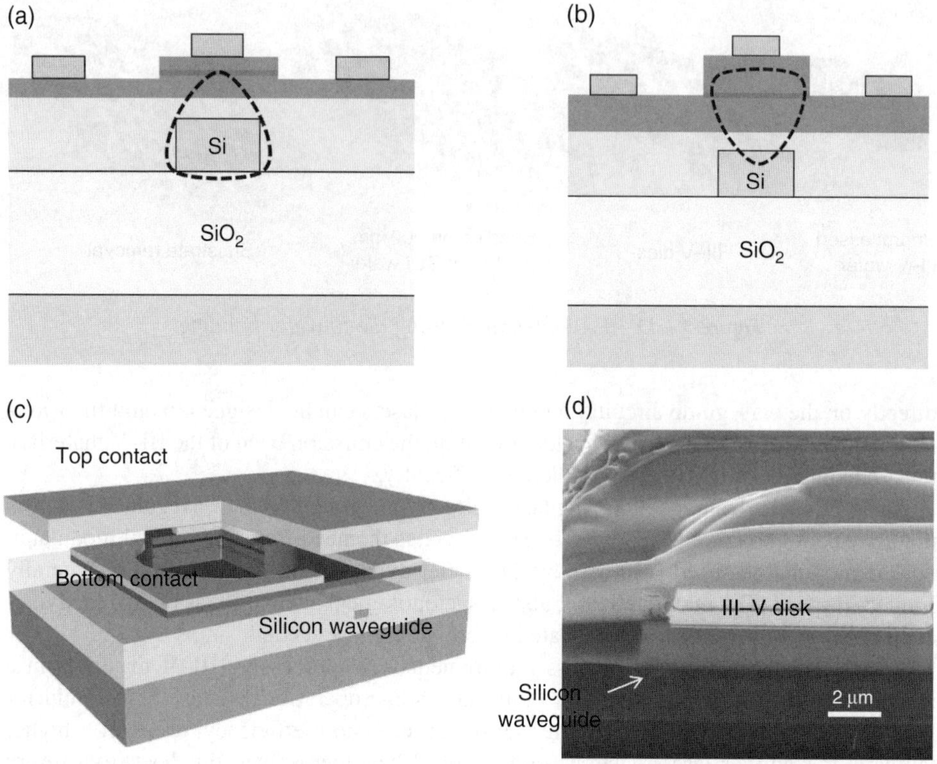


*Figure 14.18* *Silicon and III–V hybrid lasers. (a) Confinement of light largely in the silicon* [58]. *(b) Confinement of the light largely in the III–V material* [67]. *(c, d) Microdisk laser, with light in the III–V material evanescently coupled to an underlying silicon waveguide* [65]

conditions (especially temperature). Compensating for this is possible with active tuning on temperature control, but this can severely impact the power consumption of an optical link.

## 14.4 Optical Networks on a Chip

When implementing optical interconnects, the links can be set up in different topologies. The simplest topology is where communication is set up as a set of point-to-point links. While this is by far the simplest solution to manage, it is not the most scalable, and not necessarily the most efficient in terms of footprint and power consumption. Also, a large network of point-to-point links will often require a large number of waveguide intersections. While a single intersection can have quite a low loss, the number should still be kept as low as possible. An alternative to point-to-point links is the use of an *optical network on a chip* (ONoC). In an ONoC, optical signals are routed across the chip using either active switches or passive routing mechanisms. Such schemes could make more efficient use of footprint, as well as a reuse of the number of transmitters and receivers that are required. This in turn can reduce power consumption.

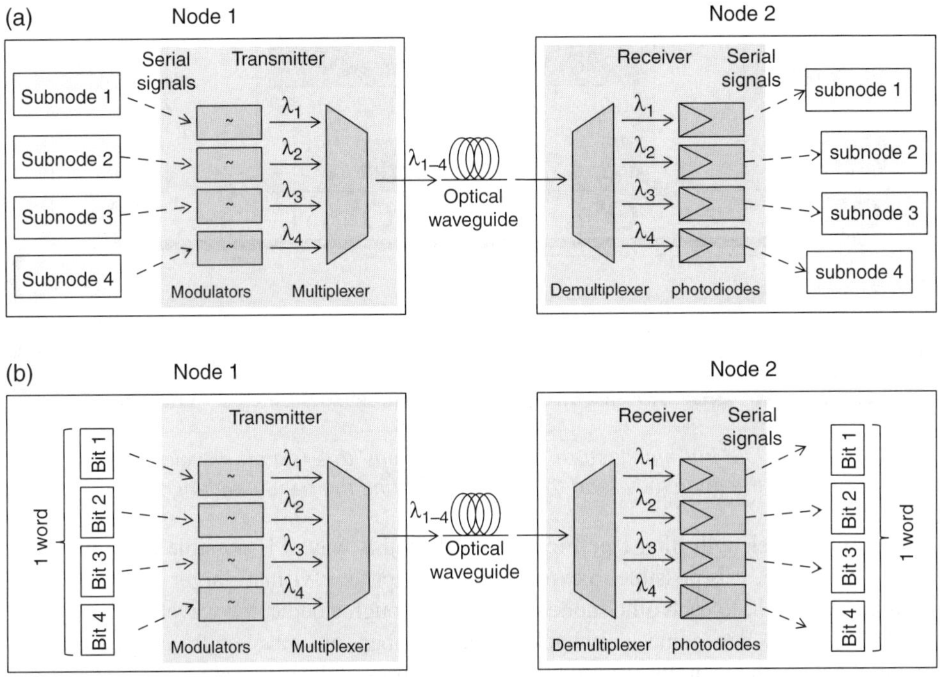


***Figure 14.19*** *Point-to-point links with WDM. (a) Two nodes communicate through a WDM link, where subnodes modulate an individual wavelength carrier. (b) The node modulates a machine word bitwise on different wavelengths*

Apart from the physical topology of the interconnect, the use of WDM introduces several additional options for the use of wavelengths. The various wavelength channels could be used to increase the available bandwidth on a single link, but the wavelengths could also multiplex multiple links on the same waveguide and be used to route light over the chip.

### 14.4.1 WDM Point-to-Point Links

The simplest way of using WDM in on-chip links is simply to use it as a multiplication of the bandwidth (see Figure 14.19). In the transmitter, multiple signals can be multiplexed on their own wavelengths and unraveled at the output, either by using (de)multiplexers or wavelength selective sources/modulators and detectors [68, 69]. This bundles a set of serial links into a single optical waveguide. A slightly different flavor of this mechanism is bit-parallel multiplexing. In this scheme, the bits of each machine word are encoded on their own wavelength, and at the output the words are reassembled [70]. In essence, this compresses a parallel link into a single line.

### 14.4.2 Bus Architecture

When a waveguide is used to carry different serial data streams on different wavelengths, it is not a strict necessity that the wavelengths are demultiplexed at the same location. Instead a bus mechanism can be used, where the nodes can extract their own wavelengths from the

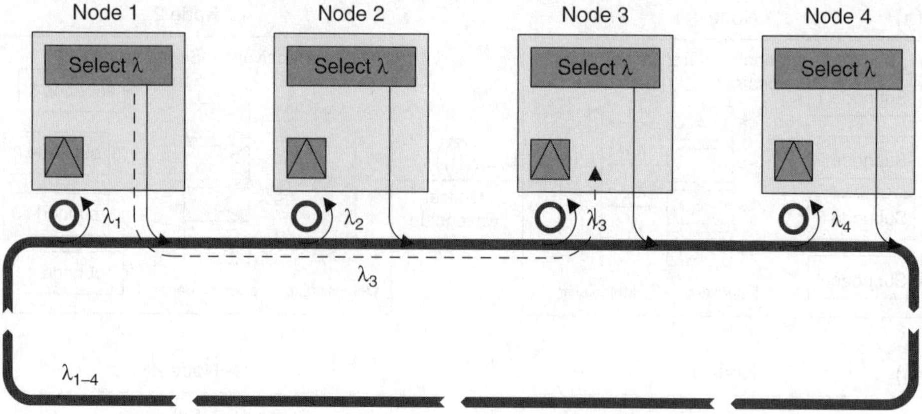


*Figure 14.20* WDM bus architecture. Each node 'listens' to a (set of) wavelengths. Other nodes can target the node they send through by selecting the right wavelength

bus using a wavelength filter (see Figure 14.20). In this way, it is possible to implement a multicast system. It is possible to expand this concept to a 'virtual' larger network, where each node can talk to each other node on the bus through a dedicated wavelength.

This scheme has some drawbacks. The number of wavelength channels scales quadratically with the number of nodes, which will impact the wavelength channel spacing. Also, as all the light traverses all the nodes, the filters in the nodes should have very high transmission and low cross-talk to mitigate compound effects.

### 14.4.3 (Reconfigurable) Networks

Instead of using a common bus to enable communication between any two nodes in a system, one can use a more scalable network topology. This can either be a dynamically switched network or a static routed network.

In a switched network, a central switching matrix changes the physical path so two network nodes become connected. Such a matrix can be made nonblocking, in the sense that it does not block any links between any two other nodes, except in some cases during the switch reconfiguration. The advantage of a physically switched network is that the full bandwidth of the waveguide remains available, including the possibility of WDM. The drawback is that an active switch matrix requires significant overhead and can induce considerable insertion loss in the link. Such a switched network requires arbitration, which is managed at the switch matrix. This is to avoid two sending nodes that try to communicate with the same receiving node.

Instead of an actively switched network, one can use WDM to perform the routing through the network in a passive circuit (see Figure 14.21). In such a system, a sender will target a specific receiver by choosing the right wavelength. In the router, the light is guided to the correct end point based on the wavelength. Such a router can be based on a set of wavelength filters or use an AWG. In such a wavelength-routed network, senders should be equipped with sources (and modulators) for all wavelengths. Microdisks, as discussed in Section 14.3, are ideally suited for this. In this way, a sender can actually communicate with different receivers at the same time, either in a one-to-one or multicast mode.

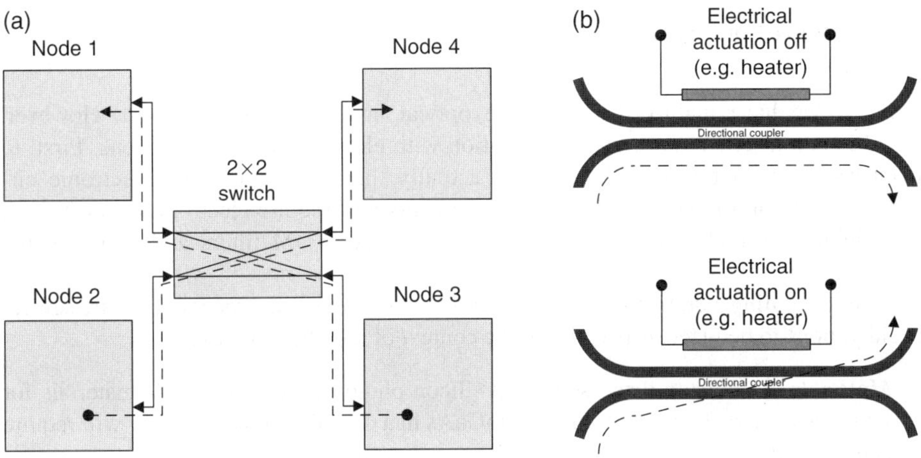


***Figure 14.21*** *Architecture of a switched 2 × 2 network. The nodes on the left can communicate with the nodes on the right using the full bandwidth. The broadband switches could be implemented like the directional coupler, as illustrated on the right*

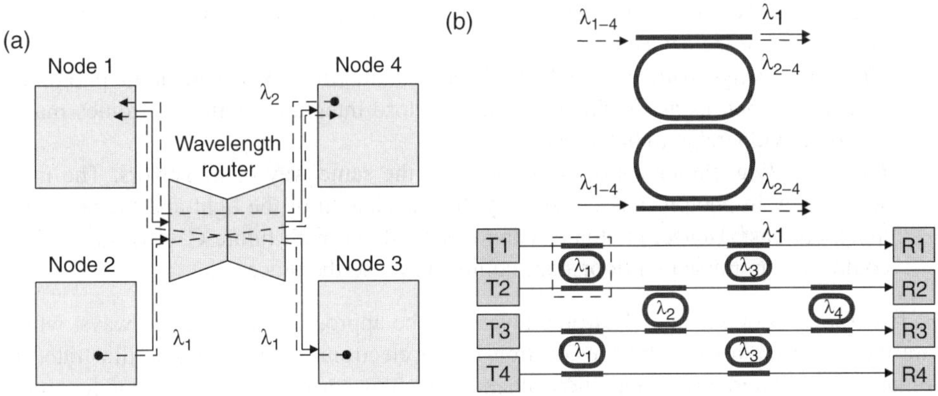


***Figure 14.22*** *Wavelength-routed network. Every node can communicate with any other node by choosing the right wavelength. Such a wavelength router can be implemented using a set of ring resonator filters or an arrayed waveguide grating*

If receivers are equipped with a single photodetector, they will have to arbitrate as to which sender is allowed to talk to them. However, if the receiver has a bank of wavelength-selective detectors (or detectors with a wavelength filter), it can in fact receive from all senders at the same time.

The advantage of a wavelength-routed network (Figure 14.22) is the simplicity of the routing/switching scheme, and that multicasting can be easily accomplished. However, the drawback is that the bandwidth between any two nodes is limited to a single wavelength channel, as WDM can no longer be used to multiply the bandwidth of a single link.

## 14.5 Integration Strategies

Currently we have mainly discussed the optical layer in an interconnect. However, optical interconnects require close integration with electronics for two reasons. First, in an interconnect context the photonics is actually there to connect the electronic circuitry. In addition to that, the photonic components will require electronic control, such as modulated current sources, feedback loops for the WDM filters and TIAs for the photodetectors.

When integrating a dramatically new technology in an existing process, there are several considerations to be made, especially in the context of a CMOS process:

1. *Materials*. As already discussed, even silicon photonics requires III–V materials for integrated lasers. Integrating InP and InGaAs in a CMOS-compatible flow will require careful contamination analysis, as well as process development.
2. *Footprint*. Depending on the integration strategy, photonics will require additional footprint on the chip, competing with the electronic circuitry, either through direct competition between photonics and transistors or between the photonic drivers and the logic.
3. *Cost*. Integrating a process for a photonic layer can quickly run into over 10 mask plates and 100 process steps. The benefits of the added bandwidth should outweigh the extra cost. Also, the additional processes can have a significant impact on the compound yield of the devices.
4. *Thermal management*. As we have discussed, thermal management in photonics, especially when using WDM, is essential. Close integration with electronics makes photonics vulnerable to hot spots.
5. *Future scaling*. Photonics does not scale in the same way of transistors. The main scaling law in photonics is dictated by the wavelength of the light and the refractive index contrast. Hence, photonics cannot scale down with future CMOS nodes. This could have an impact on floorspace competition and the processes.

Integrating photonic and electronic layers can be approached in different ways, which are most easily classified by the location in the electronics stack. This is illustrated in Figure 14.23. Alternative integration schemes, such as co-packaging, wire bonding, etc., are not really considered here, as they lack the interconnect density required for on-chip interconnects.

1. *Front-end-of-line*. Silicon photonics is generally based on SOI technology, using the top silicon layer for waveguiding. In principle, it is possible to define both transistors and waveguides in the same stack.
2. *Back-end-of-line*. By putting the optical layers on top of the metal interconnects, they become an integral part of the wiring network.
3. *Backside*. Putting the photonics on the backside partially decouples the processing of electronics and photonics.
4. *3D integration*. By using 3D integration to stack the photonics and electronics layer, the two technologies are largely decoupled, while the 3D stacking allows full access to all functionality.
5. *Flip-chip integration*. This is similar to 3D stacking, but flip-chipping implies face-to-face integration, making access and further integration difficult.

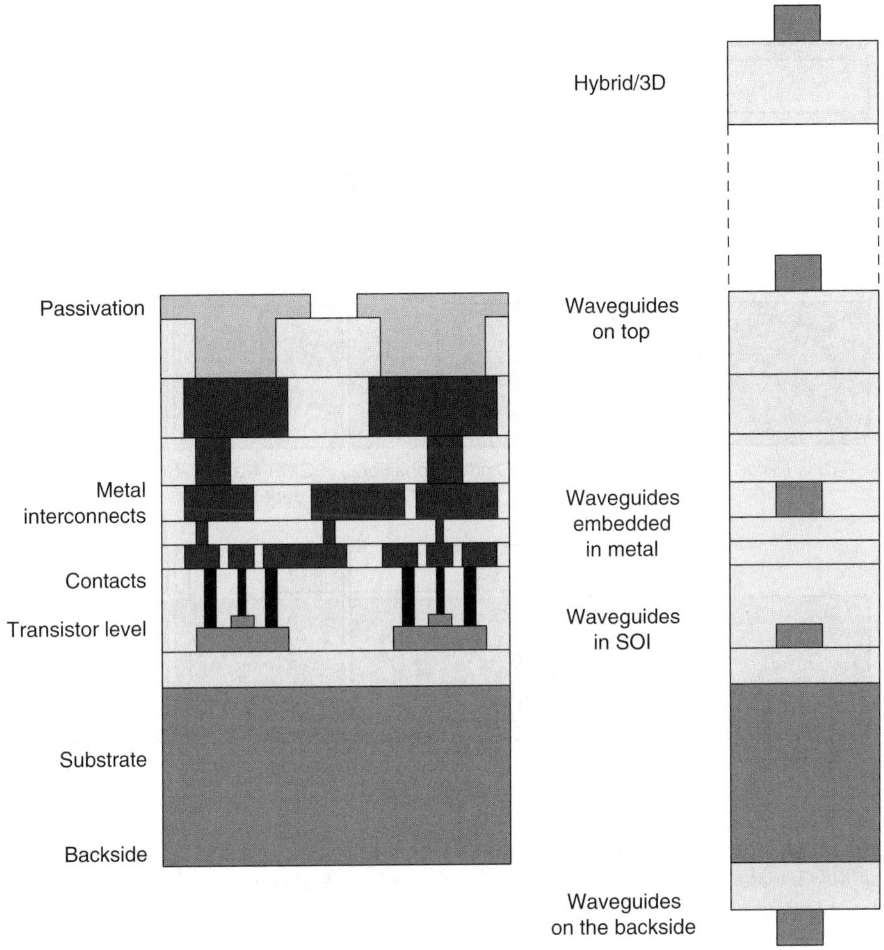


***Figure 14.23*** *Different integration options of photonic waveguides in a CMOS stack*

We will discuss the merits of the different approaches in somewhat more detail for the case of silicon photonic circuitry.

### 14.5.1 Front-End-of-Line Integration

As some silicon photonic components require steps that typically occur during transistor fabrication (e.g. implantations for modulators), it seems attractive to fabricate the photonics in the same layer as the electronics, especially when a silicon-on-insulator process is involved. This has the advantage that the photonic layers can be made with the best tools available in the fabrication and, with proper considerations, many steps could be shared between the electronics and photonics fabrication [71, 72].

However, integrating the photonics in the front-end of an electronics process faces some significant challenges. For one, the integration is only easy with an SOI electronics

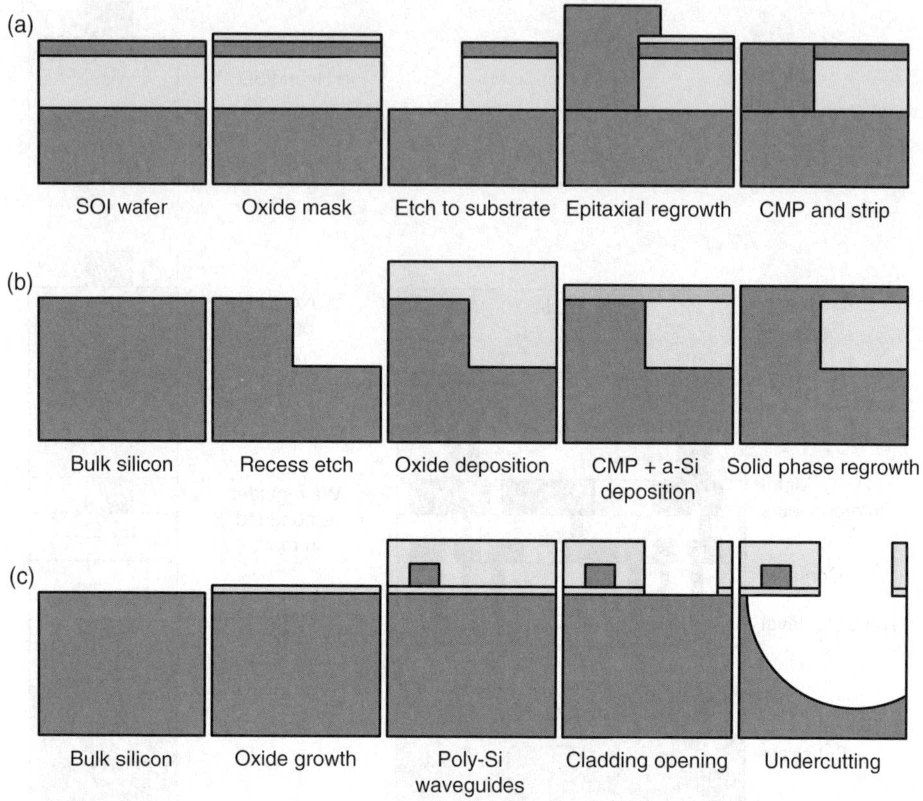


***Figure 14.24*** *Integration of silicon photonics with non-SOI electronics. (a) Building a local-bulk Si wafer from an SOI wafer. (b) Building a local-SOI wafer from a bulk Si wafer* [74]. *(c) Undercutting polysilicon waveguides to isolate them optically from the substrate* [73]

process. For bulk CMOS processes, there is no buried oxide to form the cladding. This could be overcome with deposited silicon, but the thermal anneals that occur at the end of the transistor process imply that the deposited material will be polycrystalline, which has a serious negative impact on the waveguide quality [20, 73]. However, an alternative is to construct a wafer with a local SOI for the waveguides and bulk for the electronics. Some approaches are illustrated in Figure 14.24. This can be done by starting from an SOI wafer and then backetching and regrowing. Alternatively, one can start from a bulk wafer and locally etch trenches, which can be filled with oxide. After planarization, amorphous silicon can be deposited and recrystallized using solid-phase epitaxy [74].

Even with these steps, FEOL integration of photonics is not straightforward. As the scaling of silicon waveguides only depends on the waveguide and the index contrast, they will not scale together with the CMOS. However, as the waveguides are cofabricated with the transistors, the photonic fabrication process must be redeveloped for each new node.

The strongest objection to putting the waveguides and the transistors together for on-chip interconnects is the fact that the photonic components compete for floorspace with the transistors. In this context photonics should serve the electronic functions and not the other way around. There are many applications, including off-chip interconnects, sensing, telecom, etc., where the electronics are supporting the photonics layer. For such applications, the floorspace competition is less of an issue.

### 14.5.2 Backside Integration

To avoid competition for floorspace, but still have the ability to make use of a high-quality substrate, one can make use of both sides of the wafer. As the preferred substrate for a silicon photonics layer is an SOI with a custom thickness, such a wafer could be used as the starting material to fabricate the photonics steps. All high-temperature steps are executed first, after which the wafer is flipped over for the processing of the transistors. Finally, the metal interconnect layers are made and the photonics is connected with through-silicon vias (TSV). This processing order can be slightly varied, and instead of SOI one could also use deposited layers for the waveguides.

Depending on the TSV technology, the wafer can be thinned down after the photonics process, acting as a carrier on the handling wafer for the electronics (only applicable in a bulk process). Otherwise, putting the photonics on the backside of an unthinned wafer would require relatively large TSVs [75]. Although the competition for floorspace is less than that for frontside FEOL integration, there is still a requirement for the TSVs to pass through the transistor layer, unless TSVs can be processed to contact transistors directly from the backside [75, 76].

Even though it maximizes the use of the high-quality substrate, two-sided processing brings about its own set of problems. First of all, wafer handling is much less straightforward, as backside structures have to be protected. This extends to packaging and testing, where the photonic structures should be accessible in some form. Especially in its thinned-down form, this can be a real issue.

### 14.5.3 Back-End-of-Line Integration

For on-chip interconnects, it would make sense to integrate the photonic interconnect layer inside, or directly on top of, the metal interconnect layers. However, doing this in a monolithic way is not straightforward. First of all, when considering silicon photonics, it is not possible to deposit a high-quality monocrystalline silicon layer on top of a metal interconnect. Amorphous silicon is possible, but this material has a lower optical quality. Of course, other optical waveguide materials are possible, but these in general have a lower index contrast and therefore a lower integration density as well.

The second obstacle for integrating silicon photonics in the back-end-of-line is the temperature budget. The use of metals restricts the temperature budget to around 400 °C. This again poses a challenge for depositing high-quality silicon: it is not possible to recrystallize amorphous silicon if that would be desired. However, the more serious obstacle is the integration of active photonic devices: carrier-based modulators require activated implants and good germanium photodetectors require epitaxy. Both processes require steps beyond 400 °C.

### 14.5.4 3D Integration

Instead of monolithically integrating the photonics stack, 3D integration can be used to stack on top of the electronics layer. The photonics and electronics are processed on separate wafers and only integrated near the end of the process flow. This approach has several key advantages. First of all, the photonics process can be optimized without compromising performance for the monolithic integration of transistors. Also, there is no competition for floorspace with the transistors. Finally, there is no issue with reoptimizing the photonics process for different technology nodes: as waveguides follow entirely different scaling rules, the same photonics layer could be used as a global interconnect layer for several CMOS nodes, with little or no modification. The main modification might involve optimizing the photonics for a different operating voltage. This flexibility is especially advantageous for applications other than on-chip interconnect. Low-cost optical sensors could be integrated with cheap low-power electronics, while high-bitrate transceivers could integrate with high-speed SiGe BiCMOS or even III–V electronics.

There are different approaches to 3D integration and, depending on the application, either the photonics can be placed on top of the electronics or the other way around. For on-chip interconnects, both die sizes are most likely to be the same, so both stacking orders are possible. For other applications, the die size might be different: in general, the smaller die will be stacked on the larger die.

For 3D integration, there are different processing schemes possible. Some of those are shown in Figure 14.25. For instance, one can choose to process the TSVs before bonding the wafers together. The TSV can be fabricated in the photonics wafer, which then requires no modifications to the electronics process to accommodate TSVs. In this way, the photonics is on the top of the stack, facing up, which can be relevant for applications where easy access to the photonics is needed. One of the main issues with using TSVs is that they are typically large, scaling in diameter with the thickness of the substrate, resulting in sizable series resistance and parasitic capacitance [77–79]. In a high-speed optical interconnect context, this can be a bottleneck. The thickness of the substrate is largely determined by the accuracy with which the thinning procedure can be executed and the handling of the thinned substrate.

However, because photonics uses an SOI layer stack, it can take advantage of the buried oxide to act as a selective stop layer for the thinning. This makes it possible to integrate the photonics wafer face-down on to an electronics wafer and remove the entire silicon substrate through grinding, CMP and selective etching [75, 76]. Then, vias through the oxide can connect the photonics and the electronics layer, and it is even possible to process additional metal interconnect layers on top. This approach, which used 3D stacking of very thin layers, can eliminate most of the excessive parasitics of TSVs.

One of the problems with integrating complex technologies is that of compound yield. The overall yield of the entire process is a product of the yield of the electronics and that of the photonics times the yield of the integration process. If all processes have sufficiently high yield, but if one or more has a low yield, the compound yield might make the approach unviable. 3D integration could alleviate this problem somewhat by introducing testing of electronics and photonics prior to integration, and only integrating known good dies. For this, a die-to-wafer integration approach is needed, as opposed to a wafer-to-wafer approach.

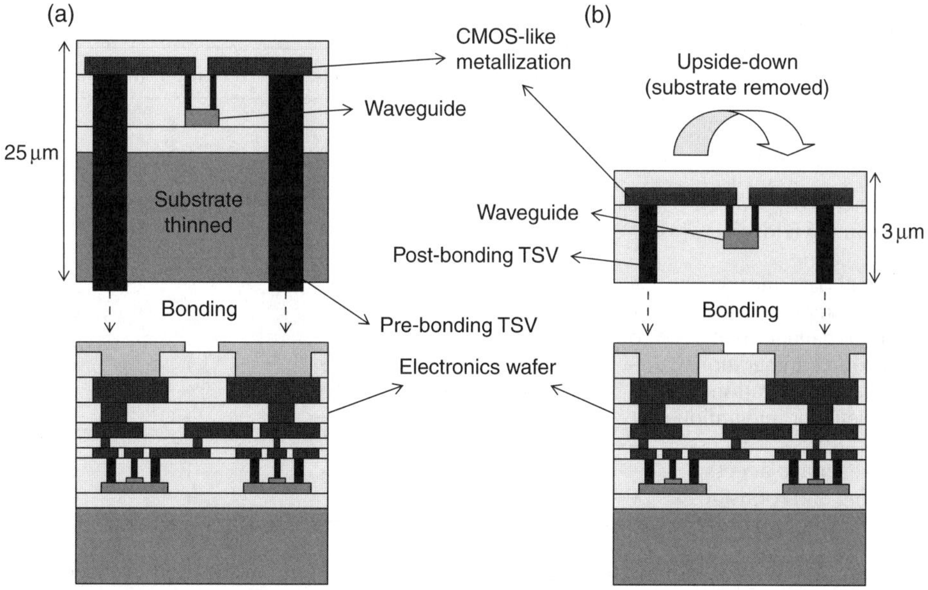


*Figure 14.25* 3D integration approaches. (a) Photonics face-up. TSVs are processed in the photonics wafer and stick out after substrate thinning. (b) Photonics face-down. The photonics wafer is bonded upside-down on the electronics wafer and metal TSV connections are processed after bonding and substrate removal

### 14.5.5 Flip-Chip Integration

Vertical stacking in 3D integration bears much similarity with simple flip-chipping [80]. However, while flip-chip technologies could be employed in the 3D integration process, there are some subtle differences. With flip-chip integration we indicate a method where the photonics and electronics face one another and no substrate-penetrating vias are used. This implies that there is no direct outside access to the 2D surface of either chip, which complicates electrical and optical I/O. Apart from this disadvantage, flip-chipping shares many advantages of 3D integration: it makes use of area-level interconnects between photonics and electronics, and microbumping technologies allow a very large density of connections. Also, the electronics and photonics layers are largely decoupled, and there is no competition for real estate.

### 14.5.6 Conclusion

These different integration strategies can have many variations and, depending on the technology, one or the other can be more suitable for optical on-chip optical interconnects. We have tried to make a qualitative comparison in Table 14.2. Given the fact that these technologies still have to prove themselves, this comparison is far from absolute.

***Table 14.2*** *Comparison between integration strategies*

| | FEOL | Backside | BEOL | 3D face-up | 3D face-down | Flip-chip |
|---|---|---|---|---|---|---|
| Floorspace competition | – | ++ | ++ | ++ | ++ | ++ |
| Thermal processing budget | ++ | + | – | ++ | ++ | ++ |
| Processing cost | ++ | – | + | – | o | o |
| Compound yield | + | – | + | ++ | – | ++ |
| Future scaling | – | + | ++ | ++ | ++ | + |
| Electrical parasitics | ++ | – | + | – | + | – |
| CMOS compatibility | + | o | + | + | + | + |
| Thermal management | o | + | – | + | – | – |
| Light source integration | – | ++ | – | ++ | + | – |
| Modulator integration | + | ++ | – | ++ | ++ | ++ |
| Photodetector integration | + | ++ | – | ++ | ++ | ++ |
| Electrical surface access | ++ | + | ++ | + | ++ | – |
| Optical surface access | – | + | ++ | + | + | – |
| Handling and robustness | ++ | – | ++ | o | + | – |

A few conclusions can be drawn from this table, depending on the metric. For on-chip optical interconnects, floorspace is essential, which rules out FEOL integration as a good solution. BEOL integration is most likely the lowest-cost technology.

If high performance is essential, the technologies that stand out most are based on 3D integration, because this allows a separate optimization of the entire photonics layer. Whether to integrate face-up or face-down then depends on the importance of the parasitics versus the thermal management.

## 14.6 Conclusion

In this chapter we have introduced the various concepts required to implement optical interconnects. Unlike in electronics, an optical link consists of a set of very heterogeneous building blocks, which typically require different material systems. The case can be argued that silicon photonics is likely to be the best candidate for future on-chip optical interconnects. It is possible to integrate all essential components to build a link. Even the light source, which still requires III–V materials, can be included using bonding technology. The processing of silicon photonics is compatible with existing CMOS manufacturing infrastructure (only recipes should be adapted) and it supports large-scale integration thanks to its high index contrast. Finally, one of the critical issues that need to be addressed is integration of the photonic layer with the electronics.

## References

[1] Krishnamoorthy, A.V., Ho, R., Zheng, X., Schwetman, H., Lexau, J., Koka, P. *et al.* (2009) Computer systems based on silicon photonic interconnects. *Proc. IEEE*, **97** (7, Special Issue SI), July, 1337–1361.

[2] Haurylau, M., Chen, G., Chen, H., Zhang, J., Nelson, N.A., Albonesi, D.H. *et al.* (2006) On-chip optical interconnect roadmap: challenges and critical directions. J. *Sel. Top Quantum Electron.*, **12** (6), 1699–1705.

[3] Barwicz, T., Watts, M.R., Popovic, M.A., Rakich, P.T., Socci, L., Kartner, F.X. *et al.* (2007) Polarization-transparent microphotonic devices in the strong confinement limit. *Nature Photonics*, **1**, January, 57–60.

[4] Lin, L.Y., Goldstein, E.L. and Tkach, R.W. (1998) Free-space micromachined optical switches with submillisecond switching time for large-scale optical crossconnects. *IEEE Photon Technol. Lett.*, **10** (4), 525–527.

[5] Lee, S.S., Huang, L.S., Kim, C.J. and Wu, M.C. (1999) Free-space fiber-optic switches based on MEMS vertical torsion mirrors. *J. Lightwave Technol.*, **17** (1), 7–13.

[6] Dragone, C. (1991) An $N \times N$ optical multiplexer using a planar arrangement of two star couplers. *IEEE Photon Technol. Lett.*, **3** (9), September, 812–814.

[7] Bogaerts, W., Dumon, P., Van Thourhout, D., Taillaert, D., Jaenen, P., Wouters, J. *et al.* (2006) Compact wavelength-selective functions in silicon-on-insulator photonic wires. *J. Sel. Top Quantum Electron.*, **12** (6), December, 1394–1401.

[8] Brouckaert, J., Bogaerts, W., Selvaraja, S., Dumon, P., Baets, R. and Van Thourhout, D. (2008) Planar concave grating demultiplexer with high reflective Bragg reflector facets. *IEEE Photon Technol. Lett.*, **20** (4), 309–311.

[9] Brouckaert, J., Bogaerts, W., Dumon, P., Van Thourhout, D. and Baets, R. (2007) Planar concave grating demultiplexer fabricated on a nanophotonic silicon-on-insulator platform. J. *Lightwave Technol.*, **25** (5), 1269–1275.

[10] Dumon, P., Bogaerts, W., Wiaux, V., Wouters, J., Beckx, S., Van Campenhout, J. *et al.* (2004) Low-loss SOI photonic wires and ring resonators fabricated with deep UV lithography. *IEEE Photon Technol. Lett.*, **16** (5), May, 1328–1330.

[11] Little, B.E., Foresi, J.S., Steinmeyer, G., Thoen, E.R., Chu, S.T., Haus, H.A. *et al.* (1998) Ultra-compact $SiSiO_2$ microring resonator optical channel dropping filters. *IEEE Photon Technol. Lett.*, **10** (4), 8212–8222.

[12] Chin, M.K. and Ho, T. (1998) Design and modeling of waveguide-coupled single-mode microring resonators. J. *Lightwave Technol.*, **16** (8), 1433–1446.

[13] Yariv, A. (2002) Critical coupling and its control in optical waveguide-ring resonator systems. *IEEE Photon Technol. Lett.*, **14** (4), April, 483–485.

[14] Xia, F, Rooks, M., Sekaric, L. and Vlasov, Y. (2007) Ultra-compact high order ring resonator filters using submicron silicon photonic wires for on-chip optical interconnects. *Opt. Express.*, **15** (19), September, 11934–11941.

[15] Colace, L., Balbi, M., Masini, G., Assanto, G., Luan, H.C. and Kimerling, L.C. (2006) Ge on Si p-i-n photodiodes operating at 10 Gbit/s. *Appl. Phys. Lett.*, **88** (10), 6 March.

[16] Ang, K.W., Liow, T.Y., Fang, Q., Yu, M.B., Ren, F.F., Zhu, S.Y. *et al.* (2009) Silicon photonics technologies for monolithic electronic–photonic integrated circuit (EPIC) applications: current progress and future outlook. IEEE International Electron Devices Meeting (IEDM), 2009, pp. 1–4.

[17] Gunn, C. (2006) Silicon photonics – poised to invade local area networks. *Photonics Spectra*, **40** (3), March, 62–68.

[18] Soref, R. (2006) The past, present, and future of silicon photonics. J. *Sel. Top Quantum Electron.*, **12** (6), 1678–1687.

[19] Bogaerts, W., Wiaux, V., Taillaert, D., Beckx, S., Luyssaert, B., Bienstman, P. *et al.* (2002) Fabrication of photonic crystals in silicon-on-insulator using 248-nm deep UV lithography. *IEEE J. Sel. Top Quantum Electron.*, **8** (4), 928–934.

[20] Foresi, J.S., Black, M.R., Agarwal, A.M. and Kimerling, L.C. (1996) Losses in polycrystalline silicon waveguides. *Appl Phys Lett.*, **68** (15), 8 April, 2052–2054.

[21] Agarwal, A.M., Liao, L., Foresi, J.S., Black, M.R., Duan, X. and Kimerling, L.C. (1996) Low-loss polycrystalline silicon waveguides for silicon photonics. J. *Appl. Phys.*, **80** (11), 6120–6123.

[22] Zhu, S., Fang, Q., Yu, M.B., Lo, G.Q. and Kwong, D.L. (2009) Propagation losses in undoped and n-doped polycrystalline silicon wire waveguides. *Opt. Express.*, **17** (23), 9 November, 20891–20899.

[23] Sparacin, D.K., Sun, R., Agarwal, A.M., Beals, M.A., Michel, J., Kimerling, L.C. *et al.* (2006) Low-loss amorphous silicon channel waveguides for integrated photonics. 3rd IEEE International Conference on Group IV Photonics, 2006, pp. 255–257.

[24] Harke, A., Krause, M. and Mueller, J. (2005) Low-loss single mode amorphous silicon waveguides. *Electron. Lett.*, **41** (25), 1377–1379.

[25] Selvaraja, S., Sleeckx, E., Schaekers, M., Bogaerts, W., Van Thourhout, D., Dumon, P. *et al.* (2009) Low-loss amorphous silicon-on-insulator technology for photonic integrated circuitry. *Opt. Commun.*, **282** (9), 1767–1770.

[26] Spector, S., Geis, M.W., Lennon, D., Williamson, R.C. and Lyszczarz, T.M. (2004) Hybrid multi-mode/single-mode waveguides for low loss, in *Optical Amplifiers and Their Applications/ Integrated Photonics Research*, Optical Society of America, p. IThE5.

[27] Selvaraja, S.K., Bogaerts, W., Absil, P., Van Thourhout, D. and Baets, R. (2010) Record low-loss hybrid rib/wire waveguides for silicon photonic circuits. 6th IEEE International Conference on Group IV Photonics, 2010, p. PD01.

[28] Prabhu, A.M., Tsay, A., Han, Z. and Van, V. (2009) Ultracompact SOI microring add-drop filter with wide bandwidth and wide FSR. *IEEE Photon Technol. Lett.*, **21** (9–12), May, 651–653.

[29] Bogaerts, W., Selvaraja, S.K., Dumon, P., Brouckaert, J., De Vos, K., Van Thourhout, D. *et al.* (2010) Silicon-on-insulator spectral filters fabricated with CMOS technology. J. *Sel. Top Quantum Electron.*, **16** (1), 33–44.

[30] Teng, J., Dumon, P., Bogaerts, W., Zhang, H., Jian, X., Han, X. *et al.* (2009) Athermal silicon-on-insulator ring resonators by overlaying a polymer cladding on narrowed waveguides. *Opt. Express.*, **17** (17), August, 14627–14633.

[31] Gunn C. (2006) CMOS photonics for high-speed interconnects. *IEEE Micro.*, **26** (2), 58–66.

[32] Jacobsen, R.S., Andersen, K.N., Borel, P.I., Fage-Pedersen, J., Frandsen, L.H., Hansen, O. *et al.* (2006) Strained silicon as a new electro-optic material. *Nature*, **441** (7090), 11 May, 199–202.

[33] Reed, G.T., Mashanovich, G., Gardes, F.Y. and Thomson, D.J. (2010) Silicon optical modulators [Review]. *Nature Photonics*, **4** (8), August, 518–526.

[34] Soref, R. and Bennett, B. (1987) Electrooptical effects in silicon. J. *Quantum Electron.*, **23** (1), 123–129.

[35] Xu, Q., Manipatruni, S., Schmidt, B., Shakya, J. and Lipson, M. (2007) 12.5 Gbit/s carrier-injection-based silicon micro-ring silicon modulators. *Opt. Express.*, **15** (2), January, 430–436.

[36] Green, W.M.J., Rooks, M.J., Sekaric, L. and Vlasov, Y.A. (2007) Ultra-compact, low RF power, 10 Gb/s silicon Mach–Zehnder modulator. *Opt. Express.*, **15** (25), December, 17106–17113.

[37] Liao, L., Liu, A., Basak, J., Nguyen, H., Paniccia, M., Rubin, D. *et al.* (2007) 40 Gbit/s silicon optical modulator for high-speed applications. *Electron. Lett.*, **43**(22).

[38] Gardes, F., Reed, G., Emerson, N. and Png, C. (2005) A sub-micron depletion-type photonic modulator in silicon on insulator. *Opt. Express.*, **13** (22), October, 8845–8854.

[39] Liu, A., Liao, L., Rubin, D., Nguyen, H., Ciftcioglu, B., Chetrit, Y. *et al.* (2007) High-speed optical modulation based on carrier depletion in a silicon waveguide. *Opt. Express.*, **15** (2), January, 660–668.

[40] Marris-Morini, D., Vivien, L., Fédéli, J.M., Cassan, E., Lyan, P. and Laval, S. (2008) Low loss and high speed silicon optical modulator based on a lateral carrier depletion structure. *Opt. Express.*, **16** (1), January, 334–339.

[41] Marris-Morini, D., Le Roux, X., Vivien, L., Cassan, E., Pascal, D., Halbwax, M. *et al.* (2006) Optical modulation by carrier depletion in a silicon PIN diode. *Opt. Express.*, **14** (22), October, 10838–10843.

[42] Liao, L., Samara-Rubio, D., Morse, M., Liu, A., Hodge, D., Rubin, D. *et al.* (2005) High speed silicon Mach–Zehnder modulator. *Opt. Express.*, **13** (8), April, 3129–3135.

[43] Alloatti, L., Korn, D., Hillerkuss, D., Vallaitis, T., Li, J., Bonk, R. *et al.* (2010) Silicon high-speed electro-optic modulator. 7th IEEE International Conference on Group IV Photonics (GFP), 2010, pp. 195–197.

[44] Ding, R., Baehr-Jones, T., Liu, Y., Bojko, R., Witzens, J., Huang, S. *et al.* (2010) Demonstration of a low *V* pi *L* modulator with GHz bandwidth based on electro-optic polymer-clad silicon slot waveguides. *Opt. Express.*, **18** (15), July, 15618–15623.

[45] Baehr-Jones, T., Hochberg, M., Wang, G.X., Lawson, R., Liao, Y., Sullivan, P.A. *et al.* (2005) Optical modulation and detection in slotted silicon waveguides. *Opt. Express.*, **13** (14), 11 July, 5216–5226.

[46] Leuthold, J., Freude, W., Brosi, J.M., Baets, R., Dumon, P., Biaggio, I. *et al.* (2009) Silicon organic hybrid technology: a platform for practical nonlinear optics. *Proc. IEEE*, **97** (7),1304–1316.

[47] Ye, T. and Cai, X. (2010) On power consumption of silicon-microring-based optical modulators. J. *Lightwave Technol.*, **28** (11), 1 June, 1615–1623.

[48] Watts, M.R., Trotter, D.C., Young, R.W., Lentine, A.L. (2008) Ultralow power silicon microdisk modulators and switches. 5th IEEE International Conference on Group IV Photonics, 2008, pp. 4–6.

[49] Geis, M.W., Spector, S.J., Williamson, R.C. and Lyszczarz, T.M. (2004) Submicrosecond submilliwatt silicon-on-insulator thermooptic switch. *IEEE Photon Technol. Lett.*, **16** (11), 2514–2516.

[50] Chu, T., Yamada, H., Ishida, S. and Arakawa, Y. (2005) Compact $1 \times N$ thermo-optic switches based on silicon photonic wire waveguides. *Opt. Express.*, **13** (25), December, 10109–10114.

[51] Vlasov, Y., Green, W.M.J. and Xia, F. (2008) High-throughput silicon nanophotonic wavelength-insensitive switch for on-chip optical networks. *Nature Photonics*, **2** (4), April, 242–246.

[52] Van Campenhout, J., Green, W.M., Assefa, S. and Vlasov, Y.A. (2009) Low-power, $2 \times 2$ silicon electro-optic switch with 110-nm bandwidth for broadband reconfigurable optical networks. *Opt. Express.*, **17** (26), December, 24020–24029.

[53] Chen, L. and Lipson, M. (2009) Ultra-low capacitance and high speed germanium photodetectors on silicon. *Opt. Express.*, **17** (10), May, 7901–7906.

[54] Vivien, L., Osmond, J., Fédéli, J.M., Marris-Morini, D., Crozat, P., Damlencourt, J.F. *et al.* (2009) 42 GHz p.i.n germanium photodetector integrated in a silicon-on-insulator waveguide. *Opt. Express.*, **17** (8), April, 6252–6257.

[55] Vivien, L., Rouvière, M., Fédéli, J.M., Marris-Morini, D., Damlencourt, J.F., Mangeney, J. *et al.* (2007) High speed and high responsivity germanium photodetector integrated in a silicon-on-insulator microwaveguide. *Opt. Express.*, **15** (15), July, 9843–9848.

[56] Assefa, S., Xia, F. and Vlasov, Y.A. (2010) Reinventing germanium avalanche photodetector for nanophotonic on-chip optical interconnects. *Nature*, **464** (7285), 4 March, 80–U91.

[57] Kang, Y., Liu, H.D., Morse, M., Paniccia, M.J., Zadka, M., Litski, S. *et al.* (2009) Monolithic germanium/silicon avalanche photodiodes with 340 GHz gain-bandwidth product. *Nature Photonics*, **3** (1), January, 59–63.

[58] Fang, A.W., Park, H., Cohen, O., Jones, R., Paniccia, M.J. and Bowers, J.E. (2006) Electrically pumped hybrid AlGaInAs–silicon evanescent laser. *Opt. Express.*, **14** (20), October, 9203–9210.

[59] Kostrzewa, M., Cioccio, L.D., Zussy, M., Roussin, J.C., Fedeli, J.M., Kernevez, N. *et al.* (2006) InP dies transferred onto silicon substrate for optical interconnects application. *Sens. Acta A: Physical*, **125** (2), 411–414.

[60] Roelkens, G., Campenhout, J.V., Brouckaert, J., Thourhout, D.V., Baets, R., Romeo, P.R. *et al.* (2007) III–V/Si photonics by die-to-wafer bonding. *Mater. Today*, **10** (7–8), October, 36–43.

[61] Roelkens, G., Van Thourhout, D. and Baets, R. (2005) Ultra-thin benzocyclobutene bonding of III–V dies onto SOI substrate. *Electron. Lett.*, **41** (9), 561–562.

[62] Roelkens, G., Brouckaert, J., Van Thourhout, D., Baets, R., Notzel, R. and Smit, M. (2006) Adhesive bonding of InP/InGaAsP dies to processed silicon-on-insulator wafers using DVS-bis-benzocyclobutene. J. *Electrochem. Soc.*, **153** (12), G1015–G1019.

[63] Hong, T., Wang, Y., Yu, H.Y., Yue, S., Chen, W.X., Liang, S. *et al.* (2010) A selective area metal bonding method for Si photonics light sources. 7th IEEE International Conference on Group IV Photonics (GFP), 2010, pp. 13–15.

[64] Romeo, P.R., Campenhout, J.V., Regreny, P., Kazmierczak, A., Seassal, C., Letartre, X. *et al.* (2006) Heterogeneous integration of electrically driven microdisk based laser sources for optical interconnects and photonic ICs. *Opt. Express.*, **14** (9), May, 3864–3871.

[65] Van Campenhout, J., Rojo Romeo, P., Regreny, P., Seassal, C., Van Thourhout, D., Verstruyft, S. *et al.* (2007) Electrically pumped InP-based microdisk lasers integrated with a nanophotonic silicon-on-insulator waveguide circuit. *Opt. Express.*, **15** (11), 6744–6749.

[66] Van Campenhout, J., Liu, L., Romeo, P.R., Van Thourhout, D., Seassal, C., Regreny, P. *et al.* (2008) A compact SOI-integrated multiwavelength laser source based on cascaded InP microdisks. *IEEE Photon Technol Lett.*, **20** (16), 1345–1347.
[67] Lamponi, M., Keyvaninia, S., Pommereau, F., Brenot, R., de Valicourt, G., Lelarge, F. *et al.* (2010) Heterogeneously integrated InP/SOI laser using double tapered single-mode waveguides through adhesive die to wafer bonding. 7th IEEE International Conference on Group IV Photonics (GFP), 2010, pp. 22–24.
[68] Xia, F., Rooks, M., Sekaric, L. and Vlasov, Y. (2007) Ultra-compact high order ring resonator filters using submicron silicon photonic wires for on-chip optical interconnects. *Opt. Express.*, **15** (19), September, 11934–11941.
[69] Shacham, A., Bergman, K., Carloni, L.P. (2008) Photonic networks-on-chip for future generations of chip multiprocessors. *IEEE Trans. Comp.*, **57** (9), 1246–1260.
[70] Patel, R.R., Bond, S.W., Pocha, M.D., Larson, M.C., Garrett, H.E., Drayton, R.F. *et al.* (2003) Multiwavelength parallel optical interconnects for massively parallel processing. J. *Sel. Top Quantum Electron.*, **9**(2), 657–666.
[71] Pinguet, T., Sadagopan, V., Mekis, A., Analui, B., Kucharski, D., Gloeckner, S. (2007) A 1550 nm, 10 Gbps optical modulator with integrated driver in 130 nm CMOS. 4th IEEE International Conference on Group IV Photonics (GFP), 2007, pp. 1–3.
[72] Pinguet, T., Analui, B., Balmater, E., Guckenberger, D., Harrison, M., Koumans, R. *et al.* (2008) Monolithically integrated high-speed CMOS photonic transceivers. 5th IEEE International Conference on Group IV Photonics (GFP), 2008, pp. 362–364.
[73] Orcutt, J.S., Khilo, A., Holzwarth, C.W., Popović, M.A., Li, H., Sun, J. *et al.* (2011) Nanophotonic integration in state-of-the-art CMOS foundries. *Opt. Express.*, **19** (3), January, 2335–2346.
[74] Shin, D.J., Lee, K.H., Ji, H.C., Na, K.W., Kim, S.G., Bok, J.K. *et al.* (2010) Mach–Zehnder silicon modulator on bulk silicon substrate; toward DRAM optical interface. 7th IEEE International Conference on Group IV Photonics (GFP), 2010, pp. 210–212.
[75] Fedeli, J.M., Augendre, E., Hartmann, J.M., Vivien, L., Grosse, P., Mazzocchi, V. *et al.* (2010) Photonics and electronics integration in the HELIOS project. 7th IEEE International Conference on Group IV Photonics (GFP), 2010, pp. 356–358.
[76] Koester, S.J., Young, A.M., Yu, R.R., Purushothaman, S., Chen, K.N., La Tulipe, D.C. *et al.* (2008) Wafer-level 3D integration technology. *IBM J. Res. Develop.*, **52** (6), 583–597.
[77] Bermond, C., Cadix, L., Farcy, A., Lacrevaz, T., Leduc, P. and Flechet, B. (2009) High frequency characterization and modeling of high density TSV in 3D integrated circuits. IEEE Workshop on Signal Propagation on Interconnects, SPI'09, 2009, pp. 1–4.
[78] Healy, M.B. and Lim, S.K. (2009) A study of stacking limit and scaling in 3D ICs: an interconnect perspective. 59th Electronic Components and Technology Conference (ECTC), 2009, pp. 1213–1220.
[79] Kim, D.H., Mukhopadhyay, S. and Lim, S.K. (2009) TSV-aware interconnect length and power prediction for 3D stacked ICs. IEEE International Interconnect Technology Conference (IITC), 2009, pp. 26–28.
[80] Zheng, X., Patil, D., Lexau, J., Liu, F., Li, G., Thacker, H. *et al.* (2011) Ultra-efficient 10 Gb/s hybrid integrated silicon photonic transmitter and receiver. *Opt. Express.*, **19** (6), March, 5172–5186.

# 15

# Wireless Interchip Interconnects

*Takamaro Kikkawa*

*Research Institute for Nanodevices and Bio Systems,
Hiroshima University, Hiroshima, Japan*

## 15.1 Introduction

According to the scaling rule of metal–oxide–semiconductor (MOS) transistors in silicon large-scale integrated circuits (LSIs), both operating frequency and power consumption can be improved by reducing the feature sizes of MOS transistors, as shown in Table 15.1 [1]. A system on a chip (SOC) has been developed because it enables the highest packing density of scaled transistors for building an ultra-large-scale integrated circuit system (ULSI). However, the scaling of metal interconnects in an SOC deteriorates the performance of the LSI system because it increases resistance–capacitance (RC) time constants in the metal interconnects, as shown in Table 15.2 [2, 3]. To solve the RC signal delay problems for advanced CMOS technology, new materials such as copper (Cu) and low dielectric constant (low-$k$) interlayer dielectric films have been introduced [4, 5], as shown in Figure 15.1 (a). Figures 15.1 (b) and (c) are schematic diagrams of complementary MOS (CMOS) inverter circuits connected with a metal interconnect line and its equivalent circuit, respectively [2, 3]. The rise time $T_{90\%}$ of the output digital signal is expressed as

$$T_{90\%} \cong 2.3(R_{tr} + R_{int})C_{int} \quad (C_L << C_{int}) \tag{15.1}$$

where $R_{int}$, $R_{tr}$, $C_{int}$ and $C_L$ are interconnect resistance, transistor output resistance, interconnect capacitance and load capacitance, respectively. Although the new materials can reduce

*Advanced Interconnects for ULSI Technology*, First Edition. Edited by Mikhail R. Baklanov, Paul S. Ho and Ehrenfried Zschech.

**Table 15.1** *Scaling rule of the MOS transistor*

| Parameter | | Constant field | Constant voltage |
|---|---|---|---|
| Gate length | $L$ | $1/k$ | $1/k$ |
| Gate width | $W$ | $1/k$ | $1/k$ |
| Area | $S$ | $1/k^2$ | $1/k^2$ |
| Gate oxide thickness | $t_{ox}$ | $1/k$ | $1/k$ |
| Voltage | $V$ | $1/k$ | 1 |
| Electric field | $E$ | 1 | $k$ |
| Current | $I$ | $1/k$ | $k$ |
| Gate capacitance | $C$ | $1/k$ | $1/k$ |
| Delay time | $T$ | $1/k$ | $1/k^2$ |
| Power consumption | $IV$ | $1/k^2$ | $k$ |
| Power delay product | $IV{\cdot}T$ | $1/k^3$ | $1/k$ |

**Table 15.2** *Scaling of interconnects*

| Parameter | | Local interconnect | Global interconnect |
|---|---|---|---|
| Interconnect length | $L$ | $1/k$ | $1/k$ |
| Interconnect width | $W$ | $1/k$ | $1/k$ |
| Interconnect height | $H$ | $1/k$ | $1/k$ |
| Chip size | $D$ | $k_D$ | $k_D$ |
| Resistance | $R$ | $k$ | $k^2k_D$ |
| Capacitance | $C$ | $1/k$ | $k_D$ |
| Voltage drop | $V$ | 1 | $kk_D$ |
| Current | $I$ | $1/k$ | $1/k$ |
| Delay time | $T$ | 1 | $k^2k_D^2$ |
| Power consumption | $IV$ | $1/k$ | $k_D$ |
| Power delay product | $IV{\cdot}T$ | $1/k$ | $k^2k_D^3$ |

the RC delay times of interconnects, they cannot be an ultimate solution for ultra-high-speed signal transmission in multiple SOCs.

Figure 15.2 shows delay time versus interconnect length [6, 7]. Delay times of a conventional metal interconnect, transmission line and electromagnetic wave transmission are $\tau = 2.3\,R_{tr}C_{int}$, $\tau = v/L$ and $\tau = c_0/L$, respectively, where $\tau$ is a signal delay time, $v$ is a phase velocity of signal propagation, $L$ is the length of an interconnect and $c_0$ is the speed of light. The system performance of an LSI is limited by the global interconnect response time. The response time of the longest global interconnect is designated by $T_c \geq T_{cs} + T_{cp}$, where $T_c$, $T_{cs}$ and $T_{cp}$ are the system cycle, the maximum clock skew in a macrocell and the critical path delay, respectively. The length of the path is designated by $L_{min}^2 = N_g A_g$, where $N_g$ and $A_g$ are the number of logic gates and gate area, respectively. Therefore, the ultimate solution for long global interconnects is electromagnetic wave transmissions ($\tau = v/L$, $\tau = c_0/L$).

System in a package (SIP) has become more and more important because of its short time to market and low nonrecurring engineering cost. However, the disadvantages of the SIP are lower integration density and longer interconnect lengths by use of printed circuit boards (PCB) in comparison with a single-chip silicon SOC. Bonding wires, which have been used for conventional two-dimensional (2D) integration of LSI chips on a PCB, have

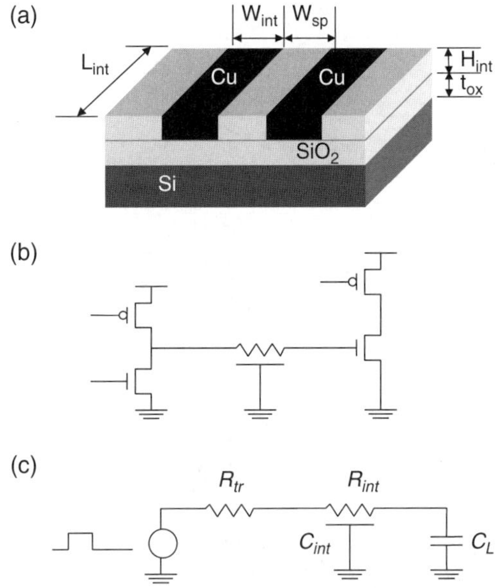


***Figure 15.1*** *Schematic diagrams of conventional metal interconnects. (a) Schematic diagram of the copper damascene structure. (b) Circuit diagram of CMOS inverters connected by a metal interconnect line. (c) Equivalent circuit diagram of the interconnect*

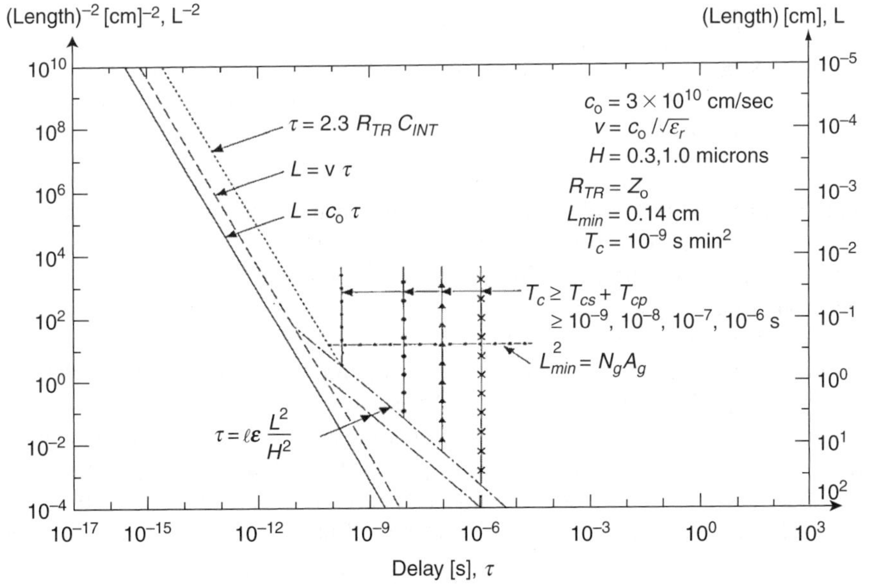


***Figure 15.2*** *Interconnect length versus signal delay time*

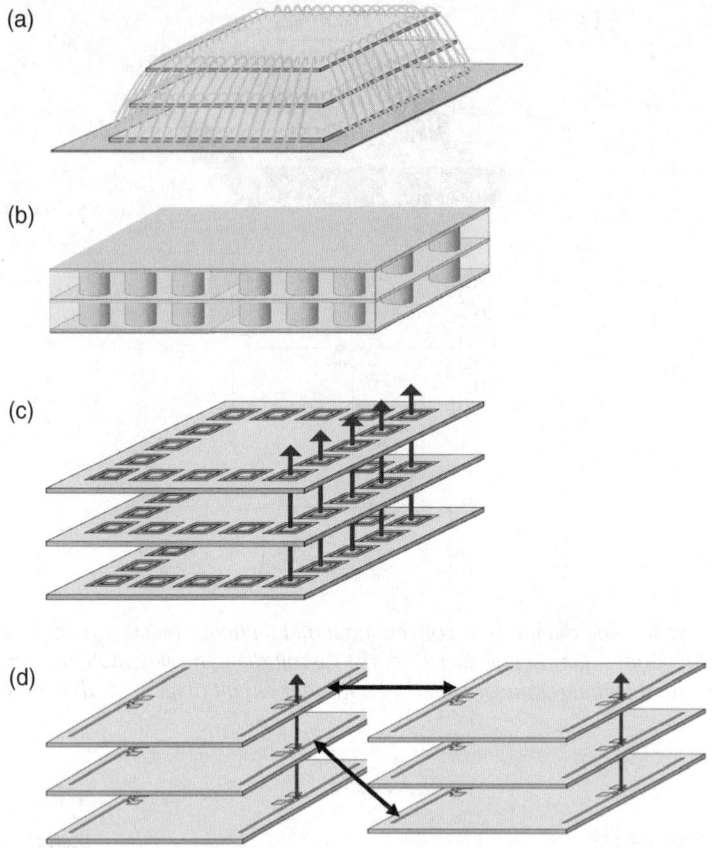


***Figure 15.3*** *Schematic diagrams of wired and wireless inter-chip interconnects: (a) wire bonding, (b) through-silicon via, (c) capacitively or inductively coupled wireless interconnects, (d) on-chip antennas*

limited the skew and the bandwidth of the signal transmission between LSI chips due to their parasitic inductances and capacitances [8]. Therefore, three-dimensional (3D) stacked LSI chip integration has been developed to shorten the interconnect lengths and increase the interconnect density [9–11].

Figure 15.3 shows schematic diagrams of interconnects for the 3D integration. In order to substitute wire bonding, shown in Figure 15.3 (a), through-silicon via (TSV) technology has been developed for the 3D integration to minimize the interconnect length and the interconnect pitch, as shown in Figure 15.3 (b). The key issues of the TSV for connecting multiple LSI chips are fabrication cost, testing, known good die and yield due to its complexity. Wireless interconnects without physical contacts are attractive alternative methods because they can eliminate the parasitic RC delays of metal interconnects and the process complexity.

Capacitive coupling [12–16] and inductive coupling [17–27], shown in Figure 15.3 (c), have the potential to solve the problems of TSV and wire bonding in terms of cost and performances at high frequencies. Electromagnetic wave propagation using on-chip antenna, shown in Figure 15.3 (d), has a potential for wireless local area network of LSI chips in the range of a few cm [28–37].

## 15.2 Wireless Interconnect Technologies

### 15.2.1 Figure of Merit for Wireless Interconnects

In order to compare the performances of wireless interconnect technologies, the figure of merit for transmission characteristics is expressed by the Friis transmission formula of the power transmission gain between the transmitter and receiver as [38]

$$G_a = \frac{|S_{21}|^2}{\left(1-|S_{11}|^2\right)\left(1-|S_{22}|^2\right)} = G_t G_r \left(\frac{\lambda}{4\pi r}\right)^2 e^{-2\alpha r} \tag{15.2}$$

where $G_a$ is the power transmission gain, $S_{21}$ and $S_{11}$ are transmission coefficient and reflection coefficient of scattering (S) parameters, respectively, $G_t$ and $G_r$ are the antenna gains of transmitter and receiver, respectively, $\lambda$ is the wavelength, $r$ is the distance between the transmitter and receiver, and $\alpha$ is the attenuation constant of the medium. $G_a$ is obtained by measuring the two-port S parameters $S_{21}$ and $S_{11}$ of a pair of transmitter and receiver, and by de-embedding the mismatch losses. The scattering matrix, or [*S*] matrix, can be measured by use of a vector network analyzer. A specific element of the [*S*] matrix can be determined as

$$S_{ij} = \left.\frac{b_i}{a_j}\right|_{a_k=0 \quad \text{for} \quad k \neq j} \tag{15.3}$$

$S_{ij}$ is determined by driving port *j* with an incident wave of $a_j$ and measuring the reflected wave amplitude, $b_i$, coming out of port *i*. The incident waves on all ports except the *j*th port are set to zero, so that all ports should be terminated in matched loads to avoid reflections. When the impedance at the interface is matched, both $S_{11}$ and $S_{22}$ are much less than 1 so that the transmission gain is nearly equal to the square of the transmission coefficient, $|S_{21}|^2$.

Figure 15.4 shows a comparison of the normalized electromagnetic field as a function of the radial distance ($R/\lambda$) of the field terms radiated by an infinitesimal dipole [38]. The electrostatic field decreases in proportion to the reciprocal cubic of the distance, while the magnetic induction field decreases in reciprocal proportion to the square of the distance. Both capacitive and inductive couplings can be used in the near field ($R/\lambda < ½\pi$). On the other hand, the radiation field decreases in reciprocal proportion to the distance so that the antenna can be used in the far field region ($R/\lambda > ½\pi$). From this consideration, the transmission ranges of capacitive and inductive couplings are limited so that they are suitable for clock and data communications between adjacent chips as local wireless interconnects. On the other hand, the electromagnetic wave transmission using on-chip antennas can reach a further distance than the capacitive and inductive couplings so that it can be applied to inter-chip reconfigurable interconnections as global wireless interconnects.

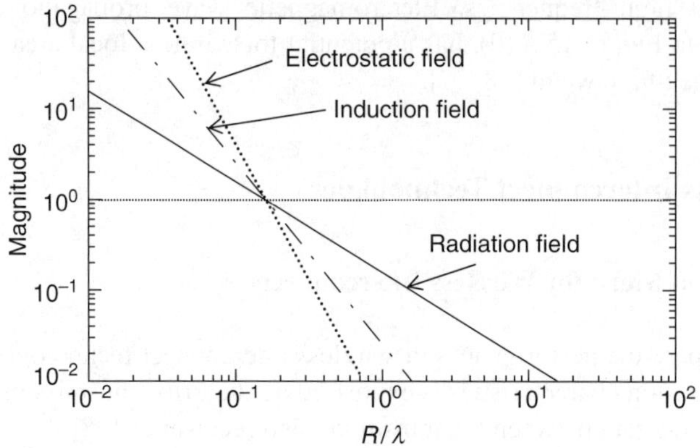


***Figure 15.4*** *Electromagnetic field versus normalized distance for the electrostatic field, induction field and radiation field*

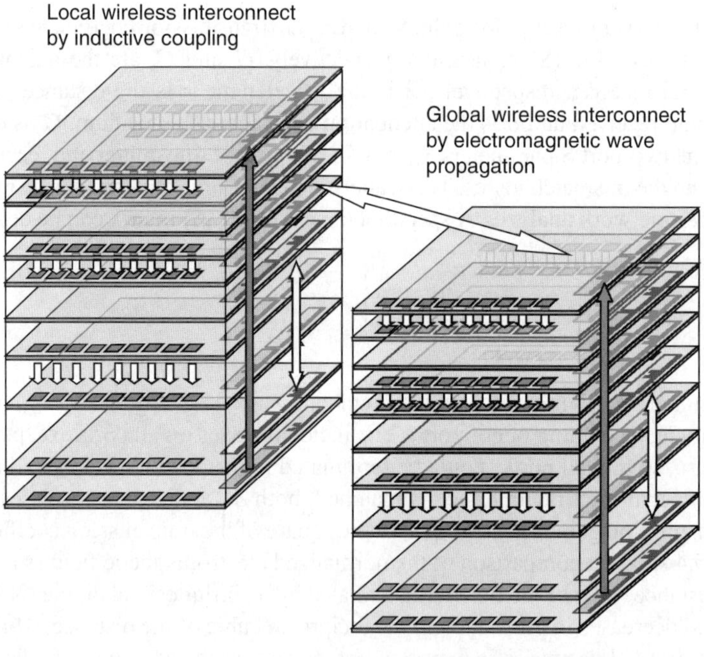


***Figure 15.5*** *Conceptual schematic diagram of local and global wireless interconnects for 3D stacked LSI chips*

A 3D wireless LSI stacked system has been proposed, as shown in Figure 15.5 [30, 31]. Local wireless interconnects (LWIs) have been developed between adjacent chips using inductive coupling and capacitive coupling. Global wireless interconnects (GWIs) have been developed among chips using electromagnetic wave propagation. The target distances of LWI and GWI are a few tens of μm and a few cm, respectively.

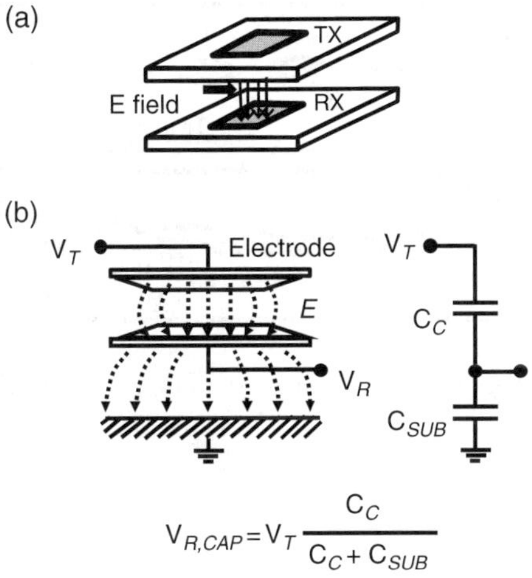


***Figure 15.6*** *Schematic diagram of capacitive coupling. (a) Schematic diagram of a capacitor. (b) Transmission model of a capacitively coupled link*

### 15.2.2 Capacitively Coupled Wireless Interconnects

Figures 15.6 (a) and (b) show schematic diagrams of capacitive coupling between chips [15, 21]. Two chips are placed face-to-face and the electrode distance is approximately a few μm. Each pair of electrodes, whose sizes are approximately 20 μm × 20 μm, forms a capacitor of the order of 10 fF. The received voltage of the capacitive-coupling link $V_{R,CAP}$ is given by $V_T C_C/(C_C + C_{SUB})$ where $V_T$ is the signal voltage in the transmitter electrode, $C_C$ is the capacitance between the electrodes and $C_{SUB}$ is the capacitance between the bottom electrode and the substrate ground. The $V_{R,CAP}$ value decreases with increasing communication distance because $C_C$ decreases while $C_{SUB}$ is constant [21].

When the chip-to-chip distance is reduced to several micrometers, the electrostatic coupling by the use of capacitors is superior due to a more confined electrical field and less cross-talk among vertical communication electrodes.

Figure 15.7 shows a capacitively coupled radio-frequency (RF) interconnect. The non-return-zero (NRZ) baseband signal is upconverted by an RF carrier using amplitude shift keying (ASK) modulation. An RF envelope detector in the RX charges/discharges the sense-amplifier input capacitor according to the NRZ input. A differential complimentary passive mixer is used in the TX to modulate NRZ data into an ASK signal and a mixer is designed as the RF envelope detector to convert the received ASK signal into NRZ data [13, 14].

Although a smaller coupling capacitor is preferred for the purpose of cost and energy savings, its size is determined according to the signal strength, receiver sensitivity and required system bit-error-rate (BER). Calculation shows that a coupling capacitor >7.3 fF is sufficient for achieving the desired $10^{-12}$ BER under a 1 V supply, which corresponds to a minimum 44 μm × 14 μm electrode area with a 3 μm chip-to-chip distance. Given the 1 V supply and the 25 GHz RF carrier, 72 μm × 38 μm capacitors are used to allow a 2 μm

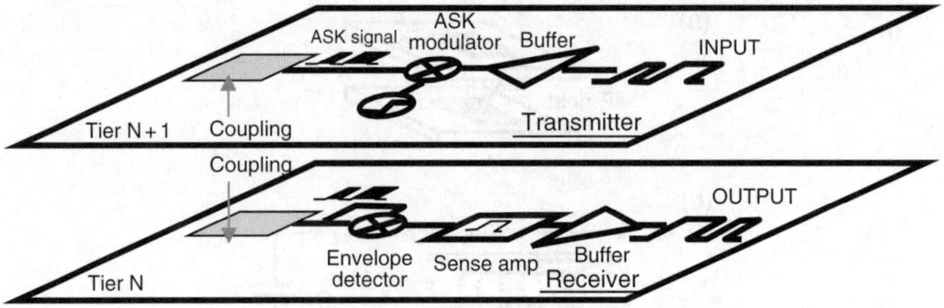


*Figure 15.7 Schematic diagram of capacitive coupling of a vertically stacked transmitter and receiver circuit chips*

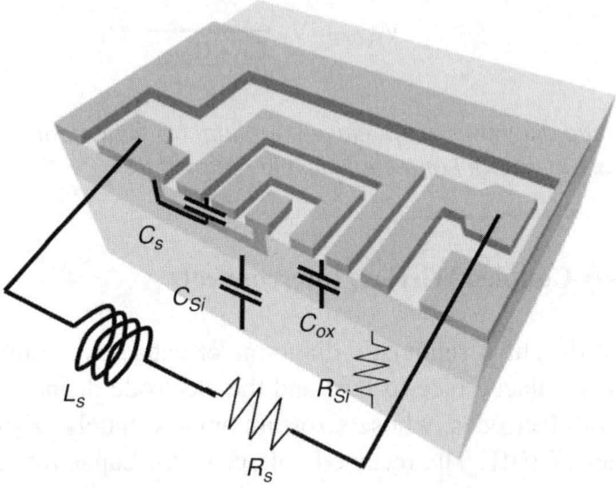


*Figure 15.8 Physical model of a spiral inductor on an Si substrate*

alignment margin on each side. The measured data rate is 11 Gb/s/pin with power consumption of 4.35 mW/pin. The root mean square (RMS) jitter is measured as 0.49 ps. The BER is measured to be $< 1.02 \times 10^{-14}$ by using a $(2^{31} - 1)$ pseudo random binary sequence (PRBS). Energy efficiency is 0.39 pJ/b. The active chip area is 0.0021 $mm^2$ [14].

### 15.2.3 Inductively Coupled Wireless Interconnects

Inductive coupling can be more efficient than capacitive coupling for longer-distance inter-chip communications, since it exploits magnetic flux that is detectable from several hundreds of micrometers away. The physical model of a spiral inductor on a silicon substrate is shown in Figure 15.8 [18]. The parasitic resistances dissipate energy through ohmic loss while the parasitic capacitances store electric energy. The inductance and resistance of the spiral and underpass are represented by the series inductance $L_s$ and series resistance $R_s$, respectively. $R_s$ represents the energy losses due to the skin effect in the spiral interconnect and the induced eddy current in conductive media. The overlap between the spiral and the underpass forms capacitive coupling $C_s$ between the two terminals of the inductor. The overlap capacitance is significant because of the relatively large potential difference between the spiral and the center-tap underpass. The oxide capacitance between the spiral and the silicon substrate is modeled by the series capacitance $C_{ox}$. The capacitance and resistance of the silicon substrate are modeled by $C_{Si}$ and $R_{Si}$, respectively. $R_{Si}$ causes the energy dissipation due to the ohmic loss in the silicon substrate. The inductor quality factor $Q$ degrades at high frequencies due to energy dissipation in the semiconducting substrate. Since the energy is stored in the inductor as a magnetic energy, the energy stored as an electric field due to parasitic capacitances has a negative effect on $Q$ so that $Q$ is proportional to the net magnetic energy stored, which is equal to the difference between the peak magnetic and electric energies.

To provide a good short to ground for the electric field, a patterned ground shield has been developed, which has slots orthogonal to the spiral for reducing the image current [18]. The slots act as an open circuit to cut off the path of the induced loop current. The slots should be sufficiently narrow such that the vertical electric field cannot leak through the patterned ground shield into the underlying silicon substrate. The shield should be strapped with the top layer metal to provide a low-impedance path to ground. To minimize the energy loss of the inductor, the shield resistance must be kept small compared to the reactance of the oxide capacitance. A typical on-chip spiral inductor has parasitic oxide capacitance between 0.25 pF and 1 pF, depending on the size and the oxide thickness. The corresponding reactance due to the oxide capacitance at 1–2 GHz is on the order of 100 Ω and, hence, shield resistance of a few ohms is sufficiently small. As the magnetic field passes through the patterned ground shield, the magnetic flux is reduced due to the skin effect. To avoid the attenuation, the shield layer must be thinner than the skin depth. Therefore, a polysilicon patterned ground shield has been developed [18].

Figure 15.9 (a) illustrates inductive coupling between spiral inductors that are fabricated on stacked silicon chips [20, 21]. The received voltage of the inductive-coupling link $V_{R,IND}$ is given by $M\,dI_T/dt$, where $M$ is the mutual inductance between coils and $I_T$ is current through the transmitter coil. By increasing the coil size, $M$ can be increased and hence $V_{R,IND}$ can be increased. Figure 15.9 (b) shows a schematic diagram of cross-talk of inductive coupling [23]. The cross-talk is attenuated with the cubic of the horizontal distance $Y$ between two spiral inductors. The magnetic flux curls around the metal line of the $T_x$ inductor, which yields the magnetic flux downward or upward in the $R_x$ inductor loop. The downward and upward fluxes cancel out at $Y_0$ so that the polarity of the cross-talk changes at the position of $Y_0$. The crosstalk is negligible at the distance $Y > Y_0$.

Figure 15.10 shows inductively coupled inter-chip signaling [19], where the chips are stacked face-up and inductively coupled by metal inductors. Since power, ground and clock can be provided by bonding wires in a face-up stacked structure, no complex mechanical process is required, whereas it is required in a face-to-face structure. Both a transmitter and a receiver have inductors and the transmitter inductor is allocated in the receiver inductor for high layout density. Since inductive coupling is a current-driven scheme, transmit power can be increased for longer distance communication even at low supply voltages in scaled devices. Transmission gain can be increased by increasing the number of turns of the inductors by exploiting an increasing level of the metal layers. The maximum data rate was 1.25 Gb/s/channel. Power dissipation was 43 mW in the transmitter and 2.6 mW in the receiver at 3.3 V [19].

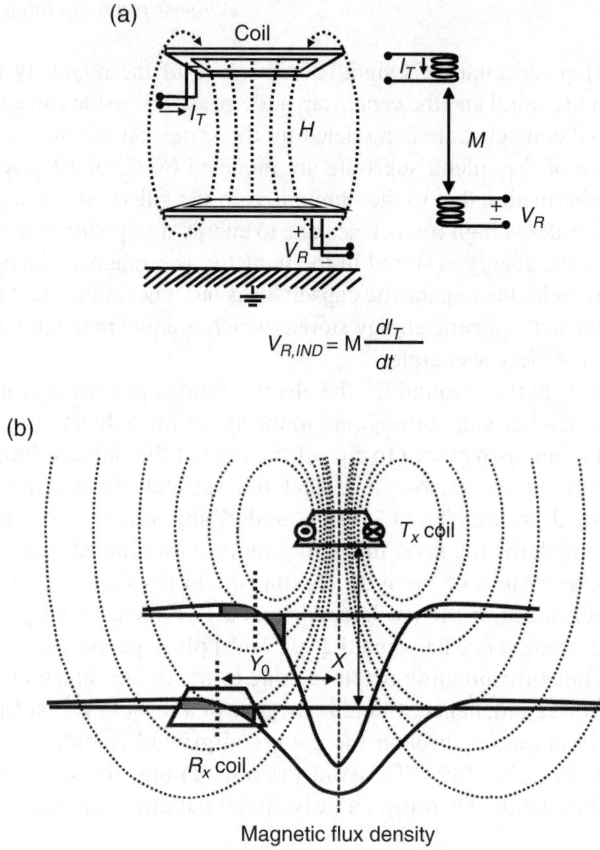


***Figure 15.9*** *Inductive coupling of stacked Si chips. (a) Inductively coupled Si chips and the recieved voltage of an inductively coupled link. (b) Schematic diagram of magnetic flux density showing cross-talk of inductive coupling*

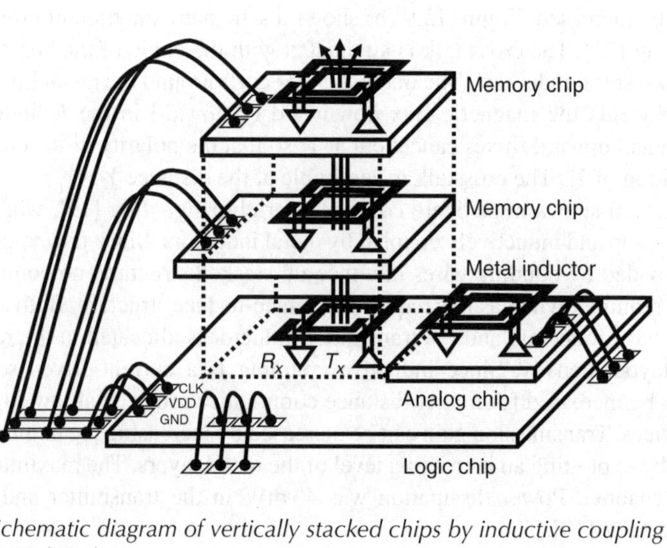


***Figure 15.10*** *Schematic diagram of vertically stacked chips by inductive coupling combined with wire bonding*

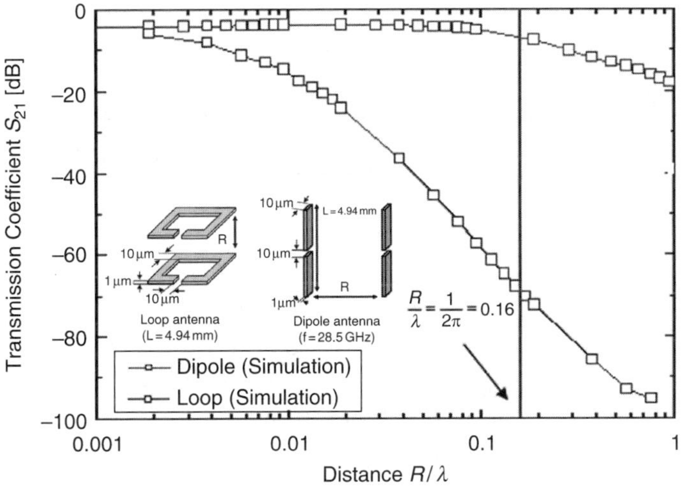


***Figure 15.11*** *Transmission coefficient $S_{21}$ versus normalized distance of dipole antennas and loop inductor*

When the diameter of a inductor is 120 µm and the number of turns is 5, the self-inductance and self-resonance frequency were 9 nH and 16 GHz, respectively. The maximum data rates of 11 Gb/s, 10.5 Gb/s and 8.5 Gb/s and the energy dissipations of 1.4 pJ/b, 1.5 pJ/b and 1.8 pJ/b were achieved with the BER of $10^{-14}$ for the distances of 15 µm, 30 µm and 45 µm, respectively [21]. The 64 stacked chips of NAND flash memories were demonstrated using both wireless interfaces for data access and bonding wires for the power supply [25]. The inductor size is 200 µm in a diameter and the shield metal size is 400 µm × 400 µm. The thicknesses of a chip and an adhesive are 50 µm and 10 µm, respectively. The average energy consumption of 15 pJ/b/chip is achieved with the BER of $10^{-12}$ at the data rate of 2 Gb/s.

### 15.2.4 Antennas and Propagation [26–37]

An electromagnetic wave can transmit signals with the highest speed between LSI chips as well as in a chip, as indicated in Figure 15.2. Figure 15.11 shows $S_{21}$ versus normalized distance ($R/\lambda$) for half-wavelength dipole antenna and loop inductor. The sizes of half-wavelength dipole antenna and loop inductor are 4.94 mm. The $S_{21}$ of the half-wavelength dipole antenna decreases in proportion to the reciprocal distance in far field ($R/\lambda > ½\pi$), while the $S_{21}$ of the loop inductor decreases in reciprocal proportion to square of the normalized distance in the near field ($R/\lambda < ½\pi$). The $S_{21}$ of antenna radiation is approximately 60 dB higher than that of the inductor coupling at $R/\lambda = ½\pi$.

When the distance of the signal path is on the order of a wavelength, impedance matching at interfaces must be taken into account according to the transmission line theory. A silicon on-chip antenna is an impedance transformer at the interface between a CMOS circuit and a transmission medium such as air [27, 28]. Schematic diagrams of half-wavelength dipole

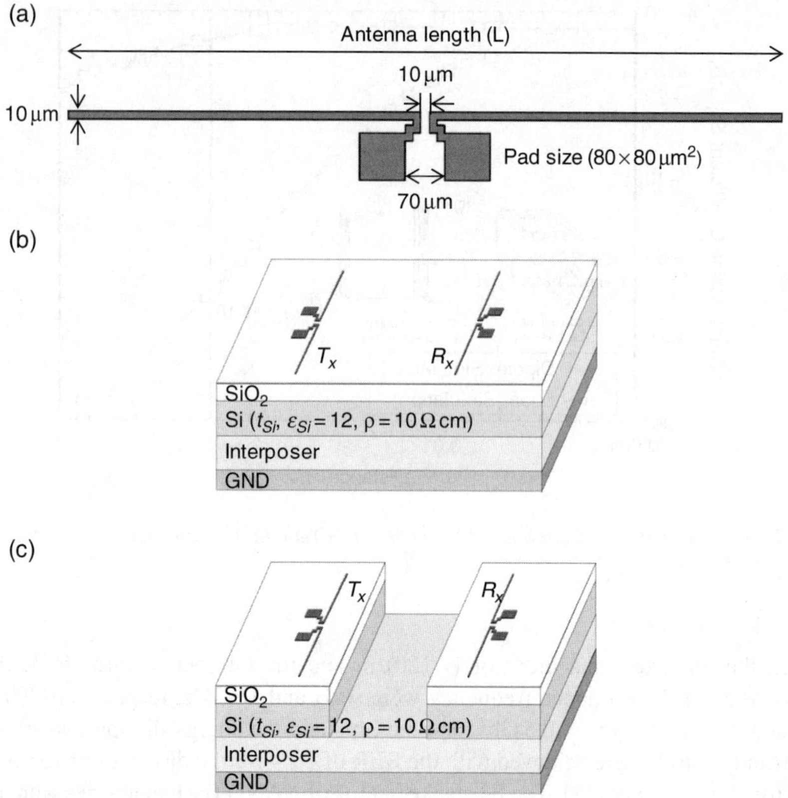


*Figure 15.12* *Schematic diagram of an on-chip antenna. (a) A plan view of a half-wavelength on-chip dipole antenna. (b) Schematic diagram of on-chip dipole antennas for intra-chip communication. (c) Schematic diagram of an inter-chip wireless interconnect using antennas*

antennas for intra-chip and inter-chip communications are shown in Figure 15.12. The on-chip antenna has some drawbacks in the occupied area due to its resonance wavelength, ranging from a few cm to a few mm at the corresponding microwave or mm-wave frequencies (1–100 GHz) and attenuation of transmission gain due to conductive loss in a standard Si substrate. The wavelength of the electromagnetic wave can be reduced by increasing the dielectric constant of a medium by a factor of $1/\sqrt{\varepsilon_r}$, where $\varepsilon_r$ is the relative dielectric constant of the medium. The wavelength as well as velocity of the electromagnetic wave decrease in an Si substrate by a factor of $1/\sqrt{\varepsilon_{Si}}$, where $\varepsilon_{Si}$ is the relative dielectric constant of Si. A practical antenna aperture size is equivalent to half of the wavelength, which corresponds to the wavelength of a resonance frequency as expressed by

$$L_A \propto \frac{\lambda_{res}}{2}\frac{1}{\sqrt{\varepsilon_{Si}}} \tag{15.4}$$

where $L_A$ is the antenna aperture size, $\lambda_{res}$ is a wavelength of a resonance frequency of the antenna, which is located in the air. The length of the dipole antenna, which is fabricated on the Si substrate, is approximately 22 mm at the frequency of 3.5 GHz by taking the factor of $1/\sqrt{\varepsilon_{Si}}$ into account. The aperture size of an on-chip dipole antenna can be scaled as the wavelength decreases and the cutoff frequency of the CMOS transistor increases. The size of the on-chip antenna can be reduced to 1.8 mm by increasing the center frequency up to 34 GHz.

Figure 15.13 (a) shows a schematic cross-section of the integrated antennas on an Si chip. There are multiple propagation paths of direct and reflected waves from a transmitter

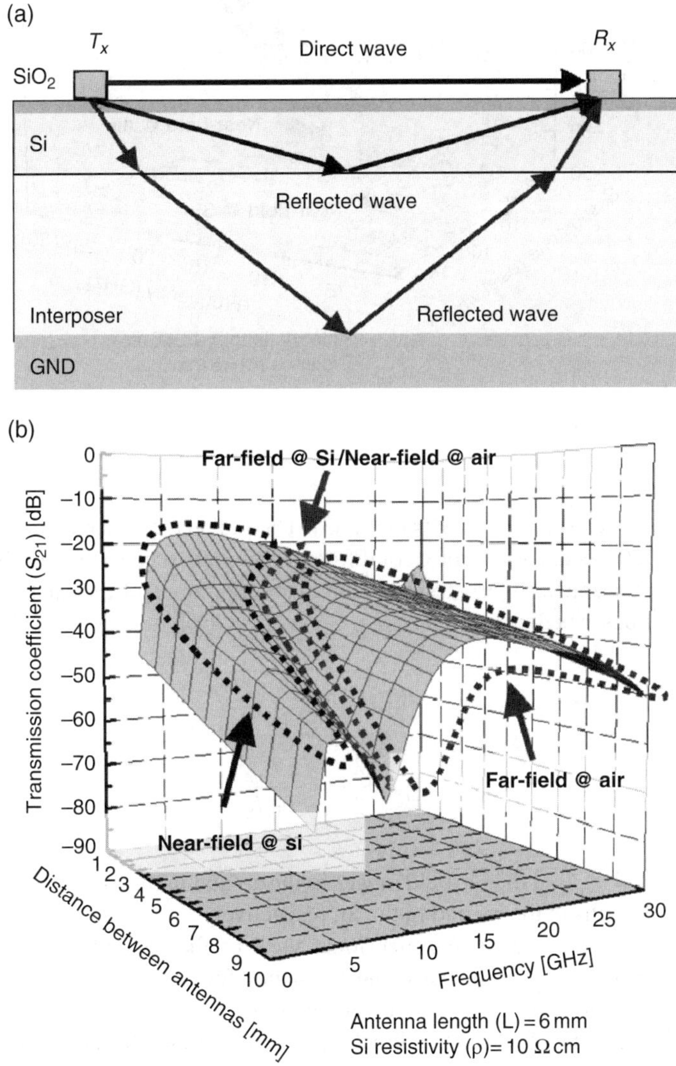


***Figure 15.13*** *Schematic diagrams of multiple propagation paths. (a) Cross-sectional diagram of on-chip antennas and multiple propagation paths. (b) Three-dimensional view of the transmission coefficient versus distance and frequency. (c) Three-dimensional view of the transmission coefficient versus Si substrate resistivity and frequency*

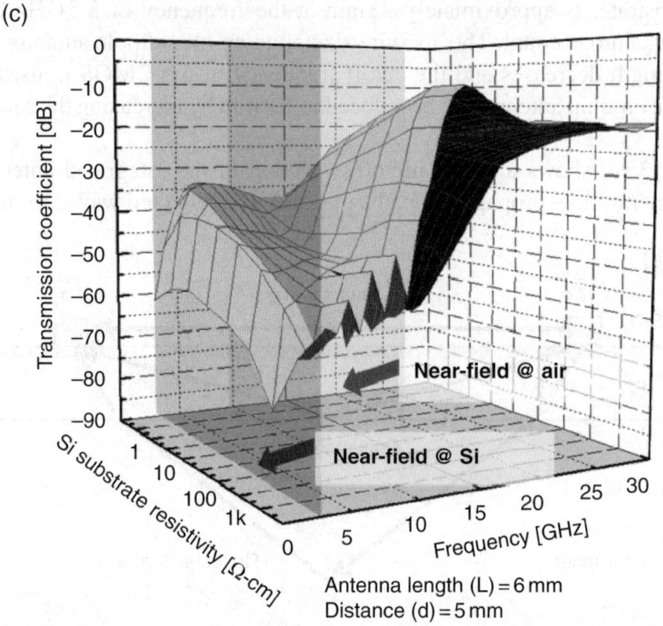


***Figure 15.13*** *(continued)*

antenna (Tx) to a receiver antenna (Rx) [32, 33]. The radiated electromagnetic wave from the Tx antenna reflects at the interfaces of different materials such as the Si substrate/interposer and interposer/metal interfaces. The boundary conditions of the far-fields in the air and in the Si are described, respectively, as

$$R_{Far(air)} > \frac{\lambda_{air}}{2\pi} \tag{15.5}$$

$$R_{Far(Si)} > \frac{\lambda_{Si}}{2\pi} = \frac{\lambda_{air}}{2\pi\sqrt{\varepsilon_{Si}}} \tag{15.6}$$

where $R_{Far(air)}$ and $R_{Far(Si)}$ are the far-field ranges in the air and in the Si, respectively; $\lambda_{air}$ and $\lambda_{Si}$ are the wavelengths in the air and in the Si, respectively.

Figure 15.13 (b) shows a three-dimensional view of the dependence of the antenna transmission coefficient $S_{21}$ on distance and frequency for the fixed antenna length fabricated on a standard Si substrate. In the far-field in the air, which corresponds to the far-field condition of Equation (15.5), both the direct wave in the air and the surface wave in the Si propagate along with the gradual slope of the ridge at the higher frequency. On the other hand, in the near-field in the Si, where the distance is shorter than the near-field condition in Si, $R_{Near(Si)} < \lambda_{air} / 2\pi\sqrt{\varepsilon_{Si}}$, the $S_{21}$ decreases more abruptly than that of the far-field with increasing distance along with the steeper slope of the ridge at the lower frequency. In the near-field, the signal is transmitted by capacitive and inductive couplings between the

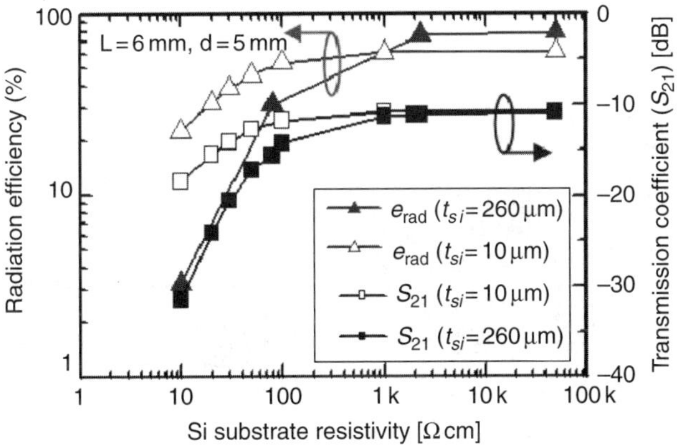


***Figure 15.14*** *Radiation efficiency and transmission coefficient $S_{21}$ of on-chip antennas as a function of Si substrate resistivity*

transmitter and the receiver antennas. In addition, the valley between the two ridges is the interface region between the far-field in Si and the near-field in air.

According to the Friis formula in Equation (15.2), one of the issues of the Si on-chip antenna is its low transmission gain due to its conductive loss in the standard Si substrate. In order to improve the transmission gain, the resistivity of the Si substrate is increased. The three-dimensional view of the dependence of $S_{21}$ on the Si substrate resistivity and frequency for the fixed distance and the antenna length is shown in Figure 15.13 (c). In the far-field, which corresponds to the far-field condition of Equation (15.6), the $S_{21}$ increases with increasing resistivity of Si from 10 Ω cm to 2.29 kΩ cm. The $S_{21}$ also increases with decreasing thickness of the Si substrate. The transmission gain of the on-chip antenna increases with decreasing Si thickness in the far-field at the high frequency. The quality factor of the antenna is improved but the bandwidth decreases with decreasing thickness of the Si substrate. In addition, the resonance frequency increases with decreasing Si thickness because the reduction of the Si thickness reduces the influence of Si on the reduction of the effective wavelength, $\lambda_{air}/\sqrt{\varepsilon_{Si}}$ .

The radiation efficiency and the transmission coefficient of the on-chip antenna increase with increasing resistivity and with increasing thickness of the Si substrate, as shown in Figure 15.14. The radiation efficiency of the antenna increases from 3 % to 70 % and the transmission gain increases from −30 dB to −10 dB with increasing resistivity of the Si substrate from 10 Ω cm to 2.29 kΩ cm. Although the high resistivity Si substrate (> 2.29 kΩ cm) can eliminate the loss in the Si substrate, the high resistivity Si substrates are not used because they are more expensive than the standard Si substrate. Consequently, the thickness of the standard Si substrate having 10 Ωcm resistivity is reduced from 260 μm to 10 μm. The corresponding radiation efficiency improved from 3 % to 25 % and the transmission gain improved from −30 dB to −20 dB.

Figure 15.15 shows electromagnetic wave propagation through the 3D stacked Si chips. A Tx antenna is located on a top Si chip and an Rx antenna is located on a bottom Si chip, as shown in Figure 15.15 (a). The electromagnetic wave penetrates through stacked Si

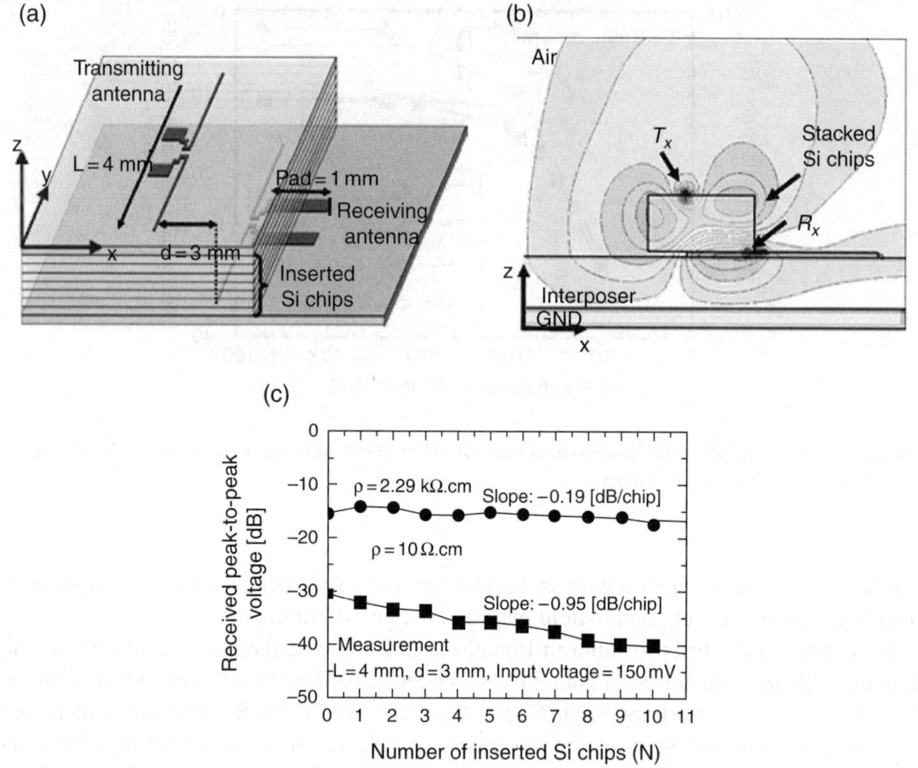


*Figure 15.15 Electromagnetic wave propagation in 3D stacked Si chips. (a) Schematic diagram of 3D-stacked LSI chips with on-chip antennas. (b) Electric field distribution of electromagnetic wave propagation through 3D-stacked Si chips. (c) Received peak-to-peak voltage versus number of stacked Si chips*

chips, as shown in Figure 15.15 (b). The received peak-to-peak voltage does not depend on the number of stacked chips when the resistivity of the Si substrate is 2.29 kΩ cm but it decreases with increasing numbers of chips when the standard Si substrate (10 Ω cm) is used, as shown in Figure 15.15 (c) [31].

Nevertheless, the transmission gain of the on-chip antenna is still low because most of the radiated power from a transmitter antenna scatters before arriving at the receiver antenna. Therefore, a low transmission gain of the on-chip antenna can be improved by use of a low-loss interposer waveguide, as shown in Figure 15.16. The radiated electromagnetic wave from the antenna is confined in the interposer when the dielectric constant of the interposer is higher than that of the air, so that the interposer acts as a medium for the local area network of Si chips. The interposer is placed under the Si chips to serve as a shared broadcasting medium of bidirectional communications. When the dielectric constant of the interposer is higher than that of Si ($\varepsilon_r = 12$), the radiated electromagnetic wave from the on-chip antenna propagates dominantly in the interposer instead of transmitting in the standard 10 Ω cm Si substrate. The electromagnetic wave from the half-wavelength dipole antenna, which is designed to have the resonance frequency in the high-*k* interposer, is confined in the interposer.

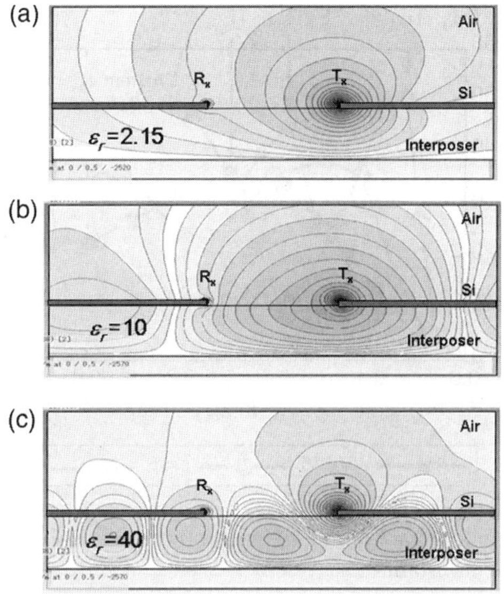


***Figure 15.16*** *Cross-sectional view of electromagnetic field radiated from Tx antenna on an Si chip. (a) Electromagnetic field distribution for the low-k interposer whose dielectric constant is 2.15. (b) Electromagnetic field distribution for the interposer whose dielectric constant is 10. (c) Electromagnetic field distribution for the high-k interposer ($\varepsilon_r = 40$)*

The electromagnetic wave, which is radiated from a transmitter antenna (Tx) on to a Si chip, propagates horizontally in the interposer and is received by a receiver antenna (Rx) on a separate Si chip, as shown in Figure 15.16. The radiated signals can be up-linked from the Tx chip to the dielectric interposer as a broadcasting medium via the on-chip Tx antenna and then down-linked to the input port of the Rx chip via the Rx antenna. The radiated electromagnetic wave propagates in both air and the interposer when the low-*k* interposer ($\varepsilon_r = 2.15$) is used, as shown in Figure 15.16 (a). On the other hand, the signal propagates mainly in the interposer when the high-*k* interposers ($\varepsilon_r = 10, 40$) are used, as shown in Figures 15.16 (b) and (c).

Figure 15.17 shows the radiation efficiency versus normalized thickness of the interposer as a parameter of the dielectric constant. The resistivity and the thickness of the Si substrate are 10 Ω cm and 10 μm, respectively. The antenna length is 6 mm. The cutoff frequencies of the transverse magnetic ($TM_n$) (for $n = 0, 2, 4, \ldots$) and transverse electric ($TE_n$) (for $n = 1, 3, 5, \ldots$) modes in the dielectric sub-waveguide are expressed as [39, 40]

$$f_c = \frac{nc_0}{4h\sqrt{\varepsilon_r - 1}} \Rightarrow \frac{h}{\lambda_c} = \frac{n}{4\sqrt{\varepsilon_r - 1}} \tag{15.7}$$

where $c_0$ is the velocity of light, $f_c$ is the cutoff frequency, $\lambda_c$ is the cutoff wavelength, $h$ is the thickness of the interposer as a waveguide and $\varepsilon_r$ is the relative dielectric constant of the interposer, respectively. The cutoff frequency depends on both the thickness and the dielectric constant of the interposer. The radiation efficiency has the local maximum values at the

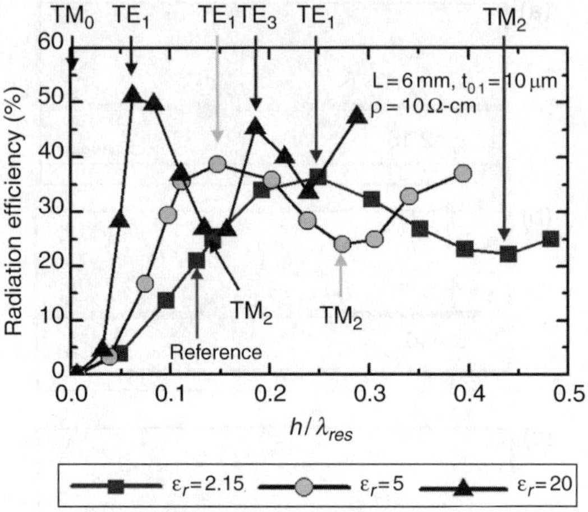


***Figure 15.17*** *Dependence of radiation efficiency on the normalized thickness of the interposers and dielectric constants*

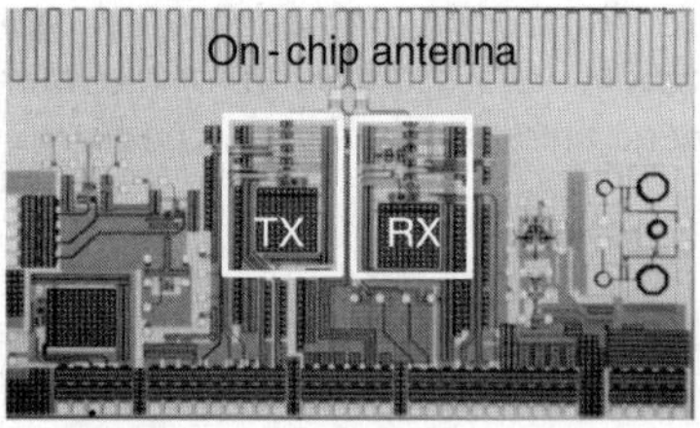


***Figure 15.18*** *Photograph of a CMOS transmitter and receiver circuit chip with on-chip antennas*

frequencies of the $TE_n$ mode. The local maximum of the radiation efficiency increases with increasing dielectric constant of the interposer. The optimum normalized thickness can be reduced by use of the higher dielectric constant material. The maximum radiation efficiency of 54% is obtained at the cutoff frequency of the $TE_1$ mode when the dielectric constant of the interposer is 38 and the normalized thickness ($h/\lambda$) is 0.06. When the thickness of the interposer decreases from the $TE_1$ mode, the radiation efficiency decreases abruptly toward zero.

The impulse radio–ultra-wideband (IR-UWB) system modulates baseband signals into impulses and the receiver conducts a timing correlation and signal recovery. A test chip of IR-UWB CMOS transmitter and receiver circuits is shown in Figure 15.18. The total antenna length is 30.58 mm, which is equivalent to the effective half-wavelength of the electromagnetic wave of 3 GHz. The BER of the IR-UWB transceiver was measured using bow-tie antennas as a function of distance. The measured BER was $6.89\times10^{-9}$ at the distance of 9 cm [41].

## 15.3 Conclusion

Inter-chip wireless interconnect technologies have been developed. Capacitive and inductive couplings are used for clock and data communications as the local wireless interconnects in the near-field whose distance is shorter than approximately 100 μm. On-chip antennas are used for the far-field transmission between LSI chips as the global wireless interconnects whose distances are longer than 1 mm.

## References


[1] Dennard, R.H., Gaensslen, F.H., Yu, H.-N., Rideout, V.L., Bassous, E. and LeBlanc, A.R. (1974) Design of ion implanted MOSFET's with very small physical dimensions. *IEEE J. Solid-State Circuits*, **SC-9**, October, 256–268.

[2] Bakoglu, H.B. and Meindl, J.D. (1985) Optimal interconnection circuits for VLSI. *IEEE Trans. Electron Devices*, **ED-32**, May, 903–909.

[3] Bakoglu, H.B. (1990) *Circuits, Interconnections, and Packaging for VLSI*, Addison-Wesley Book Express, Boston, Massachusetts.

[4] Edelstein, D., Heidenreich, J., Goldblatt, R., Cote, W., Uzoh, C., Lustig, N., Roper, P., McDevitt, T., Motsiff, W., Simon, A., Dukovic, J., Wachnik, R., Rathore, H., Schulz, R., Su, L., Luce, L.S. and Slattery, J. (1997) Full copper wiring in a sub-0.25 μm CMOS ULSI technology. IEEE International Electron Devices Meeting, Technical Digest, pp. 773–776.

[5] Kikkawa, T., Chikaki, S., Yagi, R., Shimoyama, M., Shishida, M., Fujii, N., Kohmura, K., Tanaka, H., Nakayama, T., Hishiya, S., Ono, T., Yamanishi, T., Ishikawa, A., Matsuo, H., Seino, Y., Hata, N., Yoshino, T., Takada, S., Kawahara, J. and Kinoshita, K. (2005) Advanced scalable ultra low-*k*/Cu interconnect technology for 32 nm CMOS ULSI using self-assembled porous silica and self-aligned CoWP barrier. IEEE International Electron Devices Meeting, Technical Digest, pp. 99–102.

[6] Meindl, J.D. (1995) Low power microelectronics: retrospect and prospect. *Proc. IEEE*, **83** (4), 619–635.

[7] Davis, J.A., Venkatesan, R., Kaloyeros, A., Beylansky, M., Souri, S.J., Banerjee, K., Saraswat, K.C., Rahman, A., Reif, R. and Meindl, J.D. (2001) Interconnect limits on gigascale integration (GSI) in the 21st century. *Proc. IEEE*, **89** (3), 305–324.

[8] Yeung, E. and Horowitz, M.A. (2000) A 2.4 Gb/s/pin simultaneous bidirectional parallel link with per-pin skew compensation. *IEEE J. Solid-State Circuits*, **35** (11), November, 1619–1628.

[9] Burns, J., McIlrath, L., Keast, C., Vu, D.P., Warner, K. and Wyatt, P. (2001) Three dimensional integrated circuits for low-power high-bandwidth systems on a chip. IEEE Solid-State Circuits Conference, Digest Technical Papers, February 2001, pp. 268–269.

[10] Kurita, Y., Matsui, S., Takahashi, N., Soejima, K., Komura, M., Itou, M. and Kawano, M. (2009) Vertical integration of stacked DRAM and high speed logic device using SMAFTI technology. *IEEE Trans. Advanced Packaging*, **32** (3), August, 657–665.

[11] Kang, U., Chung, H.J., Heo, S., Park, D.K., Lee, H., Kim, J.H., Ahn, S.H., Cha, S.H., Ahn, J., Kwon, D.M., Lee, J.W., Joo, H.S., Kim, W.S., Jang, D.H., Kim, N.S., Choi, J.H., Chung, T.G., Yoo, J.H., Choi, J.S., Kim, C. and Jun, Y.H. (2010) 8 Gb 3D DDR3 DRAM using through silicon via technology. *IEEE J. Solid-State Circuits*, **45** (1), January, 111–119.

[12] Chang, M.F., Roychowdhury, V.P., Zhang, L., Shin, H.C. and Qian, Y. (2001) RF/wireless interconnect for inter- and intra-chip communications. *Proc. IEEE*, **89** (4), 456–466.

[13] Chang, M.C.F., Verbauwhede, I., Chien, C., Xu, Z., Kim, J., Ko, J., Gu, Q. and Lai, B.C. (2005) Advanced RF/baseband interconnect schemes for inter- and intra-ULSI communications. *IEEE Trans. Electron Devices*, **52** (7), July, 1271–1285.

[14] Gu, Q., Xu, Z., Ko, J. and Chang, M.C.F. (2007) Two 10 Gb/s/pin low-power interconnect methods for 3D Ics. IEEE International Solid-State Circuits Conference, Digest Technical Papers, February 2007, pp. 448–449.

[15] Kanda, K., Antono, D.D., Ishida, K., Kawaguchi, H., Kuroda, T. and Sakurai, T. (2003) 1.27 Gb/s/ch 3 mW/pin wireless syperconnect interface scheme. IEEE Intrnational Solid-Statre Circuits Conference, Digest Technical Papers, February 2003, pp. 186–187.

[16] Fazzi, A., Magagni, L., Mirandola, M., Charlet, B., Di Cioccio, L., Jung, E., Canegallo, R. and Guerrieri, R. (2007) 3D capacitive interconnections for wafer-level and die-level assembly. *IEEE J. Solid-State Circuits*, **42** (10), 2270–2282.

[17] Yue, C.P. and Wong, S.S. (2000) Physical modeling of spiral inductors on silicon. *IEEE Trans. Electron Devices*, **47** (3), March, 560–568.

[18] Yue, C.P. and Wong, S.S. (1998) On-chip spiral inductors with patterned ground shields for Si-based RF ICs. *IEEE J. Solid-State Circuits*, **33** (5), May, 743–752.

[19] Miura, N., Mizokuchi, D., Yusof, Y.B., Sakurai, T. and Kuroda, T. (2005) Analysis and design of inductive coupling and transceiver circuit for inductive inter-chip wireless superconnect. *IEEE J. Solid-State Circuits*, **40** (4), April, 829–837.

[20] Mizoguchi, D., Miura, N., Yoshida, Y., Yahagishi, N. and Kuroda, T. (2006) Measurement of inductive coupling in wireless superconnect. *Jpn. J. Appl. Phys*., **45** (4B), 3286–3289.

[21] Miura, N., Kohama, Y., Sugimori, Y., Ishikuro, H., Sakurai, T. and Kuroda, T. (2009) A high-speed inductive-coupling link with burst transmission. *IEEE J. Solid-State Circuits*, **44** (3), March, 947–955.

[22] Kuroda, T. (2007) CMOS proximity inter-chip communications. IEEE 2007 CMOS Emerging Technologies Workshop Keynote.

[23] Miura, N., Sakurai, T. and Kuroda, T. (2007) Cross talk countermeasures for high density inductive-coupling channel array. *IEEE J. Solid-State Circuits*, **42** (2), February, 410–421.

[24] Ishikuro, H., Miura, N. and Kuroda, T. Wideband inductive coupling interface for high-performance portable system. IEEE Custom Integrated Circuits Conference Proceedings, 16–29 September 2007, pp. 13–20.

[25] Saito, M., Sugimori, Y., Kohama, Y., Yoshida, Y., Miura, N., Ishikuro, H., Sakurai, T. and Kuroda, T. (2010) 2 Gb/s 15 pJ/b/chip inductive-coupling programmable bus for NAND flash memory stacking. *IEEE J. Solid-State Circuits*, **45** (1), January, 134–141.

[26] Ishikuro, H. and Kuroda, T. (2010) Wireless proximity interfaces with a pulse-based inductive coupling technique. *Commun. Mag., IEEE*, **48** (10), 192–199.

[27] Sasaki, M. and Iwata, A. (2005) A 0.95 mW/1.0 Gbps spiral-inductor based wireless chip-interconnect with asynchronous communication scheme. IEEE Symposium on VLSI Circuits, Digest Technical Papers, pp. 348–351.

[28] Floyd, B.A., Hung, C.M., Kenneth, K. and O, K.K. (2002) A 15-GHz wireless interconnect implemented in a 0.18-μm CMOS technology using integrated transmitters, receivers, and antennas. *IEEE J. Solid-State Circuits*, **37** (5), May, 543–552.

[29] O, K.K., Kim, K., Floyd, B.A., Mehta, J.L., Yoon, H., Hung, C.M., Bravo, D., Dickson, T.O., Guo, X., Li, R., Trichy, N., Caserta, J., Bomstad, W.R., Branch, J., Yang, D.J., Bohorquez, J., Seok, E., Gao, L., Sugavanam, A., Lin, J.J., Chen, J. and Brewer, J.E. (2005) On-chip antennas in silicon ICs and their application. *IEEE Trans. Electron Devices*, **52** (7), July, 1312–1323.

[30] Kikkawa, T., Saha, P.K., Sasaki, N. and Kimoto, K. (2008) Gaussian monocycle pulse transmitter using 0.18 μm CMOS technology with on-chip integrated antennas for inter-chip UWB communication. *IEEE J. Solid-State Circuits*, **43** (5), May, 1303–1312.

[31] Iwata, A., Sasaki, M., Kikkawa, T., Kameda, S., Ando, H., Kimoto, K., Arizono, D. and Sunami, H. (2005) A 3D integration scheme utilizing wireless interconnections for implementing hyper brains. IEEE International Solid-State Circuits Conference, Technical Digest, San Francisco, California, February 2005, pp. 262–263.

[32] Sasaki, N., Kimoto, K., Moriyama, W. and Kikkawa, T. (2009) A single-chip ultra-wideband receiver with silicon integrated antennas for inter-chip wireless interconnection. *IEEE J. Solid-State Circuits*, **44** (2), February, 382–393.

[33] Saha, P.K., Sasaki, N. and Kikkawa, T. (2006) A single chip Gaussian monocycle pulse transmitter using 0.18 μm CMOS technology for intra/inter-chip UWB communication. Digest

of Technical Papers, IEEE Symposium on VLSI Circuits, Honolulu, Hawaii, 11–15 June 2006, pp. 252–253.
[34] Rashid, A.B.M.H., Watanabe, S. and Kikkawa, T. (2002) High transmission gain integrated antenna on extremely high resistivity Si for ULSI wireless interconnect. *IEEE Electron Device Lett.*, **23** (12), December, 731–733.
[35] Watanabe, S., Rashid, A.B.M.H. and Kikkawa, T. (2004) Effect of high resistivity Si substrate on antenna transmission gain for on-chip wireless interconnects. *Jpn. J. Appl. Phys.*, **43** (4B), April, 2297–2301.
[36] Kikkawa, T., Kimoto, K. and Watanabe, S. (2005) Ultra wideband characteristics of fractal dipole antennas integrated on Si for ULSI wireless interconnects. *IEEE Electron Device Lett.*, **26** (10), October, 767–769.
[37] Kimoto, K. and Kikkawa, T. (2005) Transmission characteristics of Gaussian monocycle pulses for inter-chip wireless interconnections using integrated antennas. *Jpn. J. Appl. Phys.*, **44** (4B), April, 2756–2760.
[38] Balanis, C.A. (2005) *Antenna Theory, Analysis and Design*, 3rd edn, John Wiley & Sons, Inc., New York.
[39] Harrington, R. (2001) *Time-Harmonic Electromagnetic Fields*, John Wiley & Sons, Inc., New York, p. 169.
[40] Pozar, D.M. (1983) Considerations for millimeter wave printed antennas. *IEEE Trans. Antennas and Propagation*, **AP-32** (5), September, 740–747.
[41] Sasaki, N. and Kikkawa, T. (2009) Bit error rate measurement of IR-UWB CMOS transmitter and receiver. Proceedings of Asia-Pacific Microwave Conference 2009, Singapore, 7–11 December 2009.

# Index

References to figures are given in italic type. References to tables are given in bold type.




*Advanced Interconnects for ULSI Technology*, First Edition. Edited by Mikhail R. Baklanov, Paul S. Ho and Ehrenfried Zschech.